T0205319

Lecture Notes in Computer Science 12464

More information about this series at http://www.springer.com/series/7409

De-Shuang Huang · Kang-Hyun Jo (Eds.)

Intelligent Computing Theories and Application

16th International Conference, ICIC 2020 Bari, Italy, October 2–5, 2020 Proceedings, Part II

Editors
De-Shuang Huang
Institute of Machine Learning
and Systems Biology
Tongji University
Shanghai, China

Kang-Hyun Jo
School of Electrical Engineering
University of Ulsan
Ulsan, Korea (Republic of)

ISSN 0302-9743 ISSN 1611-3349 (electronic)
Lecture Notes in Computer Science
ISBN 978-3-030-60801-9 ISBN 978-3-030-60802-6 (eBook)
https://doi.org/10.1007/978-3-030-60802-6

LNCS Sublibrary: SL3 – Information Systems and Applications, incl. Internet/Web, and HCI

This Springer imprint is published by the registered company Springer Nature Switzerland AG
The registered company address is: Gewerbestrasse 11, 6330 Cham, Switzerland

Preface

The International Conference on Intelligent Computing (ICIC) was started to provide an annual forum dedicated to the emerging and challenging topics in artificial intelligence, machine learning, pattern recognition, bioinformatics, and computational biology. It aims to bring together researchers and practitioners from both academia and industry to share ideas, problems, and solutions related to the multifaceted aspects of intelligent computing.

ICIC 2020, held in Bari, Italy, during October 2–5, 2020, constituted the 16th edition of this conference series. It built upon the success of ICIC 2019 (Nanchang, China), ICIC 2018 (Wuhan, China), ICIC 2017 (Liverpool, UK), ICIC 2016 (Lanzhou, China), ICIC 2015 (Fuzhou, China), ICIC 2014 (Taiyuan, China), ICIC 2013 (Nanning, China), ICIC 2012 (Huangshan, China), ICIC 2011 (Zhengzhou, China), ICIC 2010 (Changsha, China), ICIC 2009 (Ulsan, South Korea), ICIC 2008 (Shanghai, China), ICIC 2007 (Qingdao, China), ICIC 2006 (Kunming, China), and ICIC 2005 (Hefei, China).

This year, the conference concentrated mainly on the theories and methodologies as well as the emerging applications of intelligent computing. Its aim was to unify the picture of contemporary intelligent computing techniques as an integral concept that highlights the trends in advanced computational intelligence and bridges theoretical research with applications. Therefore, the theme for this conference was "Advanced Intelligent Computing Technology and Applications." Papers that focused on this theme were solicited, addressing theories, methodologies, and applications in science and technology.

ICIC 2020 received 457 submissions from 21 countries and regions. All papers went through a rigorous peer-review procedure and each paper received at least three review reports. Based on the review reports, the Program Committee finally selected 162 high-quality papers for presentation at ICIC 2020, included in three volumes of proceedings published by Springer: two volumes of *Lecture Notes in Computer Science* (LNCS), and one volume of *Lecture Notes in Artificial Intelligence* (LNAI).

This volume of LNCS includes 54 papers.

The organizers of ICIC 2020, including Tongji University, China, and Polytechnic University of Bari, Italy, made an enormous effort to ensure the success of the conference. We hereby would like to thank the members of the Program Committee and the referees for their collective effort in reviewing and soliciting the papers. We would like to thank Alfred Hofmann, executive editor from Springer, for his frank and helpful advice and guidance throughout as well as his continuous support in publishing the proceedings. In particular, we would like to thank all the authors for contributing their papers. Without the high-quality submissions from the authors, the success of the conference would not have been possible. Finally, we are especially grateful to the International Neural Network Society and the National Science Foundation of China for their sponsorship.

August 2020

De-Shuang Huang
Prashan Premaratne

Organization

General Co-chairs

De-Shuang Huang, China
Vitoantonio Bevilacqua, Italy

Program Committee Co-chairs

Eugenio Di Sciascio, Italy
Kanghyun Jo, South Korea

Organizing Committee Co-chairs

Ling Wang, China
Phalguni Gupta, India
Vincenzo Piuri, Italy
Antonio Frisoli, Italy
Eugenio Guglielmelli, Italy
Silvestro Micera, Italy
Loreto Gesualdo, Italy

Organizing Committee Members

Andrea Guerriero, Italy
Nicholas Caporusso, USA
Francesco Fontanella, Italy
Vincenzo Randazzo, Italy
Giacomo Donato Cascarano, Italy
Irio De Feudis, Italy
Cristian Camardella, Italy
Nicola Altini, Italy

Award Committee Co-chairs

Kyungsook Han, South Korea
Jair Cervantes Canales, Mexico
Leonarda Carnimeo, Italy

Tutorial Co-chairs

M. Michael Gromiha, India
Giovanni Dimauro, Italy

Publication Co-chairs

Valeriya Gribova, Russia
Antonino Staiano, Italy

Special Session Co-chairs

Abir Hussain, UK
Antonio Brunetti, Italy

Special Issue Co-chairs

Mario Cesarelli, Italy
Eros Pasero, Italy

International Liaison Co-chairs

Prashan Premaratne, Australia
Marco Gori, Italy

Workshop Co-chairs

Laurent Heutte, France
Domenico Buongiorno, Italy

Publicity Co-chairs

Giansalvo Cirrincione, France
Chun-Hou Zheng, China
Salvatore Vitabile, Italy

Exhibition Contact Co-Chairs

Michal Choras, Poland
Stefano Cagnoni, Italy

Program Committee Members

Daqi Zhu
Xinhong Hei
Yuan-Nong Ye
Abir Hussain
Khalid Aamir
Kang-Hyun Jo
Andrea Guerriero
Angelo Ciaramella
Antonino Staiano
Antonio Brunetti
Wenzheng Bao
Binhua Tang
Bin Qian
Bingqiang Liu
Bo Liu
Bin Liu
Chin-Chih Chang
Wen-Sheng Chen
Michal Choras
Xiyuan Chen
Chunmei Liu
Cristian Camardella
Zhihua Cui
Defu Zhang
Dah-Jing Jwo
Dong-Joong Kang
Domenico Buongiorno
Domenico Chiaradia
Ben Niu
Shaoyi Du
Eros Pasero
Fengfeng Zhou
Haodi Feng
Fei Guo
Francesco Fontanella
Chuleerat Jaruskulchai
Fabio Stroppa
Gai-Ge Wang
Giacomo Donato Cascarano
Giovanni Dimauro
L. J. Gong
Guoquan Liu
Wei Chen
Valeriya Gribova
Michael Gromiha
Maria Siluvay
Guoliang Li
Huiyu Zhou
Tianyong Hao
Mohd Helmy Abd Wahab
Honghuang Lin
Jian Huang
Hao Lin
Hongmin Cai
Xinguo Lu
Ho-Jin Choi
Hongjie Wu
Irio De Feudis
Dong Wang
Insoo Koo
Daowen Qiu
Jiansheng Wu
Jianbo Fan
Jair Cervantes
Junfeng Xia
Junhui Gao
Juan Carlos
Juan Carlos Figueroa García
Gangyi Jiang
Jiangning Song
Jing-Yan Wang
Yuhua Qian
Joaquín Torres-Sospedra
Ju Liu
Jinwen Ma
Ji Xiang Du
Junzhong Gu
Ka-Chun Wong
Kyungsook Han
K. R. Seeja
Yoshinori Kuno
Weiwei Kong
Laurent Heutte
Leonarda Carnimeo
Bo Li
Junqing Li
Juan Liu
Yunxia Liu
Zhendong Liu
Jungang Lou
Fei Luo
Jiawei Luo
Haiying Ma
Marzio Pennisi
Nicholas Caporusso
Nicola Altini
Giansalvo Cirrincione
Gaoxiang Ouyang
Pu-Feng Du
Shaoliang Peng
Phalguni Gupta
Ping Guo
Prashan Premaratne
Qinghua Jiang
Qingfeng Chen
Roman Neruda
Rui Wang
Stefano Squartini
Salvatore Vitabile
Wei-Chiang Hong
Jin-Xing Liu
Shen Yin
Shiliang Sun
Saiful Islam
Shulin Wang
Xiaodi Li
Zhihuan Song
Shunren Xia
Sungshin Kim
Stefano Cagnoni
Stefano Mazzoleni
Surya Prakash
Tar Veli Mumcu
Xu-Qing Tang

Vasily Aristarkhov
Vincenzo Randazzo
Vito Monaco
Vitoantonio Bevilacqua
Waqas Bangyal
Bing Wang
Wenbin Liu
Weidong Chen
Weijia Jia
Wei Jiang
Shanwen Zhang
Takashi Kuremoto

Reviewers

Wan Hussain Wan Ishak
Nureize Arbaiy
Shingo Mabu
Lianming Zhang
Xiao Yu
Shaohua Li
Yuntao Wei
Jinglong Wu
Wei-Chiang Hong
Sungshin Kim
Tianhua Guan
Shutao Mei
Yuelin Sun
Hai-Cheng Yi
Zhan-Heng Chen
Suwen Zhao
Medha Pandey
Mike Dyall-Smith
Xin Hong
Ziyi Chen
Xiwei Tang
Khanh Le
Shulin Wang
Di Zhang
Sijia Zhang
Na Cheng
Menglu Li
Zhenhao Guo
Limin Jiang
Kun Zhan
Cheng-Hsiung Chiang
Yuqi Wang
Anna Esposito
Salvatore Vitabile
Bahattin Karakaya
Tejaswini Mallavarapu
Sheng Yang
Heutte Laurent
Seeja
Pu-Feng Du
Wei Chen
Jonggeun Kim
Eun Kyeong Kim
Hansoo Lee
Yiqiao Cai
Wuritu Yang
Weitao Sun
Shou-Tao Xu
Min-You Chen
Yajuan Zhang
Guihua Tao
Jinzhong Zhang
Wenjie Yi
Miguel Gomez
Lingyun Huang
Chao Chen
Jiangping He
Jin Ma
Xiao Yang
Sotanto Sotanto
Liang Xu
Chaomin Iuo
Rohitash Chandra
Hui Ma
Lei Deng
Di Liu
María I. Giménez
Ansgar Poetsch
Dimitry Y. Sorokin
Jill F. Banfield
Can Alkan
Ji-Xiang Du
Xiao-Feng Wang
Zhong-Qiu Zhao
Bo Li
Zhong rui Zhang
Yanyun Qu
Shunlin Wang
Jin-Xing Liu
Shravan Sukumar
Long Gao
Yifei Wu
Qi Yan
Tianhua Jiang
Fangping Wan
Lixiang Hong
Sai Zhang
Tingzhong Tian
Qi Zhao
Leyi Wei
Lianrong Pu
Chong Shen
Junwei Wang
Zhe Yan
Rui Song
Xin Shao
Xinhua Tang
Claudia Guldimann
Saad Abdullah Khan Bangyal
Giansalvo Cirrincione
Bing Wang
Xiao Xiancui
X. Zheng
Vincenzo Randazzo
Huijuan Zhu
DongYuan Li
Jingbo Xia
Boya Ji
Manilo Monaco
Xiao-Hua Yu
Pierre Leblond
Zu-Guo Yu
Jun Yuan

Honglin Zhang
Yu Jie He
Benjamin Soibam
Sungroh Yoon
Mohamed Chaabane
Rong Hu
Youjie Yao
NaiKang Yu
Carlo Bianca
Giulia Russo
Dian Liu
Cheng Liang
Iyyakutti Iyappan Ganapathi
Mingon Kang
Zhang Chuanchao
Hao Dai
Geethan
Brendan Halloran
Yue Li
Qianqian Shi
Zhiqiang Tian
Yang Yang
Jalilah Arijah Mohd Kamarudin
Jun Wang
Ke Yan
Hang Wei
David A. Hendrix
Ka-Chun Wong
Yuyan Han
Hisato Fukuda
Yaning Yang
Lixiang Xu
Yuanke Zhou
Shihui Ying
Wenqiang Fan
Zhao Li
Zhe Zhang
Xiaoying Guo
Yiqi Jiang
Zhuoqun Xia
Jing Sun
Na Geng
Chen Li
Xin Ding
Balachandran Manavalan
Bingqiang Liu
Lianrong Pu
Di Wang
Fangping Wan
Guosheng Han
Renmeng Liu
Yinan Guo
Lujie Fang
Ying Zhang
Yinghao Cao
Xhize Wu
Le Zou
G. Brian Golding
Viktoriya Coneva
Alexandre Rossi Paschoal
Ambuj Srivastava
Prabakaran R.
Xingquan Zuo
Jiabin Huang
Jingwen Yang
Liu Qianying
Markus J. Ankenbrand
Jianghong Meng
Tongchi Zhou
Zhi-Ping Liu
Xinyan Liang
Xiaopeng Jin
Jun Zhang
Yumeng Liu
Junliang Shang
L. M. Xiao
Shang-han Li
Jianhua Zhang
Han-Jing Jiang
Daniele Nardi
Kunikazu
Shenglin Mu
Jing Liang
Jialing Li
Yu-Wen-Tian Sun
Zhe Sun
Wentao Fan
Wei Lan
Jiancheng Zhong
Josue Espejel Cabrera
José Sergio Ruiz Castilla
Juan de Jesus Amador
Nanxun Wang
Rencai Zhou
Moli Huang
Yong Zhang
Daniele Loiacono
Grzegorz Dudek
Joaquín Torres-Sospedra
Xingjian Chen
Saifur Rahaman
Olutomilayo Petinrin
Xiaoming Liu
Xin Xu
Zi-Qi Zhu
Punam Kumari
Pallavy Pandey
Najme Zehra
Zhenqing Ye
Hao Zhang
Zijing Wang
Lida Zhu
Lvzhou Li
Junfeng Xia
Jianguo Liu
Jia-Xiang Wang
Gongxin Peng
Junbo Liang
Linjing Liu
Xian Geng
Sheng Ding
Jun Li
Laksono Kurnianggoro
Minxia Cheng
Meiyi Li
Qizhi Zhu
Peng Chao Li
Ming Xiao
Guangdi Liu
Jing Meng
Kang Xu
Cong Feng
Arturo Yee
Yi Xiong
Fei Luo
Xionghui Zhou

Thar Baker
Haoqian Huang
Siguo Wang
Huan Liu
Jianqing Chen
Chunhui Wang
Xiaoshu Zhu
Wen Zhang
Yongchun Zuo
Dariusz Pazderski
Elif Hocaoglu
Hyunsoo Kim
Park Singu
Saeed Ahmed
Youngdoo Lee
Nathan D. Kent
Areesha Anjum
Sanjay Sharma
Shaojin Geng
Andrea Mannini
Van-Dung Hoang
He Yongqiang
Kyungsook Han
Long Chen
Jialin Lyu
Zhenyang Li
Tian Rui
Khan Alcan
Alperen Acemoglu
Duygun Erol Barkana
Juan Manuel Jacinto Villegas
Zhenishbek Zhakypov
Domenico Chiaradia
Huiyu Zhou
Yichuan Wang
Sang-Goo Jeong
Nicolò Navarin
Eray A. Baran
Jiakai Ding
Dehua Zhang
Giuseppe Pirlo
Alberto Morea
Giuseppe Mastronardi
Insoo Koo
Dah-Jing Jwo
Yudong Zhang
Zafaryab Haider
Mahreen Saleem
Quang Do
Vladimir Shakhov
Daniele Leonardis
Simona Crea
Byungkyu Park
Pau Rodr´
Alper Gün
Mehmet Fatih Demirel
Elena Battini
Radzi Ambar
Mohamad Farhan
Mohamad Mohsin
Nur Azzah Abu Bakar
Noraziah ChePa
Sasalak Tongkaw
Kumar Jana
Hafizul Fahri Hanafi
Liu Jinxing
Alex Moopenn
Liang Liang
Ling-Yun Dai
Raffaele Montella
Maratea Antonio
Xiongtao Zhang
Sobia Pervaiz Iqbal
Fang Yang
Si Liu
Natsa Kleanthous
Zhen Shen
Jing Jiang
Shamrie Sainin
Suraya Alias
Mohd Hanafi Ahmad Hijazi
Mohd Razali Tomari
Chunyan Fan
Jie Zhao
Yuchen Zhang Casimiro
Dong-Jun Yu
Jianwei Yang
Wenrui Zhao
Di Wu
Chao Wang
Alex Akinbi
Fuyi Li
Fan Xu
Guangsheng Wu
Yuchong Gong
Weitai Yang
Mohammed Aledhari
Yanan Wang
Bo Chen
Binbin Pan
Chunhou Zheng
Abir Hussain
Chen Yan
Dhanjay Singh
Bowen Song
Guojing
Weiping Liu
Yeguo Liao
Laura Jalili
Quan Zou
Xing Chen
Xiujuan Lei
Marek Pawlicki
Haiying Ma
Hao Zhu
Wang Zhanjun
Mohamed Alloghani
Yu Hu
Haya Alaskar
Baohua Wang
Hanfu Wang
Hongle Xie
Guangming Wang
Yongmei Liu
Fuchun Liu
Farid Garcia-Lamont
Yang Li
Hengyue Shi
Gao Kun
Wen Zheng Ma
Jin Sun
Xing Ruiwen
Zhong Lianxin
Zhang Hongyuan
Han Xupeng
Mon Hian Chew

Jianxun Mi
Michele Scarpiniti
Hugo Morais
Alamgir Hossain
Felipe Saraiva
Xuyang Xuyang
Yasushi Mae
Haoran Mo
Pengfei Cui
Yoshinori Kobayashi
Qing Yu Cui
Kongtao Chen
Feng Feng
Wenli Yan
Zhibo Wang
Ying Qiao
Qiyue Lu
Geethan Mendiz
Dong Li
Liu Di
Feilin Zhang
Haibin Li
Heqi Wang
Wei Wang
Tony Hao
Yingxia Pan
Chenglong Wei
My Ha Le
Yu Chen
Eren Aydemir
Naida Fetic
Bing Sun
Zhenzhong Chu
Meijing Li
Wentao Chen
Mingpeng Zheng
Zhihao Tang
Li keng Liang
Alberto Mazzoni
Domenico Buongiorno
Zhang Lifeng
Chi Yuhong
Meng-Meng Yin
Yannan Bin
Wasiq Khan
Yong Wu
Qinhu Zhang
Jiang Liu
Yuzhen Han
Pengcheng Xiao
Harry Haoxiang Wang
Fengqiang Li
Chenggang Lai
Dong Li
Shuai Liu
Cuiling Huang
Lian-Yong Qi
Qi Zhu
Wenqiang Gu
Haitao Du
Bingbo Cui
Qinghua Li
Xin Juan
Emanuele Principi
Xiaohan Sun
Inas Kadhim
Jing Feng
Xin Juan
Hongguo Zhao
Masoomeh Mirrashid
Jialiang Li
Yaping Hu
Xiangzhen Kong
Mi-Xiao Hou
Zhen Cui
Juan Wang
Na Yu
Meiyu Duan
Pavel Osinenko
Chengdong Li
Stefano Rovetta
Mingjun Zhong
Baoping Yuan
Akhilesh Mohan Srivastatva
Vivek Baghel
Umarani Jayaraman
Somnath Dey
Guanghui Li
Lihong Peng
Wei Zhang
Hailin Chen
Fabio Bellavia
Giosue' Lo Bosco
Giuseppe Salvi
Giovanni Acampora
Zhen Chen
Enrico De Santis
Xing Lining
Wu Guohua
Dong Nanjiang
Jhony Heriberto Giraldo Zuluaga
Waqas Haider Bangyal
Cong Feng
Autilia Vitiello
TingTing Dan
Haiyan Wang
Angelo Casolaro
Dandan Lu
Bin Zhang
Raul Montoliu
Sergio Trilles
Xu Yang
Fan Jiao
Li Kaiwen
Wenhua Li
Ming Mengjun
Ma Wubin
Cuco Cristanno
Chao Wu
Ghada Abdelmoumin
Han-Zhou Wu
Antonio Junior Spoleto
Zhenghao Shi
Ya Wang
Tao Li
Shuyi Zhang
Xiaoqing Li
Yajun Zou
Chuanlei Zhang
Berardino Prencipe
Feng Liu
Yongsheng Dong
Yatong Zhou
Carlo Croce
Rong Fei
Zhen Wang

Huai-Ping Jin
Mingzhe She
Sen Zhang
Yifan Zheng
Christophe Guyeux
Jun Sang
Huang Wenzhun
Jun Wu
Jing Luo
Wei Lu
Heungkyu Lee
Yinlong Qian
Hong wang
Daniele Malitesta
Fenqiang Zhao
Xinghuo Ye
Hongyi Zhang
Xuexin Yu
Guanshuo Xu
Mehdi Yedroudj
Xujun Duan
Xing-Ming Zhao
Jiayan Han
Yan Xiao
Weizhong Lu
Weiguo Shen
Hongzhen Shi
Zeng Shangyou
Zhou Yue
TaeMoon Seo
Sergio Cannata
Weiqi Luo
Feng Yanyan
Pan Bing
Jiwen Dong
Yong-Wan Kwon
Heng Chen
S. T. Veena
J. Anita Christaline
R. Ramesh
Shadrokh Samavi
Amin Khatami
Min Chen
He Huang
Qing Lei
Shuang Ye
Francesco Fontanella
Kang Jijia
Rahul Kumar
Alessandra Scotto Freca
Nicole Cilia
Alessandro Aliberti
Gabriele Ciravegna
Jacopo Ferretti
Jing Yang
Zheheng Jiang
Dan Yang
Dongxue Peng
Wenting Cui
Francescomaria Marino
Wenhao Chi
Ruobing Liang
Feixiang Zhou
Jijia Kang
Xinshao Wang
Huawei Huang
Zhi Zhou
Yanrui Ding
Peng Li
Yunfeng Zhao
Guohong Qi
Xiaoyan Hu
Li Guo
Xia-an Bi
Xiuquan Du
Ping Zhu
Young-Seob Jeong
Han-Gyu Kim
Dongkun Lee
Jonghwan Hyeon
Chae-Gyun Lim
Nicola Altini
Claudio Gallicchio
Dingna Duan
Shiqiang Ma
Mingliang Dou
Jansen Woo
Shanshan
ShanShan Hu
Hai-tao Li
Francescomaria Marino
Jiayi Ji
Jun Peng
Jie Hu
Jipeng Wu
Shirley Meng
Prashan Premaratne
Lucia Ballerini
Haifeng Hu
JianXin Zhang
Xiaoxiao Sun
Shaomin Mu
Yongyu Xu
Jingyu Hou
Zhixian Liu

Contents – Part II

Protein-Protein Interaction Prediction

Biomarker Discovery and Detection

Systems Biology

Modeling, Simulation, and Optimization of Biological Systems

Intelligent Computing in Computational Biology

Intelligent Computing in Drug Design

Computational Genomics

Artificial Intelligence in Biological and Medical Information Procession

High-Performance Modelling Methods on High-Throughput Biomedical Data

Intelligent Computing and Swarm Optimization

Construction of Large-Scale Heterogeneous Molecular Association Network and Its Application in Molecular Link Prediction

Gene Expression Array Analysis

A Novel Clustering-Framework of Gene Expression Data Based on the Combination Between Deep Learning and Self-organizing Map

Yan Cui[1(✉)], Huacheng Gao[1], Rui Zhang[1], Yuanyuan Lu[1], Yuan Xue[1], and Chun-Hou Zheng[2]

[1] National Engineering Reaearch Center of Communication and Network Technology, Nanjing University of Posts and Telecommunications, Nanjing 210003, China
cuiyan@njupt.edu.cn

[2] School of Computer Science and Technology, Anhui University, Hefei 230601, China
zhengch99@126.com

Abstract. Learning latent feature representation embedding in high-dimensional gene expression data is a crucial step for gene clustering application. Our clustering-framework method, incorporating Variational Autoencoders (VAE) into Self-Organizing Map (SOM), not only clustered gene expression data precisely, but also reduced the dimensionality of raw data effectively without any prior knowledge. The clustering results obtained from this method based on four gene datasets exhibited an impressive performance in efficiency and accuracy.

Keywords: Gene expression data · Clustering · Variational autoencoders · Self-Organizing map

1 Introduction

DNA microarrays are one of the main pathway to observe biochemical reactions in cells, protein and tissues levels of biological, which is conveyed by gene expression data [1, 2]. With the development of human observation and recording technology rapidly, the scale of gene expression data is becoming increasingly large. However, the analysis and processing on the large data of gene expression in high-dimensional and high-complexity has hampered the acquirement of accurate data analysis. Smart mathematical and powerful statistical methods based on Machine Learning (ML) can reply the issues through the accurate achievement of logical relationships and features [3]. However, although these excellent performances of ML technology are often high demanded, the theory is complex and difficult to operate.

As a popular method in ML, Deep Learning (DL) [4–8] have displayed some distinct advantage in hidden structures revealing, abstract pattern learning and multi task processing [9]. In DL family, autoencoder (AE) [10] including deep autoencoder

D.-S. Huang and K.-H. Jo (Eds.): ICIC 2020, LNCS 12464, pp. 3–13, 2020.
https://doi.org/10.1007/978-3-030-60802-6_1

architecture [11, 12], denoising autoencoders [13] and variational autoencoders (VAE) [14, 15] is an effective strategy for dimension reduction and features extraction. Introduced the reparameterization trick [14] technology, VAE as a data-driven and using unsupervised learning model shown a good performance in capturing relevant features and generating complex data in recent years.

For the novel method of this paper, the concept of DL was used to improve a shallow unsupervised clustering into a deep unsupervised clustering. The shallow unsupervised clustering we used was Self-Organizing Map (SOM) [16, 17], as an artificial neural network which developed by simulating the characteristics of the human brain for signal processing. However, missing the hidden layer, which is the most important feature of SOM, is different from DL. The SOM was proposed firstly by Kohonen [18] in 1990. The target of this clustering algorithm is to map points in high-to-low dimensional space and keep the topological structure as much as possible between points, which can insure a good clustering and visualization and can map data to two-dimensional grid. Considering the advantages of VAE mentioned above, the combining process between VAE and SOM was employed firstly to obtain more actual gene expression pattern through the accuracy enhancement of clustering.

In this paper, we proposed a novel clustering-framework method of gene expression data using VAE model to reduce the dimensions of the raw data and SOM clustering algorithm to group the gene more specific. The merge strategy we proposed has three advantages: Firstly, redundant features in data matrix can be removed effectively [19]. Secondly, the results can be draw and visualized conveniently. And the last one, speed up the calculation and reduce the time cost. After being processed by this method, we used the data after dimensionality reduction for further clustering. These consequences suggest that new insights into reducing the dimension of data sets makes contribution to improving the clustering accuracy.

The rest of this article is arranged as follows. In the next section, we not only introduce the hybrid VAE-SOM algorithm, but also describe the algorithm VAE and SOM respectively. Section 3 records and analyzes the experimental results of VAE-SOM based on four real datasets briefly. Finally, Sect. 4 concludes this paper and look forward to future research.

2 Methodology

DL model transformed the raw data at the input layer into increasingly abstract feature representations by successively combining outputs from the preceding layer and encapsulating highly complicated functions during the process, which is the key to overcome high-dimensional problems in many domains. Combined with a specific DL, a novel clustering-framework was presented by this paper to increase clustering accuracy and reduce the time cost through extending the traditional shallow clustering algorithm into the deep clustering algorithm. Figure 1 is a visual representation of the deep clustering-framework. The original information in the main data matrix has been maximized reserved by setting the parameters [20] in the VAE model. Then, SOM algorithm was used to cluster data based on VAE model in clustering module, and last, we further optimize the clustering matrix in optimization module.

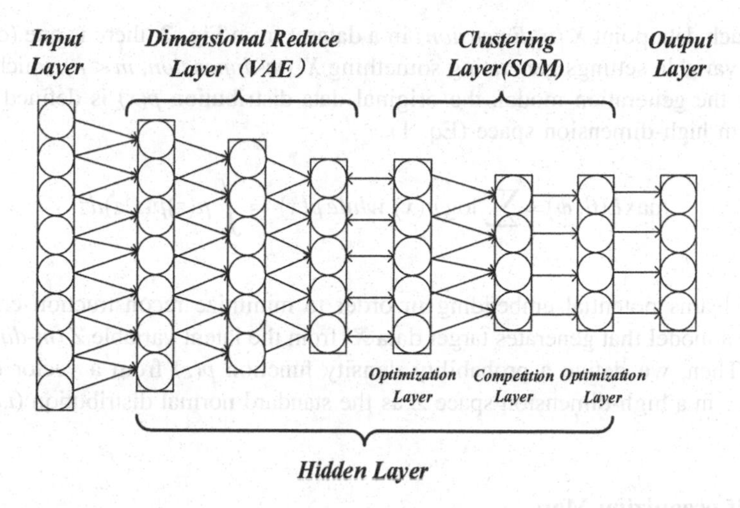

Fig. 1. Clustering-Framework based on the combination between VAE and SOM.

2.1 Variational Autoencoders

VAE is one of the derivatives of autoencoder with encoder and decoder, which uses variational inference to infer the posterior of latent embedding given input dataset. The similarity of probability distribution between the generated data and the original data is a key factor to estimate the accuracy of VAE model. VAE model based on unsupervised feature learning was employed to achieve dimensionality reduction from the unlabeled high-dimensional gene expression datasets. Specifically, the VAE model consists of three parts: encoder part, sampling part and decoder part as shown in Fig. 2.

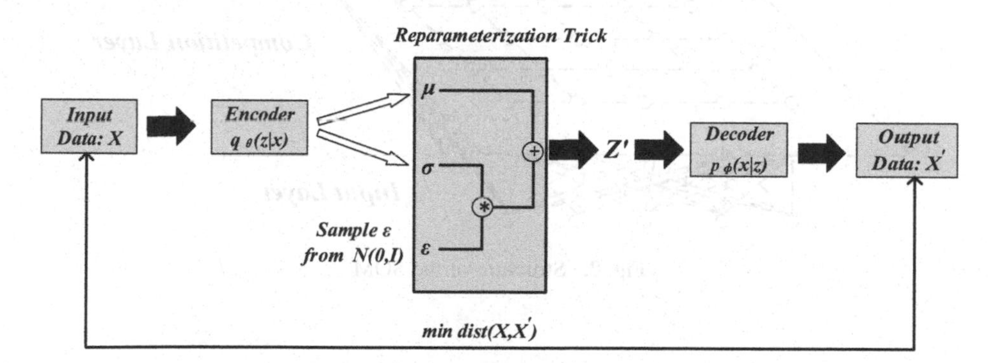

Fig. 2. Structure of the VAE

For each data point *X (n-dimension)* in a dataset from Fig. 2, there is one (or many) potential variable settings generating something *X′ (m-dimension, m < n)* which similar to *X*. For the generation model, the original data distribution *p(x)* is defined by data points *X* in high-dimension space (Eq. 1).

$$\max L(\theta, \phi) = \sum \log p(x) \; while \, p(x) = \int_z p(z) p(x|z) dz \tag{1}$$

VAE learns potential embedding in order to minimize reconstruction errors and construct a model that generates target data *X′* from the latent variable *Z (m-dimension, m < n)*. Then, we define a probability density function *p(z)* from a vector of latent variables *z* in a high-dimension space *Z* as the standard normal distribution (*i.e. p(z)* = *N(0, I)*).

2.2 Self-organizing Map

SOM is a kind of clustering algorithm from the simulation of human brain. When the neural network receives external input signal, it will be divided into different corresponding regions, each region has different response characteristics to the input signal, and the process is completed automatically. Typical characteristic of SOM networks have 2-D grid, one is the input layer and the other is competition layer, which can capture the topology of the gene expression data in high dimension. Figure 3 is a simplified form of SOM structure.

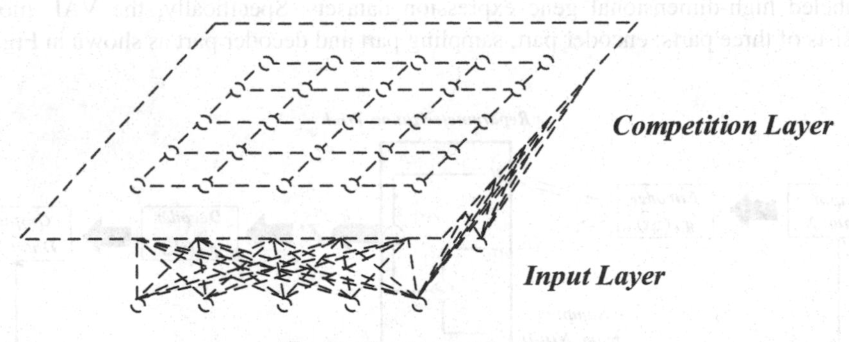

Fig. 3. Structure of the SOM

The purpose of SOM algorithm is to adaptively implement the transformation in topology and orderly manner, and turn the high-dimensional input signal pattern into 2-dimension discrete map. In other words, it is grouping the input vectors which were associated with the winner neurons on the grid.

2.3 VAE-SOM

Although SOM is an effective tool for mapping data into 2-D space, not suit for dealing with the gene expression data in high-dimensional and high-complexity. The purpose of clustering gene expression data is not only grouped genes with similar behaviors, but also revealed the potential expression patterns of different gene interactions. Hence, this paper presented a new clustering-framework based on the original idea of SOM. With VAE-SOM, it could extract some useful features from gene expression data more efficiently. In this work, the data size was further compressed by the low-dimensional manifold representation obtained from the high-dimensional data. It not only improved the clustering accuracy of SOM algorithm, but also reduced the noise effect in gene expression data.

First of all, combine them according to the similar characteristics of models VAE and SOM, which are both from the simulation of the neural connection mode of human brain. Secondly, the output layer of VAE model and the input layer of SOM model are connected to form a bridge between the two models. Finally, both the encoder and decoder in VAE model and the competition layer in SOM model become the hidden layer in the proposed model, which is the deep clustering-framework proposed in this paper (Fig. 1).

This paper took into account the complexity of the gene expression profile dataset, the encoder network used three dense layers, and the decoder network used two dense layers. The number of specific neurons in each layer was determined by the size of the dataset. The connection between layers was activated by batch normalization and rectification linear units (ReLU). The learning rate was set here to 0.0005.

The pseudocode of VAE-SOM for clustering is shown in Algorithm 1

Algorithm 1 VAE-SOM

1:**Iput:Dataset** $x_i \in \{x_1, x_2, \cdots, x_n\}$

2:**Initialization:random initialized** $\theta_0, \phi_0, \ y', w', N_{j^*}(0)$

3:**Output:** *Cluster* C_{best}

4: **repeat**

5: Sample x_i in the minibatch

6: Encoder: $q_\theta(z_i \mid x_i)$

7: Sampling: $z_i = \mu_i + \varepsilon \odot \sigma_i, \varepsilon \ \ N(0, I)$

8: Decoder: $p_\phi(x_i \mid z_i)$

9: Computer loss: $L^v_{\theta,\phi} = D_{KL}(q_\theta(z \mid x_i) \ \ p_\phi(z)) - E_{q_\theta(z|x_i)}\left[\log(p_\phi(x_i \mid z))\right]$

10: Back-propagate the gradients

12: **until** maximum iteration reached

12: **while** $\alpha(t) < \alpha(t)_{min}$ **do**

13: Using Euclidean distance to search winner neuron form y' and w_{j*}'.

14: Build winner neighborhood: $N_{j^*}(t)$

15: Adjusting the output neuron j^* and the weight of its neighboring neurons:

$$w_{ij}(t+1) = w_{ij}(t) + \alpha(t)h(d,t)\left[y_i' - w_{ij}(t)\right]$$

16: **end while**

17: Computing the output value: C_{best}

.

3 Experiments

Several experiments are executed to verify the performance of the proposed clustering-framework on various datasets. Our approach is used to compare with others clustering methods such as Order-preserving Submatrix Algorithm (OPSM) [22], Cheng and Church's algorithm (CC) [23], QUBIC [24], Flexible Overlapped bi-clustering Algorithm (FLOC) [21], BIMAX [25].

3.1 Datasets

The algorithm of VAE-SOM has been applied in four datasets: cdc_15 (D1), elutriation (D2), 1mM_menadione (D3) and constant_32nM_H2O2 (D4). The datasets respectively contain 1086 genes and 24 conditions, 935 genes and 14 conditions, 1050 genes and 9 conditions, 976 genes and 10 conditions, which is obtained from supplementary information provided in [21]. The details are shown in Table 1.

Table 1. The datasets used in experiments

Dataset name	# of gene	# of condition
cdc_15	1086	24
elutriation	935	14
1mM_meandione	1050	9
constant_32nM_H2O2	976	10

3.2 Results and Analysis

Before clustering with SOM algorithm, VAE model is used to reduce the original dimension or extract potential feature patterns of data. We use gene heat-mapping to show the similarity between the processed and the original datasets. The dimensionality reduction results can be displayed clearly in Fig. 4 through the comparison of before and after the processing by the algorithm VAE. The columns of heat-mapping which shown in Fig. 4 represents the number of conditions, meanwhile, the number of rows can be reduced from the number of genes in different datasets (~1000) to 500 uniformly.

Under this process, the clustering framework produce 36 clusters totally. First, VAE algorithm is used to reduce the dimension of high-dimensional data. Then, we use SOM algorithm to cluster and form a total of 36 clustering centers. All data points in the dataset after dimensionality reduction will be mapped to clustering neighborhoods with the smallest Euclidean distance, which are centered on winner neurons and generate 36 clusters. For example, two of them were shown in Fig. 5, where the X-axis represents the conditions and the Y-axis represents the value of gene expression.

3.3 Comparison with Traditional Methods

Several classical clustering algorithms, including FLOC, CC, QUBIC and BIMAX, were selected to evaluate with VAE-SOM algorithm. Two criteria, Average Correlation Value (ACV) and Mean Squared Residue (MSR) [25], were introduced to confirm the degree of correlation between genes in the cluster. The values of clusters were shown in Table 2.

The values of ACV can be in the interval [0, 1], where a value equal to 1, it indicates that the genes are similar. Contrary to ACV, the lower the value of MSR (→0), the stronger the coherence of genes in the cluster. In the others words, when a cluster with ACV value tending to 1 and MSR value tending to 0, that means all of the genes are more similar and coherent in describing interactions. Therefore, the two standards mentioned above can be used not only to estimate the coherence of a cluster, but also to evaluate the performance of different clustering algorithms.

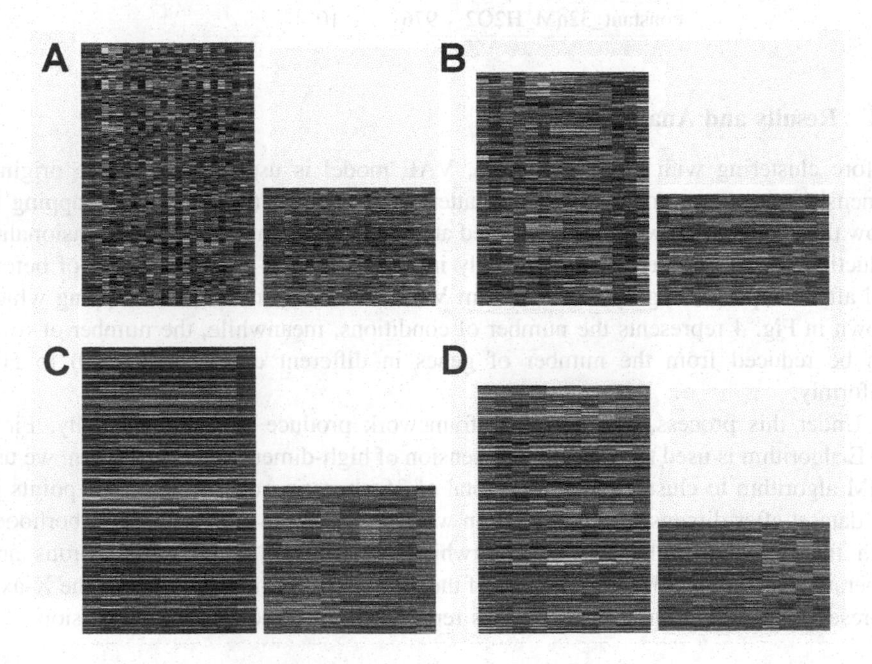

Fig. 4. Dimensionality reduction by the VAE Algorithm from four datasets, D1 (A), D2 (B), D3 (C) and D4 (D); the left side is the heat map before dimension reduction, and the right side is the heat map after dimension reduction in each groups.

From the comparison of the algorithms, we can conclude that the VAE-SOM algorithm has no obvious advantages in the value of ACV, but the value of MSR has the best performance. Excellent MSR values mean that the stronger the coherence by the clusters, and the genes under experimental conditions are fluctuating in exactly the same way. The results show that the VAE-SOM model of deep clustering-framework proposed in this paper had stronger accuracy in capturing shifting tendencies within the gene expression data.

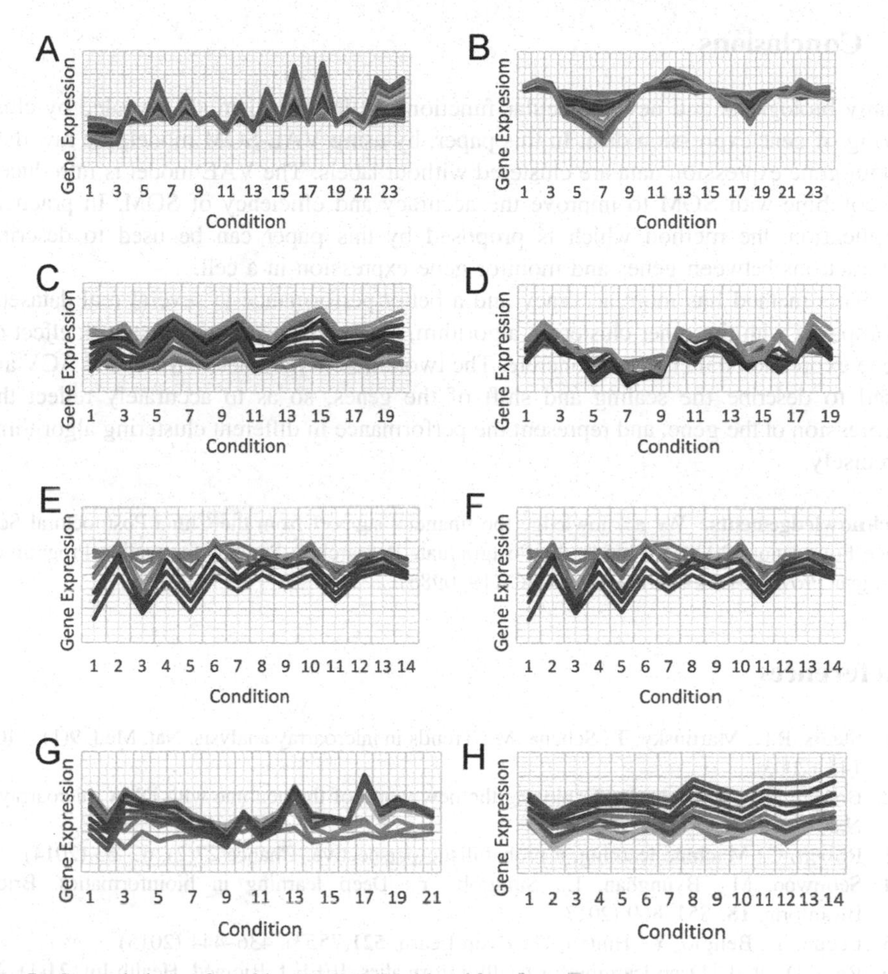

Fig. 5. Several cluster found by VAE-SOM Algorithm from Four datasets. D1 (A and B), D2(C and D), D3 (E and F), D4 (G and H).

Table 2. Comparison of clustering performance of different algorithms in four datasets

	D1		D2		D3		D4	
	ACV	MSR	ACV	MSR	ACV	MSR	ACV	MSR
VAE-SOM	0.82	0.005	0.82	0.009	0.87	0.004	0.81	0.01
OPSM	0.77	0.05	0.75	0.03	0.88	0.03	0.73	0.11
CC	0.31	0.32	0.46	0.24	0.54	0.18	0.48	0.47
QUBIC	0.71	0.41	0.46	0.24	0.67	0.27	0.71	0.64
FLOC	0.44	0.31	0.56	0.26	0.49	0.21	0.55	0.51
BIMAX	0.74	0.033	0.84	0.04	0.76	0.08	0.55	0.17

4 Conclusions

Many biologicals and new molecular functions of the organism are revealed by clustering of gene expression data. In this paper, by using VAE-SOM model, the raw data about gene expression data are clustered without labels. The VAE model is introduced to combine with SOM to improve the accuracy and efficiency of SOM. In practical application, the method which is proposed by this paper can be used to describe interactions between genes and monitor gene expression in a cell.

This method has more accuracy and a better performance in several real datasets. Compared with the other clustering algorithm, the deep clustering-framework effect of gene expression data is more concrete. The two different parameters MSR and ACV are used to describe the scaling and shift of the genes, so as to accurately reflect the expression of the gene, and represent the performance in different clustering algorithms precisely.

Acknowledgements. We acknowledge the financial support from the China Postdoctoral Science Foundation (2020M671554), the Postgraduate Research & Practice Innovation Program of Jiangsu Province (KYCX18_0921, KYCX19_0985).

References

1. Stears, R.L., Martinsky, T., Schena, M.: Trends in microarray analysis. Nat. Med. **9**(1), 140–145 (2003)
2. Brown, P.O., Botstein, D.: Exploring the new world of the genome with DNA microarrays. Nat. Genet. **21**, 33–37 (1999)
3. Robert, C.: Machine learning, a probabilistic perspective. Chance **27**(2), 62–63 (2014)
4. Seonwoo, M., Byunghan, L., Sungroh, Y.: Deep learning in bioinformatics. Brief. Bioinform. **18**, 851–869 (2017)
5. Lecun, Y., Bengio, Y., Hinton, G.: Deep Learn. **521**(7553), 436–444 (2015)
6. Ravi, D., et al.: Deep learning for health informatics. IEEE J. Biomed. Health Inf. **21**(1), 4–21 (2017)
7. Zhang, Q., Yang, L.T., Chen, Z., Li, P.: A survey on deep learning for big data. Inf. Fusion **42**, 146–157 (2018)
8. Angermueller, C., Parnamaa, T., Parts, L., Stegle, O.: Deep learning for computational biology. Mol. Syst. Biol. **12**(7), 878 (2016)
9. Bengio, Y., Courville, A., Vincent, P.: Representation learning: a review and new perspectives. IEEE Trans. Pattern Anal. Mach. Intell. **35**(8), 1798–1828 (2013)
10. Hinton, G.E., Zemel, R.S.: Autoencoders, minimum description length and Helmholtz free energy. In: Advances in neural information processing systems 6 (1994)
11. Hinton, G.E., Salakhutdinov, R.R.: Reducing the dimensionality of data with neural networks. Science **313**(5786), 504–507 (2006)
12. Weng, R., Lu, J., Tan, Y.P., Zhou, J.: Learning cascaded deep auto-encoder networks for face alignment. IEEE Trans. Multi. **18**(10), 1 (2016)
13. Vincent, P., Larochelle, H., Bengio, Y., Manzagol, P.A.: Extracting and composing robust features with denoising autoencoders. In: Proceedings of the 25th International Conference on Machine learning, pp. 1096–1103 (2008)

14. Kingma, D.P., Welling, M.: Auto-encoding variational Bayes. In: Conference Proceedings: Papers Accepted to the International Conference on Learning Representations (ICLR) Ithaca, NY: arXiv.org. (2014)
15. Doersch C.: Tutorial on variational autoencoders (2016)
16. Vesanto, J., Alhoniemi, E.: Clustering of the self-organizing map. IEEE Trans. Neural Netw. **11**(3), 586–600 (2000)
17. Zou, X., Sun, H.: Clustering analysis of micro-array data based on the SOM algorithm. In: Proceedings of the 2013 Ninth International Conference on Computational Intelligence and Security, pp. 308–312. IEEE (2013)
18. Kohonen T.: The self-organizing map. Neurocomputing 21(1/2/3), 1–6 (1990)
19. Kaski, S., Peltonen, J.: Dimensionality reduction for data visualization [applications corner]. IEEE Sig. Process. Mag. **28**(2), 100–104 (2011)
20. Hu, Q., Greene, C.S.: Parameter tuning is a key part of dimensionality reduction via deep variational autoencoders for single cell RNA transcriptomics. Pac. Symp. Biocomput. **4**, 362–373 (2019)
21. Jaskowiak, P.A., Campello, R., Costa, I.G.: Proximity measures for clustering gene expression microarray data: a validation methodology and a comparative analysis. IEEE/ACM Trans. Comput. Biol. Bioinf. **10**(4), 845–857 (2013)
22. Ben-Dor, B., Chor, R., Karp, Y.Z.: Discovering local structure in gene expression data: the order-preserving submatrix problem. In: Proceedings of the Sixth Annual International Conference on Computational Biology (RECOMB 2002), pp. 49–57 (2002)
23. Yang, J., Wang, H., Wang, W., Yu, P.: An improved biclustering method for analyzing gene expression profiles. Int. J. Artif. Intell. Tools **14**(05), 771–789 (2005)
24. Amela, P., Bleuler, S., Zimmermann, P., Wille, A., Zitzler, E.: A systematic comparison and evaluation of biclustering methods for gene expression data. Bioinformatics **22**(9), 1122–1129 (2006)
25. Saber, H., Elloumi, M.: A new study on biclustering tools, biclusters validation and evaluation functions. Int. J. Comput. Sci. Eng. Surv. **6**(1), 01–13 (2015)

Tumor Gene Selection and Prediction via Supervised Correlation Analysis Based F-Score Method

Jia-Jun Cheng[1,2] and Bo Li[1,2(✉)]

[1] School of Computer Science of Technology,
Wuhan University of Science of Technology, Wuhan 430065, Hubei, China
liberol@126.com
[2] Hubei Province Key Laboratory of Intelligent Information Processing
and Real-Time Industrial System, Wuhan 430065, Hubei, China

Abstract. It is imperative to select key genes from tumor gene expressive data for genotype prediction. In this paper, a supervised gene selection method is proposed. Firstly, all the genes are sorted as their descending F-scores, and then supervised correlation analysis is also recommended to reduce the redundancy from those selected genes. At last SVM is introduced to classify those gene subsets. Some experiments are conducted on benchmark tumor gene expressive data sets and results show the performance of the proposed method.

Keywords: Supervised correlation analysis · F-score · Gene selection · Gene microarray data

1 Introduction

Recent advances on microarray techniques bring insights on some particular cancers by making classification to their gene expressive data. Presently, a variety of methods and mathematical models have been presented to manage and interpret these high-density microarray data for gene genotype prediction, which makes it impossible for computer aided cancer diagnose [1].

Generally speaking, all these algorithms adopted for cancer prediction can be categorized into two kinds, i.e. tumor feature extraction and feature selection. Due to the fact that tumor microarray data are represented to a large number of variables (genes), some tumor feature extraction methods are often put forward, where the conventional linear models as linear discriminant analysis (LDA) [2], principal component analysis (PCA) [3], and independent component analysis (ICA) [4] are involved in. By making attempts on extracting features from gene microarray data with nonlinear techniques, kernel approaches [5] and nonlinear manifold learning models [6, 7] are also constructed, they also make their contributions on the final tumor gene expressive data genre prediction. Most of these methods show their performances at low computational expense.

Another kind aims to explore some key tumor pathogenic genes, on which the final cancer diagnosis can be determined and some treatments will be accurately offered.

D.-S. Huang and K.-H. Jo (Eds.): ICIC 2020, LNCS 12464, pp. 14–20, 2020.
https://doi.org/10.1007/978-3-030-60802-6_2

As a result, feature selection algorithms based on tumor gene expressive data have been paid many concentrations on. Tumor gene selection, also named feature subset selection, expects to find those key genes with high relations to the corresponding tumors, which can make the learning models simpler with high generalization.

In order to evaluate the performance of those selected tumor genes, some classifiers should be adopted to combine to them, which is also called the wrapper. Under such circumstances, the wrapper methods with support vector machine were developed for binary classification [8]. Furthermore, SVM-REF was also proposed to be extended to multi-class classification [9]. However, these wrapper methods can easily incur high computational expense. Another shortcoming in the wrappers is that the selection of gene subset is heavily depended on all kinds of classifiers. In other words, the performances on the same optimal gene subsets may be different by using varied classifiers.

Unlike the traditional wrapper algorithms, we will just focus on key tumor gene selection in this paper, where a new tumor gene selection approach is put forward. Firstly, an evaluation metric as F-score is introduced to sort the genes with the descending order, from which gene subsets with high scores can be chosen as the candidates. And then, the supervised correlation analysis is also recommended to reduce the redundancy existing in the candidates, by which those key genes with low correlation will be approached. At last, SVM is taken to predict the genres of the selected gene subsets.

2 Method

In this section, we will firstly introduce F-score metric, which is taken to weight the contributions of the i−th gene for discriminant learning. In order to reduce the redundancy existing in the selected subsets with F-score, the supervised correlation analysis is also described in details, after which the outline of the proposed method is concluded.

2.1 F-Score

F-score is always used to characterize the discriminative power of selected features for binary classification. For the training samples of the tumor gene expressive data points $X = [X_1, X_2, \ldots, X_n] \in \mathbb{R}^{D \times n}$, there exists n_+ positive samples and n_- negative samples, where $n_+ + n_- = n$. Thus the F - score metric for the i−th gene can be formulated to the following:

$$F_{\mathrm{i}} \text{ - score} = \frac{\left(\overline{X_i^+} - \overline{X_i}\right)^2 + \left(\overline{X_i^-} - \overline{X_i}\right)^2}{\frac{1}{n_+ - 1}\sum_{k=1}^{n_+}\left(X_{ki}^+ - X_i^+\right)^2 + \frac{1}{n_- - 1}\sum_{k=1}^{n_-}\left(X_{ki}^- - X_i^-\right)^2} \tag{1}$$

where $\overline{X_i}$, $\overline{X_i^+}$ and $\overline{X_i^-}$ denote the mean of the $i-$th gene on all the training samples, the mean of the $i-$th gene on the positive samples and the mean of the $i-$th gene on the negative samples, respectively. Moreover, X_{ki}^+ and X_{ki}^- are the corresponding values of the $i-$th gene on the k positive sample and the $i-$th gene on the k negative sample.

2.2 Supervised Correlation Analysis

Correlation analysis is often used to measure the correlations among objects. For tumor microarray data composed of a large number of genes, some of them may be of high correlations, which shows negative effects on the final prediction of tumor gene genres from the following aspects. Firstly, those genes with high correlations will result in high computational burden when making prediction because more genes are involved in. Secondly, it will be difficult in selecting key genes from those high correlated ones. Consequently, the final accuracy will also be affected. On the contrary, the gene subsets after reducing the redundancy will be more helpful for tumor gene expressive data classification. Thus, how to remove the redundant genes from the selected subsets is desired to improve the final accuracy with low computational expense. So in this subsection, a supervised correlation analysis model is also constructed for low correlated gene subsets selection.

At first, a correlation matrix will be introduced, each of which is a measurement for the similarity between any two genes f_i and f_j using the correlation coefficient ρ_{ij} defined as follows:

$$\rho_{ij} = \frac{E\big((f_i - E(f_i)) \cdot (f_j - (f_j))\big)}{\sqrt{D(f_i)} \cdot \sqrt{D(f_j)}} \tag{2}$$

where $E(\cdot)$ and $D(\cdot)$ represent the expectation and the variance of features or genes, respectively.

Thus for all the gene expressive data $X = [X_1, X_2, \ldots, X_n]$, there exists a correlation coefficient between any two of them, based on which a correlation matrix can be constructed below:

$$C = \begin{bmatrix} \rho_{11} & \rho_{12} & \cdots & \rho_{1n} \\ \rho_{21} & \rho_{22} & \cdots & \rho_{2n} \\ \cdots & \cdots & \cdots & \cdots \\ \rho_{n1} & \rho_{n2} & \cdots & \rho_{nn} \end{bmatrix} \tag{3}$$

Moreover, it takes the label information of the original tumor gene expressive data into account, where the sample matrix can be divided into A and B with the labels of 1 and -1, characterizing the positive and the negative samples, respectively. Thus according to the labeled sample matrix A and B, we can obtain the corresponding correlation matrixes as C_A and C_B.

Obviously, the correlation matrix is symmetrical, moreover, foe each element on the correlation matrix, its value will be in the interval as [0 1]. At the same time, all the diagonal elements are all with value of 1 because they reflect the correlations of themselves. In general, the larger ρ_{ij} in both the correlation matrixes C_A and C_B, the more correlated between any two genes as f_i and f_j, which also lead to a problem that how to set the coefficient threshold ρ to determine which genes are correlated. In details, those with correlation coefficient are larger than ρ, they will be highly correlated; otherwise, they will be uncorrelated.

2.3 Outline of the Proposed Method

According to the above analysis, the outline of the proposed method will be naturally concluded and displayed in the following Table 1.

Table 1. Outline of the proposed method

Input: Data samples $X=[X_1,X_2,...,X_n]\in\mathbb{R}^{D\times n}$ and their labels $C=[C_1,C_2,...,C_c]$, $C_i=\pm1, i=1,2,...,c$

Output: Selected key tumor gene subset and label prediction

Algorithm:

Step 1 Tumor gene sorting

Sort the tumor gene expressive data X according to the descending F-scores and obtain the sorted matrix X';

Step 2 Supervised correlation analysis

2.1 For the training set, divide the sorted matrix X' into A and B according to their class labels;

2.2 Compute the correlation matrixes C_A and C_B related to A and B, respectively;

2.3 Determine the optimal correlation coefficient threshold ρ using both C_A and C_B, for each gene in the test set, select genes with the correlation coefficient value larger than ρ and remove some of them

Step 3 Key tumor gene selection

Select those remained genes with high F-score as the optimal subset;

Step 4 Genres prediction

Adopt SVM classifier to classify the key tumor gene subset.

3 Experiments

In this section, experiments are conducted on three benchmark tumor gene expressive data sets including Colon, DLBCL and Leukemia. Moreover, performance comparisons are also made by using the proposed method, F-score method, Laplacian Score [10] and Fisher Score [11]. At the same time, SVM is also taken as another comparison method for tumor gene expressive data classification.

In addition, when making supervised correlation analysis on those Colon, DLBCL and Leukemia data, it occurs to the parameter as the correlation coefficient threshold ρ. In the experiment, LOO-CV (Leave One Out-Cross Validation) is also introduced to adjust parameter ρ.

Figure 1 shows the performance curves with different coefficient threshold ρ when making the supervised correlation analysis to the training sets, where the optimal

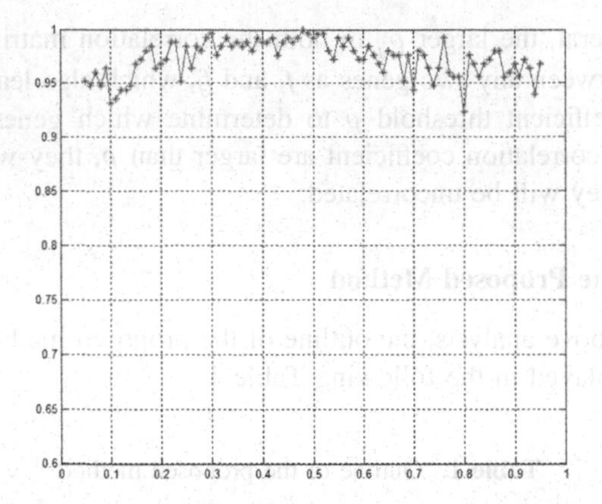

(a) Performance curves with the varied ρ on Colon data

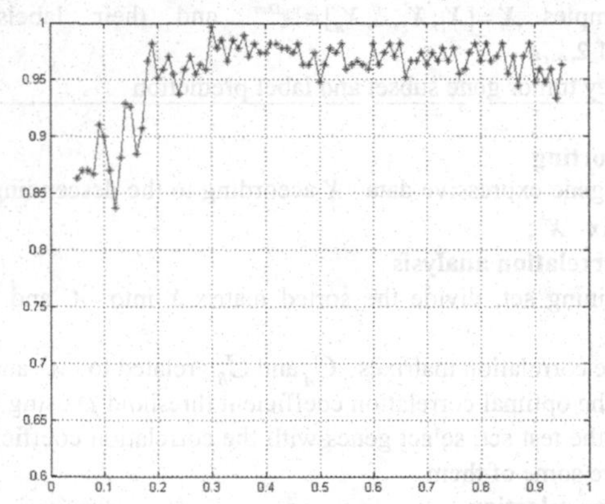

(b) Performance curves with the varied ρ on DLBCL data

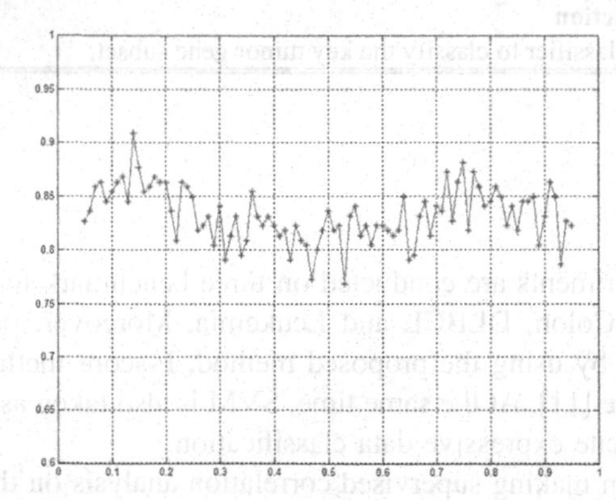

(C) Performance curves with the varied ρ on leukemia data

Fig. 1. LOO-CV performance with varied ρ for Colon, DLBCL and Acute leukemia data sets

parameter ρ will be tuned. From Fig. 1, it can be clearly found that the correlation coefficient ρ is tuned to be 0.48, 0.3 and 0.14 for Colon, DLBCL and Leukemia data set, respectively.

After tuning the coefficient threshold ρ, experiments have been conducted on these three data sets by using the proposed method, Laplacian score, Fisher score, F-score method and SVM method, respectively. The experimental results on these tumor gene expressive data sets are listed in the following Table, where it can find that the proposed method is superior to the other comparison methods Table 2.

Table 2. LOO-CV Performance comparisons on three data sets

Methods	Performance		
	Colon	DLBCL	Leukemia
Fisher score	97.12%	95.72%	90.67%
Laplacian score	96.64%	95.19%	75.11%
SVM	97.24%	95.5%	78.56%
F-score	96.63%	99.6%	89.09%
The proposed method	98.71%	100%	92.91%

4 Conclusions

In order to select the key genes from tumor gene expressive data, a feature selection method is proposed, which can be introduced for tumor genotype prediction. In the proposed method, firstly, F-score is recommended to sort the tumor expressive data with the descending order, from which those with high scores can be selected as the candidate of gene subset. And then supervised correlation analysis is also taken to reduce the redundancy existing in the candidate gene subset and the rest is the optimal gene subset. Finally, SVM is also adopted to predict the genre of them. Experiments on benchmark tumor gene expressive data sets have been conducted with some related comparison tumor gene selection methods, by which the performance of the proposed method can be validated.

Acknowledgments. This work was partly supported by China Post-doctoral Science Foundation (2016M601646).

References

1. Liu, X.H., Cai, C.Z., Yuan, Q.F., Xiao, H.G., Kong, C.Y.: Computer-aided diagnosis of breast cancer based on support vector machine. J. Chongqing Univ. (Nat. Sci. Ed.) **30**(6), 140–144 (2007)
2. Martinez, A.M., Kak, A.C.: PCA versus LDA. IEEE Trans. Pattern Anal. Mach. Intell. **23** (2), 228–233 (2001)
3. Jolliffe, I.T.: Principal component analysis (2002)

4. Huang, D.S., Zheng, C.H.: Independent component analysis based penalized discriminate method for tumor classification using gene expression data. Bioinformatics **22**, 1855–1862 (2006)
5. Ha, V.S., Nguyen, H.N.: Machine Learning and Data Mining in Pattern Recognition (2016)
6. Li, B., Tian, B.B., Zhang, X.L., Zhang, X.P.: Locally linear representation fisher criterion based tumor gene expressive data classification. Comput. Biol. Med. **44**(10), 48–54 (2014)
7. Pillati, M., Viroli, C.: Supervised locally linear embedding for classification: an application to gene expression data analysis. In: 29thAnnual Conference of the of the German Classification Society (GfKl 2005), pp. 15–18 (2005)
8. Juan, M.G.G., Juan, G.S., Pablo, E.M., Elies, F.G., Emilio, S.O.: Sparse manifold clustering and embedding to discriminant gene expression profiles of glioblastoma and meningioma tumors. Comput. Biol. Med. **43**(11), 1863–1869 (2013)
9. Mundra, P.A., Rajapakse, J.C.: SVM-RFE with MRMR filterfor gene selection. IEEE Trans. Nanobiosci. **9**(1), 31–37 (2010)
10. He, X., Cai, D., Niyogi, P.: Laplacian score for feature selection. In: Advances in Neural Information Processing Systems, pp. 507–514(2005)
11. Sun, L., Zhang, X.-Y., Qian, Y.-H., Xu, J.-C., Zhang, S.-G., Tian, Y.: Joint neighborhood entropy-based gene selection method with fisher score for tumor classification. Appl. Intell. **49**(4), 1245–1259 (2018). https://doi.org/10.1007/s10489-018-1320-1

A Machine Learning Based Method to Identify Differentially Expressed Genes

Bolin Chen[1,2], Li Gao[1,3], and Xuequn Shang[1,2(✉)]

[1] School of Computer Science,
Northwestern Polytechnical University, Xi'an, China
npu_bioinf@hotmail.com

[2] Key Laboratory of Big Data Storage and Management,
Ministry of Industry and Information Technology,
Northwestern Polytechnical University, Xi'an, China

[3] School of Software, Northwestern Polytechnical University, Xi'an, China

Abstract. Detecting differentially expressed genes (DEGs) under two biological conditions is an essential step and is one of the most common reasons for statistical analysis of Microarray and RNA-seq data. There are various methods developed to detect DEGs either originate from a sophisticated statistical model based on fold-change (FC) strategy or from an analysis of biological reasoning. In this paper, we present a machine learning based method called Fusion for identifying DEGs based on an ensemble strategy, it provides a straightforward stringent way to determine the significance level for each gene. We use the Fusion technique on two biological datasets, the results show that in each case it performs more reliably and consistently than the *Limma* as well as other methods. A validation on the Platinum Spike dataset indicates that the proposed approach is more reliable with high confidence in identifying DEGs. An analysis of the biological function of the identified genes illustrates that the designed ensemble technique is powerful for identifying biologically relevant expression changes.

Keywords: Gene expression · Differentially expressed genes · Ensemble strategy

1 Introduction

One of the main objectives in the analysis of experimental data is the identification of genes that are differentially expressed under two biological conditions. The identification of differentially expressed genes (DEGs) can be regarded as a feature selection problem that detects most significant genes, which can help us to study disease-related, cell-specific gene expression patterns [1].

The earliest approach uses a simple *fold-change* (FC) [2] criterion to detect genes of interest, but this has the obvious disadvantage that it does not provide a significance estimate for the observed changes and that the necessary *cutoff* values are essentially arbitrary [3]. Hence, replicated experiments and increasingly sophisticated statistical tests were soon suggested to achieve a more reliable identification of differentially

D.-S. Huang and K.-H. Jo (Eds.): ICIC 2020, LNCS 12464, pp. 21–31, 2020.
https://doi.org/10.1007/978-3-030-60802-6_3

regulated genes. T-test [4, 5] aims to calculate a *t-statistic* for each gene to measure the difference in gene expression under two distinct conditions, and then decide the significance level of the expression difference by calculating the p-value according to the *t*-distribution. The lower the *p-value*, the more significant the difference in expression of the gene. However, the disadvantage of this method is that genes with low expression are more susceptible to errors due to noise [6]. To be more specific, for genes with low expression levels, a small change may result in a large absolute *t-statistic* value, which can be identified as the DEGs by T-test, but in fact the average expression difference of the gene varies little between two conditions [7]. Few attempts have been made to improve upon the popular T-test. One development is the realization that variance shrinkage improves statistical power. The most popularly used *Limma* [8] integrated the linear models with empirical Bayes method, aiming to shrinkage of the estimated sample variances for more stable inference and obtain a *p-value* for each gene and choose a feasible false discovery rate (FDR) by using moderated *F*-statistics of *t* inferential approach.

Although the above methods combine sample variance with the availability of *p-value* in cope with the high level of noise of dataset, they ignore the interactions between genes and obtain a set of isolated genes in a biomolecular network, which is hard to analyze those genes' functions from a system biology point of view [9]. Thus, there are some methods were proposed to identify DEGs by taking the correlations of genes into consideration to reduce the effects of uneven distribution of the dataset instead of single genes. Min-Edge [10] detects DEGs through the interactions between genes in biomolecular networks, which attempts to facilitate the problem of biological interpretation and also increase statistical power, it has been confirmed as a useful for finding key genes related to disease. Characteristic Direction (Chdir) [11] is proposed to identify DEGs geometrically, which incorporates a regularization scheme to maximize the use of dimensional information in expression data by taking into account the variances and the correlations of the genes in a shared manner. The Chdir approach also provides an intuitive visualization of the differential expression.

In this paper, we present a machine learning based method for identifying DEGs based on an ensemble strategy, it integrates the differential characteristics of genes and perform feature learning through a random forest model. By the feature learning, the significance of the gene is analyzed according to the probability score obtained through the algorithm. The overall framework of the proposed Fusion method is illustrated in Fig. 1, where the expression data includes *n* genes with *s* samples. In the input process, the color of each row indicates the expression value of each gene in different samples. The color in feature extraction corresponds to the differential score of each gene detected by each method.

The rest of the paper is organized as follows. Section 2 describes the data sources and methods proposed in this paper. Section 3 addresses the validation results and discussions. Section 4 draws some conclusions.

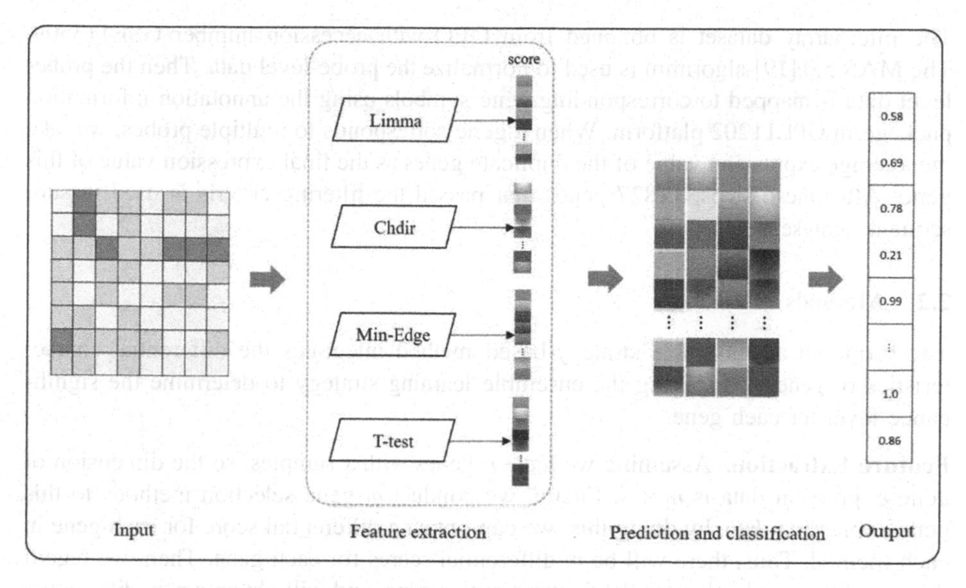

Fig. 1 The overall framework of the Fusion method. In the input process, the color of each row indicates the expression value of each gene in different samples. The color in prediction and classification process corresponds to the differential score of each gene detected by each method.

2 Materials and Method

2.1 Materials

Dataset 1. The Platinum Spike dataset is downloaded from the GEO website (accession: GSE21344) [12] consists of 18 spike-in samples (9 controls versus 9 tests). The designated FC associated file [13] contains 18952 probes, among which 1940 are known as differentially expressed probes, this ground-truth information is used for the comparison of performance of proposed method with other gene selection methods. The probe-level data is analyzed using the robust multi-array average (RMA) [14] algorithm for the determination of the chip quality, including intensity value background correction, log2 transformation, and quantile normalization [15, 16], etc. The results of interest are confirmed through the robust *R* language *affy* Bioconductor package (https://www.bioconductor.org/). By data cleaning steps, we obtained 12924 genes, among which 1690 genes are known true DEGs, which will help us to validate the sensitivity of the proposed method and also can evaluate the performance of each method.

Dataset 2. The second experiment data is generated from mouse blood cells following either x-ray or neutron irradiation, using to investigate whether these post-irradiation data can distinguish the neutron component after a mixed field x-ray/neutron exposure [17]. Mice in the experiment are either sham-irradiated or exposed to 3 Gy of mixed neutron/x-ray exposures that contained 5, 15, and 25% of neutron component, and are also exposed to 3 Gy x-rays or 0.75 Gy of neutron irradiation [18] for comparison.

The microarray dataset is obtained from GEO with accession number GSE113509. The MAS 5.0 [19] algorithm is used to normalize the probe-level data. Then the probe-level data is mapped to corresponding gene symbols using the annotation information package in GPL11202 platform. When a gene corresponds to multiple probes, we take the average expression value of the duplicate genes as the final expression value of this gene. After these steps, 11827 genes that passed the filtering criteria is used in subsequent analyses.

2.2 Methods

The proposed an ensemble strategy-based method integrates the differential characteristics of genes, combining the ensemble learning strategy to determine the significance level for each gene.

Feature Extraction. Assuming we have n genes with s samples, so the dimension of gene expression data is $n \times s$. Firstly, we conduct m gene selection methods to this gene expression data, by doing this, we can obtain a differential score for each gene in each method. Thus, there will be m differential scores for each gene. Then, we regard these m differential values as the features of this gene, and will obtain a new differential expression feature matrix with $n \times m$.

In this paper, we integrate the differential values calculated by four algorithms (Limma, Chdir, Min-Edge, T-test) as the features of each gene. We believe that these four methods separately measure gene expression characteristics in a highly integrated manner from different perspectives. Specifically, Limma is currently a widely used method, which improves the accuracy of differential genes from a linear model and Bayesian perspective. Chdir considers the expression level of gene expression in different dimensions. T-test measures the statistical differences of genes under different experimental conditions from a statistical perspective. Min-Edge starts from the interactions between genes, taking into account the differential level of the gene expression. We choose these four algorithms as the characteristics to improve the accuracy of obtained genes.

Positive Samples and Negative Samples. We firstly conduct the gene expression data to gene selection methods, such as Limma, T-test, Min-Edge, Chdir. After that, each method will obtain a gene list with its differential score. For the labeled data, we regard the truth DEGs as the positive samples and the truth none DEGs as the negative samples. For the unlabeled data, we choose the top 100 common genes of these four methods as positive samples and the last 100 common genes as negative samples.

Model Selection and Introduction. After splitting the positive and negative samples into train data and test data for training and evaluating the model. We take several machine learning methods into account, and the random forest algorithm was chosen for further analyses after 10-fold cross-validation due to its high accuracy. The random forest builds multiple decision trees and merges them together to get a more accurate and stable prediction [20]. It can be used for both classification and regression problems, which form the majority of current machine learning systems. In our method, we use it to measure the feature matrix, and predict the differential expression level by generating a probability of each gene. By this feature learning process, the significance of genes is analyzed according to the probability score obtained through the algorithm.

3 Results

Firstly, we integrate the differential values calculated by four algorithms (Limma, T-test, Min-Edge, Chdir) as the features of each gene, then we conduct the random forest method to these features to predict the probability of the genes. Finally, we compare the proposed Fusion method with those four existing gene selection methods in terms of the performance for differential expression [21].

3.1 The Results of Dataset 1

The Number of Identified DEGs. By conducting five methods on gene sets, the Limma method obtains 739 genes with p-value $\leq$ 0.05, FC $\geq$ 2 or FC $\leq$ 0.5, while 1676 DEGs are detected by Min-Edge with the same cutoffs. 1610 DEGs are detected by T-test with *p-value* $\leq$ 0.01, Chdir obtains 1695 DEGs with $b \geq 0.006$, and 1500 DEGs are obtained by our method while the probability over 0.6. Table 1 gives the number of DEGs common to each pair of methods. The diagonal elements represent the number of DEGs obtained by the method. We can conclude that the DEGs detected by Limma are contained in Fusion, Min-Edge and T-test, while there are only 464 genes in common with Chdir. The DEGs detected by Fusion method contains the most of DEGs detected by Limma, Chdir, Min-Edge and T-test methods, whereas the Chdir has few common genes with other methods.

Table 1. The number of genes common to each pair of methods.

Method	Fusion	Limma	Chdir	Min-Edge	T-test
Fusion	**1500**	739	730	1476	1369
Limma		**739**	464	739	739
Chdir			**1695**	722	758
Min-Edge				**1676**	1409
T-test					**1610**

Validation of Methods on Platinum Spike Dataset. We computed true positive rates (TPRs) as the number of true DEGs, true positive over the 1690 ground-truth DEGs, and the number of false positive genes (FPRs) over the 11234 genes that are not differentially expressed, over all of the 12924 genes. The receiver operating characteristic (ROC) curves are calculated based on the corresponding difference value obtained by each method. The ROC curves of the different methods are shown in Fig. 2 together with the corresponding AUC values. In this comparison, Fusion (AUC 0.968129) method produced the highest AUC values, followed by Min-Edge (AUC 0.943169) and Limma (AUC 0.942052). The T-test performed worse in this comparison (AUC 0.939404), while Chdir showed the overall lowest performance (AUC 0.788339). The ROC curves confirm that the proposed method is the best method over the full range of number of genes selected. The machine learning based method also is proved to be the superior method for analyzing high throughput data when the magnitude of differential expression is taken into account.

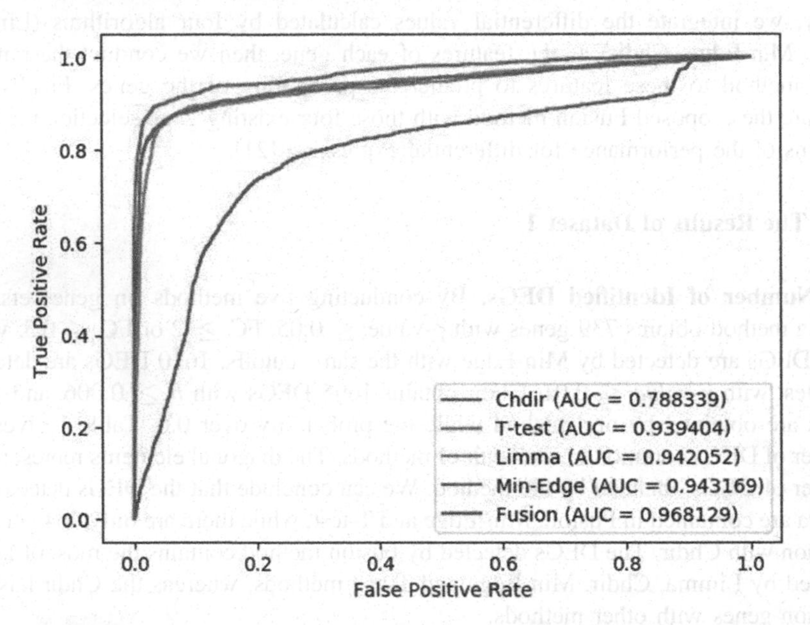

Fig. 2 ROC curves of each gene selection method on the dataset of Platinum Spike The red, green, purple, orange and blue lines depict the ROC curves of Fusion, Min-Edge, Limma, T-test and Chdir, respectively. The AUC value of Fusion is 0.968129, which is higher than that of the other four algorithms. (Color figure online)

3.2 The Results of Dataset 2

The Top Ranked Genes. Table 2 and Table 3 shows the top 15 genes from the GSE113509 with 5% neutron exposure and 25% neutron exposure, respectively. The numerical data in the table represents the significant level of the genes obtained by each method, and genes in table are ranked by Fusion method. The table includes the Chdir, the Min-Edge, the T-test and the Limma method for each gene. The Limma and Min-Edge give the absolute value of log (FC) as a result, Chdir gives the absolute value of *b*, the bigger the absolute of *b*, the more significant the gene. The result of T-test is shown with *p-value*, the smaller the *p-value*, the more significant the gene. Fusion method gives the probability for each gene, higher value represents the gene tends to be differentially expressed.

Generally speaking, top ranked genes have high confidence in analyzing the functions, the pathways enriched by them should play important roles in cellular functions. Most of the ordering of genes by Min-Edge and Limma method is the same as the result of Fusion method. For example, HAMP ranked third in Fusion, while ordered eighth in Limma and third in Min-Edge. From the confidence of genes detected by each method, we found that Fusion method is more powerful than other methods. Similarly, when the mice are exposed to 25% neutron component in Table 3, Fusion performs well than other methods in ranking genes, and Min-Edge performs second while Limma performs third.

Table 2. Top 15 genes from the GSE113509 with 5% neutron exposure.

Name	Limma	Chdir	Min-Edge	T-test	Fusion
FCER2A	7.673	0.0004	5.882	0.058	0.978
NRIP3	7.576	0.0002	5.862	0.0724	0.961
HAMP	5.613	0.039	5.852	0.024	0.956
CHST3	6.099	6.8e−06	3.578	0.053	0.949
IGHV14-2	5.921	5.0e−06	5.352	0.054	0.942
ATP2B4	5.612	0.0001	5.240	0.054	0.940
GNG13	6.422	0.002	4.990	0.049	0.938
SRPK3	5.373	0.004	4.864	0.040	0.935
EBF1	5.324	0.004	4.887	0.043	0.932
IQGAP3	6.322	9.0e−05	4.953	0.051	0.928
TNFRSF13C	5.315	3.0e−05	4.761	0.047	0.919
FAIM3	7.298	1.3e−05	4.745	0.049	0.917
EPDR1	5.132	0.002	4.724	0.066	0.915
IL4I1	6.111	2.4e−05	4.615	0.039	0.910
FCRLA	5.101	2.9e−05	4.625	0.047	0.909

Table 3. Top 15 genes from the GSE113509 with 25% neutron exposure.

Name	Limma	Chdir	Min-Edge	T-test	Fusion
NRIP3	7.505	0.0001	5.689	0.072	0.960
HAMP	6.439	0.0136	6.060	0.0039	0.956
FCER2A	6.450	4.9e−05	5.004	0.058	0.948
ALB	6.008	0.1142	3.237	0.003	0.948
MBL2	5.990	0.0487	5.410	0.009	0.937
CHST3	4.773	4.9e−05	3.364	0.053	0.935
IGHV14-2	5.759	4.3e−05	0.0	0.053	0.924
GNG13	5.913	0.0027	4.345	0.049	0.919
ANKRD36	5.621	0.0031	4.864	0.062	0.914
ATP2B4	6.545	0.0006	5.253	0.054	0.911
APOC3	5.521	0.0983	4.184	0.0028	0.907
TTR	5.712	0.1101	3.795	0.0005	0.905
HAMP2	5.461	0.0179	5.356	0.021	0.903
SRPK3	5.349	0.0089	4.166	0.040	0.897
NACAD	5.648	0.0005	5.161	0.054	0.892

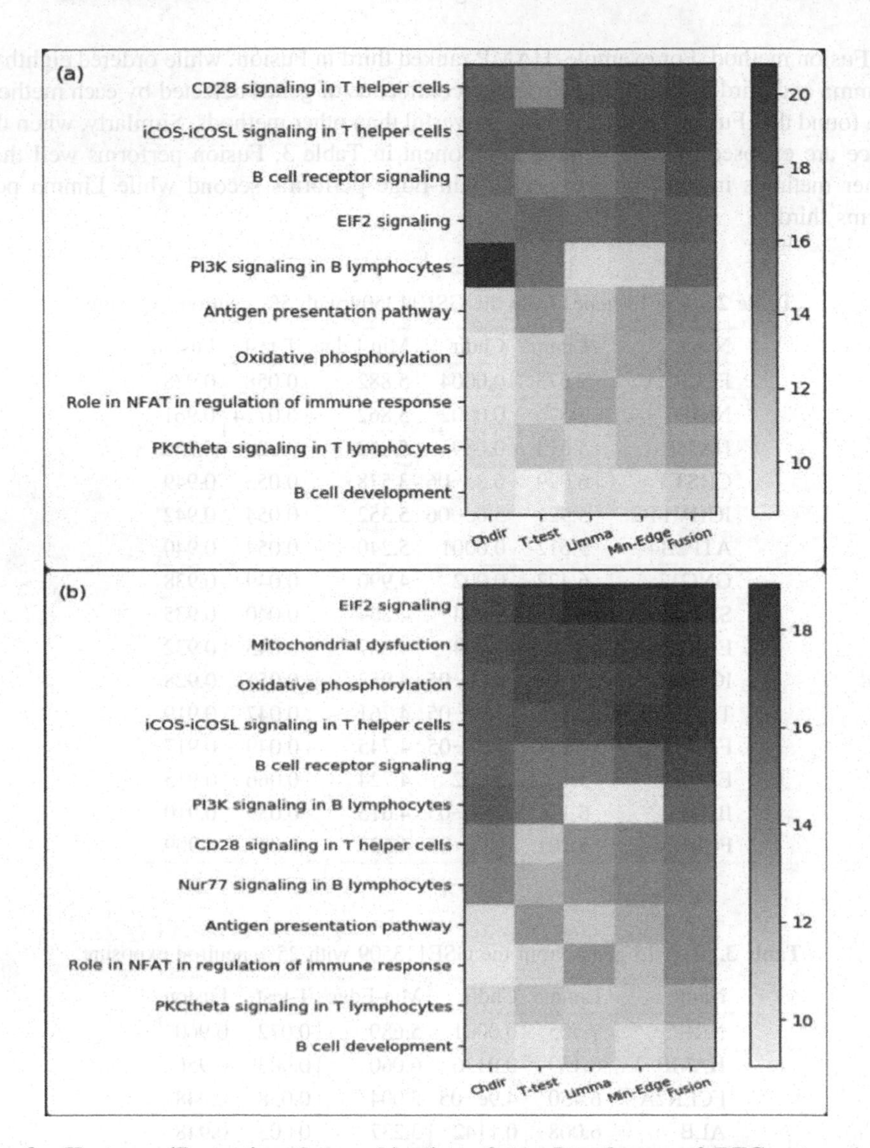

Fig. 3 Heatmap illustrating the combination of top 5 pathways of DEGs at various methods. (a) shows the combination of top 5 pathways enriched by each method, of which the data is exposed to 3 Gy of mixed neutron contained 5% neutron. (b) shows the combination of top 5 pathways enriched by each method, of which the data with 25% neutron exposure. Each row represents one pathway and each column represents a gene selection method. The number in the legend on the right indicates-log10 (p-value), which depicts the larger the number, the lower the *p-value*. Orange indicates low *p-value* of pathway, while white color indicates high *p-value*.

Pathway Enrichment Analysis. By conducting five methods on gene expression data, the Fusion, Min-Edge, Limma, T-test and Chdir methods obtained 2067, 3250, 2625, 2578 and 2486 DEGs, respectively. To further evaluate the biological significance of the DEGs, we used KEGG pathway enrichment to identify the most significantly enriched molecular functions. Biological processes and pathways over-represented among DEGs with a Benjamini-corrected *p-value* of less than 0.05 are considered significant.

The combination of top 5 pathways based on the significance of the enrichment (p-value) are shown in Fig. 3. The most statistically significant pathways identified by each method are ordered by p-value of Fusion method. The p-value of the pathway is displayed in different alpha, which means the darker the color, the smaller the p-value of the pathway. Figure 3 (a) shows the combination of top 5 pathways enriched by each method, of which the data is exposed to 3 Gy of mixed neutron contained 5% neutron. As can be seen in the figure, for most of pathways obtained by Fusion have lower p-value except PI3 K signaling in B lymphocytes pathway. For instance, CD28 signaling in T helper cells has lowest p-value in Fusion than other methods. Similarly, Fig. 3 (b) depicts the combination of top 5 pathways enriched by each method, of which the data with 25% neutron exposure. The most significant pathway in our method is the EIF2 signaling pathway (p-value = 1.16e−20). For each pathway, Fusion performs well than other methods in terms of the significance of pathways.

Furthermore, there are also some pathways significantly represented in both 5% and 25% radiation exposures. To be more specific, canonical pathways pertaining to B- and T-cell immune response are uniformly represented in both radiation exposures. And the EIF2 signaling pathway is significantly enriched in the exposures that included a neutron component. The above analyzing the results confirm that among top pathways significantly represented in the neutron exposures are those related to EIF2 signaling.

4 Conclusion

We have described a novel idea to identify DEGs by fusing the features of genes. We present a machine learning based method (Fusion) for identifying DEGs based on an ensemble strategy, aiming to provide a straightforward stringent way to determine the significance level for each gene. The method starts with known existed methods to calculate a value of each gene, and then integrates these characteristic values as the features of the gene. Then the ensemble learning method random forest is used to perform these features to distinguish the significance of the gene according to its probability.

To validate the feasibility and reliability of the proposed method, we perform two biological datasets on the fusion method. The "Platinum Spike" dataset with ground truth information was used to compare the performance of proposed method with other gene selection methods, and the microarray dataset with accession GSE113509 is used to validate the ability and interpretability of designed method through biology point of view. The results show that in each case the fusion performs more reliably and consistently than other methods. We can conclude from the experiments that the proposed method may shed new light on relevant biological mechanisms that would have remained undiscovered by the current methods. In the future, we will put more experiments on the RNA-seq data to validate the proposed method and will focus on powerful larger samples with biological interpretation in identifying differentially expressed genes.

Acknowledgements. This work was supported by the National Natural Science Foundation of China under Grant Nos. 61972320, 61772426, 61702161, 61702420, 61702421, and 61602386, the Fundamental Research Funds for the Central Universities under Grant No. 3102019DX1003, the Key Research and Development and Promotion Program of Henan Province of China under Grant 182102210213, the Key Research Fund for Higher Education of Henan Province of China under Grant 18A520003, and the Top International University Visiting Program for Outstanding Young Scholars of Northwestern Polytechnical University.

References

1. Ning, L.F., Yu, Y.Q., Guoji, E.T., Kou, C.G., Yu, Q.: Meta-analysis of differentially expressed genes in autism based on gene expression data. Genet. Mol. Res. Gmr **14**(1), 2146–2155 (2015)
2. DeRisi, J., et al.: Use of cDNA microarray to analyse gene expression patterns in human cancer. Nat. Genet. **14**, 457460 (1996)
3. Shaik, J.S., Yeasin, M.: A unified framework for finding differentially expressed genes from microarray experiments. BMC Bioinform. **8**(1), 347 (2007)
4. Goulden, C.H.: Methods of Statistical Analysis, 2nd edn., pp. 50–55. Wiley, New York (1956)
5. Baldi, P., Long, A.D.: A Bayesian framework for the analysis of microarray expression data: regularized t-test and statistical inferences of gene changes. Bioinformatics **17**, 509–519 (2001)
6. Mukherjee, S., Roberts, S.J., Laan, M.J.: Data-adaptive test statistics for microarray data. Bioinformatics **21**(2), 108–114 (2005)
7. Kvam, V.M., Liu, P., Si, Y.: A comparison of statistical methods for detecting differentially expressed genes from RNAseq data. Am. J. Bot. **99**(2), 248–256 (2012)
8. Smyth, G.K.: Linear models and empirical Bayes methods for assessing differential expression in microarray experiments. Stat. Appl. Genet. Mol. Biol. **3**(1), 1–28 (2004)
9. Zhang, Z., Zhang, G., Gao, Z., Li, S., Li, Z., Bi, J., et al.: Comprehensive analysis of differentially expressed genes associated with PLK1 in bladder cancer. BMC Cancer **17**(1), 861 (2017)
10. Chen, B., Gao, L., Shang, X.: Identifying differentially expressed genes based on differentially expressed edges. In: Intelligent Computing Theories and Application. ICIC 2019. Lecture Notes in Computer Science, vol. 11644 (2019). https://doi.org/10.1007/978-3-030-26969-2_10
11. Clark, N.R., et al.: The Characteristic direction: a geometrical approach to identify differentially expressed genes. BMC Bioinformatics **15**(1), 79 (2014)
12. Zhu, Q., Miecznikowski, J.C., Halfon, M.S.: Preferred analysis methods for Affymetrix genechips. ii. an expanded, balanced, wholly-defined spike-in dataset. BMC Bioinform. **11** (1), 285 (2010)
13. Dembele, D., Kastner, P.: Fold change rank ordering statistics: a new method for detecting differentially expressed genes. BMC Bioinform. **15**(1), 14 (2014)

14. Irizarry, R.A., et al.: Exploration, normalization, and summaries of high density oligonucleotide array probe level data. Biostatistics **4**(2), 249–264 (2003)
15. Xiao, Y., Feng, M., Ran, H., Han, X., Li, X.: Identification of key differentially expressed genes associated with nonsmall cell lung cancer by bioinformatics analyses. Mol. Med. Rep. **17**(5), 6379–6386 (2018)
16. Tang, F., et al.: Identification of differentially expressed genes and biological pathways in bladder cancer. Mol. Med. Rep. **17**(5), 6425–6434 (2018)
17. Broustas, C.G., Harken, A.D., Garty, G., Amundson, S.A.: Identification of differentially expressed genes and pathways in mice exposed to mixed field neutron/photon radiation. BMC Genom. **19**, 504 (2018). https://doi.org/10.1186/s12864-018-4884-6
18. Broustas, C.G., Xu, Y., Harken, A.D., Garty, G., Amundson, S.A.: Comparison of gene expression response to neutron and x-ray irradiation using mouse blood. BMC Genom. **18** (1), 2 (2017)
19. Pepper, S.D., Saunders, E.K., Edwards, L.E., Wilson, C.L., Miller, C.J.: The utility of MAS5 expression summary and detection call algorithms. BMC Bioinformatics **30**(8), 273 (2007)
20. Breiman, L.: Random forests. Mach. Learn. **45**(1), 5–32 (2001)
21. Hong, F., Breitling, R.: A comparison of meta-analysis methods for detecting differentially expressed genes in microarray experiments. Bioinformatics **24**(3), 374–382 (2008)

Multi-omics Classification on Kidney Samples Exploiting Uncertainty-Aware Models

Marta Lovino[1(✉)], Gianpaolo Bontempo[1], Giansalvo Cirrincione[2,3], and Elisa Ficarra[1]

[1] Department of Control and Computer Engineering, Politecnico di Torino, 10129, Turin, Italy
{marta.lovino,elisa.ficarra}@polito.it, gianpaolo.bontempo@studenti.polito.it

[2] LTI Lab, University of Picardie Jules Verne, 80025, Amiens, France
exin@u-picardie.fr

[3] University of South Pacific, Suva, Fiji

Abstract. Due to the huge amount of available omic data, classifying samples according to various omics is a complex process. One of the most common approaches consists of creating a classifier for each omic and subsequently making a consensus among the classifiers that assigns to each sample the most voted class among the outputs on the individual omics.

However, this approach does not consider the confidence in the prediction ignoring that a biological information coming from a certain omic may be more reliable than others. Therefore, it is here proposed a method consisting of a tree-based multi-layer perceptron (MLP), which estimates the class-membership probabilities for classification. In this way, it is not only possible to give relevance to all the omics, but also to label as *Unknown* those samples for which the classifier is uncertain in its prediction. The method was applied to a dataset composed of 909 kidney cancer samples for which these three omics were available: gene expression (mRNA), microRNA expression (miRNA) and methylation profiles (meth) data. The method is valid also for other tissues and on other omics (e.g. proteomics, copy number alterations data, single nucleotide polymorphism data). The accuracy and weighted average f1-score of the model are both higher than 95%. This tool can therefore be particularly useful in clinical practice, allowing physicians to focus on the most interesting and challenging samples.

Data availability: the code is freely accessible at https://github.com/Bontempogianpaolo1/Consunsus-on-multi-omics, while mRNA, miRNA and meth data can be obtained from the GDC database [2] or upon request to the authors.

Keywords: Bayesian neural networks · Gene expression · mRNA · miRNA · Methylation · Multi-Layer Perceptron (MLP) · Multi-omics · Multi-omics classification

D.-S. Huang and K.-H. Jo (Eds.): ICIC 2020, LNCS 12464, pp. 32–42, 2020.
https://doi.org/10.1007/978-3-030-60802-6_4

1 Introduction

In recent years, the reduction of costs for the sequencing of biological molecules including DNA, RNA and proteins has allowed the widespread of huge amounts of data both in the form of large structured databases and in the form of repositories specially created for the study of particular pathologies [1–3].

In this context, various omic data can be taken into account for the study and analysis of samples, either tumor or healthy: gene expression data (mRNA), microRNA expression data (miRNA), methylation data (meth), copy number alterations data (CNA), single nucleotide polymorphism data (SNV), proteomics and phosphoproteomics data.

Two large strands are typically available in multi-omics analysis: first the subdivision of the samples into its own classes [8–11] and second, the identification of specific pathways and gene patterns in the dataset [12, 13].

This work is focus exclusively on the first strand; in particular, some methods are presented for the classification of kidney cancer samples by simultaneously exploiting the information from the mRNA, miRNA, and methylation (meth) data. Although the work is focused on mRNA, miRNA and meth omics, it must be noticed that the same algorithms can be applied to a greater number of omics or other omics in place of them.

In the multi-omics classification approach, a crucial step is represented by the algorithm by which to integrate the classification results from each omic. One of the standard approaches is to make a consensus among the various omics, such that the multi-omic class is the most voted class among the outputs on the individual omics [14, 15]. However, this approach has two main limitations. At first, it is difficult to attribute to the multi-omic class in the case in which all the outputs of the individual omics are completely disjoint or more than one class is equally voted among all the omics. Secondly, each omic carries characteristics that may not be present in the other omics. For classification purposes, therefore, the contribution of a single omic should be considered according to the certainty in its classification.

This work proposes the use of a learning method that for each omic returns not only the corresponding class, but also its membership probability to that class, overcoming the main problem of standard consensus when the same sample is assigned to different classes across the omics or when there is no clear class prevalence.

In addition, the use of the class-membership probability allows to filter samples according to the class probability and consequently postpone for further analyses those samples on which there is not enough certainty in the classification across all the omics. This approach is particularly useful in creating automatic tools that, integrating different omic information, may favor clinical practice, by proposing a classification label when all the omics are enough certain in their classification and, an *Unknown* label when discrepancies are found across the omics. In this way, physicians can have a quick look at well-defined samples and focus more on the most interesting and challenging cases where human control is crucial.

2 Biological Data

Although the proposed method can be applied to any tissue and pathology, this work is deal with the study of kidney tumor samples freely available in the Genomic Data Commons (GDC) database [2]. The samples used in this study belong to three main kindney tumor subtypes: kidney renal papillary cell carcinoma (KIRP), kidney renal clear cell carcinoma (KIRCH) and kidney chromophobe (KICH). In addition, a reduced number of healthy samples is available both for KIRP and KICH subtypes (usually these tissues are healthy areas surrounding a KIRP or KICH tumor).

For KIRP, KIRCH and KICH subtypes, only samples samples available are selected for mRNA, miRNA and meth data, obtaining a final dataset of 909 samples.

The mRNA, miRNA and meth data are tabular data commonly represented as matrices, where the value in position (i, j) represents the amount of a specific biological product or the intensity of a phenomenon (mRNA, miRNA and meth respectively) in a specific sample. The mRNA, miRNA and meth matrices carry different biological information.

The mRNA expression value is strictly related to the amount of its protein (higher is the number, higher the amount of the protein) which regulates a specific pathway in the cell life cycle.

The miRNA expression value indicates the amount of a specific miRNA, a small noncoding RNA molecule which intervenes in the post-transcriptional process, regulating the amount of produced final protein.

Methylation value refers to the methylation beta value, an estimate of the methylation level computed as the ratio of intensities between methylated and unmethylated alleles. The biological effect of the methylation consists of the change of the activity of a DNA segment without changing its sequence (when methylation occurs, it reduces the DNA transcription, thus consequently reducing the amount of protein).

It must be noticed that many biological molecules act together in order to regulate the cell activity and that changes in the values of one or more omics can be correlated to a specific pathology or a tumor subtype.

2.1 Data Preprocessing

After downloading and selecting samples for which both mRNA, miRNA and meth data are available, the following preprocessing is performed:

- **mRNA**: 5000 features × 909 samples. Raw count data have originally about 60000 mRNA genes and have been normalized using the Variance Stabilizing Trasformation (VST) [16]. Then all not protein coding genes have been discarded reaching about 20000 mRNA genes and z-score transformation has been performed. In the end, the top 5000 mRNA genes with the highest standard deviation are selected.
- **miRNA**: 1200 features × 909 samples. The miRNA data have about 2000 miRNAs and have been normalized using deseq [17]. Then pseudo-counts have been computed as $log_2(count\_value + 1)$. In the end, z-score transformation has been performed and the top 1200 miRNAs with the highest standard deviation have been selected.

- **meth:** 5000 features × 909 samples. Among the 27000 features in methylation array data obtained with Illumina Human Methylation 27 platform, the top 5000 with the highest standard deviation are selected. Since original data are intrinsically normalized, no further normalization is required.

The 909 samples belong to 5 classes: **tumor KIRCH**: 509, **tumor KIRP**: 288, **tumor KICH**: 65, **healthy KIRCH**: 24, **healthy KIRP**: 23.

They have been further divided into training set (75% of the samples) and test set (25% of the samples) such that the latter includes the same proportion of samples belonging to the different classes.

In order to test the model on samples that do not belong to the kidney classes, 37 stomach samples have been obtained from GDC [2], by applying the same preprocessing steps described in 2.1. This dataset is used only as test test, without re-training the kidney model to evaluate the ability of the probabilistic approaches, such as the tree MLP classifier, to recognize unseen classes.

3 Method

An extension of the multi-layer perceptron (MLP) combining several MLPs in a tree architecture (tree MLP) is here proposed. Such an architecture has been designed to face with the classification of samples where no clear class prevalence was obtained through the consensus of the various omic-based classifiers. Moreover, it aims at identifying and filtering out samples uncertainly classified.

Since a MLP equipped with a cross-entropy loss function, with associated either logistic sigmoid (two class problem) or softmax (multiclass problem), outputs the class-membership posterior probabilities of the inputs [21], the proposed tree MLP classifier is therefore able to return the class label and the associated probability of the sample belonging to a class.

As it can be seen in Fig. 1, a tree-like architecture was created with MLP models as nodes and trained separately on subsets of the training set. For this specific problem, there are a root node (trained to recognize healthy from tumor samples) and two leaf nodes. The former is trained on healthy samples and classifies them into KIRP and KIRCH healthy tissues. The latter is trained on tumor samples and classifies them into KIRP, KIRCH, and KICH tumors.

Therefore, given a new sample S, it will be classified by the root MLP as healthy or tumor (y_{root}) with a class-membership probability P_r. After selecting the leaf node corresponding to y_{root}, it returns the subclass label y_{leaf} (tumor_KIRP, tumor_KIRCH, and tumor_KICH for tumor leaf MLP; healthy_KIRP and healthy_KIRCH for normal leaf MLP) with its class-membership probability P_{leaf}. The final class y_{pred} is equal to y_{leaf}.

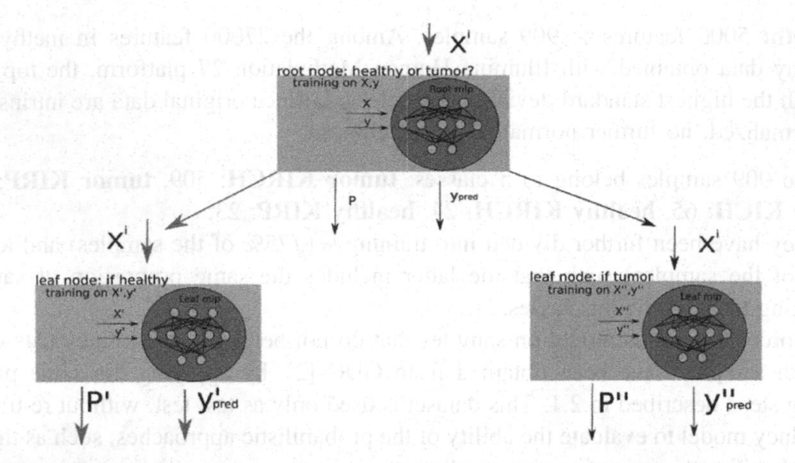

Fig. 1. Proposed tree MLP model: i) each node is trained on three different subsets of the original dataset. (X, y) aims to distinguish between healthy and tumor samples, (X′, y′) between subtypes of healthy samples and (X″, y″) between subtypes of tumor samples; ii) the output of each node consists of the predicted label y_{pred} and the class-membership probability P.

Once the classification on each individual omic is performed, the final consensus is built taking into account the final probabilities on each omic. Given:

- *n*: the number of the omics,
- *m*: the number of the classes,
- *th*: threshold on the omics, in order to filter predictions with low probabilities across all the omics,
- *tr*: threshold on the classes, in order to select only samples with a not uniform distribution of the class-membership probabilities across the *m* classes,
- P_{ij}: the class membership probability for class *i* and omic *j*,
- $S_i = sum_{j=1}^{n} P_{ij}$: the sum of the probabilities on all the omics for a single class,
- $S_a = sum_{i=1}^{m} S_i$: the sum of the probabilities on all the omics and all the samples,
- $S_m = S_i/n$: the mean of the probabilities on all the omics for a single class.

The consensus for a sample is built according to the next formula:

$$\begin{cases} Unknown,\ if\ \max\limits_{i}(S_m) < th\, or\, \max\limits_{i}(S_i/S_a) < tr \\ arg\max\limits_{i}(S_i) otherwise \end{cases}$$

In that way, a sample with a low mean probability across all the omics is labelled as *Unknown*.

In addition, when a sample receives similar S_i values for more than one class, the model is uncertain in its prediction. Therefore, a *tr* threshold is set in order to select only samples with a not uniform distribution of the class-membership probabilities across the *m* classes.

This final consensus can be applied using any number of omics as long as each omic represents different points of view of the same sample. Obviously, the larger the number of the omics, the more reliable the consensus prediction can be.

Very small architectures in the implemented neural networks were used (e.g. MLP with a single hidden layer with 20 neurons and a single activation layer) since the chosen number of main PCA components is low. This structure is the same for all the nodes of the tree MLP. Many hyper-parameters configurations have been considered. In the end, gradient descent with back propagation and the cross entropy as loss function were used. The optimizer was Adam.

In order to have a baseline for the results, a support vector machine (SVM) and a random forest (RF) classifier have been applied to the training set (with hyper-parameters optimization). Unless these models do not output a class-membership probability, they can provide valuable insights onto the data. Since they are unable to estimate the certainty of their prediction, the implementation of the consensus has been slightly modified. The final consensus for SVM and RF classifiers is given by the majority voting between the different omics.

On the other hand, to compare the tree MLP architecture with other methods that return a class-membership probability, a standard MLP classifier and a Bayesian neural network (BNN) were built.

The BNN model has the same structure as the MLP; however, it works in a completely different way. Indeed, as the loss is modified with a Bayesian regularization term, its weights are no longer deterministic like a standard MLP, but probabilistic, and each neuron learns to follow probabilistic distributions.

Therefore, it is possible to infer the level of uncertainty of the class-membership probability estimation of the input, which represents how much a sample belongs to a given class. The model is applied to the sample *n* times and the median value among all the output probabilities is selected as the final probability. For instance, if the median value is 0.95, it means that the output is highly stable and its classification uncertainty is very low.

All models have been tested both on the test set, consisting of kidney samples belonging to the 5 classes of the training set, and on the 37 stomach cancer samples.

All models were implemented in Pytorch framework [22]. In addition, the Pyro library [23] was used for the BNN to transform the parameters into random variables and to run stochastic variational inference.

4 Results

In this section, the results related to the proposed method are presented, as well as those of SVM, RF, MLP and BNN models. All performance metrics are obtained setting $th = 0.9$ and $tr = 0.25$. For tree MPL, MLP and BNN classifiers, all metrics are computed discarding *Unknown* samples.

In detail, concerning the tree MLP classifier, it is reported the confusion matrix with the classification results as well as the final consensus both on kidney test set (Fig. 2 (a)) and on 37 stomach samples (Fig. 2 (b)). Globally, the tree MLP method reached the 98% of accuracy and 97% of weighted average f1-score (Table 1). The metrics were computed disregarding *Unknown* samples as they had not been assigned to any classes. The tree MLP classifier selected as *Unknown* the 21, 49% of kidney test set samples and misclassified the 2% of the not *Unknown* samples.

Concerning the 37 stomach samples, all of them were correctly labelled as *Unknown* (Table 1).

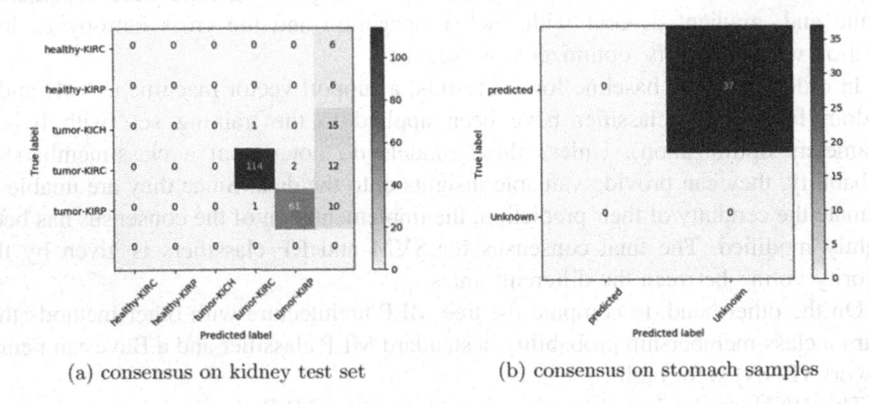

(a) consensus on kidney test set (b) consensus on stomach samples

Fig. 2. Confusion matrices for tree MLP classifier on (a) kidney test set and (b) stomach set.

Consensus confusion matrices for SVM and RF classifiers on kidney samples are reported in Fig. 3 (a-b). Both SVM and RF reached the 95% of accuracy and weighted average f1-score.

It should be noticed that the consensus creation for SVM and RF is different from that used in tree MLP, MLP and BNN models, since SVM and RF does not output the class-membership probabilities. Therefore, for SVM and RF classifiers the consensus is based on the majority voting on the three omics without considering class probabilities. As a consequence, the results for SVM and RF on stomach samples are not reported, since all the 37 stomach samples will be forced to one of the five kidney classes.

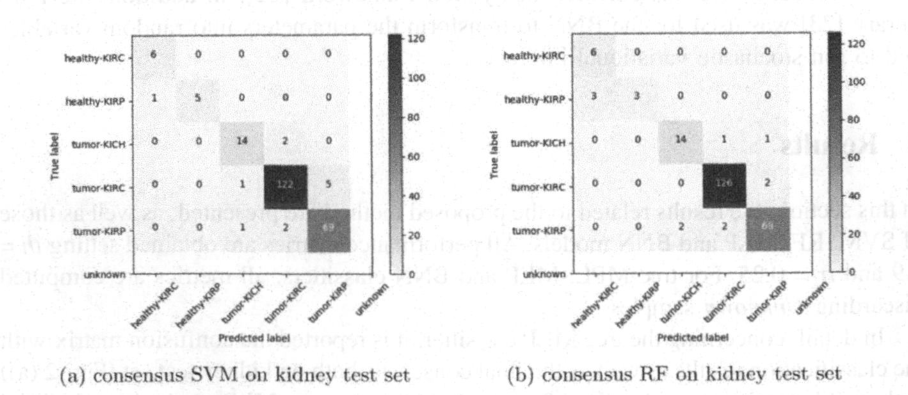

(a) consensus SVM on kidney test set (b) consensus RF on kidney test set

Fig. 3. Consensus confusion matrices on kidney test set on (a) SVM, (b) RF classifiers.

In addition, the performances with standard MLP model on the kidney test set were evaluated. After proper hyper-parameter tuning, the results obtained for the consensus are reported in Fig. 4 (a). Globally, it reached the 99% of accuracy and 99% of weighted average f1-score. All metrics were computed disregarding *Unknown* samples. Standard MLP model classified as *Unknown* the 22, 80% of kidney samples and misclassified the 1, 10% of the not *Unknown* samples. Concerning the 37 stomach samples, all of them were correctly labelled as *Unknown* (see in Fig. 5 (a)).

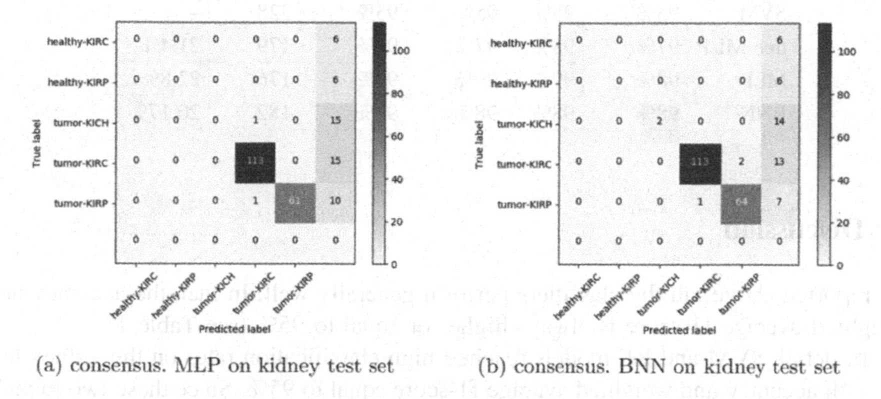

(a) consensus. MLP on kidney test set (b) consensus. BNN on kidney test set

Fig. 4. Consensus confusion matrices on kidney test set on (a) MLP, (b) BNN classifiers.

Consensus performances obtained with the BNN model are reported in Fig. 4 (b). The BNN model classified as *Unknown* the 20, 17% of kidney samples and misclassifies the 2% of the not *Unknown* samples. In addition, it achieved the 98% of accuracy and the 98% of weighted average f1-score. However, concerning the 37 stomach samples, only the 73% of them were correctly labelled as *Unknown* (see in Fig. 5 (b)).

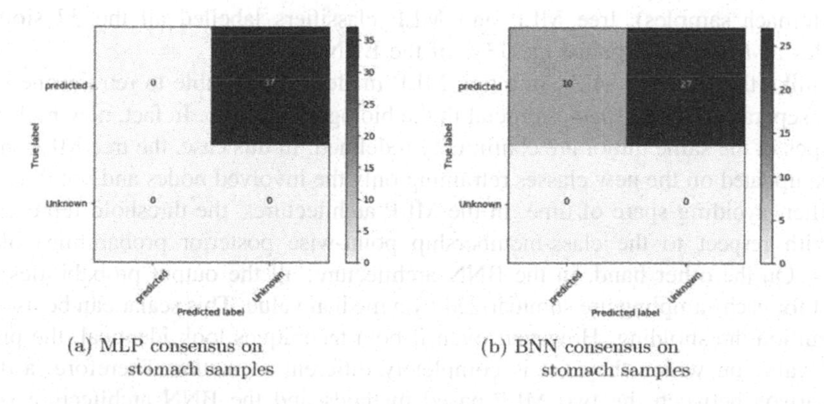

(a) MLP consensus on stomach samples (b) BNN consensus on stomach samples

Fig. 5. Confusion matrices for (a) MLP classifier and (b) BNN on 37 stomach samples.

In the end, the main results achieved for all the classifiers are reported in Table 1.

Table 1. Comparison between all the methods on the kidney test set. All the reported metrics are computed with weighted average only on not *Unknown* samples. The support metric, or the number of not *Unknown* samples, is the value on which the other metrics are based.

	Precision	Recall	F1-score	Accuracy	Support	*Unknown*
RF	95%	95%	95%	95%	228	–
SVM	95%	95%	95%	95%	228	–
tree MLP	97%	98%	97%	98%	179	21.4%
MLP	98%	99%	99%	99%	176	22.8%
BNN	98%	98%	98%	98%	182	20.17%

5 Discussion

As reported above, all the classifiers perform generally well. In fact, the accuracy and weighted average f1-score is always higher or equal to 95% (see Table 1).

In detail, SVM and RF models reached high classification rates on the kidney test set with accuracy and weighted average f1-score equal to 95%. Since these two models always force a prediction, they prevent labelling samples as *Unknown*. Although it could seem a minor issue, in real clinical practice, it is suitable to receive an *Unknown* label when the classifier is uncertain in its prediction.

Compared to the majority voting consensus used for SVM and RF classifiers, the proposed method analyses the probability values obtained on each omic and provides an integrated assessment of all the probability values. Considering tree MLP, standard MLP and BNN classifiers, they labelled as *Unknown* a similar percentage of kidney test set samples (21.49%, 22.80%, 20.17%, respectively) and had a similar weighted average f1-score (97%, 99%, 98%, respectively).

It should be noticed that, considering a tissue which the classifiers were not trained on (stomach samples), tree MLP and MLP classifiers labelled all the 37 stomach samples as *Unknown*, against the 73% of the BNN classifier.

Unlike the standard MLP, in a tree MLP model it is possible to retrain one of its nodes separately. This aspect is crucial in the biological domain. In fact, new molecular subtypes of the same tumor are continually redefined. In this case, the tree MLP model can be updated on the new classes retraining only the involved nodes and not the entire classifier, avoiding spare of time. In the MLP architectures, the threshold represents a cut with respect to the class-membership point-wise posterior probabilities of the inputs. On the other hand, in the BNN architecture, all the output probabilities estimated for each sampling are summarized by a median value. This scalar can be used for recognition thresholding. However, even if both techniques look identical, the probability value on which they act is completely different in nature. Therefore, a direct comparison between the two MLP-based methods and the BNN architecture is not completely possible. In the presented results, same *th* and *tr* values were applied to the three probabilistic approaches for the sake of comparison. This choice probably leaded the BNN to be less selective in the classification of stomach samples.

In addition, it can be noticed that a key role in the results is played by the criteria that has been used to obtain the final consensus across all the omics. In fact, the proposed consensus algorithm labels as *Unknown* samples with a low mean probability across all the omics or with similar sum probabilities on the classes (S_i). In that way, it prevents an unsafe labelling.

6 Conclusions

In the multi-omics classification task, the main limitation of the standard consensus is given by the absence of a measure to check the relevance of each individual omic in the classification.

Here, to overcome this problem, a tree MLP architecture is proposed to take into account the reliability of the classification on the individual omics exploiting uncertainty-aware models. Compared to the standard MLP and BNN architectures to classify kidney test set, the tree MLP represents a good compromise in terms of percentage of samples labelled as *Unknown*, and misclassification rate on the remaining samples (21, 49% and 2% respectively). In addition, the tree MLP model significantly outperforms the BNN model when predicting samples coming from a tissue on which the model has not been trained. This aspect is particularly relevant in clinical practice, since usually it is preferable to receive an *Unknown* label instead of a wrong prediction. Moreover, compared to a standard MLP, the tree structure is particular effective in applications where there is an ever-evolving knowledge, such as genetic complex diseases studies, preventing the classifier to be trained from scratch.

References

1. Weinstein, J.N., et al.: The cancer genome atlas pan-cancer analysis project. Nat. Genet. **45**(10), 1113 (2013)
2. Grossman, R.L., et al.: Toward a shared vision for cancer genomic data. N. Engl. J. Med. **375**(12), 1109–1112 (2016)
3. Leinonen, R., Sugawara, H., Shumway, M.: International nucleotide sequence database collaboration: the sequence read archive. Nucleic Acids Res. **39**((suppl_1)), D19–D21 (2010)
4. Pochet, N., De Smet, F., Suykens, J.A., De Moor, B.L.: Systematic benchmarking of microarray data classification: assessing the role of non-linearity and dimensionality reduction. Bioinformatics **20**(17), 3185–3195 (2004)
5. Lee, G., Rodriguez, C., Madabhushi, A.: An empirical comparison of dimensionality reduction methods for classifying gene and protein expression datasets. In: Măndoiu, I., Zelikovsky, A. (eds.) ISBRA 2007. LNCS, vol. 4463, pp. 170–181. Springer, Heidelberg (2007). https://doi.org/10.1007/978-3-540-72031-7_16
6. Kim, P.M., Tidor, B.: Subsystem identification through dimensionality reduction of large-scale gene expression data. Genome Res. **13**(7), 1706–1718 (2003)

7. Lu, M., Zhan, X.: The crucial role of multiomic approach in cancer research and clinically relevant outcomes. EPMA J. **9**(1), 77–102 (2018)
8. Wang, B., Mezlini, A.M., Demir, F., Fiume, M., Tu, Z., Brudno, M., et al.: Similarity network fusion for aggregating data types on a genomic scale. Nat. Meth. **11**(3), 333 (2014)
9. Argelaguet, R., et al.: Multi-omics factor analysis—a framework for unsupervised integration of multi-omics data sets. Mol. Syst. Biol. **14**(6), e8124 (2018)
10. Robles, A.I., Arai, E., Mathé, E.A., Okayama, H., Schetter, A.J., Brown, D., et al.: An integrated prognostic classifier for stage I lung adenocarcinoma based on mRNA, microRNA, and DNA methylation biomarkers. J. Thorac. Oncol. **10**(7), 1037–1048 (2015)
11. Tang, W., Wan, S., Yang, Z., Teschendorff, A.E., Zou, Q.: Tumor origin detection with tissue-specific miRNA and DNA methylation markers. Bioinformatics **34**(3), 398–406 (2018)
12. Cantini, L., Medico, E., Fortunato, S., Caselle, M.: Detection of gene communities in multi-networks reveals cancer drivers. Sci. Rep. **5**, 17386 (2015)
13. Barabási, A.L., Gulbahce, N., Loscalzo, J.: Network medicine: a network-based approach to human disease. Nat. Rev. Genet. **12**(1), 56–68 (2011)
14. Fuchs, M., Beißbarth, T., Wingender, E., Jung, K.: Connecting high-dimensional mRNA and miRNA expression data for binary medical classification problems. Comput. Meth. Programs Biomed. **111**(3), 592–601 (2013)
15. Mallik, S., Mukhopadhyay, A., Maulik, U., Bandyopadhyay, S.: Integrated analysis of gene expression and genome-wide DNA methylation for tumor prediction: an association rule mining-based approach. In: 2013 IEEE Symposium on Computational Intelligence in Bioinformatics and Computational Biology (CIBCB), (pp. 120–127). IEEE April 2013
16. Huber, W., Von Heydebreck, A., Sültmann, H., Poustka, A., Vingron, M.: Variance stabilization applied to microarray data calibration and to the quantification of differential expression. Bioinformatics **18**((suppl_1)), S96–S104 (2002)
17. Love, M.I., Huber, W., Anders, S.: Moderated estimation of fold change and dispersion for RNA-seq data with DESeq2. Genome Biol. **15**(12), 550 (2014)
18. Wold, S., Esbensen, K., Geladi, P.: Principal component analysis. Chemometr. Intell. Lab. Syst. **2**(1-3), 37–52 (1987)
19. Breiman, L.: Random forests. Mach. Learn. **45**(1), 5–32 (2001)
20. Ramchoun, H., Idrissi, M.A.J., Ghanou, Y., Ettaouil, M.: Multilayer perceptron: architecture optimization and training. IJIMAI **4**(1), 26–30 (2016)
21. Christopher, M.: Bishop.: Pattern Recognition and Machine Learning (Information Science and Statistics). Springer-Verlag, Berlin, Heidelberg (2006)
22. Paszke, A., et al. Automatic differentiation in pytorch (2017)
23. Bingham, E., et al.: Pyro: deep universal probabilistic programming. J. Mach. Learn. Res. **20** (1), 973–978 (2019)

Gene Regulation Modeling and Analysis

Inference Method for Reconstructing Regulatory Networks Using Statistical Path-Consistency Algorithm and Mutual Information

Yan Yan[1], Xinan Zhang[2], and Tianhai Tian[3](✉)

[1] School of Mathematics and Physics, Wuhan Institute of Technology, Wuhan, China

[2] School of Mathematics and Statistics, Central China Normal University, Wuhan, China

[3] School of Mathematics, Monash University, Melbourne, Australia
tianhai.tian@monash.edu

Abstract. The advances of high-throughout technologies havc produced huge amount of data regarding gene expressions or protein activities under various experimental conditions. The reverse-engineering of regulatory networks using these datasets is one of the top important research topics in computational biology. Although substantial efforts have been contributed to design effective inference methods, there are still a number of significant challenges to deal with the weak correlations between the observation data and the dependence of network structure on the order of variables in the systems. To address these issues, this work proposes a novel statistical approach to infer the structure of regulatory networks. Instead of using one single variable order, we generate a number of variable orders and then obtain different networks based on these orders. The weight of each edge for connecting genes/proteins is determined by the statistical measures based on the generated networks using different variable orders. Our proposed algorithm is evaluated by using the golden standard networks in Dream challenges and a cell signalling transduction pathway by using experimental data. Inference results suggest that our proposed algorithm is an effective approach for the reverse-engineering of regulatory networks with better accuracy.

Keywords: Regulatory network · Mutual information · Graphic model · Path-consistency

This work is supported by the National Natural Science Foundation of China (Grant number: 11931019, 11871238), and the Science Foundation of Wuhan Institute of Technology (Grant number K202047).

D.-S. Huang and K.-H. Jo (Eds.): ICIC 2020, LNCS 12464, pp. 45–56, 2020.
https://doi.org/10.1007/978-3-030-60802-6_5

1 Introduction

Network theory provides a systematic understanding for molecular mechanisms in biological processes [1, 2]. The groups of genes, regulatory proteins and their interactions are often referred to as regulatory networks. Recent advances in high-throughout technologies have generated huge amount of various datasets such as the miroarray gene expression data, RNA-sequence data, and proteomic data. These datasets provide unprecedented opportunities for exploring the underlying regulatory mechanisms in biological systems. The reverse-engineering of regulatory networks, which aims at reconstructing the underlying gene-gene or protein-protein interactions using the measured datasets, is considered as one of most important and challenging goals in systems biology [3, 4]. For this reason, the Dialogue for Reverse Engineering Assessments and Methods (DREAM) program is developed for providing benchmark systems to evaluate the effectiveness and efficiency of the designed algorithms [5].

Network inference methods have been widely used now in biological sciences to detect regulatory relationships between molecular components. These inference methods can be generally classified into two broad categories, namely model-free inference methods and model-based inference methods [6]. The first type methods normally are statistical inference algorithms based on certain correlation measure and thus are able to infer large scale regulatory networks. These methods mainly determine whether two components of the system have stronger relationship than the other pairs by using a statistical or information measure. The commonly used measures include the Pearson correlation coefficient, mutual information, partial correlation, rank correlation, Euclidean distance, and vector angle [7–11]. On the other hand, model-based network inference methods use a mathematical or statistical model to describe the dynamic properties of the systems. The widely used models include: Boolean networks, linear regression models, ordinary differential equations models, and chemical reaction system models [12–21].

Recently, information theory has been increasingly used for reconstructing regulatory networks due to its ability to measure nonlinear relationship between observation data [22]. Mutual information (MI) provides a natural generalization of the correlation coefficient and therefore attracts much attention in recent years [11, 23]. In addition, conditional mutual information (CMI) and part mutual information (PMI) are capable of detecting the joint regulations by exploiting the conditional dependency between genes or proteins. Methods based on both MI and CMI have been proposed to reduce the false positive rate for detecting interactions. Recently, the path-consistency algorithm (PCA) based on CMI or PMI was proposed for inferring regulatory networks [24–26]. However, the path consistency (PC) algorithms are order-dependent, namely its output of network structure is dependant on the order of variables in the dataset [27]. Although the PC-stable algorithm have been proposed to address the issue of order dependance, the relative information of PMI was not fully considered in the algorithm.

In this work, we propose a new method to construct regulatory networks using omics datasets. Instead of considering one single network, we develop a number of regulatory networks that are based on different orders of genes (or proteins) in the systems. Then a final network is constructed based on the statistical properties of these developed networks by using a weighted measure for each edge that connects two variables. The method is validated on the benchmark networks in Dream challenges and the mitogen-activated protein (MAP) kinase pathway. The cross-validation results confirmed the effectiveness of our proposed statistical method, which outperforms over the existing methods.

2 Methods

This section first introduce the definitions of information theory for calculating MI, CMI and PMI. After a brief discussion of the published PCA-PMI algorithm and PC-stable algorithm, we will introduce the proposed statistical path-consistent algorithm (SPCA) with PMI for inferring regulatory networks.

2.1 Mutual Information and Conditional Mutual Information

Compared with correlation coefficient, mutual information (MI) from information theory is able to describe the nonlinear dependence between two random variables. Let X be a random variable with density function $p(x)$. The entropy $H(X)$ of the random variable X is defined by

$$H(X) = -\sum\nolimits_{x \in X} p(x)\log p(x) = -\int_{\Omega} p(x)\log p(x)dx. \tag{1}$$

The above definitions are designed for discrete and continuous random variables, respectively. In addition, for two random variables X and Y, the joint entropy $H(X, Y)$ is defined by

$$H(X,Y) = -\sum\nolimits_{x \in X, y \in Y} p(x,y)\log p(x,y) = -\int_{\Omega} p(x,y)\log p(x,y)dxdy \tag{2}$$

where $p(x, y)$ is the joint probability of random variables X and Y.

MI measures the nonlinear dependency between two random variables. For discrete variables X and Y, MI can be calculated from

$$MI(X,Y) = -\sum\nolimits_{x \in X, y \in Y} p(x,y)\log\frac{p(x,y)}{p(x)p(y)}. \tag{3}$$

where $p(x)$ and $p(y)$ are marginal density function of variable X and Y, respectively. In addition, MI can be measured in terms of entropies as

$$MI(X,Y) = H(X) + H(Y) - H(X,Y). \tag{4}$$

If the MI value is zero, two random variables are independent to each other. However, a larger value of MI generally suggests closer relationship between the two random variables. For a system with more random variables, the strong dependence relationship of two random variables may be caused by the third random variable.

To address this issue, conditional mutual information (CMI) measures conditional dependency between two random variables under the condition of the third variable. The value of CMI between variables X and Y given Z can be calculated from

$$\mathrm{CMI}(X, Y|Z) = -\sum\nolimits_{x \in X, y \in Y, z \in Z} p(x, y, z) \log \frac{p(x, y|z)}{p(x|z)p(y|z)}, \quad (5)$$

where $p(x|z)$ and $p(y|z)$ are the density functions of random variables X and Y under the given third random variable Z, respectively, and $p(x, y|z)$ is the joint probability of the random variables X and Y under the condition of given Z. If variables X and Y are independent of each other under the condition of variable Z, then $\mathrm{CMI}(X, Y|Z) = 0$.

For a regulatory network of m genes, the activity of gene X_i is measured by the expression levels at different time points $(x_{i1}, \ldots, x_{in})$. We can calculate the frequency of the expression data and then use the frequency to approximate the mutual information [23]. To this end, we first uniformly divide the interval $\left[\min_j(x_{ij}), \max_j(x_{ij})\right]$ into k subintervals, and then compute the frequency of the expression data falling into the subinterval q, and then approximate the probability by using

$$f_{iq} \approx \frac{f_{iq}}{q}, \quad i = 1, \ldots n, q = 1, \ldots, k.$$

The mutual information can be calculated by using the approximated probability. Similar formulas can be derived for the joint probability. We can also calculate MI by using the assumed probability density functions. A particular case is the Gaussian kernel probability density function. The detailed formulas can be found in [24].

2.2 Path-Consistency Algorithm

Now we consider the problem of estimating regulatory networks underlying a fixed set of genes. Let $G = (V, E)$ be the graph with node (i.e., gene) set V of size $|V| = p$ and edge set E. Let $D = \{x_t\}_{t=1}^{T}$ be of the set of samples of node states. The path-consistency (PC) algorithm construct a graphic network by removing edges that have independent correlation from the fully connected network. The PC algorithm [30] was first designed for learning directed acyclic graphs under the assumption of causal sufficiency. To construct a network with adequate sparse ϵ is selected. In the first stage, for any adjacent genes i and j that are connected by an edge $e(i,j)$, if the value of $MI(i,j)$ is zero or less than the threshold value, we regard these two genes are independent, and the edge $e(i,j)$ is removed from the network. In the second stage, for adjacent genes i and j, find a gene k that is adjacent to both genes i and j. If gene k does not exist, keep edge $e(i,j)$ in the network. Otherwise, calculate $\mathrm{PMI}(i,j|k)$, and remove edge $e(i,j)$ from the network if $\mathrm{PMI}(i,j|k) < \epsilon$. Similar procedure is also applied to the PMI with more genes as the condition. The detailed process of PCA-PMI is given in the following Algorithm 1.

Algorithm 1 PCA-PMI algorithm

1: Input the genes expression data D. Set θ for deciding the independence. Generate the complete network G for all genes.
Set $L = -1$.
2: For $L = L + 1$. For a non-zero edges $G(i,j) \neq 0$, search for adjacent genes connected with both genes i and j. Compute the number T of these adjacent genes (not including genes i and j) .
3: Step 2: if $T < L$, stop. If $T \geq L$, select L genes from these T genes and let them as $K = [k_1, \ldots, k_L]$. The number of all selections for K is C_T^L. Compute the L-order PMI$(i,j|K)$ for all C_T^L selections, and choose the maximal one denoting as PMI$_{max}(i.j|K)$. If PMI$_{max}(i.j|K) < \theta$, set $G(i,j) = 0$. Return to Step 1.

One of the key issues of the PC algorithm is that the inferred network is dependent on the order of genes in the network. For example, if the value of PMI$(i,j|k)$ is smaller than the threshold value, the edge $e(i,j)$ should be removed. However, if the edge connecting k and i or that connecting k and j was removed from the network before computing PMI$(i,j|k)$, this PMI does not exist and edge $e(i,j)$ cannot be removed. To address this, the PC-stable algorithm was proposed to check all the possible adjacent gene pairs first before computing the values of PMI [27]. After calculating the PMI values of one edge, we make the decision to keep or remove this edge. Unlike the PCA-PMI, the edge is not removed immediately but we just save the edge index which should be removed. After all PMI values are determined at one stage, we then remove the edges from the index set.

Although the PC-stable algorithm has solved the problem for the dependence of gene order, it raises the issue of possible false positive regulations. Note that in the PC algorithm, an edge remains if the maximal value of all the related PMI values is greater than the threshold. This consideration is reasonable since one genetic regulation exists if there is at least one PMI supports this regulation. However, this consideration may lead to false positive regulations if this supportive PMI actually does not exist. If there are more values of PMI for a given edge, it is more possible to keep this edge.

2.3 Statistical Path-Consistency Algorithm

In these section, we introduce a new statistical method to construct regulatory network. Since we do not know the order of genes in the considered network (or such an optimal order actually does not exist), we consider a large number of different orders, and then develop the corresponding network for each order using PCI-PMI. Assume that N networks are developed by using N different orders of variables. For the s-th developed network, we label the edge connecting genes i and j with weight $w_{ij}^s = 0$ (if the edge does not appear in the developed network) or $w_{ij}^s = 1$ (if the edge exists). The average weight of this edge is calculated by

$$AW_{ij} = \frac{1}{N}\sum_{s=1}^{N} w_{ij}^s, \quad (6)$$

If the value of AW_{ij} is zero or one, we remove or select this edge in the final developed network. If this value is between zero and one, we assume that the probability for selecting this edge is larger if the value of AW_{ij} is large.

To further use the values of PMI, we define the weighted PMI by

$$\mathrm{AWT}_{ij} = \frac{1}{N}\sum_{s=1}^{N} \mathrm{PMI}_{ij}^s w_{ij}^s, \quad (7)$$

where PMI_{ij}^s is the maximal PMI value for edge $e(i,j)$ in the s-th network. The advantage of the weighted PMI is the ability to rank all edges whose weighted PMI is not zero. Note that, for a given edge, it may have MI values only (i.e. $L = 0$), or the PMI value (i.e. $L = 1$), or the value of higher order PMI (i.e. $L > 1$). Thus the final value of the weighted PMI is the value of the highest possible order. Based on the above discussion, we formulate the statistical path-consistency algorithm in Algorithm 2.

Algorithm 2 Statistical path-consistent Algorithm (SPCA)

Input: Molecular (gene or protein) activity dataset $D_{n \times m}$ with n variables and m observations. The node set is $V = \{v_1, \ldots, v_n\}$. N is the number of different variable orders. Threshold value is ϵ. M is the number of edges in the output network.
Output: Network $G(V,E)$. E is the set of selected edges.

1: for id=1:N do
2: Generate a sample order $\{k_1, \ldots, k_n\}$ for $\{1, \ldots, n\}$ and reorganize the dataset based on the new order of variables $\{x_{k_1}, \ldots, x_{k_n}\}$.
3: Use PCA-PMI algorithm (Algorithm 1) to construct a network by the new order.
4: For each edge (x_i, x_j), find edge weight w_{ij}^{id}, and PMI value of the highest or der PMI_{ij}^{id}.
5: **end for**
6: Calculate the average weight AW_{ij} (6) or the weighted PMI AWT_{ij} (7). Sort all edges according to the average weight or weighted PMI.
7: Select the top M edges with the highest values of average weight or weighted PMI.
8: Export the generated network $G(V, VE)$.

2.4 Accuracy of Inference Algorithms

To measure the accuracy of inference methods, we use the sensitivity and accuracy (ACC), given by

$$\mathrm{ACC} = \frac{\mathrm{TP} + \mathrm{TN}}{\mathrm{TP} + \mathrm{FT} + \mathrm{TN} + \mathrm{PN}}$$

where TP, FP, TN, and PN are the numbers of the true positive, false positive, true negative and false negative regulations predicted by the proposed method, respectively. In addition, we also use the following definitions

$$\mathrm{TPR} = \frac{\mathrm{TP}}{\mathrm{TP} + \mathrm{PN}}$$

$$\mathrm{FPR} = \frac{\mathrm{FP}}{\mathrm{FP} + \mathrm{PN}}$$

$$\mathrm{PPV} = \frac{\mathrm{TP}}{\mathrm{TP} + \mathrm{FP}}$$

to measure the true positive rate (TPR), false positive rate (FPR) and positive predictive value (PPV), respectively. Similar to other approaches, the receiver operating characteristic curve (ROC) and the area under the ROC curve (AUC) are used as the key measure to assess the accuracy of the inference method.

3 Results

We first show the dependence of network structure on the order of variables in the system. We test our method on a DREAM3 datasets about the SOS DNA repair gene regulatory networks [29]. The dataset has 100 genes and 101 samples, and the standard network has 166 edges. We first test the influence of gene orders on the inferred network structure. We generate 1000 different orders of these 100 genes and then use PCA-PMI to infer a network based on each generated variable order. Table 1 gives that the maximal, minimal and average edge numbers of the networks using four different threshold values. This result clearly suggests the dependence of the inferred networks on the variable orders. In addition, the variation percentage is defined by

$$\text{Variation ratio} = \frac{\text{max edge number} - \text{min edge number}}{\text{average edge number}}.$$

The variation ratio is more than 10% for all these four tests, and the largest value is 17.14%. When the edge number of the network is smaller, there are larger variations in the edge numbers in different networks determined by the same threshold value.

We classify all edges in the network into three types. The first type of edges does not appear in any networks, the second one appears in part of the networks but not in

Table 1. Network inference results using four different threshold values based on 1000 different variables orders for the DREAM3 network with 100 genes

Threshold value	Average number of edges	Max number of edge	Min number of edge	Variation ratio
0.05	105	95	113	17.14%
0.03	155	136	161	16.13%
0.02	197	186	209	11.68%
0.016	250	234	262	11.12%
0.05	105	95	113	17.14%

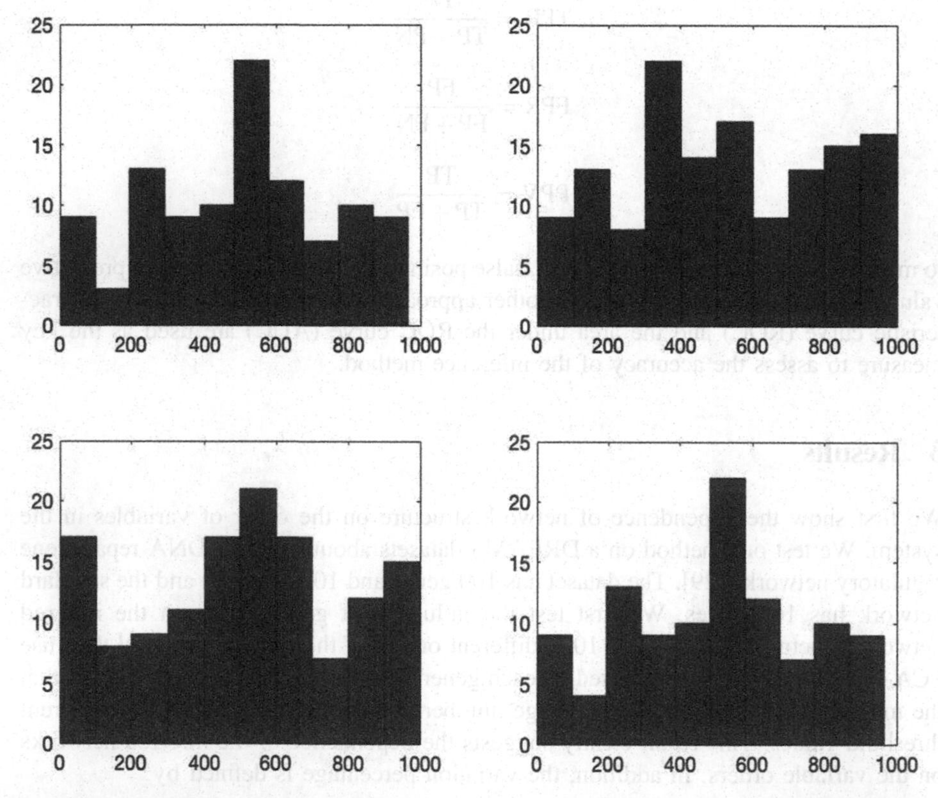

Fig. 1. Distribution of edge frequency for the second type of edges (namely edges appearing in part of the inferred network but not in all networks). Top-left: networks with average 105 edges. Top-right: networks with average 155 edges. Bottom-left: networks with average 197 edges. Bottom-right: networks with average 250 edges.

all networks, and the third one appears in all the networks. Figure 1 shows the frequency distribution for the second type of edges in the generated 1000 networks. These distributions close to the uniform distribution, though the frequency around 500 is relatively larger than the other frequency numbers. In addition, the edge numbers of the second and third types increase gradually in accordance with the increase of edge numbers in the networks by using smaller threshold values. For example, the numbers of type two and three edges are 104 and 53 when the threshold value is 0.05, while these numbers increase to 155 and 166 when the threshold value is decreased to 0.016. The number of type three edges increases substantially while that of type two just increases moderately.

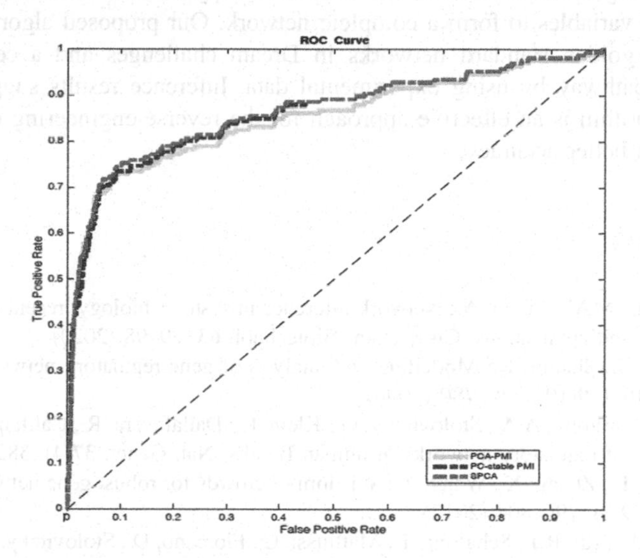

Fig. 2. ROC curves of the inferred SOS DNA repair genetic regulatory network from Dream3 challenge by using PCA-PMI (green line), PC-stable-PMI (blue line) and our proposed method SPCA (red line). (Color figure online)

We next examine the effectiveness of our proposed method by comparing the the accuracy of three methods, namely PCA-PMI, PC-stable with PMI, and the proposed algorithm SPCA. We use the same SOS DNA repair gene dataset for reconstructing gene regulatory networks [29]. The PCA-PMI algorithm is obtained from the published programs [24] and the PC-stable method is implemented in an open-source software with the R-package pcalg [32]. The performance of these three algorithms is evaluated by true-positive rate (TPR) and false positive rate (FPR). The area under ROC curve (AUC) is calculated to indicate the accuracy of the inference methods. Figure 2 shows that our proposed method has AUC value 0.8649 which is larger than the value of PC-stable with PMI (i.e. 0.8605) and that of PCA-PMI (i.e. 0.8571). These results suggest that our proposed algorithm has better accuracy than these two algorithms by using the real gene expression data.

4 Conclusions

In this paper, we propose a new algorithm to infer regulatory networks using omics datasets. To address the issue of independence of network structure on the order of variables, we propose a new statistical method that determine the network structure by using a large number of networks based on different variable orders. Instead of using one single variable order, we generate a number of variable orders and then obtain different networks based on these orders. The weight of each edge for connecting genes/proteins is determined by the statistical measures based on the generated networks using different variable orders. This new method has successfully addressed the issues in existing methods such as the various entropy values in different setting and connection of variables to form a complete network. Our proposed algorithm is evaluated by the golden standard networks in Dream challenges and a cell signalling transduction pathway by using experimental data. Inference results suggest that our proposed algorithm is an effective approach for the reverse-engineering of regulatory networks with better accuracy.

References

1. Saintantoine, M.M., Singh, A.: Network inference in systems biology: recent developments, challenges, and applications. Curr. Opin. Biotechnol. **63**, 89–98 (2020)
2. Karlebach, G., Shamir, R.: Modelling and analysis of gene regulatory networks. Nat. Rev. Mol. Cell Biol. **9**(10), 770–780 (2008)
3. Basso, K., Margolin, A.A., Stolovitzky, G., Klein, U., Dallafavera, R., Califano, A.: Reverse engineering of regulatory networks in human B cells. Nat. Genet. **37**(4), 382–390 (2005)
4. Li, H., Xie, L., Zhang, X., Wang, Y.: Wisdom of crowds for robust gene network inference. Nat. Meth. **9**(8), 796–804 (2012)
5. Marbach, D., Prill, R.J., Schaffter, T., Mattiussi, C., Floreano, D., Stolovitzky, G.: Revealing strengths and weaknesses of methods for gene network inference. Proc. Natl. Acad. Sci. **107** (14), 6286–6291 (2010)
6. Huynh-Thu, V.A., Sanguinetti, G.: Gene regulatory network inference: an introductory survey. In: Sanguinetti, G., Huynh-Thu, V.A. (eds.) Gene Regulatory Networks. MMB, vol. 1883, pp. 1–23. Springer, New York (2019). https://doi.org/10.1007/978-1-4939-8882-2_1
7. Omranian, N., Eloundoumbeb Stuart, J.M., Segal, E., Koller, D., Kim, S.M.: A gene-coexpression network for global discovery of conserved genetic modules. Science **302**(5643), 249–255 (2003)
8. Farahmand, S., Oconnor, C., Macoska, J., Zarringhalam, K.: Causal inference engine: a platform for directional gene set enrichment analysis and inference of active transcriptional regulators. Nucleic Acids Res. **47**(22), 698852 (2019)
9. Omranian, N., Eloundou-Mbebi, J.M., Mueller-Roeber, B., Nikoloski, Z.: Gene regulatory network inference using fused LASSO on multiple data sets. Sci. Rep. **6**(1), 20533 (2016)
10. Casadiego, J., Nitzan, M., Hallerberg, S., Timme, M.: Model-free inference of direct network interactions from nonlinear collective dynamics. Nat. Commun. **8**(1), 2192 (2017)
11. Liu, Z.: Quantifying gene regulatory relationships with association measures: a comparative study. Front. Genet. **8**, 96 (2017)
12. Li, H., Xie, L., Zhang, X., Wang, Y.: Output regulation of Boolean control networks. IEEE Trans. Autom. Control **62**(6), 2993–2998 (2017)

13. Cantone, I., et al.: A yeast synthetic network for in vivo assessment of reverse-engineering and modeling approaches. Cell **137**(1), 172–181 (2009)
14. Chan, T.E., Stumpf, M.P.H., Babtie, A.C.: Gene regulatory network inference from single-cell data using multivariate information measures. Cell Syst. **5**(3), 251 (2017)
15. Barman, S., Kwon, Y.K.: A Boolean network inference from time-series gene expression data using a genetic algorithm. Bioinformatics **34**(17), i927–i933 (2018)
16. Kishan, K.C., Li, R., Cui, F., Haake, A.R.: GNE: a deep learning framework for gene network inference by aggregating biological information. BMC Syst. Biol. **13**(2), 1–14 (2019)
17. Yuan, L., et al.: Integration of multi-omics data for gene regulatory network inference and application to breast cancer. IEEE/ACM Trans. Comput. Biol. Bioinf. **16**(3), 782–791 (2019)
18. Wang, J., Wu, Q., Hu, X.T., Tian, T.: An integrated platform for reverse-engineering protein-gene interaction network. Methods **110**, 3–13 (2016)
19. Wei, J., Hu, X., Zou, X., Tian, T.: Reverse-engineering of gene networks for regulating early blood development from single-cell measurements. BMC Med. Genomics **10**(5), 72 (2017)
20. Yang, B., Bao, W.: RNDEtree: regulatory network with differential equation based on flexible neural tree with novel criterion function. IEEE Access **7**, 58255–58263 (2019)
21. Yang, B., Bao, W., Huang, D.-S., Chen, Y.: Inference of large-scale time-delayed gene regulatory network with parallel mapreduce cloud platform. Sci. Rep. **8**(1), 17787 (2018)
22. Meyer, P.E., Kontos, K., Lafitte, F., Bontempi, G.: Information-theoretic inference of large transcriptional regulatory networks. EURASIP J. Bioinf. Syst. Biol. **2007**(1), 8 (2007)
23. Guo, X., Zhang, H., Tian, T.: Development of stock correlation networks using mutual information and financial big data. PLoS ONE **13**(4), e0195941 (2018)
24. Zhang, X., et al.: Inferring gene regulatory networks from gene expression data by path consistency algorithm based on conditional mutual information. Bioinformatics **28**(1), 98–104 (2012)
25. Zhang, X., Zhao, J., Hao, J., Zhao, X., Chen, L.: Conditional mutual inclusive information enables accurate quantification of associations in gene regulatory networks. Nuc Aci Res. **43**(5), e31 (2015)
26. Zhao, J., Zhou, Y., Zhang, X., Chen, L.: Part mutual information for quantifying direct associations in networks. Proc. Natl. Acad. Sci. **113**(18), 5130–5135 (2016)
27. Colombo, D., Maathuis, M.H.: Order-independent constraint-based causal structure learning. J. Mach. Learn. Res. **15**(1), 3741–3782 (2014)
28. Janzing, D., Balduzzi, D., Grosse-Wentrup, M., Schölko pf, B.: Quantifying causal influences. Ann. Stat. **41**(5), 2324–2358 (2013)
29. Ronen, M., Rosenberg, R., Shraiman, B.I., Alon, U.: Assigning numbers to the arrows: parameterizing a gene regulation network by using accurate expression kinetics. Proc. Natl. Acad. Sci. **99**(16), 10555–10560 (2002)
30. Spirtes, P., Glymour, C., Scheines, R.: The MIT Press (1993)
31. Cargnello, M., Roux, P.P.: Activation and function of the mapks and their substrates the mapk-activated protein kinases. Microbiol. Mol. Biol. Rev. **75**(1), 50–83 (2011)
32. Kalisch, M., Maechler, M., Colombo, D.: Causal inference using graphical models with the r package pcalg. J. Stat. Softw. **047**(1), 1–26 (2012)
33. Spirtes, P.: Causation, prediction, and search, 45(3), 272–273 (1996)
34. Tian, T., Song, J.: Mathematical modelling of the MAP kinase pathway based on proteomics dataset. PLoS ONE **7**(8), e42230 (2012)
35. Greenfield, A., Madar, A., Ostrer, H., Bonneau, R.: DREAM4: combining genetic and dynamic information to identify biological networks and dynamical models. PLoS ONE **5**(10), e13397 (2014)

36. Lawrence, R.T., Perez, E.M., Hernandez, D., et al.: The proteomic landscape of triple-negative breast cancer. Cell Rep. **11**(4), 630–644 (2015)
37. Zhang, W., Liu, H.T.: Mapk signal pathways in the regulation of cell proliferation in mammalian cells. Cell Res. **12**(1), 9 (2002)
38. Kanehisa, M., Goto, S., Kawashima, S., et al.: The KEGG resource for deciphering the genome. Nucleic Acids Res. **32**(suppl 1), D277–D280 (2004)

Exploring lncRNA-MRNA Regulatory Modules Based on lncRNA Similarity in Breast Cancer

Lei Tian, Shu-Lin Wang(✉), and Xing Zhong

College of Computer Science and Electronic Engineering, Hunan University, Changsha 410082, Hunan, China
Smartforesting@163.com

Abstract. The latest study has shown that long non-coding RNA (lncRNA) is involved in many fundamental biological processes and is involved in various diseases including breast cancer. The development of computational methods to identify lncRNA-related modules is a powerful tool to reveal the role of lncRNA in diseases. Here, we proposed a novel computational method called BCLMM to identify lncRNA-mRNA modules of breast cancer based on lncRNA similarity. First, BCLMM calculated the Pearson correlation matrix between any two lncRNAs. Second, BCLMM formed the lncRNA similarity matrix based on the Pearson correlation matrix and collected lncRNA functional similarity. Third, BCLMM applied BCPlaid to the lncRNA similarity matrix to obtain subclasses. Finally, the lncRNA-mRNA modules were identified by querying lncRNA-mRNA pairs from the validated lncRNA-mRNA interactions according to the lncRNAs involved in each subclass. Each module was analyzed by biological enrichment, classification performance and survival performance and also was validated by miRspongR. The experimental results indicate that BCLMM has excellent performance to explore regulatory modules.

Keywords: Regulatory modules · lncRNA similarity · Breast cancer · Biological enrichment · Clustering algorithm

1 Introduction

The study reveals that less than 2% of sequences in the human genome are protein-coding genes. More than 98% of the whole genome does not decode into protein [1], which are called non-coding RNA [2] (ncRNA). Long non-coding RNA [3] (lncRNA), whose length is usually longer than 200 nucleotides, accounts for 80% to 90% of all ncRNA. The function of lncRNA involves almost all processes of physiology and pathology [4]. LncRNA can not only regulate physiological processes such as cell proliferation, differentiation and metabolism, but also participates in regulating diverse pathological processes, such as cancer, diabetes, and immune diseases [5]. Abnormal expression of lncRNA in breast cancer plays a key role in the development of tumors. Identifying lncRNA-related regulatory modules and exploring the role of lncRNA in the development of breast cancer provides research and theoretical basis for the diagnosis and treatment of patients.

D.-S. Huang and K.-H. Jo (Eds.): ICIC 2020, LNCS 12464, pp. 57–66, 2020.
https://doi.org/10.1007/978-3-030-60802-6_6

Various computational methods have been proposed to identify regulatory modules. For example, Zhang et al. [6] proposed a miRSM method to explore miRNA sponge modules by expression data and putative miRNA-target interactions. First, he calculated the miRNA-mRNA regulatory score matrix *W* with the Pearson correlation method. The context ++ score matrix *T* was obtained by querying binding information from TargetScan [7]. The miRNA-mRNA regulatory score matrix *S* was generated by combining *W* and *T*. Second, the bicluster [8] method was applied to *S* to get various biclusters. Third, according to mRNA involved in each bicluster, miRNA-mRNA pair, which met given conditions, were considered to form a module. Xiao et al. [9] proposed JONMF method to identify lncRNA and mRNA co-expression modules in ovarian cancer. This method combined large-scale genomic data and a model to characterize the lncRNA-related modules through network-regularized constraints.

Here, we proposed a novel computational method to identify lncRNA-mRNA modules of breast cancer based on lncRNA similarity, and our method was called BCLMM. First, BCLMM calculated the Pearson correlation matrix between lncRNAs based on the corresponding lncRNA expression data of breast cancer. Second, BCLMM calculated the lncRNA similarity matrix, whose values relied on the average value of the Pearson correlation matrix and the collected functional similarity matrix. Third, BCLMM applied BCPlaid [8] to the lncRNA similarity matrix. Subsequently, various subclasses were generated. The lncRNAs in each subclass were regarded as having a strong similarity. Then the collected lncRNAs of breast cancer were utilized to filter each subclass to get a subclass that only included lncRNAs of breast cancer. Finally, lncRNA-mRNA pairs were searched from validated lncRNA-mRNA interactions based on the lncRNA in each subclass to form the corresponding lncRNA-mRNA modules.

For each module, biological analysis and survival analysis were conducted. The biological analysis shows that each module is enriched in GO terms [10] and KEGG enrichment [11], indicating that each module is closely associated with breast cancer. The survival analysis shows that each module can act as a marker of breast cancer. Simultaneously SVM [12] with 10-fold cross-validation was applied to each module to distinguish normal and tumor samples, the results also show the power classification performance of each module.

2 Materials and Methods

2.1 Data Collection

LncRNA Functional Similarity. Chen et al. [13] calculated the Spearman correlation coefficient between any two lncRNAs. With the point that similar lncRNAs tend to show similar interaction and non-interaction patterns with diseases, the Gaussian interaction profile kernel similarity of lncRNA was calculated. Based on the Spearman correlation coefficient and Gaussian interaction profile kernel similarity of lncRNA, the functional similarity of 104 lncRNAs was generated.

LncRNA and mRNA Expression Profiles. Zhang et al. [14] collected the lncRNA and mRNA expression profiles of breast cancer from Paci et al. [15]. He utilized R packages to handle the data and finally obtained 470 lncRNAs and 11157 mRNAs expression profiles. Those expression profiles contain 72 normal samples and 72 tumor samples. Besides, 58 lncRNAs of breast cancer were gathered from his processed data.

Validated lncRNA-mRNA Interactions. Zhang et al. [14] collected 195,257 experimentally verified lncRNA-target interactions from NPInter v3.0 [16], LncRNA2Target v1.2 [17] and LncRNADisease v2015 [18]. 978, 56, 746, and 1,444 experimentally validated lncRNA-mRNA interactions were collected from LncACTdb 2.0 [19], LncCeRBase [20], LncRNA2Target v2.0 [17], and starBase v2.0 [21], respectively. All collected lncRNA-mRNA interactions were combined into integrity.

2.2 Method

As shown in Fig. 1, BCLMM includes the following five steps:

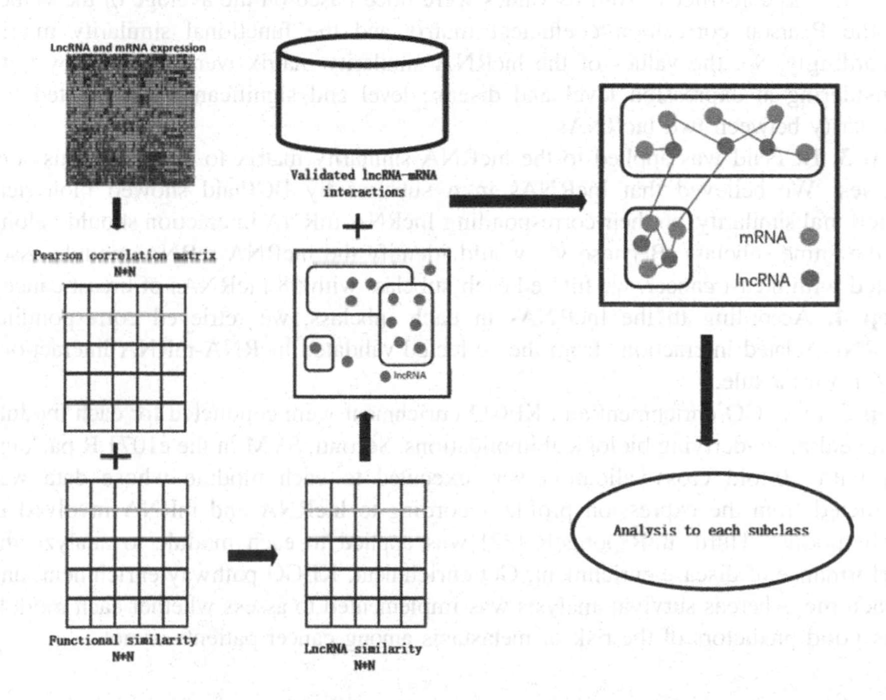

Fig. 1. The workflow of BCLMM to explore lncRNA-mRNA modules. First, BCLMM calculated the Pearson correlation matrix between any two lncRNAs. Second, BCLMM formed the lncRNA similarity matrix based on the Pearson correlation matrix and the lncRNA functional similarity. Third, we applied BCPlaid to the lncRNA similarity matrix to obtain subclasses. Four, lncRNA-mRNA modules were identified by querying lncRNA-mRNA pairs from validated lncRNA-mRNA interactions according to lncRNAs involved in each subclass. Finally, each module was analyzed.

Step 1. There co-existed 29 lncRNAs among the 470 lncRNAs collected from Zhang and 104 lncRNAs collected from Chen. So, a Pearson correlation coefficient matrix and a functional similarity matrix whose size both were 29 * 29 were constructed. Each row or column corresponded to one of 29 lncRNAs. The values of the Pearson correlation matrix were calculated based on the two corresponding lncRNA expression data from the expression profile. The values of the functional similarity matrix were filled based on the matched value querying from the lncRNA functional similarity designed by Chen.

Step 2. Zhang et al. [6] constructed miRNA-mRNA regulatory score matrix S based on the formula $S = a * W + b * T$, W was the miRNA-mRNA correlation matrix and was calculated based on Pearson correlation between miRNA and mRNA, and T was context ++ score matrix and was obtained by querying putative miRNA-target binding information, a and b were tuning parameters with default values of 0.5. In our paper, the Pearson correlation matrix was constructed by considering lncRNA similarity at the expression level and the functional similarity matrix was obtained by considering lncRNA similarity at the disease level. A lncRNA similarity matrix whose size was 29 * 29 was constructed. And its values were filled based on the average of the values of the Pearson correlation coefficient matrix and the functional similarity matrix accordingly. So, the values of the lncRNA similarity matrix were obtained by both considering at expression level and disease level and significantly represented the similarity between two lncRNAs.

Step 3. BCPlaid was applied to the lncRNA similarity matrix to obtain various subclasses. We believed that lncRNAs in a subclass by BCPlaid showed biological functional similarity, so their corresponding lncRNA-mRNA interaction should belong to the same subclass. Because we would identify the lncRNA-mRNA module associated with breast cancer, we filtered each subclass with 58 lncRNAs of breast cancer.

Step 4. According to the lncRNAs in each subclass, we retrieved corresponding lncRNA-related interactions from the collected validated lncRNA-mRNA interactions to form a module.

Step 5. First, GO enrichment and KEGG enrichment were conducted for each module to reveal the underlying biological implications. Second, SVM in the e1071 R package [8] with 10-fold cross-validation was executed to each module whose data was extracted from the expression profile according to lncRNA and mRNA involved in each module. Third, miRspongeR [22] was applied to each module to analyze the performance of disease enrichment, GO enrichment, KEGG pathway enrichment, and Reactome, whereas survival analysis was implemented to assess whether each module was good predictors of the risk of metastasis among cancer patients or not.

3 Results

As shown in Fig. 2. Four lncRNA-mRNA regulatory modules (M1, M2, M3, M4) of breast cancer were identified. Four modules included 4,970 lncRNA-mRNA interactions. 2,395 lncRNA-mRNA interactions co-existed between module 3 and module 4, almost all lncRNA-mRNA interactions in module 1 were included in module 3, and ever there was no overlap among the four modules, which indicated that lncRNA-mRNA interactions tended to be module conserved across two modules and a small number of lncRNA-mRNA interactions were module-specific.

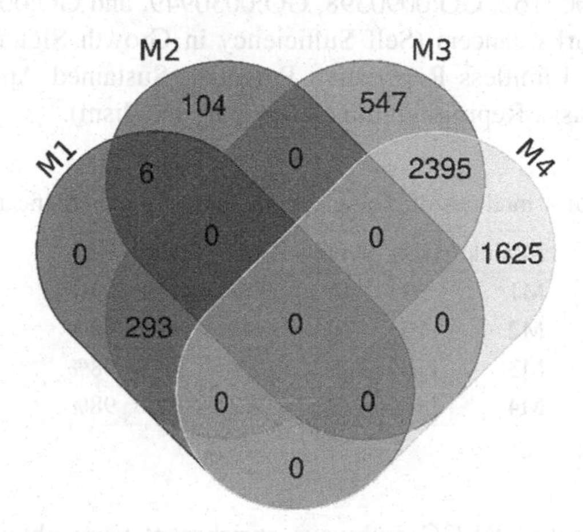

Fig. 2. The overlap between 4 modules

There were a large amount of common lncRNA-mRNA interactions between module 3 and module 4. And KEGG enrichment and Go enrichment were made to those common interactions, the results showed that those interactions contained ample biological enrichment associated with the disease. Hub analysis of those interactions indicated the top 10 hub genes participated in numerous pathways and may be potential cancer drivers. Survival analysis of common interactions between modules 3 and module 4 indicated those interactions had the powerful ability to separate normal samples and cancer samples and also to predict the metastasis risks for patients. Those facts showed common interactions between module 3 and module 4 required more focus to study.

3.1 KEGG Pathways and GO Analysis

GO enrichment and KEGG analysis at a significant level were conducted to the four modules, respectively. As shown in Table 1 module 1 included 393 GO terms, among which 5 GO terms (GO:0045787, GO:0071456, GO:0007162, GO:0090398, and GO:0045765) were associated with 5 hallmarks cancers (Self Sufficiency in Growth Signals, Limitless Replicative Potential, Sustained Angiogenesis, Tissue Invasion and Metastasis, Reprogramming Energy Metabolism). Module 2 included 197 GO terms, among which 2 GO terms (GO:0001837 and GO:0007162) were associated with tissue invasion and metastasis. Module 3 included 1,174 GO terms, among which 12 GO terms (GO:0045787, GO:0006096, GO:0071456, GO:0001837, GO:0007162, GO:0032206, GO:0030308, GO:0001570, GO:0002367, GO:0045765, GO:0030307, and GO:0045 005) were associated with 8 hallmarks cancers (Self Sufficiency in Growth Signals, Insensitivity to Antigrowth Signals, Limitless Replicative Potential, Sustained Angiogenesis, Tissue Invasion and Metastasis, Genome Instability and Mutation, Tumor

Promoting Inflammation, Reprogramming Energy Metabolism, Evading Immune Detection). Module 4 included 1,184 GO terms, among which 11 GO terms (GO:0045787, GO:0006096, GO:0001837, GO:0030308, GO:0071456, GO:0032206, GO:0001570, GO:0007162, GO:0090398, GO:0030949, and GO:0045765) were associated with 6 hallmarks cancers (Self Sufficiency in Growth Signals, Insensitivity to Antigrowth Signals, Limitless Replicative Potential, Sustained Angiogenesis, Tissue Invasion and Metastasis, Reprogramming Energy Metabolism).

Table 1. Details of 4 modules of biological enrichment and classification performance

Module	GOs	KEGG	ACC	AUC	OPI
M1	393	41	99%	100%	99%
M2	197	30	99%	99%	99%
M3	1184	108	98%	99%	98%
M4	1174	112	98%	99%	98%

41, 30, 108, and 112 KEGG pathways enrichment were obtained from the four modules, respectively, among those many KEGG pathways participated in many biological processes (such as apoptosis, cell cycle, DNA replication, signaling pathway and so on) and were closely related to the development of various diseases (such as Gastric cancer, Bladder cancer, Prostate cancer Colorectal cancer and so on). Especially, hsa04110 shared by the four modules were related to cell cycle, and hsa05224 shared by the four modules were associated closely with breast cancer. Several KEGG pathways (hsa04115, hsa04151, and hsa04010) were experimentally verified as playing a crucial role in breast cancer.

3.2 Classification Analysis

The expression profile of each gene was comprised of 144 samples, including 72 normal samples and 72 cancer samples. SVM with 10-fold cross-validation was conducted to each module whose data was extracted from the expression profile according to lncRNAs and mRNAs involved in each module. The classification performance to classify normal and tumor samples in each module was evaluated by ACC, AUC and OPI [14]. As shown in Table 1, 4 modules displayed higher 98% classification performance.

3.3 Validation by MiRspongeR

MiRspongeR, an R/Bioconductor package, provides seven methods and an integrative approach to identify and explore miRNA sponge interactions and its modules. It also includes tools to analyze biological enrichment for modules. Function enrichment was performed to the four modules. First, a disease enrichment analysis was evaluated by DO numbers (DOs), DGN numbers (DGNs), and NCG numbers (NCGs). Second, the function to analyze GO, KEGG pathway, and Reactome enrichment analyses for the four modules were evaluated by GO numbers (GOs), KEGG numbers (KEGGs), and Reactome number (Reactomes). As shown in Table 2, the four modules were abundant in Ds, DGNs, NCGs, GOS, KEEGs, and Reactomes.

Table 2. DOs, DGNs, NCGs, GOS, KEEGs, and Reactomes of four modules

Module	DOs	DGNs	NCGs	GOs	KEGGs	Reactomes
M1	215	1025	4	390	85	55
M2	262	954	0	200	30	160
M3	304	1861	12	1859	110	202
M4	251	1178	9	1180	112	87

MiRspongeR also provides a function to make survival analysis to identified module. All samples are divided into the high risk and low risk groups according to their risk scores with the Cox model. As shown in Fig. 3. Four modules were divided into high risk and the low risk groups.

4 Comparison with Other Methods

MiRSM, miRCoPPI [14] and EPMSIs [23] were the latest best methods to explore modules of breast cancer. In this section, BCLMM was compared with three other methods. MiRSM and miRCoPPI firstly identified modules from a large amount of data which included breast cancer genes and non-breast cancer genes. Then the modules were obtained by filtering the breast cancer gene. BCLMM and EPMSIs firstly utilized the breast cancer gene to identify the module effectively.

Because the number of modules identified by these four methods was different, the average number of GO enrichment and KEGG enrichment for four methods were used to compare. As shown in Fig. 4. Four modules identified by BCLMM were more enriched than the other three methods in terms of biological enrichment. BCLMM evaluated classification performance and survival analysis, and the results also demonstrated the great potential to identify biological significance modules. The results validated by miRspongeR showed the modules identified by BCLMM associated more closely with the disease.

5 Conclusion and Discussion

Increasing evidence shows that lncRNAs are important regulators. They play a very important role in the body's physiological and pathological processes and are particularly relevant to diverse diseases. The study of their mechanism can provide diagnosis and solutions to the prevention of disease. In this paper, we proposed a novel computation method named BCLMM to identify lncRNA-mRNA modules of breast cancer based on lncRNA similarity. First, BCLMM calculated the lncRNA similarity matrix

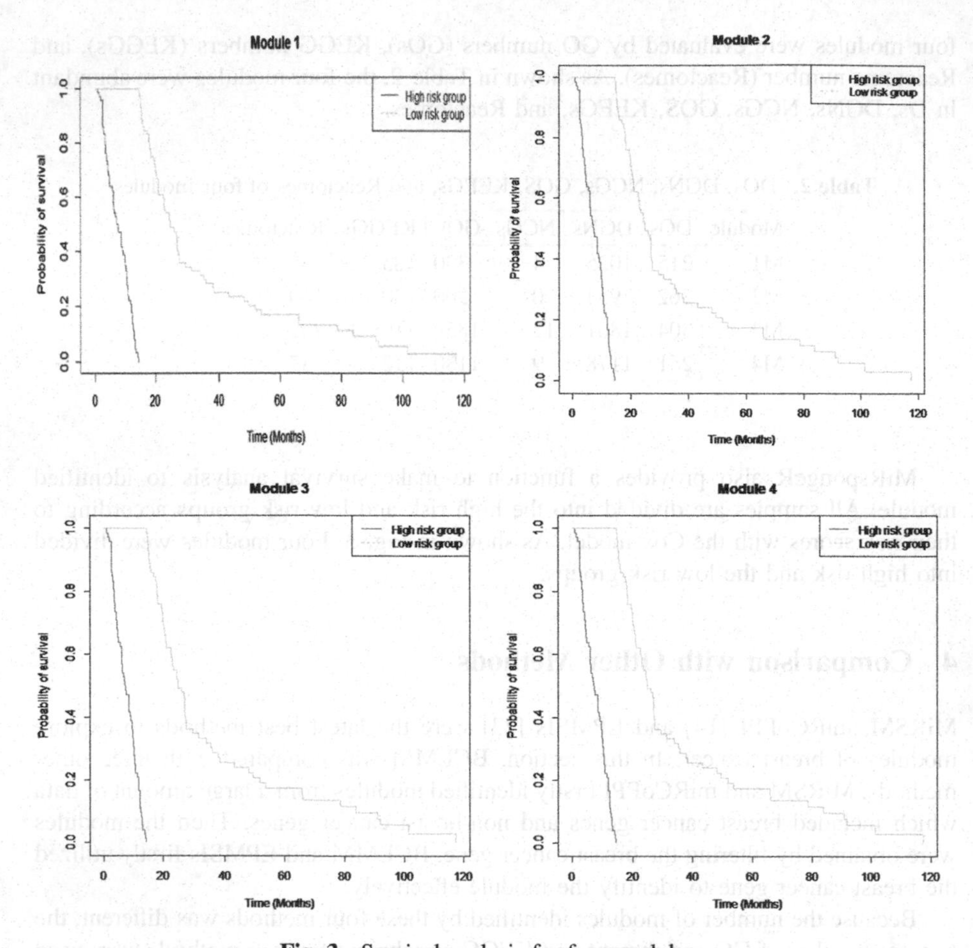

Fig. 3. Survival analysis for four modules.

based on the Pearson correlation matrix and the collected functional similarity matrix. Then BCLMM applied BCPlaid to the lncRNA similarity matrix to identify subclasses. Then the lncRNA of breast cancer in each class was used to search for lncRNA-mRNA pairs from validated lncRNA-mRNA interactions to form the corresponding lncRNA-mRNA modules. Consequently, 4 modules were identified. Biological analysis and survival analysis were conducted to 4 modules, the result showed BCLMM had excellent in identifying biological significant modules associated with breast cancer and significantly indicated that each module was biologically significant and associated closely with the disease, and could act as good predictors of the metastasis risks of breast cancer patients. Simultaneously SVM with 10-fold cross-validation was applied to each module to distinguish normal and tumor samples, which indicated four modules could act as module signatures for prognostication of human breast cancer for excellent classification performance. Compared with other methods, the modules identified by BCLMM showed a stronger correlation with biological significance, which indicated that four modules identified were biologically significant and also could act as a

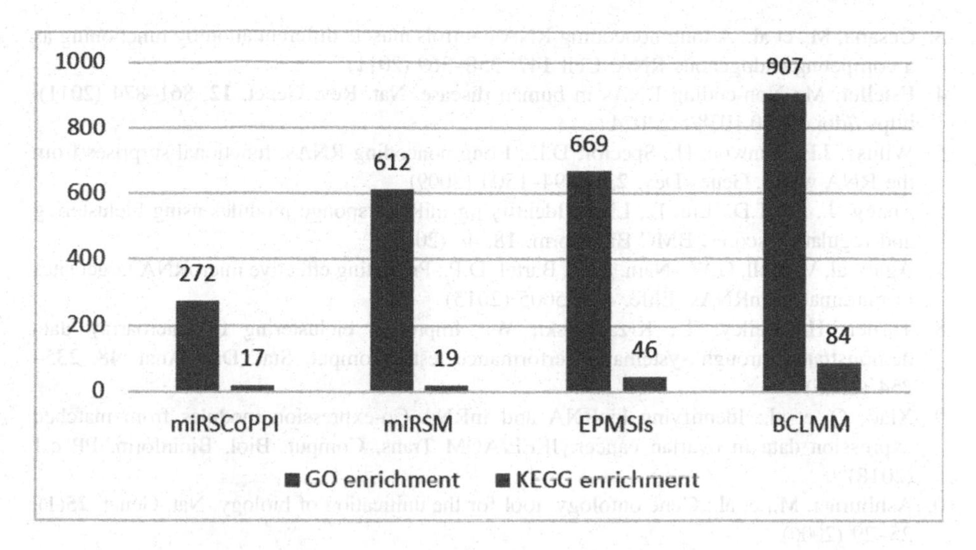

Fig. 4. Comparison between four methods in GO and KEGG enrichment

functional unit related to breast cancer. All results demonstrated that BCLMM was a promising way to identify modules significantly enriched in biological enrichment and associated with breast cancer, and maybe it could provide a new way to explore and study physiological and pathological processes of other diseases.

Although BCLMM has achieved good performance, there are still many disadvantages that need to be resolved in the future. The first thing to be solved is to build a larger data set of lncRNA. BCLMM only constructed a similar matrix of 29 lncRNAs, which may make it hard to have a landscape of regulations for lncRNA. The second thing to be solved is the coefficient ratio between the Pearson correlation coefficient matrix and the functional similarity matrix. In this paper, we calculated the average value simply by reference to the method of Zhang. Perhaps the different coefficient ratios between the two matrixes may generate a different result. The third thing is that we need to conceive more ways for validation of identified modules, not just limited to function enrichment analysis.

Acknowledgments. This work was supported by the National Nature Science Foundation of China (Grant Nos. 61672011 and 61472467), and the National Key Research and Development Program (Grant Nos. 2017YFC1311003).

References

1. Bertone, P., et al.: Global identification of human transcribed sequences with genome tiling arrays. Science **306**(5705), 2242–2246 (2004)
2. Birney, E., et al.: Identification and analysis of functional elements in 1% of the human genome by the ENCODE pilot project. Nature **447**, 799–816 (2007)

3. Cesana, M., et al.: A long noncoding RNA controls muscle differentiation by functioning as a competing endogenous RNA. Cell **147**, 358–369 (2011)
4. Esteller, M.: Non-coding RNAs in human disease. Nat. Rev. Genet. **12**, 861–874 (2011). https://doi.org/10.1038/nrg3074
5. Wilusz, J.E., Sunwoo, H., Spector, D.L.: Long noncoding RNAs: functional surprises from the RNA world. Genes Dev. **23**, 1494–1504 (2009)
6. Zhang, J., Le, T.D., Liu, L., Li, J.: Identifying miRNA sponge modules using biclustering and regulatory scores. BMC Bioinform. **18**, 44 (2017)
7. Agarwal, V., Bell, G.W., Nam, J.W., Bartel, D.P.: Predicting effective microRNA target sites in mammalian mRNAs. Elife. **4**, e05005 (2015)
8. Turner, H., Bailey, T., Krzanowski, W.: Improved biclustering of microarray data demonstrated through systematic performance tests. Comput. Stat. Data Anal. **48**, 235–254 (2005)
9. Xiao, Q, et al.: Identifying lncRNA and mRNA Co-expression modules from matched expression data in ovarian cancer. IEEE/ACM Trans. Comput. Biol. Bioinform. PP c:1 (2018)
10. Ashburner, M., et al.: Gene ontology: tool for the unification of biology. Nat. Genet. **25**(1), 25–29 (2000)
11. Ogata, H., Goto, S., Sato, K., Fujibuchi, W., Bono, H., Kanehisa, M.: KEGG: kyoto encyclopedia of genes and genomes. Nucleic Acids Res. **28**(1), 27–30 (1999)
12. Chang, C., Lin, C., Tieleman, T.: LIBSVM: a library for support vector machines. ACM Trans Intell Syst Technol. **2**, 27 (2008)
13. Chen, X., Clarence Yan, C., Luo, C., Ji, W., Zhang, Y., Dai, Q.: Constructing lncRNA functional similarity network based on lncRNA-disease associations and disease semantic similarity. Sci. Rep. **5**(11338), 11338 (2015). https://doi.org/10.1038/srep11338
14. Zhang, J., Le, T.D., Liu, L., Li, J.: Inferring miRNA sponge co-regulation of protein-protein interactions in human breast cancer. BMC Bioinform. **18**, 235 (2017)
15. Paci, P., Colombo, T., Farina, L.: Computational analysis identifies a sponge interaction network between long non-coding RNAs and messenger RNAs in human breast cancer. BMC Syst. Biol. **8**, 83 (2014)
16. Hao, Y., et al.: NPInter v3.0: an upgraded database of noncoding RNA-associated interactions. Database: baw057–baw057 (2016)
17. Cheng, L., et al.: Lncrna2target v2.0: a comprehensive database for target genes of lncrnas in human and mouse. Nucleic Acids Res. **47**, D140–D144 (2019)
18. Chen, G., et al.: LncRNADisease: a database for long-non-coding RNA-associated diseases. Nucleic Acids Res. **41**, 983–986 (2013)
19. Wang, P., et al.: Lncactdb 2.0: an updated database of experimentally supported cerna interactions curated from low- and high-throughput experiments. Nucleic Acids Res. **47**, D121–D127 (2019)
20. Pian, C., Zhang, G., Tu, T., Ma, X., Li, F.: LncCeRBase: a database of experimentally validated human competing endogenous long non-coding RNAs. Database **2018**, 1–4 (2018)
21. Li, J.H., Liu, S., Zhou, H., Qu, L.H., Yang, J.H.: StarBase v2.0: decoding miRNA-ceRNA, miRNA-ncRNA and protein-RNA interaction networks from large-scale CLIP-Seq data. Nucleic Acids Res. **42**, D92–D97 (2014)
22. Zhang, J., et al.: MiRspongeR: an R/Bioconductor package for the identification and analysis of miRNA sponge interaction networks and modules. BMC Bioinform. **20**, 1–12 (2019)
23. Tian, L., Wang, S.-L.: Exploring the potential microrna sponge interactions of breast cancer based on some known interactions. J. Bioinform. Comput. Biol. 6 October: 1–8 (2019). https://doi.org/10.1038/srep36625

Three-Layer Dynamic Transfer Learning Language Model for E. Coli Promoter Classification

Ying He[1(✉)], Zhen Shen[1], Qinhu Zhang[1], Siguo Wang[1], Changan Yuan[2], Xiao Qin[3], Hongjie Wu[4], and Xingming Zhao[5]

[1] Institute of Machine Learning and Systems Biology, Department of College of Electronics and Information Engineering, Tongji University, Shanghai 201804, China
heyingzy@163.com

[2] Guangxi Academy of Science, Nanning 530025, China

[3] School of Computer and Information Engineering, Nanning Normal University, Nanning 530299, China

[4] School of Electronic and Information Engineering, Suzhou University of Science and Technology, Suzhou 215009, China

[5] Institute of Science and Technology for Brain Inspired Intelligence (ISTBI), Fudan University, Shanghai 200433, China

Abstract. Classification of functional genomic regions (such as promoters or enhancers) based on sequence data alone is a very important problem. Various data mining algorithms can be used well to apply to predict the promoter region. For example, association and clustering algorithms like Classification And Regression Tree (CART), machine learning algorithms like Simple Logistic, BayesNet, Random forest, or the most popular deep learning like Recurrent Neural Network (RNN), Convolutional Neural Networks (CNN). However, due to large amount of genetic data are unlabeled, these methods cannot directly solve this challenge. Therefore, we present a three-layer dynamic transfer learning language model (TLDTLL) for E. coli promoter classification problems. TLDTLL is an effective algorithm for inductive transfer learning that utilizes pre-training on large unlabeled genomic corpuses. This is particularly advantageous in the context of genomics data, which tends to contain significant volumes of unlabeled data. TLDTLL shows improved results over existing methods for classification of E. coli promoters using only sequence data.

Keywords: Pre-training model · Transfer learning · Data mining · E. coli promoter · Classification

1 Introduction

Motif plays a key role in the gene-expression regulating on both transcriptional and post-transcriptional levels. DNA-Protein/RNA-Protein binding motifs are involved in many biological processes including transcription, translation, and alternative splicing. As an important part of gene expression, the main task of translational regulation is to

D.-S. Huang and K.-H. Jo (Eds.): ICIC 2020, LNCS 12464, pp. 67–78, 2020.
https://doi.org/10.1007/978-3-030-60802-6_7

regulate processes related to protein synthesis [1]. From the late 1990s to the early 21st century, researchers through biological experiments gradually identified a large number of proteins with binding functions and their corresponding binding sites on the genome sequence, because the binding sites of the same protein have certain conservative, so people initially used conservative sequences to describe protein binding sites [2].

Advances in genomics research, such as high-throughput sequencing, have driven the explosive growth of genomic data. Large amounts of data require powerful algorithms to analyze and interpret the data. Genome classification-assigning labels to genome sequences based on function or some other attribute is the basis for processing raw genomic data. There are algorithms for classifying sequences based on manual engineering features, but these methods are difficult to solve the structural heterogeneity of genomic sequences [3]. The structure of genome regulatory features (such as promoters) is well known to be complex and diverse. A large amount of experimental data is provided for a more systematic study of the primitive binding mode. For example, ChIP-seq technology [4] combines chromatin immunoprecipitation technology and high-throughput sequencing technology (High-throughput sequencing).

Protein is the final product of gene expression and it is deeply involved in various complex life activities in the human body. Abnormalities in the stage of protein synthesis can cause disturbances in cell function, which in turn disrupts normal life activities in the human body and leads to the development of complex malignant diseases [5]. Therefore, the systematic study of DNA-Protein/RNA-Protein binding can help us understand the translational regulation mechanism in gene expression [6]. DNA-Protein/RNA-Protein binding sites is also known as Motif Discovery [7]. With the development of high-throughput sequencing technology [8], a large number of biologically meaningful experimental data has been provided for motif discovery [9], which greatly promoted research related to DNA-Protein/RNA-Protein binding [10] and gives researchers the opportunity to further explore the nature of translation regulation and complex life phenomena [11]. The use of traditional machine learning algorithms for motif discovery can overcome the huge consumption of human, material and financial resources by biological experiments [12], but it also faces the defects of high time complexity [13], noise sensitivity and poor generalization performance [14, 15]. The motif discovery model based on deep neural network overcomes the shortcomings of traditional motif discovery methods and achieves better prediction results [16–18], which brings a new direction for the design and development of efficient motif discovery algorithms [19].

Recently, deep learning methods have attempted to use large genomic datasets to train end-to-end models to address the complex structural diversity of genomic sequences. In the context of classification problems, deep learning has been applied to promoter classification [3, 20], enhancer classification [21–24], enhancer-promoter interaction [25], CRISPR guide score [26], transcription factor binding site [19], metagenomics classification [27], delitrious mutation classification [28] and long non-coding RNA [29] classification. The limitation of the implementation of these methods is that they all rely on marked data. Even if techniques such as pre-training and transfer learning are used, such steps are only applied to labeled categorical corpora [19, 24, 25]. This poses a problem because labeled genomic data sets are often very small, and deep learning models tend to overfit on small data sets [30].

Pre-trained models [31] and transfer learning [32] are widely used in algorithms such as NLP and CV, and are used to train text classification models using transfer learning. The pre-training model uses unsupervised pre-training on large unlabeled domain corpora to improve the classification of smaller labeled datasets. Compared with training from scratch, this method has shown to reduce the error rate of ordinary text classification tasks by 18–24%, while also reducing data requirements by 100 times [1]. In the context of genomic data, a pre-trained model allows us to pre-train a large amount of unlabeled sequence data, and then use the pre-trained model to classify smaller labeled data. This helps to solve the limitation caused by labeled sequence availability.

In order to solve the limitation of labeled sequence availability, we utilized applications of pre-trained models and transfer learning in NLP and CV to obtain TLDTLL models. TLDTLL adapt pre-trained models and transfer learning to genomics data with k-mer based tokenization.

2 Method

2.1 Data Preparation

All genomes are downloaded from NCBI. For example, E. coli genomic data and TSS locations are taken from the E. coli reference genome genbank file available at NCBI Escherichia coli.

To train a sequence model on genomic data, we first need to figure out how to process the data into a form that can be used by neural networks. We need to convert the sequence data into a digital form that can be processed mathematically. We proceed in two steps: tokenization and digitization.

Tokenization is the process of breaking the sequence down into sub-units or tokens. Numericalization is the process of mapping tokens to integer values.

A common genomic tokenization method is to label with nucleotides [33, 34]. Single nucleotide labeling will treat the sequence “ATCGCGTACG” as “A T C G C G T A C G”. This is feasible, but it provides a very limited representation of the genomic subunit. This allows each nucleotide to be viewed in a vacuum. Essentially, no matter where or in what context “A” appears, it should be considered the same. The more nuanced representation method is to use the k-mers instead [35–37]. Such as, we can tokenize the sequence “ATCGCGTACGATCCG” as Table 1.

Table 1. Operation process parameters.

k-mers	Strides	Tokenized
3	3	ATC GCG TAC GAT CCG
4	4	ATCG CGTA CGAT
5	5	ATCGC GTACG ATCCG
4	2	ATCG CGCG CGTA TACG CGAT ATCC
4	3	ATCG GCGT TACG GATC

Notice how the stride parameter affects the number of tokens created per length of input sequence. The impact of the choice of k-mer and stride values is discussed more in Sect. 3. For now, understand that k-mer and stride values are hyperparameters that must be decided before training begins, and that the choice of k-mer and stride has an effect on compute time and performance.

2.2 TLDTLL Model

After data preparation, we can process the genomic data into a form that can be fed into the TLDTLL model, we need to determine the strategy for training the TLDTLL model. Let's start by defining the ultimate goal: we want to train a sequence model to classify E. coli promoter using only sequence input. In E. coli promoter classification, sequence models often require large amounts of data to be effectively trained, and labeled promoter classification datasets may be small. The TLDTLL method provides a solution for this. TLDTLL divides the training into three stages:

First, we use an unsupervised training general domain language model on a large unlabeled corpus.

Secondly, We fine-tune the general language model on the classified corpus to create task-specific language models.

Finally, We fine-tune the task-specific language model for promoter classification.

Before going further, let us define two types of models that will be processed. A language model is a model that absorbs a series of k-mer sequence tokens and predicts the next token in the sequence. A classification model is a model that accepts a labeled sequence and predicts the category or category to which the sequence belongs.

Language models are trained without supervision, which means that no labeled data is required. Since the goal of the language model is to predict the next k-mer in the sequence, each k-mer will become the correct output prediction for the sequence before the sequence. This means that we can generate a large amount of pairing data from any unlabeled genomic sequence (input sequence + next k-mer).

A classification model in a supervised fashion, requires paired labeled data. For example, in this task is a promoter classification requires for data labeled as non-promoters or promoters, all sequences in the classification dataset must be marked as 0 or 1.

The main structure of TLDTLL is composed of classification model and language model follows a similar structure-consisting of embedding, encoder and linear header (see Fig. 1).

At a higher level, these layers function in the following ways:

1. Embedding Layer: convert the digital markers of the input sequence into a vector representation

2. Encoder Layer: processing vectorized sequences into hidden states

3. Decision Layer: Use hidden states to make classification decisions.

Notice that, as we move between stages, we transfer the learned weights from one model to another. When training a language model, we will transfer all three parts of the model. When we moved to the classification model, we only transferred the embedding and encoder, because the classification model requires different decision layer. In specific language model, the output of the decision layer is the predicted next

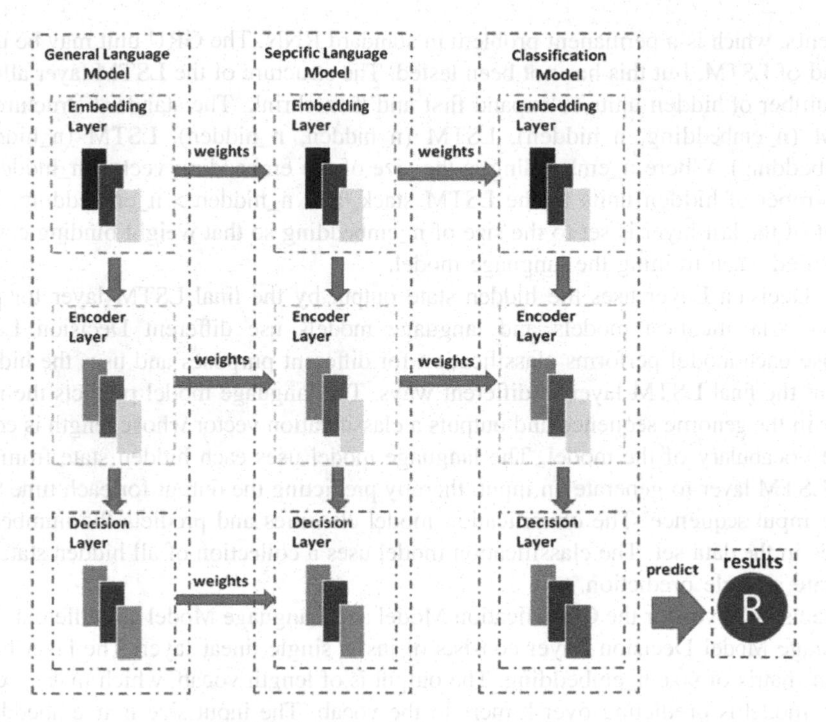

Fig. 1. The structure of TLDTLL

k-mer. The more accurate the prediction, the better the embedding layer. Next we introduce these three layers in detail.

2.3 TLDTLL Model Parameters

In Embedding Layer, the final input of the model is an integer vector. Each integer value represents k-mer in our data set vocabulary. The integer sequence represents the true sequence of the k-mer. The first step in processing the input sequence is to convert the integer value to some vector representation. A common way to accomplish this in the literature is to convert each token into a hot-coded vector. This is possible, but it is not a very rich representation. All vectors are mutually orthogonal and do not contain any information about the relationship between k-mers. By using embedding to express k-mers as learning vectors, the model can understand more meaningful relationships between k-mers. This is also functionally the same as using a hot representation and passing them through the learned weight matrix. The size of the embedding weight matrix is vocab x n_embedding, where vocab is the length of the model vocabulary and n_embedding is the length of the embedding vector.

In Encoder Layer part consists of three stacked LSTM layers. The structure comes from the AWD-LSTM model [13]. LSTM is used on standard RNN because the updated structure of LSTM allows the model to retain information on longer sequences and filter the information at each time step [14]. LSTM is also not easy to disappear gradients, which is a permanent problem in standard RNN. The GRU unit may be used instead of LSTM, but this has not been tested. The structure of the LSTM layer allows the number of hidden units to expand first and then shrink. The standard structure is: LSTM (n_embedding, n_hidden), LSTM (n_hidden, n_hidden), LSTM (n_hidden, n_embedding). Where n_embedding is the size of the embedding vector, n_hidden is the number of hidden units in the LSTM stack, and n_hidden > n_embedding. The output of the last layer is set to the size of n_embedding so that weight binding can be performed when training the language model.

In Decision Layer uses the hidden state output by the final LSTM layer for prediction. Classification models and language models use different Decision Layer because each model performs classification for different purposes and uses the hidden state in the final LSTM layer in different ways. The language model predicts the next k-mer in the genome sequence and outputs a classification vector whose length is equal to the vocabulary of the model. The language model uses each hidden state from the final LSTM layer to generate an input, thereby predicting the output for each time step in the input sequence. The classification model classifies and predicts the number of classes in the data set. The classification model uses a collection of all hidden states to generate a single prediction.

Decision Layer for the Classification Model and Language Model are different. The Language Model Decision Layer consists of just a single linear layer. The layer has a weight matrix of size n_embedding. The output is of length vocab, which makes sense as the model is predicting over k-mers in the vocab. The input size is n_embedding, which makes the size of the output weight matrix identical to the input embedding matrix, just with the dimensions flipped. This is intentional, as it allows us to tie the weights of the embedding to the softmax layer. This technique is motivated by [15, 16] and found by [13] to lead to significantly improved language modeling performance. It also has the nice effect of reducing the number of parameters in the model.

The Language Model Decision Layer uses hidden states from all time steps to output predictions at every time step. So if a sequence section is 200 k-mers long, the model outputs a matrix of 200 x vocab of next-k-mer predictions at every time step. This allows us to massively expand the amount of usable data we have for the language model. Each k-mer serves as the output value for the previous k-mer, and the input value for the subsequent k-mer. An important caveat to note is that this training approach is not compatible with bidirectional LSTMs. If a bidirectional LSTM is used, the model will be predicting over k-mers it has already seen. In practice, the linear layer is implemented using Pytorch's nn. Linear module. The weights for the softmax layer are tied to the weights of the embedding layer. Typically, bias is also included in the final layer, so strictly speaking there is not 100% weight tying between the embedding and the softmax layer.

The Decision Layer for the Classification Model is more complex than the linear head for the Language Model. The Classification Model head consists of two linear layers with batchnorm layers in between. The classification head also uses the LSTM hidden states differently. Typically, the final hidden state from the last LSTM layer is used for classification, but the most important parts of the sequence might be buried in the middle. Following the methods used in [1], we take the three vectors - the final hidden state, a maxpooling over all hidden states, and an average pooling over all hidden states - and concatenate them together. So for a vector containing hidden states at all time steps, we create and use the vector as input to the linear head. It will have a length of n_embedding*3. Since we transfer the learned wights from the embedding and encoder of the language model, we still use the LSTMs from the language model which output hidden states of length n_embedding.

3 Results and Discussion

3.1 TLDTLL Practical Model Parameters

First of all, we introduce the parameters used in practice as shown in Fig. 2.

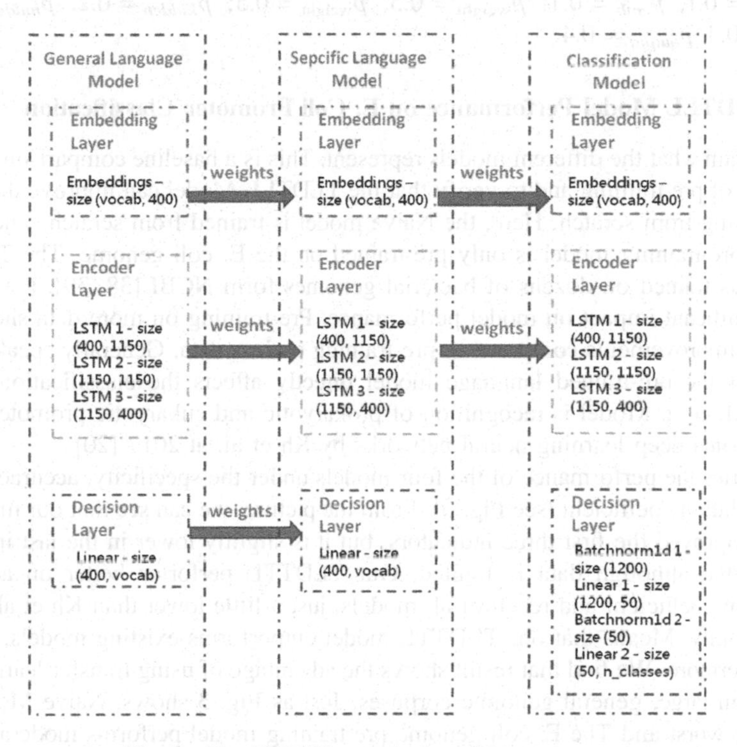

Fig. 2. The TLDTLL practical model parameters

Just as the Fig. 2 shows, For Embeddings Layer: size (vocab, 400), LSTM 1-size (400, 1150), LSTM 2-size (1150, 1150), LSTM 3-size (1150, 400). For the Language Model Decision Layer: Linear-size (400, vocab). For the Classification Model Decision Layer: Batchnorm1d 1-size (1200), Linear 1-size (1200, 50), Batch norm1d 2 - size (50), Linear 2 - size (50, n_classes).

Beside four dropout hyperparameters to set:

1. p_{emb} - dropout for the Embedding Layer
2. p_{weight} - dropout for the weights of the Encoder layers

3. p_{hidden} - variational dropout on the activations of the Encoder layer
4. p_{output} - standard dropout applied to activations in the Decision Layer

It is important to consider both the magnitude of each dropout parameter and the relative ratios between dropout parameters. In practice we find it best to set values for all dropout parameters and tune a parameter that scales all four dropout parameters while keeping the same ratio. Please note that we have not optimized the parameters, we just used.

For language models:

$p_{emb} = 0.02, p_{emb} = 0.02; p_{\text{weight}} = 0.15, p_{\text{weight}} = 0.15; p_{hidden} = 0.1\ p_{hidden} = 0.1;$ $p_{output} = 0.25\ p_{output} = 0.25$

For classification models:

$p_{emb} = 0.1,\ p_{emb} = 0.1;\ p_{\text{weight}} = 0.5,\ p_{\text{weight}} = 0.5;\ p_{hidden} = 0.2,\ p_{hidden} = 0.2;$ $p_{output} = 0.4, p_{output} = 0.4$

3.2 TLDTLL Model Performance on E. Coli Promoter Classification

First explain what the different models represent. This is a baseline comparison to show the effect of pre-training and to verify that the TLDTLL Model can improve the results than training from scratch. Here, the Naive model is trained from scratch. The E. coli genome pre-training model is only pre-trained on the E. coli genome. The TLDTTL model was trained on dozens of bacterial genomes form NCBI [38, 39]. Pre-training has a significant impact on model performance. Pre-training on more data shows that there are improvements compared to pre-training on less data. Generally speaking, the quality of the pre-trained language model directly affects the classification performance. Kh et al. Model is recognition of prokaryotic and eukaryotic promoters using convolutional deep learning neural networks by Kh et al. at 2017 [20].

Statistics the performance of the four models under the specificity, accuracy, recall and correlation coefficient (see Fig. 3). From the picture, we can see that our model has greatly improved the first three indicators, but it is slightly lower in the last indicator. We see that although data is limited, Our TLDTTL performs better on accuracy, correlation coefficient and recall of all models, just a little lower than Kh et al. Model on Specificity. Means that our TLDTTL model outperforms existing models.

Furthermore, We find that result shows the advantage of using transfer learning and training on large, general genomic corpuses. Just as Fig. 3 shows, Naive Model performs the worst and The E. coli genome pre-training model performs moderately and TLDTTL model performs best. When the corpus of pre-trained models is gradually increased, the performance of the model is getting better and better.

In other words, the performance from naive model to the TLDTLL model gradually gets better shows that the unlabeled corpus learned through the transfer learning model is helpful to the model and the recognition ability of the model is gradually enhanced as the model expects to increase.

As the Table 2 shows that performance have a close linear up relationship with labeled examples. Note that the TLDTLL model using only a small amount of data (20 labeled exmples) has almost the same accuracy as the Naive model using a lot of data (more than 6,000 labeled exmples). This means that our model has learned a lot of

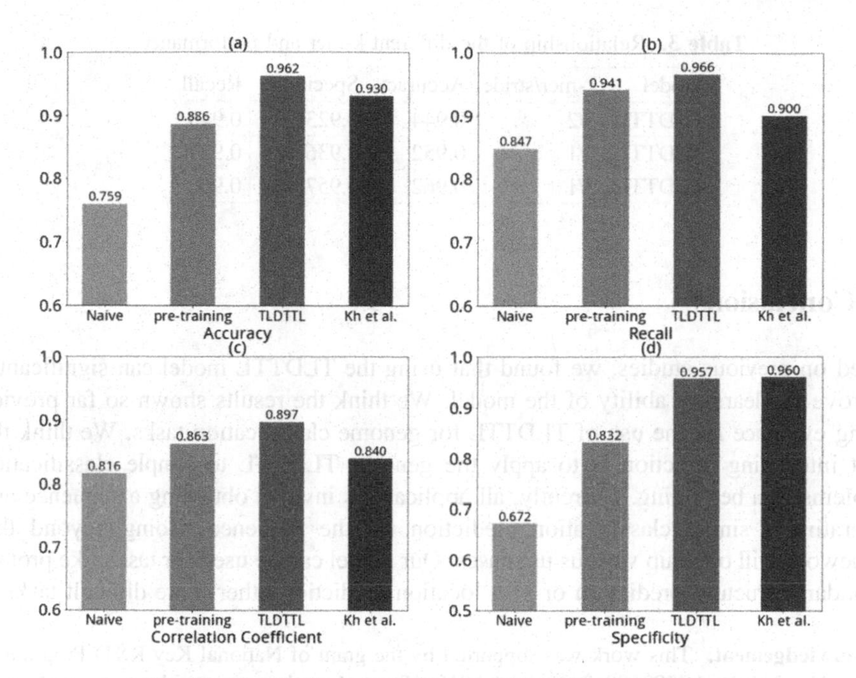

Fig. 3. Four indicators evaluation of the different models

knowledge in pre-training so that the model still has good recognition ability when using only small sample data sets.

Table 2. Relationship of the different examples and performance.

Examples	Accuracy	Recall	Correlation coefficient	Specificity
20	0.843	0.794	0.622	0.675
400	0.886	0.858	0.833	0.854
3500	0.951	0.937	0.873	0.945
6791	0.962	0.966	0.897	0.957

Just as the Table 3 shows that tokenizing genomic sequences with different k-mer and stride values impact classification performance. In this experiment we taked k-mer = 3 and stride = 1 through experimental comparison. There is no fixed method for the k-mer and stride value. We can divide the DNA sequence into k-mer sequences by specifying the k-mer length and sliding window size, and use the word embedding (Word Embedding) method to encode the k-mer sequence. Through the conversion of the embedded layer in the model, each k-mer will be converted into a vector, which contains the association information between the k-mer and the context. Different datasets respond differently to changes in k-mer and stride.

Table 3. Relationship of the different k-mer and performance.

Model	k-mer/stride	Accuracy	Specificity	Recall
TLDTTL	5/2	0.944	0.923	0.948
TLDTTL	5/1	0.952	0.936	0.957
TLDTTL	3/1	0.962	0.957	0.966

4 Conclusions

Based on previous studies, we found that using the TLDTTL model can significantly improve the learning ability of the model. We think the results shown so far provide strong evidence for the use of TLDTTL for genome classification tasks. We think the most interesting direction is to apply the genome TLDTTL to simple classification problems as a beginning. Currently, all applications involve obtaining a sequence and generating a single classification prediction for the sequence. Going beyond this framework will open up various use cases. Our model can be used for tasks like protein secondary structure prediction or TSS location prediction other more difficult tasks.

Acknowledgement. This work was supported by the grant of National Key R&D Program of China (No. 2018YFA0902600 & 2018AAA0100100) and partly supported by National Natural Science Foundation of China (Grant nos. 61520106006, 61861146002, 61702371, 61932008, 61732012, 61772370, 61532008, 61672382, 61772357, and 61672203) and China Postdoctoral Science Foundation (Grant no. 2017M611619) and supported by "BAGUI Scholar" Program and the Scientific & Technological Base and Talent Special Program, GuiKe AD18126015 of the Guangxi Zhuang Autonomous Region of China and supported by Shanghai Municipal Science and Technology Major Project (No.2018SHZDZX01), LCNBI and ZJLab.

References

1. Huang, D.-S.: Systematic Theory of Neural Networks for Pattern Recognition. Publishing House of Electronic Industry of China, Beijing, vol. 201 (1996)
2. Huang, D.-S., Zhao, X.-M., Huang, G.-B., Cheung, Y.-M.: Classifying protein sequences using hydropathy blocks. Pattern Recogn. **39**, 2293–2300 (2006)
3. Umarov, R., Kuwahara, H., Li, Y., Gao, X., Solovyev, V.: Promid: human promoter prediction by deep learning (2018). arXiv preprint arXiv:1810.01414
4. Zhu, L., Guo, W.-L., Deng, S.-P., Huang, D.-S.: ChIP-PIT: Enhancing the analysis of ChIP-Seq data using convex-relaxed pair-wise interaction tensor decomposition. IEEE/ACM Trans. Comput. Biol. Bioinf. **13**, 55–63 (2015)
5. Huang, D.-S.: Radial basis probabilistic neural networks: Model and application. Int. J. Pattern Recogn. Artif. Intell. **13**, 1083–1101 (1999)
6. Huang, D.-S., Huang, X.: Improved performance in protein secondary structure prediction by combining multiple predictions. Protein Pept. Lett. **13**, 985–991 (2006)
7. Huang, D.-S., Zheng, C.-H.: Independent component analysis-based penalized discriminant method for tumor classification using gene expression data. Bioinformatics **22**, 1855–1862 (2006)

8. Huang, D.-S., Du, J.-X.: A constructive hybrid structure optimization methodology for radial basis probabilistic neural networks. IEEE Trans. Neural Networks **19**, 2099–2115 (2008)
9. Zheng, C.-H., Huang, D.-S., Zhang, L., Kong, X.-Z.: Tumor clustering using nonnegative matrix factorization with gene selection. IEEE Trans. Inf. Technol. Biomed. **13**, 599–607 (2009)
10. Xia, J.-F., Zhao, X.-M., Song, J., Huang, D.-S.: APIS: accurate prediction of hot spots in protein interfaces by combining protrusion index with solvent accessibility. BMC Bioinf. **11**, 174 (2010)
11. Zheng, C.-H., Zhang, L., Ng, V.T.-Y., Shiu, C.K., Huang, D.-S.: Molecular pattern discovery based on penalized matrix decomposition. IEEE/ACM Trans. Comput. Biol. Bioinf. **8**, 1592–1603 (2011)
12. Huang, D.-S., Jiang, W.: A general CPL-AdS methodology for fixing dynamic parameters in dual environments. IEEE Trans. Syst. Man Cybern. Part B (Cybern.) **42**, 1489–1500 (2012)
13. Deng, S.-P., Huang, D.-S.: SFAPS: An R package for structure/function analysis of protein sequences based on informational spectrum method. Methods **69**, 207–212 (2014)
14. Huang, D.-S., Yu, H.-J.: Normalized feature vectors: a novel alignment-free sequence comparison method based on the numbers of adjacent amino acids. IEEE/ACM Trans. Comput. Biol. Bioinf. **10**, 457–467 (2013)
15. Zhu, L., You, Z.-H., Huang, D.-S., Wang, B.: t-LSE: a novel robust geometric approach for modeling protein-protein interaction networks. PLoS One **8**, e58368 (2013)
16. Deng, S.-P., Zhu, L., Huang, D.-S.: Mining the bladder cancer-associated genes by an integrated strategy for the construction and analysis of differential co-expression networks. BMC Genomics **16**(Suppl 3), S4 (2015)
17. Deng, S.-P., Zhu, L., Huang, D.-S.: Predicting hub genes associated with cervical cancer through gene co-expression networks. IEEE/ACM Trans. Comput. Biol. Bioinf. **13**, 27–35 (2015)
18. Zhu, L., Deng, S.-P., Huang, D.-S.: A two-stage geometric method for pruning unreliable links in protein-protein networks. IEEE Trans. Nanobiosci. **14**, 528–534 (2015)
19. Shen, Z., Bao, W., Huang, D.-S.: Recurrent neural network for predicting transcription factor binding sites. Sci. Rep. **8**, 1–10 (2018)
20. Umarov, R.K., Solovyev, V.V.: Recognition of prokaryotic and eukaryotic promoters using convolutional deep learning neural networks. PLoS One **12**, e0171410 (2017)
21. Min, X., Zeng, W., Chen, S., Chen, N., Chen, T., Jiang, R.: Predicting enhancers with deep convolutional neural networks. BMC Bioinf. **18**, 478 (2017)
22. Yang, B., et al.: BiRen: predicting enhancers with a deep-learning-based model using the DNA sequence alone. Bioinformatics **33**, 1930–1936 (2017)
23. Cohn, D., Zuk, O., Kaplan, T.: Enhancer identification using transfer and adversarial deep learning of DNA sequences. BioRxiv 264200 (2018)
24. Liu, F., Li, H., Ren, C., Bo, X., Shu, W.: PEDLA: predicting enhancers with a deep learning-based algorithmic framework. Sci. Rep. **6**, 28517 (2016)
25. Zeng, W., Wu, M., Jiang, R.: Prediction of enhancer-promoter interactions via natural language processing. BMC Genom. **19**, 84 (2018)
26. Chuai, G., et al.: DeepCRISPR: optimized CRISPR guide RNA design by deep learning. Genome Biol. **19**, 80 (2018)
27. Fiannaca, A., et al.: Deep learning models for bacteria taxonomic classification of metagenomic data. BMC Bioinf. **19**, 198 (2018)
28. Plekhanova, E., Nuzhdin, S.V., Utkin, L.V., Samsonova, M.G.: Prediction of deleterious mutations in coding regions of mammals with transfer learning. Evol. Appl. **12**, 18–28 (2019)

29. Baek, J., Lee, B., Kwon, S., Yoon, S.: Lncrnanet: long non-coding rna identification using deep learning. Bioinformatics **34**, 3889–3897 (2018)
30. Trabelsi, A., Chaabane, M., Ben-Hur, A.: Comprehensive evaluation of deep learning architectures for prediction of DNA/RNA sequence binding specificities. Bioinformatics **35**, i269–i277 (2019)
31. Dahl, G.E., Yu, D., Deng, L., Acero, A.: Context-dependent pre-trained deep neural networks for large-vocabulary speech recognition. IEEE Trans. Audio Speech Lang. Process. **20**, 30–42 (2011)
32. Pan, S.J., Yang, Q.: A survey on transfer learning. IEEE Trans. Knowl. Data Eng. **22**, 1345–1359 (2009)
33. Trieschnigg, D., Kraaij, W., de Jong, F.: The influence of basic tokenization on biomedical document retrieval. In: Proceedings of the 30th Annual International ACM SIGIR Conference on Research and Development in Information Retrieval, pp. 803–804 (2007)
34. Jiang, J., Zhai, C.: An empirical study of tokenization strategies for biomedical information retrieval. Inf. Retrieval **10**, 341–363 (2007)
35. Chikhi, R., Medvedev, P.: Informed and automated k-mer size selection for genome assembly. Bioinformatics **30**, 31–37 (2014)
36. Ghandi, M., Lee, D., Mohammad-Noori, M., Beer, M.A.: Enhanced regulatory sequence prediction using gapped k-mer features. PLoS Comput. Biol. **10**, e1003711 (2014)
37. Koren, S., Walenz, B.P., Berlin, K., Miller, J.R., Bergman, N.H., Phillippy, A.M.: Canu: scalable and accurate long-read assembly via adaptive k-mer weighting and repeat separation. Genome Res. **27**, 722–736 (2017)
38. Sherry, S.T., et al.: dbSNP: the NCBI database of genetic variation. Nucleic Acids Res. **29**, 308–311 (2001)
39. www.ncbi.nlm.nih.gov/genome/?term=escherichia%20coli

A New Method Combining DNA Shape Features to Improve the Prediction Accuracy of Transcription Factor Binding Sites

Siguo Wang[1], Zhen Shen[1], Ying He[1], Qinhu Zhang[1](✉), Changan Yuan[2], Xiao Qin[3], Hongjie Wu[4], and Xingming Zhao[5]

[1] The Institute of Machine Learning and Systems Biology, School of Electronics and Information Engineering, Tongji University, Shanghai 201804, China
20310031@tongji.edu.cn
[2] Guangxi Academy of Science, Nanning 530025, China
[3] School of Computer and Information Engineering, Nanning Normal University, Nanning 530299, China
[4] School of Electronic and Information Engineering, Suzhou University of Science and Technology, Suzhou 215009, China
[5] Institute of Science and Technology for Brain Inspired Intelligence (ISTBI), Fudan University, Shanghai 200433, China

Abstract. Identifying transcription factor (TF) binding sites (TFBSs) has play an important role in the computational inference of gene regulation. With the development of high-throughput technologies, there have been many conventional methods and deep learning models used in the identification of TFBSs. However, most methods are designed to predict TFBSs only based on raw DNA sequence leads to low accuracy. Therefore, we propose a Dual-channel Convolutional neural network (CNN) model combining DNA sequences and DNA Shape features to predict TFBSs, named DCDS. In the DCDS model, the convolution layer captures low-level features from input data and parallel pooling operations are used to find the most significant activation signal in a sequence for each filter to improve the prediction accuracy of TFBSs. We conduct a series of experiments on 66 *in vitro* datasets and experimental results show that proposed model DCDS is superior to some state-of-the-art methods.

Keywords: Convolutional neural network · Transcription factor binding sites · DNA shape features · DNA sequences · Pooling operations

1 Introduction

Transcription is the first step in the process of gene expression and plays a vital role in the process of gene expression. The regulation of the transcription process is usually effected by some special transcription factors (TF) [1–4], which cooperate with others to regulate the specific expression of genes. Transcription factors regulate the transcription process of genes by specifically binding the DNA regions which is called the transcription factor binding sites (TFBSs) and a transcription factor can simultaneously

D.-S. Huang and K.-H. Jo (Eds.): ICIC 2020, LNCS 12464, pp. 79–89, 2020.
https://doi.org/10.1007/978-3-030-60802-6_8

regulate multiple genes. Therefore, systematically studying the characteristics of transcription factors becomes particularly important for the study of human gene regulatory relationships. The traditional experimental methods that identify TFBSs and solve other biological problems [5–10] are time-consuming and expensive, making false positives and false negatives in the experimental results [11].With the development of high-throughput technology, especially chromatin immunoprecipitation assay (ChIP) and protein binding microarrays (PBM), a large amount of *in vivo* and *in vitro* experimental data has been provided for researchers. Subsequently, many computational methods were developed to predict transcription factor binding sites.

First of all, PWM-based methods have achieved great success in modeling DNA-binding proteins compared with experimental methods, but one of the main drawbacks using these methods is high false positives [12]. Later, the experimental results showed gkm-SVM has a greater advantage than the PWM-based which converted the DNA sequence into k-mer features, and then distinguished whether it was TFBS in the DNA sequence by SVM [13]. Due to neural networks can mine feature information of different types of data from massive input data and find potential correlations between different features [14]. Deep neural networks have been widely used in many fields and have achieved excellent results, such as computer vision, natural language processing, image classification [15–18] and other intractable issues [19–25]. DeepBind [26] was the earliest and successful methods for applying convolutional neural networks (CNN) to genomics, which were used to model sequence specificity of protein binding. And DeepSEA [27] were proposed to predict the noncoding-variant effects de novo from sequence based on multilayer convolutional neural network. Whereafter, the deep learning method Basset [28] was used to learn the functional activities of DNA sequences and determine cell specificity and DNA accessibility. Later, some researchers suppose that the capability of CNNs to extract dependencies between sequences sequence is limited, a series of hybrid CNN and recurrent neural network (RNN) models were proposed. DanQ [29] and DeeperBind [30] are two typical hybrid deep learning models to predict TFBSs, and their performance outperformers the traditional methods. Multiple studies have shown that the readout of 3D DNA structure is an important part of TF binding specificity [31, 32] and an initial database on the four DNA shape has been provided for researchers, four DNA shape features, including helix twist (HelT), minor groove width (MGW), propeller twist (ProT) and Roll [33]. Some existing methods have used the DNA shape features in the study of predicting TFBSs and improve the accuracy. For example, a kernel-based framework leverages DNA sequence and shape information to better understand protein-DNA binding preference and affinity [34]. And, the shared model proposed in our recently research work was far superior to the kernel-based method combining DNA sequence and DNA shape information [35]. However, how to effectively use DNA information and DNA shape information to improve the recognition accuracy of TFBSs is still a challenge.

In this paper, we present a dual-channel convolutional network model that uses DNA sequence data and DNA shape features to improve the prediction accuracy of transcription factor binding sites, named DCDS. In particular, our model mainly includes three parts: data input, model training and output. In the data input stage, the

DNA sequence is converted into a matrix similar to an image using one-hot encoding. In the model training stage, the DNA sequence information and DNA shape information after data processing are input into the convolutional layer to extract low-level information. Then input the extract low-level information into the parallel max pooling layer and average pooling layer at the same time to obtain find the most significant activation signal in a sequence for each filter. Then the features obtained by the pooling layer are fused into a feature vector and input to the model output stage. In the output stage, the fused feature vectors are sequentially input into the batch normalization layer, fully connected layer and the dropout layer to obtain the final binding value. Finally, we conducted a set of experiments on 66 vitro datasets, the experimental results show that DCDS outperformed some state-of-the-art TFBS detection techniques.

The outline of this paper is as follows: Sect. 2 represents the details of our proposed model, Sect. 3 shows the experimental results and analysis, and Sect. 4 concludes the paper.

2 Methods

In this part, we mainly elaborate on the proposed model DCDS using DNA sequence data and DNA shape features based a dual-channel convolutional network model, including the following parts.: data processing, model training, output stage and model implementation and hyper parameter setting. Figure 1 shows the overall framework of the proposed method.

2.1 Data Processing

Two kinds of data are used in the data input stage, one is the original DNA sequence data of nucleotides, and the other is the DNA shape data representing an alternative approach for encoding nucleotide dependencies implicitly [36]. The DNA sequence is converted into a matrix using one-hot encoding. The DNA sequence input is a 4 * L matrix where L is the length of the sequence (35 bp in our experiments) and 4 represents bases ATCG. And four bases are expressed as A = [1,0,0,0], T = [0,1,0,0], C = [0,0,1,0] and G = [0,0,0,1] by one-hot encoding respectively. According to the survey [37], they proposed a high-throughput method to derive four DNA shape features: minor groove width (MGW), roll, propeller twist (MGW) and helix twist (HelT) from large amounts of sequence data by Monte Carlo simulations and the pentanucleotides can be used to describe the degeneracy of the sequence structure of the double helix with sufficient accuracy using a sliding window approach. A pentamer contributes one MGW value, one ProT value, two Roll values and two HelT, and we take the average of the two Roll and HelT values, one MGW value and ProT value as 4 * L input matrix of DNA shape features.

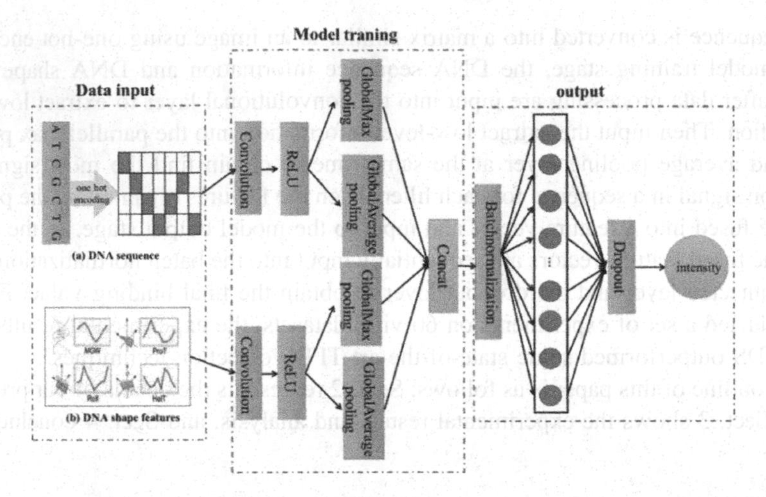

Fig. 1. The framework of the proposed DCDS. The data input of DCDS include DNA sequence and DNA shape features; model training include convolution layer and parallel pooling layer; out stage include batch normalization layer, fully connected layer and dropout.

2.2 Model Training

At this stage, firstly input the obtained matrix of one-hot encoding DNA sequence and DNA structure of the pentamer data into the convolutional layer. We used the widely used linear rectification activation function (ReLU), which not only makes the network sparse and reduces the interdependence between parameters, alleviating the occurrence of overfitting. This layer has 16 hidden neurons and used the ReLU activation function as above. Both low-level features of DNA sequence and DNA shape features are obtained by the convolutional layer, and the low-level features are input into the subsequent pooling operation. Both the global max pooling and the global average pooling are applied in the pooling layer. Applying global max pooling is used to obtain the most significant activation signal for each filter in the sequence. Taking sequence context into account, the average pooling is adopted because it can take the average value by scanning the filter at each position in the sequence. All the merged results of the DNA sequence and DNA shape features are combined in one vector and input to the output stage.

2.3 Output Stage

In the output stage, we firstly input the feature vectors representing high-level features of the DNA sequence and DNA shape features into the batch normalization layer for feature normalization. The normalized features are input to the subsequent fully connected layer and dropout layer. We use two fully connected layers, and the number of neurons is 16 and 1 respectively. In the output stage, we add batch normalization layer and dropout layer to enhance the generalization ability of the model, and also solve the problem of gradient disappearance in the process of back propagation. After the fully connected layer with the number of neurons is 1, we get the classification results of transcription factors binding sites.

Table 1. Hyper-parameter sets tested

Hyper-parameter	Choices
Dropout	0.2, 0.5
Momentum in AdaDelta	0.9, 0.99, 0.999
Delta in AdaDelta	1e−8, 1e−6, 1e−4

2.4 Model Implementation and Hyper Parameter Setting

Note that, our implementation is written in Python, utilizing the Keras 1.2.2 library (https://github.com/fchollet/keras) with the Tensorflow [38] backend. We used a Linux machine with 32 GB of memory and an NVIDIA Titan X Pascal GPU for training. For each *in vitro* experiment, 30 hyper-parameter settings were randomly sampled from all possible combinations, which is the same as the number used by DeepBind. For each potential hyper-parameter, the epochs of training are set 100. In order to accelerate convergence, an early stopping strategy is used in the training process and the hyper-parameter set with the best accuracy was used to train the model on the training set and test set. The detailed settings of the two models are shown in Table 1.

3 Experimental Results and Discussion

We firstly comprehensively collected data used this model, including 66 universal PBM (uPBM) data in *vitro* datasets and corresponding DNA shape data, and then conduct a series of experiments and analyze the experimental results. We use five-fold cross-validation and use the R^2 evaluation index which used in [2] to evaluate the accuracy of TFBSs in each data set.

3.1 Experimental Datasets

In this paper, we mainly used two data, including the original DNA sequence and DNA shape data. In order to verify the effectiveness of our proposed model, we conducted a series of experiments on 66 universal PBM (uPBM) data in *vitro* data sets which are obtained from the DREAM5 [20]. Each uPBM subset contains more than 40,000 arrays containing all patterns of 66TF unaligned 35-mer probes from various protein families.

Four DNA shape features including MGW, ProT, HelT and Roll used in this paper were obtained from [37] and the pentanucleotides can be used to describe the degeneracy of the sequence structure of the double helix with sufficient accuracy using a sliding window approach.

3.2 Compared with Other Methods to Identify Transcription Factor Binding Sites

In this section, we compare the proposed method with the four methods to systematically evaluate the performance, including Deepbind [26], Spectrum + shape kernel, Di-mismatch + shape kernel [34] and DLBSS [35]. Deepbind [26] only uses DNA sequences based on the traditional multi-convolutional layer deep learning model. The two kernel-based method Spectrum + shape kernel was derived from classic k-mer spectrum kernel [39] and Di-mismatch + shape kernel was developed on the basis of [40], which adopts DNA sequences and DNA shape information. Another method is DLBSS [15], which uses DNA sequences and DNA shape features to identify TFBSs based on a shared deep convolution model, the performance of this method is far superior to the kernel-based method Spectrum + shape kernel and Di-mismatch + shape kernel [34] also uses DNA sequences and DNA shape information. In order to comprehensively evaluate the performance of the proposed method, the R^2 evaluation metric are used to analyze the prediction accuracy of TFBSs on proposed model and compared method.

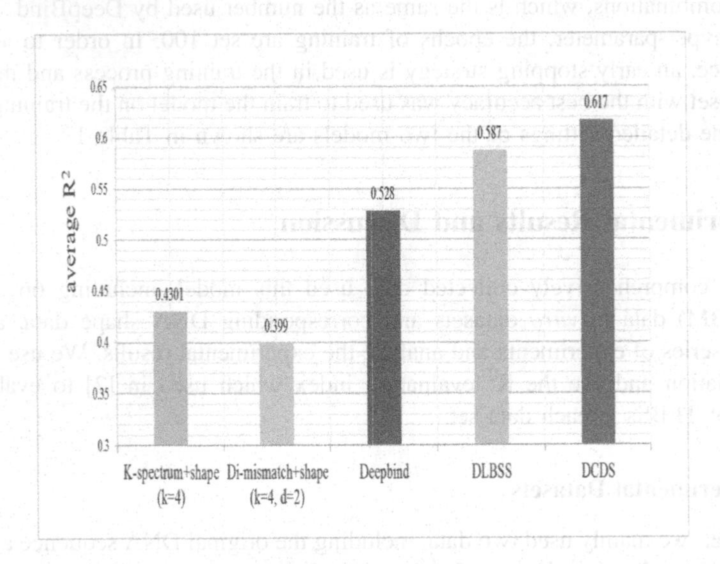

Fig. 2. The average R2 value of DCDS and the compared methods

As shown in Fig. 2, we plot histogram about average R^2 value of DCDS and the compared methods. We can see from Fig. 2, our proposed method DCDS clearly performed better than other methods about R^2, which indicates that DCDS was effective to predict TFBSs. It is noteworthy that the proposed DCDS has achieved excellent results compared with the two kernel-based method. It indicts that the deep learning method is superior to the traditional machine learning method in the identification of TFBSs, which overcomes the limitations of kernel calculation. Especially, the average R^2 of proposed DCDS is increased by 9% compared with Deepbind. It can be concluded from this that DNA shape data play an important pole in predicting in

TFBSs. When both DNA sequence data and DNA shape data are input into the model, the average R^2 of DCDS is improved by 3% compared to DLBSS. As far as I am concerned, the model design of DCDS is more suitable for extracting important information from DNA sequence data and DNA shape data to predict TFBSs. The superiority of DCDS is mainly reflected in the use of parallel pooling layers to extract high-dimensional feature information from convolutional layers, and add batch normalization to avoid the gradient disappearance of the model training.

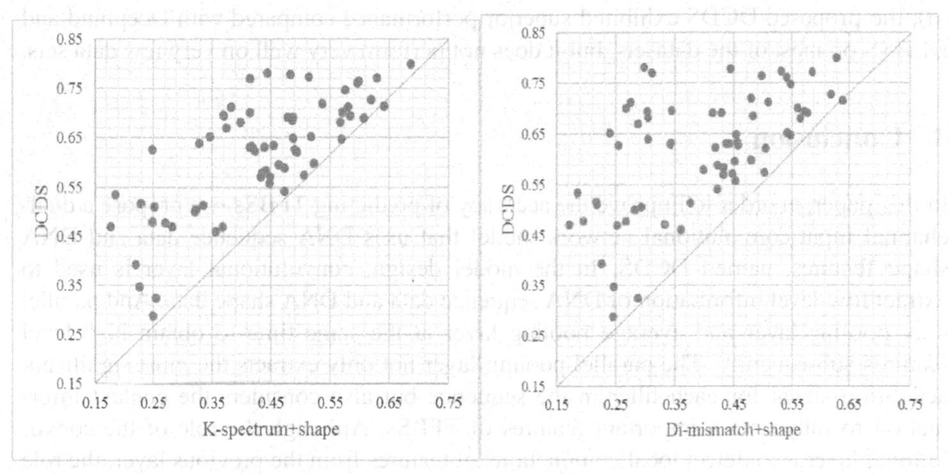

(a) DCDS compared with K-spectrum+shape (b) DCDS compared with Di-mismatch+shape

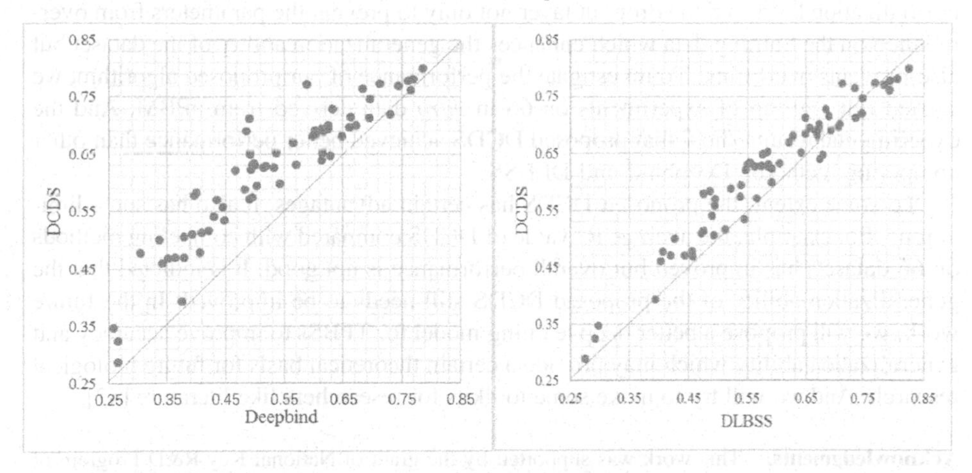

(c) DCDS compared with Deepbind (d) DCDS compared with DLBSS

Fig. 3. An overall performance comparison about R2 of DCDS and the compared methods on 66 vitro datasets. (a) represents the performance of DCDS compared with K-spectrum + shape (k = 4), (b) represents the performance of DCDS compared with Di-mismatch + shape (k = 4, d = 2), (a) represents the performance of DCDS compared with Deepbind, (c) represents the performance of DCDS compared with DLBSS

For an overall performance comparison about R^2 of DCDS and the compared methods on 66 *vitro* datasets shown in Fig. 3. As can be seen from the Fig. 3 (a) and 3 (b), the proposed outperforms two kernel-based method on every dataset. As we can see from the Fig. 3 (c), only two data sets of method Deepbind have a slightly lower R^2 value than proposed DCDS in the 66 in *vitro* datasets. It can be seen from Fig. 2 that the average R^2 value of proposed DCDS is 3% higher than method DLBSS, but as we can be seen from Fig. 3 (d), there are about ten datasets where the R^2 value of proposed DCDS is slightly lower than method DLBSS. From the perspective of Fig. 3 (c) and 3 (d), the proposed DCDS exhibited superior performance compared with Deepbind and DLBSS on most of the datasets, but it does not perform very well on very few data sets.

4 Conclusion

In this paper, in order to improve the accuracy of predicting TFBSs, we propose a dual-channel input convolutional network model that uses DNA sequence data and DNA shape features, named DCDS. In the model design, convolutional layer is used to extract low-level information of DNA sequence data and DNA shape data. And parallel max pooling layer and average pooling layer at the same time to obtain high-level features subsequently. The parallel pooling layer not only extracts the most significant activation signal for each filter in the sequence but also considers the context information to fully obtain important features of TFBSs. Although the role of the convolutional layer is to detect local conjunctions of features from the previous layer, the role of the pooling layer is to merge semantically similar features into one. The complexity of the deep learning model makes it predisposed to overfitting [41]. We add batch normalization layer and the dropout layer not only to prevent the parameters from over-reliance on the training data which enhances the generalization ability of the dataset but also prevents overfitting. To investigate the performance of our proposed algorithm, we carried out a group of experiments on 66 in *vitro* data derived from uPBM. And the experimental results show that proposed DCDS achieved better performance than other competing methods: Deepbind and DLBSS.

To some extent, the proposed DCDS has certain advantages, it also has some limitations. For example, the average R^2 value of DCDS compared with competing methods on 66 datasets has improved but overall performance is not good. It is indicted that the generalization ability of the proposed DCDS still needs to be improved. In the future work, we will propose a better deep learning model for TFBSs to improve accuracy and generalization ability which may provide a certain theoretical basis for future biological research. And we will try to make some toolkits for researchers like literature [42].

Acknowledgments. This work was supported by the grant of National Key R&D Program of China (No. 2018YFA0902600 & 2018AAA0100100) and partly supported by National Natural Science Foundation of China (Grant nos. 61861146002, 61520106006, 61772370, 61702371, 61732012, 61932008, 61532008, 61672382, 61772357, and 61672203) and China Postdoctoral Science Foundation (Grant no. 2017M611619) and supported by "BAGUI Scholar" Program and the Scientific & Technological Base and Talent Special Program, GuiKe AD18126015 of the Guangxi Zhuang Autonomous Region of China and supported by Shanghai Municipal Science and Technology Major Project (No.2018SHZDZX01), LCNBI and ZJLab.

References

1. Brand, L.H., Fischer, N.M., Harter, K., Kohlbacher, O., Wanke, D.: Elucidating the evolutionary conserved DNA-binding specificities of WRKY transcription factors by molecular dynamics and in vitro binding assays. Nucleic Acids Res. **41**, 9764–9778 (2013)
2. Weirauch, M.T., et al.: Evaluation of methods for modeling transcription factor sequence specificity. Nat. Biotechnol. **31**, 126–134 (2013)
3. Zheng, C.-H., Zhang, L., Ng, V.T., Shiu, C.K., Huang, D.S.: Molecular pattern discovery based on penalized matrix decomposition. IEEE/ACM Trans. Comput. Biol. Bioinf. **8**, 1592–1603 (2011)
4. Zheng, C., Huang, D.S., Zhang, L., Kong, X.: Tumor clustering using nonnegative matrix factorization with gene selection. IEEE Trans. Inf. Technol. Biomed. **13**, 599–607 (2009)
5. Huang, D.S., Zhang, L., Han, K., Deng, S.P., Yang, K., Zhang, H.B.: Prediction of protein-protein interactions based on protein-protein correlation using least squares regression. Curr. Protein Pept. Sci. **15**, 553–560 (2014)
6. Huang, D.S., Yu, H.-J.: Normalized feature vectors: a novel alignment-free sequence comparison method based on the numbers of adjacent amino acids. IEEE/ACM Trans. Comput. Biol. Bioinf. **10**, 457–467 (2013)
7. Huang, D.S., Zhao, X.M., Huang, G.B., Cheung, Y.M.: Classifying protein sequences using hydropathy blocks. Pattern Recogn. **39**, 2293–2300 (2006)
8. Huang, D.S., Huang, X.: Improved performance in protein secondary structure prediction by combining multiple predictions. Protein Pept. Lett. **13**, 985–991 (2006)
9. Huang, D.S., Zheng, C.: Independent component analysis-based penalized discriminant method for tumor classification using gene expression data. Bioinf. (Oxford, Engl.) **22**, 1855–1862 (2006)
10. Deng, S.P., Zhu, L., Huang, D.S.: Mining the bladder cancer-associated genes by an integrated strategy for the construction and analysis of differential co-expression networks. BMC Genomics **16**(Suppl 3), S4 (2015)
11. Coppe, A., et al.: Motif discovery in promoters of genes co-localized and co-expressed during myeloid cells differentiation. Nucleic Acids Res. **37**, 533–549 (2009)
12. Stormo, G.D.: Modeling the specificity of protein-DNA interactions. Quant. Biol. (Beijing, China) **1**, 115–130 (2013)
13. Fletez-Brant, C., Lee, D., McCallion, A.S., Beer, M.A.: kmer-SVM: a web server for identifying predictive regulatory sequence features in genomic data sets. Nucleic Acids Res. **41**, 544–556 (2013)
14. Huang, D.S., Songde, M.: A new radial basis probabilistic neural network model (1996)
15. Karpathy, A., Toderici, G., Shetty, S., Leung, T., Li, F.F.: Large-scale video classification with convolutional neural networks. In: 2014 IEEE Conference on Computer Vision and Pattern Recognition (2014)
16. Deng, L.: Deep learning for natural language processing and related applications (Tutorial at ICASSP) (2014)
17. Krizhevsky, A., Sutskever, I., Hinton, G.: ImageNet classification with deep convolutional neural networks. Adv. Neural. Inf. Process. Syst. **25**, 125–223 (2012)
18. Chen, C., Seff, A., Kornhauser, A., Xiao, J.: DeepDriving: Learning Affordance for Direct Perception in Autonomous Driving (2015)
19. Deng, S.P., Lin, Z., Huang, D.S.: Predicting hub genes associated with cervical cancer through gene co-expression networks. IEEE/ACM Trans. Comput. Biol. Bioinf. **13**, 27–35 (2016)

20. Zhu, L., Deng, S.P., Huang, D.S.: A two-stage geometric method for pruning unreliable links in protein-protein networks. IEEE Trans. Nanobiosci. **14**, 528–534 (2015)
21. Lin, Z., You, Z.H., Huang, D.S., Bing, W., Deane, C.M.: t-LSE: a novel robust geometric approach for modeling protein-protein interaction networks. PLoS One **8**, e58368 (2013)
22. Huang, D.S., Jiang, W.: A general cpl-ads methodology for fixing dynamic parameters in dual environments. IEEE Trans. Syst. Man Cybern. Part B Cybern. Publ. IEEE Syst. Man Cybern. Soc. **42**, 1489–1500 (2012)
23. Zhu, L., Guo, W.-L., Deng, S.P., Huang, D.S.: ChIP-PIT: enhancing the analysis of ChIP-Seq data using convex-relaxed pair-wise tensor decomposition. IEEE/ACM Trans. Comput. Biol. Bioinf. **13**, 55–63 (2015)
24. Huang, D.S., Du, J.-X.: A constructive hybrid structure optimization methodology for radial basis probabilistic neural networks. IEEE Trans. Neural Networks **19**, 2099–2115 (2008)
25. Xia, J.F., Zhao, X.M., Song, J., Huang, D.S.: APIS: accurate prediction of hot spots in protein interfaces by combining protrusion index with solvent accessibility. BMC Bioinf. **11**, 174 (2010)
26. Alipanahi, B., Delong, A., Weirauch, M.T., Frey, B.J.: Predicting the sequence specificities of DNA-and RNA-binding proteins by deep learning. Nat. Biotechnol. **33**, 831–838 (2015)
27. Zhou, J., Troyanskaya, O.G.: Predicting effects of noncoding variants with deep learning-based sequence model. Nat. Methods **12**, 931–934 (2015)
28. Kelley, D.R., Snoek, J., Rinn, J.L.: Basset: learning the regulatory code of the accessible genome with deep convolutional neural networks. Genome Res. **26**, 990–999 (2016)
29. Quang, D., Xie, X.: DanQ: a hybrid convolutional and recurrent deep neural network for quantifying the function of DNA sequences. Nucleic Acids Res. **44**, e107 (2016)
30. Hassanzadeh, H.R., Wang, M.D.: DeeperBind: Enhancing prediction of sequence specificities of DNA binding proteins. In: 2016 IEEE International Conference on Bioinformatics and Biomedicine (BIBM), pp. 178–183 (2016)
31. Gordân, R., et al.: Genomic regions flanking E-box binding sites influence DNA binding specificity of bHLH transcription factors through DNA shape. Cell Rep. **3**, 1093–1104 (2013)
32. Zhou, T., et al.: Quantitative modeling of transcription factor binding specificities using DNA shape. Proc. Natl. Acad. Sci. U.S.A. **112**, 4654–4659 (2015)
33. Yang, L., et al.: TFBSshape: a motif database for DNA shape features of transcription factor binding sites. Nucleic Acids Res. **42**, 148–155 (2014)
34. Ma, W., Yang, L., Rohs, R., Noble, W.S.: DNA sequence + shape kernel enables alignment-free modeling of transcription factor binding. Bioinf. (Oxford, Engl.) **33**, 3003–3010 (2017)
35. Zhang, Q., Shen, Z., Huang, D.S.: Predicting in-vitro transcription factor binding sites using DNA sequence + shape. In: IEEE/ACM Transactions on Computational Biology and Bioinformatics (2019). https://doi.org/10.1109/tcbb.2019.2947461
36. Rohs, R., West, S.M., Sosinsky, A., Peng, L., Honig, B.: The role of DNA shape in protein-DNA recognition. Nature **461**, 1248–1253 (2009)
37. Zhou, T., Yang, L., Lu, Y., Dror, I., Dantas Machado, A.C., Ghane, T., Di Felice, R., Rohs, R.: DNAshape: a method for the high-throughput prediction of DNA structural features on a genomic scale. Nucleic Acids Res. **41**, 56–62 (2013)
38. Abadi, M.: TensorFlow: Large-Scale Machine Learning on Heterogeneous Distributed Systems (2016) arxiv 1603
39. Leslie, C., Eskin, E., Noble, W.S.: The spectrum kernel: a string kernel for svm protein classification. Pacif. Symp. Biocomput. Pacif. Symp. Biocomput. **7**, 564–575 (2002)
40. Agius, P., Arvey, A., Chang, W., Noble, W.S., Leslie, C.: High resolution models of transcription factor-DNA affinities improve in vitro and in vivo binding predictions. PLoS Comput. Biol. **6**, e1000916 (2010)

41. Lecun, Y., Bengio, Y., Hinton, G.: Deep learning. Nature **521**, 436 (2015)
42. Deng, S., Yuan, J., Huang, D.S., Zhen, W.: SFAPS: An R package for structure/function analysis of protein sequences based on informational spectrum method. IEEE Int. Conf. Bioinf. Biomed. **3**, 207–212 (2013)

Predicting *in-Vitro* Transcription Factor Binding Sites with Deep Embedding Convolution Network

Yindong Zhang[1(✉)], Qinhu Zhang[1], Changan Yuan[2], Xiao Qin[3], Hongjie Wu[4], and Xingming Zhao[5]

[1] Institute of Machine Learning and Systems Biology, School of Electronics and Information Engineering, Tongji University, Shanghai, China
ydzhang@tongji.edu.cn
[2] Guangxi Academy of Science, Nanning 530025, China
[3] School of Computer and Information Engineering, Nanning Normal University, Nanning 530299, China
[4] School of Electronic and Information Engineering, Suzhou University of Science and Technology, Suzhou 215009, China
[5] Institute of Science and Technology for Brain Inspired Intelligence (ISTBI), Fudan University, Shanghai 200433, China

Abstract. With the rapid development of deep learning, convolution neural network achieve great success in predicting DNA-transcription factor binding, aka motif discovery, In this paper, we propose a novel neural network based architecture i.e. eDeepCNN, combining multi-layer convolution network and embedding layer for predicting in-vitro DNA protein binding sequence. Our model fully utilize fitting capacity of deep convolution neural network and is well designed to capture the interaction pattern between motifs in neighboring sequence. Meanwhile continuous embedding vector serves as a better description of nucleotides than one-hot image-like representation owing to its superior expressive ability. We verify the effectiveness of our model on 20 motif datasets from in-vitro protein binding microarray data (PBMs) and present promising results compared with well-established DeepBind model. In addition, we emphasis the significance of dropout strategy in our model to fight against the overfitting problem along with growing model complexity.

Keywords: Convolution neural network · Continuous embedding vector · One-hot representation · Motif discovery · DNA protein binding · Transcription factor

1 Introduction

DNA binding proteins [1], also known as transcription factor, play an important role in cell biological processes including transcription, translation, repair, and replication machinery [2–4]. It has also been reported that some genomic variants in TFBSs are associated with serious diseases [5]. Therefore, discovering transcription factor binding

D.-S. Huang and K.-H. Jo (Eds.): ICIC 2020, LNCS 12464, pp. 90–100, 2020.
https://doi.org/10.1007/978-3-030-60802-6_9

site (TFBS), i.e. motif discovery, is crucial for further understanding of the transcriptional regulation mechanism in gene expression.

In the past decade, with the development of high-throughput sequencing technology [5–9], a variety of experimental methods for extracting binding regions have been proposed. In particular, protein binding microarrays (PBMs) [10] is to measure the in-vitro binding preferences of TFs to thousands of DNA sequences, which provides a large amount of available data for better studying in-vitro protein-DNA binding.

Traditionally, transcription factors binding sites are presented by positional weight matrices (PWMs) derived from aligned DNA motif sequences [11, 12], whose elements represent a probability distribution over DNA alphabet {A, C, G, and T} for each position in motif sequence. However, such method suffer the obvious defect that a position within a binding site contributes to binding affinity independent of other positions, especially its neighboring sequence. Dependencies between nucleotides in a TF binding site can be explicitly encoded by kmers [13, 14], which partly covers the shortage of PWM method and shows great superiority in performance. However kmer-based method can barely capture the sequence ordinal information between kmers. With the rapid development of deep learning in recent years, new computational methods such as convolutional neural network (CNN) and recurrent neural network (RNN) have been applied to many fields such as natural language processing and computer graphic and made great achievement [15–17]. DeepBind [19] is one of the earliest attempts to apply deep learning to the motif discovery task and has proved to be an effective model. DeepBind took motif discovery as a binary classification /regression task, and designed a single-layer convolutional neural network to predict transcription factor binding, by converting the one-dimensional genomic sequence with four nucleotides {A, C, G, T} into a one-hot encoded two-dimensional image-like format. However, the one-hot image-like representation suffer its inherent limitation of expressiveness, especially lack of interaction between neighboring nucleotides. DeepBind, along with many other deep learning methods [20–25] defeat traditional methods in both in-vitro and in-vivo motif discovery tasks and improve the accuracy to a new height.

In this paper, we first focus on in-depth exploitation of deep convolution neural network with application on in-vitro motif discovery task. To tackle the overfitting problem along with the growing scale of convolution network, we propose adding dropout layer [17] between convolutional layers, helping achieve considerable improvement compare to shallow CNN model, deepBind for example. Taking advantage of the expressiveness of continuous embedding vector, we try continuous embedding vectors in embedding layer as novel data representation format and successfully obtain further improvement.

2 Materials and Methods

In this section, we first introduce the relevant in-vitro DNA protein binding dataset and its data preprocessing procedure. Second, architecture of our deep convolution network namely eDeepCNN is presented in detail. Third, we give a briefing of evaluation metric and training hyper-parameters in our experiment.

2.1 Dataset and Preprocessing

We downloaded 20 universal protein binding microarrays (uPBMs) datasets from the DREAM5 project [19], which comes from a variety of protein families. Each TF dataset, consisting of ~40,000 unaligned 35-mer probe sequences, comprises a complete set of PBM probe intensities from two distinct microarray designs named HK and ME. These datasets have been normalized according to the total signal intensity.

To accurately evaluate the performance of our proposed method, five-fold cross-validation strategy was adopted in this paper. Five-fold cross-validation strategy repeated five times in total. Within each time, TF dataset was randomly divided into 5 folds of roughly equal size, and four of them were used as the training data while the rest was used as the test data. During training, we randomly sampled 1/8 of the training set as the validation set.

After that, we prepared each DNA sequence in two different representation format for further experiment. On the one hand, DNA sequences were transformed into one-hot image-like inputs for further experiment. Given a DNA sequence $s = (s_1, s_2, \ldots, s_n)$ of length n, where s_i comes from {A, C, G, and T}, it is encoded into a matrix of size $n \times 4$, each column of which corresponds to a one-hot vector. On the other hand, DNA sequences was encoded according to its label ordinal in nucleotides. For example, A was encoded into 0, C to 1, G to 2 and T to 3, which suits for further input into embedding layer in neural network [27] Fig 1.

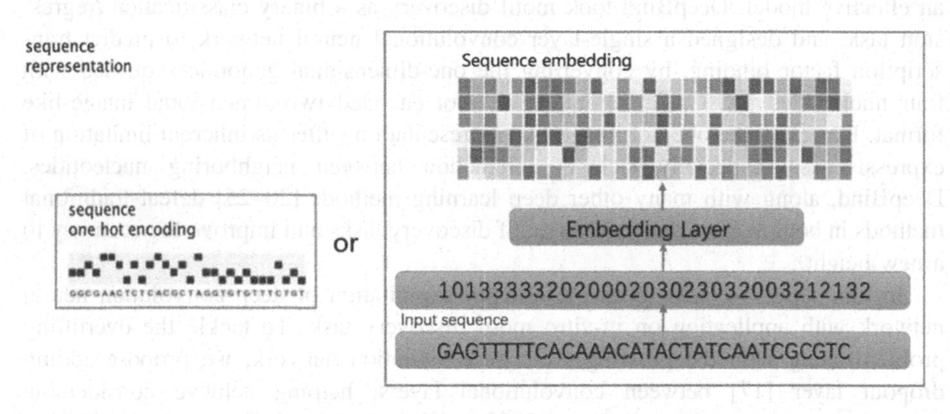

Fig. 1. A graphical demonstration of the difference between the two sequence encoding method, one-hot image-like encoding and continuous embedding vector encoding.

2.2 Network Architecture

DeepBind [19] introduced a single layer convolution neural network followed by a max global pooling layer to extract sequence features in motif discovery, which proved to be a great success. Hereby, we proposed a deeper neural network model composed of three convolution layers accompanied by dropout and local pooling strategies, namely DeepCNN. The first convolutional layer computes a score for all potential local motif, which is the same as deepBind. And we design the second and third convolutional layers, in the hope that it can capture the interaction pattern in neighboring sequence. The second convolution layer takes the motif score sequence computed by the first convolution layer as input and recognizes the distribution pattern of the motif score sequence, which, in other words, takes the interaction of the local motifs into consideration. With the same logic, the third convolution layer with bigger receptive field is capable of capturing the local motif interaction in a wider scope. Combining multiply Convolution layer improves the receptive field of DeepCNN model and allows an overall pattern recognition of the candidate sequence. Each convolution layer is followed by a local max pooling layer and a dropout layer. It should be noticed that dropout strategy plays an important role in our model, in the light of the overfitting risk accompanied by the expanding parameter size and model complexity. A global max pooling layer is used to capture the global context information of DNA sequences and feeds it into a two layer fully connected neural network to obtain final prediction.

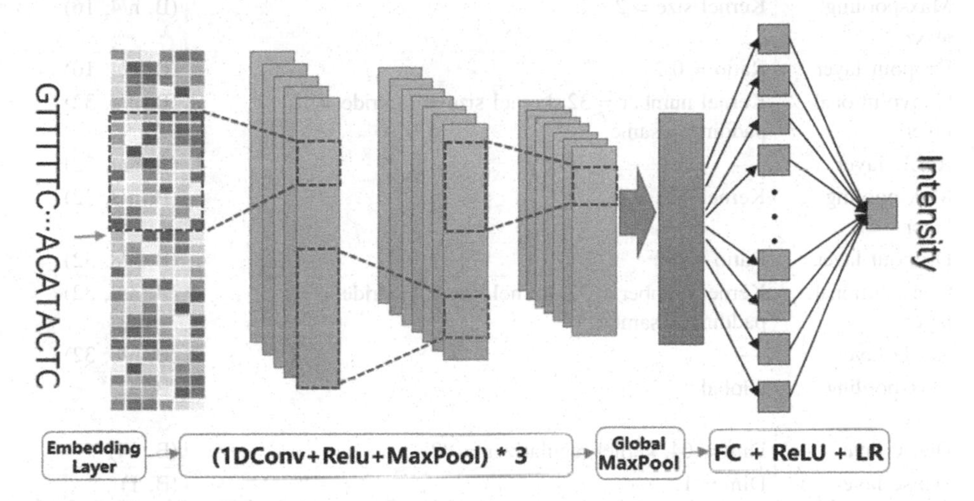

Fig. 2. A graphical illustration of eDeepCNN Model. DNA sequences are first fed into embedding layer followed by three convolution module. Each convolution module is composed of 1D convolution layer, Relu activation layer and maxPooling layer. GlobalMaxPooling layer extract the global sequence information and feed into fully connected layer and linear regression model.

The continuous embedding vector is widely used in various fields including natural language processing, information retrieval and recommendation system [27, 28]. Compared with one-hot image like representation, continuous embedding vector have an advantage for its expressive power in high dimensional latent vector space. The relative position of four nucleotides in high dimensional space are supposed to contain more information of underlying interaction pattern between nucleotides. In the meantime, continuous embedding vector profits from its flexibility in embedding size and cost efficiency considering its size of parameter. We add an optional embedding layer at the entrance of DeepCNN model and rename the alternative model as eDeepCNN.

Figure 2 plots a graphical illustration of eDeepCNN and the detailed parameter settings including convolution kernel size and number of filters in each layer are listed in Table 1. It should be mentioned that part of our hyper-parameter settings inherent from classic deep learning methods in motif discovery like deepBind, which have proved to be optimal choices, while some other parts were chosen from hyper-parameter grid search in training procedure.

Table 1. Parameter setting of eDeepCNN Model in detail.

Architectures	Settings	Output shape
Embedding layer	Input dim = 4, output dim = 32	(B, n, 32)
Convolutional layer	Kernel number = 16, kernel size = 13, stride = 1, padding = same	(B, n, 16)
ReLU layer	—	(B, n, 16)
Max-pooling layer	Kernel size = 2	(B, n/4, 16)
Dropout layer	Ratio = 0.2	(B, n/4, 16)
Convolutional layer	Kernel number = 32, kernel size = 7, stride = 1, padding = same	(B, n/4, 32)
ReLU layer	—	(B, n/4, 32)
Max-pooling layer	Kernel size = 2	(B, n/8, 32)
Dropout layer	Ratio = 0.2	(B, n/8, 32)
Convolutional layer	Kernel number = 32, kernel size = 5, stride = 1, padding = same	(B, n/8, 32)
ReLU layer	—	(B, n/8, 32)
Max-pooling layer	Global	(B, 32)
Dense layer	Dim = 64, kernel regularizer = '12'	(B, 64)
Dense layer	Dim = 1	(B, 1)

2.3 Evaluation Metric

We used the coefficient of determination R^2 to measure the correlation between the predicted intensities and the actual intensities. The R^2 measurement has been used previously to evaluate regression performance for PBM data [18]. In general, $1–R^2$ compute the proportion of mean squared error of regression model and inherent variance of ground truth of the dataset. The closer R^2 to the value 1, the smaller prediction error compared to dataset's inherent variance, which naturally indicates better predictive ability of our model.

$$R^2 = 1 - \frac{\sum_i \left(\hat{y}_i - y_i\right)^2}{\sum_i (y_i - \bar{y})^2} \tag{1}$$

2.4 Experiment Setting

The learnable parameters (e.g. weights and bias) in neural network were initialized by Glorot uniform initializer [29], and optimized by AdaDelta [30] algorithm with a mini-batchsize of 300. We implemented grid search strategy over some sensitive hyper-parameters, i.e. dropout ratio, L2 weight decay, and momentum in AdaDelta optimizer. An early stopping strategy was also adopted to fight against overfitting problem in our model. Detailed hyper-parameter setting is listed in Table 2.

Table 2. A list of sensitive hyper-parameters and grid search space in experiment.

Hyper-parameters	Settings
Dropout ratio	0.2, 0.5
Learning rate	1
Momentum in AdaDelta	0.999, 0.99, 0.9
Weight decay	5E−4, 1E−3, 5E−3
Early stopping tolerance	20
Mini-batch size	300
Loss function	L2 loss

3 Results and Analysis

In order to verify the effectiveness of our model, we conducted extensive experiments on 20 in-vitro uPBM datasets mentioned above. As mentioned above, deepBind, a typical representative of deep learning model in motif discovery, achieve high performance among all kinds of methods. We use DeepBind [19] as the competing method and evaluate the performance with the average R^2 over the 5-fold cross validation on each dataset.

Figure 3 (a) plots an overall performance comparison of DeepCNN and the competing methods DeepBind on 20 uPBM datasets. We can clearly observe that DeepCNN outperforms the shallow DeepBind Model by a large margin in all 20 datasets. DeepCNN achieves an average R^2 of 0.565 compared to 0.481 in DeepBind Model with an increase of 0.084 or a relative improvement in 17%. The prominent increase can be attributed to additional consideration of motif dependency and interaction in neighboring sequence as well as the improved receptive field in DeepCNN.

By switching the one-hot image-like representation to continuous embedding vector in embedding layer, we observed another wave of improvement. The embedding size is set to 32 and we will discuss the influence of embedding vector size latter. As show in Fig. 3(b), eDeepCNN obtains better results than DeepCNN in all the 20 uPBM dataset and achieves an average R^2 of 0.590 with margin gain of 0.025 and relative improvement of 4%, which can be explained by the simple fact that continuous

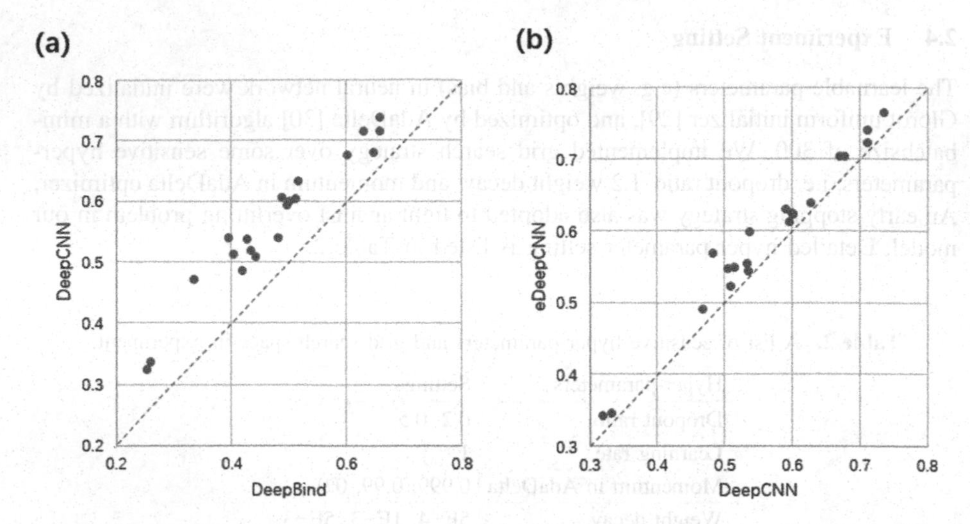

Fig. 3. (a) A comparison of DeepBind and DeepCNN experimented on 20 uPBM datasets. (b) Comparison of the performance of DeepCNN and eDeepCNN.

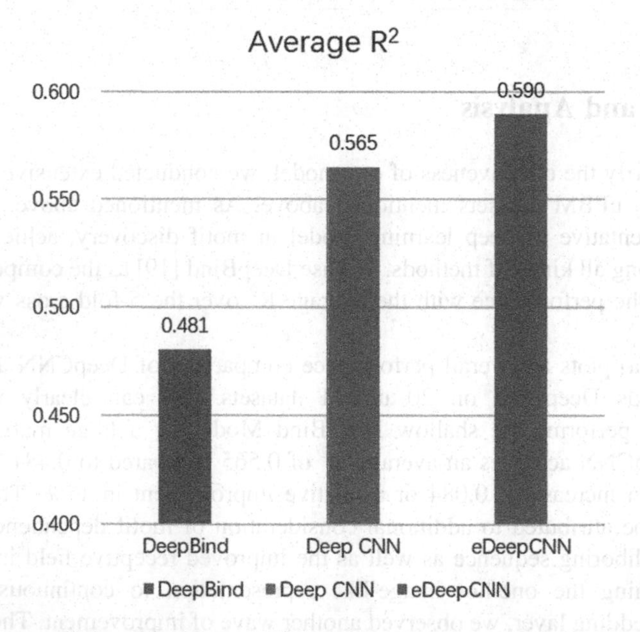

Fig. 4. Comparison of DeepBind, DeepCNN and eDeepCNN with their average performance on 20 datasets.

embedding vectors, i.e., latent variables in high dimensional space, serve as a superior description of nucleotides. Figure 4 compares the average performance of DeepBind, DeepCNN and eDeepCNN on 20 uPBM datasets and clearly illustrates the superiority of our model owing to the continuous improvement.

3.1 Effect of Dropout Layer

It has been shown in many areas that deep learning methods are always haunted with overfitting risk along with the increasing of parameter scale [17]. Zeng et al. [31] conducted a systematic exploration of the performance of different CNN architectures. The author designed a deep CNN architecture, which is composed of three convolutional layers each followed by a ReLU layer and a max-pooling layer, similar to our model, but the performance of it was worse than that of a shallow one.

However, our result clearly verify the superiority of our multi-layer DeepCNN model on in-vitro uPBM dataset than shallow one. We ascribe our success to the anti-overfitting strategy in experiment, including the dropout and regularization strategy as well as the corresponding hyper-parameter search. We performed the same experiment with DeepCNN model except that we remove the dropout strategy and Table 3 compare the results on part of the datasets. It is clearly shown that using dropout strategy make evident improvement in most of uPBM motif dataset and achieve an average R^2 of 0.563 in DeepCNN, in comparison with 0.515 in that without dropout. We can also observe that DeepCNN without dropout show better performance than DeepBind with an average improvement of 0.032, which confirm our argument that multi-layer DeepCNN is capable of capturing the neighboring motif dependency and obtains superior results as a consequence.

It should be noticed that Zeng *et al.* [31] conducted the experiment on in-vivo ChIP-seq data in contrast to our in-vitro uPBM data, and suggest that multi-layer convolution network have a good effect on in-vitro dataset, not necessarily on in-vivo dataset, which may be a result of the distinct sequence length and other data characteristics in two kind of datasets. This may be another reason for the different model behavior in their paper.

Table 3. Performance of DeepCNN without dropout strategy, in comparison with DeepBind and DeepCNN with dropout.

	DeepBind	DeepCNN without dropout	DeepCNN
TF_1_Ar_pTH1739_HK	0.265	0.275	0.334
TF_3_Foxo6_pTH3477_HK	0.632	0.671	0.713
TF_4_Klf12_pTH0977_HK	0.406	0.427	0.511
TF_59_Zbtb1_pTH2366_ME	0.504	0.558	0.598
TF_6_Klf9_pTH2353_HK	0.661	0.705	0.713
Average	0.483	0.515	0.563

3.2 Effect of Embedding Size

In this section, we discuss the influence of embedding size in eDeepCNN. To demonstrate the effect of embedding vector size, we alter the embedding vector size in a list of 8, 16 32 and 64 on three in-vitro datasets and record the performance of eDeepCNN in Table 4.

Table 4. Performance of eDeepCNN when altering embedding size.

	8	16	32	64
TF_1_Ar_pTH1739_HK	0.321	0.335	0.345	0.343
TF_3_Foxo6_pTH3477_HK	0.711	0.730	0.739	0.728
TF_6_Klf9_pTH2353_HK	0.709	0.714	0.721	0.719
Average	0.580	0.593	0.602	0.597

It is no surprise to see that we got reasonable improvement in model performance when increasing embedding size from 8 to 32. But further increasing of embedding vector dimension to 64 results in a slight performance loss in all 3 datasets. The former improvement can be explained by growing fitting ability and expressive power along with increasing embedding size and we ascribe the latter performance loss to the overfitting problem resulting from the unnecessarily large embedding vector. Therefore, we set the embedding layer dimension to 32.

4 Conclusion

In this paper, we propose a deep embedding convolution neural network model named eDeepCNN for predicting in-vitro DNA-protein binding site. By adding additional convolution layers, our model takes the interaction between local motifs into consideration and makes full advantage of the fitting capacity of deep convolution network. Improved receptive field of DeepCNN model also allows an overall pattern recognition of the input DNA sequence. Experiment results on 20 in-vitro uPBMs datasets validate the superiority of our model than competing method. Moreover, we analysis the underlying reason for our success and make an in-depth exploration. By comparing the performance of our model with one-hot image-like input and embedding vector input, we confirm the continuous embedding vector as a better description of nucleotides. And we emphasis the importance of dropout and other anti-overfitting strategies adopted in our method through contrast experiment. On the other hand, large dropout rate may lead to relative unstable training procedure and require careful training hyper-parameter settings.

Growing evidence [32, 33] suggest that shape in local DNA sequence plays an important role in DNA protein binding process. In other words, transcription factor show clear DNA shape preference in addition to the local nucleotides sequence specificity. Therefore, incorporating the DNA shape information into deep convolution neural network would be a promising method to improve DNA binding site prediction, which would be our future work direction.

Acknowledgements. This work was supported by the grant of National Key R&D Program of China (Nos. 2018AAA0100100 & 2018YFA0902600) and partly supported by National Natural Science Foundation of China (Grant nos. 61861146002, 61520106006, 61772370, 61702371, 61732012, 61932008, 61532008, 61672382, 61772357, and 61672203) and China Postdoctoral Science Foundation (Grant no. 2017M611619) and supported by "BAGUI Scholar" Program and the Scientific & Technological Base and Talent Special Program, GuiKe AD18126015 of the Guangxi Zhuang Autonomous Region of China and supported by Shanghai Municipal Science and Technology Major Project (No.2018SHZDZX01), LCNBI and ZJLab.

References

1. Lambert, S.A., et al.: The human transcription factors. Cell **172**, 650–665 (2018)
2. Vaquerizas, J.M., Kummerfeld, S.K., Teichmann, S.A., Luscombe, N.M.: A census of human transcription factors: function, expression and evolution. Nat. Rev. Genet, vol. 10, p. 252 (2009)
3. Stormo, G.D.J.B.: DNA binding sites: representation and discovery. Bioinformatics **16**, 16–23 (2000)
4. Lee, T.I., Young, R.A.: Transcriptional regulation and its misregulation in disease. Cell **152**, 1237–1251 (2013)
5. Zhu, L., Zhang, H.-B., Huang, D.-S.: Direct AUC optimization of regulatory motifs. Bioinformatics **33**, i243–i251 (2017)
6. Zhang, H., Zhu, L., Huang, D.: WSMD: weakly-supervised motif discovery in transcription factor ChIP-seq data. Sci. Rep. **7**, 3217 (2017)
7. Zhang, H., Zhu, L., Huang, D.S.: DiscMLA: An efficient discriminative motif learning algorithm over high-throughput datasets. IEEE/ACM Trans. Comput. Biol. Bioinf. **15**(6), 1810–1820 (2018)
8. Shen, Z., Zhang, Y.-H., Han, K., Nandi, A.K., Honig, B., Huang, D.-S.: miRNA-disease association prediction with collaborative matrix factorization. Complexity **2017**, 9 (2017)
9. Zhu, L., Guo, W.-L., Deng, S.-P., Huang, D.-S.: ChIP-PIT: enhancing the analysis of ChIP-Seq data using convex-relaxed pair-wise interaction tensor decomposition. IEEE/ACM Trans. Comput. Biol. Bioinform. **13**, 55–63 (2016)
10. Berger, M.F., Philippakis, A.A., Qureshi, A.M., He, F.S., Estep III, P.W., Bulyk, L.: Compact, universal DNA microarrays to comprehensively determine transcription-factor binding site specificities. Nat. Biotechnol. **24**, 1429 (2006)
11. Stormo, G.D.: Consensus patterns in DNA. Methods Enzymol. **183**, 211–221 (1990)
12. Stormo, G.D.: DNA binding sites: representation and discovery. Bioinformatics **16**, 16–23 (2000)
13. Gordân, R., et al.: Genomic regions flanking e-box binding sites influence DNA binding specificity of bHLH transcription factors through DNA shape. Cell Reports **3**, 1093–1104 (2013)
14. Fletezbrant, C., Lee, D., Mccallion, A.S., Beer, M.: kmer-SVM: a web server for identifying predictive regulatory sequence features in genomic data sets. Nucleic Acids Res. **41**, 544–556 (2013)
15. Shen, Z., Bao, W., Huang, D.: Recurrent Neural Network for Predicting Transcription Factor Binding Sites. Sci. Rep. **8**, 15270 (2018)

16. Krizhevsky, A., Sutskever, I., Hinton, G.E.: ImageNet classification with deep convolutional neural networks. In: Neural Information Processing Systems, pp. 1097–1105 (2012)
17. Srivastava, N., Hinton, G.E., Krizhevsky, A., Sutskever, I., Salakhutdinov, R.: Dropout: a simple way to prevent neural networks from overfitting. J. Mach. Learn. Res. **15**, 1929–1958 (2014)
18. Weirauch, M.T., et al.: Evaluation of methods for modeling transcription factor sequence specificity. Nat. Biotechnol. **31**, 126–134 (2013)
19. Alipanahi, B., Delong, A., Weirauch, M.T., Frey, B.J.: Predicting the sequence specificities of DNA-and RNA-binding proteins by deep learning. Nat. Biotechnol. **33**, 831–838 (2015)
20. Zhen, S., Qinhu Z., Kyungsook, H., Huang, D.S.: A deep learning model for RNA-protein binding preference prediction based on hierarchical LSTM and attention network. IEEE/ACM Transactions on Computational Biology and Bioinformatics
21. Zhen, S., Su-Ping, D., Huang, D.S.: Capsule network for predicting RNA-Protein binding preferences using hybrid feature. IEEE/ACM Transactions on Computational Biology and Bioinformatics (TCBBSI-2019-04-0191)
22. Shen, Z., Deng, S., Huang, D.: RNA-protein binding sites prediction via multi scale convolutional gated recurrent unit networks. IEEE/ACM Transactions on Computational Biology and Bioinformatics, pp. 1–1 (2019)
23. Zhang, Q., Zhu, L., Bao, W., Huang, D.: Weakly-supervised convolutional neural network architecture for predicting protein-dna binding. IEEE/ACM Transactions on Computational Biology and Bioinformatics, pp. 1–1 (2019)
24. Zhang, Q., Zhu, L., Huang, D.: High-Order convolutional neural network architecture for predicting DNA-protein binding sites. IEEE/ACM Trans. Comput. Biol. Bioinf. **16**, 1184–1192 (2019)
25. Zhang, Q., Shen, Z., Huang, D.: Modeling in-vivo protein-DNA binding by combining multiple-instance learning with a hybrid deep neural network. Scientific Reports **9**, 8484 (2019)
26. Zhang, Q., Shen, Z., Huang, D.: Predicting in-vitro transcription factor binding sites using DNA sequence + shape. IEEE/ACM Transactions on Computational Biology and Bioinformatics, pp. 1–1 (2019)
27. Tsatsaronis, G., Panagiotopoulou, V.: A generalized vector space model for text retrieval based on semantic relatedness. In: Conference of the European Chapter of the Association for Computational Linguistics, pp. 70–78 (2009)
28. Wang, J., Huang, P., Zhao, H., Zhang, Z., Zhao, B., Lee, D.L.: Billion-scale Commodity Embedding for E-commerce Recommendation in Alibaba. In: Knowledge Discovery and Data Mining, pp. 839–848 (2018)
29. Glorot, X., Bengio, Y.: Understanding the difficulty of training deep feedforward neural networks. In: International Conference on Artificial Intelligence and Statistics, pp. 249–256 (2010)
30. Zeiler, M.D.: ADADELTA: An Adaptive Learning Rate Method (2012). arXiv abs/1212.5701
31. Zeng, H., Edwards, M.D., Liu, G., Gifford, D.K.: Convolutional neural network architectures for predicting DNA–protein binding. Bioinformatics **32**, 121–127 (2016)
32. Rohs, R., West, S.M., Sosinsky, A., Liu, P., Mann, R.S., Honig, B.: The role of DNA shape in protein–DNA recognition. Nature **461**, 1248–1253 (2009)
33. Zhou, T., et al.: Quantitative modeling of transcription factor binding specificities using DNA shape. Proc. Natl. Acad. Sci. U.S.A. **112**, 4654–4659 (2015)

Protein-Protein Interaction Prediction

Prediction of Membrane Protein Interaction Based on Deep Residual Learning

Tengsheng Jiang, Hongjie Wu(✉), Yuhui Chen, Haiou Li, Jin Qiu, Weizhong Lu, and Qiming Fu

School of Electronic and Information Engineering, Suzhou University of Science and Technology, Suzhou 215009, China
Hongjie.wu@qq.com

Abstract. The interaction of membrane proteins in organisms is directly related to their functions. Accurate prediction of membrane protein interaction can better predict its spatial structure. Therefore, it is of great significance to study the interaction of membrane proteins. Currently, there is no contact method specifically for membrane protein prediction. In this paper, a membrane protein prediction tool based on deep residual learning is established. Combined with the transformation of the covariance matrix, it can well predict the interaction of membrane proteins. Compared with other methods, the experimental data and results of this model are more accurate.

Keywords: Membrane protein · Deep learning · Contact map

1 Introduction

Membrane proteins play a very important role in many life activities of organisms. Such as cell proliferation and differentiation, signal transduction [1] and so on. It is generally believed that the interaction of membrane proteins directly affects their function [2]. In structural biology, the research on predicting the interaction of proteins has previously been studied using Mutation to predict contact [3], use of co-evolution [4], Bayesian network methods [5]. But all are limited by the inability to effectively deal with noise interference. With the development of machine learning, the use of machine learning methods for contact prediction greatly improves the accuracy rate. CNN convolutional neural [6] network and other machine learning methods to predict contact graphs. The covariance matrix has been widely used in the top-level methods of casp13 [7], such as DMP, Shen-Cdeep, Yang-Server, etc. However, there is currently no method applied in the field of membrane protein interaction prediction. Therefore, in this paper, starting from the prediction of membrane protein interaction, combined with the characteristics of membrane protein, from the perspective of computational biology, a tool suitable for the prediction of membrane protein interaction is proposed.

D.-S. Huang and K.-H. Jo (Eds.): ICIC 2020, LNCS 12464, pp. 103–108, 2020.
https://doi.org/10.1007/978-3-030-60802-6_10

2 Data Sources and Data Set

2.1 Data Sources

In this paper, all the experimental data comes from SCOPE, which is a database for structural classification-expansion of proteins developed at Berkeley Laboratories and the University of California, Berkeley. Website: https://scop.berkeley.edu/. Then use the hhblits [8] tool to calculate MSA.

2.2 Data Set

In this paper, the data of membrane protein classification in the SCOPe database—Lineage for Class f: Membrane and cell surface proteins [9] and peptides [10] are used. The classification includes 60 flods, 119 superfamilies, and 173 Number of families. We choose 4784 of them for model training. Table 1 lists representative partial membrane protein.

Table 1. Part of the Protein used in this paper.

#	Class	Folds	Superfamilies	Protein	Quantity
1	f	f.1	Toxins' membrane translocation domains	d1ddta3	324
2	f	f.3	Light-harvesting complex subunits	1dx7	144
3	f	f.4	Transmembrane beta-barrels	1bxw	347
4	f	f.17	Rotary ATPase ring subunits2	d1c99a_	229
5	f	f.6	C-X-C chemokine receptor type 4	d3m2la_	153
6	f	f.26	Bacterial photosystem II	d1rzhl_	229
7	f	f.21	Baker's yeast	d1kb9c2	185
8	f	f.13	G protein-coupled receptor (GPCR)	d1uazb_	334
9	f	F.23	Bacterial aa3 type cytochrome c oxidase subunit IV	d5b3sq_	1247

The PDB file containing sequence coordinate information is provided in the SCOPE database. After selecting the sequence, the PDB file of each sequence can be used to calculate the true contact status of the residue pair. The definition and categorization of contact predictions follow the conventional criterions in CASP, i.e., a residue pair, among which the Euclidean distance between two Cb (Ca for Glycine) atoms is smaller than 8 Å, is considered as in contact.

3 Methods

In this paper, the flow chart of the research method is shown in Fig. 1.

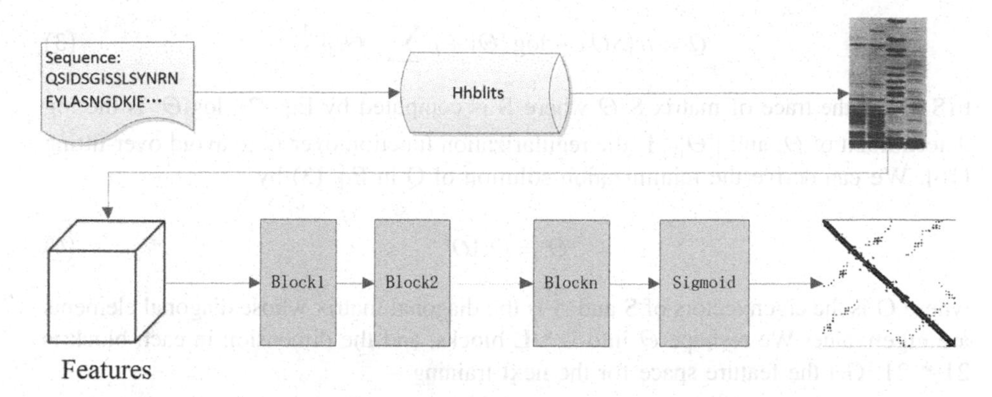

Fig. 1. Flowchart of this method

3.1 Multiple Sequence Alignment Generation

First, we use one sequence to generate the corresponding multiple sequence alignment (MSA). Information-rich MSA is essential for contact map prediction. In this paper, MSA is generated by HHblits and the coverage threshold of the query sequence is 60%. Through two iterations, the latest comparison database UniRef30_2020_02 is used for comparison. Then, one-hot encoding [11] is performed for each position in the MSA. Each position becomes a 21-dimensional vector.

3.2 Constructing the Characteristic Space of Membrane Proteins

First, we use MSA to calculate the covariance matrix [12]. The covariance matrix $\sum$ as the input feature is proposed by DeepCov [13]. Considering the MSA of N rows and L columns, we can calculate the sample covariance matrix with dimensions 21 * L * 21 * L as follows:

$$S_{ij}^{ab} = f_{i,j} = (a, b) - f_i(a)f_j(b) \tag{1}$$

where $f_{i,j}$ (a, b) is the relative frequency of residue pairs a and b at positions i and j, $f_i(a)$ is the frequency of occurrence of residual type a at position i, and $f_j(b)$ is at position j the frequency of occurrence of residual type b. There are 21 residue types (20 standard amino acid types plus one gap type).

But the covariance matrix cannot effectively eliminate the interference between variables. Therefore, the Mahalanobis distance transformation matrix [14] $D_M(x)$ is further adopted as the feature extraction to achieve the purpose of eliminating the influence of variable metrics.

$$D_M(x) = \sqrt{(x-\mu)^T \sum{}^{-1} (x-\mu)} \tag{2}$$

Then minimize the objective function to get the matrix estimate Θ for further training by minimizing the objective function [15]:

$$Q = tr(S\Theta) - \log|\Theta| + \rho \sum ||\Theta_{i,j}||_2^2 \quad (3)$$

tr(S Θ) is the trace of matrix S Θ where S is computed by Eq. (2). log$|\Theta|$ is the log determinant of Θ; and $||\Theta||_2^2$ is the regularization function over Θ to avoid over-fitting [16]. We can derive the minimization solution of Q in Eq. (3) by

$$\widehat{\Theta} = O\Lambda O^T \quad (4)$$

where O is the eigenvectors of S and Λ is the diagonal matrix whose diagonal elements are eigenvalues.We reshape Θ into L * L blocks, and the dimension in each block is 21 * 21. Get the feature space for the next training.

3.3 Deep Residual Learning Based Model Training

In the field of computer image processing, the effect of using CNN to extract features of pictures is very significant. Here, we also use CNN to process the extracted feature space. Each pixel represents a residue pair. When the length of a sequence is L, its input feature is composed of residue pairs in the sequence, and its dimension is L * 21 * L * 21. Full convolutional network architecture [17] has been shown to be an effective solution for training end-to-end, pixel-to pixel models on semantic segmentation, i.e., pixel-wise labeling.

With the increase of training depth, the traditional deep learning is prone to gradient explosion or gradient disappearance. The residual network [18] proposed by He Kaiming can effectively solve this problem by using the characteristics of skip-connection. He proposed the residual network (ResNets). As shown in followed, The output of each residual block [19] x_t is:

$$x_t = f(x_{t-1} + F(x_{t-1}, w_t)) \quad (5)$$

where w_t is a set of weights, and the final residual output is obtained through the activation function ReLU [20]. In addition, the dimension of the convolution kernel is 3 * 3, padding is 1 in each block.

4 Experiments and Discussion

We rank the probability value of the prediction result of the sequence of length L from high to low. In this paper, the experimental results are evaluated, some of which are shown in Table 2. Sort all results from high to low. Assuming the sequence length is L, the accuracy of L/2, L/5, L/10 is calculated in sequence. Residue pairs in contact and separated by at least 24 residues in the sequence are considered as long-range contacts, where those with a sequence separation between 12 and 23 or 6 and 11 are considered as medium- or short-range contacts, respectively.

Table 2. Performance comparisons between covariance (Cov) and Mahalanobis-matrix (Dmx) based features on the dataset

		L/10	L/5	L/2	L
Short	Cov	0.809	0.712	0.432	0.231
	Dmx	0.827	0.732	0.453	0.298
Medium	Cov	0.812	0.721	0.489	0.241
	Dmx	0.832	0.721	0.493	0.250
Long	Cov	0.843	0.742	0.523	0.42
	Dmx	0.852	0.753	0.532	0.43

We can see that Dmx has higher accuracy than Cov in predicting contact in Table 2. From the perspective of short, medium, and long, the results are better than using Cov alone. We believe that Dmx can effectively improve the prediction accuracy.

5 Conclusion

In this paper, a membrane protein prediction tool based on deep learning is established. This paper tested the membrane protein family in the SCOPe database and obtained a large amount of experimental data. The results show that the tool can accurately reflect the contact of membrane proteins.

Acknowledgement. This paper is supported by the National Natural Science Foundation of China (61772357, 61502329, 61672371, and 61876217), Jiangsu Province 333 Talent Project, Top Talent Project (DZXX-010), Suzhou Foresight Research Project (SYG201704, SNG201610, and SZS201609).

References

1. Heijne, G.V.: Membrane proteins. Eur. J. Biochem. **120**, 12 (1981)
2. Kubota, Y., Nash, R.A., Klungland, A., et al.: Reconstitution of DNA base excision-repair with purified human proteins: interaction between DNA polymerase beta and the XRCC1 protein. EMBO J. **15**(23), 6662–6670 (1997)
3. GoBel, U., Sander, C., Schneider, R., et al.: Correlated mutations and residue contacts in proteins. Proteins Struct. Funct. Genet. **18**(4), 309–317 (1994)
4. Martin, L.C., Gloor, G.B., Dunn, S.D., et al.: Using information theory to search for co-evolving residues in proteins. Bioinformatics **21**(22), 4116–4124 (2005)
5. Feizi, S., et al.: Network deconvolution as a general method to distinguish direct dependencies in networks. Nat. Biotechnol. **31**, 726 (2013)
6. Krizhevsky, A., Sutskever, I., Hinton, G.: ImageNet classification with deep convolutional neural networks. Adv. Neural Inf. Process. Syst. **25**(2), 1097–1105 (2012)
7. Buchan, D.W.A., Jones, D.T.: Improved protein contact predictions with the MetaPSICOV2 server in CASP12. Proteins Struct. Funct. Bioinf. **86**(Suppl 1), 17 (2017)

8. Remmert, M., Biegert, A., Hauser, A., et al.: HHblits: Lightning-fast iterative protein sequence searching by HMM-HMM alignment. Nat. Methods **9**(2), 173–175 (2011)
9. Ushkaryov, Y., Petrenko, A., Geppert, M., et al.: Neurexins: synaptic cell surface proteins related to the alpha-latrotoxin receptor and laminin. Science **257**(5066), 50–56 (1992)
10. Udenfriend, S., Stein, S., Bohlen, P., et al.: Fluorescamine: a reagent for assay of amino acids, peptides, proteins, and primary amines in the picomole range. Science **178**(4063), 871–872 (1972)
11. Knapp, S.K.: Accelerate FPGA macros with one-hot approach. Electron. Des. **38**, 71–78 (1990)
12. Fan, J., et al.: An overview of the estimation of large covariance and precision matrices. Econ. J. **19**, C1–C32 (2016)
13. Jones, D.T., et al.: High precision in protein contact prediction using fully convolutional neural networks and minimal sequence features. Bioinformatics **34**, 3308–3315 (2018)
14. Maesschalck, R.D., Jouan-Rimbaud, D., Massart, D.L.: The Mahalanobis distance. Chemometr. Intell. Lab. Syst. **50**(1), 1–18 (2000)
15. Guu, S.M., Wu, Y.K.: Minimizing a linear objective function with fuzzy relation equation constraints. Fuzzy Optim. Decis. Making **1**(4), 347–360 (2002)
16. Cawley, G.C.: On over-fitting in model selection and subsequent selection bias in performance evaluation. J. Mach. Learn. Res. **11**, 2079–2107 (2010)
17. Long, J., Shelhamer, 等.: Fully convolutional networks for semantic segmentation. In: IEEE Transactions on Pattern Analysis & Machine Intelligence (2017)
18. He, K., Zhang, X., Ren, S., et al.: Deep Residual learning for image recognition. In: IEEE Conference on Computer Vision & Pattern Recognition. IEEE Computer Society (2016)
19. Carlos, V.A., et al.: The significance of quantifiable residual normal karyotype hematopoietic cells for toxicity and outcome. Leuk. Lymphoma **52**(6), 943 (2011)
20. Chizat, L., Bach, F.: Implicit Bias of Gradient Descent for Wide Two-layer Neural Networks (2020)

GCNSP: A Novel Prediction Method of Self-Interacting Proteins Based on Graph Convolutional Networks

Lei Wang[1,2], Zhu-Hong You[2(✉)], Xin Yan[3,4(✉)], Kai Zheng[5], and Zheng-Wei Li[2]

[1] College of Information Science and Engineering, Zaozhuang University, Zaozhuang, China
[2] Xinjiang Technical Institute of Physics and Chemistry, Chinese Academy of Sciences, Urumqi, China
zhuhongyou@ms.xjb.ac.cn
[3] School of Computer Science and Technology, China University of Mining and Technology, Xuzhou, China
xinyanuzz@gmail.com
[4] School of Foreign Languages, Zaozhuang University, Zaozhuang, China
[5] School of Computer Science and Engineering, Central South University, Changsha, China

Abstract. As an essential protein interaction, self-interacting proteins (SIPs) plays a vital role in biological processes. Identifying and confirming SIPs is of great significance for the exploration of new gene functions, protein function research and proteomics research. Although a large number of SIPs have been confirmed with the rapid development of high-throughput technology, the biological experimental method is still limited by blindness and high cost, and has a high false-positive rate. Therefore, the use of computational techniques to accurately and efficiently predict SIPs has become an urgent need. In this study, a novel SIPs prediction method GCNSP based on Graph Convolutional Networks (GCN) is proposed. Firstly, the evolution information of protein is described by Position-Specific Scoring Matrix (PSSM). Then the feature information is extracted by GCN, and finally fed into Random Forest (RF) classifier for accurate classification. In the five-fold cross-validation on *Human* and *Yeast* data sets, GCNSP achieved 93.65% and 90.69% prediction accuracy with 99.64% and 99.08% specificity, respectively. In comparison with different classifier models and other existing methods, GCNSP shows strong competitiveness. The excellent results show that the proposed method is very suitable for SIPs prediction and can provide highly reliable candidates for biological experiments.

Keywords: Self-interacting protein · Graph Convolutional Networks · Protein-protein interactions · Position-specific scoring matrix · Random forest

D.-S. Huang and K.-H. Jo (Eds.): ICIC 2020, LNCS 12464, pp. 109–120, 2020.
https://doi.org/10.1007/978-3-030-60802-6_11

1 Introduction

Proteins are the primary component of an organism, and almost participate in all biological processes in cells. Exploring protein structure and function is of great significance for understanding life activities, disease treatment, and new drug development. Studies have shown that proteins usually do not perform functions alone, but interact with other proteins (protein-protein interactions, PPIs) to complete a specific function [1, 2]. With the rapid development of high-throughput biological experiment technologies, such as mass spectrometry, yeast two-hybrid, protein chip technology, and chromosomal co-immunoprecipitation, etc., protein interaction data has emerged in large numbers [3]. How to mine biological knowledge from massive protein interaction data to reveal the reaction pathways, regulatory mechanisms, and molecular components involved in life processes are facing new challenges.

In protein interactions, self-interacting proteins (SIPs) occupies a prominent position. A pair of interacting SIP proteins is translated by the same gene to form a homodimer. Evidence suggests that homo-oligomerization plays a crucial role in critical biological processes including enzyme activation, gene expression regulation, signal transduction and immune response. Based on the analysis of the origin and evolution of large-scale biological protein complexes, Pereira-Leal *et al.* [4] found that the evolution of many protein complexes was initially established by self-interactions. Ispolatov *et al.* [5] found that there are a large number of SIPs in the protein interaction network, which may play an important role in the cellular system. Moreover, SIPs can regulate protein functions through its interactions without increasing the size of the genome, thereby expanding their functional diversity.

In recent years, researchers have proposed some computational models for predicting PPIs [6–10]. For example, Wang *et al.* [11] proposed a new method CNN-FSRF based on a convolutional neural network to predict PPIs. This method can extract the deep features of protein effectively by using convolutional neural network, thereby achieving excellent results. You *et al.* [12] designed a new hierarchical PCA-EELM model, which only uses protein sequence information to predict PPIs, and achieved high accuracy in the benchmark data sets. Jia *et al.* [13] predicted PPIs by integrating chaos game representation information into Pseudo Amino Acid Composition (PseAAC), so that some key sequence-order or sequence-pattern information could be effectively combined during the treatment of the protein pair samples. Wang *et al.* [14] proposed a pure biological language processing model to predict PPIs. The model uses natural language processing to extract the semantic features of protein sequences, thereby effectively predicting the potential interaction between proteins.

In this paper, we design the GCNSP model to predict SIPs based on the evolutionary information of protein sequences and the latest deep learning algorithm Graph Convolutional Network (GCN). Specifically, we build the model through the following three steps. Firstly, the Position-Specific Scoring Matrix (PSSM) is used to describe the evolution information of protein sequence; secondly, GCN with excellent characterization ability is used to deeply extract the features of protein; finally, the random forest classifier is used to accurately predict potential SIPs. In the five-fold cross-validation experiment, GCNSP achieved 90.69% and 93.65% prediction accuracy with 99.08%

and 99.64% specificity on the benchmark data sets *yeast* and *human*, respectively. To further evaluate the performance of GCNSP, we compared it with other existing methods on these two benchmark datasets. Competitive results indicate that GCNSP can effectively predict potential SIPs from a large amount of data, and provide highly reliable candidates for biological experiments.

2 Materials and Methods

2.1 Data Sources

We collected human protein sequences that can interact with themselves from databases including UniProt [15], InnateDB [16], BioGRID [17], DIP [18] and MatrixDB [19] to construct the experimental data set. These data have been processed as follows: (1) Delete protein sequences with length longer than 5000 or less than 50 residues from the human proteome; (2) The protein data selected as positive samples meet at least one of the following conditions: (a) in UniProt database, the protein has been defined as homooligomer; (b) at least two related publications reported on them; (c) at least verified by two kinds of large-scale or one kind of small-scale experiments; (3) All known self-interacting proteins were removed from the negative data set. Finally, 1441 *human* SIPs and 15938 *human* non-SIPs were selected as positive and negative sample sets, respectively. In addition, in order to further evaluate the model, the same strategy was used to create the *yeast* data set, including 710 positive SIPs and 5511 negative non-SIPs.

2.2 Position-Specific Scoring Matrix

The Position-specific scoring matrix (PSSM) is a kind of sequence matrix introduced by Gribskov *et al.* [20] to detect distantly related proteins, and a scoring matrix is generated from a group of sequences pre-arranged by sequence or structural similarity [21–25]. Each protein sequence can be converted into a $N \times 20$ PSSM matrix $PM(i,j)$, which is expressed as follows:

$$PM = \begin{bmatrix} \sigma_{1,1} & \sigma_{1,2} & \cdots & \sigma_{1,20} \\ \sigma_{2,1} & \sigma_{2,2} & \cdots & \sigma_{2,20} \\ \vdots & \vdots & \vdots & \vdots \\ \sigma_{N,1} & \sigma_{N,2} & \cdots & \sigma_{N,20} \end{bmatrix} \quad (1)$$

Here $\sigma_{i,j}$ represents the probability that the *ith* residue being mutated into the *jth* naive amino acid during the evolutionary process of protein multiple sequence alignment. In the experiment, we use the Position Specific Iterated BLAST (PSI-BLAST) tool to extract the evolutionary information of protein sequence and generate PSSM matrix by comparing the homologous proteins in *SwissProt* database.

2.3 Fast Learning with Graph Convolutional Networks

In this study, we use the Fast learning with Graph Convolutional Networks (FastGCN) [26] to extract the information describing the deep features of proteins to classify and predict them more clearly. FastGCN is a newly proposed deep learning algorithm, which can interpret graph convolution as the integral transformation of embedded function under probability measure.

Assume that the vertex set V' of the graph G' is associated with the probability space (V', F, P). For sub graph G of graph G', whose vertices are i.i.d. sample of V' generated by probability measure P. The function generalization can be described as follows:

$$\tilde{h}^{(l+1)}(v) = \int \hat{A}(v,u)h^{(l)}(u)W^{(l)}dP(u), h^{(l+1)}(v) = \sigma\left(\tilde{h}^{(l+1)}(v)\right), l = 0, \ldots, M-1 \tag{2}$$

Here, function $h^{(l)}$ is an embedded function from the lth layer, u and v are independent random variables of P. The loss L is the expected value of $g\left(h^{(M)}\right)$ and is embedded in $h^{(M)}$, which is described as:

$$L = E_{v \sim P}\left[g\left(h^{(M)}\right)(v)\right] = \int g\left(h^{(M)}\right)(v)dP(v) \tag{3}$$

Therefore, the i.i.d. sample $u_1^{(l)}, \ldots, u_{t_l}^{(l)} \sim P$ of t_l can be used to approximate the estimation of the integral transformation in the lth layer, which is expressed as follows:

$$\tilde{h}_{t_{l+1}}^{(l+1)}(v) := \frac{1}{t}\sum\nolimits_{j=1}^{t_l} \hat{A}\left(v, u_j^{(l)}\right)h_{t_l}^{(l)}\left(u_j^{(l)}\right)W^{(l)}, \quad h_{t_{l+1}}^{(l+1)}(v) := \sigma\left(\tilde{h}_{t_{l+1}}^{(l+1)}(v)\right) \tag{4}$$

where $h_{t_0}^{(0)}$ is $h^{(0)}$. Such loss L can be converted into:

$$L_{t_0,t_1,\ldots,t_M} := \frac{1}{t_M}\sum\nolimits_{i=1}^{t_M} g\left(h_{t_M}^{(M)}\left(u_i^{(M)}\right)\right) \tag{5}$$

2.4 Random Forest

We used the random forest (RF) [27–29] classifier to classify SIPs data in the experiment accurately. Random forest is a classifier that contains multiple decision trees and has a prominent classification function. The random forest algorithm can be divided into two stages, one stage is to create a random forest, and the other is to make a prediction based on the random forest classifier created in the previous stage. The entire process is as follows:

The first is the random forest creation stage:

(1) Randomly select K features from all m features, where $k << m$.
(2) Among the K features, calculate the node d with the best split point.
(3) Split the node into child nodes with the best split.
(4) Repeat the previous three steps until you get L number of nodes.
(5) Repeat steps 1 to 4 n times to create n trees to form a forest.

In the second stage, RF makes the final prediction based on the random forest classifier created in the previous stage. The process is as follows:

(1) Select the test features, use the rules of each randomly created decision tree to predict the results, and save the predicted results.
(2) Settle the number of votes obtained for each prediction result.
(3) The prediction result with the most votes is used as the final prediction of the random forest algorithm.

3 Results

3.1 Evaluation Criteria

In the experiment, we use Accuracy (Acc.), Specificity (Spe.), F1-score (F1), Matthews correlation coefficient (MCC) and area under the receiver operating characteristic curve (AUC) as evaluation parameters to evaluate the performance of the model [30–33]. Their calculation formula is as follows:

$$Acc. = \frac{TP + TN}{TP + TN + FP + FN} \tag{6}$$

$$Spe. = \frac{TN}{TN + FP} \tag{7}$$

$$F1 = \frac{2TP}{2TP + FP + FN} \tag{8}$$

$$MCC = \frac{TP \times TN - FP \times FN}{\sqrt{(TP + FP) \times (TN + FN) \times (TP + FN) \times (TN + FP)}} \tag{9}$$

Here TP represents the number of true positive samples, TN represents the number of true negative samples, FP represents the number of false negative samples and FN represents the number of false negative samples. In addition, the five-fold cross-validation is used to verify the predictive ability of the model [34–37].

3.2 Performance on the Benchmark Dataset

We implement five-fold cross-validation on benchmark data sets *human* and *yeast* to evaluate the performance of GCNSP model. Table 1 summarizes the five-fold

Table 1. The five-fold cross-validation results performed by GCNSP on *human* data set

Testing set	Acc.	Spe.	F1	MCC	AUC
1	92.78%	99.03%	34.81%	37.20%	58.41%
2	93.41%	99.84%	33.62%	41.77%	61.98%
3	94.10%	99.85%	22.64%	32.09%	54.22%
4	93.53%	99.68%	44.72%	49.63%	61.08%
5	94.42%	99.78%	49.74%	54.36%	67.73%
Average	93.65 ± 0.64%	99.64 ± 0.35%	37.11 ± 10.54%	43.01 ± 9.04%	60.68 ± 4.96%

Table 2. The five-fold cross-validation results performed by GCNSP on *yeast* data set

Testing set	Acc.	Spe.	F1	MCC	AUC
1	91.32%	99.82%	44.33%	49.87%	65.99%
2	91.08%	98.84%	35.09%	37.04%	64.13%
3	90.35%	98.99%	41.75%	43.85%	68.38%
4	90.11%	98.72%	37.56%	39.07%	61.81%
5	90.60%	99.01%	33.14%	36.13%	61.22%
Average	90.69 ± 0.50%	99.08 ± 0.43%	38.37 ± 4.63%	41.19 ± 5.69%	64.30 ± 2.97%

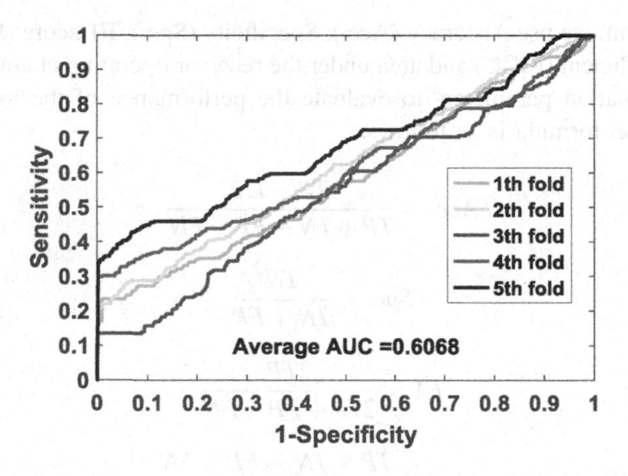

Fig. 1. ROC curves of five-fold cross-validated performed by GCNSP on *human* data set

cross-validation experimental results generated by GCNSP on *human* data. We can see from the table that GCNSP achieved the prediction accuracy of 93.65%, the specificity of 99.64%, the F1-score of 37.11%, the Matthews correlation coefficient of 43.01% and the AUC of 60.68%. The standard deviations of these evaluation criteria are 0.64%, 0.35%, 10.54%, 9.04% and 4.96%, respectively. Table 2 lists the five-fold cross-validation experimental results generated by GCNSP on *yeast* data. As can be seen in Table 2, the average accuracy of GCNSP is 90.69%, the specificity is 99.08%, the F1-score is 38.37%, the Matthews correlation coefficient is 41.19% and the AUC is 64.30%. The standard deviations of these evaluation criteria are 0.50%, 0.43%, 4.63%,

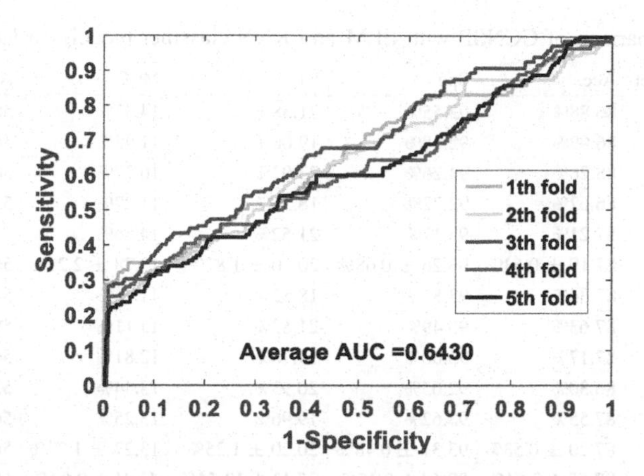

Fig. 2. ROC curves of five-fold cross-validated performed by GCNSP on *yeast* data set

5.69% and 2.97%, respectively. The ROC curve of five-fold cross-validated generated by GCNSP model on *human* and *yeast* datasets is shown in Fig. 1 and Fig. 2.

3.3 Comparison of Different Classifier Models

Although GCNSP achieved good performance from the experimental results, in order to further evaluate the impact of the classifier on the overall performance of the model, we compared it with the excellent ELM [38] and KNN [39] classifier models. We use the same feature extraction method in the experiment, only replacing the classifier in the model to verify on *human* and *yeast* data sets.

Tables 3 and 4 summarize the results of the five-fold cross-validation of the ELM and KNN classifier models on the *human* and *yeast* data sets, respectively. It can be seen from Table 3 that the ELM classifier model achieved the prediction accuracy of 87.19% on *human* data set, and the results of its five experiments were 86.88%, 86.99%, 88.26%, 86.62%, and 87.21%, respectively; the KNN classifier model achieved the prediction accuracy of 87.20%, and the results of its five experiments were 87.34%, 87.63%, 87.17%, 86.30%, and 87.55%, respectively. However, the GCNSP model achieved the prediction accuracy of 93.65%, which was higher than them by 6.46% and 6.45%, respectively.

From Table 4, we can see that the prediction accuracy of ELM and KNN classifier models on the *yeast* data set is 79.68% and 82.86%, respectively, which is 11.01% and 7.83% lower than that of GCNSP model. In other evaluation criteria, our model also achieves the optimal experimental results. In order to facilitate intuitive comparison, we will display the results of all the evaluation criteria in the form of histogram, and the results are shown in Figs. 3 and 4.

Table 3. Comparison of GCNSP with ELM and KNN classifier models on *human* data set

Model	Testing set	Acc.	Spe.	F1	MCC	AUC
ELM	1	86.88%	93.55%	21.38%	14.33%	58.32%
	2	86.99%	92.58%	19.00%	11.95%	56.66%
	3	88.26%	94.26%	23.02%	16.77%	56.21%
	4	86.62%	92.72%	18.56%	11.27%	55.46%
	5	87.21%	93.17%	21.52%	14.56%	54.59%
	Average	87.19 ± 0.63%	93.26 ± 0.68%	20.70 ± 1.87%	13.78 ± 2.21%	56.25 ± 1.40%
KNN	1	87.34%	93.81%	18.52%	11.76%	54.53%
	2	87.63%	93.49%	21.82%	15.11%	59.00%
	3	87.17%	93.00%	19.78%	12.81%	56.21%
	4	86.30%	92.65%	20.93%	13.44%	53.86%
	5	87.55%	93.62%	19.96%	13.25%	56.66%
	Average	87.20 ± 0.53%	93.31 ± 0.48%	20.20 ± 1.25%	13.27 ± 1.22%	56.05 ± 2.01%
GCNSP	Average	**93.65 ± 0.64%**	**99.64 ± 0.35%**	**37.11 ± 10.54%**	**43.01 ± 9.04%**	**60.68 ± 4.96%**

Table 4. Comparison of GCNSP with ELM and KNN classifier models on *yeast* data set

Model	Testing set	Acc.	Spe.	F1	MCC	AUC
ELM	1	79.18%	87.01%	20.80%	8.93%	55.50%
	2	79.82%	85.93%	21.32%	10.53%	56.27%
	3	80.14%	86.47%	23.53%	12.70%	55.02%
	4	80.87%	86.96%	27.88%	17.41%	59.00%
	5	78.39%	86.04%	22.48%	10.14%	51.64%
	Average	79.68 ± 0.94%	86.48 ± 0.50%	23.20 ± 2.82%	11.94 ± 3.35%	55.49 ± 2.64%
KNN	1	82.32%	90.59%	22.54%	12.60%	59.12%
	2	83.44%	90.68%	20.16%	10.92%	52.72%
	3	81.75%	90.11%	22.53%	12.21%	54.78%
	4	82.88%	91.35%	20.82%	11.35%	53.99%
	5	83.94%	92.07%	18.70%	9.94%	54.18%
	Average	82.86 ± 0.87%	90.96 ± 0.76%	20.95 ± 1.64%	11.40 ± 1.06%	54.96 ± 2.44%
GCNSP	Average	**90.69 ± 0.50%**	**99.08 ± 0.43%**	**38.37 ± 4.63%**	**41.19 ± 5.69%**	**64.30 ± 2.97%**

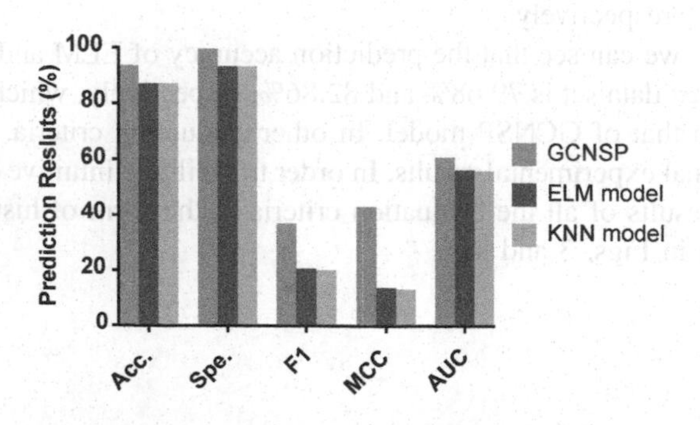

Fig. 3. Comparison of different classifier models on *human* dataset

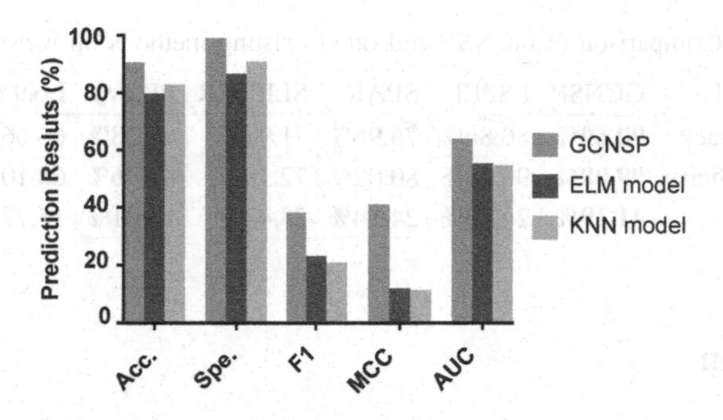

Fig. 4. Comparison of different classifier models on *yeast* dataset

3.4 Comparison with Other Existing Methods

In recent years, some excellent methods including PSPEL [40], SPAR [41], SLIPPER [42], PPIevo [43] and LocFuse [44] have been proposed to predict SIPs. In order to understand the ability of GCNSP to predict SIPs more clearly, we compare it with these methods on *human* and *yeast* data sets. Table 5 and Table 6 summarize the results generated by the above methods on *human* and *yeast* data sets. It can be seen from Table 5 that the GCNSP model achieved the highest prediction accuracy on the *human* dataset, which were 1.56% higher than the second highest SPAR method and 5.84% higher than the average result. On the evaluation criteria specificity and MCC, GCNSP also achieved the best results, which were 20.31% and 12.76% higher than the average values of the other five methods.

Table 5. Comparison of GCNSP and other existing methods on *human* data set

Model	GCNSP	PSPEL	SPAR	SLIPPER	PPIevo	LocFuse
Accuracy	**93.65%**	91.30%	92.09%	91.10%	78.04%	80.66%
Specificity	**99.64%**	97.89%	97.40%	95.06%	25.82%	80.50%
MCC	**43.01%**	29.84%	38.36%	41.97%	20.82%	20.26%

Similarity, from Table 6 we can see that the GCNSP model also achieved the best prediction accuracy on the *yeast* data set, and the result is 16.96% higher than the average of the other five methods. The comparison results show that our proposed GCNSP model has good prediction performance and can provide higher accuracy than the currently existing methods. GCNSP also achieved the best results on the evaluation criteria specificity and MCC, which were 18.27% and 18.91% higher than the average of the other five methods, respectively.

Table 6. Comparison of GCNSP and other existing methods on *yeast* data set

Model	GCNSP	PSPEL	SPAR	SLIPPER	PPIevo	LocFuse
Accuracy	**90.69%**	86.86%	76.96%	71.90%	66.28%	66.66%
Specificity	**99.08%**	96.28%	80.02%	72.18%	87.46%	68.10%
MCC	**41.19%**	24.36%	24.84%	28.42%	18.01%	15.77%

4 Conclusion

With the continuous accumulation of protein sequence data and the rapid development of computing technology, the prediction of SIPs based on computing has made significant progress. In this study, we proposed a computational method GCNSP that uses GCN to predict SIPs based on protein sequence information. This method can make full use of the evolutionary information of proteins, and accurately and efficiently predict potential SIPs through deep learning algorithm and ensemble classifier. GCNSP has achieved excellent results on both *human* and *yeast* data sets. In comparison with other classifier models and existing methods, this method showed strong competitiveness. These results indicated that GCNSP can be used as a promising tool for predicting SIPs to provide highly reliable candidates for further biological experiments.

Acknowledgements. This work is supported is supported in part by the National Natural Science Foundation of China, under Grants 61702444, in part by the West Light Foundation of The Chinese Academy of Sciences, under Grant 2018-XBQNXZ-B-008, in part by the Chinese Postdoctoral Science Foundation, under Grant 2019M653804, in part by the Tianshan youth - Excellent Youth, under Grant 2019Q029, in part by the Qingtan scholar talent project of Zaozhuang University. The authors would like to thank all anonymous reviewers for their constructive advices.

References

1. Giot, L., et al.: A protein interaction map of Drosophila melanogaster. Science **302**, 1727–1736 (2003)
2. Li, S., et al.: A map of the interactome network of the metazoan C. elegans. Science **303**, 540–543 (2004)
3. Gavin, A.C., et al.: Functional organization of the yeast proteome by systematic analysis of protein complexes. Nature **415**, 141–147 (2002)
4. Pereira-Leal, J.B., et al.: Evolution of protein complexes by duplication of homometric interactions. Genome Biol. **8**, 51 (2007). https://doi.org/10.1186/gb-2007-8-4-r51
5. Ispolatov, I., et al.: Binding properties and evolution of homodimers in protein-protein interaction networks. Nucleic Acids Res. **33**, 3629–3635 (2005)
6. Wang, Y.-B., et al.: Predicting protein–protein interactions from protein sequences by a stacked sparse autoencoder deep neural network. Mol. BioSyst. **13**, 1336–1344 (2017)
7. Wang, L., et al.: Advancing the prediction accuracy of protein-protein interactions by utilizing evolutionary information from position-specific scoring matrix and ensemble classifier. J. Theor. Biol. **418**, 105–110 (2017)

8. Wang, L., et al.: An ensemble approach for large-scale identification of protein-protein interactions using the alignments of multiple sequences. Oncotarget **8**, 5149 (2017)
9. Zhu, L., You, Z.H., Huang, D.S.: Increasing the reliability of protein–protein interaction networks via non-convex semantic embedding. Neurocomputing **121**, 99–107 (2013)
10. You, Z.H., et al.: Using manifold embedding for assessing and predicting protein interactions from high-throughput experimental data. Bioinformatics **26**, 2744–2751 (2010)
11. Wang, L., et al.: Predicting Protein-Protein Interactions from Matrix-Based Protein Sequence Using Convolution Neural Network and Feature-Selective Rotation Forest. Sci. Rep. **9**, 9848 (2019)
12. You, Z.-H., et al.: Prediction of protein-protein interactions from amino acid sequences with ensemble extreme learning machines and principal component analysis. BMC Bioinf. **14**, S10 (2013). https://doi.org/10.1186/1471-2105-14-S8-S10
13. Jia, J., et al.: iPPI-PseAAC (CGR): Identify protein-protein interactions by incorporating chaos game representation into PseAAC. J. Theor. Biol. **460**, 195–203 (2019)
14. Wang, Y., et al.: A high efficient biological language model for predicting protein–protein interactions. Cells **8**, 122 (2019)
15. Bateman, A., et al.: UniProt: a hub for protein information. Nucleic Acids Res. **43**, D204–D212 (2015)
16. Breuer, K., et al.: InnateDB: systems biology of innate immunity and beyond-recent updates and continuing curation. Nucleic Acids Res. **41**, D1228–D1233 (2013)
17. Chatr-aryamontri, A., et al.: The BioGRID interaction database: 2015 update. Nucleic Acids Res. **43**, D470–D478 (2015)
18. Salwinski, L., Miller, C.S., Smith, A.J., et al.: The database of interacting proteins: 2004 update. Nucleic Acids Res. **32**, D449–D451 (2004)
19. Launay, G., et al.: MatrixDB, the extracellular matrix interaction database: updated content, a new navigator and expanded functionalities. Nucleic Acids Res. **43**, D321–D327 (2015)
20. Gribskov, M., McLachlan, A.D., Eisenberg, D.: Profile analysis: detection of distantly related proteins. Proc. Natl. Acad. Sci. U.S.A. **84**, 4355–4358 (1987)
21. Zheng, K., et al.: Dbmda: A unified embedding for sequence-based mirna similarity measure with applications to predict and validate mirna-disease associations. Molecular Therapy-Nucleic Acids **19**, 602–611 (2020)
22. Deng, S.-P., Zhu, L., Huang, D.-S.: Predicting hub genes associated with cervical cancer through gene co-expression networks. IEEE/ACM Trans. Comput. Biol. Bioinf. **13**, 27–35 (2015)
23. Wang, L., et al.: Computational methods for the prediction of drug-target interactions from drug fingerprints and protein sequences by stacked auto-encoder deep neural network. In: Cai, Z., Daescu, O., Li, M. (eds.) ISBRA 2017. LNCS, vol. 10330, pp. 46–58. Springer, Cham (2017). https://doi.org/10.1007/978-3-319-59575-7_5
24. Wang, L., et al.: Prediction of RNA-protein interactions by combining deep convolutional neural network with feature selection ensemble method. J. Theor. Biol. **461**, 230–238 (2019)
25. Chen, Z.-H., et al.: Identification of self-interacting proteins by integrating random projection classifier and finite impulse response filter. Bmc Genomics **20**, 1 (2019)
26. Jie, C., Ma, T., Cao, X.: FastGCN: Fast learning with graph convolutional networks via importance sampling. In: International Conference on Learning Representations (2018)
27. Wang, L., et al.: An efficient approach based on multi-sources information to predict circrna-disease associations using deep convolutional neural network. Bioinformatics **36**, 4038–4046 (2019)
28. You, Z-H., et al.: Large-scale protein-protein interactions detection by integrating big biosensing data with computational model. Biomed. Res. Int (2014

29. Zheng, K., Wang, L., You, Z.-H.: CGMDA: an approach to predict and validate microRNA-disease associations by utilizing chaos game representation and LightGBM. IEEE Access **7**, 133314–133323 (2019)
30. Wang, L., et al.: GCNCDA: A new method for predicting circRNA-disease associations based on Graph Convolutional Network Algorithm. PLoS Comput. Biol. **16**, e1007568 (2020)
31. Zheng, K., et al.: iCDA-CGR: identification of circrna-disease associations based on chaos game representation. PLoS Comput. Biol. **16**, e1007872 (2020)
32. Chen, Z.-H., et al.: An improved deep forest model for predicting self-interacting proteins from protein sequence using wavelet transformation. Front. Genet. **10**, 90 (2019)
33. Chen, Z.-H., et al.: Prediction of self-interacting proteins from protein sequence information based on random projection model and fast fourier transform. Int. J. Mol. Sci. **20**, 930 (2019)
34. Wang, L., et al.: Incorporating chemical sub-structures and protein evolutionary information for inferring drug-target interactions. Sci. Rep. **10**, 1–11 (2020)
35. Wang, M-N., et al.: LDGRNMF: LncRNA-Disease Associations Prediction based on Graph Regularized Non-Negative Matrix Factorization. Neurocomputing (2020)
36. Xia, J.-F., Han, K., Huang, D.-S.: Sequence-based prediction of protein-protein interactions by means of rotation forest and autocorrelation descriptor. Protein Pept. Lett. **17**, 137–145 (2010)
37. You, Z.-H., et al.: A MapReduce based parallel SVM for large-scale predicting protein–protein interactions. Neurocomputing **145**, 37–43 (2014)
38. Wang, L., et al.: Combining high speed elm learning with a deep convolutional neural network feature encoding for predicting protein-RNA interactions. IEEE/ACM Trans. Comput. Biol. Bioinf. **1**, 1 (2018)
39. Zheng, K., et al.: MLMDA: a machine learning approach to predict and validate MicroRNA–disease associations by integrating of heterogenous information sources. J. Translational Med. **17**, 260 (2019)
40. Li, J.Q., et al.: PSPEL: In silico prediction of self-interacting proteins from amino acids sequences using ensemble learning. IEEE/ACM Trans. Comput. Biol. Bioinf. pp. 1 (2016)
41. Liu, X., Yang, S., Li, C., Zhang, Z., Song, J.: SPAR: a random forest-based predictor for self-interacting proteins with fine-grained domain information. Amino Acids **48**(7), 1655–1665 (2016). https://doi.org/10.1007/s00726-016-2226-z
42. Liu, Z., et al.: Proteome-wide prediction of self-interacting proteins based on multiple properties. Mol. Cell. Proteomics **12**, 1689–1700 (2013)
43. Zahiri, J., et al.: PPIevo: Protein-protein interaction prediction from PSSM based evolutionary information. Genomics **102**, 237–242 (2013)
44. Zahiri, J., et al.: LocFuse: human protein–protein interaction prediction via classifier fusion using protein localization information. Genomics **104**, 496–503 (2014)

Predicting Protein-Protein Interactions from Protein Sequence Using Locality Preserving Projections and Rotation Forest

Xinke Zhan, Zhuhong You(✉), Changqing Yu, Jie Pan, and Ruiyang Li

School of Information Engineering, Xijing University, Xi'an 710123, China
zhuhongyou@gmail.com

Abstract. Protein-protein interactions (PPIs) play an important role in nearly every aspect of the cell function in biological system. A number of high-throughput technologics have been proposed to detect the PPIs in past decades. However, they have some drawbacks such as time-consuming and high cost, and at the same time, a high rate of false positive is also unavoidable. Hence, developing an efficient computational method for predicting PPIs is very necessary and urgent. In this paper, we propose a novel computational method for predicting PPIs from protein sequence using Locality Preserving Projections (LPP) and Rotation Forest (RF) model. Specifically, the protein sequence is firstly transformed into Position Specific Scoring Matrix (PSSM) generated by multiple sequences alignments. Then, the LPP descriptor is applied to extract protein evolutionary information from. Finally, the RF classifier is adopted to predict whether the given protein pair is interacting or not. When the proposed method performed on *Yeast* and *H. pylori* PPIs datasets, it achieved the results with an average accuracy of 92.52% and 91.46%, respectively. To further verify the performance of the proposed method, we compare the proposed method with the state-of-the-art support vector machine (SVM) and get good results. The promising results indicated the proposed method is stable and robust for predicting PPIs.

Keywords: Locality preserving projections · Rotation forest · PSSM

1 Introduction

Protein-protein interaction (PPI) plays a crucial role in almost all biological cellular process [1, 2] such as gene expression, cell cycle regulation, signal transduction, cell growth and so on. Therefore, it is significant to detect the potential properties of biological processes by studying PPIs. Recently, a number of high-throughput biological experimental technologies have been proposed to detect the interaction, such as yeast two-hybrid screen (Y2H), immunoprecipitation, and tandem affinity purification tagging (TAP) [3]. Meanwhile, with the rapid accumulate of PPI datasets, a variety of databases like the Biomolecular Interaction Network Database (BIND) [4], the Database of Interacting Proteins (DIP) [5], and the Molecular Interaction database (MINT) [6] have been constructed to store different types of PPIs dataset. However, these traditional high-throughput methods are high cost, labor-intensive and time-consuming.

D.-S. Huang and K.-H. Jo (Eds.): ICIC 2020, LNCS 12464, pp. 121–131, 2020.
https://doi.org/10.1007/978-3-030-60802-6_12

And only a small portion of the whole PPI network has been validated through biological experiment methods. Moreover, a high rate of false positive is also an inevitable disadvantage of these methods. Thus, developing a novel computational method for predicting PPIs has great practical significance.

Recently, a number of approaches have been proposed to predict the interactions among proteins which contain different types, such as structure information [7–9], evolutionary information [10, 11], and genomic information. In order to fully utilize the potential information of protein data, many machine learning methods have been widely developed to predict the PPIs [12–14]. For example, Wang *et al.* [15] proposed a computational method combining probabilistic classification vector machine (PCVM) and Zernike moment (ZM) to detect the interaction of protein pairs. You *et al.* [16] split the length of protein sequence for obtaining multi-scale local information and achieved a better result in S. cerevisiae dataset. Huang *et al.* [17] adopt discrete cosine transform (DCT) and weighted sparse representation classifier (WSRC) to predicted PPIs. Qi *et al.* [18] compared with several classifiers performance such as logistic regression (LR), decision tree (DT), SVM, and random forest (RF). Zhao *et al.* [19] proposed an ensemble ELM based on kernel principal component analysis (KPCA), which made a great progress in reliability and accuracy. These previous work illustrate that effective feature extraction methods make a great contribution to improving the prediction performance of the classifier.

In this work, we proposed a new computational method which is using evolutionary information of protein sequence. We combined position specific scoring matrix (PSSM), locality preserving projections (LPP) and rotation forest (RF) for predicting protein-protein interactions. Specifically, we convert the protein amino acid sequence into PSSM which can retain the evolutionary information. We then use LPP to extract the feature descriptor from each PSSM. Finally, the RF classifier is used to predict the PPIs. We apply our proposed method on *Yeast* and *H. pylori* PPIs datasets, the results of average accuracy are 92.52% and 91.46%, respectively. In order to better evaluate our method, we compare our method with the superior support vector machine (SVM) and we obtain good results. The promising results indicate that the proposed approach can be a useful tool to identify PPIs.

2 Materials and Methods

2.1 Golden Standard Datasets

In order to evaluate the prediction performance of our proposed method, the highly reliable *saccharomyces cerevisiae* PPI dataset is collected from the DIP database [5]. We would directly remove protein pairs which length is less than 50 residues or more than forty percent sequence identity protein, the golden standard positive dataset consists of the remaining 5594 protein pairs. With regards to the golden standard negative dataset, the additional 5594 protein pairs which have different subcellular localization were selected. Thus, the whole golden standard dataset of *Yeast* is made up of 11188 protein pairs that positive and negative samples are equally divided. We also validate our method on *H. pylori* dataset. In the *H. pylori* PPIs dataset, the whole dataset is constructed by 2916 protein pairs (1458 interacting pairs and 1458 non-interacting pairs) as described by Martin *et al.*

2.2 Position Specific Scoring Matrix

Position Specific Scoring Matrix (PSSM) [20] is a representation pattern which is widely used in biological sequence, which can be employed to detect distantly related protein [21]. It has been widely used in various biological experiments. Given protein sequence with length *N*, we would obtain a matrix of 20 columns and *N* rows. The matrix can be interpreted as $D = \{\alpha_{i,j} : i = 1 \cdots N, and\ j = 1 \cdots 20\}$, where 20 denotes the number of amino acids. The PSSM can be represented as follows:

$$D = \begin{bmatrix} \alpha_{1,1} & \alpha_{1,2} & \cdots & \alpha_{1,20} \\ \alpha_{2,1} & \alpha_{2,2} & \cdots & \alpha_{2,20} \\ \vdots & \vdots & \vdots & \vdots \\ \alpha_{N,1} & \alpha_{N,2} & \cdots & \alpha_{N,20} \end{bmatrix} \tag{1}$$

where $\alpha_{i,j}$ means the probability of the *i*th residue being mutated into type *j* of 20 native amino acid during the evolutionary process of in the protein from multiple sequence alignments. In our study, we convert each protein sequence into PSSM by employing the Position Specific Iterated Basic Local Alignment Search Tool (PSI-BLAST) [22]. For obtaining a highly broad homologous sequence, the parameters of iterations and e-value are set to be 3 and 0.001, respectively, and the default of the other parameters.

2.3 Locality Preserving Projections (LPP)

Locality Preserving Projections (LPP) algorithm [23] is a dimensionality reduction method that aims to extracting feature information of the data. The main consideration of LPP is to retain the structure between the neighboring points by using a linear approximation to the Laplace characteristic map. In LPP algorithm, given a training sample $X = [x_1, x_2, \ldots, x_m]^T \in R^D$, where *m* denotes the feature vectors of each sample and *D* is the feature dimension. Then seeking a projection matrix *W*, which aims to project high-dimensional input dataset *X* into a low-dimensional dataset $Y = [y_1, y_2, \ldots, y_m]$. The objective function of LPP can be defined as follows:

$$\arg\min \sum_{i,j=1}^{m} (y_i - y_j)^2 P_{ij} \tag{2}$$

where

$$P_{ij} = \begin{cases} \exp(-\frac{\|x_i - x_j\|^2}{t}), x_i\ and\ x_j\ is\ linked \\ 0, x_i\ and\ x_j\ is\ not\ linked \end{cases} \tag{3}$$

$$y_i = w^T x_i \tag{4}$$

where w is a transformation vector, t represent a scale parameter and P_{ij} denotes the heat kernel which is generated by the nearest-neighbor graph. The distance formula is shown as follows:

$$d(x_i, x_j) = \|x_i - x_j\| \tag{5}$$

Meanwhile, minimizing the projection matrix W is to make sure that if x_i and x_j are close neighbors and y_i and y_j are close as well. The steps are as follows:

$$\begin{aligned} \frac{1}{2}\sum_{i,j}(y_i - y_j)^2 P_{ij} &= \frac{1}{2}\sum_{i,j}(w^T x_i - w^T x_j)^2 P_{ij} \\ &= w^T X(D - W)X^T w = w^T XLX^T w \end{aligned} \tag{6}$$

where D denotes a diagonal matrix, $D_{ii} = \sum_{j=1}^{n} W_{ij}$ and $L = D - W$ represent the Laplacian matrix. The constraints are as follows:

$$w^T XDX^T w = 1 \tag{7}$$

The above mentioned problem of finding a transfer matrix w that minimizes the objective function can be translated into the following generalized eigenvalue problem:

$$XLX^T w = \lambda XDX^T w \tag{8}$$

Find all the eigenvalues and eigenvectors corresponding to Eq. (8), and sort the eigenvalues from small to large to obtain k eigenvalues: $\lambda_0, \lambda_1, \ldots, \lambda_{k-1}$, the corresponding characteristic vectors is $\{w_0, w_1, \ldots, w_{k-1}\}$. We choose the first l eigenvectors to form the projection matrix $W = [w_0, w_1, \ldots, w_{l-1}]$. Thus, the final embedding is as follows:

$$x_i \rightarrow y_i = W^T x_i \tag{9}$$

In our work, the final LPP feature descriptors that represent a protein sequence are constructed of 140 coefficients.

2.4 Rotation Forest

Rotation Forest (RF) proposed by Rodriguez *et al.* [24] is a typical ensemble learning algorithm, which is an integrated optimization algorithm developed on the basis of a random forest. Supposing that the training sample set S, which is a $M \times m$ matrix. Each training sample M is made up of m feature vectors. Let X be the feature set and the corresponding lables $Y = (y_1, y_2, \cdots, y_M)^T$. Assuming that there are L decision tree in rotation forest totally, which can be denotes as $Q_1, Q_2, \cdots, Q_L$, respectively. And the whole feature set is randomly divided into K subsets equally. The preprocessing steps for a single classifier Q_i are shown as follows:

(1) Select the appropriate parameters K which is a factor of M, and X randomly divided into K disjoint subsets, each subsets contains the number of features is $C = {}^{n}/_{K}$;
(2) Let $X_{i,j}$ represent the jth subset of features, which is trained by the classifier Q_i. And a new matrix $S'_{i,j}$ is generated by applying a bootstrap sampling technique from 75% of the original training dataset S.
(3) Adopt PCA method in matrix $S'_{i,j}$ which only using the C features in $S_{i,j}$. The principal components coefficients are stored in $T_{i,j}$ which can be represented as $\gamma_{ij}^{(1)}, \ldots, \gamma_{ij}^{(C_j)}$.
(4) The coefficients obtained in the matrix $T_{i,j}$ are constructed a sparse rotation matrix F_i, which is shown as follows:

$$F_i = \begin{bmatrix} \gamma_{i,1}^{(1)}, \ldots, \gamma_{i,1}^{(C_1)} & 0 & \cdots & 0 \\ 0 & \gamma_{i,2}^{(1)}, \ldots, \gamma_{i,2}^{(C_2)} & \cdots & 0 \\ \vdots & \vdots & \vdots & \vdots \\ 0 & 0 & \cdots & \gamma_{i,k}^{(1)}, \ldots, \gamma_{i,k}^{(C_k)} \end{bmatrix} \quad (10)$$

Given a test sample x in the prediction phase, Let $d_{i,j}(xF_i^{\gamma})$ be the probability which is generated by the classifier Q_i. The average combined method can compute the confidence of a class that the formula is shown as follows:

$$\mu_j = \frac{1}{L}\sum\nolimits_{i=1}^{L} d_{i,j}(xF_i^{\gamma}) \quad (11)$$

Thus, the test sample x is easily assigned to the classes with greatest possible.

3 Results and Discussion

3.1 Evaluation Criteria

In order to measure the performance of our approach, five standard criteria were employed including accuracy (Accu.), precision (Prec.), sensitivity (Sen.) and Matthews correlation coefficient (MCC). They are defined as follows:

$$Accu. = \frac{TP + TN}{TP + TN + FP + FN} \quad (12)$$

$$Prec. = \frac{TN}{TN + FP} \quad (13)$$

$$Sen. = \frac{TP}{TP + FN} \quad (14)$$

$$MCC = \frac{TP \times TN - FP \times FN}{\sqrt{(TP + FN) \times (TN + FP) \times (TN + FN) \times (TP + FP)}} \quad (15)$$

where true-positive (TP) represents the number of interacting protein pairs that identified correctly. False-positive (FP) represent the number of interacting protein pairs that identified incorrectly. False-negative (FN) is the number of non-interacting protein pairs that identified incorrectly. True-negative (TN) is the number of non-interacting protein pairs which identified correctly. Moreover, the receiver operating characteristic (ROC) curve is adopted and the area under ROC curve (AUC) is also computed for which can summarize ROC curve in a numerical way.

3.2 Assessment of Prediction Ability

For evaluating the effectiveness of the proposed model, experiments are performed on two PPIs datasets of *Yeast*, and *H. pylori*, and five-fold cross-validation method is adopted for better accessing the proposed method. Specially, the whole dataset is divided into five parts equally which one fifth is used as a testing set and four fifths are used as a training set. Meanwhile, the corresponding parameters *K* and *L* of the rotation forest classifier were set the same for ensuring the fairness of experiment. After optimizing the parameters, we select the $K = 5$ and $L = 5$, respectively. Here, the parameter of *L* denotes the number of decision trees and *K* represents the number of feature subsets. The results of three datasets were shown in Tables 1 and 2.

Table 1. 5-fold cross-validation results achieved by using proposed model on *Yeast* dataset.

Testing set	Accuracy	Precision	Sensitivity	MCC	AUC
1	92.58%	95.47%	89.28%	86.22%	0.9531
2	92.71%	96.03%	88.73%	86.42%	0.9532
3	92.45%	95.99%	88.78%	86.00%	0.9507
4	93.07%	96.35%	89.56%	87.07%	0.9606
5	91.77%	95.95%	87.51%	84.86%	0.9526
Average	**92.52 ± 0.48%**	**95.96 ± 0.32%**	**88.77 ± 0.79%**	**86.12 ± 0.81%**	**0.9541 ± 0.0038**

The results of our proposed method on PPIs of *Yeast* dataset are list in Table 1, the values of average accuracy, precision, sensitivity, and MCC are 92.52%, 95.96%, 88.77%, and 86.12%, respectively. The standard deviations are 0.48%, 0.32%, 0.79%, and 0.81%, respectively. When exploring the *H. pylori* PPIs dataset, the proposed method yielded results of average accuracy, precision, sensitivity, and MCC of 91.46%, 92.74%, 89.89%, and 84.34%, respectively. The standard deviations of above criteria values are 1.09%, 2.54%, 1.84%, and 1.84%, respectively. Meanwhile, The AUC values of *Yeast* and *H. pylori* datasets are 0.9541, and 0.9407, respectively. The ROC curves of three datasets are shown in Figs. 1 and 2.

Table 2. 5-fold cross-validation results achieved by using proposed model on *H. pylori* dataset.

Testing set	Accuracy	Precision	Sensitivity	MCC	AUC
1	90.74%	93.03%	88.70%	83.18%	0.9404
2	92.28%	92.25%	91.93%	85.75%	0.9488
3	90.57%	89.58%	91.17%	82.90%	0.9377
4	92.97%	96.63%	90.25%	86.87%	0.9412
5	90.74%	92.19%	87.41%	83.02%	0.9394
Average	**91.46 ± 1.09%**	**92.74 ± 2.54%**	**89.89 ± 1.84%**	**84.34 ± 1.84%**	0.9407 ± 0.0027

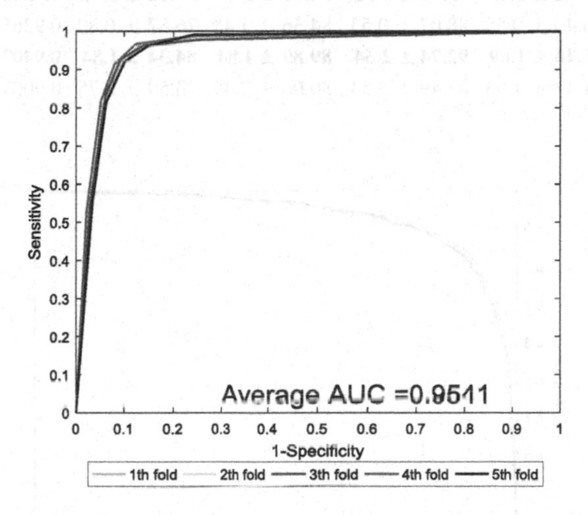

Fig. 1. ROC curves performed on *Yeast* PPIs dataset by the proposed method.

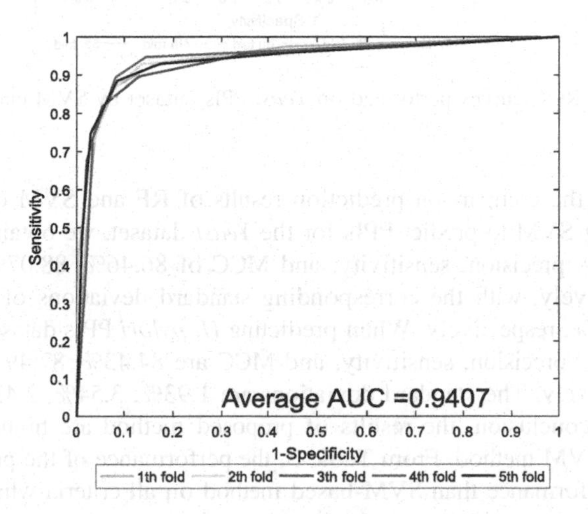

Fig. 2. ROC curves performed on *H. pylori* PPIs dataset by the proposed method.

3.3 Performance Comparison Between RF and SVM Model

Nowadays, a number of machine learning models have been proposed for predicting PPIs. In order to better evaluate the proposed method. We make a comparison between RF and the superior support vector machine (SVM) classifier. It is noteworthy that the same feature descriptors are used as input data in classifiers. In the experiment, we set the optimized corresponding parameters $c = 1$, $g = 4$ of *Yeast* PPIs datasets. When detecting on PPIs of *H. pylori* dataset, we set $c = 5$, $g = 0.1$, respectively.

Table 3. Comparison with SVM on *Yeast* and *H. pylori* datasets.

Dataset	Model	Accuracy	Precision	Sensitivity	MCC	AUC
Yeast	**RF**	**92.52 ± 0.48**	**95.96 ± 0.32**	**88.77 ± 0.79**	**86.12 ± 0.81**	**0.9541 ± 0.0038**
	SVM	86.46 ± 0.55	88.07 ± 0.53	84.36 ± 1.48	76.57 ± 0.80	0.9268 ± 0.0039
H.pylori	**RF**	**91.46 ± 1.09**	**92.74 ± 2.54**	**89.89 ± 1.84**	**84.34 ± 1.84**	**0.9407 ± 0.0027**
	SVM	84.43 ± 1.93	87.49 ± 3.54	80.06 ± 2.45	73.59 ± 2.75	0.9007 ± 0.0149

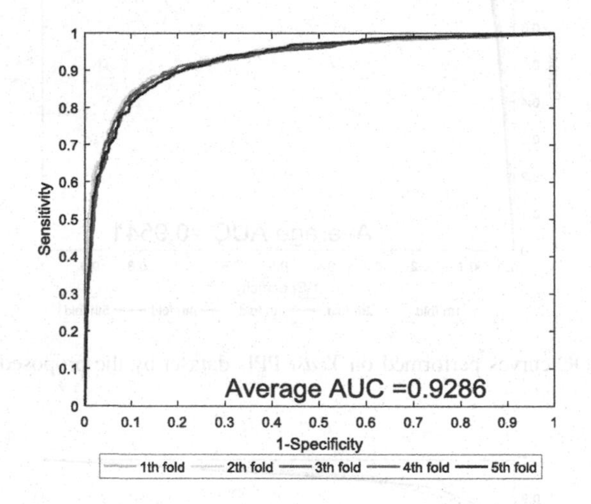

Fig. 3. ROC curves performed on *Yeast* PPIs dataset by SVM classifier.

Table 3 lists the comparison prediction results of RF and SVM on two datasets. When employing SVM to predict PPIs for the *Yeast* dataset, we obtained results with average accuracy, precision, sensitivity, and MCC of 86.46%, 88.07%, 84.36%, and 76.57%, respectively, with the corresponding standard deviations of 0.55%, 0.53%, 1.48%, and 0.80%, respectively. When predicting *H. pylori* PPIs dataset, the values of average accuracy, precision, sensitivity, and MCC are 84.43%, 87.49%, 80.06%, and 73.59%, respectively. The standard deviations are 1.93%, 3.54%, 2.45%, and 2.75%, respectively. In conclusion, the results of proposed method are higher than average attained by the SVM method. From Table 3, the performance of the proposed method have a better performance than SVM-based method on all criteria which illustrate the

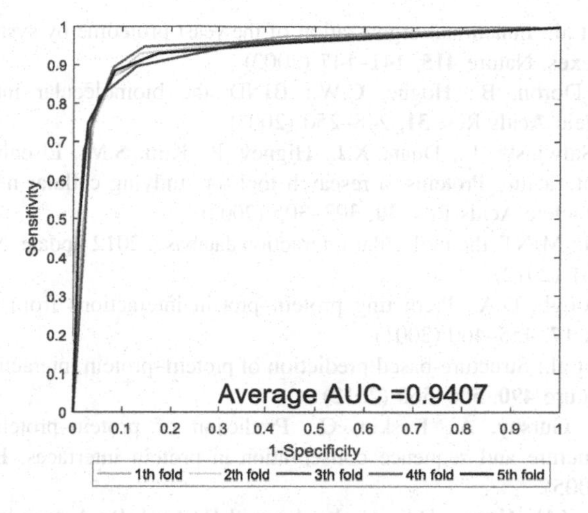

Fig. 4. ROC curves performed on *H. pylori* PPIs dataset by SVM classifier.

stability and robust of the method. Moreover, we calculated the corresponding AUC values on *Yeast* and *H. pylori* datasets and the ROC curves were shown in Figs. 3 and 4, respectively. Therefore, it is reliable to use the proposed method for predicting PPIs.

4 Conclusions

In this work, we proposed a novel computational method for predicting potential protein-protein interactions by using locality preserving projections and rotation forest model. It is well known that extracting effective feature descriptors is the key to predicting PPIs, so the main improvement of this work is to maintain an internally fixed local structure from feature descriptors by employing locality preserving projections. Then, adopting RF to improve the diversity of each base classifier and get the final prediction result by using 5-fold cross-validation. The proposed method yielded average prediction accuracies of 92.52% and 91.46% on *Yeast* and *H. pylori* PPIs datasets, respectively. Moreover, we further compared the proposed method with the SVM model and found that our method still has a significant improvement over SVM-based method. Promising results illustrated that the proposed method achieved well performance in predicting PPIs and we hope that the method can be a useful tool to predict PPIs.

References

1. Wang, L., et al.: An ensemble approach for large-scale identification of protein-protein interactions using the alignments of multiple sequences. Oncotarget **8**, 5149–5159 (2017)
2. Braun, P., Gingras, A.C.: History of protein-protein interactions from egg-white to complex networks. Proteomics **12**, 1478–1498 (2012)

3. Gavin, A.C., et al.: Functional organization of the yeast proteome by systematic analysis of protein complexes. Nature **415**, 141–147 (2002)
4. Bader, G.D., Doron, B., Hogue, C.W.: BIND: the biomolecular interaction network database. Nucleic Acids Res. **31**, 248–250 (2003)
5. Xenarios, I., Salwinski, L., Duan, X.J., Higney, P., Kim, S.M., Eisenberg, D.: DIP, the Database of Interacting Proteins: a research tool for studying cellular networks of protein interactions. Nucleic Acids Res. **30**, 303–305 (2002)
6. Licata, L., et al.: MINT, the molecular interaction database: 2012 update. Nucleic Acids Res. **40**, D857–D861 (2012)
7. Bock, J.R., Gough, D.A.: Predicting protein–protein interactions from primary structure. Bioinformatics **17**, 455–460 (2001)
8. Zhang, Q.C., et al.: Structure-based prediction of protein–protein interactions on a genome-wide scale. Nature **490**, 556–560 (2012)
9. Aytuna, A.S., Gursoy, A., Keskin, O.: Prediction of protein–protein interactions by combining structure and sequence conservation in protein interfaces. Bioinformatics **21**, 2850–2855 (2005)
10. Yi, H.C., You, Z.H., Huang, D.S., Li, X., Jiang, T.H., Li, L.P.: A deep learning framework for robust and accurate prediction of ncRNA-protein interactions using evolutionary information. Mol. Ther. Nucleic Acids **11**, 337–344 (2018)
11. Valencia, A., Pazos, F.: Prediction of protein-protein interactions from evolutionary information. Methods Biochem. Anal. **44**, 411–426 (2003)
12. Wang, Y., You, Z., Li, L., Chen, Z.: A survey of current trends in computational predictions of protein-protein interactions. Front. Comput. Sci. **14**(4), 1–12 (2020). https://doi.org/10.1007/s11704-019-8232-z
13. An, J.Y., You, Z.H., Zhou, Y., Wang, D.F.: Sequence-based Prediction of Protein-Protein Interactions Using Gray Wolf Optimizer-Based Relevance Vector Machine. Evol. Bioinform. **15**, 1176934319844522 (2019)
14. You, Z.H., Huang, W.Z., Huang, Y.A., Yu, C.Q., Li, L.P.: An efficient ensemble learning approach for predicting protein-protein interactions by integrating protein primary sequence and evolutionary information. IEEE/ACM Trans. Comput. Biol. Bioinform. **16**, 809–817 (2018)
15. Wang, Y., You, Z.H., Li, X., Chen, X., Jiang, T., Zhang, J.: PCVMZM: using the probabilistic classification vector machines model combined with a zernike moments descriptor to predict protein–protein interactions from protein sequences. Int. J. Mol. Sci. **18**, 1029 (2017)
16. You, Z.H., Chan, K.C., Hu, P.: Predicting protein-protein interactions from primary protein sequences using a novel multi-scale local feature representation scheme and the random forest. PLoS ONE **10**, e0125811 (2015)
17. Huang, Y.A., You, Z.H., Gao, X., Wong, L., Wang, L.: Using weighted sparse representation model combined with discrete cosine transformation to predict protein-protein interactions from protein sequence. Biomed Res. Int. (2015)
18. Zhao, L.J., Chai, T.Y., Yuan, D.C.: Selective ensemble extreme learning machine modeling of effluent quality in wastewater treatment plants. Int. J. Autom. Comput. **9**, 627–633 (2012)
19. Qi, Y., Bar, J.Z., Klein, S.J.: Evaluation of different biological data and computational classification methods for use in protein interaction prediction. Proteins: Struct., Funct., Bioinf. **63**, 490–500 (2006)
20. Gribskov, M., McLachlan, A.D., Eisenberg, D.: Profile analysis: detection of distantly related proteins. Proc. Natl. Acad. Sci. **84**, 4355–4358 (1987)

21. Huang, C., Yuan, J.: Using radial basis function on the general form of Chou's pseudo amino acid composition and PSSM to predict subcellular locations of proteins with both single and multiple sites. Biosystems. **113**, 50–57 (2013)
22. Altschul, S.F., et al.: Gapped BLAST and PSI-BLAST: a new generation of protein database search programs. Nucleic Acids Res. **25**, 3389–3402 (1997)
23. He, X., Niyogi, P.: Locality preserving projections. Adv. Neural Inf Process. Syst. **2004**, 153–160 (2004)
24. Rodriguez, J.J., Kuncheva, L.I., Alonso, C.J.: Rotation forest: a new classifier ensemble method. IEEE Trans. Pattern Anal. Mach. Intell. **28**, 1619–1630 (2006)

Predicting Protein-Protein Interactions from Protein Sequence Information Using Dual-Tree Complex Wavelet Transform

Jie Pan, Zhu-Hong You(✉), Chang-Qing Yu, Li-Ping Li, and Xin-ke Zhan

School of Information Engineering, Xijing University, Xi'an 710123, China
zhuhongyou@gmail.com

Abstract. Protein-protein interactions (PPIs) play major roles in most biological processes. Although a number of high-throughput technologies have been established for generating PPIs, it still has unavoidable problems such as time-consuming and labor intensive. In this paper, we develop a novel computational method for predicting PPIs by combining dual-tree complex wavelet transform (DTCWT) on substitution matrix representation (SMR) and weighted sparse representation-based classifier (WSRC). When predicting PPIs of *Yeast* and *Human* datasets, the proposed method obtained remarkable results with average accuracies as high as 97.12% and 97.56%, respectively. The performance of the proposed method is obviously better than the existing methods. Furthermore, we compare it with the superior support vector machine (SVM) classifier for further evaluating the prediction performance of our method. The promising results illustrate that our method is robust and stable for predicting PPIs, and it is anticipated that it would be a useful tool to predict PPIs in a large-scale.

Keywords: Protein-protein interactions · Protein sequence · Dual-tree complex wavelet transform · Weighted sparse representation

1 Introduction

Identification of protein–protein interactions (PPIs) is crucial for studying protein function and deep understanding of biological processes in a cell. In recent years, plenty of high-throughput technologies, such as *Yeast two-hybrid (Y2H) screens* [1, 2], *tandem affinity purification* (TAP) [3] and *mass spectrometric protein complex identification* (MS-PCI) [4], have been developed for the large-scale PPIs detection. However, these previous methods are expensive and require a great deal of human effort. In addition, only a small part of the whole PPIs have been identified. Thus, we can draw a conclusion that only using experimental methods is difficult to identify unknown PPIs.

In recent years, researchers have developed different types of computational methods for the prediction of PPIs [5–16]. For example, Li *et al.* proposed a novel computational model combining Position Weight Matrix (PWM) and Scale-Invariant Feature Transform (SIFT) algorithm [17]. An *et al.* proposed an effective algorithm that using Gray Wolf Optimizer–Based Relevance Vector Machine to identify PPIs [18].

D.-S. Huang and K.-H. Jo (Eds.): ICIC 2020, LNCS 12464, pp. 132–142, 2020.
https://doi.org/10.1007/978-3-030-60802-6_13

Zhu *et al.* proposed a useful tool which used the Position-Specific Scoring Matrices (PSSMs) and ensemble learning algorithm Rotation Forest (RF) [19]. You *et al.* developed a novel method for detecting PPIs by integrating a new protein sequence substitution matrix feature representation and ensemble weighted sparse representation model classifier [20]. Huang *et al.* proposed a sequence-based method based on the combination of weighted sparse representation based classifier (WSRC) and global encoding (GE) of amino acid sequence [21]. Wen *et al.* proposed an effective method based on similar network fusion (SNF) model to integrate the physical and chemical properties of proteins [22]. Leon Wong *et al.* presented a novel computational approach that combining a Rotation Forest Model with a novel PR-LPQ Descriptor [23]. Huang *et al.* developed an effective algorithm, which is based on Extreme Learning Machine (ELM) and combined the concept of Chou's Pseudo-Amino Acid Composition (PseAAC) composition [24].

In this paper, we propose a novel computational method for predicting PPIs, which combines weighted sparse representation based classifier (WSRC) and the dual-tree complex wavelet transform (DTCWT). Specifically, we first select substitute matrix representation (SMR) based on BLOSUM62 to represent protein sequences. Then, we adopt the DTCWT to extract feature vectors from each SMR matrix. Finally, we utilize WSRC to predict PPIs on two different biological datasets: *Yeast* and *Human*. Our model achieves excellent performance results which obtain average accuracies of 97.12% and 97.56%, respectively. In order to further evaluating the proposed method, we compared the WSRC with the state-of-the-art support vector machine (SVM) classifier. The promising results demonstrated that the proposed method is robust and stable for the prediction of PPIs.

2 Materials and Methodology

2.1 Godden Standard Datasets

The first dataset that we chose in this paper is gathered from publicly available database of interacting proteins (DIP). We removed the protein pairs whose length are less than 50 residues because these might be fragments. The pairs with $\geq 40\%$ sequence identity have been deleted too. In this way, the positive dataset is constructed by the remaining 5594 protein pairs. Moreover, we selected 5594 additional protein pairs of different subcellular localizations to build the negative dataset. Consequently, the whole dataset is made up of 11188 protein pairs.

In order to demonstrate the generality of the proposed method, we validated our method on another PPI dataset. We collected the dataset from the Human Protein References Database (HPRD). Those protein pairs which have $\geq 25\%$ sequence identity have been removed. Finally, we used the remaining 3899 protein-protein pairs of experimentally verified PPIs from 2502 different human proteins, so that we can comprise the golden standard positive dataset. Following the previous work [25], we assume that the proteins in different subcellular compartments will not interact with each other and finally obtained 4262 protein pairs from 661 different human proteins as the negative dataset. As a result, the Human dataset is constructed by 8161 protein pairs.

2.2 Substitution Matrix Representation

Substitution matrix representation (SMR) is a modified version of representation method reported by [26]. In this novel matrix representation for proteins, we generated a $N \times 20$ matrix to represent the N-length protein sequence, which are based on a substitution matrix. BLOSUM62 matrix is a powerful substitution matrix and has been utilized in this work for the sequence alignment of proteins. SMR can be defined as follows:

$$SMR(i.j) = B(P(i),j)\, i = 1 \cdots N, j = 1 \cdots 20 \tag{1}$$

In this formula, B means the BLOSUM62 matrix, it is a 20×20 substitution matrix and $B(i,j)$ represents the value in row i and column j of BLOSUM62 matrix, this value represents the probability rate of amino acid i converting to amino acid j in the evolution process; $P = (p1, p2 \cdots pN)$ is the given protein sequence constructed by N amino acids.

2.3 The Dual-Tree Complex Wavelet Transform

The dual-tree complex wavelet transform (DTCWT) [27] is a variant of the traditional complex wavelet transform (DWT). It inherited the characteristics of the Multi-scale and Multi-resolution of discrete wavelet transform. At the same time, it makes up for the deficiencies of complex wavelet transform with large amount of calculation and high complexity. Different with the traditional DWT, DTCWT utilized two real DWTs to form a complex transform [28]. The first part symbolizes the real component and the second part represents the imaginary component of this transform.

The DTCWT settled the matters about the "shift-invariant problems" and "directional selectivity in two or more dimensions," which are both weak points of conventional DWT [29]. It acquired directional selectivity by using approximate analytic wavelets. It also has the skills to generate a total of six directionally discriminating subbands oriented in the $\pm 15°$, $\pm 45°$ and $\pm 75°$ directions, for both the real (R) and imaginary (I) parts. Let $h_i(n)$ and $g_i(n)$ be the filters in the first stage. Let the new stage response of the first filter bank be $H_{new}^{(k)}(e^{jw})$ and second filter bank be $H_{new}^{'(k)}(e^{jw})$; we now have the following result.

Suppose one is provided with CQF pairs $\{h_0(n), h_1(n)\}, \{h_0^{'}(n), h_1^{'}(n)\}$. For $k > 1$.

$$H_{new}^{(k)}(e^{jw}) = H\left\{H_{new}^{'(k)}(e^{jw})\right\} \tag{2}$$

if and only if

$$h_0^{'(1)}(n) = h_0^{(1)}(n-1) \tag{3}$$

A 2D image $f(x,y)$ can be decomposed by 2D DTCWT over a series of dilations and translations of a complicated scaling function and six complex wavelet function $\varphi^{\theta}_{j,l}$; that is:

$$f(x,y) = \sum_{l \in Z^2} s_{j_0,l}\phi_{j_0}l^{(x,y)} + \sum_{\theta \in \Theta}\sum_{j \geq j_0}\sum_{l \in Z^2} c^{\theta}_{j},l\varphi^{\theta}_{j},l^{(x,y)} \tag{4}$$

where $\theta \in \Theta = \{\pm 15°, \pm 45°, \pm 75°\}$ gives the directionality of the complex wavelet function.

2.4 Weighted Sparse Representation-Based Classification

In the past twenty years, sparse representation based classifier (SRC) [30, 31] has earned considerable attention in the field of signal processing, pattern recognition and computer vision because of the great development of linear representation methods (LRBM) and compressed sensing (CS) theory. Sparse representation attempts to optimize matrix to reveal the relationship between any given test sample and the training set. Therefore, it would be a good trial to use it for building a prediction system for PPIs. In this work, we build a computational model by employing weighted sparse representation-based classifier (WSRC).

Given a training sample matrix $x \in R^{m \times n}$ which is made up of n samples of m dimensions. If there are sufficient training samples belonging to the kth class, then the sub-matrix constructed by the samples of the kth class can be symbolized as $X_k[l_{k1}, l_{k2} \cdots l_{kn_k}]$, where l_i denotes the class of ith sample and n_k is the number of samples belonging to kth class. Thus, X can be further rewritten as $X = [X_1, X_2 \cdots X_K]$, where K is the class number of the whole samples. Given a test sample $y \in R^m$ and it can be represented as

$$y = \alpha_{k,1}l_{k,1} + \alpha_{k,2}l_{k,2} + \cdots + \alpha_{k,n_k}l_{k,n_k} \tag{5}$$

when considering the whole training set representation, Eq. (5) can be further symbolized as

$$y = X\alpha_0 \tag{6}$$

where $\alpha_0 = \left[0,\ldots 0, \alpha_{k,2}, \cdots, \alpha_{k,n_k}, 0, \cdots 0\right]^T$. For these reason that the nonzero entries in α_0 are only associated with the kth class, so if class number of samples become large, the α_0 would come to be sparse. The key question of SRC algorithm is searching the α vector which can subject to Eq. (6) and minimize the ℓ_0-norm of itself:

$$\begin{aligned} \widehat{\alpha_0} &= \arg\min \|\alpha\|_0 \\ \text{subject to } y &= X\alpha \end{aligned} \tag{7}$$

Problem (7) is an NP-hard problem and it can be achieved but difficultly to be solved precisely. According to the theory of compressive sensing [32, 33] show, if α is sparse enough, we can solve the related problem convex l_1-minimization problem instead of solving the l_0-minimization problem directly.

$$\begin{gathered} \widehat{\alpha_1} = \arg\min\|\alpha\|_1 \\ \text{subject to } y = X\alpha \end{gathered} \tag{8}$$

When dealing with occlusion, we should extend Eq. (8) to the stable ℓ_1-minimization problem

$$\begin{gathered} \widehat{\alpha_1} = \arg\min\|\alpha\|_1 \\ \text{subject to } \|y - X\alpha\| \leq \varepsilon \end{gathered} \tag{9}$$

where $\varepsilon > 0$ represent the tolerance of reconstruction error. Given the solution from Eq. (9), the SRC algorithm assigns the label of test sample y to class c with the reconstruction residual:

$$\min_c r_c(y) = \|y - X\widehat{\alpha_1}^c\|, \; c = 1 \cdots K \tag{10}$$

Besides sparse representation, Nearest Neighbor (NN) is another popular classifier which only considering the influence of the Nearest Neighbor in training data to classify the test sample and SRC uses the linearity structure of data and overcomes the drawback of NN. Some researches shows that locality is more essential than sparsity in some cases [34, 35]. Lu *et al.* [36] have proposed a modified version of traditional sparse representation based classifier called weighted sparse representation based classifier (WSRC), it integrates the locality structure of data into basic sparse representation. Specifically, Gaussian distance between single sample and the whole training samples will be first computed and WSRC can use it as the weights of each training sample. The Gaussian distance between two samples, s_1 and s_2 can be described as follow:

$$d_G(s_1, s_2) = e^{-\|s_1 - s_2\|^2 / 2\alpha^2} \tag{11}$$

where σ is the Gaussian kernel width. In this way, weighted sparse representation based classifier can retain the locality structure of data and then it turned to solve the following problem:

$$\begin{gathered} \widehat{\alpha_1} = \arg\min\|W\alpha\|_1 \\ \text{subject to } y = X\alpha \end{gathered} \tag{12}$$

specifically,

$$diag(W) = \left[d_G(y, x_1^1), \ldots, d_G(y, x_{n_k}^k)\right]^T \tag{13}$$

where W is a block-diagonal matrix of locality adaptor and n_k is the sample number of training set in class k. Dealing with this occlusion, we can eventually solve the following stable ℓ_1-minimization problem:

$$\widehat{\alpha_1} = \arg\min \|W\alpha\|_1 \quad \text{subject to } \|y - X\alpha\| \leq \varepsilon \tag{14}$$

where $\varepsilon > 0$ is the tolerance value.

In summary, the WSRC algorithm can be summarized by the following steps:

Algorithm1. Weighted Sparse Representation Based Classifier (WSRC)

(1). Input : training samples matrix $X \in R^{m \times n}$ and any test sample $y \in R^d$.
(2). Normalize the columns of X to have unit ℓ_2 -norm.
(3). Calculate the Gaussian distances between y and each sample in X and make up matrix W.
(4). Solve the stable ℓ_1 -minimization problem defined in Eq.(12)
(5). Compute each residual of K classes: $r_c(y) = \left\| y - X\widehat{\alpha_1}^c \right\|$ (c=1,2,...,K)
(6). Output: assign to the class c by the rule: $identity(y) = \arg\min_c (r_c(y))$.

3 Results and Discussion

In order to evaluate the performance of the proposed method, the overall prediction accuracy (Acc.), sensitivity (Sen.), precision (PR.), and Matthews's correlation coefficient (MCC.) were calculated. They are defined as follows:

$$Acc. = \frac{TP + TN}{TP + FP + TN + FN} \tag{15}$$

$$Sen. = \frac{TP}{TP + FN} \tag{16}$$

$$PR. = \frac{TP}{TP + FP} \tag{17}$$

$$MCC. = \frac{TP \times TN - FP \times FN}{\sqrt{(TP + TN) \times (TN + FN) \times (TP + FP) \times (TN + FN)}} \tag{18}$$

In these algorithm, true positive (TP) represents the number of true samples which are predicted correctly; false negative (FN) is the number of samples predicted to be non-interacting pairs incorrectly; false positive (FP) is the number of true non-interacting pairs predicted to be PPIs falsely; and true negative (TN) is the number of true noninteracting pairs predicted correctly. What's more, the receiver operating characteristic (ROC) curves are also calculated to evaluate the performance of proposed method. In order to summarize ROC curve in a numerical way, the area under an ROC curve (AUC) was computed.

3.1 Assessment of Prediction Ability

For the sake of fairness, when predicting PPIs of *Yeast* and *Human*, we set the same corresponding parameters of weighted sparse representation-based classifier. We set the $\sigma = 1.5$ and $\varepsilon = 0.00005$. In addition, 5-fold cross-validation [37] was employed in our experiments in order to avoid overfitting and get a stable and reliable model from the proposed method. Specifically, we divided the whole dataset into five subsets. Four of the subsets are used for training and the last part was used for testing. By this way, the results of these experiments in which we used the proposed model to predict PPIs of *Yeast* and *Human* datasets are shown in Tables 1 and 2.

Table 1. 5-fold cross-validation results obtained by using proposed method on *Yeast* dataset.

Testing set	Acc. (%)	PR. (%)	Sen. (%)	MCC (%)	AUC (%)
1	96.65	100	93.24	93.50	96.59
2	97.63	100	95.14	95.36	98.01
3	97.18	100	94.43	94.52	96.92
4	97.00	100	94.02	94.18	97.04
5	97.14	100	94.38	94.44	97.16
Average	**97.12 ± 0.35**	**100 ± 0.00**	**94.24 ± 0.69**	**94.40 ± 0.67**	**97.14 ± 0.53**

Table 2. 5-fold cross-validation results obtained by using proposed method on *Human* dataset.

Testing set	Acc. (%)	PR. (%)	Sen. (%)	MCC (%)	AUC (%)
1	98.28	99.86	96.44	96.60	97.89
2	97.43	99.32	95.19	94.95	96.73
3	96.63	99.74	93.45	93.47	96.20
4	97.49	99.05	95.57	95.07	97.61
5	97.98	99.48	96.35	96.03	98.10
Average	**97.56 ± 0.63**	**99.49 ± 0.32**	**95.40 ± 1.21**	**95.23 ± 1.19**	**97.31 ± 0.81**

When using the proposed method to predict PPIs of the *Yeast* dataset, we obtained the prediction results with average accuracy, precision, sensitivity, and MCC of 97.12%, 100%, 94.24% and 94.40%. The standard deviations of these criteria values are relatively low, which of accuracy, precision, sensitivity and MCC are 0.35%, 0.00%, 0.69%, and 0.67%, respectively. Form Table 2, when exploring the *Human* dataset, the proposed method yielded results of average accuracy, precision, sensitivity and MCC of 97.56%, 99.49%, 95.40%, and 95.23%. The standard deviations are 0.63%, 0.32%, 1.21%, and 1.19%, respectively. The ROC curves performed on these two datasets are shown in Fig. 1. To better evaluate the prediction performance of the proposed method, we computed the AUC values of *Yeast* and *Human* datasets, which are 97.14% and 97.31%, respectively.

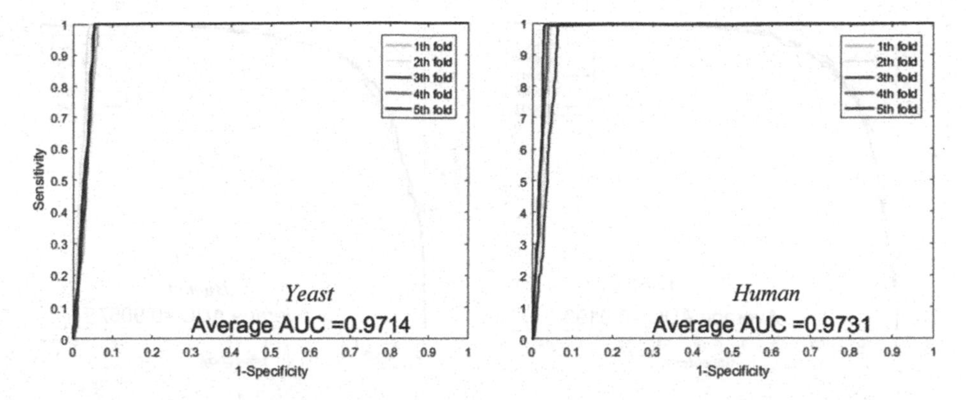

Fig. 1. ROC from proposed method result for *Yeast* and *Human* PPIs dataset.

The high accuracies show that WSRC based model combining the SMR-DTCWT descriptors is feasible and effective for predicting PPIs. All experimental results demonstrate the feasibility, effectiveness and robustness of the proposed method.

3.2 Comparison with SVM-Based Method

Table 3. Comparison with support vector machine on *Yeast* and *Human* datasets

Dataset	Classifier	Acc. (%)	PR. (%)	Sen. (%)	MCC (%)	AUC (%)
Yeast	SVM	84.74 ± 0.78	83.44 ± 0.96	86.71 ± 1.61	74.11 ± 1.08	91.99 ± 0.74
	WSRC	**97.12 ± 0.35**	**100.00 ± 0.00**	**94.24 ± 0.69**	**94.40 ± 0.67**	**97.14 ± 0.53**
Human	SVM	83.14 ± 0.83	81.92 ± 1.19	83.08 ± 2.17	71.92 ± 1.09	90.57 ± 0.41
	WSRC	**97.56 ± 0.63**	**99.49 ± 0.32**	**95.40 ± 1.21**	**95.23 ± 1.19**	**97.31 ± 0.81**

In order to further evaluate the performance of the proposed method, we compared the prediction performance of the proposed method with the state-of-the-art SVM classifier on the dataset of *Human* and *Yeast*. We utilized the same feature extraction method and a grid search method to optimize two corresponding parameters of SVM c and g. Here, we set the $c = 0.3$ and $g = 0.3$.

Table 3 shows that when using SVM to predict PPIs of *Yeast* dataset, we obtained relatively poor results with the average accuracy, precision, sensitivity, MCC, and AUC of 84.74%, 83.44%, 86.71%, 74.11%, and 91.99%, respectively. When exploring the *Human* dataset with the SVM-based method yielded relatively low results with the average accuracy, precision, sensitivity, MCC, and AUC of 83.14%, 81.92%, 83.08%, 71.92%, and 90.57%, respectively. Considering the comparison result and higher values for criteria and lower standard deviations, the prediction performance of SVM-based method is lower than that of WSRC. The ROC curves performed by SVM classifier on the two datasets are shown in Fig. 2.

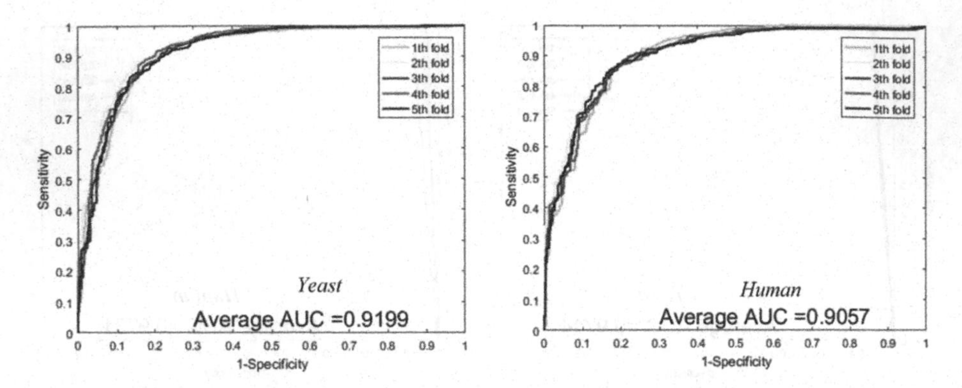

Fig. 2. ROC from SVM-based method result for *Yeast* and *Human* PPIs dataset.

4 Conclusions and Discussion

It is becoming more and more important to develop an effective and accurate method for predicting PPIs. In this work, we explore a novel computation model for predicting PPIs by combing weighted sparse representation-based classifier and the dual-tree complex wavelet transform. In the step of feature extraction, it has been proven that it is effective to combine the SMR matrix and dual-tree complex wavelet transform. Compared with the previous methods, the main improvement of the proposed method is to adopt a novel protein feature representation and utilizing a powerful classifier. In addition, good experiment results indicate that the proposed method performs well in PPIs prediction and has great generalization ability.

Competing Interests. The authors declare that they have no competing interests.

Acknowledgements. This work is supported by the National Science Foundation of China under Grants 61873212.

References

1. Nishihara, T., Nishikawa, J., Kanayama, T., Dakeyama, F., Saito, K., Imagawa, M., et al.: Estrogenic activities of 517 chemicals by yeast two-hybrid assay. J. Health Sci. **46**, 282–298 (2000)
2. Sato, T., Hanada, M., Bodrug, S., et al.: Interactions among members of the Bcl-2 protein family analyzed with a yeast two-hybrid system. Proc. Natl. Acad. Sci. **91**, 9238–9242 (1994)
3. Puig, O., Caspary, F., Rigaut, G., et al.: The tandem affinity purification (TAP) method: a general procedure of protein complex purification. Methods **24**, 218–229 (2001)
4. Ho, Y., Gruhler, A., Heilbut, A., Bader, G.D., Moore, L., Adams, S.-L., et al.: Systematic identification of protein complexes in Saccharomyces cerevisiae by mass spectrometry. Nature **415**, 180–183 (2002)

5. You, Z.H., Zhou, M.C., Luo, X., Li, S.: Highly efficient framework for predicting interactions between protein. IEEE Trans. Cybernet. **47**, 731–743 (2016)
6. Wang, L., et al.: Using two-dimensional principal component analysis and rotation forest for prediction of protein-protein interactions. Scientific Reports **8**, 1–10 (2018)
7. Wang, L., You, Z.H., Xia, S.X., Liu, F., Chen, X., Yan, X., Zhou, Y.: Advancing the prediction accuracy of protein-protein interactions by utilizing evolutionary information from position-specific scoring matrix and ensemble classifier. J. Theor. Biol. **418**, 105–110 (2017)
8. Li, Z.W., et al.: Accurate prediction of protein-protein interactions by integrating potential evolutionary information embedded in PSSM profile and discriminative vector machine classifier. Oncotarget **8**(14), 23638 (2017)
9. You, Z.H., Li, X., Chan, K.C.C.: An Improved sequence-based prediction protocol for protein-protein interactions using amino acids substitution matrix and rotation forest ensemble classifiers. Neurocomputing **228**, 277–282 (2017)
10. Wang, L., et al.: An ensemble approach for large-scale identification of protein-protein interactions using the alignments of multiple sequences. Oncotarget **8**, 5149 (2017)
11. Wang, Y.B., et al.: Predicting protein-protein interactions from protein sequences by stacked sparse auto-encoder deep neural network. Molecular BioSyst. **13**, 1336–1344 (2017)
12. An, J.Y., You, Z.H., Chen, X., Huang, D.S., Yan, G.Y.: Robust and accurate prediction of protein self-interactions from amino acids sequence using evolutionary information. Molecular BioSyst. **12**, 3702–3710 (2016)
13. Huang, Y.A., et al.: Construction of reliable protein-protein interaction networks using weighted sparse representation based classifier with pseudo substitution matrix representation. Neurocomputing **218**, 131–138 (2016)
14. An, J.Y., Meng, F.R., You, Z.H., Chen, X.: Improving protein-protein interactions prediction accuracy using protein evolutionary information and relevance vector machine model. Protein Sci. **25**, 1825–1833 (2016)
15. You, Z.H., Chan, K.C.C.: Prediction of protein-protein interactions from primary protein sequence using random forest model with a novel multi-scale local feature representation. PLoS ONE **10**, e0131091 (2015)
16. You, Z.H., Zhu, L., Zheng, C.H., Yu, H.J., Deng, S.P., Ji, Z.: Prediction of protein-protein interactions from amino acid sequences using a novel multi-scale continuous and discontinuous feature set. BMC Bioinform. **15**(15), s9 (2014)
17. Li, J., Shi X., You, Z.H., Yi, H.C., Chen, Z., Lin, Q., et al.: Using weighted extreme learning machine combined with scale-invariant feature transform to predict protein-protein interactions from protein evolutionary information. IEEE/ACM Trans. Comput. Biol. Bioinform. (2020)
18. An, J.-Y., You, Z.H., Zhou, Y., Wang, D.F.: Sequence-based prediction of protein-protein interactions using gray wolf optimizer-based relevance vector machine. Evol. Bioinform. **15**, 1176934319844522 (2019)
19. Zhu, H.J., You, Z.H., Shi, W.L., Xu, S.K., Jiang, T.H., Zhuang, L.H.: Improved prediction of protein-protein interactions using descriptors derived from PSSM via Gray Level Co-Occurrence Matrix. IEEE Access **7**, 49456–49465 (2019)
20. You, Z.H., Huang, W.Z., Zhang, S., Huang, Y.A., Yu, C.Q., Li, L.P.: An efficient ensemble learning approach for predicting protein-protein interactions by integrating protein primary sequence and evolutionary information. IEEE/ACM Trans. Comput. Biol. Bioinform. **16**, 809–817 (2018)
21. Huang, Y.A., You, Z.H., Chen, X., Chan, K., Luo, X.: Sequence-based prediction of protein-protein interactions using weighted sparse representation model combined with global encoding. BMC Bioinform. **17**, 184 (2016)

22. Wen, Y.T., Lei, H.J., You, Z.H., Lei, B.Y., Chen, X., Li, L.P.: Prediction of protein-protein interactions by label propagation with protein evolutionary and chemical information derived from heterogeneous network. J. Theor. Biol. **430**, 9–20 (2017)
23. Wong, L., You, Z.-H., Li, S., Huang, Y.-A., Liu, G.: Detection of protein-protein interactions from amino acid sequences using a rotation forest model with a novel PR-LPQ descriptor. In: Huang, D.-S., Han, K. (eds.) ICIC 2015. LNCS (LNAI), vol. 9227, pp. 713–720. Springer, Cham (2015). https://doi.org/10.1007/978-3-319-22053-6_75
24. Huang, Q.Y., You, Z.H., Li, S., Zhu, Z.X.: Using Chou's amphiphilic pseudo-amino acid composition and extreme learning machine for prediction of protein-protein interactions. In: IEEE, pp. 2952–2956 (2014)
25. You, Z.H., Yu, J.Z., Zhu, L., Li, S., Wen, Z.K.: A MapReduce based parallel SVM for large-scale predicting protein–protein interactions. Neurocomputing **145**, 37–43 (2014)
26. Yu, X., Zheng, X., Liu, T., Dou, Y., Wang, J.: Predicting subcellular location of apoptosis proteins with pseudo amino acid composition: approach from amino acid substitution matrix and auto covariance transformation. Amino Acids **42**, 1619–1625 (2012)
27. Selesnick, I.W., Baraniuk, R.G., Kingsbury, N.G.: The dual-tree complex wavelet transform. IEEE Signal Process. Mag. **22**, 123–151 (2005)
28. Kingsbury, N.: Complex wavelets for shift invariant analysis and filtering of signals. Appl. Comput. Harmon. Anal. **10**, 234–253 (2001)
29. Barri, A., Dooms, A., Schelkens, P.: The near shift-invariance of the dual-tree complex wavelet transform revisited. J. Math. Anal. Appl. **389**, 1303–1314 (2012)
30. Yin, J., Liu, Z., Jin, Z., Yang, W.: Kernel sparse representation based classification. Neurocomputing **77**, 120–128 (2012)
31. Gao, Y., Ma, J., Yuille, A.L.: Semi-supervised sparse representation based classification for face recognition with insufficient labeled samples. IEEE Trans. Image Process. **26**, 2545–2560 (2017)
32. Candes, E.J., Tao, T.: Near-optimal signal recovery from random projections: universal encoding strategies? IEEE Trans. Inf. Theory **52**, 5406–5425 (2006)
33. Chen, S., Donoho, D.L., Saunders, M.A.: Atomic decomposition by basis pursuit. SIAM J. Sci. Comput. **20**, 33–61 (1998)
34. Liu, C.H.W.: Gabor feature based classification using the enhanced fisher linear discriminant model for face recognition. IEEE Trans. Image Process. **11**, 467–476 (2002)
35. Wang, J., Yang, J., Yu, K., Lv, F., Huang, T., Gong, Y.: Locality-constrained linear coding for image classification. In: IEEE, pp. 3360–3367 (2010)
36. Lu, C.Y., Min, H., Gui, J., Zhu, L., Lei, Y.K.: Face recognition via weighted sparse representation. J. Vis. Commun. Image Rep. **24**, 111–116 (2013)
37. Zhang, P.: Model selection via multifold cross validation. Ann. Stat. **21**, 299–313 (1993)

Biomarker Discovery and Detection

Biomarkers Selection of Abnormal Functional Connections in Schizophrenia with $\ell_{2,1-2}$-Norm Based Sparse Regularization Feature Selection Method

Na Gao[1], Chen Qiao[1(✉)], Shun Qi[1,2], Kai Ren[2], Jian Chen[3], and Hanfeng Fang[4]

[1] Xi'an Jiaotong University, Xi'an 710049, China
qiaochen@xjtu.edu.cn
[2] Xijing Hospital, Fourth Military Medical University, Xi'an 710032, China
[3] Huashan Hospital, Fudan University, Shanghai 200040, China
[4] Suzhou Hanlin Information Technology Development Co., Ltd., Suzhou 215138, China

Abstract. Schizophrenia (SZ) is a serious mental disease involving multiple symptoms such as sensation, cognition, emotion and memory. Recent studies have shown that schizophrenia is related to abnormal functional connections in brain regions. Functional magnetic resonance imaging (fMRI) has become a powerful tool to examine the abnormal connectivity of brain networks in SZ, which provides an approach to bring psychiatry from subjective description into objective and tangible brain-based measures. However, there are many irrelevant and redundant features in fMRI. How to eliminate redundant and irrelevant features and locate accurately the biomarkers for schizophrenia from fMRI data are very important for diagnosis and further study of SZ. In this paper, the $\ell_{2,1-2}$-norm, which is a kind of nonconvex and Lipschitz continuous norm, is introduced as a regularization term to the embedded feature selection method. The $\ell_{2,1-2}$-norm sparse regularization feature selection method can be used to select features that are crucial for discrimination, and then explore the abnormal functional connections of schizophrenia with fMRI data. Our results showed that the abnormal functional connections are related to superior parietal gyrus, parahippocampal gyrus, caudate nucleus and middle occipital gyrus. The changes of functional connections involved in these brain regions may result to incorrect information processing in the brain, which provide further evidence for the cognitive disorder of schizophrenia.

Keywords: Schizophrenia · Abnormal functional connections · Feature selection · $\ell_{2,1-2}$ norm

1 Introduction

Schizophrenia is a severe chronic and highly disabling mental disorder involving sensory, cognitive, emotional, memorial and other symptoms. The global lifetime prevalence rate is about 1%, which brings a heavy burden to the families and society.

D.-S. Huang and K.-H. Jo (Eds.): ICIC 2020, LNCS 12464, pp. 145–158, 2020.
https://doi.org/10.1007/978-3-030-60802-6_14

It is an urgent problem to reveal the pathogenic mechanism and find out biomarkers of schizophrenia for the prevention, diagnosis and treatment of SZ. In recent decades, researches on brain structure and function have shed some light on some understanding of the neurobiological mechanisms behind its symptoms. The findings from functional brain imaging studies support a main hypothesis that SZ stems from disconnectivity, namely abnormal interactions between widely distributed brain networks. At present, many researchers have studied functional magnetic resonance imaging (fMRI) data to look for different functional connections between people with schizophrenia and normal people. Shine et al. [1] studied the fMRI data of SZ and found that the dynamic changes of brain functional network play an important role in cognitive information processing. Du et al. [2] used a novel group information guided method to process rs-fMRI data and explored the intrinsic dynamic functional connectivity disorder in schizophrenia. They found that the abnormalities of SZ were mainly distributed in the cerebellum, frontal cortex, thalamus and temporal cortex. Rosenberg et al. [3] found that the strength of connections in the whole brain functional network can be used as a biological marker to distinguish patients with mental illness from healthy controls. These studies on developing biomarkers have pushed the field of imaging analysis and psychiatry forward.

Given that SZ is often accompanied by cognitive decline, the thorough investigation of brain functional connections in SZ seems important in order to better understand the underlying neural mechanism. Recently, the use of machine learning algorithms to analyze fMRI data to achieve accurate diagnosis of the disease and explore the mechanism of schizophrenia has attracted much attention. In the process of model building, it is possible to mistake the irrelevant features as the beneficial features for identifying the data, which leads to the over-fitting of the model, so that the diagnostic accuracy of schizophrenia cannot achieve the desired results, the computational efficiency and the model of interpretability are also greatly affected [4]. Feature selection methods are usually used to eliminate redundancy and irrelevance by selecting a group of features that have the greatest contribution to distinguish different objects from the original feature set according to the a certain criteria. The criteria could be either of the following items. The feature subset is the optimal choice with some evaluation indexes for classifiers; the dimension of the feature subset is the lowest for a given accuracy; the conditional probability distribution function of the data is consistent with that of the selected features; the error rate of the classifier would not be reduced by not increasing or decreasing the number of features. By such a selection process we can effectively reduce the overfitting, enhance the generalization ability of the model, improve the computational speed and strengthen the interpretability which is crucial for exploring the mechanisms of why things are different.

The embedded method based on penalty term is a widely used feature selection method, which adds regularization term to the loss function and produces row sparse solutions by constraining the norm of the transformation matrix, that is, the norms for most rows of the transformed matrix are zero or close to zero, only a small part of the features appear in the final model, so as to achieve the goal of feature selection. Since the convex optimization problem has been studied for many years, there are many mature algorithms, and it has been proved that the local optimal solution is the global optimal solution, so most sparse regularization methods adopt convex function as the regularization term, such as the ℓ_1 norm and its variants for binary classification and regression [5–7] and the $\ell_{2,1}$ norm for multiclass classification [8–10]. However, researches have shown that non-convex regularization performs better in some cases [11–13]. For example, smoothly clipped absolute deviation [11] and minimax concave penalty [12] could produce unbiased estimation from the statistic point of view. $\ell_{2,p}$ norm is a nonconvex norm, which had been proved to perform better than $\ell_{2,1}$ norm [14]. However, it is very difficult to calculate due to its non-Lipschitz continuity and unknown hyperparameter p. Therefore, for embedded feature selection, it is an urgent problem to choose suitable and appropriate norm for sparse regularization terms, which has a great influence on the implementation of the algorithm and the recognition ability of the selected features.

In this paper, we introduce $\ell_{2,1-2}$ norm as sparse regularization term to loss function in the study of fMRI data of patients with schizophrenia and health controls. As the difference of $\ell_{2,1}$ norm and $\ell_{2,2}$ norm, $\ell_{2,1-2}$ norm is a kind of nonconvex regularization measure and Lipschitz continuous. In the process of solving the nonconvex problem, we introduce an iterative algorithm in the framework of Concave-Convex Procedure (CCCP) [15]. With this new embedded method, we studied the pathogenesis of schizophrenia according to the abnormal functional connections related to schizophrenia corresponding to the selected features, which provides a direction for the diagnosis and early prevention of schizophrenia.

In our study, based on the $\ell_{2,1-2}$-norm sparse regularization feature selection method, we found that there are 27 abnormal functional connections associated with schizophrenia, and they are mainly concentrated in the superior parietal gyrus (SPG. L), parahippocampal gyrus (PHG. L), middle occipital gyrus (MOG. R) and caudate nucleus (CAU. L). The superior parietal gyrus mainly corresponds to the sensory area of the human body. The parahippocampal gyrus plays a key role in memory, emotion and cognition, which is an important emotion regulating center. The middle occipital gyrus is the primary visual processing center, its impairment can not only lead to visual disorder, but also result in memory impairment and motion perception impairment. The abnormalities in the motor system are related to the functional connections associated with the caudate nucleus.

2 Methodology

Adding sparse regularization into the loss function can be considered as an embedded feature selection method, which has been widely used in many fields of feature learning due to its good performance and interpretability. From the perspective of sparsity, $\ell_{2,0}$ norm might be the most desirable. However, it will result in the NP-hard and is very difficult to solve. Therefore, several proxies have been studied to approximate the $\ell_{2,0}$ norm such as $\ell_{2,1}$ norm (LASSO), $\ell_{2,2}$ norm (ridge regression), which are both convex functions. However, nonconvex regularization performs better than convex regularization in some cases. $\ell_{2,p}$ norm ($0 < p < 1$) is a nonconvex regularization measure and experiments have empirically demonstrated that $\ell_{2,p}$ outperforms $\ell_{2,1}$ [16], but the $\ell_{2,p}$ norm is more difficult to compute. Firstly, there is an unknown hyperparameter p, which controls the effect of feature selection. The smaller the p, the better the performance. Secondly, $\ell_{2,p}$ is non-Lipschitz continuous, which makes it difficult to solve.

Recently, a nonconvex and Lipschitz continuous sparse metric, $\ell_{2,1-2}$ norm is proposed, which is the difference of the $\ell_{2,1}$ norm and the $\ell_{2,2}$ norm [17]. The feature selection method which adds $\ell_{2,1-2}$ norm as a regularization can be expressed as

$$\begin{aligned}\min_{W} F(W) &= \min_{W} \| Y - XW \|_{2,1} + \alpha \| W \|_{2,1-2} \\ &= \min_{W} (\| Y - XW \|_{2,1} + \alpha \| W \|_{2,1}) - \alpha \| W \|_{2,2}\end{aligned} \tag{1}$$

Where $X \in \mathbb{R}^{n \times d}$ is a data matrix with n samples of d features, and $Y \in \mathbb{R}^{n \times c}$ is a tag matrix with n samples of c classes. If the *i*th sample belongs to the *j*th class, $y_{i,j} = 1$, otherwise $y_{i,j} = 0$.

From (1), we can know that F(W) is a nonconvex function and can be split into the difference between two convex functions. CCCP (concave - convex process) is a monotone decreasing and global optimization method for solving nonconvex form, the form can be represented as the difference between two convex functions or the sum of convex function and concave function [17]. Therefore, (1) can use an iterative algorithm in the framework of CCCP. The iterative steps are as follows: To start with, linearize the second term $\| W \|_{2,2}$ in the objective function at the current solution W_k.

$$\begin{aligned}\| W \|_{2,2} &\geq \| W_k \|_{2,2} + \langle W - W_k, A_k \rangle = \| W_k \|_{2,2} + \langle W, A_k \rangle - \langle W_k, A_k \rangle \\ &= \langle W, A_k \rangle\end{aligned}$$

Where A_k is the subgradient of $\| W \|_{2,2}$ at W_k.

$$A_k = \begin{cases} \frac{W_k}{\| W_k \|_{2,2}}, & W_k \neq 0 \\ 0, & W_k = 0 \end{cases}$$

Then a new one W_{k+1} is obtained by solving the following subproblem:

$$\min_{W} \| Y - XW \|_{2,1} + \alpha \| W \|_{2,1} - \alpha \langle W, A_k \rangle \tag{2}$$

Based on the fact that W_{k+1} minimizes (2) and the definition of subgradient, we have

$$\begin{aligned}F(W_{k+1}) &\leq (\| Y - XW_{k+1} \|_{2,1} + \alpha \| W_{k+1} \|_{2,1}) - \alpha \langle W_{k+1}, A_k \rangle \\ &\leq (\| Y - XW_k \|_{2,1} + \alpha \| W_k \|_{2,1}) - \alpha \langle W_k, A_k \rangle \\ &= (\| Y - XW_k \|_{2,1} + \alpha \| W_k \|_{2,1}) - \alpha \| W_k \|_{2,2} = F(W_k)\end{aligned}$$

Repeat the above process until the CCCP stop criterion is satisfied, a monotonically decreasing sequence of F(W) is obtained. The sequence of iterative points $\{W_k\}_{k=0}^{\infty}$ generated by CCCP is proved to be strong global convergent.

Then ADMM algorithm is used to solve the subproblem (2). The ADMM combines the weak conditional convergence of multiplier method and the decomposability of dual ascent method, so it is a widely used and efficient algorithm [18]. The problems solved with ADMM have the following basic forms, some other problems can also be used after deformation.

$$\min_{x,z} f(x) + g(z)$$
$$\text{s.t. } Ax + Bz = C \tag{3}$$

The augmented Lagrange expression is

$$l_\rho(x, z, y) = f(x) + g(z) + y^T(Ax + Bz - C) + \frac{\rho}{2} \| Ax + Bz - C \|_2^2$$

Where y is Lagrange multipliers corresponding to the equality constraints, and $\rho > 0$ is a penalty parameter which trades off the bias and variance. The iterative solution process is as follows

$$x^{k+1} = \arg\min_x L_\rho(x, z^k, y^k)$$
$$z^{k+1} = \arg\min_z L_\rho(x^{k+1}, z, y^k)$$
$$y^{k+1} = y^k + \rho(Ax^{k+1} + Bz^{k+1} - C) \tag{4}$$

Repeat the above process until the ADMM stop criterion is satisfied. In general, the stop criterion is that the norm of dual residual s^{k+1} and the initial residual r^{k+1} are less than the set threshold. Where $s^{k+1} = \rho A^T B(z^{k+1} - z^k)$ and $r^{k+1} = Ax^{k+1} + Bz^{k+1} - C$.To meet the formula of ADMM, two auxiliary variables $M = Y - XW$ and $N = W$ are introduced to construct the standard form of ADMM, and the above subproblem (2) is converted into the following form:

$$\min_W(\| M \|_{2,1} + \alpha \| N \|_{2,1}) - \alpha\langle W, A^k \rangle$$
$$\text{s.t. } M = Y - XW, N = W \tag{5}$$

The augmented Lagrange expression is: $\min_{W,M,N} \| M \|_{2,1} + \alpha \| N \|_{2,1} - \alpha\langle W, A^k \rangle + \langle \Phi, Y - XW - M \rangle + \langle \psi, W - N \rangle + \frac{\rho}{2}(\| Y - XW - M \|_F^2 + \| W - N \|_F^2)$. Where Φ and Ψ are two Lagrange multipliers corresponding to two equality constraints, respectively. The specific iterative process is as follows.

ADMM for solving $\ell_{2,1-2}$ norm based sparse regularization

---

1. **Input:** data matrix X, label matrix Y, regularization parameter α and penalty parameter ρ
2. **Initialize:** outer loop k=0, $W_0 = \mathbf{0}$, $W_{-1} = \mathbf{0}$, ε
3. **While** $F(W_{k-1}) - F(W_k) > \varepsilon$ **or** $k = 0$:
   3.1 Calculate the subgradient of current W_k, and named it A^k
   Initialize the inner loop s=0, and $N^0 = \mathbf{0}$, $M^0 = \mathbf{0}$ and $\Psi^0 = \mathbf{0}$, $\Phi^0 = \mathbf{0}$
   3.2 $W = \rho^{-1}(X^TX + I_d)^{-1}\left(\alpha A^k - \Psi^s + \rho N^s + X^T(\Phi^s + \rho Y - \rho M^s)\right)$
   3.3 Update N: Let $V^s = W + \frac{\Psi^s}{\rho}$, v_i represents the ith row of V^s, n_i represents the ith row of N^{s+1}.
   If $\alpha < \rho\|v_i\|_2$, then $n_i = \left(1 - \frac{\alpha}{\rho\|v_i\|_2}\right)v_i$; otherwise $n_i = \mathbf{0}$
   3.4 Update M: Let $U^s = Y - XW + \frac{\Phi^s}{\rho}$, u_i represents the ith row of U^s, m_i represents the ith row of M^{s+1}.
   If $1 < \rho\|u_i\|_2$, then $m_i = \left(1 - \frac{1}{\rho\|u_i\|_2}\right)u_i$; otherwise $m_i = \mathbf{0}$
   3.5 Update Ψ and Φ: $\Psi^{s+1} = \Psi^s + \rho(W - N^{s+1})$
   $\Phi^{s+1} = \Phi^s + \rho(Y - XW - M^{s+1})$
   3.6 Update $s := s + 1$
   3.7 Repeat 3.2~3.6 until the ADMM stop criterion is satisfied
   3.8 $k := k + 1$, $W_k = W$
4. **end while**
5. **Output:** W_k

---

3 Experiments

3.1 Data Collection and Preprocessing

In this study, the Machine Learning for Signal Processing Schizophrenia classification challenge data were used (https://www.kaggle.com/c/mlsp-2014-mri). Image preprocessing was performed using statistical parametric mapping software. Further feature extraction was done by the GIFT Toolbox (http://mialab.mrn.org/software/gift/), yielding FNC features for resting state functional MRI. The data consisted of 40 patients with SZ and 46 HCs. Each sample had 378 FNC features. FNC features were the pair-wise correlation values between the time-courses of 28 brain regions and can be seen as a functional modality feature describing the subjects' overall level of synchronicity between brain areas [18]. These 28 brain regions were selected according to the anatomical automatic labeling template.

$\ell_{2,1-2}$ algorithm has two hyperparameters, α and ρ. α is the penalty parameter of the sparse regularization model, which controls the generalization ability of the model, ρ is the augmented Lagrange constant in ADMM algorithm, which controls the convergence speed of ADMM and the accuracy of CCCP subproblem. According to the grid search, we set $\alpha = 0.01$ and $\rho = 2$ in the experiment.

The original data is randomly split into two parts: 70% for training and 30% for testing. On the training set, the sorting of all features is obtained according to the feature selection algorithm. Thus, features that contribute the most to identifying the classes are represented with the highest rankings, and features with low rankings are generally considered as noise, irrelevant or redundant to discriminate data. On the test set, the classification accuracy of each feature subset is obtained. In order to avoid the influence of random factors of data on feature ordering and obtain better classification accuracy on the test set, we used cross validation on the training set. First, the training set is divided into 10 parts on average, nine of which are selected for feature selection each time, and the remaining one is used as the validation set. To get more stable results, we repeatedly run random split 30 times, and counted the frequency of each feature appearing in the optimal feature subset obtained from each split, and then determined the final feature ranking and optimal feature subset.

3.2 Results

To test the performance of $\ell_{2,1-2}$ regularization, we compare it with Relief and Max-Relevance and Min-Redundancy (mRMR). Relief is a typical feature selection method, using distance measurement as its evaluation function. The running time of Relief algorithm increases linearly with the number of samples and the number of original features, the high efficiency and satisfactory results makes the algorithm widely used. Max-Relevance and Min-Redundancy (mRMR), which adopts mutual information to measure redundancy between features and correlations between features and labels, is also often used because it can eliminate redundancy. All of the three algorithms use linear support vector machine (SVM) as the classifier and take the default parameter

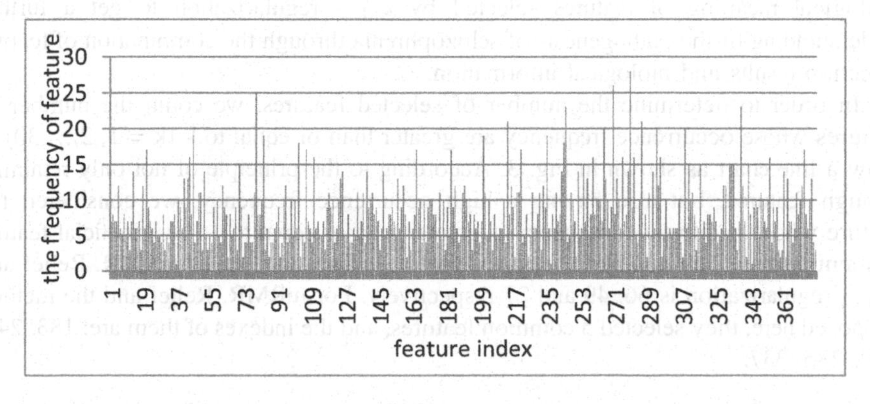

Fig. 1. The frequency of 378 FNC features obtained by $\ell_{2,1\text{-}2}$ regularization.

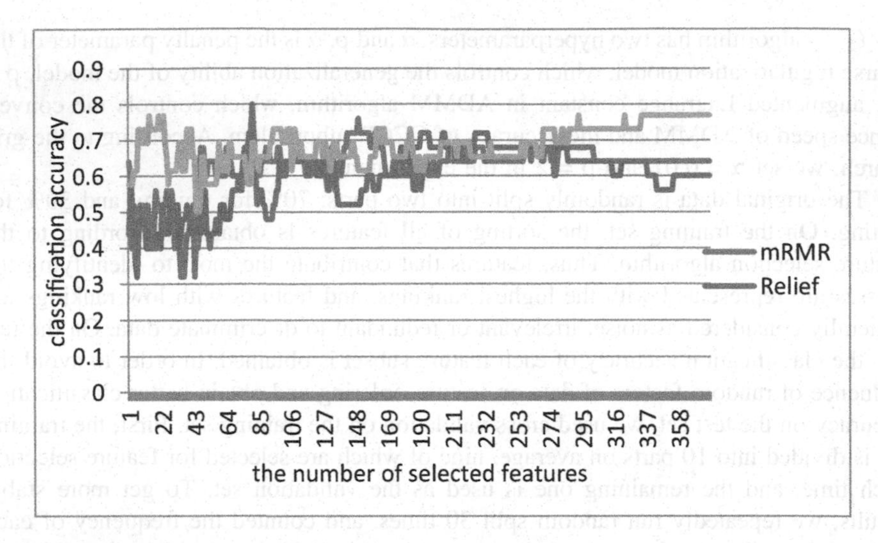

Fig. 2. Classification accuracy with the number of selected features.

C = 1. The frequency of each feature obtained by $\ell_{2,1-2}$ regularization is shown in Fig. 1. Figure 2 shows the variation of classification accuracy on the test set with the number of selected features.

In the classification accuracy curve of $\ell_{2,1-2}$ regularization, it can be found that the classification accuracy reaches the maximum when the number of selected features is around 20, and then with the addition of irrelevant features and noise, the classification accuracy declines and shows an oscillating trend. When the number of features increases to a certain degree, the obtained classification accuracy is the same as the accuracy when the number of features is around 20. Therefore, the $\ell_{2,1-2}$ regularization can eliminate redundant and irrelevant features, simplify model effectively and save computation time while ensuring classification accuracy at the same time. Although the Relief algorithm and the mRMR algorithm obtained the highest test accuracy under a certain feature subset, the result is unstable. So in the next section, we mainly study the biological meaning of features selected by $\ell_{2,1-2}$ regularization to get a further understanding of the pathogenesis of schizophrenia through the combination of feature selection results and biological information.

In order to determine the number of selected features, we count the number of features whose occurrence frequency are greater than or equal to k (k = 1, 2,…,30) to draw a line chart as shown in Fig. 3. According to the principle of not only retaining enough features, but also having a high occurrence frequency, we considered the feature whose occurrence frequency is greater than or equal to 16 as beneficial feature to identify data. The number of selected features corresponding to mRMR, Relief and $\ell_{2,1-2}$ regularization is 66, 49 and 27 respectively. For mRMR, Relief and the method proposed here, they selected 5 common features, and the indexes of them are: 183, 243, 279, 285, 337.

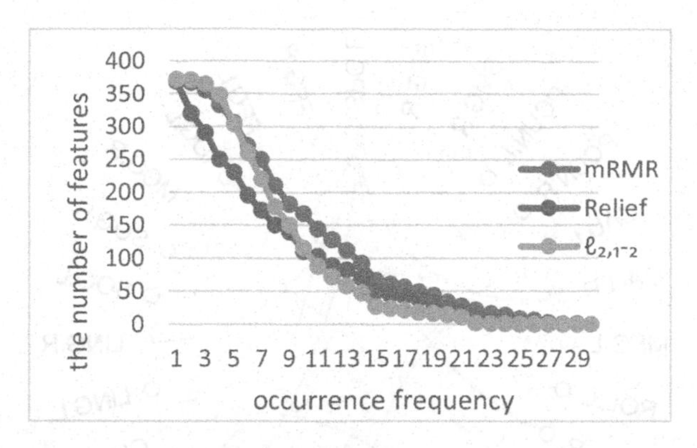

Fig. 3. The number of features whose occurrence frequency are greater than or equal to k (k = 1, 2, 3,…, 30).

Combining the results in Fig. 2 and Fig. 3, high classification accuracy can be obtained by using the 27 features selected from $\ell_{2,1-2}$ regularization. We also use other metrics to measure classification performance of the 27 features selected by $\ell_{2,1-2}$ regularization, and the precision score, recall score, F1 score and ROC are 0.8182, 0.6923, 0.75 and 0.7692 respectively. Statistical significance of classification performance is also considered, the result of permutation test is that the classification score is 0.79 and the p-value = 0.009 < 0.05, which indicates that the classification performance passed the significance test. Additionally, the classification accuracy of feature selection method using ℓ_1-norm and ℓ_2-norm as regularizations is 0.73 and 0.69 respectively, and the classification accuracy by using $\ell_{2,1-2}$ regularization is 0.79.

We can apply this algorithm to the diagnosis of schizophrenia with the help of practical observation, family feedback and other clinical means, and then provide technical support for the early detection and treatment of the diseases. In addition, except with the accurate diagnosis of schizophrenia, we also care about abnormal brain regions and abnormal function connections associated with the disease, so as to reveal the inner pathogenesis of schizophrenia, provide theoretical basis for the development of effective drugs and the creation of medical techniques. The Table 1 lists the 27 abnormal functional connections of the regions of interest selected by $\ell_{2,1-2}$ algorithm. The feature index in Table 1 represents the functional connections between two of the 28 brain regions selected according to the anatomical automatic labeling template. For example, the feature with index number of 279 represents the functional connection between brain region 59 (Superior parietal gyrus) and brain region 60 (Superior parietal gyrus). To test whether the 27 features have a significant difference in normal and schizophrenia patients, we used the single factor multivariate analysis of variance and obtained p value = 3.7384e–11 < 0.05, so the 27 feature defined as abnormal functional connections have a significant difference in normal and schizophrenia patients. The schematic map of the selected functional connections is also shown in Fig. 4 as circular connectivity graph.

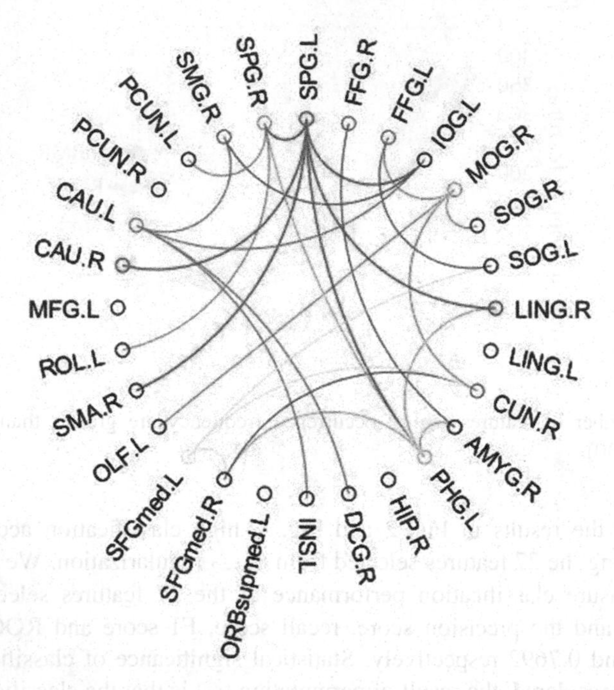

Fig. 4. Circular connectivity graph.

Among these abnormal functional connections associated with schizophrenia shown in Figs. 4 and 5, there are six functional connections related with the superior parietal gyrus (SPG.L): inferior occipital gyrus (IOG.L), parahippocampal gyrus (PHG. L), supplementary motor area (SMA.R), caudate nucleus (CAU.R), superior parietal gyrus (SPG.R) and Lingual gyrus (LING.R). In addition, there also existed five abnormal functional connections related to the PHG.L, including Amygdala (AMYG. R), lingual gyrus (LING.R), SPG.L, SPG.R and Superior frontal gyrus, medial (SFGmed.R), four aberrant functional connection in middle occipital gyrus (i.e., fusiform gyrus, olfactory cortex, cuneus and superior occipital gyrus) and four aberrant functional connection in CAU.L (i.e., supramarginal gyrus, inferior occipital gyrus, median cingulate and paracingulate gyri and Insula).

The superior parietal gyrus mainly corresponds to the sensory area of the human body, the highest center of somatosensory sensation. The area is determined by stimulation and excision and can perceive various sensory stimuli. Recent studies on schizophrenia have found that the onset of schizophrenia may be related to the abnormal functional connections of the somatosensory center. It can be found in the repeat tentacles stimulation experiments of mice that, comparing with the normal control mice, the number of activated cells in the sensory cortex of mice with schizophrenia risk gene is greatly reduced, which shows that the mutated gene can lead to somatosensory sensory dysfunction, the occurrence of schizophrenia may be associated with abnormal function in the sensory area. The study of Pawełhas found that the abnormal connect between superior parietal gyrus and prefrontal lobe may lead to abnormalities in the brain's cognitive processing speed, Mayer's researches have shown that some connectivity related to superior parietal gyrus had significantly decreased in patients with schizophrenia, including posterior cingulate cortex, thalamus, cerebellum, and visual cortex.

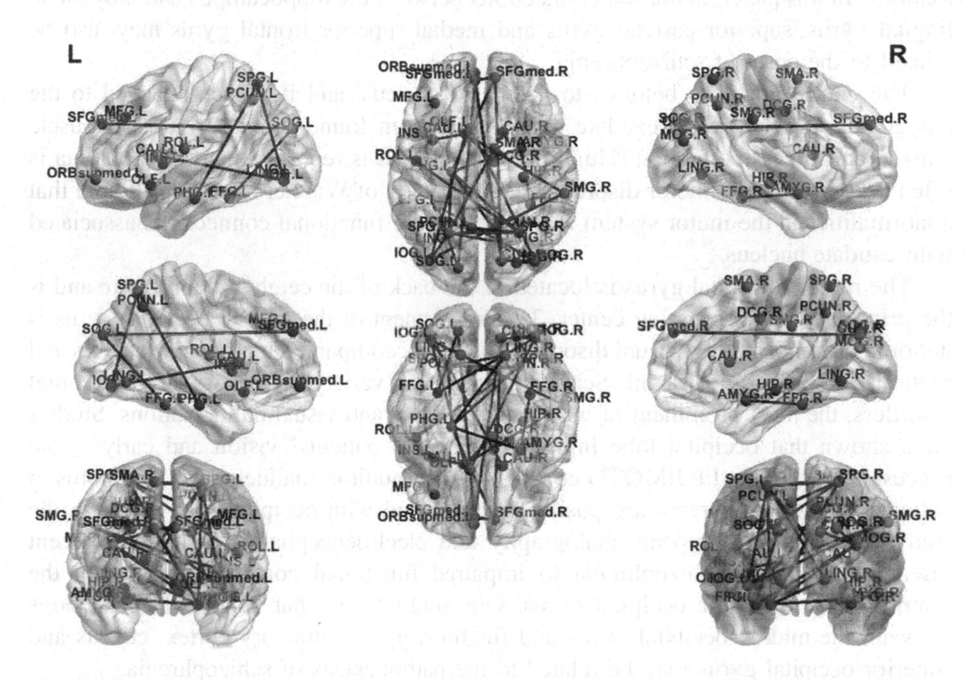

Fig. 5. The locations and their connections of the selected brain regions by the BrainNet Viewer toolbox.

Table 1. Connections

Feature index	Connectivity	Feature index	Connectivity	Feature index	Connectivity
279	59—60	265	39—60	183	29—71
270	39—42	200	46—52	20	21—52
79	23—24	45	17—60	372	55—49
337	34—71	234	67—60	165	56—42
244	48—59	83	23—46	110	24—39
213	64—53	243	48—39	275	59—53
285	59—20	102	23—49	259	39—59
293	50—52	106	24—46	353	52—55
308	53—71	220	64—71	281	59—72

The parahippocampal gyrus, which is located on the medial side of the occipital lobe and the lower temporal lobe, is considered to be an important emotion regulating center and plays a key role in memory, emotion and cognition. It has been found that the structural damage of the hippocampus can cause cognitive and emotional abnormalities. In this paper, abnormal connections between the hippocampus and amygdala, lingual gyrus, superior parietal gyrus and medial superior frontal gyrus may also be related to the onset of schizophrenia.

The caudate nucleus belongs to the basal nucleus, and its tail is attached to the amygdala. Atrophy of the caudate nucleus has been found to cause reduced muscle tension and excessive motion (Huntington chorea). It is reported that schizophrenia is often accompanied by motor disorders, and the study of Walther S et al. has found that abnormalities in the motor system are related to the functional connections associated with caudate nucleus.

The middle occipital gyrus is located in the back of the cerebral hemisphere and is the primary visual processing center. The impairment of the middle occipital gyrus is not only accompanied by visual disorder, but also accompanied by memory deficits and motion perception impairment. Schizophrenia has a variety of sensory and perceptual disorders, the most prominent of which are auditory and visual hallucinations. Studies have shown that occipital lobe injury is related to patients' vision and early visual processing disorders. FUJIMOTO et al. found that auditory hallucination symptoms in patients with schizophrenia are positively correlated with occipital lobe injury in the study based on Magnetoencephalography and electroencephalography data. Recent research has linked schizophrenia to impaired functional connections between the brain's bilateral middle occipital gyrus. Our study found that abnormal connections between the middle occipital gyrus and fusiform gyrus, olfactory cortex, cuneus and superior occipital gyrus may be related to the pathogenesis of schizophrenia.

Existing studies have shown that schizophrenia is associated with abnormal functional connections in brain [19, 20], there are multiple abnormal functional connections between brain regions in patients with schizophrenics, which are either higher or lower than the normal values, hinder the interactions of each regions and lead to a variety of clinical symptoms, involving sensory and perceptual disorders, thinking disorders, cognitive disorders, emotional disorders, will and behavior disorders. Our study of functional connections between 28 brain regions showed that, the superior parietal gyrus, parahippocampal gyrus, caudate nucleus and middle occipital gyrus of SZ had abnormal connections with other brain regions. These abnormal connections in the brain may be related to schizophrenia, medical workers could do further verification and exploration refer to this result.

4 Conclusions

In this paper, 27 abnormal functional connections in brains of SZ were discovered by the $\ell_{2,1-2}$ norm based sparse regularization. The selected features showed that most of abnormal functional connections were related with superior parietal gyrus, parahippocampal gyrus, caudate nucleus and middle occipital gyrus. These four networks are closely related to somatosensory dysfunctions, cognitive deficits, emotional abnormalities, memory deficits and motion perception impairments. Our results provide new ideas for the application of fMRI and the study of schizophrenia.

Acknowledgement. This work was supported by NSFC (No. 11471006 and No. 81601456), Science and Technology Innovation Plan of Xi'an (No. 2019421315KYPT004JC006), the Fundamental Research Funds for the Central Universities (No. xjj2017126) and the HPC Platform, Xi'an Jiaotong University.

References

1. Shine, J.M., et al.: The dynamics of functional brain networks: integrated network states during cognitive task performance. Neuron **92**, 544–554 (2016)
2. DU, Y., et al.: Dynamic functional connectivity impairments in early schizophrenia and clinical high-risk for psychosis. NeuroImage **180**, 632–645 (2018)
3. Rosenberg, M.D., et al.: A neuromarker of sustained attention from whole-brain functional connectivity. Nat. Neurosci. **19**, 165–171 (2016)
4. Suk, H.I., Lee, S.W., Shen, D.: Deep ensemble learning of sparse regression models for brain disease diagnosis. Med. Image Anal. **37**, 101–113 (2017)
5. Tibshirani, R.: Regression shrinkage and selection via the lasso. J. Royal Stat. Soc. **58**(1), 267–288 (1996)
6. Bradley, P.S., Mangasarian, O.L.: Feature selection via concave minimization and support vector machines. In: Proceedings of 13th International Conference Machine Learning, pp. 82–90 (1998)
7. Yuan, M., Lin, Y.: Model selection and estimation in regression with grouped variables. J. Royal Stat. Soc. **68**(1), 49–67 (2006)
8. Nie, F., Huang, H., Cai, X., Ding, C. H.: Efficient and robust feature selection via joint $\ell_{2,1}$-norms minimization. In: Proceedings of Advance Neural Information Processing System, pp. 1813–1821 (2010)
9. Cai, X., Nie, F., Huang, H., Ding, C.: Multi-class $\ell_{2,1}$-norm supportvector machine. In: Proceedings of IEEE 11th International Conference Data Mining, pp. 91–100 (2011)
10. Xiang, S., Nie, F., Meng, G., Pan, C., Zhang, C.: Discriminative least squares regression for multiclass classification and feature selection. IEEE Trans. Neural Netwk. Learn. Syst. **23** (11), 1738–1754 (2012)
11. Fan, J., Li, R.: Variable selection via nonconcave penalized likelihoodand its oracle properties. J. Amer. Statist. Assoc. **96**(456), 1348–1360 (2001)
12. Zhang, C.H.: Nearly unbiased variable selection under minimax concave penalty. Ann. Statist. **38**(2), 894–942 (2010)
13. Xu, Z., Chang, X., Xu, F., Zhang, H.: ℓ2,1-2 regularization: A thresholding representation theory and a fast solver. IEEE Trans. NeuralNetw. Learn. Syst. **23**(7), 1013–1027 (2012)
14. Zhang, M., Ding, C., Zhang, Y., Nie, F.: Feature selection at the discrete limit. In: Proceedings of AAAI Conference Artificial Intelligence, pp. 1355–1361 (2014)
15. Shi, Y., Miao, J., Wang, Z., Zhang, P., Niu, L.: Feature selection with ℓ2,1-2 regularization [J]. IEEE Trans. Neural Netk. Learn. Syst. **29**, 4967–4981 (2018)

16. Du, X., Yan, Y., Pan, P., Long, G., Zhao, L.: Multiple graph unsupervised feature selection. Signal Process **120**, 754–760 (2016)
17. Boyd, S., Parikh, N., Chu, E., Peleato, B., Eckstein, J.: Distributed optimization and statistical learning via the alternating direction method of multipliers. Found. Trends Mach. Learn. **3**(1), 1–122 (2011)
18. Allen, E.A., et al.: A baseline for the multivariate comparison of resting-state networks. Front. Syst. Neurosci. **5**(5), 2 (2011)
19. Liang, M., et al.: Widespread functional disconnectivity in schizophrenia with resting-state functional magnetic resonance imaging. NeuroReport **17**, 209–213 (2006)
20. Xu, Y., et al.: Selective functional disconnection of the orbitofrontal subregions in schizophrenia. Psychol. Med. **47**, 1637–1646 (2017)

Systems Biology

Identification and Analysis of Genes Involved in Stages of Colon Cancer

Bolin Chen[1,2], Teng Wang[3], and Xuequn Shang[1,2(✉)]

[1] School of Computer Science, Northwestern Polytechnical University, Xi'an, China
npu_bioinf@hotmail.com

[2] Key Laboratory of Big Data Storage and Management Ministry of Industry and Information Technology, Northwestern Polytechnical University, Xi'an, China

[3] School of Computer and Information Technology, Shanxi University, Taiyuan, China

Abstract. Colon cancer is a common malignant tumor that occurs in the colon, with high morbidity and mortality. In addition, most cancers are caused by genetic mutations. With the continuous development of sequencing technology, the amount of gene expression data increases dramatically, so we can study biological data from the perspective of data mining. In this study, we mainly through data analysis and network analysis to explore the colon cancer related genes and its biological functions. First, we obtained the gene expression data and some other data, then did the data preprocessing. Second, we grouped the data by sample disease stage, and differential expression analysis was performed for each group. Then, the PPI network and the functional interaction network were constructed by differentially expressed genes. Finally, we conducted Newman clustering algorithm on PPI network and functional interaction network, then carried out graph topology analysis and functional pathway analysis respectively. The result showed that CTP3A4, FLNC, CNTN2, MEP18 and MAOA might be colon cancer related genes. Besides, cAMP signaling pathway, chemokine signaling pathway and neuroactive ligand-receptor interaction were KEGG pathways with significant differences in four stages of colon cancer. And some pathways are closely related to other pathways at adjacent stages, such as chemokine signaling pathway and cytokine-cytokine receptor interaction.

Keywords: Colon cancer · Cancer stage · Protein-protein interaction · Functional interaction · Differentially expressed gene · Functional enrichment analysis

1 Introduction

Colon cancer is a common malignant tumor of the digestive tract that occurs in the colon and the incidence rate is the third in gastrointestinal tumors. In China, the incidence of colon cancer has been on the rise in recent years, and the incidence increases significantly after the age of 35, which increases with the age. The incidence is higher in urban areas than in rural areas, and the male to female ratio is about 3:1.

D.-S. Huang and K.-H. Jo (Eds.): ICIC 2020, LNCS 12464, pp. 161–172, 2020.
https://doi.org/10.1007/978-3-030-60802-6_15

For the colon cancer, the disease usually has no special symptoms in the early stage and is easily overlooked. However, once the symptoms are discovered, it is in the middle or later stage. The cancer stage of cancer patients is extremely important to determine the clinical diagnosis and treatment plan [1]. If the cancer is at an early stage, the cure rate is higher, but if the patient is in the middle or later stages of cancer, cancer cell proliferation or even metastasis will occur, then the general medical diagnosis and treatment methods are beyond help. Although there are many effective screening methods like enteroscopy, the molecular mechanism of colon cancer development has not completely studied. Further understanding of its mechanism will promote the further development of innovative screening methods, prognostic indicators and treatment methods.

As the material basis of inheritance, genes are specific nucleotide sequences with genetic information. There are about tens of thousands of genes in human, which contain answers to unknown puzzles in the process of birth, growth and dead of life. Therefore, analysis of gene sequencing data is an effective way to interpret life, moreover, some colon cancer related genes have been discovered. For instance, CRYAB [2] promoted the invasion and metastasis of (colorectal cancer) CRC cells via epithelial-mesenchymal transition (EMT), HMGCS2 [3] enhance cell invasion and metastasis during CRC by directly binding to peroxisome proliferator-activated receptor alpha (PPARα) to promoted Src activity, MEP1A [4] promote the cancer progression by block cell proliferation. However, the current molecular mechanism of colon cancer and its stage related genes have not been fully clarified.

With the advent of next-generation sequencing sets and automated analysis of high-through sequencing technology (also known as next-generation sequencing technology) supported by high-performance computer cluster technology, it makes it possible to conduct a comprehensive and in-depth analysis of the transcriptome and genome of a species. The progressive development of sequencing technology makes researchers unsatisfactory to study only one single gene in the face of complex life activities, thus they turning to use the result of high-through sequencing technology data for a more comprehensive study of omics data, which at entirety, network, and dynamic levels. In this study, we aimed to use the method of data analysis, that with different from the perspective of medicine and biology, to explore the colon cancer stage related genes and biological function.

In this study, we used the TCGA database to obtain gene expression data, colon cancer clinical information, and other relevant data, then integrated the expression data of colon cancer samples in stages. We analyzed the gene expression data at the data level and the network level respectively. For gene expression data, we have done the differential expression analysis and obtained the differentially expressed (DE) genes at each group. Next, we used DE genes to construct the PPI (Protein-protein interaction) network and functional interaction network, then applied graph clustering algorithm and pathway enrichment analysis, in order to analyze the results of colon cancer related genes and biological function. The result of this study would provide some new information about colon cancer at different stages, and it is hoped that it will be helpful for the treatment and prognosis of colon cancer.

2 Materials and Method

2.1 Materials

Gene expression data. In this study, gene expression data of colon cancer were collected from The Cancer Genome Atlas (TCGA), and the project name was TCGA-COAD. This database not only contain HTSeq-Counts data but also HTSeq-FPKM and HT-FPKM-UQ data, but only the row expression data that HTSeq-Counts data were obtained and investigated.

The data form Ensemble database [5] annotated the types of all genes in the gene expression data which obtained from TCGA. And only selected 19587 protein-coding genes for subsequent analysis. This database also provides corresponding information of gene coding, which is used to unify various data files.

Clinical data. In this study, clinical data of colon cancer were also obtained from TCGA. The original clinical data contained variety of clinical records of the samples, and only the information of sample number and cancer stage were selected.

After correlated the gene expression data with the corresponding clinical data, all samples could be divided into 5 categories: healthy human(normal), stage I, Stage II, Stage III, Stage IV. The normal samples were combined with each stage, and there are151, 223, 193, 138 samples respectively. We defined these four differentially expressed sets as DE sets.

Known colon cancer related genes. The Comparative Toxicogenomics Database (CTD) [6] stores genes which are associated with colonic neoplasms or its descendants, and these genes has either a curated association to the disease (M marker/mechanism and/or T therapeutic) or an inferred association via a curated chemical interaction. Because they played important role in colon cancer, so we considered these genes to be known colon cancer related genes in this study.

2.2 Data Preprocessing

Detecting differentially expressed (DE) genes was based on the analysis of gene expression quantification. But in RNA-seq studies, variability in measurement can be attributed to technical factors. Sources of technical variation include difference in library preparation across samples, gene length, sequencing depth and so on [7]. These noises would not only significantly affect the accuracy of statistical inferences from RNA-seq, but also prevent researchers from better modeling the group-specific changes. Hence, it was necessary to preprocess the data to minimize noise.

In order to filter low counts genes, each group of data should be calculated CPM (counts per million) value, and filtered the genes which more than 10 samples meet CPM < 1. Then, for the purpose of comparing gene expression data in different samples effectively, the gene expression data should be normalized. For each group of data, we considered the TMM (Trimmed Mean of M-values) algorithm [8] to reduce the impact of sample preparation and sequencing.

2.3 Differential Expression Analysis

The study of differential expression analysis, from statistical point of view, is the hypothesis testing. For the data which already known about the distribution, the false positive rate of analysis result will be relatively low. The normal distribution is often used to normalize microarray data, however, the RNA-seq technology produces the dispersion measures of gene expression data, so we considered the negative binomial distribution of the disperse probability distribution.

In this study, the exact test method that based on negative binomial distribution in edgeR package [9] in R language was used to identify the differentially expressed genes for each DE set. We calculated the p-value, which was further adjusted into false discovery (FDR) by using Benjamini-Hochberg method. We also calculated the FC value for each group, and only the genes with FDR < 0.05 and $|\log^{FC}| \geq 1$ could be regarded as differentially expressed (DE) genes.

2.4 Functional Interaction Network

For the functional interaction network, the nodes represented the genes, and the edge represented the functional associations and interaction associations between genes. Genes in 4 DE sets identified above were merged (all DE genes), and the genes which were used to construct the FI network were all DE genes. The data of edge was obtained from many databases that describe the biological function or metabolic pathway of homo sapiens, in this way, the edge contained the significance of biological function, which provided basis for the functional analysis. The FI network was construct by using a Cytoscape software [10], ReactomeFIViz [11].

After constructed the functional interaction network, we used the Newman graph clustering algorithm to divide multiple modules. And then, mapped the genes of each module with known colon cancer related genes and extracted the mapped modules from functional interaction network. For each mapped module, should do the pathway enrichment analysis (criterial: DE genes counts > 4 and p-value < 0.005).

2.5 Pathway Analysis Method

Because there were a great variety of pathways obtained by pathway enrichment, and the results of pathway enrichment in 4 sets were mixed, it was not conductive to the staged biological function analysis. Hence, we considered using the method of representing the relationship between pathways, that enriched in different stages, by graph, so as to obtain the results of functional analysis in stages.

In this study, we considered 2 relationships of pathways between adjacent stages, one of which was to investigate which pathways were significantly different at each stage, and the other was to investigate which pathways were associated with adjacent pathways at different stages. For the former, we constructed a network that takes the pathway as the node and takes whether the genes included in the pathways at adjacent DE sets overlapped as the basis of the edge. In particular, all nodes (pathways) in the network were staged, and there was no edge within the stages, only the same pathways in different stages may be connected. For the latter, we constructed a network which nodes are the same with the former one, but the edges depended on whether the genes involved in adjacent DE sets of the different pathway overlap. We defined the overlapping genes of the pathway in adjacent DE sets as overlap genes.

2.6 Protein-Protein Interaction Network

Protein-protein interaction (PPI) network was also involved in this study. Within cells, proteins rarely act as independent material to executive functions, and the study have shown that proteins involved in the same cellular processes often interact with each other, so protein-protein interactions (PPIs) are the basis of almost all biological processes. With the development of high-throughput technology, there was a surge in the data of PPIs, so that PPIs data can be represented in the form of networks, i.e., PPI network.

The Search Tool for the Retrieval of Interacting Genes/Proteins (STRING) [12] is a database that covers a wide range of species and stores an enormous amount of information about PPIs. The PPI network was constructed based on STRING database in this study, nodes represent proteins and edges represent PPIs. Due to the variable shearing and post-transcription modification of eukaryotes, a protein-coding gene will produce multiple proteins. However, Different isoforms generated by the same gene, which are merged as nodes in the PPI network. In this study, the nodes in PPI network were composed of the intersection of 4 DE sets.

2.7 Analysis of Protein-Protein Interaction Network

After constructed the PPI network, we used the Newman graph clustering algorithm to divide multiple interaction modules, and we considered that gene connections within same modules were tighter and functional associations were stronger. For each interaction module of PPI network, the module which contained known genes were extracted, and then the first neighbor nodes of known genes were extracted from these modules to form sub-networks (criteria: DE genes count > 4). Next, mapping nodes of FI network to each sub-network, then calculated betweenness value for each mapped node and arranged it in descending order to find the node with the highest betweenness value except the nodes that represented known colon cancer related genes.

3 Results

3.1 Different Expression Genes

A total of 1644,1573,1414,1665 significant DE genes (FDR 0.05 and $|\log^{FC}| \geq 1$) were found between normal and colon cancer stage I, II, III and IV, respectively. Clustering analysis have shown that most of expression of these 4 DE sets were different between normal and colon cancer samples.

For 4 DE sets, there were 1299 DE genes in total, and the number of intersections of DE genes is 880. Comparing the value of these 4 DE sets (Fig. 1) by VennPlex [13], the number of up-regulated genes were much more than down-regulated, and there were not much difference between the number of up-regulated genes or down-regulated genes in each DE set. There were still few contra-regulated genes, about 2.3% of the total genes, and suggested that the molecular function at different stages acted in much the same way.

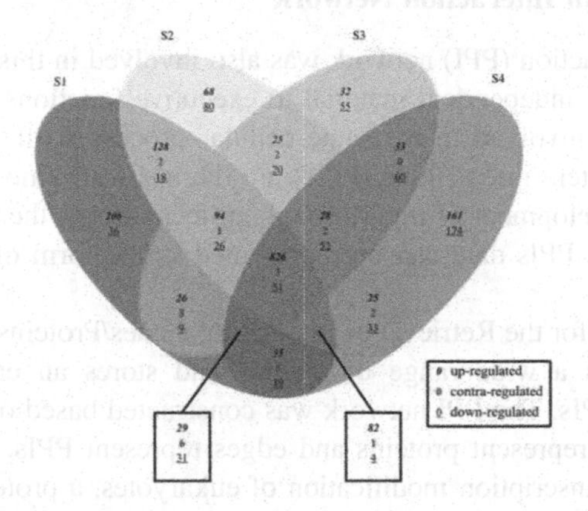

Fig. 1. Comparison between 4 DE 1sets. S1: DE set I (stage I and normal), S2: DE set II (stage II and normal), S3: DE set III (stage III and normal), S4: DE set IV (stage IV and normal).

3.2 Functional Interaction Network

The functional interactions between genes were obtained by analyzing all DE genes, and the functional interaction network was constructed. The network had 1153 nodes and 3491 edges, and 30 modules (DE genes counts > 4) were obtained by Newman clustering (Fig. 2). The functional interaction network also contained 32 known colon cancer related genes and 510 intersection DE genes. In the network, only 12 modules (DE genes counts > 4) contained known colon cancer related genes, and the pathway enrichment analysis have done to these modules.

3.3 Pathway Analysis of Functional Interaction Network

By above pathway enrichment analysis, 277 pathways were obtained. And through further analysis of the staged genes involved in the pathway, the correlation between 4 DE sets of pathways was obtained and expressed in the form of graph, called pathway interaction network. For the pathway interaction network (I) (Fig. 3-A), there were 1108 nodes and 831 edges in the network. Only overlap genes counts > 19 were screened as the significant difference pathway (Fig. 3-B). There were 9 pathways were obtained, and only 5 pathways were all significantly different in 4 DE sets (Fig. 3-B). For the pathway interaction network (II) (Fig. 4-A), there were 412 nodes and 3050 edges in the network, and only overlap genes counts > 14 could be extracted to analyze the relationship between different pathways in adjacent DE sets (Fig. 4-B). There were 7 pathways were obtained, and the result shown that multiple pathways in the previous

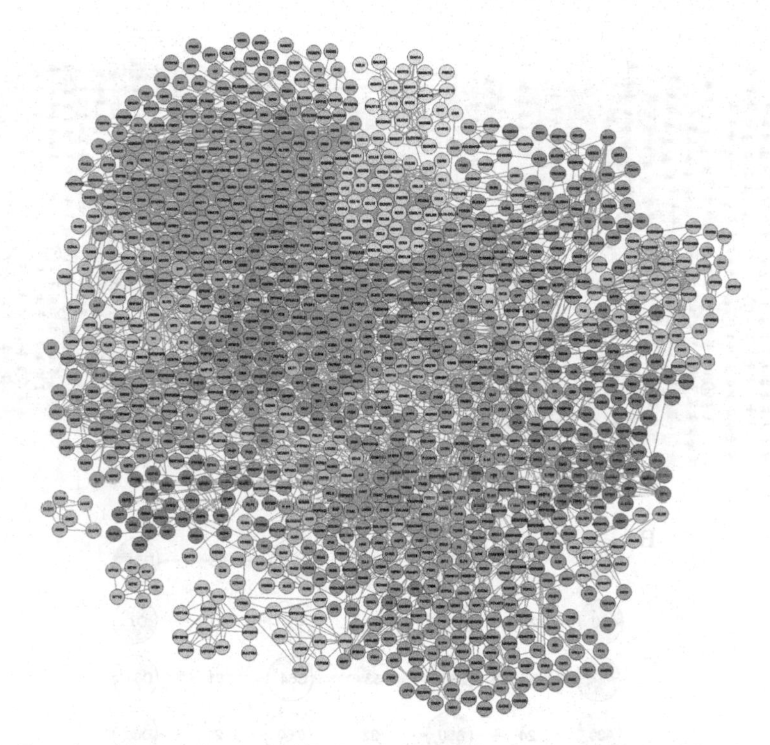

Fig. 2. Functional interaction network. Nodes with different represent the different modules.

stage were enriched in only one pathway in the next stage, or one pathway in the previous stage was enriched in multiple pathways in the next stage.

3.4 Protein-Protein Interaction Network

PPIs between intersection DE genes were investigated and a protein-protein interaction (PPI) network was constructed. The network involved 803 nodes (intersection DE genes) and 3840 edges (PPIs), and 17 modules (DE genes > 4) were obtained by Newman clustering (Fig. 5). For PPI network, there were 406 nodes appeared in the functional interaction network, and 11 nodes were known colon cancer related genes, such as MMP7, HMGCS2, FOLR1, SULT1A1, MET1A, GAP43, MAOB, TAGLN, CRYAB, ACTG2, GSTM5.

3.5 Protein-Protein Interaction Sub-network

Trough the further analysis of PPI network, 5 sub-networks that met the requirements were obtained (Fig. 6). These sub-networks contained 31 nodes (140 PPIs), 26 nodes (104 PPIs), 22 nodes (86 PPIS), 9 nodes (17 PPIs), 6 nodes (8 PPIs), respectively. Finally, we found 5 genes (i.e., CTP3A4, FLNC, CNTN2, MEP18, MAOA) as candidate colon cancer related genes.

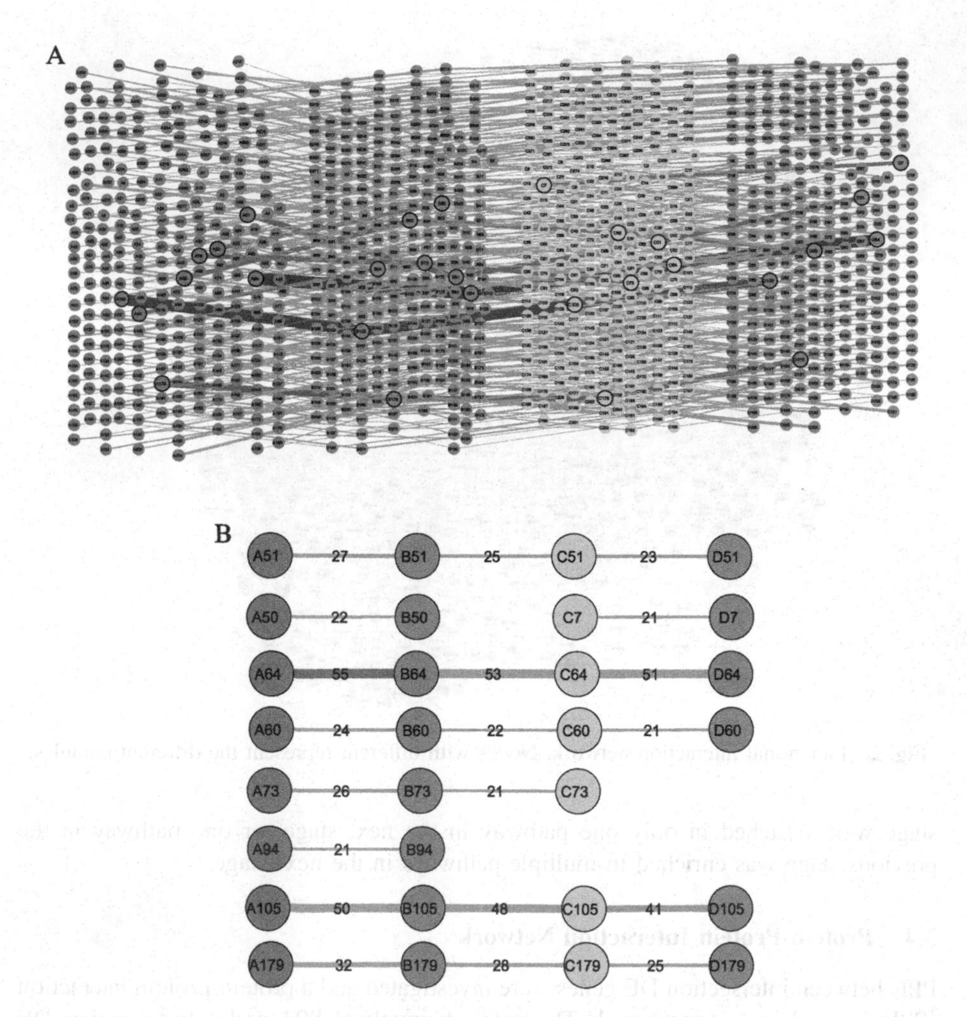

Fig. 3. Pathway interaction network (I). Red nodes or "A": DE set I, blue nodes or "B": DE set II, yellow nodes or "C": DE set III, green nodes or "D": DE set IV. (A) the whole network. Thick edges were select by overlap genes counts (B), 7: Adrenergic signaling in cardiomyocytes(K), 50: Calcium signaling pathway(K), 51: cAMP signaling pathway(K), 60: Chemokine signaling pathway(K), 64: Class A/1 (Rhodopsin-like receptors)(R), 73: Cytokine-cytokine receptor interaction(K), 94: Extracellular matrix organization(R), 105: G alpha (i) signaling events(R), 179: Neuroactive ligand-receptor interaction(K). The edge table is overlap genes count between connected nodes (Color figure online).

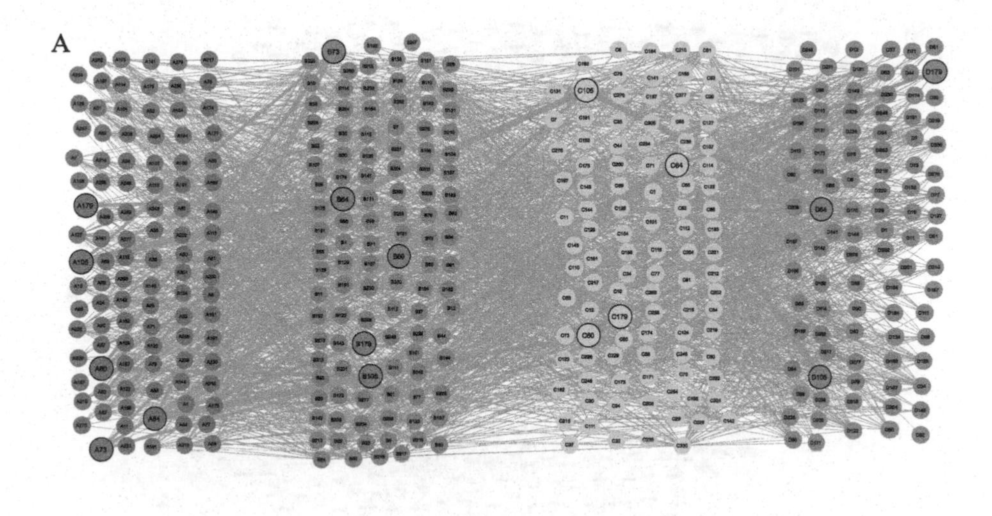

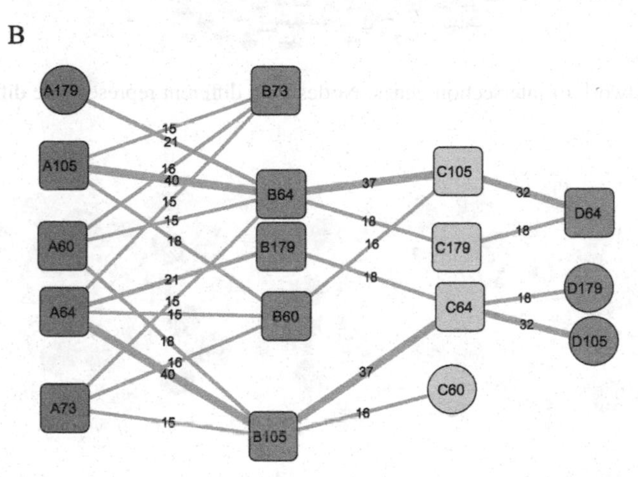

Fig. 4. Pathway interaction network (II). Red nodes or "A": DE set I, blue nodes or "B": DE set II, yellow nodes or "C": DE set III, green nodes or "D": DE set IV. (A) the whole network. Thick edges were select by overlap genes counts. (B) 50: Calcium signaling pathway(K), 51: cAMP signaling pathway(K), 60: Chemokine signaling pathway(K), 64: Class A/1 (Rhodopsin-like receptors)(R), 73: Cytokine-cytokine receptor interaction(K), 105: G alpha (i) signaling events (R), 179: Neuroactive ligand-receptor interaction(K). The edge table is overlap genes count between connected nodes. Square nodes: nodes with degree equal to 1, circle nodes: nodes with degree greater than 1 (Color figure online).

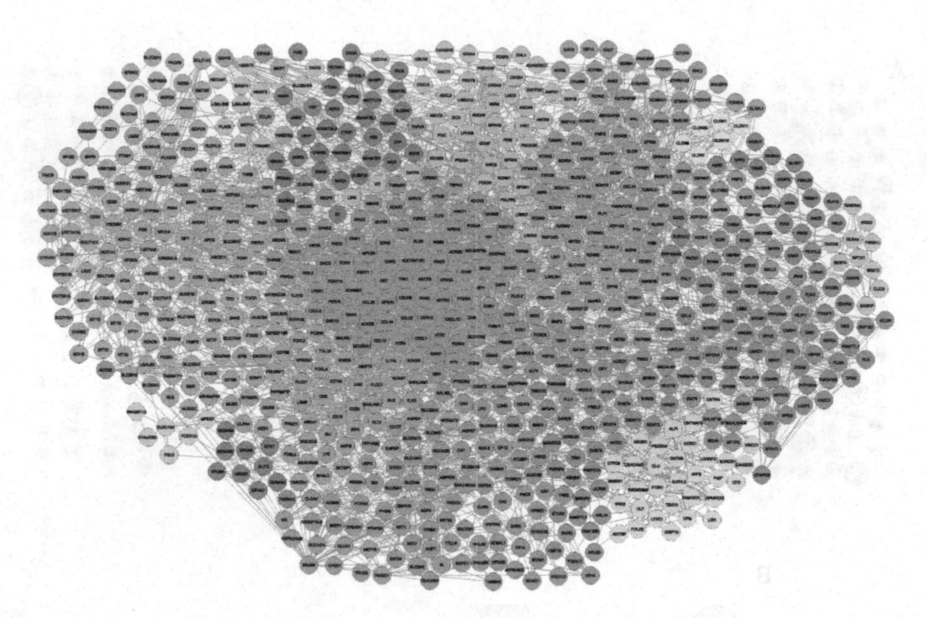

Fig. 5. PPI network of intersection genes. Nodes with different represent the different modules.

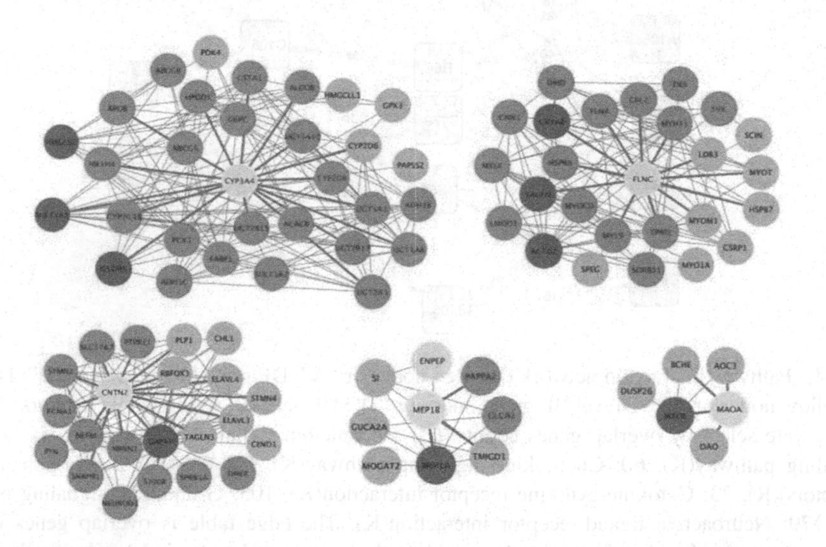

Fig. 6. PPI sub-network. Blue nodes: nodes of PPI network, Green nodes: nodes appeared in PPI network and functional interaction network, red nodes: known colon cancer related genes, yellow nodes: potential colon cancer stage related genes (Color figure online).

4 Conclusion

In this study, we used a series of methods to identify the colon cancer related genes and analyze the biological functions. First, we chosen the TCGA as the source of gene expression data and clinical information. Considered the technical noise of sequencing data, such as sequencing depth, library difference, gene length and other factors, so we do the data preprocessing to reduce the noise as possible. Second, in order to better describe the data distribution, the negative binomial distribution was used, and for this study, negative binomial distribution exact test was used within each DE set, then calculated the $\log^{FC}$, p-value, FDR, etc. Only genes which FDR 0.05 and $|\log^{FC}| \geq 1$ could be chosen as the differentially expressed (DE) genes. Then, we constructed the functional interaction network, and performed the Newman clustering algorithm on it and extracted some modules based on the known colon cancer genes. For these modules, performed the pathway enrichment analysis, and for the sake of analysis better, we constructed 2 networks that described relationship between pathways in each stage. Finally, we constructed the PPI network, and extracted the sub-network to narrow the scope of analysis. Through calculated the betweenness of nodes to choose the potential genes from sub-networks, then got the potential colon cancer stage related genes CTP3A4, FLNC, CNTN2, MEP18 and MAOA.

In order to analyze the staged biological functions directly, we compared the relationship between the same pathway and different pathways in adjacent DE sets. For analyzing the same pathway at adjacent stages, the KEGG pathway that significant different at all stages were cAMP signaling pathway, chemokine signaling pathway and neuroactive ligand-receptor interaction. Among them, cAMP signaling pathway [14] regulates key physiological processes, include metabolism, secretion, calcium homeostasis, muscle contraction, cell fate and gene transcription, and the primary function of chemokine signaling pathway [15] is to recruit leukocyte subsets under both homeostasis and pathological conditions. And there were also have 2 pathways from Reactome database which is Class A/1 (Rhodopsin-like receptors) and G alpha (i) signaling events. They're all about the G protein pathway which is an important GTP binding protein in intracellular signaling pathway and indirectly promotes cAMP production and signaling [16]. For analyzing the different pathway at adjacent stages, the pathway Class A/1 (Rhodopsin-like receptors) and G alpha (i) signaling events were closely connected. Besides these two pathways, chemokine signaling pathway and cytokine-cytokine receptor interaction are all significantly connected by multiple pathways of previous stages.

In conclusion, we identified 5 potential colon cancer stage related genes and obtained some biological information about the colon cancer. The result would provide some useful information about stages of the colon cancer, and the genes might be the staged biomarkers of colon cancer. However, we still need to do some biological experiments to verify the accuracy of the results.

Acknowledgements. This work was supported by the National Natural Science Foundation of China under Grant Nos. 61972320, 61772426, 61702161, 61702420, 61702421, and 61602386, the Fundamental Research Funds for the Central Universities under Grant No. 3102019DX1003, the education and teaching reform research project of Northwestern Polytechnical University under Grant No 2020JGY23, the Key Research and Development and Promotion Program of Henan Province of China under Grant 182102210213, the Key Research Fund for Higher Education of Henan Province of China under Grant 18A520003, and the Top International University Visiting Program for Outstanding Young Scholars of Northwestern Polytechnical University.

References

1. Stephen, M., Jane, E.M.: Racial differences in colorectal cancer mortality: the importance of stage and socioeconomic status. J. Clin. Epidemiol. **54**(4), 359–366 (2001)
2. Shi, C., Yang, X., Bu, X., Hou, N., Chen, P.: Alpha B-crystallin promotes the invasion and metastasis of colorectal cancer via epithelial-mesenchymal transition. Biochem. Biophys. Res. Commun. **489**(4), 369–374 (2017)
3. Chen, S.W., Chou, C.T., Chang, C.C., et al.: HMGCS2 enhances invasion and metastasis via direct interaction with PPARα to activate Src signaling in colorectal cancer and oral cancer. Oncotarget **8**(14), 22460 (2017)
4. Wang, X., Chen, J., Wang, J., et al.: Metalloproteases meprin-α (MEP1A) is a prognostic biomarker and promotes proliferation and invasion of colorectal cancer. BMC Cancer **16**(1), 383 (2016)
5. Yates, A.D., Achuthan, P., Akanni, W., et al.: Ensembl 2020. Nucleic Acids Res. **48**(D1), D682–D688 (2020)
6. Davis, A.P., Grondin, C.J., Johnson, R.J., et al.: The comparative toxicogenomics database: update 2019. Nucleic Acids Res. **47**(D1), D948–D954 (2019)
7. Abbas-Aghababazadeh, F., Li, Q., Fridley, B.L.: Comparison of normalization approaches for gene expression studies completed with high-throughput sequencing. PLoS ONE **13**(10), e0206312 (2018)
8. Robinson, M.D., Oshlack, A.: A scaling normalization method for differential expression analysis of RNA-seq data. Genome Biol. **11**(3), R25 (2010)
9. Robinson, M.D., McCarthy, D.J., Smyth, G.K.: edgeR: a Bioconductor package for differential expression analysis of digital gene expression data. Bioinformatics **26**(1), 139–140 (2010)
10. Shannon, P., Markiel, A., Ozier, O., et al.: Cytoscape: a software environment for integrated models of biomolecular interaction networks. Genome Res. **13**(11), 2498–2504 (2003)
11. Wu, G., Dawson, E., Duong, A., et al.: Reactomefiviz: a cytoscape app for pathway and network-based data analysis. F1000Res **3**, 146 (2014)
12. Szklarczyk, D., et al.: STRING v11: protein–protein association networks with increased coverage, supporting functional discovery in genome-wide experimental datasets. Nucleic Acids Res. **47**(1), 607–613 (2018)
13. Cai, H., Chen, H., Yi, T., et al.: VennPlex–a novel venn diagram program for comparing and visualizing datasets with differentially regulated datapoints. PLoS ONE **8**(1), e53388 (2013)
14. Cho-Chung, Y.S., Nesterova, M., Becker, K.G., et al.: Dissecting the circuitry of protein kinase A and cAMP signaling in cancer genesis: antisense, microarray, gene overexpression, and transcription factor decoy. Ann. N. Y. Acad. Sci. **968**(1), 22–36 (2002)
15. Mellado, M., Rodríguez-Frade, J.M., Mañes, S., et al.: Chemokine signaling and functional responses: the role of receptor dimerization and TK pathway activation. Ann. Rev. Immunol. **19**(1), 397–421 (2001)
16. Vossler, M.R., Yao, H., York, R.D., et al.: cAMP activates MAP kinase and Elk-1 through a B-Raf-and Rap1-dependent pathway. Cell **89**(1), 73–82 (1997)

Modeling, Simulation, and Optimization of Biological Systems

Take It or Leave It:

A Computational Model for Flexibility in Decision-Making in Downregulating Negative Emotions

Nimat Ullah[1(✉)], Sander L. Koole[2], and Jan Treur[1]

[1] Social AI Group, Department of Computer Science, Vrije Universiteit Amsterdam, Amsterdam, The Netherlands
nimatullah09@gmail.com, j.treur@vu.nl

[2] Amsterdam Emotion Regulation Lab, Department of Clinical Psychology, Vrije Universiteit Amsterdam, Amsterdam, The Netherlands
s.l.koole@vu.nl

Abstract. Flexibility in emotion regulation strategies is one of the properties associated to healthy minds. Emotion regulation strategies are context dependent and the adaptivity of those strategies is solely subjected to the context. Flexibility, therefore, plays a key role in the use of these emotion regulation strategies. The computational model presented in this paper, models flexibility in emotion regulation strategies that are dependent on context. Simulation results of the model are presented which provides insight of four emotion regulation strategies and highlights the role of context which activates them respectively.

Keywords: Emotion regulation · Strategies · Context · Flexibility · Adaptivity · Psychopathology

1 Introduction

Imagine that you are an office worker who gets upset when your colleague criticizes you unfairly. You can choose to walk away, distract yourself with chores, hide your negative reaction, or mentally distance yourself from your colleague. What would you do? Which of these strategies is optimal, depends on the situation. For instance, if your colleague is from a different department, you may find it easy to walk away. However, if the colleague is your boss, walking away may not be of the question, so you may be forced to distract yourself with chores. This simple example illustrates how different emotion-regulation strategies tend to yield different outcomes in different situations [1]. It thus follows that people should be able to choose flexibly between different emotion-regulation strategies in the face of different situational demands. The latter capacity is known as emotion regulation flexibility [2, 3].

Within the area of emotion regulation which is our long-term research focus, we recently proposed a computational model of emotion regulation flexibility which models a person who can switch between two strategies- expressive suppression and attention modulation - in managing anger in different work situations [4]. Building on and extending this work, the present model is inspired by the vast body of literature on

D.-S. Huang and K.-H. Jo (Eds.): ICIC 2020, LNCS 12464, pp. 175–187, 2020.
https://doi.org/10.1007/978-3-030-60802-6_16

flexibility in emotion regulation strategies and, therefore, addresses the challenge to design a computational model of emotion regulation flexibility in which a person can switch between four different emotion-regulation strategies: situation modification, attention modulation, expressive suppression, and cognitive reappraisal.

In what follows, in Sect. 2, we discuss the theoretical background of our work, which is grounded in the psychological literature on emotion regulation [5–7]. Next, in Sect. 3, we present the computational model. In Sect. 4, we present the results of simulations of the model using four different scenarios. Finally, in Sect. 5, we review the main conclusions and implications of the present work.

2 Background

Emotion regulation has been defined as the set of processes whereby people control and redirect the spontaneous flow of their emotions [8]. A large body of research has implicated difficulties in emotion regulation as a transdiagnostic factor that is central to the development and maintenance of psychopathology [9, 10]. Accordingly, clinical psychologists are increasingly moving toward unified treatments that target emotion regulation for individuals with multiple disorders [11]. In addition, emotion regulation is seen as a vital positive contributor to psychological health and wellbeing [12].

The influential process model of emotion regulation [13–15] has distinguished five families of emotion regulation strategies. The first family of strategies is situation selection, and consists of taking steps to influence which situation one will be exposed to. The second family of strategies is situation modification, and consists of changing one or more relevant aspects of the situation. The third family of strategies is attentional deployment, and consists of influencing which portions of the situation are attended to. The fourth family of strategies is cognitive change, and consists of altering the way the situation is cognitively represented. Finally, the fifth family of strategies is response modulation, and consists of directly modifying emotion-related actions.

Originally, the process model [14, 16] proposed that some emotion-regulation strategies are inherently more effective than other emotion-regulation strategies. The main evidence for this notion came from studies showing that cognitive reappraisal–a prototypical cognitive change strategy- is less effortful and more effective than emotional suppression–a prototypical response modulation strategy [16]. Subsequent research, however, has shown that the effectiveness of emotion-regulation strategies is highly dependent on situational context. For instance, there are situations in which cognitive reappraisal is ineffective, or may even backfire [17–19]. Conversely, expressive suppression may be less problematic when this strategy can be flexibly applied [20].

Evidence for context-dependent effects of emotion-regulation strategies has inspired a new generation of theories that emphasize emotion regulation flexibility [1, 3, 7, 15, 21]. Although these theories differ in their particulars, they converge on the notion that, in healthy emotion regulation, strategies have to be adjusted to the demands of the situation. Consequently, emotion regulation flexibility is key to successful emotion regulation.

Empirical research on emotion regulation flexibility has so far been limited. The available research to date has focused on switching between two strategies. This two-strategy approach is presumably derived from the limitations of experimentally examining alternations between a greater number of emotion regulation strategies. For instance, Sheppes and colleagues [22] have studied the choice between distraction (an attentional deployment strategy) and reappraisal (a cognitive change strategy). Likewise, Bonanno and colleagues have operationalized emotion-regulatory flexibility as the ability to both up-regulate and down-regulate negative emotion [23].

In line with the two-strategy approach, we recently proposed a computational model of emotion regulation flexibility in which the person switches between expressive suppression and attention modulation in managing anger in different work situations [4]. Simulation results illustrated the capacity of the model to display adaptivity in emotion regulation across different contexts.

An important strength of our computational approach is that it can afford insight into the interplay between a large number of variables simultaneously, larger than it is practical to study experimentally. In the present work, we therefore sought to extend our earlier computational model to a model in which we could examine flexible, context-dependent switching between four emotion-regulation strategies.

3 The Computational Model

The computational model, presented in this paper, is hereby thoroughly elaborated. This model simulates four emotion regulation strategies as per criteria presented in Table 3. The basic concepts of the modeling approach used for this model, called Network-Oriented modeling approach, can be consulted in Table 1 from [24]. In Network-Oriented modeling approach, a phenomenon is represented in a network form consisting of nodes that varies over time. The nodes are interpreted as states and the connections between these states are interpreted as relations and it defines the impact of one state on another state over time. This type of network is, therefore, referred to as temporal- network. Conceptually, a model in temporal–network can be represented as labelled graph in which:

- Each connection carries some *connection weight* from one state to another called *impact* represent by $\boldsymbol{\omega}_{X,Y}$.
- There's some way to *aggregate multiple impacts* on a state (combination function $\mathbf{c}_Y(..)$).
- There's a notion of *speed of change* of each state to define how faster a state changes because of the incoming impact (speed factor $\boldsymbol{\eta}_Y$).

A temporal–network is defined by these three notions, see Table 1 for more explanation of the terms and for the numerical representation of the concepts.

Table 1. Basics of Network oriented modeling approach.

Concept	Conceptual Representation	Explanation
States and connections	$X, Y, X \rightarrow Y$	Describes the nodes and links of a network structure (e.g., in graphical or matrix form)
Connection weight	$\boldsymbol{\omega}_{X,Y}$	The *connection weight* $\boldsymbol{\omega}_{X,Y}$ usually in [−1, 1] represents the strength of the causal impact of state X on state Y through connection $X \rightarrow Y$
Aggregating multiple impacts on a state	$\mathbf{c}_Y(..)$	For each state Y a *combination function* $\mathbf{c}_Y(..)$ is chosen to combine the causal impacts of other states on state Y
Timing of the effect of causal impact	$\boldsymbol{\eta}_Y$	For each state Y a *speed factor* $\boldsymbol{\eta}_Y \geq 0$ is used to represent how fast a state is changing upon causal impact
Concept	**Numerical representation**	**Explanation**
State values over time t	$Y(t)$	At each time point t each state Y in the model has a real number value, usually in [0, 1]
Single causal impact	$\mathbf{impact}_{X,Y}(t) = \boldsymbol{\omega}_{X,Y} X(t)$	At t state X with a connection to state Y has impact on Y, using connection weight $\omega_{X,Y}$
Aggregating multiple causal impacts	$\mathbf{aggimpact}_Y(t) = \mathbf{c}_Y(\mathbf{impact}_{X1,Y}(t), \ldots, \mathbf{impact}_{Xk,Y}(t)) = \mathbf{c}_Y(\boldsymbol{\omega}_{X1,Y}X_1(t), \ldots, \boldsymbol{\omega}_{Xk,Y}X_k(t))$	The aggregated causal impact of multiple states X_i on Y at t, is determind using combination function $c_Y(..)$
Timing of the causal effect	$Y(t + \Delta t) = Y(t) + \boldsymbol{\eta}_Y [\mathbf{aggimpact}_Y(t) - Y(t)] \Delta t = Y(t) + \boldsymbol{\eta}_Y [\mathbf{c}_Y(\boldsymbol{\omega}_{X1,Y}X_1(t), \ldots, \boldsymbol{\omega}_{Xk,Y}X_k(t)) - Y(t)] \Delta t$	The causal impact on Y is exerted over time gradually, using speed factor η_Y; here the X_i are all states with outgoing connections to state Y

For aggregation of multiple incoming impacts, Network-Oriented modeling approach provides a library of over 35 combination functions. Besides that, own-defined functions can also be added for better flexibility.

Conceptual representation of the model presented in this paper is given in Fig. 1, and the nomenclature of the states is provided in Table 2.

The model presented in Fig. 1, presents various courses of action of one person in different contexts. It switches among four different emotion regulation strategies depending on the context in which emotion has been felt and on the level of emotions felt.

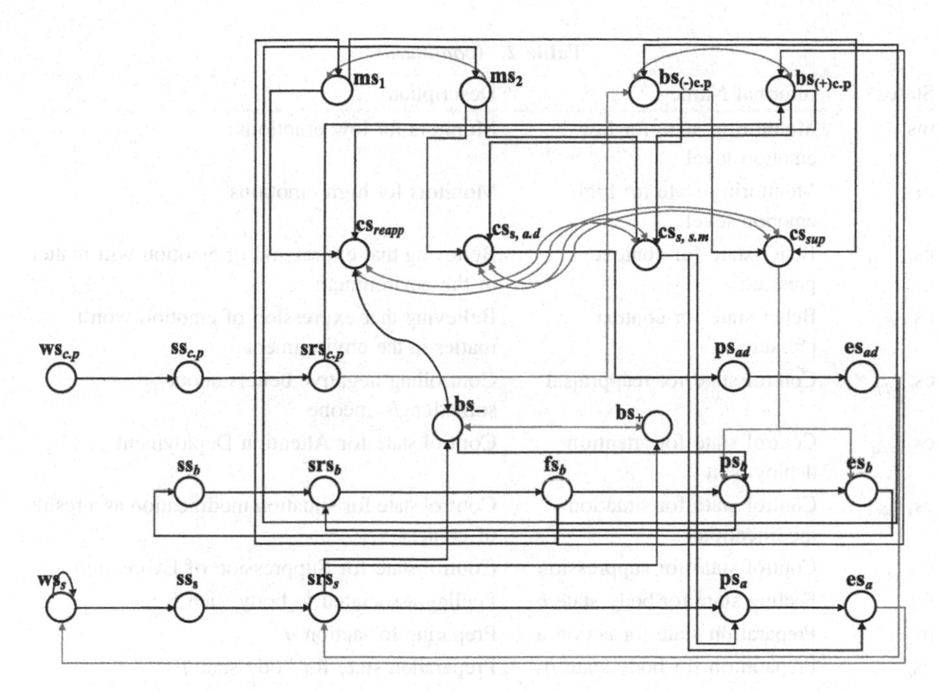

Fig. 1. Conceptual representation of the computational model as a temporal-causal network

Table 2. Nomenclature of the states of the proposed model (in connection to Fig. 1)

States	Informal Name	Description
ws_s	World state for stimulus *s*	The situation in the real world that triggers emotion
$ws_{c.p}$	World state for context pressure	A real-world situation which decides expression of emotion
ss_s	Sensor state for stimulus *s*	Sensor state for the stimulus *s* in the real world
$ss_{c.p}$	Sensor state for context pressure	Senses state for context pressure
ss_b	Sensor state for body	Sensor state for body
srs_s	Sensory representation state for stimulus *s*	Internal representation of the emotion triggering situation
$srs_{c.p}$	Sensory rep: state for context pressure	Internal representation of the context pressure in the real world
srs_b	Sensory representation state for body	Internal body representation state
bs_-	Negative believe state	The negative believe that the person has about something/someone
bs_+	Positive believe state	The positive believe that the person has about something/someone

(*continued*)

Table 2. (*continued*)

States	Informal Name	Description
ms_1	Monitoring state for low emotion level	Monitors for low emotions
ms_2	Monitoring state for high emotion level	Monitors for high emotions
$bs_{(+)c.p}$	Belief state for context pressure	Believing that expression of emotion will matter in the environment
$bs_{(-)c.p}$	Belief state for context pressure	Believing that expression of emotion won't matter in the environment
cs_{reapp}	Control state for reappraisal	Controlling negative beliefs about something/someone
$cs_{s,\ a.d}$	Control state for attention deployment	Control state for Attention Deployment
$cs_{s,s.m}$	Control state for situation modification	Control state for situation modification as a result of context
cs_{sup}	Control state for suppression	Control state for Suppression of Expression
fs_b	Feeling state for body state *b*	Feeling associated to body state *b*
ps_a	Preparation state for action a	Preparing for action *a*
ps_b	Preparation for body state *b*	Preparation state for body state *b*
ps_{ad}	Preparation state for attention deployment	Preparation for the Attention deployment action
es_a	Execution state for action *a*	Execution station for action *a*
es_b	Execution state for body state *b*	Execution state for body state *b*
es_{ad}	Execution state for attention deployment	Execution state for the Attentional Deployment action

Table 3. Choice of strategies under high/low intensity of emotions and +/− belief about context pressure

Flexibility Parameters		Repertoire of Strategies			
Emotion Strength	Context Pressure (CP)	Situation Modification	Attention Deployment	Reappraisal	Expressive Suppression
+	+	✓			
+	−		✓		
−	+			✓	
−	−				✓

The emotions can either be of high or low intensity and about the Context Pressure (CP), belief can either be positive or negative. Belief about CP refers to the person's belief about presence or absence of any environmental factor in which expression of emotion matters or doesn't matter. In other words, the belief refers to one's prediction of such factor. Selection of each strategy is subjected to two conditions i.e. the intensity of emotion and CP. The '+' symbol represents high intensity under emotion strength and positive belief/prediction about presence of an environment/factor where expression of emotion will matter. For instance, if a person feels high intensity of negative emotions and he's expecting a factor i.e. boss etc. due to which he believes his expression of emotions can have bad consequences for him, he would prefer situation modification. There are four combinations of the primary and secondary stimulus as described in Table 1, which leads into four different emotion regulation strategies interpreted below:

1. High intensity of emotion (+) and (+) belief about CP leads to situation modification "cs_{sm}".
2. High intensity of emotion (+) and (−) belief about CP activates attention deployment "$cs_{a.d}$".
3. Low intensity of emotion (−) and (+) belief about CP triggers reappraisal "cs_{reapp}".
4. Low intensity of emotion (−) and (−) belief about CP initiates expressive suppression "cs_{sup}".

4 Scenarios and Simulation Results

The computational model presented in this paper is loosely based on an example discussed by [2] with flexibility in emotion regulation strategies as modeled in [4] and decision making among various emotion regulation strategies as [25]. It's worth mentioning here that [25] is the only model, so far, considering decision making for three ER strategies up till now, which the current model extends to a repertoire of four strategies.

> *"An employee A feels angry every time a particular obnoxious coworker B starts talking. Next week the organization has monthly review meeting where presence of all the employees is mandatory unless emergency. Employee A doesn't want anyone, especially his boss to come to know about his attitude towards employee B. Employee A has four options to handle the situation, all depending upon the combination of his intensity of emotions and the chances of presences or absence of their boss in the meeting as shown in Table 3."*

The parameter values given in Tables 4 and 5, in the absence of availability of quantitative data, qualitatively validates the proposed model against the findings from social sciences and psychology that serve as qualitative evaluation indicators. These parameter values make the model reproducible; they give the simulation results as shown in Figs. 2, 3, 4 and 5. In Table 4 each state can either have value of scaling factor (λ) for which scale sum function has been used or it can have values for steepness (σ) and threshold (τ) for which alogistic combination function has been used.

Table 4. Values used for alogistic, scaled-sum combination functions and speed factor

State	λ	τ	σ	η	State	τ	σ	η
ws_s	0.94	0	0	0.1	ms_2	0.5	50	0.5
ss_s	0	0	0	0.5	$bs_{(-)c.p}$	0.1	50	0.5
ss_b	0	0	0	0.5	$bs_{(+)c.p}$	0.5	17	0.5
srs_s	1	0	0	0.5	cs_{reapp}	0.5	8	0.15
srs_b	1.4	0	0	0.5	$cs_{a.d}$	0.85	12	0.2
bs_-	0.91	0	0	0.5	$cs_{s.m}$	0.85	12	0.3
bs_+	0	0.1	10	0.5	cs_{sup}	0.5	6	0.15
ps_b	1.8	0	0	0.5	ps_a	0.6	5	0.5
es_b	0.98	0	0	0.5	$ps_{a.d}$	0	0	0.3
fs_b	1	0	0	0.5	es_a	0.5	3	0.5
ms_1	0	0.1	5	0.5	$es_{a.d}$	0	0	0.3

Table 5. Values used for connection weights

Connection	Weight	Connection	Weight	Connection	Weight	Connection	Weight
$\omega_{wss, wss}$	0.95	$\omega_{bs+, bs-}$	−0.4	$\omega_{csreapp, css.m}$	−1	$\omega_{fsb, ms1}$	0.5
$\omega_{wss, sss}$	1	$\omega_{ms1, csreapp}$	0.2	$\omega_{csreapp, cssup}$	−1	$\omega_{fsb, ms2}$	0.8
$\omega_{sss, srss}$	1	$\omega_{ms1, cssup}$	0.4	$\omega_{csa.d, psa.d}$	1	$\omega_{fsb, bs(-)c.p}$	0.5
$\omega_{ssb, srsb}$	0.7	$\omega_{ms2, ms1}$	−1	$\omega_{csa.d, css.m}$	−1	$\omega_{fsb, bs(+)c.p}$	0.5
$\omega_{srss, bsc-}$	0.9	$\omega_{ms2, csa.d}$	0.35	$\omega_{csa.d, cssup}$	−1	$\omega_{fsb, psb}$	0.9
$\omega_{srss, bsc+}$	0.4	$\omega_{ms2, css.m}$	0.5	$\omega_{css.m, psa}$	0.8	$\omega_{psa, esa}$	0.5
$\omega_{srss, psa}$	0.3	$\omega_{bs(-)c.p, bs(+)c.p}$	−1	$\omega_{css.m, esa}$	0.8	$\omega_{psb, srsb}$	0.75
$\omega_{srsc.p, bs(-)c.p}$	−1	$\omega_{bs(-)c.p, cssup}$	0.3	$\omega_{css.m, csreapp}$	−1	$\omega_{psb, esb}$	1
$\omega_{srsc.p, bs(+)c.p}$	1	$\omega_{bs(-)c.p, csa.d}$	0.6	$\omega_{css.m, csa.d}$	−1	$\omega_{psa.d, esa.d}$	1
$\omega_{srsb, fsb}$	1	$\omega_{bs(+)c.p, bs(-)c.p}$	−1	$\omega_{cssup, psb}$	−1	$\omega_{esa, wss}$	−0.5
$\omega_{bs-, bs+}$	−0.4	$\omega_{bs(+)c.p, css.m}$	0.5	$\omega_{cssup, esb}$	−0.2	$\omega_{esb, ssb}$	1
$\omega_{bs-, csreapp}$	0.05	$\omega_{bs(+)c.p, csreapp}$	0.33	$\omega_{cssup, csreapp}$	−1	$\omega_{esa.d, srss}$	0.63
$\omega_{bs-, psb}$	1	$\omega_{csreapp, bs-}$	−0.35	$\omega_{cssup, csa.d}$	−1		

All the simulation results, given below, only have the most essential states for the explanation of the results.

Figure 2 depicts a context with low intensity of emotions and positive belief about CP. This means that the person knows that his emotional expression will have consequences for him and the intensity of negative emotions that he/she is feeling is of moderate/low level. So, as the person already anticipates presence of the CP, therefore, he's reappraising his belief about the stimuli. In the figure, it can be seen that initially negative belief bs_- is quite high but it decreases as control state for reappraisal cs_{reapp} gets activated. As a result, the positive belief bs_+ increases and the feeling state fs_b also gets lower as the negative belief gets weaker and the positive belief gets stronger.

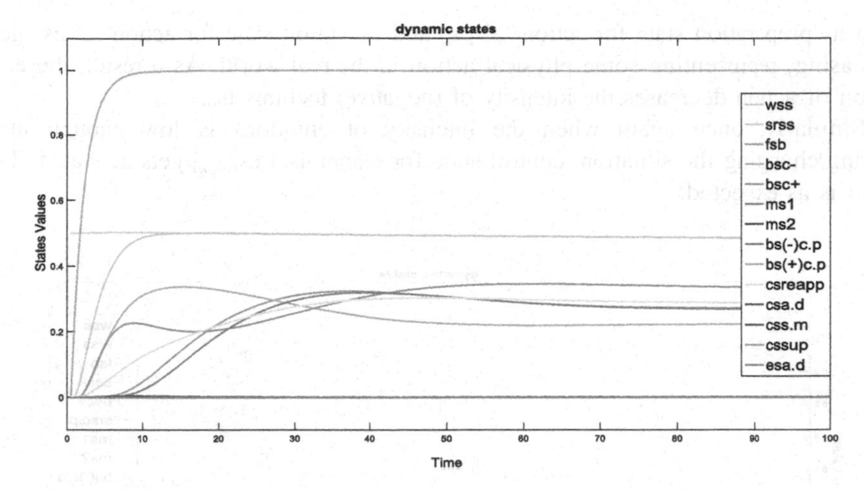

Fig. 2. Cognitive Reappraisal: low intensity of emotions and positive belief/prediction about the context pressure

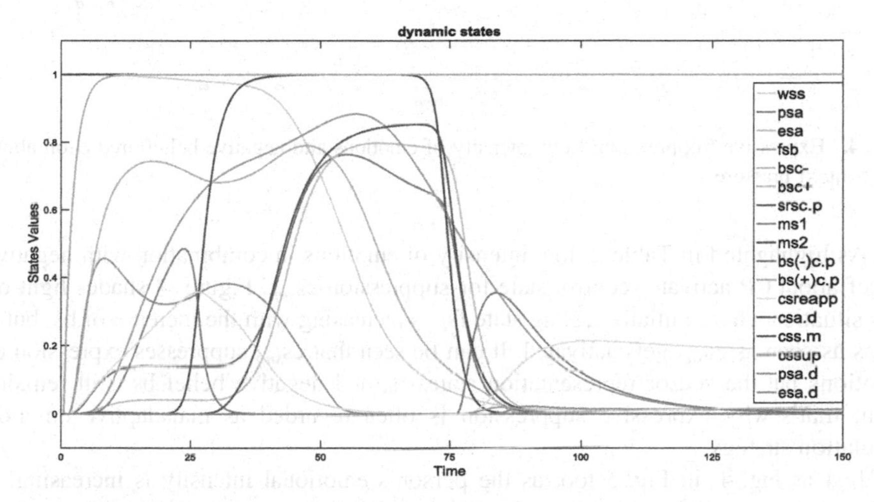

Fig. 3. Situation Modification: as a result of high intensity of emotions and positive belief/prediction about the context pressure

Figure 3 depicts the activation of situation modification as a strategy, which gets activated when the person is feeling high intensity of emotions while having positive belief about the CP (i.e. he is predicting an environment where he can't afford if his emotions are observed). In the figure it can be seen that initially control state for reappraisal cs_{reapp} gets activated. It's because the person tries to reappraise initially when the intensity of his emotions is not yet high. Later on, as the intensity of negative emotions increases, control state for situation modification $cs_{s.m}$ gets activated. As situation modification means that the person is leaving/changing the situation, therefore, the world state ws_s where the emotional event is/will take place, gets decreased as soon as preparation state for action 'a' ps_a and execution state for action 'a' es_a gets increasing, representing some physical action in the real world. As a result, the execution of action decreases the intensity of (negative) feelings fs_b.

Similarly, once again when the intensity of emotions is low enough after leaving/changing the situation, control state for reappraisal cs_{reapp} gets activated. The result is as expected.

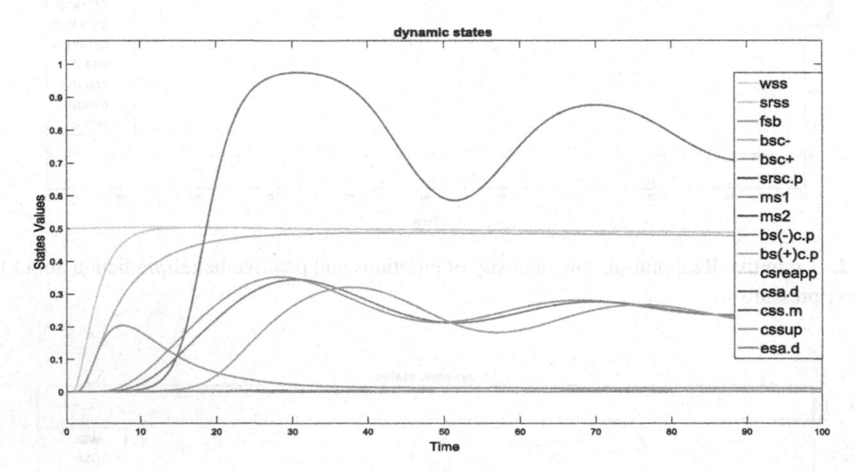

Fig. 4. Expressive Suppression: Low intensity of emotions and negative belief/prediction about the context pressure

As highlighted in Table 2, low intensity of emotions in combination with negative belief about CP activates control state for suppression cs_{sup}. Figure 4 shades light on this situation where initially feeling state fs_b is increasing with the increase of bs_- but it stops as soon as cs_{sup} gets activated. It can be seen that cs_{sup} suppresses expression of emotions but the sensor representation state srs_s and negative belief bs_- still remains high, that's why expressive suppression is often regarded as maladaptive emotion regulation strategy.

Just as Fig. 4, in Fig. 5 too, as the person's emotional intensity is increasing, it activates two strategies. Initially, when the intensity of negative emotions is yet low and as belief about the CP is already negative (i.e. he is predicting an environment where he can afford if his expression of emotions is observed), control state for suppression cs_{sup} gets activated. Its affect can also be seen in fs_b.

Later on, as the intensity of emotions gets higher and as the belief about CP is already negative, control state for attention deployment $cs_{a.d}$ gets activated. Activation of $cs_{a.d}$ decreases intensity of the stimuli and therefore, the bs_- also decreases, resulting in decrease of fs_b.

The results obtained from the model in Fig. 1 are in line with the literature from psychology and social sciences and best describe the working of emotion regulation strategies as described in the aforementioned literature.

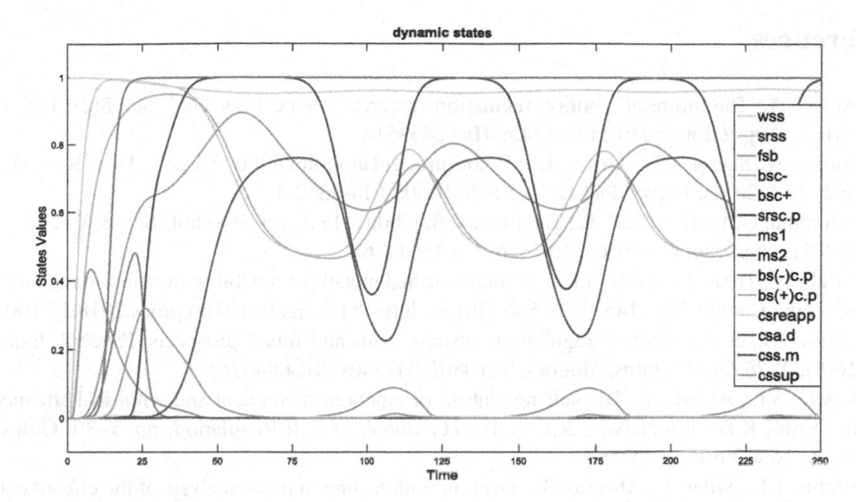

Fig. 5. Attention Deployment: high intensity of emotions and negative belief/prediction about the context pressure

5 Conclusion

This network-oriented temporal-causal network model models four different emotion regulation strategies with four different contexts. Each strategy depends on a specific context which activates it. The model not only acknowledges the ongoing debate about impact of context on various emotion regulation strategies, it rather efficiently and with computational clarity highlights the flexibility of the emotion regulation strategies dependent on a specific context.

Context plays a very profound impact in selection of a strategy. A strategy may not be as efficient in one context as it could be in another context. Therefore, a strategy can't be termed as maladaptive just because it's not adaptive in one context. Similarly, flexibility is referred to as a practice of healthy minds [3, 26] which this model has highlighted by being able to switch between different strategies as per demand of the context. Moreover, this model also gives hint to the possibility of modeling simultaneous activation of multiple strategies as found by as described in [27] For instance, in case of situation modification or attention deployment, it's possible that reappraisal or suppression is also activated at the same time, respectively. This phenomenon can be considered for further study.

This model, apart from giving insight into the phenomenon of flexibility of emotion regulation strategies, also acknowledges the strength of network oriented temporal-causal modeling [24, 28] of being able to effectively model such problems and give a clear insight into its working mechanisms.

To carry on with flexibility, in future, some other emotion regulation strategies with explicit decision-making ability maybe modeled to give insight into their working mechanism. Moreover, simultaneous activation of multiple emotion regulation strategies, from a broader repertoire, can also be considered as future project.

References

1. Aldao, A.: The future of emotion regulation research. Perspect. Psychol. Sci. **8**(2), 155–172 (2013). https://doi.org/10.1177/1745691612459518
2. Aldao, A., Sheppes, G., Gross, J.J.: Emotion regulation flexibility. Cognit. Ther. Res. **39**(3), 263–278 (2014). https://doi.org/10.1007/s10608-014-9662-4
3. Bonanno, G.A., Burton, C.L.: Regulatory flexibility. Perspect. Psychol. Sci. **8**(6), 591–612 (2013). https://doi.org/10.1177/1745691613504116
4. Ullah, N., Treur, J., Koole, S.L.: A computational model for flexibility in emotion regulation. Procedia Compu. Sci. **145**, 572–580 (2018). https://doi.org/10.1016/j.procs.2018.11.100
5. Gross, J.J., et al.: Emotion regulation: current status and future prospects. Psychol. Inquiry **26**(1), 1–26 (2015). https://doi.org/10.1080/1047840x.2014.940781
6. Koole, S.L., Aldao, A.: The self-regulation of emotion: theoretical and empirical advances. In: Vohs, K.D., Baumeister, R.F. (eds.) Handbook of Self-Regulation, pp. 1–36. Guilford Press, New York (2015)
7. Webb, T.L., Miles, E., Sheeran, P.: Dealing with feeling: a meta-analysis of the effectiveness of strategies derived from the process model of emotion regulation. Psychol. Bull. **138**(4), 775–808 (2012). https://doi.org/10.1037/a0027600
8. Koole, S.L.: The psychology of emotion regulation: An integrative review. Cognit. Emotion **23**(1), 4–41 (2009). https://doi.org/10.1080/02699930802619031
9. Aldao, A., Nolen-Hoeksema, S.: When are adaptive strategies most predictive of psychopathology? J. Abnorm. Psychol. **121**(1), 276–281 (2012). https://doi.org/10.1037/a0023598
10. Sloan, E., Hall, K., Moulding, R., Bryce, S., Mildred, H., Staiger, P.K.: Emotion regulation as a transdiagnostic treatment construct across anxiety, depression, substance, eating and borderline personality disorders: a systematic review. Clinical Psychol. Rev. **57**, 141–163 (2017). https://doi.org/10.1016/j.cpr.2017.09.002
11. Moses, E.B., Barlow, D.H.: A new unified treatment approach for emotional disorders based on emotion science. Curr. Dir. Psychol. Sci. **15**(3), 146–150 (2006). https://doi.org/10.1111/j.0963-7214.2006.00425.x
12. Buruck, G., Dorfel, D., Kugler, J., Brom, S.S.: Enhancing well-being at work: The role of emotion regulation skills as personal resources. J. Occup. Health Psychol. **21**(4), 480–493 (2016). https://doi.org/10.1037/ocp0000023
13. Gross, J.J.: The emerging field of emotion regulation: an integrative review. Rev. Gen. Psychol. **2**(3), 271–299 (1988). https://doi.org/10.1037/1089-2680.2.3.271
14. Richards, J.M., Gross, J.J.: Emotion regulation and memory: The cognitive costs of keeping one's cool. J. Pers. Soc. Psychol. **79**(3), 410–424 (2000). https://doi.org/10.1037/0022-3514.79.3.410
15. Gross, J.J.: The extended process model of emotion regulation: elaborations, applications, and future directions. Psychol. Inquiry **26**(1), 130–137 (2015). https://doi.org/10.1080/1047840x.2015.989751
16. Gross, J.J.: Antecedent- and response-focused emotion regulation: Divergent consequences for experience, expression, and physiology. J. Pers. Soc. Psychol. **74**(1), 224–237 (1998). https://doi.org/10.1037/0022-3514.74.1.224
17. Ford, B.Q., Karnilowicz, H.R., Mauss, I.B.: Understanding reappraisal as a multicomponent process: the psychological health benefits of attempting to use reappraisal depend on reappraisal success. Emotion **17**(6), 905–911 (2017). https://doi.org/10.1037/emo0000310

18. Troy, A.S., Shallcross, A.J., Mauss, I.B.: A person-by-situation approach to emotion regulation: cognitive reappraisal can either help or hurt, depending on the context. Psychol. Sci. **24**(12), 2505–2514 (2013). https://doi.org/10.1177/0956797613496434
19. Veenstra, L., Schneider, I.K., Koole, S.L.: Embodied mood regulation: the impact of body posture on mood recovery, negative thoughts, and mood-congruent recall. Cognit. Emotion **31**(7), 1361–1376 (2017). https://doi.org/10.1080/02699931.2016.1225003
20. Dworkin, J.D., Zimmerman, V., Waldinger, R.J., Schulz, M.S.: Capturing naturally occurring emotional suppression as it unfolds in couple interactions. Emotion **19**(7), 1224–1235 (2019). https://doi.org/10.1037/emo0000524
21. Sheppes, G.: Emotion regulation choice: theory and findings. In: Handbook of Emotion Regulation, 2nd edn., pp. 126–139. Guilford Press, New York (2014)
22. Sheppes, G., Scheibe, S., Suri, G., Gross, J.J.: Emotion-regulation choice. Psychol. Sci. **22**(11), 1391–1396 (2011). https://doi.org/10.1177/0956797611418350
23. Bonanno, G.A., Wortman, C.B., Nesse, R.M.: Prospective patterns of resilience and maladjustment during widowhood. Psychol. Aging **19**(2), 260–271 (2004). https://doi.org/10.1037/0882-7974.19.2.260
24. Treur, J.: Network-oriented modeling and its conceptual foundations. Network-Oriented Modeling. UCS, pp. 3–33. Springer, Cham (2016). https://doi.org/10.1007/978-3-319-45213-5_1
25. Manzoor, A., Abro, A.H., Treur, J.: Monitoring the impact of negative events and deciding about emotion regulation strategies. In: Criado Pacheco, N., Carrascosa, C., Osman, N., Julián Inglada, V. (eds.) EUMAS/AT -2016. LNCS (LNAI), vol. 10207, pp. 350–363. Springer, Cham (2017). https://doi.org/10.1007/978-3-319-59294-7_30
26. Kashdan, T.B., Rottenberg, J.: Psychological flexibility as a fundamental aspect of health. Clinical Psychol. Rev. **30**(7), 865–878 (2010). https://doi.org/10.1016/j.cpr.2010.03.001
27. Dixon-Gordon, K.L., Aldao, A., De Los Reyes, A.: Emotion regulation in context: examining the spontaneous use of strategies across emotional intensity and type of emotion. Pers. Individ. Differ. **86**, 271–276 (2015). https://doi.org/10.1016/j.paid.2015.06.011
28. Treur, J.: Network-Oriented Modeling for Adaptive Networks: Designing Higher-Order Adaptive Biological, Mental and Social Network Models. SSDC, vol. 251. Springer, Cham (2020). https://doi.org/10.1007/978-3-030-31445-3

A Network-Driven Approach for LncRNA-Disease Association Mapping

Lin Yuan(✉), Tao Sun, Jing Zhao, Song Liu, Ai-Min Li, Qin Lu, Yu-Shui Geng, and Xin-Gang Wang(✉)

School of Computer Science and Technology, Qilu University of Technology (Shandong Academy of Sciences), Jinan 250353, Shandong, China
yuanlindc@126.com

Abstract. Aberrant lncRNA may contributes to development of cancer. It remains, however, a challenge to understand associations between lncRNA and cancer because of complex mechanisms involved in associations and insufficient sample sizes. With unprecedented wealth of lncRNA data, gene expression data and disease status data give us a new opportunity to design machine learning method to investigate underlying association mechanisms. In this paper, we propose a network-driven approach named NLDAM which for lncRNA-disease association mapping. NLDAM detects associations between lncRNA and genes, genes and disease. NLDAM constructs an association network, where nodes represent lncRNA, genes or disease status, and weighted edges represent significance of associations between nodes. NLDAM identifies significant paths from lncRNA to disease based on the weighted scores. The experimental results on synthetic datasets show the advantage of NLDAM in terms of lncRNA selection accuracy than traditional methylation sites search methods (including NTSDMHN and IDHI-MIRW) under false positive control. Furthermore, we applied NLDAM on ovarian cancer data from The Cancer Genome Atlas database and identified significant lncRNA-gene-disease path associations, among which we analyzed top 10 paths associated with oncogenes. We also provide hypothetical biological path associations to explain our findings.

Keywords: LncRNA-disease association · Machine learning · Similarity measure · Biological network science

1 Introduction

Non-coding RNAs do not encode proteins and are usually biologically identified as transcriptional noise and have no biological function [1]. However, more and more studies have shown that non-coding RNAs play a key role in many biological processes [2, 3]. Among them, long noncoding RNAs (lncRNAs) contain the largest subclass of noncoding RNAs, which is a class of noncoding RNAs with a length of more than 200 nucleotides [4, 5]. Many studies have confirmed that the human genome contains many thousands of lncRNAs [6]. A large quantity of lncRNAs play significant roles in many important biological processes including chromatin modification, transcriptional and posttranscriptional regulation, genomic splicing, differentiation, immune responses and

D.-S. Huang and K.-H. Jo (Eds.): ICIC 2020, LNCS 12464, pp. 188–197, 2020.
https://doi.org/10.1007/978-3-030-60802-6_17

so on [7, 8]. In addition, mutations and disorders of lncRNAs are closely related to the occurrence of breast cancer, ovarian cancer and other diseases [9]. LncRNAs have been novel candidate biomarkers for cancers. Therefore, lncRNAs are considered to be important factors for disease diagnosis and treatment [10].

Predicting the relationship between lncRNAs and disease is not only beneficial for disease diagnosis, but also helps to understand biological processes [11]. Unfortunately, identifying those unknown associations is still a major challenge. Traditional biological experiments and clinical methods will consume a lot of time and effort. Therefore, researchers need to develop effective calculation methods to solve this problem.

A large number of methods to predict the association between lncRNAs and diseases have been proposed [12]. IDHI-MIRW infers disease-related lncRNA candidates by integrating diverse heterogeneous information sources with positive pointwise mutual information and random walk with restart algorithm [13]. NTSDMHN infers potential lncRNA-disease associations based on geometric matrix completion [14]. IDLDA is an improved diffusion potential lncRNA–disease associations prediction model [15]. However, due to the particularity of the prediction problems associated with lncRNAs and diseases, the accuracy of current prediction methods needs to be further improved [16]. Experimental scientists urgently need lncRNAs-disease association prediction methods that can give more reference value. Against this background of needs, research and development of high-precision prediction methods are particularly important [17].

In this paper, to predict complex relationship between lncRNA and disease, we present a novel prediction method based on random walk with restart algorithm, named NLDAM. We integrate diverse heterogeneous information sources to construct a Network which is a heterogeneous network containing three types nodes which generated from bioinformatics linked datasets and use random walk with restart algorithm method to extract topological structure features of the nodes in the linked tripartite network for calculating similarities. We exploit different data sources including LncRNADisease [18], Lnc2Cancer [19], MNDR [20], DisGeNET [21] and LncACTdb [22]. Our proposed method can be separated into the following steps: Firstly, we integrate heterogeneous data to construct a network: containing the interactions of known lncRNA-disease, lncRNA-gene, lncRNA-miRNA and gene-disease. Secondly, the topological structure features of the nodes are extracted based on random walk with restart algorithm. Thirdly, similarity scores of disease-disease pairs and lncRNA-lncRNA pairs are computed based on the topology of this network. Finally, new lncRNA and disease associations are discovered with lncRNA-lncRNA similarities. Our proposed method shows superior predictive performance for prediction of lncRNA-disease associations based on topological similarity from heterogenous network. The AUC value is used to show the performance of our method. The similarity measurement using network topology based on random walk with restart algorithm provide a novel perspective which is different from the similarity derived from sequence or structure information [23, 24].

Our contributions could be summarized as follows:

- We propose a new machine learning framework for lncRNA-disease association by combining GNN and inductive matrix completion.
- We construct similarity graphs for lncRNAs and diseases respectively, based on the domain knowledge from various data sources.
- We analyze different factors that influence the performance of our method.
- We have conducted extensive experiments on simulated and real data sets, and the experimental results show that the method outperforms other representative algorithms by a large margin in most evaluation indicators.

2 Materials and Methods

2.1 Method Overview

Our proposed method can be separated into the following steps: Firstly, we integrate diverse heterogeneous information sources to construct a network: containing the interactions of lncRNA-disease, lncRNA-miRNA-gene, lncRNA-gene and gene-disease. The interaction of lncRNA-miRNA-gene can be defined as lncRNA related triplet, we can identify lncRNA active in cancer through this triplet. Secondly, the topological structural features of the nodes are extracted based on random walk with restart algorithm. Thirdly, similarity scores of disease-disease pairs and lncRNA-lncRNA pairs are inferred based on the topology feature vectors of the nodes. Finally, new lncRNA and disease associations are discovered by means of lncRNA-lncRNA similarities. We extract lncRNA and disease information from multi-omics data such as lncRNA, miRNA, gene and disease, and predict unknown lncRNA-disease associations.

2.2 Network Construction

A diverse heterogeneous network we constructed and called tripartite network consists of three types of nodes, lncRNAs, genes and diseases. Tripartite network contains three kinds of relationships, including lncRNA-disease associations, lncRNA-gene associations and gene-disease associations. For example given n lncRNA sets $U=\{u_1, u_2, \ldots u_m\}$ and r disease sets $D=\{d_1, d_2, \ldots, d_r\}$, the correlation matrix is $M \in R^{n \times r}$. $M_{ij}=1$ indicates that u_i and v_j are related, $M_{ij}=0$ indicating that u_i and v_j are unrelated or unknown, and the relationship between the two needs to be predicted.

The relationships in the tripartite network are obtained as follows: lncRNA-disease associations were downloaded from LncRNADisease database, Lnc2Cancer database and MNDR datasets. LncRNA-disease associations dataset from lncRNADisease database is experimentally supported, which integrated 2900 lncRNA-disease associations and 470 lncRNA interaction entries, including 900 lncRNAs and 320 diseases. 1470 lncRNA-cancer associations are obtained from lnc2Cancer database, including 660 lncRNAs and 90 human cancers. After getting rid of duplicate lncRNA-disease associations, 1440 lncRNA-disease associations including 800 lncRNAs and 270 diseases were obtained. LncRNA-gene associations from LncACTdb database, 5300 high-quality experimentally verified human lncRNA-gene associations about 550 lncRNAs and 1208 genes are downloaded. And gene-disease associations were downloaded from DisGeNET database, which provided the most comprehensive experimentally confirmed gene-disease associations based on large scale experimental data. After getting rid of duplicate interactions, 2198 gene-disease associations were obtained, including 1077 genes and 227 diseases.

Kernel functions have proven to be efficient and useful methods in machine learning and in many bioinformatics classifications. Gaussian kernel function is a commonly used radial basis function, and it is a symmetric scalar function along the radial direction. It is usually a monotonic function of the euclidean distance between any point x in space and a certain center x_c. The form of the Gaussian kernel function is:

$$K(x, x_c) = \exp(-\frac{\|x - x_c\|^2}{2h^2}) \tag{1}$$

where x_c is the center of the Gaussian kernel function, and h is the bandwidth parameter of the Gaussian kernel function, used to control the radial range of the kernel function.

In order to obtain the similarity information between the biological information, the Gaussian kernel function is applied to the topology relationship network between the biological information nodes, and the kernel method is used to establish the kernel function from the feature vector to extract the relevant features of biological information, named Gaussian Interaction Profile (GIP) kernels function. We can calculate the Gaussian interaction attribute between each biological factor pair:

$$KL(l_i, l_j) = \exp(-\gamma_l \|IP(l_i) - IP(l_j)\|^2) \tag{2}$$

$$\gamma_l = \gamma_l(\frac{1}{n_l}\sum_{i=1}^{n_l} \|IP(l_i)\|^2) \tag{3}$$

where n_l represents the number of lncRNA, KL represents the Gaussian interaction attribute kernel similarity matrix of lncRNA, element $KL(l_i, l_j)$ means Gaussian interaction properties between lncRNA l_i and lncRNA l_j.

The basic idea of the random walk algorithm is to traverse a graph starting from one or a series of vertices. At any vertex, the traversor will walk to the neighboring vertex of this vertex with probability 1-*a*, and randomly jump to any vertex in the graph with probability *a*, called a probability of jump occurrence, which is obtained after each walk A probability distribution that describes the probability that each vertex in the graph is visited. Use this probability distribution as the input for the next walk and iterate the process repeatedly. When certain prerequisites are met, this probability distribution will tend to converge. After convergence, a stable probability distribution can be obtained. random walk model is widely used in data mining and Internet field, pagerank algorithm can be seen as an example of random walk model.

Random walk with restart algorithm is an improvement on the basis of the random walk algorithm. Starting from a certain node in the graph, each step faces two choices, randomly selects adjacent nodes, or returns to the starting node. The algorithm contains a parameter a as the restart probability, and 1-a represents the probability of moving to the neighboring node. After iteration, the probability reaches stability, and the probability distribution obtained after the stability can be regarded as the distribution affected by the starting node. Restarting random walk can capture the multi-faceted relationship between two nodes and capture the overall structure information of the graph.

First, we perform feature extraction on lncRNA and disease-related data. On the one hand, lncRNA related data, including lncRNA-disease related data, lncRNA similarity data, and lncRNA-miRNA related data; on the other hand, disease related data, including disease similarity data, disease-miRNA related data, and disease-lncRNA related data. The data is expressed in the form of a matrix, for example, lncRNA-disease matrix, each row represents an lncRNA, and each column represents a disease. We plan to use the kernel principal component analysis (KPCA) method for feature extraction. For each lncRNA correlation matrix, we construct the lncRNA feature matrix by combining the singular vectors containing the main features method to generate disease feature matrix $Q \in R^{r \times k_v}$, k_u and k_v represent the dimension of feature vector.

We can calculate the similarity matrix of lncRNA and disease based on feature matrix separately:

$$Corr(u_i, u_j) = p_i^{\mathrm{T}} p_j \tag{4}$$

$$Corr(d_i, d_j) = q_i^{\mathrm{T}} q_j \tag{5}$$

$$G(u)_{ij} = Corr(u_i, u_j) \tag{6}$$

$$G(d)_{ij} = Corr(d_i, d_j) \tag{7}$$

where p_i and q_i are the row vectors of P and Q, respectively. $G(u) \in R^{n \times n}$ and $G(v) \in R^{r \times r}$ are the similarity matrix of lncRNA and disease, respectively. We obtain the adjacency matrix through the similarity matrix:

$$A(u) = I + \mathrm{Norm}(G(u)) \tag{8}$$

$$A(v) = I + \mathrm{Norm}(G(v)) \tag{9}$$

We can get the new information matrix of lncRNA and disease and the prediction matrix of lncRNA-disease association based on graph neural network to train the network and its parameters:

$$P = \sigma(A(u)PW_u) \tag{10}$$

$$Q = \sigma(A(v)PW_v) \tag{11}$$

where W_u and W_v are parameters in the graph neural network.

2.3 Nonconvex Induction Matrix Completion

In the real data obtained by the experimental observations, commonly exist missing data. In order to operate missing data effectively, we extend the proposed basic model as following. Ω is used to represent an observed subset of entries in lncRNA-disease association data. And $\Omega^{\dagger}$ represent the unobserved entries. For mathematical concept of linear operator, we define an orthogonal projection operator ρ that keeps the entries in Ω of a matrix unchanged and those entries in $\Omega^{\dagger}$ of matrix, i.e. projecting matrix onto the linear space of matrices supported by Ω:

$$\Pr(W)(i,j) = \begin{cases} W_{ij}, if(i,j) \in \Omega \\ 0, if(i,j) \notin \Omega \end{cases} \tag{12}$$

And $\Pr_{\Omega^{\dagger}}(W)$ is its complementary projection. Because we want to find a sparse coefficient matrix and a low-rank matrix based on the observed data, we propose to solve the following nonconvex inductive matrix completion problem:

$$L = \left\| S - \widetilde{S} \right\|_F^2 + \lambda \sum_{i=1}^{d} h(\beta_i) \tag{13}$$

Where λ is the parameter of the balanced regular term and $h(\cdot)$ represents the non-convex penalty function. We intend to use the better performing smooth truncated absolute deviation (SCAD) function and the minimum maximum concave penalty (MCP) function. β_i represent the number of the tandem matrix $[Wu; Wv]$ i column.

3 The Proposed Method NLDAM

In this part we will introduce our experimental process. The learning procedure of NLDAM is constructed on the random walk with restart algorithm. The steps of NLDAM is elaborated as follow.

1. Initialize $t = 0$, the maximum number of iterations *itermax*, the position and velocity of each P_i;
2. Compute the β_i, and evaluate P_i by three objective functions;
3. Initialize P_{ib} and determine the dominance relationship between all the particles;
4. Update the external achieve $\mathcal{REP}$;
5. Update each particle in the population;
6. Compute the β_i to the new P_i, evaluate P_i and determine the dominance relationship between all the particles;
7. $t = t + 1$, go to Step 4 until $t \geq$ *itermax*;
8. Using Bagging method to ensemble the IMCs in $\mathcal{REP}$.

4 Experimental Results and Discussion

In this section, we evaluate the NLDAM and compare it with NTSDMHN and IDHIMIRW on several datasets. With an increased complexity and sophistication of a proposed method, it is very important to compare its performance with existing methods in an objective way. To achieve this end, we conduct extensive simulation studies to compare the proposed methods for joint analysis of multiple variables. Any simulation experiment is necessarily in complete and does not represent real lncRNA-disease experiments. Nevertheless, we try to simulate a relatively 'realistic' lncRNA-disease association model and evaluate the performance with different sample sizes and correlation structures. We consider a backcross population with sample sizes of 200, 400, and 800 to represent very small and large sample sizes. We simulate a genome with 19 chromosomes, each of length 100 cm with 11 equally spaced markers (markers placed 10 cM apart) on each chromosome. Ten percent of the genotypes of these markers were assumed to be randomly missing in all cases. For each of the three sample sizes, we consider two correlation structures, namely, low and high. Therefore, we have six cases with three samples sizes and two correlation structures. For each of these six cases, we simulate six lncRNAs that control the disease: E1 and E2 (E3 and E4) are nonpleitropic lncRNAs, influencing only the disease y1 (y2) with moderate-sized and weak effects, respectively; E5 is a moderate-sized pleiotropic lncRNAs affecting both y1 and y2; while E6 is a weak pleiotropic lncRNAs affecting both y1 and y2. Table 1 presents the simulated positions of six lncRNAs, their effect values, and their heritabilities (proportion of the phenotypic variation explained by a lncRNAs).

For each of the six cases, we generate 100 replicated data sets, resulting in 600 total data sets. For each of these 600 data sets we perform analysis using three methods, namely, the NTSDMHN, joint analysis using a IDHIMIRW model, and our new algorithm NLDAM.

Table 1. Six lncRNAs in the simulation data set

	E_1	E_2	E_3	E_4	E_5	E_6
Chromosome	2	2	4	4	5	6
Position(cM)	30	46	30	56	56	67
y_1	0.9	0.75	0	0	0.9	0.75
y_2	0.15	0.1	−0.8	−0.6	0.7	0.6
y_1 (%)	9.1	5.1	0	1.0	8.2	5.2
y_2 (%)	0.1	0.2	9.0	6.2	8.9	4.9

To illustrate the advantages of using a more complex method of analysis it is important to have an objective and reproducible plan of evaluation. However, in the model selection framework of multiple lncRNAs mapping this assessment becomes a little more complicated as one has to account for model uncertainty. Nonetheless, to assess the performance of different methods we adopt a simple approach. For all six cases we simulate 100 null data sets and compute the genomewide maximum 2 $\log_e$BF (twice the natural logarithm of Bayes factors) for each disease. The 95th percentile of the max 2 $\log_e$BF empirical distribution is considered as the threshold value above which a lncRNA-disease association would be deemed 'significant'. At each replication, the number of correctly identified lncRNA-disease association and the number of incorrectly identified or extraneous lncRNA-disease are recorded. A peak in the 2 logBF profile is considered a lncRNA-disease association if it crosses the significance threshold. It is deemed correct if it is within 10 cM (Broman and Speed 2002) of a true lncRNA-disease association. If there is more than one peak within 10 cM of the true lncRNA-disease association, only one is considered correct. Table 2 represents the average correct and extraneous (incorrect) lncRNA-disease detections for the six situations and for all three methods for y1 and y2, respectively.

Table 2. Average correct and incorrect lncRNA-disease detected for disease y_1 (first row) and y_2 (second row)

$(n, \rho_{y_1 y_2})$	Correct			Extraneous		
	NTSDMHN	IDHIMIRW	NLDAM	NTSDMHN	IDHIMIRW	NLDAM
(200, 0.5)	1.25	1.58	1.14	1.53	2.32	1.10
	1.58	1.66	1.32	1.58	2.32	1.12
(200, 0.8)	0.72	2.02	2.04	0.48	3.72	1.63
	1.36	2.14	2.54	1.42	3.44	1.68
(400, 0.5)	3.21	4.89	4.98	2.12	5.12	1.62
	3.52	4.32	4.32	1.26	5.12	1.68
(400, 0.8)	3.02	5.21	4.89	1.26	6.03	1.48
	3.42	5.13	5.15	1.98	6.23	1.88
(800, 0.5)	6.12	6.45	6.51	2.12	6.21	1.88
	7.21	7.42	7.53	1.52	6.24	1.52
(800, 0.8)	7.35	7.99	7.62	2.41	6.38	2.03
	7.21	8.02	7.42	2.36	6.25	1.42

5 Conclusion

In this work, we propose a new machine learning framework for lncRNA-disease association task by integrating diverse heterogeneous information sources with positive pointwise mutual information and random walk with restart algorithm, named NLDAM. We integrate diverse heterogeneous information sources to construct a Network which is a heterogeneous network containing three types nodes which generated from bioinformatics linked datasets and use random walk with restart algorithm method to extract topological structure features of the nodes in the linked tripartite network for calculating similarities. In addition, NLDAM is inductive–it can generalize to lncRNAs/diseases unseen during the training (given that their associations exist), and can even transfer to new tasks. We concatenate all the features as the input feature. Dense similarities matrices are constructed by the inner product of features. Once we got association values of genes and diseases by GIP, we use the inner product of them to get the preference score matrix. An ϵ-Insensitive loss is proposed to reduce the effect of data sparsity. We use nonconvex inductive matrix completion approach to optimize this loss function. From our experiments, we could see that NLDAM outperforms all other state-of-the-art methods a lot on most of metrics.

In the analysis of our results, we could see that extra information is the key point to improve the performance in disease-gene association task. In the future, we would try to mine more accurate correlations by learning method. Moreover, we will apply our method on more applications that include similarity information and contain sparse data, such as POI recommendation.

Acknowledgement. This work was supported by the National Key R&D Program of China [No. 2019YFB1404700], the Natural Science Foundation of Shandong Province [No. ZR2017LF019].

References

1. Hobert, O.: Gene regulation by transcription factors and microRNAs. Science **319**(5871), 1785–1786 (2008)
2. Swami, M.: Transcription factors: MYC matters. Nat. Rev. Cancer **10**(12), 812 (2010)
3. Collins, F.S., Morgan, M., Patrinos, A.: The human genome project: lessons from large-scale biology. Science **300**(5617), 286–290 (2003)
4. Yuan, L., Guo, L.H., Yuan, C.A., et al.: Integration of multi-omics data for gene regulatory network inference and application to breast cancer. IEEE/ACM Trans. Comput. Biol. Bioinf. **16**(3), 782–791 (2019)
5. International Human Genome Sequencing Consortium: Initial sequencing and analysis of the human genome. Nature **409**(6822), 860 (2001)
6. Louro, R., Smirnova, A.S., Verjovski-Almeida, S.: Long intronic noncoding RNA transcription: expression noise or expression choice? Genomics **93**(4), 291–298 (2009)
7. Yuan, L., Zhu, L., Guo, W.L., Huang, D.S.: Nonconvex penalty based low-rank representation and sparse regression for eQTL mapping. IEEE/ACM Trans. Comput. Biol. Bioinf. **14**(5), 1154–1164 (2017)
8. Geisler, S., Coller, J.: RNA in unexpected places: long non-coding RNA functions in diverse cellular contexts. Nat. Rev. Mol. Cell Biol. **14**(11), 699–712 (2013)
9. Xing, Z., Lin, A., Li, C., et al.: lncRNA directs cooperative epigenetic regulation downstream of chemokine signals. Cell **159**(5), 1110–1125 (2014)
10. Yuan, L., Yuan, C.A., Huang, D.S.: FAACOSE: a fast adaptive ant colony optimization algorithm for detecting SNP epistasis. Complexity **1**, 1–10 (2017)
11. Yuan, L., Huang, D.S.: A network-guided association mapping approach from DNA methylation to disease. Sci. Rep. **9**(1), 1–16 (2019)
12. Chen, X., Yan, C.C., Zhang, X., et al.: Long non-coding RNAs and complex diseases: from experimental results to computational models. Brief. Bioinform. **18**(4), 558–576 (2016)
13. Gao, Y., Meng, H., Liu, S., et al.: LncRNA-HOST2 regulates cell biological behaviors in epithelial ovarian cancer through a mechanism involving microRNA let-7. Hum. Mol. Genet. **24**(3), 841–852 (2014)
14. Yuan, L., Zheng, C.H., Xia, J.F., Huang, D.S.: Module based differential coexpression analysis method for type 2 diabetes. Biomed. Res. Int. **1**, 1–8 (2015)

15. Chen, G., Wang, Z., Wang, D., et al.: LncRNADisease: a database for long-non-coding RNA-associated diseases. Nucleic Acids Res. **41**(D1), D983–D986 (2012)
16. Lan, W., Li, M., Zhao, K., et al.: LDAP: a web server for lncRNA-disease association prediction. Bioinformatics **33**(3), 458–460 (2016)
17. Wang, J., Ma, R., Ma, W., et al.: LncDisease: a sequence based bioinformatics tool for predicting lncRNA-disease associations. Nucleic Acids Res. **44**(9), e90–e90 (2016)
18. Zhou, M., Wang, X., Li, J., et al.: Prioritizing candidate disease-related long non-coding RNAs by walking on the heterogeneous lncRNA and disease network. Mol. BioSyst. **11**(3), 760–769 (2015)
19. Zhang, J., Zhang, Z., Chen, Z., et al.: Integrating multiple heterogeneous networks for novel lncRNA-disease association inference. IEEE/ACM Trans. Comput. Biol. Bioinf. **16**(2), 396–406 (2017)
20. Yao, Q., Wu, L., Li, J., et al.: Global prioritizing disease candidate lncRNAs via a multi-level composite network. Sci. Rep. **7**, 39516 (2017)
21. Ganegoda, G.U., Li, M., Wang, W., et al.: Heterogeneous network model to infer human disease-long intergenic non-coding RNA associations. IEEE Trans. Nanobiosci. **14**(2), 175–183 (2015)
22. Chen, X., Yan, G.Y.: Novel human lncRNA–disease association inference based on lncRNA expression profiles. Bioinformatics **29**(20), 2617–2624 (2013)
23. Zheng, C.H., Yuan, L., Sha, W., et al.: Gene differential coexpression analysis based on biweight correlation and maximum clique. BMC Bioinform. **15**(S15), S3 (2014)
24. Yuan, L., Han, K., Huang, D.-S.: Novel algorithm for multiple quantitative trait loci mapping by using bayesian variable selection regression. In: Huang, D.-S., Han, K., Hussain, A. (eds.) ICIC 2016. LNCS (LNAI), vol. 9773, pp. 862–868. Springer, Cham (2016). https://doi.org/10.1007/978-3-319-42297-8_80

Intelligent Computing in Computational Biology

A Graph Convolutional Matrix Completion Method for miRNA-Disease Association Prediction

Wei Wang[1], Jiawei Luo[1(✉)], Cong Shen[1], and Nguye Hoang Tu[2]

[1] College of Computer Science and Electronic Engineering, Hunan University, Changsha 410083, China
luojiawei@hnu.edu.cn

[2] Faculty of Information and Technology, Hanoi University of Industry, Hanoi 100803, Vietnam

Abstract. MicroRNAs (miRNAs) play a key role in various biological processes associated with human diseases. Identification of miRNA-disease relationships can help to understand disease pathogenesis. Experimentally verifying substantial associations between miRNAs and diseases is the most convincing but time-consuming, while in silico methods can provide efficient alternatives. However, existing computational methods still have room for improvement in considering topology and prior information of network nodes. In this paper, we presented a novel model called GCMCAP, in which we referred to the prediction of potential miRNA-disease associations as a recommendation problem. In our framework, we integrated graph convolution networks as feature extractors into a matrix completion to predict diseases related miRNAs. We tested GCMCAP and other three methods on the same dataset. The results indicate that GCMCAP outperforms other methods with respect to average AUC value. In addition, case studies show that GCMCAP has a great capability to discover novel miRNA-disease associations.

Keywords: MicroRNA · Disease · Association · Graph convolution networks · Matrix completion

1 Introduction

MicroRNAs (miRNAs) are small non-coding RNAs that can regulate gene expression during post-transcription and influence the output of protein-coding genes [1]. Existing researches have shown that miRNAs are involved in many biological processes, such as differentiation [2], development [3], immune reaction [4], apoptosis [5], and pathogenesis. There are already compelling evidences that human miRNAs are associated with complex diseases [6]. For example, studies by Zare et al. have shown that miRNA expression is a regulator of tumorigenesis [7]. In addition, the Human MicroRNA Disease Database (HMDD v2.0) has shown that more than 10,000 associations between miRNA and disease have been identified [8], involving 378 diseases and 572 miRNAs. Therefore, it is necessary to discover other potential associations between miRNA and disease in order to more fully understand pathogenesis of human

D.-S. Huang and K.-H. Jo (Eds.): ICIC 2020, LNCS 12464, pp. 201–215, 2020.
https://doi.org/10.1007/978-3-030-60802-6_18

diseases associated with miRNAs, thereby facilitating disease diagnosis, treatment, and prevention. Identifying the associations between miRNA and disease through biological experiments is the most convincing but time-consuming, while in silico methods can provide efficient alternatives [9–11].

The most of these proposed methods are based on the assumption that similar miRNAs are related to similar diseases [12, 13]. For example, Zhao et al. discovered disease related miRNA candidates using gene expression data and miRNA-gene regulation [14]. Chen et al. proposed a method to identify disease-related miRNAs by random walks with restart on the miRNA similarity network [15]. Xuan et al. evaluated the *k* most functionally similar neighbors by considering the disease terms and phenotype similarity [16]. You et al. constructed a heterogeneous network by integrating different types of heterogeneous biodata sets and proposed a path-based approach to calculate the association score between miRNA and disease [17]. Zou et al. predicted miRNA-disease association based on social network analysis [18]. Xiao et al. used graph regularization non-negative matrix factorization to identify microRNA-disease associations [19]. Luo et al. predicted Small Molecule-microRNA Associations based on Non-negative Matrix Factorization [20]. All these methods make good use of the prior information in the miRNA/disease interaction network.

On the other hand, machine learning-based methods are proposed as many known associations are confirmed by biological experiments. Based on transduction learning, Luo et al. developed a collective prediction method of disease-associated miRNAs [21]. Chen et al. presented a computational model named Laplacian Regularized Sparse Subspace Learning, which projected miRNAs/diseases' graph feature profile to a common subspace [22]. These results show that machine learning based methods improve the performance of association prediction. Based on inductive matrix completion, Ding, X et al. exploited an improved computation method. Li et al. released an effective computational model of Matrix Completion for MiRNA-Disease Association prediction (MCMDA) [23]. Recently, methods combined neural network have been used for association prediction. For example, Hou et al. combined neural network model and induction matrix completion (NIMC) to predict disease-gene association [24]. Xuan, P et al. proposed a method to predict disease related miRNAs based on network representation learning and convolutional neural networks [25]. Han, P et al. predicted disease-gene association by integrating graph convolutional network and matrix factorization [26].

The methods based on similarity measure can effectively integrate heterogeneous data, but it is limited by known associations. On the other side, machine learning-based methods have impressive results for miRNA-disease association predictions, but there is still room for improvement in using miRNAs and disease prior information. In this paper, we referred to the prediction of potential miRNA-disease associations as a recommendation problem and proposed a novel computational framework named GCMCAP. We integrated graph convolutional neural networks into matrix completion to predict the associations between miRNA and disease. With the help of graph convolutional networks, non-linear and high-order neighborhood information can be captured. In addition, the problem of no negative sample in training progress can be circumvented. We tested GCMCAP and other three methods on the same dataset. The results suggest that GCMCAP outperforms other methods in terms of average AUC values. Moreover, case studies show that GCMCAP has a great capability to discover novel miRNA-disease associations.

2 Materials and Methods

2.1 Datasets

We downloaded miRNA-gene interactions from experimentally verified databases, including miRecords v4.0 [27], TarBase v6.0 [28] and miRTarBase v4.5 [29]. After unioning and removing duplicates, we got 38,089 interactions between miRNAs and genes, involving 477 miRNAs and 12,422 genes. We downloaded gene-gene interaction network from HumanNet which contains 476,399 interactions among 16,243 genes [30]. We downloaded validated miRNA-disease associations datasets from the HMDD database v2.0. As done in Xuan et al. [31], we regarded multiple miRNA transcripts as the same mature miRNA. So we acquired 5424 associations involving 378 diseases and 495 miRNAs from HMDD v2.0. We downloaded the disease hierarchical directed acyclic graphs (DAGs) from MeSH (https://www.nlm.nih.gov/mesh/). To ensure consistency of miRNAs, diseases and associations, we removed some irregularly named diseases in MeSH and eliminated miRNAs that are missing from the three miRNA target databases mentioned above. 4887 experimentally validated miRNA-disease associations involving 327 diseases and 351 miRNAs were retained. Consequently, miRNA matrix $S_m \in R^{m \times m}$, disease matrix $S_d \in R^{d \times d}$ and miRNA-disease association matrix $S \in R^{m \times d}$ were formed for prediction task.

2.2 Method Overview

We regarded the prediction of miRNA-disease associations as a recommendation problem with four main steps. In the first step, we calculated miRNA-miRNA and disease-disease pairwise similarities and compiled them into two adjacency matrices. In the second step, we respectively performed a multi-layer graph convolutional networks on the adjacency matrices to assign node feature to each miRNA and disease. In such way, the high-order neighborhood information of each miRNA and disease node is encoded into embeddings. In the third step, we modeled association ratings as the inner product of the embeddings projected onto a latent space. In the last step, we used matrix completion principle to obtain the ratings for each miRNA-disease association. In this stage, the embeddings of miRNA/disease are adjusted by minimizing the difference between the reconstruction matrix and the initial matrix. Figure 1 depicts the whole framework of the proposed method.

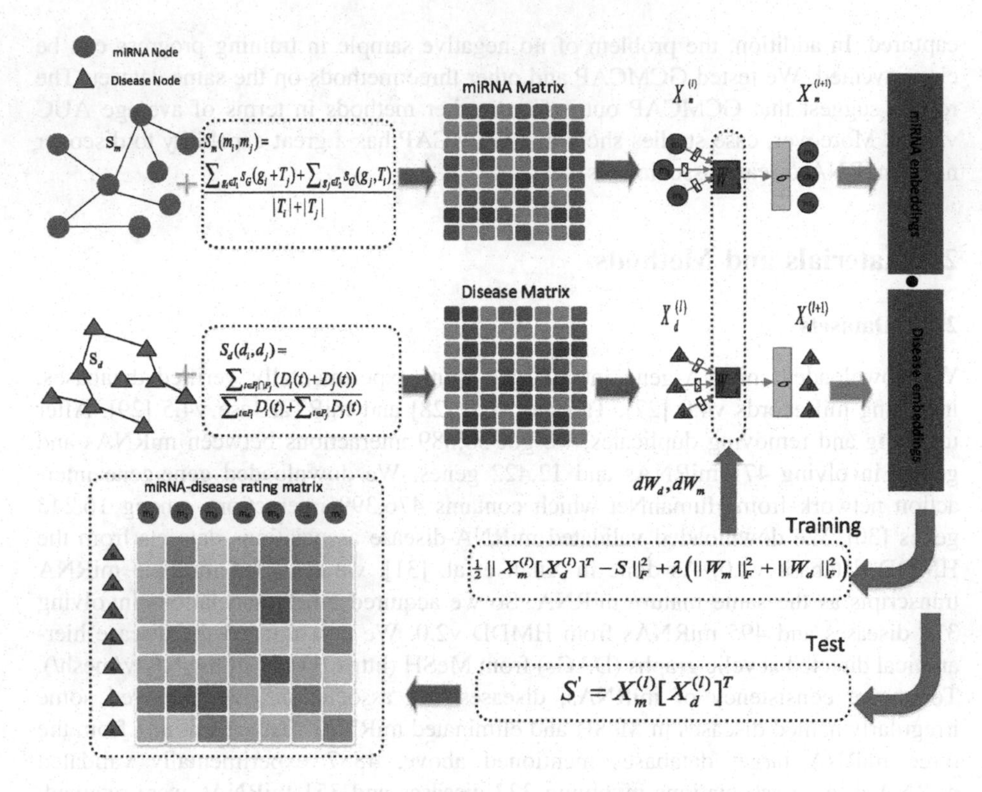

Fig. 1. Workflow of GCMCAP

2.3 Diseases Semantic Similarity and MiRNA Functional Similarity

The MeSH database provides a valuable reference system in the form of DAGs (http://www.ncbi.nlm.nih.gov/). Disease similarity can be calculated using DAGs according to Wang's method [32]. In DAGs, nodes and edges represent diseases and the associations between diseases respectively. We calculated the similarity between the two diseases through the disease hierarchical relationship in DAGs. The semantic value of disease d is calculated by the following formula:

$$\mathrm{D(t)} = \begin{cases} \max\{\Delta * \mathrm{D(t')} | \in \mathrm{childrenof(t)}\} & t \neq d \\ D(\mathrm{d}) = 1 & t = d \end{cases} \quad (1)$$

where Δ represents the semantic contribution factor. We set Δ as 0.5 as suggested by Wang et al.. According to the assumption that diseases that share a large part of the DAG tend to have higher semantic similarity, the following calculation methods are available:

$$S_d(d_i, d_j) = \frac{\sum_{t \in P_i \cap P_j} (D_i(t) + D_j(t))}{\sum_{t \in P_i} D_i(t) + \sum_{t \in P_j} D_j(t)} \tag{2}$$

$D_i(t)$ and $D_j(t)$ are the semantic values related to disease d_i and disease d_j, respectively. The semantic similarity of each two diseases is calculated based on the position of the two diseases in the DAG and their semantic associations with ancestor diseases.

To avoid reliance on existing associations between miRNA and disease, inspired by Xiao et al., we estimated the similarity between miRNAs using a weighted gene interaction network and an experimentally validated miRNA-target regulatory relationship. Initially the gene function interaction network is downloaded from the HumanNet. Interaction score between two genes indicates the strength of the association between genes. We used the following normalization technique on the gene interaction network to get the pairwise similarity score S_g:

$$S_g(g_i, g_j) = \frac{S_G(g_i, g_j) - S_{min}}{S_{max} - S_{min}} \tag{3}$$

where $S_G(g_i, g_j)$ denotes similarity score before normalization. We obtained the miRNA-target regulatory relationship from collated miRNA-gene data. Then the similarity between the gene and the target gene set is defined as the maximum similarity between the gene and others. It can be described by the following formula:

$$S_G(g_t, T) = \max_{g_{ti} \in T} (S_g(g_t, g_{ti})) \tag{4}$$

Based on the assumption that the greater the number of common target genes, the greater the similarity between miRNAs, the functional similarity $S_m(m_i, m_j)$ between miRNA m_i and m_j can be calculated by the following BMA method [33]:

$$S_m(m_i, m_j) = \frac{\sum_{g_i \in T_1} S_G(g_i, T_j) + \sum_{g_j \in T_2} S_G(g_j, T_i)}{|T_i| + |T_j|} \tag{5}$$

where T_i and T_j denote target gene sets of m_i and m_j respectively.

2.4 Graph Convolutional Feature Extractor

The graph convolutional network (GCN) is a kind of multilayer neural network with graph as input. It represents neighbor node features and network information as output vectors. GCNs have been successfully applied in areas such as recommendation systems and drug interactions [34, 35]. In this paper, we resorted to GCNs to convert miRNA and disease networks into embeddings and map the learned embeddings into latent space. The outputs will be applied to the matrix completion in the downstream framework.

Based on the assumption that node features are associated with all of its neighbors, we employed a simple solution to integrate network information. Let miRNA and disease similarity networks (adjacent matrices) be S_m and S_d, respectively. We defined the multiplication of S_m and X_m to obtain the neighborhood information of the current miRNA node. So we obtained the following neural network propagation rule for miRNA nodes.

$$f(X_m^{(l)}, S_m) = \sigma(S_m X_m^{(l)} W_m^{(l)}) \tag{6}$$

where σ denotes the non-linear activation function, $X_m^{(l)}$ denotes the miRNA features output by the l-layer neural network, and $W_m^{(l)}$ represents the weight of the miRNA features. However, this method of multiplication modeling can easily lead to overfitting the local neighborhood structure of a graph with a wide node degree distribution. Therefore, we treated each initial node feature X_m as a graph signal according to spectral graph theory, and use the spectral convolution operation $S_m \bigstar X_m$ on the graph S_m to replace the multiplication. Referring to the convolution theorem that convolution operation is equivalent to the product after Fourier transform, the following equation can be obtained:

$$S_m \bigstar X_m = U_S((U_S^T g) \odot (U_S^T X_m)) \tag{7}$$

where U_S is the eigenvector matrix of S_m, g represents the convolution kernel, and $\odot$ represents the hardmard product. To simplify formula (2), we treat diagonal matrix $g_\theta = diag\left(U_S^T g\right)$ as parameterized convolution kernel. Then the hardmard product can be changed to the form of matrix multiplication. However, Eq. (7) is still limited by 3 problems. First, the information of miRNA nodes is very important for network feature extraction, but the adjacency matrix S_m does not contain node information. Second, S_m is not normalized, which will lead to a larger feature values extracted by nodes with more neighbors. Third, Eq. (7) requires the eigendecomposition of S_m, which has high computational complexity for a large miRNA/disease networks.

To solve the first two limitations, we used $\tilde{S}_m = S_m + I$ to add the identity matrix I to the miRNA matrix to form a self-loop, and used the normalized Laplacian matrix $L_m = I - D^{-1/2} S_m D^{-1/2}$ to represent the network structure. The normalized Laplacian matrix $\tilde{L}_m$ with self-loop is obtained, i.e.

$$\tilde{L}_m = I - \tilde{D}_m^{-\frac{1}{2}} \tilde{S}_m \tilde{D}_m^{-\frac{1}{2}} \tag{8}$$

where $\tilde{D}_m$ denotes the diagonal matrix $\left[\tilde{D}_m\right]_i = \sum_{j=1}^n \left[\tilde{S}_m\right]_{ij}$. In this way, we can perform eigendecomposition on $\tilde{L}_m$, and replace the U_S with the eigenvector matrix U_L obtained from the feature decomposition. Equation (7) can be transformed into the following form.

$$S_m \bigstar X_m = U_L g_\theta U_L^T X_m \tag{9}$$

To circumvent the third limitation, we used Kipf's method to avoid eigendecomposition [36]. Based on Eq. (9), the first-order Chebyshev polynomial is used as the convolution kernel, and the maximum eigenvalue λ of L_m is approximated to 2. The propagation rule of miRNA nodes on the graph S_m can be written as:

$$X_m^{(l+1)} = \sigma(\tilde{D}_m^{-\frac{1}{2}}\tilde{S}_m\tilde{D}_m^{-\frac{1}{2}}X_m^{(l)}W_m^{(l)}) \tag{10}$$

where $X_m^{(l)} \in R^{m\times e}$ denotes the miRNA features output by l-layer network, m is the number of miRNA nodes, and e is the embedding size. The multi-layer networks can be achieved by stacking Eq. (10). Multi-layer networks imply the use of higher-order neighborhood information. Similarly, the propagation rule of disease nodes on the graph S_d can be written as:

$$X_d^{(l+1)} = \sigma(\tilde{D}_d^{-\frac{1}{2}}\tilde{S}_d\tilde{D}_d^{-\frac{1}{2}}X_d^{(l)}W_d^{(l)}) \tag{11}$$

where $X_d^{(l)} \in R^{d\times e}$ denotes the disease features output by l-layer network, d is the number of disease nodes. Disease and miRNA embeddings are trained simultaneously.

2.5 Matrix Completion

Matrix completion has been well used in recommendation system [37]. Similarly, we considered association prediction as a recommendation problem and use known miRNA-disease associations to recover missing entries in association matrix $S \in R^{m\times d}$. Where m is the number of miRNAs. For each entry in matrix S, $S_{ij} = 1$ if the miRNA is associated with the disease, otherwise, $S_{ij} = 0$. Let S' be the matrix to be completed. We modeled association ratings as the inner product of the features of miRNAs and diseases projected onto a latent space. i.e. $S' = KH$, where $K \in R^{m\times j}$ and $H \in R^{d\times j}$ denote miRNA and disease space, respectively. According to low-rank assumption, j satisfies $j << m, d$. The following formula can be defined:

$$\min_{K,H}\frac{1}{2}\|KH^T - S\|_F^2 + \lambda\left(\|K\|_F^2 + \|H\|_F^2\right) \tag{12}$$

where $\| * \|_F$ denotes Frobenius norm and λ denotes the regularization coefficient. However, the classical matrix completion cannot directly take advantage of the embeddings of known miRNA and disease. Inspired by Natarajan et al., we set T_1 and T_2 as projection matrices [38]. The features of miRNA and disease can be mapped into latent spaces with the same dimensions by $K = X_mT_1$ and $H = X_dT_2$, respectively. Therefore, the known associations and the features of miRNA and disease can be simultaneously used by the following improved matrix completion.

$$\min_{T_1,T_2}\frac{1}{2}\|X_mT_1T_2^TX_d^T - S\|_F^2 + \lambda\left(\|T_1\|_F^2 + \|T_2\|_F^2\right) \tag{13}$$

2.6 Training and Evaluation

A classic matrix completion for miRNA-disease association prediction considered the miRNA matrix X_m and disease matrix X_d as inputs. T_1 and T_2 represent the low rank decomposition of the projection matrix T. In proposed method, the similarity matrices X_m and X_d are encoded by graph convolutional networks, respectively. The linear projection $X_m T_1 T_2 X_d$ is replaced by the nonlinear rating $X_m^{(l)}\left[X_d^{(l)}\right]^T$ generated by the graph convolution network. According to the above annotation, the loss function of the graph convolutional matrix completion model can be defined as:

$$loss = \frac{1}{2}\left\|X_m^{(l)}\left[X_d^{(l)}\right]^T - S\right\|_F^2 + \lambda\left(\|W_m\|_F^2 + \|W_d\|_F^2\right) \tag{14}$$

where $X_m^{(l)} = \sigma\left(\tilde{D}_m^{-\frac{1}{2}}\tilde{S}_m\tilde{D}_m^{-\frac{1}{2}}X_m W_m^{(l)}\right)$ and $X_d^{(l)} = \sigma\left(\tilde{D}_d^{-\frac{1}{2}}\tilde{S}_d\tilde{D}_d^{-\frac{1}{2}}X_d W_d^{(l)}\right)$ denote the outputs of graph convolution networks. Considering the advantage that all operations are differentiable, the proposed model can be train in an end-to-end workflow by gradient descent algorithm. Algorithm 1 shows the overall prediction procedure.

Algorithm 1. GCMCAP Algorithm

Input: association matrix $S \in R^{m \times n}$; similarity matrix $S_m \in R^{m \times m}$, $S_d \in R^{d \times d}$; learning rate γ; iterations t; embedding size e; regularization parameters λ;

1: **for p=0 : t do**
2: **Apply the GCNs on S_d to produce embeddings $X_d^{(l)} \in R^{d \times e}$**
3: **Apply the GCNs on S_m to produce embeddings $X_m^{(l)} \in R^{m \times e}$**
4: **for i = 1 : m do**
5: **for j=1:d do**
6: **Apply embeddings $\left[X_m^{(l)}\right]_i$ and $\left[X_d^{(l)}\right]_j$ to generate predicted values S'_{ij}**
7: $loss = \frac{1}{2}\|S' - S\|_F^2 + \lambda\left(\|W_m\|_F^2 + \|W_d\|_F^2\right)$
8: **end for**
9: **end for**
10: **Update the GCNs parameters W_m and W_d**
11: **end for**
12: **Apply $S' = X_m^{(l)}\left[X_d^{(l)}\right]^T$ to produce ratings**

Output: rating matrix S'

3 Results

In this part, we verified the practicality of the proposed method GCMCAP through experiments. Firstly, the evaluation indicators for the performance of all methods are introduced. Then the performance of the GCMCAP and several other common miRNA-disease prediction methods are compared. Next, we carried out parameter analysis to verify the reliability and robustness of the model. Finally, case studies are arranged to explore the capacity of GCMCAP to discover novel disease-associated miRNAs.

3.1 Experiment Settings

To evaluate the performance of the GCMCAP algorithm, we performed 5-fold cross validation (CV) and 10-fold CV. Taking 5-fold CV as an example, in each round, the known miRNA-disease associations are randomly divided into five disjoint subsets. One subset is used as the validation set and the remaining subsets are utilized as the training set. For each fold, embedding size e is set from $\{32, 64, 128, 256\}$, learning rate γ is set from $\{0.001, 0.01, 0.1\}$, the number of GCN layers l is set from $\{2, 3, 4, 5\}$ and probability of dropout is set from $\{0.5, 1\}$. All parameters are considered based on grid search. All experiments are repeated for 5 times, the reports are average of the 5 runs. Refer to Kipf's method, we used the identity matrix as the initial feature matrix to feed GCNs. All methods processes are implemented using Tensorflow framework (v1.9) and trained using the Adam stochastic optimization algorithm [39].

3.2 Evaluation Metrics

After the training progress completed, we obtained all association ratings from the rating matrix. The four values of true positive (TP), false positive (FP), false negative (FN), and true negative (TN) are calculated. The true positive rate(TPR),false positive rate (FPR), Precision, and Recall were calculated as following formulas:

$$\mathrm{TPR} = \frac{\mathrm{TP}}{\mathrm{TP} + \mathrm{FN}}, \mathrm{FPR} = \frac{\mathrm{FP}}{\mathrm{FP} + \mathrm{TN}} \tag{15}$$

$$\mathrm{Precision} = \frac{\mathrm{TP}}{\mathrm{TP} + \mathrm{TP}}, \mathrm{Recall} = \frac{\mathrm{TP}}{\mathrm{TP} + \mathrm{FN}} \tag{16}$$

we used the TPRs and FPRs to plot the receiver operating characteristic (ROC) curves. Then the area under the ROC curves (AUCs) are used to measure the global performance of models.

To get indicators that reflects global performance in the event of class imbalance, the mean Average Precision (mAP) is used to solve the single-point value limitation of the Precision. We focused more on the set of top K miRNA related to a disease, the sequence of the whole recommendation list may be not important. Therefore, we utilized Precision@N and Recall@N to evaluate the quality of the recommendation list. Where N denotes the percentage of sorted rating results. We set N = $\{10, 20, 30\}$ in all experiments.

3.3 Performance Evaluation

To assess the performance of GCMCAP, we compared it with the other three methods: RWRMDA [15], KATZ [18], MIDPE [31]. RWRMDA is a plain baseline and utilizes random walk algorithm to identify potential unknown association. KATZ combines social network analysis methods with machine learning and predicts unknown miRNA-disease relationships. MIDPE completely integrates various ranges of topologies around the different categories of nodes. We reproduced the three methods based on the formulas in the papers, respectively. We used 5-fold CV and 10-fold CV to compare our method with existing methods based on the same dataset. Table 1 and Table 2 list the results of all methods. GCMCAP outperforms other methods with respect to the most indicators. In 5-fold CV, our method yields an average AUC value of 0.894, which is better than RWRMDA (0.804), MIDPE (0.702) and KATZ (0.864). Table 2 shows that the results of 10-fold CV are similar to the 5-fold CV, and GCMCAP is outperform other methods. Furthermore, P@10 and R@10 of GCMCAP are significantly higher than existing methods, which shows a better performance in top *K* prediction ability. Figure 2 shows the ROC curves of various comparison methods, which suggests the overall performance of the models. σ in the figure denotes the standard deviation. The ROC curve generated by GCMCAP is above all the curves and closer to the upper left corner of the coordinate system. The results suggest that the associations in the similarity matrix maybe not directly reflect the relationship between miRNA or disease. Since GCMCAP integrates graph convolution networks into matrix completion method, non-linear neighborhood information can be abstracted from miRNA network and disease network. Furthermore, GCMCAP does not rely on negative samples for training, which effectively circumvent the problem of no negative sample.

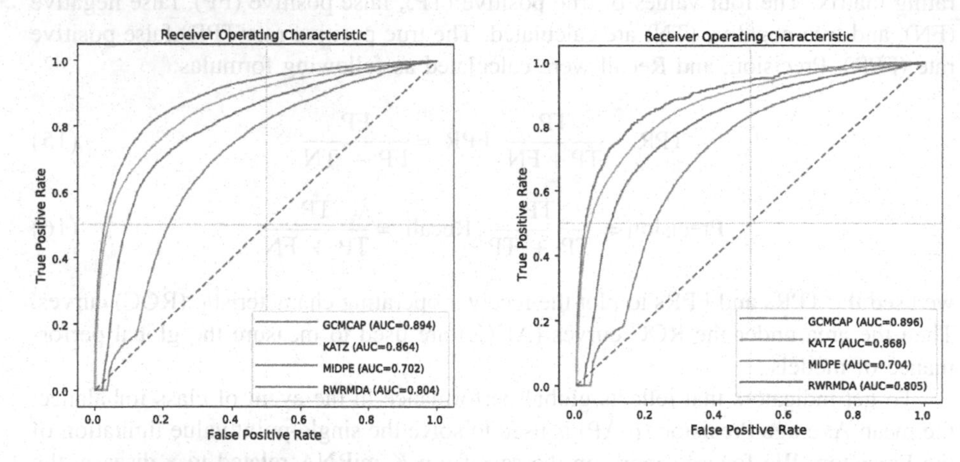

Fig. 2. (a) ROC curves of 5-fold CV; (b) ROC curves of 10-fold CV.

Table 1. Results on 5-fold cross validation

Method	MAP	AUC	P@10	P@20	P@30	R@10	R@20	R@30
RWRMDA	0.032	0.804	0.026	0.039	0.046	0.102	0.207	0.309
MIDPE	0.016	0.702	0.015	0.014	0.018	0.114	0.219	0.325
KATZ	0.14	0.864	0.267	0.206	**0.163**	0.134	0.245	0.346
GCMCAP	**0.197**	**0.894**	**0.347**	**0.223**	0.153	**0.189**	**0.319**	**0.449**

Table 2. Results on 10-fold cross validation

Method	MAP	AUC	P@10	P@20	P@30	R@10	R@20	R@30
RWRMDA	0.016	0.805	0.012	0.018	0.022	0.102	0.204	0.306
MIDPE	0.008	0.704	0.008	0.007	0.009	0.123	0.241	0.358
KATZ	0.083	0.868	0.17	0.12	**0.099**	0.119	0.223	0.327
GCMCAP	**0.153**	**0.896**	**0.238**	**0.142**	0.093	**0.225**	**0.368**	**0.501**

3.4 Parameter Analysis

In this section, we analyzed the effects of parameters through the parameter adjustment. Because AUC is an indicator that can evaluate the comprehensive performance of the model, the influence of parameter changes on the AUC value is analyzed. Learning rate is an important parameter for the optimization model using gradient descent algorithm. We fixed the other parameters and set the learning rate from $\{0.001, 0.01, 0.1\}$. Figure 3(a) shows that there may be an optimal value for the initial learning rate. A small learning rate will cause the model to converge slowly, conversely a large learning rate may make it difficult to converge. Figure 3(b) shows that the embedding size does not affect convergence of GCMCAP within an appropriate size range, which indicates that the proposed method has the ability to obtain prior information steadily. However, when the embedding size is too large, too many parameters make the model difficult to train or even overfitting. This pattern is consistent with other related studies [40].

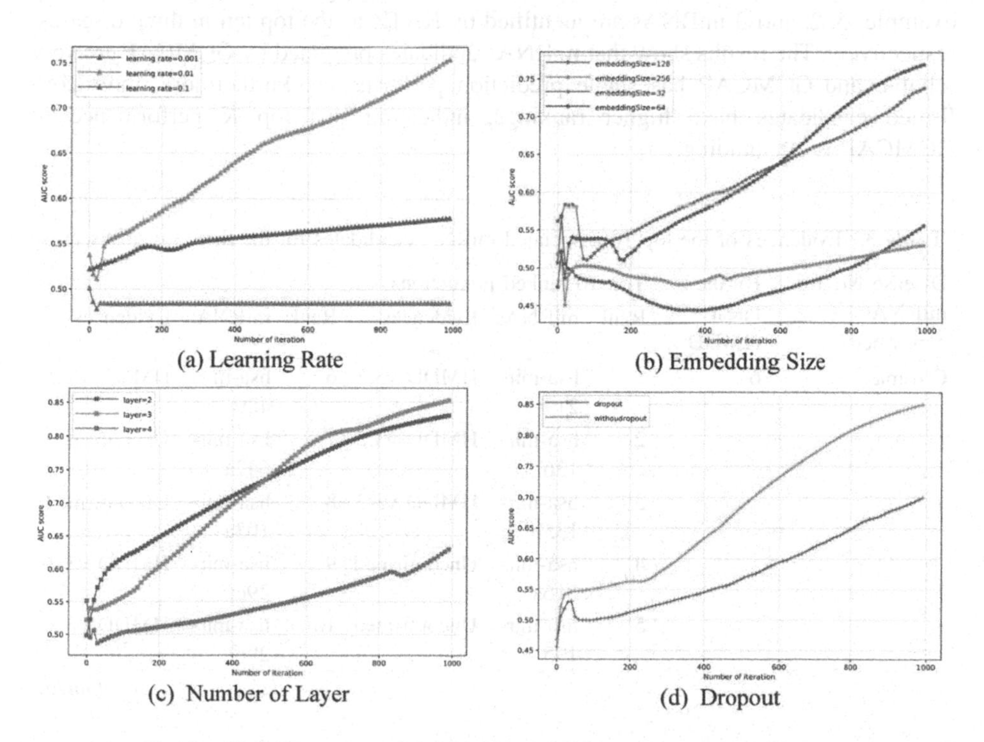

(a) Learning Rate (b) Embedding Size

(c) Number of Layer (d) Dropout

Fig. 3. (a) ROC curves of 5-fold CV; (b) ROC curves of 10-fold CV.

The model with a small number of GCN layers l perform well, and the performance decreases rapidly when $l >= 4$ according to Fig. 3(c). The increase of layer may capture the more global information, but also the more noise is captured. Dropout is a commonly used method to improve performance by avoiding overfitting. However, there is no obvious effect of the dropout on the performance of GCMCAP according to Fig. 3(d). The possible reason is that the effect of dropout is not obvious due to sparse data.

3.5 Case Studies

To further explore the capacity of GCMCAP to discover potential miRNAs associated with disease, 3 given diseases of Glioma, Carcinoma Hepatocellular and Ovarian Neoplasms are analyzed. All known associations in HMDD v2.0 are used to train and unknown associations are used for validation. For each disease, the candidate miRNAs are ranked based on the ratings. The top 10 miRNA candidates are obtained from prediction results.

The latest human microRNA disease database (HMDD v3.2) is used to confirm miRNA candidates for given 3 diseases. HMDD V3.2 provides extensive experimentally supported evidence for human microRNA (miRNA) and disease associations. HMDD v3.2 contains twice as many human miRNA disease associations as previous HMDD v2.0. As shown in Table 3, HMDD v3.2 confirms 6,6 and 5 miRNA candidates are associated with Glioma, Carcinoma Hepatocellular and Ovarian Neoplasms, respectively. In addition, some candidates also rank higher in other methods. For example, 3, 2, and 3 miRNAs are identified by KATZ as the top ten in three diseases, respectively. The results show that miRNA candidates predicted by GCMCAP are very reliable and GCMCAP has stable prediction performance. Furthermore, most confirmed candidates have higher rankings, indicating that top K performance of GCMCAP is outstanding.

Table 3. Evidences of the top 10 associated miRNA candidates for the three given diseases

Disease No. of miRNAs confirmed	By the latest HMDD	Top 10 ranked predictions					
		Rank	miRNAs	Evidences	Rank	miRNAs	Evidences
Glioma	6	1	hsa-mir-219	HMDD v3.2	6	hsa-mir-429	HMDD v3.2
		2	hsa-mir-136	HMDD v3.2	7	hsa-mir-642a	Unconfirmed
		3	hsa-mir-135b	HMDD v3.2	8	hsa-mir-103a	Unconfirmed
		4	hsa-mir-216a	Unconfirmed	9	hsa-mir-29c	HMDD v3.2
		5	hsa-mir-195	Unconfirmed	10	hsa-mir-296	HMDD v3.2

(*continued*)

Table 3. (*continued*)

Disease No. of miRNAs confirmed	By the latest HMDD	Top 10 ranked predictions					
		Rank	miRNAs	Evidences	Rank	miRNAs	Evidences
Carcinoma, Hepatocellular	6	1	hsa-mir-1296	HMDD v3.2	6	hsa-mir-519a	HMDD v3.2
		2	hsa-mir-632	Unconfirmed	7	hsa-mir-219	Unconfirmed
		3	hsa-mir-17	HMDD v3.2	8	hsa-mir-490	HMDD v3.2
		4	hsa-mir-379	HMDD v3.2	9	hsa-mir-545	Unconfirmed
		5	hsa-mir-518a	HMDD v3.2	10	hsa-mir-659	Unconfirmed
Ovarian Neoplasms	5	1	hsa-mir-134	HMDD v3.2	6	hsa-mir-379	Unconfirmed
		2	hsa-mir-1296	Unconfirmed	7	hsa-mir-518a	Unconfirmed
		3	hsa-mir-632	Unconfirmed	8	hsa-mir-519a	HMDD v3.2
		4	hsa-mir-17	HMDD v3.2	9	hsa-mir-219	HMDD v3.2
		5	hsa-mir-362	Unconfirmed	10	hsa-mir-490	HMDD v3.2

4 Discussion and Conclusion

Identifying miRNA-disease relationships helps to understand disease pathogenesis, diagnosis and treatment. We make observation that existing computational methods still have room for improvement in considering the topology and prior information of network nodes. To adapt to the graph structured disease network and miRNA network, we use graph convolutional networks to model miRNA and disease node embeddings. In addition, GCMCAP also takes advantage of matrix completion theory to circumvent the problem of no negative sample in training progress. Cross validations are implemented to evaluate the performance of GCMCAP. The proposed method GCMCAP obtained an average AUC value of 0.894 and 0.896. In comparison with several other methods, GCMCAP outperforms other baselines in terms of most indictors. The result shows that associations between disease and miRNA can be more accurately identified by GCMCAP. The process of parameter analysis shows that our model is less affected by parameters and has stable prediction ability. Case studies show the ability of discovering new disease associated miRNAs. For future work, we may explore obtaining miRNA and disease node information from other data sources for more prior information. Additionally, graph convolution networks with attention mechanism may further improve the predictive ability for disease-related miRNAs.

Acknowledgements. This work has been supported by the National Natural Science Foundation of China (Grant no. 61873089).

References

1. Ambros, V.: microRNAs: tiny regulators with great potential. Cell **107**(7), 823–826 (2001)
2. Chen, C.-Z., Li, L., Lodish, H.F., Bartel, D.P.: MicroRNAs modulate hematopoietic lineage differentiation. Science **303**(5654), 83–86 (2004)
3. Ambros, V.J.C.: MicroRNA pathways in flies and worms: growth, death, fat, stress, and timing. Cell **113**(6), 673–676 (2003)
4. Taganov, K.D., Boldin, M.P., Chang, K.-J., Baltimore, D.J.P.: NF-κB-dependent induction of microRNA miR-146, an inhibitor targeted to signaling proteins of innate immune responses. Proc. Natl Acad. Sci. **103**(33), 12481–12486 (2006)
5. Petrocca, F., et al.: E2F1-regulated microRNAs impair TGFβ-dependent cell-cycle arrest and apoptosis in gastric cancer. Cancer cell **13**(3), 272–286 (2008)
6. Shi, B., Sepp-Lorenzino, L., Prisco, M., Linsley, P., DeAngelis, T., Baserga, R.: Micro RNA 145 targets the insulin receptor substrate-1 and inhibits the growth of colon cancer cells. J. Biol. Chem. **282**(45), 32582–32590 (2007)
7. Zare, M., Bastami, M., Solali, S., Alivand, M.R.: Aberrant miRNA promoter methylation and EMT-involving miRNAs in breast cancer metastasis: diagnosis and therapeutic implications. J. Cell. Physiol. **233**(5), 3729–3744 (2018)
8. Li, Y., et al.: HMDD v2.0: a database for experimentally supported human microRNA and disease associations. Nucleic Acids Res. **42**(D1), D1070–D1074 (2013)
9. Luo, J., Xiao, Q.: A novel approach for predicting microRNA-disease associations by unbalanced bi-random walk on heterogeneous network. J. Biomed. Inform. **66**, 194–203 (2017)
10. Zou, Q., Li, J., Song, L., Zeng, X., Wang, G.: Similarity computation strategies in the microRNA-disease network: a survey. Briefings Funct. Genomics **15**(1), 55–64 (2015)
11. Chen, X.J.M.B.: miREFRWR: a novel disease-related microRNA-environmental factor interactions prediction method. Mol. BioSyst. **12**(2), 624–633 (2016)
12. Zeng, X., Zhang, X., Zou, Q.: Integrative approaches for predicting microRNA function and prioritizing disease-related microRNA using biological interaction networks. Briefings Bioinform. **17**(2), 193–203 (2015)
13. Ding, P., Luo, J., Xiao, Q., Chen, X.: A path-based measurement for human miRNA functional similarities using miRNA-disease associations. Sci. Rep. **6**, 32533 (2016)
14. Zhao, X.-M., et al.: Identifying cancer-related microRNAs based on gene expression data. Bioinformatics **31**(8), 1226–1234 (2014)
15. Chen, X., Liu, M.-X., Yan, G.-Y.: RWRMDA: predicting novel human microRNA–disease associations. Mol. BioSyst. **8**(10), 2792–2798 (2012)
16. Xuan, P., et al.: Prediction of microRNAs associated with human diseases based on weighted k most similar neighbors. PLoS ONE **8**(8), e70204 (2013)
17. You, Z.-H., et al.: PBMDA: a novel and effective path-based computational model for miRNA-disease association prediction. PLoS Comput. Biol. **13**(3), e1005455 (2017)
18. Zou, Q., et al.: Prediction of microRNA-disease associations based on social network analysis methods. BioMed Res. Int. **2015** (2015)
19. Xiao, Q., Luo, J., Liang, C., Cai, J., Ding, P.: A graph regularized non-negative matrix factorization method for identifying microRNA-disease associations. Bioinformatics **34**(2), 239–248 (2018)

20. Luo, J., Shen, C., Lai, Z., Cai, J., Ding, P.: Incorporating clinical, chemical and biological information for predicting small molecule-microRNA associations based on non-negative matrix factorization. IEEE/ACM Trans. Comput. Biol. Bioinform. 1 (2020)
21. Luo, J., Ding, P., Liang, C., Cao, B., Chen, X.: Collective prediction of disease-associated miRNAs based on transduction learning. IEEE/ACM Trans. Comput. Biol. Bioinf. **14**(6), 1468–1475 (2016)
22. Chen, X., Huang, L.: LRSSLMDA: Laplacian regularized sparse subspace learning for MiRNA-disease association prediction. PLoS Comput. Biol. **13**(12), e1005912 (2017)
23. Li, J.-Q., Rong, Z.-H., Chen, X., Yan, G.-Y., You, Z.-H.: MCMDA: matrix completion for MiRNA-disease association prediction. Oncotarget **8**(13), 21187 (2017)
24. Hou, S.: Neural Inductive Matrix Completion for Predicting Disease-Gene Associations (2018)
25. Xuan, P., Sun, H., Wang, X., Zhang, T., Pan, S.: Inferring the disease-associated miRNAs based on network representation learning and convolutional neural networks. Int. J. Mol. Sci. **20**(15), 3648 (2019)
26. Han, P., et al.: GCN-MF: disease-gene association identification by graph convolutional networks and matrix factorization. In: Proceedings of the 25th ACM SIGKDD International Conference on Knowledge Discovery & Data Mining, pp. 705–713 (2019)
27. Xiao, F., Zuo, Z., Cai, G., Kang, S., Gao, X., Li, T.: miRecords: an integrated resource for microRNA–target interactions. Nucleic Acids Res. **37**(Suppl. 1), D105–D110 (2008)
28. Vergoulis, T., et al.: TarBase 6.0: capturing the exponential growth of miRNA targets with experimental support. Nucleic Acids Res. **40**(D1), D222–D229 (2011)
29. Chou, C.-H., et al.: miRTarBase 2016: updates to the experimentally validated miRNA-target interactions database. Nucleic Acids Res. **44**(D1), D239–D247 (2016)
30. Lee, I., Blom, U.M., Wang, P.I., Shim, J.E., Marcotte, E.M.: Prioritizing candidate disease genes by network-based boosting of genome-wide association data. Genome Res. **21**(7), 1109–1121 (2011)
31. Xuan, P., et al.: Prediction of potential disease-associated microRNAs based on random walk. Bioinformatics **31**(11), 1805–1815 (2015)
32. Wang, D., Wang, J., Lu, M., Song, F., Cui, Q.: Inferring the human microRNA functional similarity and functional network based on microRNA-associated diseases. Bioinformatics **26**(13), 1644–1650 (2010)
33. Wang, J.Z., Du, Z., Payattakool, R., Yu, P.S., Chen, C.-F.: A new method to measure the semantic similarity of GO terms. Bioinformatics **23**(10), 1274–1281 (2007)
34. Berg, R., Kipf, T.N., Welling, M.: Graph convolutional matrix completion, arXiv preprint arXiv:1706.02263 (2017)
35. Zitnik, M., Agrawal, M., Leskovec, J.: Modeling polypharmacy side effects with graph convolutional networks. Bioinformatics **34**(13), i457–i466 (2018)
36. Kipf, T.N., Welling, M.: Semi-supervised classification with graph convolutional networks (2016)
37. Koren, Y., Bell, R., Volinsky, C.: Matrix factorization techniques for recommender systems. Computer **42**(8), 30–37 (2009)
38. Natarajan, N., Dhillon, I.S.: Inductive matrix completion for predicting gene–disease associations. Bioinformatics **30**(12), i60–i68 (2014)
39. Kingma, D.P., Ba, J.: Adam: a method for stochastic optimization, arXiv preprint arXiv: 1412.6980 (2014)
40. Gui, H., Liu, J., Tao, F., Jiang, M., Norick, B., Han, J.: Large-scale embedding learning in heterogeneous event data. In: 2016 IEEE 16th International Conference on Data Mining (ICDM), pp. 907–912. IEEE (2016)

An Efficient Computational Method to Predict Drug-Target Interactions Utilizing Structural Perturbation Method

Xinguo Lu(✉), Fang Liu, Li Ding, Xinyu Wang, Jinxin Li, and Yue Yuan

College of Computer Science and Electronic Engineering, Hunan University, Changsha, China
hnluxinguo@126.com

Abstract. Accurately and quickly identifying potential drug candidates for therapeutic targets (i.e., drug-target interactions, DTIs) is a basic step in the early drug discovery process. However, the experimental determination of drug-target interactions is expensive and time-consuming. Therefore, a continuous demand that the effective calculation algorithm is developed with a more accurate prediction of drug-target interactions. Some algorithms have been designed to infer new interactions but ignored the link generation mechanism in the DTI network, and the missing information can be completed by generalizing the observed DTI network structure according to some consistency rules. We propose a calculation model named SPDTI, which is based on the structural perturbation method and uses the explicit and implicit relations of the drug-target adjacency to predict novel DTIs. In our framework, we first construct the implicit relationship between nodes of the same type through the display relationship of the DTI adjacency matrix. Then a bilayer network is designed to integrate these two relationships, and finally, the structure perturbation method is used to predict the potential interactions between drugs and targets. To evaluate the performance of SPDTI, we carried out a lot of experiments on four benchmark datasets. The experimental results show that SPM can be used as an effective computing method to improve the identification accuracy of drug-target interactions.

Keywords: Drug-target interaction · Structural perturbation · Implicit relations

1 Introduction

Discovering to determine the potential beneficial therapeutic effects or medical uses of a new drug is a lengthy process. Including target identification and verification, compound leads to be identified, verified, and optimized, as well as the initial stages of different types of preclinical and clinical trials until the final approval by the US Food and Drug Administration (FDA) [1]. Drug treat diseases by acting on their targets [2–5]. From the perspective of cost and time, the introduction of new drugs from abstract concepts to the market is a costly and long process. Despite the increasing investment in drug development and design, in fact drugs rarely only bind to the intended target, as well as non-target effects are very common and may even cause unnecessary side

D.-S. Huang and K.-H. Jo (Eds.): ICIC 2020, LNCS 12464, pp. 216–226, 2020.
https://doi.org/10.1007/978-3-030-60802-6_19

effects [6], which is another important reason why the number of new drugs put on the market every year shows a downward trend. This phenomenon also promotes research in the field of drug reuse, that is, the application of developed drugs to treat new diseases. With the continuous improvement of public biomedical databases, it has become possible to develop a practical computing framework, overcoming the limitations of traditional experimental methods, to help find new associations between existing drugs and off-target effects [7]. Nevertheless, owing to the potentially unknown complexity of pharmacology and biology, drug reuse is full of challenges.

Computational prediction methods are very useful for prioritizing candidate drug-target interactions by utilizing known drug-target information [8]. How to use the observed information to predict potential DTIs? This problem is actually a matter of link prediction for drug-target networks. Kevin et al. proposed a supervised learning calculation model called BLM [9]. In this study, the authors used a bipartite local model (RLSavg [10]) as the basic algorithm to predict unknown drug-target interactions by converting edge prediction problems into binary classification problems. In 2013, Mei et al. combined the neighbor-based interaction-profile inferring (NII) with the existing BLM model to further expand the BLM model [11]. The idea for this method came from the BLM model could not be applied without any known target of the drug. Xia et al. formulated a semi-supervised prediction model, namely, Laplacian regularized least square (LapRLS) [12] and augments it to include the kernel constitute from the observed drug-target interaction network [13]. Wang et al. focused on using the topology information in the drug-target network, which integrated the observed DTIs, drug-drug similarity, and target-target similarity into a heterogeneous graph, modeling the DTIs as stable information flow between them across the heterogeneous graph, and making full use of the hidden relationship between drug and target to predict the DTIs [14]. To further improve their method, Wang et al. extended the two-layer heterogeneous graph model (HGBI) to the three-layer heterogeneous graph model (TL_HGBI), by capturing the internal relationship among diseases-diseases, drugs-drugs and targets-targets, as well as the relationship among diseases, drugs and targets to predict the use of the new drug [15]. Laarhoven et al. introduced a simple weighted nearest neighbor algorithm (WNN) and employed RLS with the WNN algorithm for DTIs prediction problems [16]. Moreover, Zheng et al. proposed a multiple similarities collaborative matrix factorization method (MSCMF), which combined multiple kinds of target similarities and drug similarities to improve the drug-target interactions prediction accuracy [17]. Luo et al. used the local structure of the interaction data and developed a method based on logistic matrix factorization (NRLMF) to predict the DTIs [18]. The DNILMF model combines the profile information of the drug and target to enhance the NRLMF method [19]. Meanwhile, NeoDTI is a model that integrates neural information passing and aggregation technology into an end-to-end learning framework, extracting potential features of drugs and targets from heterogeneous networks for DTI prediction. The method used a neural network algorithm and utilized the information about nodes such as drugs, targets, side effects, and diseases [20]. Moreover, TriModel is an effective drug target association prediction algorithm based on knowledge graph embedding. The model transform the DTIs prediction problem into representation learning and link prediction problems in the knowledge graph. It integrated existing biological knowledge bases such as KEGG, UniProt, and Drugbank to generate knowledge maps related to drugs and targets, and then learn effective vector representations of drugs and targets through the TriModel model [21].

The existing methods can be divided into four categories: (i) network-based models, such as HGBI [14] and TL_HGBI [15]; (ii) machine learning-based models, such as BLM [9], NetLapRLS [13] and TriModel [21]; (iii) matrix factorization-based models, such as MSCMF [17], NRLMF [18], DNILMF [19]; (iii) Deep learning-based models, such as NeoDTI [20].

The existing researches ignore the link generation mechanism in the drug-target network. The DTIs prediction problem can be modeled as the link prediction problem in the bipartite networks and the observed DTI network structure can be generalized according to some consistency rules to correct the missing information [22]. If the added new link has a little effect on the consistency of the original network structure, the more likely it is a missing link [23]. In this study, we propose a novel approach, named SPDTI, for DTI prediction, and we apply the SPDTI algorithm to Yamanishi_08 dataset. In order to make the structure perturbation method [23] applicable to bipartite networks, we construct the implicit relationship between drugs and the implicit relationship between targets and targets through the explicit relationship of DTI adjacency matrix. Then integrate these two relationships to design a bilayer network. We use perturbation sets on the constructed bilayer network to stir the remaining links through first-order approximation, and then obtain the perturbed adjacency matrix. The unobserved links are sorted according to their scores in the perturbation matrix, and a new association that does not seriously affect the consistency of the network structure is selected, that is, the interaction with the highest ranking is selected as the prediction result. The experimental results show that our method can effectively predict the drug-target interactions in the 5-fold cross-validation and obtain better performance than many previous methods.

2 Materials and Methods

2.1 Datasets

The observed DTIs was downloaded from [24] to evaluate the DTI prediction algorithm. The Yamanishi_08 dataset was originally compiled by [24] and was used as a reference in many studies [11, 25, 26]. This dataset contains known drug-target interactions as retrieved from DrugBank [27], SuperTarget [28], BRENDA [29] and KEGG BRITE databases [30]. DTIs information in the dataset is divided into four groups according to different target protein class: (i) enzymes (E); (ii) ion-channels (IC); (iii) G-protein-coupled receptors (GPCR) and (iv) nuclear receptors (NR). Table 1 shows the statistics of the four data types.

Table 1. Summary of the DTI datasets in this study.

Datasets	Drugs	Targets	DTIs	Sparsity of the interaction matrix
IC	210	204	1476	96.55%
GPCR	223	95	635	97.00%
NR	54	26	90	93.59%
E	445	664	2926	99.01%

2.2 Measurement of Implicit Relationship Between Nodes of the Same Type

In a bipartite graph network, two nodes of the same type are not directly connected, but this does not mean that they are irrelevant. The effectiveness of implicit relations (higher-order information) in network embedded prompting us to explore the implicit relations of link prediction in bipartite graph networks [31]. Based on the idea that similar drug exhibit similar interaction profiles with target, we used the Gaussian Interaction Profile (GIP) kernel to express the implicit relationship between drug nodes. Let $D = \{d_1, d_2, \ldots, d_m\}$ is the set of drugs, $T = \{t_1, t_2, \ldots, t_m\}$ is the set of targets, and $Y \in R^{m \times n}$ is the adjacency matrix of the bipartite graph network to describe known drug-target interactions with m drugs as rows and n targets as columns. If drug d_i and target t_i have a known interaction, the y_{ij} is 1, otherwise 0. The implicit relationship between drug d_1 and d_2 can be calculated as follows.

$$S_d(d_1, d_2) = \exp(-\gamma_d \parallel yd_1 - yd_2 \parallel^2) \tag{1}$$

where $y_{d_1} = \{y_{11}, y_{12}, \ldots, y_{1m}\}$ and $y_{d_2} = \{y_{21}, y_{22}, \ldots, y_{2m}\}$ are the interaction profiles of drug d_1 and drug d_2, respectively. The parameter γ_d is used to regulate the kernel bandwidth, which can be obtained from the standardization of new bandwidth γ_d' by the average number of targets associated with each drug. Thus, γ_d can be computed as follows.

$$\gamma_d = \gamma_d' / \left(\frac{1}{m} \sum_{i=1}^{m} \parallel yd_i \parallel^2 \right) \tag{2}$$

the parameter γ_d' can be set by cross-validation. In this study, we set the parameter γ_d' to 1 according to the previous studies [16].

The kernel S_t for the similarity between target t_1 and target t_2 and the kernel bandwidth γ_t between target t_1 and target t_2 are defined analogously.

$$S_t(t_1, t_2) = \exp(-\gamma_t \parallel yt_1 - yt_2 \parallel^2) \tag{3}$$

$$\gamma_t = \gamma_t' / \left(\frac{1}{n} \sum_{i=1}^{n} \parallel yt_i \parallel^2 \right) \tag{4}$$

2.3 Construction of the Bilayer Network

Since the structural perturbation method can only be applied to real symmetric matrices, we construct a bilayer network using A_{dt}, A_{dd} and A_{tt}. We define matrix A_{dd} as an implicit network of drug, the weights are S_d values. And A_{tt} as the an implicit network of target, the weights are S_t values. We define matrix A_{dt} as the known DTIs network, if drug i and target j are connected, the element $A_{dt}(i,j) = 1$; otherwise, $A_{dt}(i,j) = 0$. Therefore, drug-target bilayer network can be expressed by a $\mathrm{K} \times \mathrm{K}$ (K = M+N is the total number of drugs and targets in the network) undirected adjacency matrix $B_{K \times K}$.

$$B = \begin{bmatrix} A_{dd} & A_{dt} \\ \mathrm{A}_{dt}^{T} & A_{tt} \end{bmatrix} \tag{5}$$

2.4 Structural Perturbation Method for DTI Prediction

When the network changes dynamically due to a certain evolution or disturbance process, the information is represented by a certain regularity or consistency in the network structure [22]. The structural perturbation method first-order approximation agitates the rest of the links by randomly deleting a small number of links. The purpose is to recover the missing links considered as unknown information by perturbing the network with another set of observed links. It is worth noting that the original SPM [23] can only be used for monopartite link prediction. When the network is an asymmetric matrix that cannot be diagonalized, the original SPM is not available. In this article, we extend this method to bipartite link prediction on the bilayer network.

The drug-target adjacency matrix B can be represented as a bipartite graph network $\mathrm{G(U, V, E)}$. U is the set of drug nodes, V is the set of target nodes and E is the set of the weighted edges. Then, we randomly remove a small part of links from E to construct the perturbation set ΔE, and the rest of the links define as $E^R = E - \Delta E$. Therefore, we can get a new network $\mathrm{G}^R(\mathrm{U}, \mathrm{V}, \mathrm{E}^R)$ with the adjacency matrix $\mathrm{B}^R = B - \Delta B$, where ΔB denote the adjacency matrix for removed links. B^R is real symmetric, thus it can be defined as follows.

$$B^R = \sum_{k=1}^{N} \lambda_k x_k x_k^T \tag{6}$$

where x_k are the orthogonal and normalized eigenvectors for adjacency matrix B^R, and λ_k are the corresponding eigenvalues.

We used the ΔE as a perturbation to the B^R, then, the perturbed matrix is constructed by the first-order approximation which allows the eigenvalue to change but keep the eigenvector unchanged. After perturbation, the eigenvalue λ_k is change to be $\lambda_k + \Delta\lambda_k$, and its corresponding eigenvalue will change to be $x_k + \Delta x_k$. Therefore we have

$$(B^R + \Delta B)(x_k + \Delta x_k) = (\lambda_k + \Delta\lambda_k)(x_k + \Delta x_k), k = 1, 2, \ldots, n. \tag{7}$$

By left-multiplying eigenfunction x_k^T in Eq. 4 and neglecting the second-order terms $x_k^T \Delta B \Delta x_k$ and $\Delta\lambda_k x_k^T \Delta x_k$, we can obtain

$$\Delta\lambda_k \approx \frac{x_k^T \Delta B x_k}{x_k^T x_k} \tag{8}$$

By fixing the eigenvectors, the perturbed matrix B' can be obtained as follows according to the perturbed eigenvalue,

$$B' = \sum_{k=1}^{N} (\lambda_k + \Delta\lambda_k) x_k x_k^T \tag{9}$$

which can be regarded as a linear approximation of the given bilayer network B, if the expansion is based on B^R.

Matrix eigenvectors can reflect the structural characteristics of the network [23]. If the eigenvectors of matrix B' and matrix B are almost the same, it means that the perturbation set has no significant effect on the structural characteristics. The smaller the effect of the restored new link on the original network structure, the more likely it is the missing link. We choose the upper right DT' of B" to get the final perturbed matrix, B" is obtained by averaging over 5 independent selections of perturbation set. All unobserved links are sorted in descending order according to their corresponding scores in the perturbed matrix. The higher the score of these links, the greater the possibility of existence.

3 Result

3.1 Experimental Setting and Evaluation Metrics

We used 5-fold cross validation (5-CV) based on the observed drug-target interactions in Yamanishi_08 dataset to evaluate the predictive performance of the SPDTI model. The drug-target pairs in A_{dt} are randomly divided into 5 equal sized subsets. In each fold, a single subset is used as the probe set for testing the model, and the remaining 4 subsets together with the A_{dd} and A_{tt} are used as training sets. We randomly select a fraction of links from training sets to form the perturbation set. In this work, 10% of links in training sets are selected at random to form the perturbation set. To avoid the bias in data split, the cross validation process is run independently 5 times, and each of the 4 subsets is only used once as the probe set. Then, we average the calculated 5 results to obtain the final estimate.

Here, for each prediction algorithm, at each in case of cross-validation, we adopt four evaluation metrics to measure performances of prediction models, i.e., precision, recall and area under the ROC curve (AUC) and the area under the precision-recall curve (AUPR). In our task, DTIs take a small proportion of all drug pairs, and thus AUPR, which takes into account both recall and precision, is used as the primary evaluation metric.

3.2 Prediction Results on the Bilayer Network

To demonstrate the effectiveness of our method, we searched the highly ranked interactions in the ChEMBL, DrugBank, and KEGG databases, and found that the top ten of the four databases were reported in at least one of the three databases, which shows that our method is reliable. Table 2 shows the top ten DTIs predicted by our developed SPDTI model on IC, GPCR, NR, and E datasets.

Table 2. Top ten predicted interactions by SPDTI on NR, GPCR, IC, and E datasets under 5-CV

	Rank	Drug	Target		Rank	Drug	Target
IC	1	D02356	hsa23704	NR	1	D00954	hsa2099
	2	D00332	hsa774		2	D04066	hsa2099
	3	D02356	hsa6263		3	D00094	hsa6258
	4	D00358	hsa6331		4	D00067	hsa2099
	5	D00437	hsa785		5	D00312	hsa2099
	6	D00358	hsa6328		6	D00129	hsa7421
	7	D00629	hsa785		7	D00105	hsa2099
	8	D00294	hsa3788		8	D00316	hsa6256
	9	D00629	hsa782		9	D00930	hsa7421
	10	D02356	hsa3743		10	D00316	hsa5914
GPCR	1	D00454	hsa3356	E	1	D03784	hsa1545
	2	D00607	hsa146		2	D03784	hsa1548
	3	D00454	hsa148		3	D00225	hsa1551
	4	D00509	hsa146		4	D00964	hsa1545
	5	D01603	hsa146		5	D03784	hsa1551
	6	D00454	hsa146		6	D01425	hsa1545
	7	D00283	hsa1813		7	D01198	hsa5151
	8	D00283	hsa3356		8	D01133	hsa5151
	9	D00513	hsa146		9	D00528	hsa1559
	10	D00283	hsa3354		10	D00964	hsa1548

3.3 Comparison with Other Existing Methods

We evaluate our algorithm on the Yamanishi_08 database, and compare the SPDTI with five widely applied drug-target interactions prediction method: NetLapRLS [13], BLM-NII [11], WNN-GIP [16], MSCMF [17] and NRLMF [18] under the same conditions for all model, i.e. under the 5-fold cross-validation based on same datasets, we set the parameters of these methods as the same of these best setting provided in original paper.

We show that SPDTI, using 5-fold cross validation, achieved better performance than other methods. Figure 1 and Fig. 2 shows the overall results of AUC values and AUPR values for all compared methods. The results show that the SPDTI algorithm is superior to all other algorithms in the terms of AUC and AUPR on all benchmarking dataset. It is worth noting that our model achieves a better AUPR value with a margin of 2%, 8%, 7%, 2% on IC, GPCR, NR and E dataset, respectively.

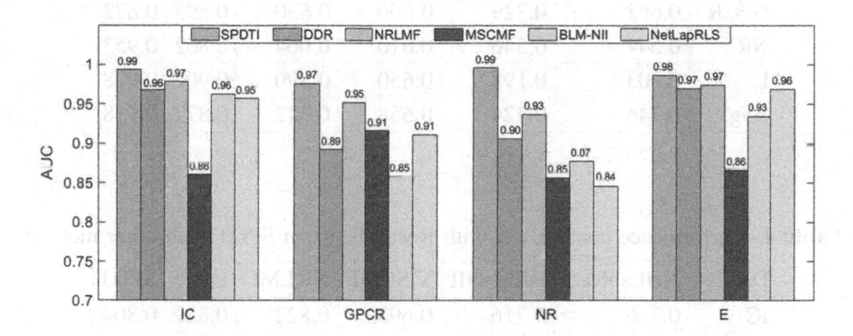

Fig. 1. The AUCs averaged over 10 runs by different methods.

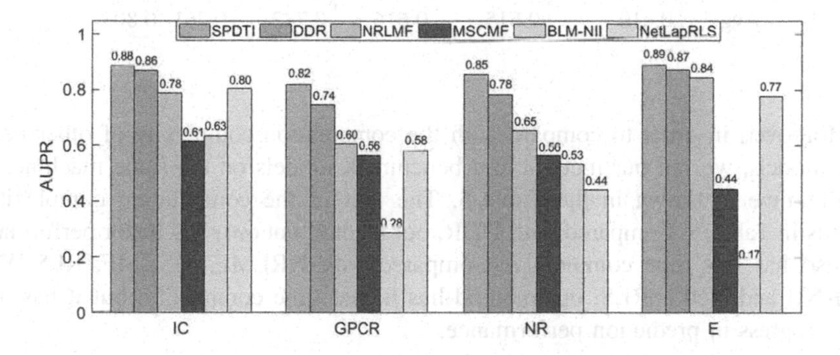

Fig. 2. The AUPRs averaged over 10 runs by different methods.

In addition, Table 3 and 4 shows the Precision and Recall. The average Precision scores by SPDTI is 0.918, which is 4.1% higher than the second method DDR. Moreover, our approach obtains the best Recall values over three datasets (i.e., GPCR, NR, E) and achieves the third-best Recall scores on the IC dataset, where NRLMF and DDR outperforms SPDTI. The average Recall obtained by SPDTI is 0.803, which is 4.2% higher than that achieves by the second-best model DDR. It is worth noting that different from other methods, our method is also very useful for small data, and it has also achieved good performance on the NR data set. This may be because our method considers the structural characteristics of the network, so its predictive ability is not limited to the size of the network. In summary, based on our prediction results (Fig. 1, Table 3 and 4), the SPDTI achieves superior results by using the universal structural feature information.

Table 3. Performance comparison with Precision between SPDTI and other methods in 5-CV based on IC, GPCR,NR and E datasets.

Dataset	NetLapRLS	BLM-NII	MSCMF	NRLMF	DRR	SPDTI
IC	0.873	0.621	0.776	0.785	0.911	0.918
GPCR	0.663	0.329	0.596	0.630	0.893	0.872
NR	0.544	0.546	0.610	0.664	0.802	0.953
E	0.903	0.198	0.650	0.890	0.902	0.928
Avg.	0.746	0.424	0.658	0.742	0.877	0.918

Table 4. Performance comparison with Recall between SPDTI and other methods.

Dataset	NetLapRLS	BLM-NII	MSCMF	NRLMF	DDR	SPDTI
IC	0.716	0.716	0.693	0.822	0.829	0.804
GPCR	0.567	0.494	0.608	0.694	0.703	0.754
NR	0.441	0.600	0.534	0.657	0.702	0.820
E	0.714	0.650	0.630	0.794	0.811	0.831
Avg.	0.610	0.615	0.616	0.742	0.761	0.803

Moreover, in order to compare with the computation complexity of other benchmark models, we ran our method and benchmark models on the same machine. The running time is shown in the Table 5. The unit of the computation complexity is seconds in Table 5. Compared with DDR, our method not only has better performance, but also has less time complexity. Compared with NRLMF, MSCMF, RLS-WNN, BLM-NII and NetLapRLS, our method has higher time complexity, but it has made great progress in prediction performance.

Table 5. Computation complexity comparison between SPDTI and other methods.

Dataset	NetLapRLS	BLM-NII	MSCMF	NRLMF	DDR	SPDTI
IC	3.20	1.10	349.47	22.90	11399.18	1990.12
GPCR	2.12	0.72	219.27	13.58	3396.13	910.31
NR	0.24	0.19	41.24	1.62	337.18	19.21
E	28.30	14.12	815.48	183.27	847454.55	49378.74

4 Conclusions

In this article, we propose a novel framework to predict drug-target interactions. Our method constructs a bilayer network to integrate the explicit and implicit information of the DTI adjacency matrix, and then uses the idea of matrix perturbation to predict the potential drug-target interactions, SPDTI considers the link generation mechanism in the DTI network, and uses the general structural feature information to generalize the observed DTI network structure according to the consistency rules to recover the lost link. We compared SPDTI with NetLapRLS, BLM-NII, RLSWNN, MSCMF, and NRLMF. The results show that our model is powerful in discovering potential DTIs. In addition, the high ranking interactions predicted by our method were reported in the existing database, which further proves the reliability of our model. In the future, we will consider how to design the network to improve the consistency of the network structure to further improve the predictive performance of drug targets.

Acknowledgements. This work was supported by Natural Science Foundation of China (Grant No. 61972141) and Natural Science Foundation of Hunan Province, China (Grant No. 2018JJ2053).

References

1. Olayan, R.S., Ashoor, H., Bajic, V.B.: DDR: efficient computational method to predict drug–target interactions using graph mining and machine learning approaches. Bioinformatics **34** (7), 1164–1173 (2017)
2. Gao, M.-M., Cui, Z., Gao, Y.-L., Liu, J.-X., Zheng, C.-H.: Dual-network sparse graph regularized matrix factorization for predicting miRNA-disease associations. Mol. Omics **15** (2), 130–137 (2019)
3. Yin, M.-M., Cui, Z., Gao, M.-M., Liu, J.-X., Gao, Y.-L.: LWPCMF: logistic weighted profile-based collaborative matrix factorization for predicting MiRNA-disease associations. IEEE/ACM Trans. Comput. Biol. Bioinform. **PP**(99), 1 (2019)
4. Cui, Z., Liu, J.-X., Gao, Y.-L., Zheng, C.-H., Wang, J.: RCMF: a robust collaborative matrix factorization method to predict miRNA-disease associations. BMC Bioinform. **20**(S25), 686 (2019)
5. Gao, Y.-L., Cui, Z., Liu, J.-X., Wang, J., Zheng, C.-H.: NPCMF: nearest profile-based collaborative matrix factorization method for predicting miRNA-disease associations. BMC Bioinform. **20**, 353 (2019)
6. Bowes, J., et al.: Reducing safety-related drug attrition: the use of in vitro pharmacological profiling. Nat. Rev. Drug Discov. **11**(12), 909–922 (2012)
7. Vilar, S., Hripcsak, G.: The role of drug profiles as similarity metrics: applications to repurposing, adverse effects detection and drug-drug interactions. Brief. Bioinform. **18**, 670–681 (2016)
8. Yuan, Q., et al.: DrugE-Rank: improving drug–target interaction prediction of new candidate drugs or targets by ensemble learning to rank. Bioinformatics **32**(12), i18–i30 (2016)
9. Bleakley, K., Yamanishi, Y.: Supervised prediction of drug–target interactions using bipartite local models. Bioinformatics **25**(18), 2397–2403 (2009)
10. Suykens, J.A.K., Vandewalle, J.: Least squares support vector machine classifiers. Neural Process. Lett. **9**(3), 293–300 (1999)

11. Mei, J.-P., Kwoh, C.-K., Yang, P., Li, X.-L., Zheng, J.: Drug–target interaction prediction by learning from local information and neighbors. Bioinformatics **29**(2), 238–245 (2013)
12. Belkin, M., Niyogi, P., Sindhwani, V.: Manifold regularization: a geometric framework for learning from labeled and unlabeled examples. J. Mach. Learn. Res. **7**(1), 2399–2434 (2006)
13. Xia, Z., Wu, L.Y., Zhou, X., Wong, S.T.: Semi-supervised drug-protein interaction prediction from heterogeneous biological spaces. BMC Syst. Biol. **4**(Suppl. 2), S6 (2010)
14. Wang, W., et al.: Drug target predictions based on heterogeneous graph inference. In: Pacific Symposium on Biocomputing, pp. 53–64 (2013)
15. Wang, W., Yang, S., Zhang, X., Li, J.: Drug repositioning by integrating target information through a heterogeneous network model. Bioinformatics **30**(20), 2923–2930 (2014)
16. van Laarhoven, T., Marchiori, E.: Predicting drug-target interactions for new drug compounds using a weighted nearest neighbor profile. PLoS ONE. **8**(6), e66952 (2013)
17. Zheng, X., Ding, H., Mamitsuka, H., Zhu, S.: Collaborative matrix factorization with multiple similarities for predicting drug-target interactions. In: KDD'13: Proceedings of the 19th ACM SIGKDD International Conference on Knowledge Discovery and Data Mining, pp. 1025–1033 (2013)
18. Liu, Y., et al.: Neighborhood regularized logistic matrix factorization for drug-target interaction prediction. PLoS Comput. Biol. **12**, e1004760 (2016)
19. Hao, M., et al.: Predicting drug-target interactions by dual-network integrated logistic matrix factorization. Sci. Rep. **7**, 40376 (2017)
20. Wan, F., Hong, L., Xiao, A., Jiang, T., Zeng, J.: NeoDTI: neural integration of neighbor information from a heterogeneous network for discovering new drug–target interactions. Bioinformatics **35**(1), 104–111 (2019)
21. Mohamed, S.K., Nováček, V., Nounu, A.: Discovering protein drug targets using knowledge graph embeddings. Bioinformatics **36**(2), 603–610 (2020)
22. Xu, X., Liu, B., Wu, J., et al.: Link prediction in complex networks via matrix perturbation and decomposition. Sci. Rep. **7**(1), 14724 (2017)
23. Lü, L., et al.: Toward link predictability of complex networks. Proc. Natl. Acad. Sci. **112**, 2325–2330 (2015)
24. Yamanishi, Y., Araki, M., Gutteridge, A., Honda, W., Kanehisa, M.: Prediction of drug–target interaction networks from the integration of chemical and genomic spaces. Bioinformatics **24**(13), i232–i240 (2008)
25. Ba-Alawi, W., et al.: DASPfind: new efficient method to predict drug-target interactions. Cheminformatics **8**(1), 15 (2016)
26. Daminelli, S., et al.: Common neighbours and the local-community-paradigm for topological link prediction in bipartite networks. New J. Phys. **17**(11), 113037 (2015)
27. Wishart, D.S., et al.: DrugBank: a knowledgebase for drugs, drug actions and drug targets. Nucleic Acids Res. **36**, D901–D906 (2008)
28. Gunther, S., et al.: SuperTarget and Matador: resources for exploring drug-target relationships. Nucleic Acids Res. **36**, D919–D922 (2008)
29. Schomburg, I.: BRENDA, the enzyme database: updates and major new developments. Nucleic Acids Res. **32**, D431–D433 (2004)
30. Kanehisa, M., et al.: From genomics to chemical genomics: new developments in KEGG. Nucleic Acids Res. **34**, D354–D357 (2006)
31. Chen, X., Jiao, P., Yu, Y., Li, X., Tang, M.: Toward link predictability of bipartite networks based on structural enhancement and structural perturbation. Physica A: Stat. Mech. Appl. **530**(121072), 0378–4371 (2019)

Predicting Human Disease-Associated piRNAs Based on Multi-source Information and Random Forest

Kai Zheng[1], Zhu-Hong You[2(✉)], Lei Wang[2(✉)], Hao-Yuan Li[2], and Bo-Ya Ji[2]

[1] School of Computer Science and Engineering, Central South University, Changsha, China
[2] Xinjiang Technical Institute of Physics and Chemistry, Chinese Academy of Sciences, Urumqi, China
{zhuhongyou,leiwang}@ms.xjb.ac.cn

Abstract. Whole genome analysis studies have shown that Piwi-interacting RNA (piRNA) play a crucial role in disease progression, diagnosis, and therapeutic target. However, traditional biological experiments are expensive and time-consuming. Thus, computational models could serve as a complementary means to provide potential disease-related piRNA candidates. In this study, we propose a novel computational model called APDA to identify piRNA-disease associations. The proposed method integrates disease semantic similarity and piRNA sequence information to construct feature vectors, and maps them to the optimal feature subspace through the stacked autoencoder to obtain the final feature vector. Finally, random forest classifier is used to infer disease-related piRNA. In five-fold cross-validation, the APDA achieved an average AUC of 0.9088 and standard deviation of 0.0126, which is significantly better than the compared method. Therefore, the proposed APDA method is a powerful and necessary tool for predicting human disease-associated piRNAs and provide new impetus to reveal the underlying causes of human disease.

Keywords: PIWI-interacting RNA · Disease · piRNA-disease associations · Heterogenous information · Multi-source information

1 Introduction

PIWI-interacting RNA (piRNA) is an important branch of non-coding RNA [1–6]. Compared to microRNA (miRNA) and Small interfering RNA (siRNA), the maturation of piRNA does not require the participation of the enzyme Dicer [7–10]. In addition, piRNA exists only in the animal kingdom. However, miRNAs and siRNAs are involved in the regulation of animals, plants, and protists [11]. It has been found in the past two decades that piRNAs, combined with Piwi proteins in the Argonaute family, inhibit insertional mutations caused by self-crossing DNA strands called transposons in animal germline cells [12]. This molecular defense is likened to the immune system of the genome and plays a vital role in maintaining the integrity of the genome [7, 13, 14]. piRNAs primarily express functioning in spermatogenesis, particularly, in germ cell [15].

D.-S. Huang and K.-H. Jo (Eds.): ICIC 2020, LNCS 12464, pp. 227–238, 2020.
https://doi.org/10.1007/978-3-030-60802-6_20

Since piRNAs are involved in transcriptional regulation of genes and epigenetic silencing of transposable elements, the role of them in human diseases is receiving increasing attention [16–21]. In principle, piRNA targeting mRNA transcripts affects tumor growth by inhibiting tumor suppressor genes or oncogenes [22–25]. On the other hand, ectopic expression of the piRNA/PIWI complex promotes genomic instability leading to a more aggressive cancer phenotype [11]. From concrete examples, piR-823 promotes tumor growth by promoting phosphorylation and its phosphorylation at Ser326 [26]. In the same year, Zhang et al. found that piR-932 /PIWIL2 complex leads to hypermethylation of Latexin, which plays a driving role in breast cancer [27]. Meanwhile, Law et al. found that the expression of PIWIL2 was positively correlated with piR-Hep1 in hepatocellular carcinoma, indicating that the PIWIL2 and piR-Hep1 complexes have a driving effect on tumors [28]. Subsequently, Yan et al. found that inhibition of piR-823 reduced global methylation in multiple myeloma cells, suggesting its role in tumor growth [29].

With in-depth study, several databases have been built to provide basic information related to piRNAs, such as piRNABank [30], piRBase [31], piRNAQuest [32]. They provide a variety of narratives for piRNAs, including pseudogenes and synchronization information, especially sequence and positional data. In addition, an experimentally supported piRNA-disease association database was proposed as more and more piRNAs were confirmed to be strongly associated with multiple diseases, piRDisease [12]. The introduction of piRDisease brings an opportunity to design computational tools to identify potential disease-associated piRNAs. However, the predictor for piRNA-disease association have not been proposed so far.

In this study, we simply designed a method to verify the feasibility of predicting potential associations, namely APDA (attribute-based descriptor), based on attribute characteristics. In APDA, attribute information contains sequence information and disease semantic information for describing RNA and disease. In order to verify the effectiveness of the method, we applied APDA on the large-scale benchmark datasets piRDisease. The experimental results show that the proposed method have good prediction performance. Among them, the superior performance of APDA is due to the quantification of piRNA and disease characteristics. Through APDA, we can predict not only the unknown piRNA-disease associations but also the piRNA-disease associations of the undefined attribute, thus expanding the scope of the model prediction and making them more applicable. In short, APDA are powerful and necessary tools to promote the improvement of human interaction groups, providing new impetus to promote the perfection of the completion of the human interactome, and making the first attempt in the field of large-scale prediction of the potential association between piRNA and diseases. We hope that this work will stimulate more research on piRNA-disease association prediction.

2 Materials and Methods

2.1 Data Set

The recently established experimentally validated piRDisease v1.0 is a comprehensive and unique disease-related piRNA database resource that collects 7939 associations from over 2500 articles involving 4796 piRNAs and 28 diseases [12]. Since the piRDisease v1.0 benchmark dataset has not been used to develop computational methods for predicting piRNA-disease associations, how to process data in piRDisease v1.0 to obtain more accurate predictions needs to be explored (Table 1).

Due to the redundancy of the raw data, we merged the same associations in the data. After removal of the replicates, the final baseline data set used in our experiments contained 5214 experimentally confirmed piRNA-disease associations involving 4503 piRNAs and 27 diseases.

Table 1. The details of benchmark dataset of piRNA-disease associations.

Benchmark dataset	piRNA	Disease	Association
piRDisease	4503	27	5214

2.2 Construct piRNA and Disease Attribute-Based Representation

Representation of piRNA Sequences. Sequence motifs, as entities of genetic information, can essentially describe the functional information of RNA. In order to describe the characteristics of a piRNA with attribute information, the sequence is mapped to a vector to form an attribute descriptor of the piRNA. The most common sequence mapping method is k-mer, where k is the length of the short motif [33, 34]. For example, 3-mers of RNA sequence can be represented as AAA, AAC, …, UUU. In this study, a vector consisting of 3-mer frequencies is used to deconstruct piRNA function information and construct the piRNA sequence descriptor $Attribute(p_i)$ where p_i is the ith piRNA.

Representation of Disease Semantic. To quantify the relationship between diseases, disease semantic information is used to construct a directed acyclic graph (DAG). The disease semantics comes from the Medical Subject Headings (MeSH) database (https://www.nlm.nih.gov/), which is a strict disease classification and enable more in-deep disease semantic studies to be carried out [35, 36]. Each disease in the MeSH database has a corresponding MeSH ID. For example, the MeSH ID of "Digestive system neoplasms" is "C04.588.274; C06.301". Meanwhile, "Digestive system neoplasms" has parent nodes, "Neoplasms by site" and "Digestive system disease" whose MeSH IDs are "C04.588" and "C06" respectively as Fig. 1. According to the relationship between diseases in the MeSH database, the disease d_i is described as $DAG_{d_i} = (d_i, T_{d_i}, E_{d_i})$. Where T_{d_i} is the node in the DAG that contains the disease and its parent nodes, such as "Neoplasms by site" and "Digestive system disease", and E_{d_i} is an edge set containing all the associations between diseases within T_{d_i}. According to previous study [37–41], the semantic contribution of disease t to disease d_i is defined as follows:

$$\begin{cases} Sem_{d_i}(t) = 1 & if\ t = d_i \\ Sem_{d_i}(t) = max(\kappa * Sem_{d_i}(t')|t' \in children\ of\ t) & if\ t \neq d_i \end{cases} \quad (1)$$

In the above formula, the semantic contribution κ is set to 0.5 here. If the relationship between t and d_i is closer, the higher the semantic contribution. Thus, the semantic value of the disease d_i is defined as follows:

$$SV(d_i) = \sum\nolimits_{t=T_{d_i}} Sem_{d_i}(t) \quad (2)$$

According to the hypothesis that if the two diseases have closer common ancestors, the two diseases are more similar in semantics [42–47]. Therefore, the semantic similarity score of disease d_i and d_j can be calculated as follows:

$$Sim(d_i, d_j) = \frac{\sum\nolimits_{t \in N_{d_i} \cap N_{d_j}} \left(Sem_{d_i}(t) + Sem_{d_j}(t)\right)}{SV(d_i) + SV(d_j)} \quad (3)$$

In $Sim(d_i, d_j)$, the hierarchical relationship between the two diseases in DAG is quantified [50–57]. However, in DAG of other diseases, the frequency of occurrence of each disease is different, and the contribution of diseases with low occurrence frequency should be higher than that of diseases with high occurrence frequency [58–63]. Therefore, a second model to calculate disease contribution value is introduced to quantify this relationship:

$$Sem'_{d_i}(t) = -log\left(\frac{number\ of\ DAGs\ including\ t}{num(disease)}\right) \quad (4)$$

Therefore, the new semantic similarity formula can be defined as follows:

$$Sim'(d_i, d_j) = \frac{\sum\nolimits_{t \in N_{d_i} \cap N_{d_j}} \left(Sem'_{d_i}(t) + (Sem'_{d_j}(t)\right)}{SV(d_i) + SV(d_j)} \quad (5)$$

No matter Sim or Sim', the description of semantic similarity is one-sided, so the two similarities are integrated to obtain a more comprehensive semantic similarity, which is defined as follows:

$$SIM(d_i, d_j) = max\left(Sim(d_i, d_j), Sim'(d_i, d_j)\right) \quad (6)$$

In this study, the vector composed of the similarities between disease d_i and other diseases is constructed as the attribute descriptor $Attribute(d_i)$ of the disease.

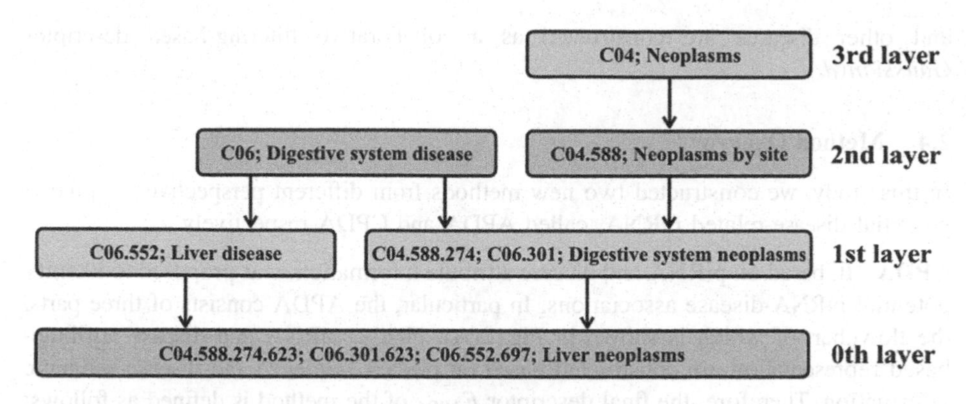

Fig. 1. The disease DAGs of liver neoplasms

2.3 Construct piRNA and Disease Collaborative Filtering-Based Representation

According to the previous research and the idea of collaborative filtering, Gaussian interaction profile kernel similarity is calculated to quantify the degree of correlation between diseases [58, 59]. Specifically, a binary vector $V(p_i)$ composed of the association between piRNA p_i and all diseases is used to describe the network characteristics of piRNA p_i in piRNA-disease association network. Therefore, the Gaussian interaction profile kernel similarity between piRNA p_i and piRNA p_j can be calculated according to their corresponding binary vectors $V(p_i)$ and $V(p_j)$:

$$GP(p_i, p_j) = exp\left(-\chi_p \|V(p_i) - V(p_j)\|^2\right) \tag{7}$$

According to the above formula, χ_p is the kernel width coefficient which can be calculated by the following formula:

$$\chi_p = \frac{1}{\frac{1}{np}\sum_{k=1}^{np}\|V(p_k)\|^2} \tag{8}$$

where np is the number of piRNAs. Similar to $GP(p_i, p_j)$, the Gaussian interaction profile kernel similarity of disease can be defined as follows:

$$GD(d_i, d_i) = exp\left(-\chi_d \|V(d_i) - V(d_j)\|^2\right) \tag{9}$$

$$\chi_d = \frac{1}{\frac{1}{nd}\sum_{k=1}^{nd}\|V(d_k)\|^2} \tag{10}$$

where nd is the number of diseases. In this study, a numerical vector consisting of the similarities between piRNA p_i and other piRNAs are constructed as a collaborative filtering-based descriptor $Gaussian(p_i)$. Similarly, the similarities between disease d_i

and other diseases are constructed as a collaborative filtering-based descriptor $Gaussian(d_i)$.

2.4 Method Overview

In this study, we constructed two new methods from different perspectives to predict potential disease-related piRNA, called APDA and CPDA respectively.

APDA. It, based on piRNA and disease attribute information, was proposed to identify potential piRNA-disease associations. In particular, the APDA consists of three parts, the flowchart of which is shown in Fig. 2(A). Firstly, piRNA and disease attribute-based representation are constructed based on piRNA sequence and disease semantic information. Therefore, the final descriptor F_{APDA} of the method is defined as follows:

$$F_{APDA}(p_i, d_j) = (Attribute(p_i), Attribute(d_j)) \tag{11}$$

Secondly, the classifier is trained according to the constructed descriptor $F_{APDA}(p_i, d_j)$ to obtain the predictor. Finally, the predictor is used to quantify the correlation degree of the unknown associated piRNA-disease pairs. It is worth noting that the higher the predictive score of piRNA-disease pairs, the greater the possibility of correlation between them.

CPDA. It, based on piRNA-disease association network to obtain the similarity between piRNA and diseases through collaborative filtering to assist in identifying piRNA-disease association with biological significance but not yet mapped. In particular, the CPDA consists of three parts, the flowchart of which is shown in Fig. 2(B). First of all, we construct a collaborative filtering-based representation of piRNA and disease, that is, descriptor construction, based on the original adjacency matrix of association between piRNA and disease. Therefore, the final descriptor F_{CPDA} of the method is defined as follows:

$$F_{CPDA}(p_i, d_j) = (Gaussian(p_i), Gaussian(d_j)) \tag{12}$$

Secondly, the classifier is trained according to the constructed descriptor $F_{CPDA}(p_i, d_j)$ to obtain the predictor. Finally, the predictor is used to quantify the correlation degree of the unknown associated piRNA-disease pairs. Similarly, the higher the predictive score of piRNA-disease pairs, the greater the likelihood of correlation between them.

Positive and Negative Samples. Negative samples are constructed by downsampling. Specifically, negative samples with the same number as positive samples are randomly selected from unproven piRNA-disease associations. The proportion of positive and negative samples in the data set used in the experiment is half each.

The Classifier. Both methods use Random Forest as the classifier to predict potential associations, and the classifier utilizes default parameters. The random forest model is trained by final descriptors F_{APDA} and F_{CPDA} to predict potential associations. The higher the associated prediction score, the greater the likelihood that piRNA will be associated with the disease.

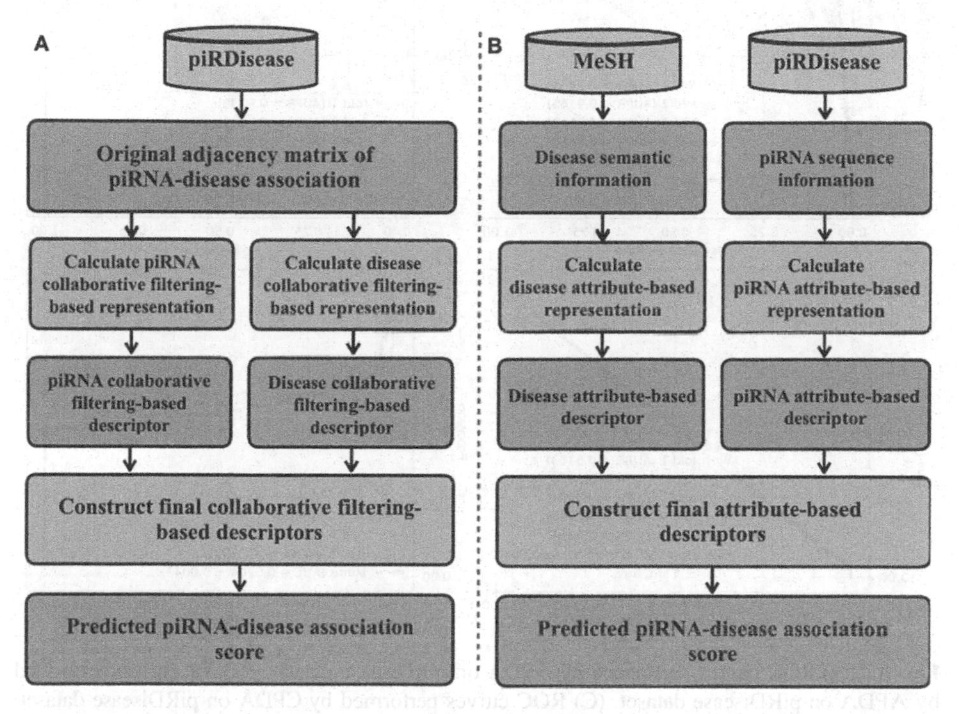

Fig. 2. The flowchart includes three steps: descriptor construction; Training predictor; Prediction of association between potential piRNA and disease. (A) The flowchart of CPDA. (B) The flowchart of APDA.

3 Results

We first tested APDA and CPDA in the piRDisease benchmark data set. As can be seen in Table 2, the APDA's average Accuracy of five-fold cross-validation is 0.8564, the recall is 0.8634, the Precision is 0.8571, and the F1-value is 0.8580. As can be seen in Table 3, the CPDA's average Accuracy of five-fold cross-validation is 0.5157, the recall is 0.7947, the Precision is 0.0432, and the F1-value is 0.0818. Figure 3 shows the ROC curves generated by APDA and CPDA on the piRDisease benchmark dataset, respectively.

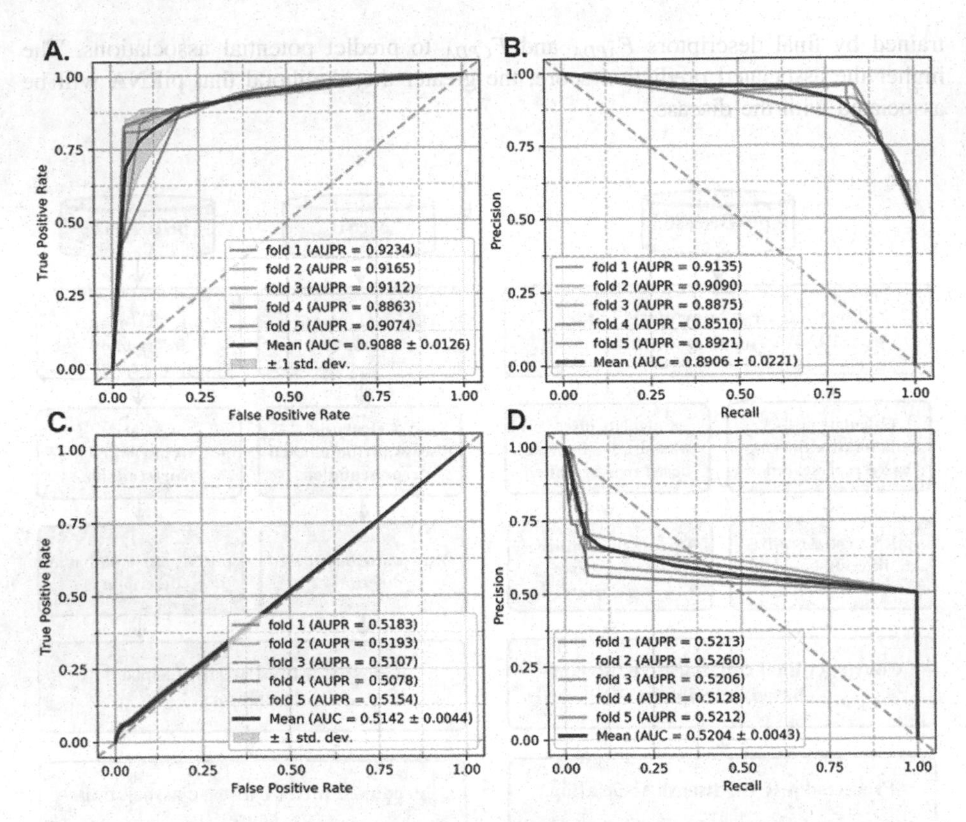

Fig. 3. (A) ROC curves performed by APDA on piRDisease dataset. (B) PR curves performed by APDA on piRDisease dataset. (C) ROC curves performed by CPDA on piRDisease dataset. (D) PR curves performed by CPDA on piRDisease dataset

Table 2. The five-fold cross-validation results performed by APDA on piRDisease dataset

	Accuracy	Sensitivity	Precision	F1-score
Average	0.8564 ± 0.0192	0.8571 ± 0.0572	0.8634 ± 0.0317	0.8580 ± 0.0123

Table 3. The five-fold cross-validation results performed by CPDA on piRDisease dataset

Testing set	Accuracy	Sensitivity	Precision	F1-score
Average	0.5157 ± 0.0024	0.7947 ± 0.0673	0.0432 ± 0.0027	0.0818 ± 0.0047

4 Conclusion

In recent years, piRNAs have gradually been confirmed to be closely related to the development of various diseases. To date, no predictors specific to piRNA-disease association have been proposed. Therefore, it is necessary to develop computational methods to identify piRNA-disease associations that are biologically significant but not yet mapped.

In this study, a new computational method called APDA were proposed to identify the associations between potential piRNAs and diseases, based on attribute characteristics. The experimental results show that our proposed method have excellent prediction performance. APDA's outstanding success stems from its ability to capture piRNA and quantitative attributes of diseases. However, it also stems from the fact that although the interaction verified by experiments provides relatively incomplete coverage of the complete interaction group, it has reached the coverage and accuracy required by the predictor, making accurate prediction possible in the future. Through APDA, the applicability of model prediction piRNA-disease associations has been greatly improved. In short, the APDA is powerful and necessary tools to promote the perfection the completion of the human interactome, providing new impetus to reveal the root causes of human diseases.

Acknowledgments. This work is supported by the Xinjiang Natural Science Foundation under Grant 2017D01A78.

References

1. Yin, H., Lin, H.: An epigenetic activation role of Piwi and a Piwi-associated piRNA in Drosophila melanogaster. Nature **450**, 304 (2007)
2. Siomi, M.C., Sato, K., Pezic, D., Aravin, A.A.: PIWI-interacting small RNAs: the vanguard of genome defence. Nat. Rev. Mol. Cell Biol. **12**, 246 (2011)
3. Iwasaki, Y.W., Siomi, M.C., Siomi, H.: PIWI-interacting RNA: its biogenesis and functions. Ann. Rev. Biochem. **84**, 405–433 (2015)
4. Grimson, A., et al.: Early origins and evolution of microRNAs and Piwi-interacting RNAs in animals. Nature **455**, 1193 (2008)
5. Aravin, A.A., Hannon, G.J., Brennecke, J.: The Piwi-piRNA pathway provides an adaptive defense in the transposon arms race. Science **318**, 761–764 (2007)
6. Malone, C.D., et al.: Specialized piRNA pathways act in germline and somatic tissues of the Drosophila ovary. Cell **137**, 522–535 (2009)
7. Leslie, M.: The immune system's compact genomic counterpart. American Association for the Advancement of Science (2013)
8. Pall, G.S., Codony-Servat, C., Byrne, J., Ritchie, L., Hamilton, A.: Carbodiimide-mediated cross-linking of RNA to nylon membranes improves the detection of siRNA, miRNA and piRNA by northern blot. Nucleic Acids Res. **35**, e60 (2007)
9. Marcon, E., Babak, T., Chua, G., Hughes, T., Moens, P.: miRNA and piRNA localization in the male mammalian meiotic nucleus. Chromosome Res. **16**, 243–260 (2008)
10. Armisen, J., Gilchrist, M.J., Wilczynska, A., Standart, N., Miska, E.A.: Abundant and dynamically expressed miRNAs, piRNAs, and other small RNAs in the vertebrate Xenopus tropicalis. Genome Res. **19**, 1766–1775 (2009)
11. Moyano, M., Stefani, G.: piRNA involvement in genome stability and human cancer. J. Hematol. Oncol. **8**, 38 (2015)
12. Muhammad, A., Waheed, R., Khan, N.A., Jiang, H., Song, X.: piRDisease v1. 0: a manually curated database for piRNA associated diseases. Database 2019 (2019)

13. Rajasethupathy, P., et al.: A role for neuronal piRNAs in the epigenetic control of memory-related synaptic plasticity. Cell **149**, 693–707 (2012)
14. Houwing, S., et al.: A role for Piwi and piRNAs in germ cell maintenance and transposon silencing in Zebrafish. Cell **129**, 69–82 (2007)
15. Gou, L.-T., et al.: Pachytene piRNAs instruct massive mRNA elimination during late spermiogenesis. Cell Res. **24**, 680 (2014)
16. Zou, A.E., et al.: The non-coding landscape of head and neck squamous cell carcinoma. Oncotarget **7**, 51211 (2016)
17. Chu, H., et al.: Identification of novel piRNAs in bladder cancer. Cancer Lett. **356**, 561–567 (2015)
18. Cheng, J., et al.: piRNA, the new non-coding RNA, is aberrantly expressed in human cancer cells. Clinica Chimica Acta **412**, 1621–1625 (2011)
19. Assumpcao, C.B., et al.: The role of piRNA and its potential clinical implications in cancer. Epigenomics **7**, 975–984 (2015)
20. Li, Y., et al.: Piwi-interacting RNAs (piRNAs) are dysregulated in renal cell carcinoma and associated with tumor metastasis and cancer-specific survival. Mol. Med. **21**, 381–388 (2015)
21. Romano, G., Veneziano, D., Acunzo, M., Croce, C.M.: Small non-coding RNA and cancer. Carcinogenesis **38**, 485–491 (2017)
22. Simon, B., et al.: Recognition of 2′-O-methylated 3′-end of piRNA by the PAZ domain of a Piwi protein. Structure **19**, 172–180 (2011)
23. Rouget, C., et al.: Maternal mRNA deadenylation and decay by the piRNA pathway in the early Drosophila embryo. Nature **467**, 1128 (2010)
24. Ghildiyal, M., et al.: Endogenous siRNAs derived from transposons and mRNAs in Drosophila somatic cells. Science **320**, 1077–1081 (2008)
25. Vourekas, A., Alexiou, P., Vrettos, N., Maragkakis, M., Mourelatos, Z.: Sequence-dependent but not sequence-specific piRNA adhesion traps mRNAs to the germ plasm. Nature **531**, 390 (2016)
26. Yin, J., et al.: piR-823 contributes to colorectal tumorigenesis by enhancing the transcriptional activity of HSF 1. Cancer Sci. **108**, 1746–1756 (2017)
27. Zhang, H., Ren, Y., Xu, H., Pang, D., Duan, C., Liu, C.: The expression of stem cell protein Piwil2 and piR-932 in breast cancer. Surg. Oncol. **22**, 217–223 (2013)
28. Lee, J.H., et al.: Stem-cell protein Piwil2 is widely expressed in tumors and inhibits apoptosis through activation of Stat3/Bcl-XL pathway. Hum. Mol. Genetics **15**, 201–211 (2005)
29. Yan, H., et al.: piRNA-823 contributes to tumorigenesis by regulating de novo DNA methylation and angiogenesis in multiple myeloma. Leukemia **29**, 196 (2015)
30. Sai Lakshmi, S., Agrawal, S.: piRNABank: a web resource on classified and clustered Piwi-interacting RNAs. Nucleic Acids Res. **36**, D173–D177 (2007)
31. Wang, J., et al.: piRBase: a comprehensive database of piRNA sequences. Nucleic Acids Res. **47**, D175–D180 (2018)
32. Sarkar, A., Maji, R.K., Saha, S., Ghosh, Z.: piRNAQuest: searching the piRNAome for silencers. BMC Genomics **15**, 555 (2014)
33. Kirk, J.M., et al.: Functional classification of long non-coding RNAs by k-mer content. Nat. Genetics **50**, 1474 (2018)
34. Wong, L., Huang, Y.A., You, Z.H., Chen, Z.H., Cao, M.Y.: LNRLMI: linear neighbour representation for predicting lncRNA-miRNA interactions. J. Cell. Mol. Med. **24**(1), 79–87 (2019)
35. Xiang, Z., Qin, T., Qin, Z.S., He, Y.: A genome-wide MeSH-based literature mining system predicts implicit gene-to-gene relationships and networks. BMC Syst. Biol. **7**, S9 (2013)

36. Lipscomb, C.E.: Medical subject headings (MeSH). Bull. Med. Libr. Assoc. **88**, 265 (2000)
37. Ji, B.-Y., You, Z.-H., Cheng, L., Zhou, J.-R., Alghazzawi, D., Li, L.-P.: Predicting miRNA-disease association from heterogeneous information network with GraRep embedding model. Sci. Rep. **10**, 1–12 (2020)
38. Zheng, K., You, Z.-H., Wong, L., Chen, Z.-H., Jiang, H.-J.: Inferring Disease-Associated Piwi-Interacting RNAs via Graph Attention Networks. bioRxiv (2020)
39. Chen, X., et al.: WBSMDA: within and between score for MiRNA-disease association prediction. Sci. Rep. **6**, 21106 (2016)
40. Chen, Z.-H., Li, L.-P., He, Z., Zhou, J.-R., Li, Y., Wong, L.: An improved deep forest model for predicting self-interacting proteins from protein sequence using wavelet transformation. Front. Genetics **10**, 90 (2019)
41. Chen, X., Wang, C.-C., Yin, J., You, Z.-H.: Novel human miRNA-disease association inference based on random forest. Mol. Therapy-Nucleic Acids **13**, 568–579 (2018)
42. Zheng, K., You, Z.-H., Li, J.-Q., Wang, L., Guo, Z.-H., Huang, Y.-A.: iCDA-CGR: Identification of circRNA-disease associations based on Chaos Game Representation. PLOS Comput. Biol. **16**, e1007872 (2020)
43. Wang, L., et al.: LMTRDA: Using logistic model tree to predict MiRNA-disease associations by fusing multi-source information of sequences and similarities. PLoS Comput. Biol. **15**, e1006865 (2019)
44. Li, J.-Q., Rong, Z.-H., Chen, X., Yan, G.-Y., You, Z.-H.: MCMDA: matrix completion for MiRNA-disease association prediction. Oncotarget **8**, 21187 (2017)
45. Chen, Z.-H., You, Z.-H., Li, L.-P., Wang, Y.-B., Li, X.: RP-FIRF: prediction of self-interacting proteins using random projection classifier combining with finite impulse response filter. In: Huang, D.-S., Jo, K.-H., Zhang, X.-L. (eds.) ICIC 2018. LNCS, vol. 10955, pp. 232–240. Springer, Cham (2018). https://doi.org/10.1007/978-3-319-95933-7_29
46. Zheng, K., You, Z.-H.: iMDA-BN: Identification of miRNA-Disease Associations based on the Biological Network and Graph Embedding Algorithm. bioRxiv (2020)
47. Wang, L., You, Z.-H., Li, L.-P., Zheng, K., Wang, Y.-B.: Predicting circRNA-disease associations using deep generative adversarial network based on multi-source fusion information. In: 2019 IEEE International Conference on Bioinformatics and Biomedicine (BIBM), pp. 145-152. IEEE (2019)
48. Zheng, K., You, Z.-H., Wang, L., Wong, L., Zhan, Z.-H.: SPRDA: a matrix completion approach based on the structural perturbation to infer disease-associated Piwi-Interacting RNAs. bioRxiv (2020)
49. Chen, X., et al.: A novel computational model based on super-disease and miRNA for potential miRNA–disease association prediction. Mol. bioSyst. **13**, 1202–1212 (2017)
50. Zheng, K., You, Z.-H., Wang, L., Li, Y.-R., Wang, Y.-B., Jiang, H.-J.: MISSIM: improved miRNA-disease association prediction model based on chaos game representation and broad learning system. In: Huang, D.-S., Huang, Z.-K., Hussain, A. (eds.) ICIC 2019. LNCS (LNAI), vol. 11645, pp. 392–398. Springer, Cham (2019). https://doi.org/10.1007/978-3-030-26766-7_36
51. Wang, M.-N., You, Z.-H., Wang, L., Li, L.-P., Zheng, K.: LDGRNMF: LncRNA-disease associations prediction based on graph regularized non-negative matrix factorization. Neurocomputing (2020)
52. Ma, L., et al.: Multi-neighborhood learning for global alignment in biological networks. IEEE/ACM Trans. Comput. Biol. Bioinform. (2020)
53. Chen, X., Huang, Y.-A., You, Z.-H., Yan, G.-Y., Wang, X.-S.: A novel approach based on KATZ measure to predict associations of human microbiota with non-infectious diseases. Bioinformatics **33**, 733–739 (2017)

54. Wang, L., You, Z., Li, Y., Zheng, K., Huang, Y.: GCNCDA: A New Method for Predicting CircRNA-Disease Associations Based on Graph Convolutional Network Algorithm. bioRxiv 858837 (2019)
55. Wang, Y.-B., You, Z.-H., Yang, S., Yi, H.-C., Chen, Z.-H., Zheng, K.: A deep learning-based method for drug-target interaction prediction based on long short-term memory neural network. BMC Med. Inf. Decis. Making **20**, 1–9 (2020)
56. Lei, Y.-K., You, Z.-H., Ji, Z., Zhu, L., Huang, D.-S.: Assessing and predicting protein interactions by combining manifold embedding with multiple information integration. BMC Bioinform. **13**, S3 (2012)
57. Zheng, K., You, Z.-H., Wang, L., Zhou, Y., Li, L.-P., Li, Z.-W.: DBMDA: a unified embedding for sequence-based miRNA similarity measure with applications to predict and validate miRNA-disease associations. Mol. Therapy-Nucleic Acids **19**, 602–611 (2020)
58. Chen, X., Yan, C.C., Zhang, X., You, Z.-H.: Long non-coding RNAs and complex diseases: from experimental results to computational models. Briefings Bioinform. **18**, 558–576 (2016)
59. You, Z.-H., Zhan, Z.-H., Li, L.-P., Zhou, Y., Yi, H.-C.: Accurate prediction of ncRNA-protein interactions from the integration of sequence and evolutionary information. Front. Genetics **9**, 458 (2018)
60. Zheng, K., You, Z.-H., Wang, L., Zhou, Y., Li, L.-P., Li, Z.-W.: MLMDA: a machine learning approach to predict and validate MicroRNA–disease associations by integrating of heterogenous information sources. J. Transl. Med. **17**, 1–14 (2019)
61. You, Z.-H., Zhou, M., Luo, X., Li, S.: Highly efficient framework for predicting interactions between proteins. IEEE Trans. Cybern. **47**, 731–743 (2017)
62. Zheng, K., Wang, L., You, Z.-H.: CGMDA: an approach to predict and validate MicroRNA-disease associations by utilizing chaos game representation and LightGBM. IEEE Access **7**, 133314–133323 (2019)
63. Zhu, L., You, Z.-H., Huang, D.-S., Wang, B.: t-LSE: a novel robust geometric approach for modeling protein-protein interaction networks. PLoS ONE **8**, e58368 (2013)

Inferring Disease-Associated Piwi-Interacting RNAs via Graph Attention Networks

Kai Zheng[1], Zhu-Hong You[2(✉)], Lei Wang[2(✉)], Leon Wong[2], and Zhan-Heng Chen[2]

[1] School of Computer Science and Engineering, Central South University, Changsha, China

[2] Xinjiang Technical Institute of Physics and Chemistry, Chinese Academy of Sciences, Urumqi, China

{zhuhongyou,leiwang}@ms.xjb.ac.cn

Abstract. Piwi proteins and Piwi-Interacting RNAs (piRNAs) are commonly detected in human cancers. However, it is time-consuming and costly to detect piRNA-disease associations (PDAs) by traditional experimental methods. In this study, we present a computational method GAPDA to identify potential and biologically significant PDAs based on graph attention network. Specifically, we combined piRNA sequence information, disease semantic similarity, and piRNA-disease association network to construct a new attribute network. Then, the network embedding in node-level is learned via the attention-based graph neural network. Finally, potential piRNA-disease associations are scored.To be our knowledge, this is the first time that the attention-based Graph Neural Networks is introduced to the field of ncRNA-related association prediction. In the experiment, the proposed GAPDA method achieved AUC of 0.9038 using five-fold cross-validation. The experimental results show that the GAPDA approach ensures the prospect of the graph neural network on such problems and can be an excellent supplement for future biomedical research.

Keywords: PIWI-interacting RNA · Disease · piRNA-disease association · Graph attention network · Self-attention mechanism

1 Introduction

Piwi-interacting RNA (piRNA) is a small, non-coding RNA that clusters at transposon loci in the genome and is typically 24–32 nucleotides in length. Its discovery has greatly expanded the RNA world [1–5]. Since the discovery and formal definition of piRNA in 2006, the PIWI–piRNA field has been developed rapidly, and its functions in developmental regulation, transposon silencing, epigenetic regulation, and genomic rearrangement are being revealed gradually [6–9]. piRNA interact exclusively with PIWI proteins which belong to germline-specific subclade of the Argonaute family [10]. The best-known role of it is to repress transposons and maintain germline genome integrity through DNA methylation, as the depletion of PIWI leads to a sharp increase in transposon messenger RNA expression [11, 12]. Compared with microRNA (miRNA) and small interfering RNA (siRNA) that are small RNAs, (1) longer than

D.-S. Huang and K.-H. Jo (Eds.): ICIC 2020, LNCS 12464, pp. 239–250, 2020.
https://doi.org/10.1007/978-3-030-60802-6_21

miRNA or siRNA; (2) only present in animals; (3) more diverse sequences and constitute the largest class of noncoding RNA; (4) testes-specific [6, 12–15].

Recently, emerging evidence suggests that piRNA and PIWI proteins are abnormally expressed in various cancers [16–21]. Therefore, the function and potential mechanism of piRNA in cancer become one of the important research directions in tumor diagnosis and treatment. For example, Fu *et al.* found that abnormal expression of piR-021285 promoted methylation of ARHGAP11A at the 5'-UTR/first exon CpG site, thereby pro-moting mRNA apoptosis and inhibiting apoptosis of Breast cancer cells [22]. Subsequently, Tan *et al.* found that down-regulation of piRNA-36712 in breast cancer increases SLUG levels, while P21 and E-cadherin levels were reduced, thereby promoting the malignant phenotype of cancer [23]. piR-30188 binds to OIP5-AS1 to inhibit glioma cell progression while low expression of OIP5-AS1 reduces CEBPA levels and pro-motes the malignant phenotype of glioma cells which discovered by liu *et al.* [24]. Also glioblastoma, Jacobs *et al.* found that piR-8041 can inhibit the expression of the tumor cell marker ALCAM/CD166, with the clinical role of targeted therapy [25]. In addition, piRNA is directly or indirectly involved in the formation of liver cancer. In 2016, Rizzo *et al.* found that hsa_piR_013306 accumulates only in hepatocellular carcinomas [26].

piRNA is gaining enormous attention, and tens of thousands of them have been identified in mammals and are rapidly accumulating. In order to accelerate research in this field and provide access to piRNA data and annotations, multiple databases such as piRNABank [27], piRBase [28], piRNAQuest [29] have been successively established. Subsequently, the role of piRNA and PIWI proteins in the epigenetics of cancer were discovered, and some of them can serve as novel biomarkers and therapeutic targets. Taking this as an opportunity, an experimentally supported piRNA-disease association database called piRDisease [30] was proposed, which made it possible to predict potential associations on a large scale. Although many disease-related ncRNA prediction models have been proposed, predictors for disease-related piRNA is relatively unexplored [31–35].

In this paper, we propose a piRNA-disease association predictor based on attention-based graph neural network, called GAPDA. On the association dataset piRDisease, GAPDA achieves an AUC of 91.45% with an accuracy of 84.49%. In general, the proposed method can provide new impetus for cancer mechanism research, provide new research ideas for small sample data sets, and determine the prospects of attention-based Graph Neural Networks on such issues. In addition, we hope that this work will stimulate more association prediction research based on graph neural network.

2 Materials and Methods

2.1 Benchmark Dataset

With the rapid increase of PIWI-interacting RNA (piRNA) related research, the contribution of piRNA in disease diagnosis and prognosis gradually emerges. These manually managed, complex and heterogeneous information may lead to data inconsistency and inefficiency, it put data analysis into a dilemma. To this end, the piRDisease database, which integrates experimentally supported association between piRNAs and disease, was proposed in 2019 [30]. Azhar *et al*. developed piRDisease v1.0 by searching more than 2,500 articles, which provided 7939 piRNA-disease associations with 4,796 piRNAs and 28 diseases. The baseline set by simple filtering is named GPRD, as shown in Table 1.

Table 1. The details of benchmark dataset of piRNA-disease associations.

Benchmark dataset	piRNA	Disease	Association
GPRD	501	22	1212

GPRD. Currently, research on the relationship between piRNA and disease is in the ascendant, so the degree of some piRNAs are only 1 in the association network. Too many nodes with the degree = 1 affect the performance of the network-based approach. Therefore, in GPRD, we only retained 501 piRNAs with the degree greater than 1 and constituted 1212 associations. The training dataset T can be defined as:

$$T = T^p \cup T^n \tag{1}$$

where T^p is a positive subset of the piRNA-disease association construct in GPRD, and T^n is a negative subset containing 1212 associations which were randomly extracted from all 11022 unconfirmed associations between piRNA and disease.

2.2 Construct New piRNA-Disease Association Network

The structure of the network. At present, ncRNA-related associations with experimental verification are very limited, so the network-based method is difficult to achieve satisfactory prediction results. In addition, it's difficult to get the desirable accuracy by attribute-based methods. In the meanwhile, biological data is often complex, and network representations computed from sparse networks cannot cover all real-world behavior information. Therefore, a method of enriching the hidden representations contained in sparse network is urgently needed. To this end, we propose a simple network construction method with association as a node. The new association adjacency matrix A based on n associations is calculated as follows:

$$A = \begin{bmatrix} a_{1,1} & \cdots & a_{1,n} \\ \vdots & \ddots & \vdots \\ a_{n,1} & \cdots & a_{n,n} \end{bmatrix} \tag{2}$$

where n is the number of associations in the training dataset T. The element $a_{i,j}$ is set to 1 if the i-th association is related with the j-th association, otherwise 0. In particular, the links between associations is various. In detail, the dark blue line represents the confirmed piRNA-disease association, the light blue line is the similarity between piRNAs or diseases, and the dotted line is the unproven piRNA-disease association. In this paper, we utilize piRNA and disease as link vectors, respectively, and define them as follows:

$$\alpha_{i,j}^{R} = \begin{cases} 1 & if\, association(i).piRNA = association(j).piRNA \\ 0 & otherwise \end{cases} \tag{3}$$

$$\alpha_{i,j}^{D} = \begin{cases} 1 & if\, association(i).disease = association(j).disease \\ 0 & otherwise \end{cases} \tag{4}$$

According to the above formula, a plurality of superimposable adjacency matrices can be obtained to enrich the structural information of the network, like A^R composed of $\alpha_{i,j}^{R}$ and A^D composed of $\alpha_{i,j}^{D}$. Since the size of the abstracted adjacency matrix is uniform, they can be stacked by weighting. For the sake of simplicity, we only performed an addition operation on the adjacency matrix A^R and the adjacency matrix A^D. Therefore, the element $\alpha_{i,j}^{RD}$ of the adjacency matrix A^{RD} is calculated as follows:

$$\alpha_{i,j}^{RD} = \begin{cases} 1 & if\, \alpha_{i,j}^{R} = 1\, or\, \alpha_{i,j}^{D} = 1 \\ 0 & otherwise \end{cases} \tag{5}$$

Node Attributes. The attribute of the node is mainly composed of two parts: piRNA sequence features and disease semantic features. These two kinds of attribute information are described in detail below. The specific structure and function of RNA is determined by the sequence carrying the genetic information, so describing the sequence as a descriptor is an effective way to characterize its function. k-mers is a common alignment algorithm that the basic principle is to divide a sequence into subsequences of length k and count its frequency. Recent studies show that ncRNAs of related function often have related k-mer contents [36]. For example, 3-mer of piRNA can be expressed as CCC, CCG,..., GGG. Herein, the k-mer deconstructs and reconstructs the piRNA functional features to obtain piRNA descriptor $Feature(p_a)$ where p_a is piRNA with serial number a.

It is still an urgent and tough problem to characterizate disease attributes. So far, methods for constructing directed acyclic graphs (DAG) by the Medical Subject Headings (MeSH) to quantify the relationship between diseases are commonly used [37, 38]. The semantic contribution C of disease w to disease d is calculated:

$$\begin{array}{ll} C_d(w) = 1 & if\; w = d \\ C_d(w) = max(\nabla * C_d(w')|w' \in children\, of\, w) & if\; w \neq d \end{array} \tag{6}$$

Here we set the seismic contribution decay factor ∇ to 0.5. w' is the child node of w. If the disease d is farther apart from the disease w in the DAG, the contribution of the disease w to the disease d is lower. For example, “Neoplasms” contributes less to “Lip Neoplasms” than “Mouth Neoplasms”. According to the semantic contribution C, the semantic value V of disease d is calculated:

$$V(d) = \sum\nolimits_{w \in T_d} C_d(w) \tag{7}$$

If the two diseases share more DAGs and near common ancestors, the two diseases are more semantically similar. Under that assumption, the semantic similarity scores *SS* for disease *a* and disease *b* can be defined as follows:

$$SS(a,b) = \frac{\sum_{w \in N_a \cap N_b} (C_a(w) + C_b(w))}{V(a) + V(b)} \tag{8}$$

The semantic similarity score SS considers the existence of common ancestors between diseases. However, its performance is not unlimited. For example, "Neoplasms by site" appears in the DAGs of many diseases, while the "Stomatognathic Disease" of the same layer appears less frequently. Since "Stomatognathic Disease" has a higher specificity for "Lip Neoplasms", its weight should also be higher. To quantify such differences in weight, the second semantic contribution is designed:

$$C'_d(w) = -log\left(\frac{num(DAGs\ including\ w)}{num(disease)}\right) \tag{9}$$

Similarly, the second semantic similarity scores SS' for disease *a* and disease *b* can be defined as follows:

$$SS'^{(a,b)} = \frac{\sum_{w \in N_a \cap N_b} (C'_a(w) + C'_b(w))}{V(a) + V(b)} \tag{10}$$

Both *SS* and SS' are unilateral in principle. In order to combine the advantages of two semantic similarity scores, the comprehensive semantic similarity *S* is calculated:

$$S(a,b) = max(SS(a,b), SS'(u,b)) \tag{11}$$

In this study, the degree of semantic association between disease d_b and other diseases was used as the descriptor $Feature(d_b)$ for the disease.

2.3 Gaussian Interaction Profile Kernel Similarity

The Gaussian interaction profile kernel similarity (GIP) is a commonly used collaborative filtering algorithm [47–54]. According to previous studies, the method can calculate the similarity matrix between ncRNAs and diseases from the known adjacency matrix [39, 55–60]. In detail, the similarity between piRNA p_a and piRNA p_b can be defined as follows:

$$G_p(p_a, p_b) = exp\left(-\psi_p \|V(p_a) - V(p_b)\|^2\right) \tag{12}$$

where $V(p_a)$ is a two-dimensional vector composed of the relationship between piRNA and all diseases, as is $V(p_b)$. In addition, ψ_p as kernel width coefficient is defined as follows:

$$\psi_p = \frac{1}{\frac{1}{num_p}\sum_{k=1}^{num_p}||V(p_k)||^2} \tag{13}$$

num_p is the number of piRNAs. Similarly, the similarity between piRNA d_a and piRNA d_b can also be calculated by this algorithm:

$$\mathrm{G}_d(d_a, d_b) = exp\left(-\psi_d||V(d_a) - V(d_b)||^2\right) \tag{14}$$

$$\psi_d = \frac{1}{\frac{1}{num_d}\sum_{k=1}^{num_d}||V(p_k)||^2} \tag{15}$$

num_d is the number of diseases. In this paper, it is compared as a traditional method with the proposed method. In this study, the degree of Gaussian association between piRNA p_a and other piRNAs was used as the descriptor $Feature'(p_a)$ for the disease. And, the degree of Gaussian association between disease d_b and other diseases was used as the descriptor $Feature'(d_b)$ for the disease.

2.4 Graph Attention Networks

Graph Attention Network (GAT) is a graph neural network based on self-attention mechanism proposed in 2018 [62]. The main contribution is to construct a hidden self-attention layer to specify different weights to different nodes in a neighborhood without any time-consuming matrix operations (such as inversion) or a priori knowledge of the graph structure. The input to the graph attention layer is n node features of length H, $f = \left\{\vec{f}_1, \vec{f}_2, \vec{f}_3, \ldots, \vec{f}_n\right\}, \vec{f}_i \in R^H$. $\vec{f}_i$ is the initial feature of the i-th node. And, the output of the layer is produced as $f' = \left\{\vec{f}_1', \vec{f}_2', \vec{f}_3', \ldots, \vec{f}_n'\right\}, \vec{f}_i' \in R^{H'}$, where H and H' have different dimensions. $\vec{f}_i'$ is the projected feature of the i-th node. In order to implement self-attention mechanism, a shared linear transformation parameter matrix $W \in R^{H' \times H}$ is designed to be applied to each node. Therefore, the attention coefficient $e_x(y)$ of node x to node y can be calculated as follows:

$$e_x(y) = att\left(W\vec{f}_x, W\vec{f}_y\right) \tag{16}$$

Here att denotes a mapping, $R^{H'} \times R^{H'} \rightarrow R$. It converts two vectors of length F' into a scalar as the attention coefficient. In addition, self-attention assigns attention to all nodes in the graph, which obviously loses structural information. Therefore, a method called masked attention is proposed:

$$\theta_x(y) = softmax_x(e_x(y)) = \frac{exp(e_x(y))}{\sum_{t \in N_x} exp(e_x(t))}$$
$$= \frac{exp(leakyReLU(\vec{\lambda}^T[W\vec{f_x}||W\vec{f_y}]))}{\sum_{t \in N_x} exp\left(leakyReLU(\vec{\lambda}^T[W\vec{f_x}||W\vec{f_t}])\right)} \quad (17)$$

Where N_x is the set of neighbor nodes of node x. $softmax_x$ is utilized to normalize the attention coefficient $e_x(y)$ to obtain the weight coefficient $\theta_x(y)$. $\vec{\lambda}$ is the weight coefficient vector of the graph attentional layer, and the length is $2F'$. *LeakyReLU* is the activation function. T represents transposition and $||$ represents connection operation. Therefore, the embedding of node x can be fused by the projected node features of neighbors with different weights, as follows:

$$\vec{f_x'} = \partial\left(\sum_{t \in N_x} \theta_x(t) \cdot W\vec{f_y}\right) \quad (18)$$

In order to solve the problem of large variance of the graph data caused by the scale-free of the heterogeneous graph, multi-head attention is performed to make the training process more stable. Specifically, features of m independent attention mechanisms are integrated to achieve specific embedding:

$$\vec{f_x'} = \partial\left(\frac{1}{K}\sum_{k=1}^{K} \sum_{t \in N_x} \theta_x(t) \cdot W\vec{f_y}\right) \quad (19)$$

2.5 Method Overview

In this study, we propose a novel method called GAPDA to predict biologically significant, yet unmapped associations between piRNA and disease on a large scale. GAPDA is generally composed of five components. First, we construct piRNA and disease feature descriptors based on sequence information, disease semantic information, and Gaussian interaction profile kernel similarity information. Therefore, the final feature $\vec{f}$ is defined as follows:

$$\vec{f}(p_a, d_b) = (Feature(p_a), Feature'(p_a), Feature(d_b), Feature'(d_b)) \quad (20)$$

Second, based on the existing associated network, an abstract network topology is constructed to expand the information contained in the network. Third, the reconstructed abstract network topology is combined with the final descriptor $\vec{f}$ to obtain a new piRNA-disease association attribute network. Fourth, the network embedding in node-level is learned via the attention-based graph neural network. Finally, the degree of association between piRNA and disease pairs is scored. In particular, the predicted scores of piRNA and disease pairs are directly proportional to the probability of association.

3 Results

In this part, we choose $\alpha^{R}_{i,j}$ as an element for abstract network topology. In order to evaluate the performance of the proposed method, it is applied to the benchmark database GPRD. Figure 1 depicts the ROC curve generated on the baseline data and the average AUC of five-fold cross-validation is 0.9038. In addition, Table 2 lists the results of the detailed evaluation criteria, with the average accuracy (Acc.) of 0.8569, the precision (Pre.) is 0.8550, the Recall (Rec.) is 0.8638 and the F1-score is 0.8577. Their standard deviations are 0.92%, 3.56%, 4.16%, 0.92%, respectively. From the results, the lowest accuracy in the five experiments reached 0.8395, and the highest accuracy reached 0.8642. Meanwhile, this experiment relies on the network structure to make predictions, and the prediction results obtained by different attribute networks have error. Overall, our approach yielded convincing results, suggesting that GAPDA can provide powerful candidates for piRNA as a biomarker and has the potential to drive disease diagnosis and to identify disease mechanisms.

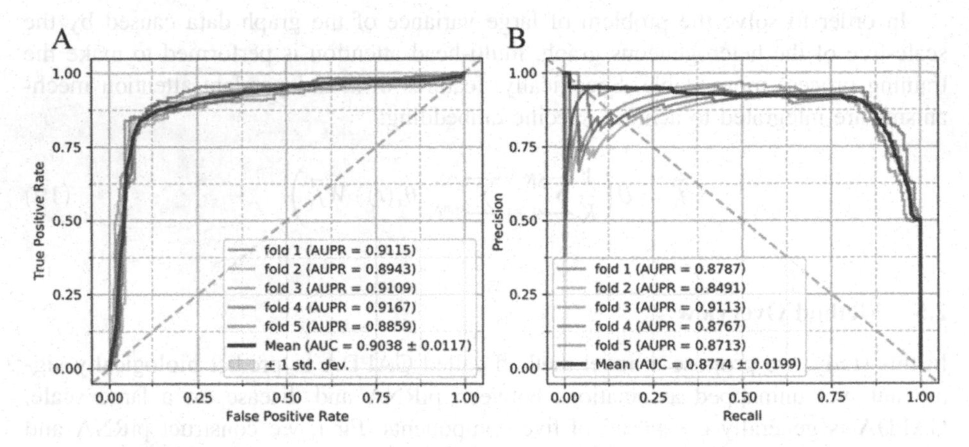

Fig. 1. (A) ROC curves performed by GAPDA on GPRD dataset. (B) PR curves performed by GAPDA on GPRD dataset.

Table 2. The five-fold cross-validation results performed by GAPDA on GPRD dataset

	Accuracy	Sensitivity	Precision	F1-score
Average	0.8569 ± 0.0092	0.8550 ± 0.0356	0.8638 ± 0.0416	0.8577 ± 0.0091

4 Conclusion

In this study, we proposed a piRNA-disease association prediction framework based on the graph attention network. In particular, we introduced attention-based graph neural networks into the field of ncRNA-related association prediction for the first time, and proposed an abstract network topology suitable for small samples. Supported by these two novel methods, GAPDA showed encouraging results in predicting piRNA-disease associations. In detail, in five-fold cross-validation, GAPDA got an AUC of 0.9038, AUPR of 0.8774, and accuracy of 0.8569. In addition, we compared it with two traditional methods and methods based on different abstract network topology. Experiments showed that GAPDA can be an excellent complement to future biomedical research and has determined the prospect of the graph neural grid on such problems. We hope that the proposed method can provide a powerful candidate for piRNA biomarkers and can be extended to other graph-based tasks.

Acknowledgments. This work is supported by the Xinjiang Natural Science Foundation under Grant 2017D01A78.

References

1. Yin, H., Lin, H.: An epigenetic activation role of Piwi and a Piwi-associated piRNA in Drosophila melanogaster. Nature **450**, 304 (2007)
2. Iwasaki, Y.W., Siomi, M.C., Siomi, H.: PIWI-interacting RNA: its bio-genesis and functions. Ann. Rev. Biochem. **84**, 405–433 (2015)
3. Grimson, A., et al.: Early origins and evolution of microRNAs and Piwi-interacting RNAs in animals. Nature **455**, 1193 (2008)
4. Aravin, A.A., Hannon, G.J., Brennecke, J.: The Piwi-piRNA pathway provides an adaptive defense in the transposon arms race. Science **318**, 761–764 (2007)
5. Malone, C.D., et al.: Specialized piRNA pathways act in germline and somatic tissues of the Drosophila ovary. Cell **137**, 522–535 (2009)
6. Leslie, M.: The Immune System's Compact Genomic Counterpart. American Association for the Advancement of Science (2013)
7. Pall, G.S., Codony-Servat, C., Byrne, J., Ritchie, L., Hamilton, A.: Carbodiimide-mediated cross-linking of RNA to nylon membranes improves the detection of siRNA, miRNA and piRNA by northern blot. Nucl. Acids Res. **35**, e60 (2007)
8. Marcon, E., Babak, T., Chua, G., Hughes, T., Moens, P.: miRNA and piRNA localization in the male mammalian meiotic nucleus. Chromos. Res. **16**, 243–260 (2008)
9. Armisen, J., Gilchrist, M.J., Wilczynska, A., Standart, N., Miska, E.A.: Abundant and dynamically expressed miRNAs, piRNAs, and other small RNAs in the vertebrate Xenopus tropicalis. Genom. Res. **19**, 1766–1775 (2009)
10. Moyano, M., Stefani, G.: piRNA involvement in genome stability and human cancer. J. Hematol. Oncol. **8**, 38 (2015)
11. Brennecke, J., et al.: Discrete small RNA-generating loci as master regulators of transposon activity in Drosophila. Cell **128**, 1089–1103 (2007)
12. Siomi, M.C., Sato, K., Pezic, D., Aravin, A.A.: PIWI-interacting small RNAs: the vanguard of genome defence. Nat. Rev. Mol. Cell Biol. **12**, 246 (2011)
13. Rajasethupathy, P., et al.: A role for neuronal piRNAs in the epigenetic control of memory-related synaptic plasticity. Cell **149**, 693–707 (2012)
14. Houwing, S., et al.: A role for Piwi and piRNAs in germ cell maintenance and transposon silencing in Zebrafish. Cell **129**, 69–82 (2007)
15. Moazed, D.: Small RNAs in transcriptional gene silencing and genome defence. Nature **457**, 413 (2009)

16. Zou, A.E., et al.: The non-coding landscape of head and neck squamous cell carcinoma. Oncotarget **7**, 51211 (2016)
17. Chu, H., et al.: Identification of novel piRNAs in bladder cancer. Cancer Lett. **356**, 561–567 (2015)
18. Cheng, J., et al.: piRNA, the new non-coding RNA, is aberrantly expressed in human cancer cells. Clin. Chimica Acta **412**, 1621–1625 (2011)
19. Assumpcao, C.B., et al.: The role of piRNA and its potential clinical implications in cancer. Epigenomics **7**, 975–984 (2015)
20. Ng, K.W., et al.: Piwi-interacting RNAs in cancer: emerging functions and clinical utility. Mol. Cancer **15**, 5 (2016)
21. Romano, G., Veneziano, D., Acunzo, M., Croce, C.M.: Small non-coding RNA and cancer. Carcinogenesis **38**, 485–491 (2017)
22. Fu, A., Jacobs, D.I., Hoffman, A.E., Zheng, T., Zhu, Y.: PIWI-interacting RNA 021285 is involved in breast tumorigenesis possibly by remodeling the cancer epigenome. Carcinogenesis **36**, 1094–1102 (2015)
23. Tan, L., et al.: PIWI-interacting RNA-36712 restrains breast cancer progression and chemoresistance by interaction with SEPW1 pseudogene SEPW1P RNA. Mol. Cancer **18**, 9 (2019)
24. Liu, X., et al.: PIWIL3/OIP5-AS1/miR-367-3p/CEBPA feedback loop regulates the biological behavior of glioma cells. Theranostics **8**, 1084 (2018)
25. Jacobs, D.I., Qin, Q., Fu, A., Chen, Z., Zhou, J., Zhu, Y.: piRNA-8041 is downregulated in human glioblastoma and suppresses tumor growth in vitro and in vivo. Oncotarget **9**, 37616 (2018)
26. Rizzo, F., et al.: Specific patterns of PIWI-interacting small noncoding RNA expression in dysplastic liver nodules and hepatocellular carcinoma. Oncotarget **7**, 54650 (2016)
27. Sai Lakshmi, S., Agrawal, S.: piRNABank: a web resource on classified and clustered Piwi-interacting RNAs. Nucl. Acids Res. **36**, D173–D177 (2007)
28. Wang, J., et al.: piRBase: a comprehensive database of piRNA sequences. Nucl. Acids Res. **47**, D175–D180 (2018)
29. Sarkar, A., Maji, R.K., Saha, S., Ghosh, Z.: piRNAQuest: searching the piRNAome for silencers. BMC Genom. **15**, 555 (2014)
30. Muhammad, A., Waheed, R., Khan, N.A., Jiang, H., Song, X.: piRDisease v1. 0: a manually curated database for piRNA associated diseases. In: Database 2019 (2019)
31. Wang, L., Wang, H.-F., Liu, S.-R., Yan, X., Song, K.-J.: Predicting protein-protein interactions from matrix-based protein sequence using convolution neural network and feature-selective rotation forest. Sci. Rep. **9**, 9848 (2019)
32. Zheng, K., You, Z.-H., Wang, L., Li, Y.-R., Wang, Y.-B., Jiang, H.-J.: MISSIM: improved miRNA-disease association prediction model based on chaos game representation and broad learning system. In: Huang, D.-S., Huang, Z.-K., Hussain, A. (eds.) ICIC 2019. LNCS (LNAI), vol. 11645, pp. 392–398. Springer, Cham (2019). https://doi.org/10.1007/978-3-030-26766-7_36
33. Zheng, K., You, Z.-H., Wang, L., Zhou, Y., Li, L.-P., Li, Z.-W.: MLMDA: a machine learning approach to predict and validate MicroRNA–disease associations by integrating of heterogenous information sources. J. Transl. Med. **17**, 1–14 (2019)
34. Wang, L., et al.: LMTRDA: Using logistic model tree to predict MiRNA-disease associations by fusing multi-source information of sequences and similarities. PLoS Comput. Biol. **15**, e1006865 (2019)
35. Li, Y., Li, L.-P., Wang, L., Yu, C.-Q., Wang, Z., You, Z.-H.: An ensemble classifier to predict protein-protein interactions by combining PSSM-based evolutionary information with local binary pattern model. Int. J. Mol. Sci. **20**, 3511 (2019)

36. Kirk, J.M., et al.: Functional classification of long non-coding RNAs by k-mer content. Nat. Genet. **50**, 1474 (2018)
37. Xiang, Z., Qin, T., Qin, Z.S., He, Y.: A genome-wide MeSH-based literature mining system predicts implicit gene-to-gene relationships and networks. BMC Syst. Biol. **7**, S9 (2013)
38. Xuan, P., et al.: Prediction of microRNAs associated with human diseases based on weighted k most similar neighbors. PLoS ONE **8**, e70204 (2013)
39. van Laarhoven, T., Nabuurs, S.B., Marchiori, E.: Gaussian interaction pro-file kernels for predicting drug–target interaction. Bioinformatics **27**, 3036–3043 (2011)
40. Zheng, K., You, Z.-H., Li, J.-Q., Wang, L., Guo, Z.-H., Huang, Y.-A.: iCDA-CGR: Identification of circRNA-disease associations based on chaos game representation. PLoS Comput. Biol. **16**, e1007872 (2020)
41. Wang, Y.-B., You, Z.-H., Yang, S., Yi, H.-C., Chen, Z.-H., Zheng, K.: A deep learning-based method for drug-target interaction prediction based on long short-term memory neural network. BMC Med. Inf. Decis. Mak. **20**, 1–9 (2020)
42. Wang, M.-N., You, Z.-H., Wang, L., Li, L.-P., Zheng, K.: LDGRNMF: LncRNA-disease associations prediction based on graph regularized non-negative matrix factorization. Neurocomputing (2020)
43. Zheng, K., You, Z.-H., Wang, L., Zhou, Y., Li, L.-P., Li, Z.-W.: DBMDA: a unified embedding for sequence-based miRNA similarity measure with applications to predict and validate miRNA-disease associations. Mol. Therap. Nucleic Acids **19**, 602–611 (2020)
44. Fan, C., Lei, X., Wu, F.-X.: Prediction of CircRNA-disease associations using KATZ model based on heterogeneous networks. Int. J. Biol. Sci. **14**, 1950 (2018)
45. Zheng, K., You, Z.-H., Wang, L., Wong, L., Zhan, Z.-h.: SPRDA: a matrix completion approach based on the structural perturbation to infer disease-associated Piwi-Interacting RNAs. bioRxiv (2020)
46. Wang, L., You, Z.-H., Li, L.-P., Zheng, K., Wang, Y.-B.: Predicting circRNA-disease associations using deep generative adversarial network based on multi-source fusion information. In: 2019 IEEE International Conference on Bioinformatics and Biomedicine (BIBM), pp. 145–152. IEEE (2019)
47. Zheng, K., You, Z.-H., Wang, L., Li, Y.-R., Zhou, J.-R., Zeng, H.-T.: MISSIM: an incremental learning-based model with applications to the prediction of miRNA-disease association. IEEE/ACM Trans. Comput. Biol. Bioinf
48. You, Z.-H., Li, X., Chan, K.C.: An improved sequence-based prediction protocol for protein-protein interactions using amino acids substitution matrix and rotation forest ensemble classifiers. Neurocomputing **228**, 277–282 (2017)
49. Wang, Y.-B., et al.: Predicting protein–protein interactions from protein sequences by a stacked sparse autoencoder deep neural network. Mol. BioSyst. **13**, 1336–1344 (2017)
50. Zheng, K., You, Z.-H.: iMDA-BN: Identification of miRNA-Disease Associations based on the Biological Network and Graph Embedding Algorithm. bioRxiv (2020)
51. Chen, X., Huang, Y.-A., You, Z.-H., Yan, G.-Y., Wang, X.-S.: A novel approach based on KATZ measure to predict associations of human microbiota with non-infectious diseases. Bioinformatics **33**, 733–739 (2017)
52. Sun, X., Bao, J., You, Z., Chen, X., Cui, J.: Modeling of signaling cross-talk-mediated drug resistance and its implications on drug combination. Oncotarget **7**, 63995 (2016)
53. Dong, Y., Sun, Y., Qin, C., Zhu, W.: EPMDA: Edge perturbation based method for miRNA-disease association prediction. IEEE/ACM Trans. Comput. Biol. Bioinf. (2019)
54. Guo, Z.-H., et al.: MeSHHeading2vec: a new method for representing MeSH headings as vectors based on graph embedding algorithm. Brief. Bioinf. (2020)

55. Zhu, H.-J., You, Z.-H., Zhu, Z.-X., Shi, W.-L., Chen, X., Cheng, L.: DroidDet: effective and robust detection of android malware using static analysis along with rotation forest model. Neurocomputing **272**, 638–646 (2018)
56. Luo, X., You, Z., Zhou, M., Li, S., Leung, H., Xia, Y., Zhu, Q.: A highly efficient approach to protein interactome mapping based on collaborative filtering framework. Sci. Rep. **5**, 7702 (2015)
57. Zheng, K., Wang, L., You, Z.-H.: CGMDA: an approach to predict and validate MicroRNA-disease associations by utilizing chaos game representation and LightGBM. IEEE Access **7**, 133314–133323 (2019)
58. Lei, Y.-K., You, Z.-H., Ji, Z., Zhu, L., Huang, D.-S.: Assessing and predicting protein interactions by combining manifold embedding with multiple information integration. In: BMC bioinformatics, vol. 13, p. S3 (2012)
59. Li, S., You, Z.-H., Guo, H., Luo, X., Zhao, Z.-Q.: Inverse-free extreme learning machine with optimal information updating. IEEE Trans. Cybern. **46**, 1229–1241 (2015)
60. Ji, B.-Y., You, Z.-H., Jiang, H.-J., Guo, Z.-H., Zheng, K.: Prediction of Drug-target Interactions from Heterogeneous Information Network Based on LINE Embedding Model (2020)
61. Wang, L., et al.: An improved efficient rotation forest algorithm to predict the interactions among proteins. Soft. Comput. **22**, 3373–3381 (2018)
62. Veličković, P., Cucurull, G., Casanova, A., Romero, A., Lio, P., Bengio, Y.: Graph attention networks. arXiv preprint arXiv:1710.10903 (2017)

A Novel Improved Algorithm for Protein Classification Through a Graph Similarity Approach

Hsin-Hung Chou[1], Ching-Tien Hsu[2], Hao-Ching Wang[3], and Sun-Yuan Hsieh[2(✉)]

[1] Department of Information Management, Chang Jung Christian University, Tainan 71101, Taiwan
chouhh@mail.cjcu.edu.tw

[2] Department of Computer Science and Information Engineering and Institute of Medical Information, National Cheng Kung University, No. 1, University Road, Tainan 70101, Taiwan
hondyshu@gmail.com, hsiehsy@mail.ncku.edu.tw

[3] Graduate Institute of Translational Medicine, College of Medical Science and Technology, Taipei Medical University, Taipei 110, Taiwan
wanghc@tmu.edu.tw

Abstract. The protein classification problem is considered in this paper. In the proposed algorithm, we use graphs to represent proteins, whereby every amino acid in a protein corresponds to every vertex in a graph, and the links between the amino acids correspond to the edges between the vertices in the graph. We then classify the proteins according to similarities in their corresponding graph structures.

Keywords: B-factor · Bioinformatic algorithms · Graph similarity · Protein classification · Protein structures

1 Introduction

Proteins are large biomolecules comprising one or more long chain of amino acid residues and are ubiquitous in living cells. Research on protein classification have used graphs to represent proteins, whereby every amino acid in a protein corresponds to one vertex in a graph, and each link between the amino acids in the protein corresponds to one edge between the vertices in the graph [5].

Several methods use graphs to denote proteins, such as Unsubstituted Patterns (UnSubPatt) [6], Family Frequent Subgraph Mining (FFSM) [9], Functional Neighbors [1]. Hsieh's algorithm [8] uses a specific data structure CAM (Canonical Adjacency Matrix) as patterns to classify proteins.

Because proteins have a complex structure, three-dimensional approaches require considerable computation and are time-consuming. In the proposed algorithm, we use graphs to represent proteins, and we analyze some benchmarks selected from the Protein Data Bank (PDB) [2, 17], and use some novel techniaues to control the time consumed and accuracy.

D.-S. Huang and K.-H. Jo (Eds.): ICIC 2020, LNCS 12464, pp. 251–261, 2020.
https://doi.org/10.1007/978-3-030-60802-6_22

The remainder of this paper is organized as follows. In Sect. 2, we discuss the protein classification problem and introduce the format of the PDB file and some data structures that will be used in our proposed algorithm. We present our proposed algorithm in Sect. 3 and discuss the experimental results in Sect. 4. Section 5 presents our concluding remarks.

2 Preliminary

2.1 PDB File

The PDB is a crystallographic database comprising the three-dimensional structural data of large biological molecules that were established at Brookhaven National Laboratory in 1971 [3, 17]. The PDB is now overseen by the Worldwide Protein Data Bank. Data of the PDB are submitted by scientists from all over the world and can be downloaded for free from the homepage of the Research Collaboratory for Structural Bioinformatics (https://www.rcsb.org), which is where we obtained the inputs for our experiments.

Table 1. PDB file of protein 1AF3 belonging to the bcl-xl family.

Line	ATOM	Amino acid	Chain	No.	Coord. (x, y, z)	Occup.	B-factor	Element
1	N	SER	A	2	35.882 -3.709 36.253	1.00	110.33	N
2	CA	SER	A	2	34.644 -3.764 35.404	1.00	110.75	C
3	C	SER	A	2	33.397 -3.882 36.258	1.00	107.88	C
4	O	SER	A	2	32.485 -3.058 36.202	1.00	105.56	O
5	CB	SER	A	2	34.815 -4.872 34.362	1.00	111.21	C
6	OG	SER	A	2	35.710 -4.443 33.340	1.00	112.07	O
7	N	GLN	A	3	33.371 -4.933 37.114	1.00	107.81	N
8	CA	GLN	A	3	32.259 -5.049 38.088	1.00	106.02	C
9	C	GLN	A	3	32.655 -4.588 39.472	1.00	97.19	C
10	O	GLN	A	3	31.831 -4.147 40.244	1.00	90.63	O
11	CB	GLN	A	3	31.785 -6.502 38.125	1.00	112.76	C
12	CG	GLN	A	3	30.492 -6.717 38.911	1.00	115.28	C
13	CD	GLN	A	3	29.319 -6.111 38.170	1.00	120.06	C
14	OE1	GLN	A	3	29.373 -5.073 37.528	1.00	120.62	O
15	NE2	GLN	A	3	28.232 -6.863 38.250	1.00	122.25	N
16	N	SER	A	4	33.980 -4.641 39.716	1.00	82.20	N
17	CA	SER	A	4	34.513 -3.870 40.841	1.00	82.48	C
18	C	SER	A	4	34.095 -2.395 40.746	1.00	78.95	C
19	O	SER	A	4	33.797 -1.739 41.721	1.00	83.75	O
20	CB	SER	A	4	36.050 -4.092 40.905	1.00	82.39	C
21	OG	SER	A	4	36.641 -3.476 39.734	1.00	90.52	O
22	N	ASN	A	5	34.042 -1.951 39.469	1.00	73.25	N
23	CA	ASN	A	5	33.556 -0.599 39.166	1.00	64.41	C
24	C	ASN	A	5	32.146 -0.313 39.662	1.00	59.23	C
25	O	ASN	A	5	31.895 0.716 40.189	1.00	56.99	O

Table 1 is an example of a PDB file [16]. The first column in Table 1 is the number of lines, the second is the types of atoms, and the third is the type of amino acids. The sixth column contains three floating-point numbers that are X, Y, and Z coordinates. The seventh, eighth, and ninth columns are the occupancy, B-factor [11, 15], and element name, respectively. In our algorithm, we use one label to denote each amino acid and X, Y, and Z coordinates to construct a three-dimensional structure. Moreover, we use the B-factor to control time complexity.

2.2 Distance Threshold

We construct three-dimensional structures based on PDB files, and we use graphs to denote proteins. The contents in the third column of Table 2 represent types of amino acids, whereby each type of element is denoted by a label, and we take each element in this column as a vertex. As shown in Tabel 2, every type of amino acid corresponds to one label.

Table 2. Amino acids corresponding to labels.

No	Amino acid	Label
1	ALA	A
2	CYS	C
3	ASP	D
4	GLU	E
5	PHE	F
6	GLY	G
7	HIS	H
8	ILE	I
9	LYS	K
10	LEU	L
11	MET	M
12	ASN	N
13	PRO	P
14	GLN	Q
15	ARG	R
16	SER	S
17	THR	T
18	VAL	V
19	TRP	W
20	TYR	Y
21	Other	X

After determining the vertices, we then consider how to construct edges between the vertices. The sixth column of Table 1 presents three floating-point numbers that are the coordinates of elements. We compute the Euclidean distance between any pair of elements. We then set a threshold to determine whether there is an edge between two vertices. If the distance between two vertices is less than or equal to the threshold, we set an edge between them [15]. Otherwise, there is no edge between them.

We provide an example in Table 3, whereby we assume that the threshold is 1. First, we use four vertices with labels corresponding to types of amino acids to denote each element in Table 3. We then calculate the distance between each pair of elements; because there are three distances less than or equal to the threshold of 1, there are three edges in the structure: one links the ALA and GLY elements, and the other two link the GLY element and two SER elements. Next, we construct a graph, as shown in Fig. 1.

Table 3. Example of the PDB file.

Line	ATOM	Amino acid	Chain	No.	Coord. (x, y, z)	Occup.	B-factor	Element
1	N	ALA	A	2	0.000 1.000 0.000	1.00	110.33	N
2	CA	GLY	A	2	0.000 0.000 0.000	1.00	110.75	C
3	C	SER	A	2	0.000 0.000 1.000	1.00	107.88	C
4	O	SER	A	2	1.000 0.000 0.000	1.00	105.56	O

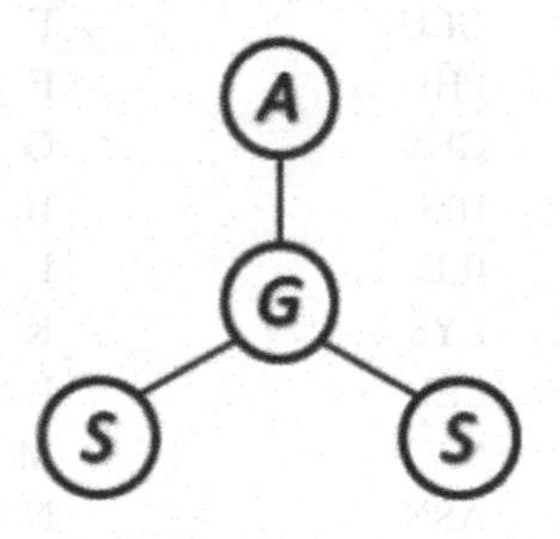

Fig. 1. The graph used to denote the PDB file in Table 3.

The distance threshold can be used to control time complexity: the larger the threshold, the longer the computation time.

2.3 B-Factor

The eighth column of Table 1 is the B-factor, which we use to reduce the complexity of the graph. The B-factor contributes to the thermal stability of proteins substitutions [11, 14]. The elements with relatively high mobility have relatively large B-factor, which means that the atom may have larger fluctuations in the space. If an atom has very large fluctuations in a space, it will not diffract well, reducing the accuracy of the atomic coordinates. To prevent unstable atomic coordinates and reduce time complexity, we eliminate atoms that have high B-factor ~\cite{HSI18}. In our algorithm, we set a

B-factor value, and if the B-factor of an atom exceeds the value, we eliminate the atom from the protein's structural data.

2.4 Protein Adjacency Matrix

In this paper, we define the *protein adjacency matrix*, which is the adjacency matrix of the graph corresponding to a protein, and the diagonal entries of the protein adjacency matrix are filled with amino acid labels [8]. In the graph corresponding to a protein structure, there is only one link between amino acids, and an element cannot have links to itself. Therefore, the graph is simple.

For example, in Fig. 2, we present a graph with four vertices, 1, 2, 3, and 4, which correspond to labels A, G, left S, and right S. First, we construct a 4 × 4 square matrix M and fill the diagonal entries $M_{i,i}$ with amino acid labels. The entry $M_{i,j}$, $i < j$, is determined when there is an edge connecting vertex i and vertex j. Because vertices 1 and 2 are connected, we fill $M_{1,2}$ with 1. Because vertices 1 and 3 are not connected, we fill $M_{1,3}$ with 0.

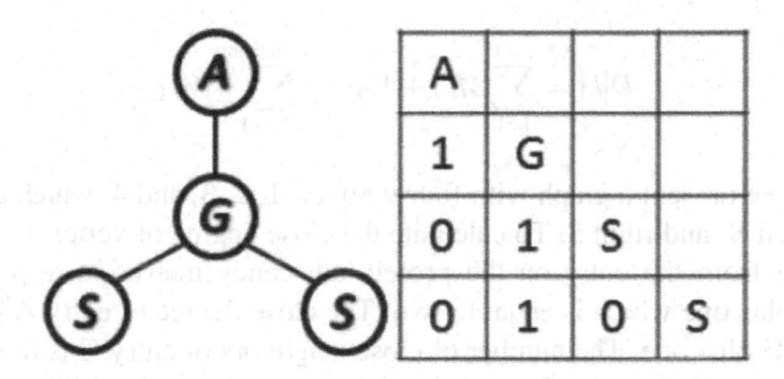

Fig. 2. Example graph with the protein adjacency matrix.

3 Proposed Algorithm

In this section, we discuss the main algorithm. Our algorithm classifies proteins into the families they belong to, and we use PDB files as inputs. In our experiments, every protein family set has proteins that belong to it; we use these proteins for structural classification. The main algorithm compares the structure of an input file with the proteins of all families, and the output will be a string denoting the family the input protein belongs to.

We use an input PDB file to construct a graph and present the graph with the protein adjacency matrix. We use the threshold and B-factor to control the size of the protein adjacency matrix.

3.1 Degree Vector

We use the degrees of the vertices in a graph to classify the corresponding proteins, whereby the degree of a vertex is equal to the number of its neighboring vertices, which can be obtained from the protein adjacency matrix.

We construct a vector called the degree vector, which can be used to show the structural features of a graph. Each entry of the degree vector is equal to the sum of the degrees of the vertices with the same label. In our experiment, there are 21 types of amino acids; therefore, we set the size of the degree vector to 21, and 21 entries of the degree vector correspond to 21 labels, as presented in Table 2. There are 21 entries in the degree vector, and some types of amino acids do not exist in a protein. To distinguish those elements, we use the number of close neighbors to calculate the degree of entry. The close neighbor of a vertex v is a vertex set that includes the neighboring vertices of v and itself, and we use $D[k]$ to denote the number of close neighbors of the kth element. We define the *close degree* of a vertex as the number of elements in close neighbors. We calculate the close degree of an element through (1), in which we sum the close degrees of vertices with the same label and fill it into the entry corresponding to a label.

$$D[k] = \sum_{i=1}^{k-1} M_{k,i} + 1 + \sum_{i=k+1}^{\#\text{of elements}} M_{i,k} \tag{1}$$

In Fig. 3, we present a graph with four vertices, 1, 2, 3, and 4, which correspond to labels A, G, left S, and right S. To calculate the close degree of vertex 1, we add to the left and down from the entry on the protein adjacency matrix corresponding to the vertex 1 and plus one which is equal to two. The close degree of entry A is two, which is one link to G plus one. The number of close neighbors of entry G is four: one link to A and two links to S plus one. The number of close degrees of the third entry S is two: one link to G plus one. There are two entries S, and their close degrees are both equal to two; we combine them to obtain the close degrees of entries S in the degree vector. After the calculation, we can construct a degree vector. We fill the entry in the degree vector corresponding to labels A, G, and S with two, four, and four, respectively, and fill the other entries with zero.

3.2 Normalized Degree Vector

We obtain the degree vector in the previous step. Because the sizes of PDB files and the number of amino acids in a protein vary, directly comparing degree vectors creates bias. Accordingly, we normalize the content of a degree vector. In a vector, all entries are converted to a number between 0 and 1 according to the proportion of the entries in the vector, as in (2).

$$V_{[i]} = \frac{V_{[i]}}{\sum_{i=1}^{21} V_{[i]}} \tag{2}$$

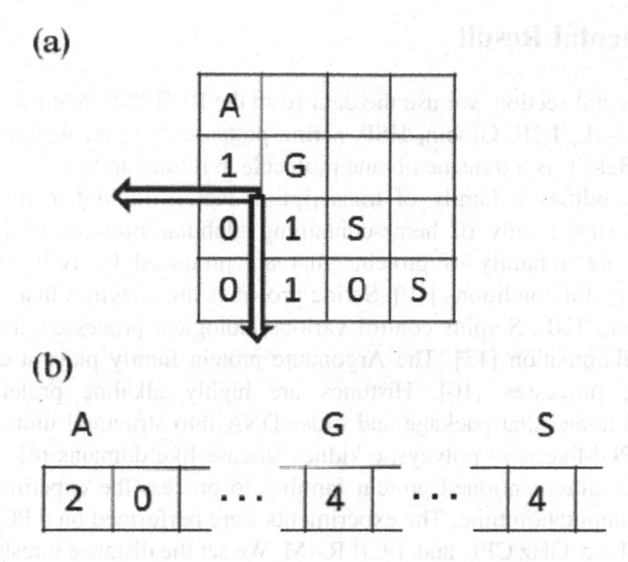

Fig. 3. (a) Protein adjacency matrix and (b) the degree vector converted from the adjacency matrix.

3.3 Family Characteristic Vector

We construct a vector for each protein family, which we call the family characteristic vector, to represent the characteristics of families. The family characteristic vector is generated as follows: we convert proteins belonging to a family to normalized degree vectors, and we average those vectors to obtain the family characteristic vector, as in (3):

$$V_{family} = \frac{\sum_{i=1}^{\#\text{ of protein in family}} V_i}{\#\text{ of protein in family}} \tag{3}$$

3.4 Cosine Similarity

Cosine similarity is a measure of similarity between two nonzero vectors of an inner product space that measures the cosine of the angle between them; we use it to compare the similarity between normalized degree vectors and family characteristics vectors. We compute the similarity between the normalized degree vector converted from an input protein and all the family characteristic vectors, and we use it to classify input proteins.

3.5 Main Algorithm

After calculating the similarity between the family characteristic vectors and the normalized degree vector of an input protein, we choose the family with the highest similarity as the output.

4 Experimental Result

In the experimental section, we use the data from the PDB [17]. We use the PDB files of families, Bcl-xL, E2F, Globin, HSP, serine proteases, serpin, Argonaute, Histone, and pkd-like. Bcl-xL is a transmembrane molecule in mitochondria [7]. E2F is a group of genes that codifies a family of transcription factors in higher eukaryotes [18]. Globins are a superfamily of heme-containing globular proteins [21]. Heat shock proteins(HSP) are a family of proteins that are produced by cells in response to exposure to stressful conditions [19]. Serine proteases are enzymes that cleave peptide bonds in proteins [20]. Serpins control various biological processes, including coagulation and inflammation [13]. The Argonaute protein family plays a central role in RNA-silencing processes [10]. Histones are highly alkaline proteins found in eukaryotic cell nuclei that package and order DNA into structural units called nucleosomes [12]. Pkd-like is in polycystic kidney disease-like domains [4].

We use the aforementioned protein families to process the experiment. First, we examine the computation time. The experiments were performed on a PC with an Intel Core i5-5210M 2.5 GHz CPU and 4 GB RAM. We set the distance threshold to 10, 15, and B-factor to 50, 60, the threshold, and B-factor parameters are from [8]. The computation time is presented in Table 4. The experiments that have a relatively high distance threshold and B-factor require more computation time. The difference between the average computation time of different families is low.

Table 4. Average execution time of the proposed method when B-factor = 50, the threshold = 10, B-factor = 60, and the threshold = 15.

Data set	B-factor	Threshold	Time consumed
bc1-x1	50	10	79.249 s
E2F	50	10	82.395 s
Globin	50	10	84.829 s
HSP	50	10	83.385 s
serine proteases	50	10	80.116 s
serpin	50	10	83.579 s
Argonaute	50	10	85.34 s
Histone	50	10	93.03 s
pkd-like	50	10	84.16 s
bc1-x1	60	15	87.484 s
E2F	60	15	94.269 s
Globin	60	15	90.379 s
HSP	60	15	90.419 s
serine proteases	60	15	91.657 s
serpin	60	15	90.719 s
Argonaute	60	15	88.807 s
Histone	60	15	88.876 s
pkd-like	60	15	95.862 s

Table 5. Accuracy of protein classification when B-factor = 50 and the threshold = 10.

Data set	B-factor	Threshold	Proposed algorithm	DALI	ProtNN	Hsieh
bcl-x1	50	10	100%*	94%	100%*	94%
E2F	50	10	100%*	100%*	89%	100%*
Globin	50	10	100%*	100%*	100%*	100%*
HSP	50	10	80%*	80%*	80%*	60%
serine proteases	50	10	100%*	67%	67%	83%
serpin	50	10	100%*	91%	73%	73%
Argonaute	50	10	71%	86%	93%*	75%
Histone	50	10	78%	78%	56%	88%*
pkd-like	50	10	80%	80%	0%	100%*

Table 6. Accuracy of protein classification when B-factor = 60 and the threshold = 15.

Data set	B-factor	Threshold	Proposed algorithm	DALI	ProtNN	Hsieh
bcl-x1	60	15	100%*	94%	100%*	94%
E2F	60	15	100%*	100%*	89%	100%*
Globin	60	15	100%*	100%*	100%*	100%*
HSP	60	15	80%	80%	80%	82%*
serine proteases	60	15	100%*	67%	67%	83%
serpin	60	15	100%*	91%	73%	90%
Argonaute	60	15	79%	86%	93%	86%
Histone	60	15	67%	78%	56%	88%*
pkd-like	60	15	80%	80%	0%	100%*

Tables 5 and 6 compare the precision of our algorithm with two other programs. In the tables, the best values are marked with *. In our algorithm, the proteins from bcl-xl, E2F, Globin, serine proteases, and serpin are precisely classified, reaching an accuracy rate of 100%. The accuracy rate for classifying other proteins also exceeded 67%, which means that our algorithm can classify proteins with high accuracy.

5 Conclusion

In this paper, we classified proteins according to similarities in their 3D graph structures. The accuracy of our method was good; it achieved 100% accuracy in the five selected protein families.

References

1. Bandyopadhyay, D., et al.: Functional neighbors: inferring relationships between nonhomologous protein families using family-specific packing motifs. IEEE Trans. Inf. Technol. Biomed. **14**(5) (2010)
2. Berman, H.M., et al.: The protein data bank. Nucleic Acids Res. **28**(1), 235–242 (2000)
3. Bernstein, F.C., Koetzle, T.F., Williams, G.J.B., Meyer Jr., E.: The protein data bank: a computerbased archival file for macromolecular structures. Eur. J. Biochem. **80**, 319–324 (1977)
4. Bycroft, M., et al.: The structure of a PKD domain from polycystin-1: implications for polycystic kidney disease. EMBO J. **18**(2), 297–305 (1999)
5. Dhifli, W., Diallo, A.B.: ProtNN: fast and accurate protein 3D-structure classification in structural and topological space. BioData Min. **9**(30) (2016)
6. Dhifli, W., Saidi, R., Nguifo, E.M.: Smoothing 3D protein structure motifs through graph mining and amino acid similarities. J. Comput. Biol. **21**(2) (2014)
7. Fiebig, A.A., Zhu, W., Hollerbach, C., Leber, B., Andrews, D.W.: Bcl-XL is qualitatively different from and ten times more effective than BCL-2 when expressed in a breast cancer cell line. BMC Cancer **6,** 213 (2006). https://doi.org/10.1186/1471-2407-6-213
8. Hsieh, S.-Y., et al.: Classifying protein specific residue structures based on graph mining. IEEE Access **6**, 55828–55837 (2018)
9. Huan, J., Wang, W., Bandyopadhyay, D., Snoeyink, J., Prins, J., Tropsha, A.: Mining protein family specific residue packing patterns from protein structure graphs. In: Proceedings of the 8th Annual International Conference on Research in Computational Molecular Biology (RECOMB), pp. 308–315 (2004)
10. Hutvagner, G., Simard, M.J.: Argonaute proteins: key players in RNA silencing. Nat. Rev. Mol. Cell Biol. **9**, 22–32 (2008)
11. Parthasarathy, S., Murthy, M.R.N.: Protein thermal stability: insights from atomic displacement parameters (B values). Protein Eng. **13**(1), 9–13 (2000)
12. Redon, C., Pilch, D., Rogakou, E., Sedelnikova, O., Newrock, K., Bonner, W.: Histone H2A variants H2AX and H2AZ. Curr. Opin. Genet. Dev. **12**(2), 162–169 (2002)
13. Silverman, G.A., et al.: The serpins are an expanding superfamily of structurally similar but functionally diverse proteins evolution, mechanism of inhibition, novel functions, and a revised nomenclature. J. Biol. Chem. **276**(36), 33293–33296 (2001)
14. Tronrud, D.E.: Knowledge-based b-factor restraints for the refinement of proteins. J. Appl. Cryst. **29**, 100–104 (1996)
15. Weskamp, N., Kuhn, D., Hllermeier, E., Klebe, G.: Efficient similarity search in protein structure databases by k - clique hashing. Bioinformatics **20**(10), 1522–1526 (2005)
16. Pdb file format. http://www.wwpdb.org/documentation/fileformat-content/format33/v3.3.html
17. Protein Data Bank. https://www.rcsb.org/pdb
18. Zheng, N., Fraenkel, E., Pabo, C.O., Pavletich, N. P.: Structural basis of DNA recognition by the heterodimeric cell cycle transcription factor E2F–DP. Howard Hughes Medical Institute, Cellular Biochemistry, and Biophysics Program, Memorial Sloan-Kettering Cancer Center, New York, New York 10021 USA; 2 Howard Hughes Medical Institute, Department of Biology, Massachusetts Institute of Technology, Cambridge, Massachusetts 02139 USA (1999)

19. Matz, M., Blake, M.J., Tatelman, H.M., Lavoi, K.P., Holbrook, N.J.: Characterization and regulation of coldinduced heat shock protein expression in mouse brown adipose tissue. Am. J. Physiol. **269**(1 Pt 2), R38–R47 (1995)
20. Hedstrom, L.: Serine protease mechanism and specificity. Chem. Rev. **102**(12), 4501–4524 (2002)
21. Vinogradov, S.N., et al.: A model of globin evolution. Received 6 December 2006; received in revised form 20 February 2007; accepted 21 February 2007 Available online 4 May (2007)

LncRNA-Disease Association Prediction Based on Graph Neural Networks and Inductive Matrix Completion

Lin Yuan(✉), Jing Zhao, Tao Sun, Xue-Song Jiang, Zhen-Yu Yang, Xin-Gang Wang, and Yu-Shui Geng(✉)

School of Computer Science and Technology, Qilu University of Technology (Shandong Academy of Sciences), Jinan 250353, Shandong, China
yuanlindc@126.com

Abstract. Emerging evidence indicates that long non-coding RNA (lncRNA) plays a crucial role in human disease. Discovering disease-gene association is a fundamental and critical biomedical task, which assists biologists and physicians to discover complex pathogenic mechanisms under the phenotype. With high-throughput sequencing technology and various clinical biomarkers to measure the similarities between lncRNA and disease phenotype, network-based semi-supervised learning has been commonly utilized by these studies to address this class imbalanced large-scale data issue. However, most existing approaches are based on linear models and suffer from two major limitations: 1) They implicitly consider a local-structure representation for each candidate; 2) They are unable to capture nonlinear associations between lncRNAs and diseases. In this paper, we propose a new framework for lncRNA-disease association task by combining Graph Neural Network (GNN) and inductive matrix completion, named GNN-IMC. With the help of GNN, we could generate subgraphs based on (lncRNA, disease) pairs from the observed association matrix and maps these subgraphs to their corresponding associations. In addition, GNN-IMC is inductive–it can generalize to lncRNAs/diseases unseen during the training (given that their associations exist), and can even transfer to new tasks. Empirical results demonstrate that the proposed deep learning algorithm outperforms all other state-of-the-art methods on most of metrics.

Keywords: Disease-lncRNA association · Graph neural network · Inductive matrix completion · Non-convex penalty function · Deep learning

1 Introduction

The Human Genome Project (HGP) determined the nucleotide sequence of 3 billion base pairs in the chromosome [1]. However, how to encode and regulate genes in the genome needs further study [2, 3]. As a follow-up project, the ENCODE (Encyclopedia of DNA elements) project aims to study the functional information of each gene on the human genome [4]. With the completion of the project, the study found that genes with protein coding functions account for only 1.5% of the genome, meaning that more than 80% of the human genome is not yet a coding protein sequence. These genomic regions

D.-S. Huang and K.-H. Jo (Eds.): ICIC 2020, LNCS 12464, pp. 262–269, 2020.
https://doi.org/10.1007/978-3-030-60802-6_23

are usually transcribed into non-coding RNAs (non-coding RNA, ncRNAs). Since non-coding RNAs do not encode proteins, they are generally regarded as transcriptional noise [5]. However, more and more studies have shown that non-coding RNAs play a key role in many biological processes. Among them, the largest subclass of non-coding RNA is long noncoding RNA, which is a class of noncoding RNAs with a length of more than 200 nucleotides. Studies have shown that lncRNAs play an important role in a variety of biological mechanisms, such as cell differentiation, proliferation and apoptosis [6, 7]. In addition, mutations and disorders of lncRNAs are closely related to the occurrence of breast cancer, ovarian cancer and other diseases [8, 9]. Therefore, lncRNAs are considered to be important factors for disease diagnosis and treatment. Predicting the relationship between lncRNAs and disease is not only beneficial for disease diagnosis, but also helps to understand biological processes. Unfortunately, identifying those unknown associations is still a major challenge. Traditional biological experiments and clinical methods will consume a lot of time and effort. Therefore, researchers need to develop effective calculation methods to solve this problem.

A large number of methods to predict the association between lncRNAs and diseases have been proposed [10–12]. GCNLDA infers disease-related lncRNA candidates based on graph convolutional network and convolutional neural network [13]. GMCLDA infers potential lncRNA-disease associations based on geometric matrix completion [14]. IDLDA is an improved diffusion potential lncRNA–disease associations prediction model [15]. However, due to the particularity of the prediction problems associated with lncRNAs and diseases, the accuracy of current prediction methods needs to be further improved [16]. Experimental scientists urgently need lncRNAs-disease association prediction methods that can give more reference value. Against this background of needs, research and development of high-precision prediction methods are particularly important [17–19].

In this paper, to predict complex relationship between lncRNA and disease, we combine GNN and inductive matrix completion to generate a new machine learning framework for lncRNA-disease association task, named GNN-IMC. We exploit different data sources to construct the lncRNA-lncRNA similarity and disease-disease similarity. Then we extract an h-hop enclosing subgraph for each training lncRNA-disease pair and train a graph regression model. Our resulting algorithm is inductive, given a trained GNN, we can also apply it to unseen lncRNAs/diseases without retraining. Moreover, we define a margin control loss function to reduce the effect of sparsity. Note that GNN-IMC does not address the extreme cold-start problem, as it still requires an unseen lncRNA-disease pair's enclosing subgraph (i.e., the lncRNA and disease should at least have some interactions with neighbors so that the enclosing subgraph is not empty).

Our contributions could be summarized as follows:

- We propose a new machine learning framework for lncRNA-disease association by combining GNN and inductive matrix completion.
- We construct similarity graphs for lncRNAs and diseases respectively, based on the domain knowledge from various data sources.

- We analyze different factors that influence the performance of our method. The most important factor of proposed method is the hyperparameter λ (Eq. 10). We denote the second largest singular value of matrix S as a $\lambda_{\max}$.
- We have conducted extensive experiments on simulated and real data sets, and the experimental results show that the method outperforms other representative algorithms by a large margin in most evaluation indicators.

2 Preliminaries

We extract lncRNA and disease information from multi-omics datasets including lncRNA, miRNA and disease data, construct characteristic matrix for lncRNA and disease, and learn the non-linear association information and similarity information in lncRNA-disease association network. We use graph neural network Inductive matrix completion methods to improve the accuracy of the model, and predict unknown lncRNA-disease associations.

Given n lncRNA sets $U = \{u_1, u_2, \ldots u_m\}$ and r disease sets $D = \{d_1, d_2, \ldots, d_r\}$, the correlation matrix is $M \in R^{n \times r}$. $M_{ij} = 1$ indicates that u_i and v_j are related, $M_{ij} = 0$ indicating that u_i and v_j are unrelated or unknown, and the relationship between the two needs to be predicted.

2.1 Graph Neural Network Training

First, we perform feature extraction on lncRNA and disease-related data. On the one hand, lncRNA related data, including lncRNA-disease related data, lncRNA similarity data, and lncRNA-miRNA related data; on the other hand, disease related data, including disease similarity data, disease-miRNA related data, and disease-lncRNA related data. The data is expressed in the form of a matrix, for example, lncRNA-miRNA matrix, each row represents an lncRNA, and each column represents a miRNA. We plan to use the kernel principal component analysis (KPCA) method for feature extraction. For each lncRNA correlation matrix, we construct the lncRNA feature matrix by combining the singular vectors containing the main features method to generate disease feature matrix $Q \in R^{r \times k_v}$, k_u and k_v represent the dimension of feature vector.

We can calculate the similarity matrix of circRNA and disease based on feature matrix separately:

$$Corr(u_i, u_j) = p_i^{\mathrm{T}} p_j \tag{1}$$

$$Corr(d_i, d_j) = q_i^{\mathrm{T}} q_j \tag{2}$$

$$G(u)_{ij} = Corr(u_i, u_j) \tag{3}$$

$$G(d)_{ij} = Corr(d_i, d_j) \tag{4}$$

where p_i and q_i are the row vectors of P and Q, respectively. $G(u) \in R^{n \times n}$ and $G(v) \in R^{r \times r}$ are the similarity matrix of lncRNA and disease, respectively. We obtain the adjacency matrix through the similarity matrix:

$$A(u) = I + \text{Norm}(G(u)) \tag{5}$$

$$A(v) = I + \text{Norm}(G(v)) \tag{6}$$

We can get the new information matrix of lncRNA and disease and the prediction matrix of lncRNA-disease association based on graph neural network to train the network and its parameters:

$$P = \sigma(A(u)PW_u) \tag{7}$$

$$Q = \sigma(A(v)PW_v) \tag{8}$$

where W_u and W_v are parameters in the graph neural network.

2.2 Nonconvex Induction Matrix Completion

In the real data obtained by the experimental observations, commonly exist missing data. In order to operate missing data effectively, we extend the proposed basic model as following. Ω is used to represent an observed subset of entries in lncRNA-disease association data. And $\Omega^\dagger$ represent the unobserved entries. For mathematical concept of linear operator, we define an orthogonal projection operator ρ that keeps the entries in Ω of a matrix unchanged and those entries in $\Omega^\dagger$ of matrix, i.e. projecting matrix onto the linear space of matrices supported by Ω:

$$\Pr(W)(i,j) = \begin{cases} W_{ij}, if\,(i,j) \in \Omega \\ 0, if\,(i,j) \notin \Omega \end{cases} \tag{9}$$

And $\Pr_{\Omega^\dagger}(W)$ is its complementary projection. Because we want to find a sparse coefficient matrix and a low-rank matrix based on the observed data, we propose to solve the following nonconvex inductive matrix completion problem:

$$L = \left\| S - \widetilde{S} \right\|_F^2 + \lambda \sum_{i=1}^{d} h(\beta_i) \tag{10}$$

Where λ is the parameter of the balanced regular term and $h(\cdot)$ represents the non-convex penalty function. We intend to use the better performing smooth truncated absolute deviation (SCAD) function and the minimum maximum concave penalty (MCP) function. β_i represent the number of the tandem matrix [*Wu*; *Wv*] i column.

3 The Proposed Method GNN-IMC

In this part we will introduce our experimental process. The learning procedure of GNN-IMC is constructed on the framework of GNN and IMC. We now present our Graph neural network and inductive matrix completion (GNN-IMC) framework. The steps of GNN-IMC is elaborated as follow.

1. Initialize $t = 0$, the maximum number of iterations *itermax*, the position and velocity of each P_i;
2. Compute the β_i, and evaluate P_i by three objective functions;
3. Initialize P_{ib} and determine the dominance relationship between all the particles;
4. Update the external achieve $\mathcal{REP}$;
5. Update each particle in the population;
6. Compute the β_i to the new P_i, evaluate P_i and determine the dominance relationship between all the particles;
7. $t = t + 1$, go to Step 4 until $t \geq itermax$;
8. Using Bagging method to ensemble the IMCs in $\mathcal{REP}$.

4 Experimental Results and Discussion

In this section, we evaluate the GNN-IMC and compare it with GCNLDA [13], GMCLDA [14] on several datasets. With an increased complexity and sophistication of a proposed method, it is very important to compare its performance with existing methods in an objective way. To achieve this end, we conduct extensive simulation studies to compare the proposed methods for joint analysis of multiple variables. Any simulation experiment is necessarily in complete and does not represent real lncRNA-disease experiments. Nevertheless, we try to simulate a relatively 'realistic' lncRNA-disease association model and evaluate the performance with different sample sizes and correlation structures. We consider a backcross population with sample sizes of 200, 400, and 800 to represent very small and large sample sizes. We simulate a genome with 19 chromosomes, each of length 100 cm with 11 equally spaced markers (markers placed 10 cM apart) on each chromosome. Ten percent of the genotypes of these markers were assumed to be randomly missing in all cases. For each of the three sample sizes, we consider two correlation structures, namely, low and high. Therefore, we have six cases with three samples sizes and two correlation structures. For each of these six cases, we simulate six lncRNAs that control the disease: E1 and E2 (E3 and E4) are nonpleitropic lncRNAs, influencing only the disease y1 (y2) with moderate-sized and weak effects, respectively; E5 is a moderate-sized pleiotropic lncRNAs affecting both y1 and y2; while E6 is a weak pleiotropic lncRNAs affecting both y1 and y2. Table 1 presents the simulated positions of six lncRNAs, their effect values, and their heritabilities (proportion of the phenotypic variation explained by a lncRNAs).

LncRNA expression profiles are obtained from the ratios of background-corrected probe intensities measured by RNAseq platform, downloaded from TCGA. Disease data also downloaded from TCGA.

Table 1. Six lncRNAs in the simulation data set

	E_1	E_2	E_3	E_4	E_5	E_6
Chromosome	2	2	4	4	5	6
Position (cM)	24	56	24	65	65	45
y_1	0.8	0.6	0	0	0.8	0.6
y_2	0	0	−0.8	−0.6	0.8	0.6
y_1 (%)	8.9	4.8	0	0	8.9	4.8
y_2 (%)	0	0	9.2	5.4	9.2	5.4

For each of the six cases, we generate 100 replicated data sets, resulting in 600 total data sets. For each of these 600 data sets we perform analysis using three methods, namely, the GCNLDA, joint analysis using a GMCLDA model, and our new algorithm GNN-IMC.

To illustrate the advantages of using a more complex method of analysis it is important to have an objective and reproducible plan of evaluation. However, in the model selection framework of multiple lncRNAs mapping this assessment becomes a little more complicated as one has to account for model uncertainty. Nonetheless, to assess the performance of different methods we adopt a simple approach. For all six cases we simulate 100 null data sets and compute the genomewide maximum 2 $\log_e$BF (twice the natural logarithm of Bayes factors) for each disease. The 95th percentile of the max 2 $\log_e$BF empirical distribution is considered as the threshold value above which a lncRNA-disease association would be deemed 'significant'. At each replication, the number of correctly identified lncRNA-disease association and the number of incorrectly identified or extraneous lncRNA-disease are recorded. A peak in the 2 logBF profile is considered a lncRNA-disease association if it crosses the significance threshold. It is deemed correct if it is within 10 cM (Broman and Speed 2002) of a true

Table 2. Average correct and incorrect lncRNA-disease detected for disease y_1 (first row) and y_2 (second row)

	Correct			Extraneous		
$(n, \rho_{y_1 y_2})$	GCNLDA	GMCLDA	GNN-IMC	GCNLDA	GMCLDA	GNN-IMC
(200,0.5)	1.32	**1.61**	1.34	1.40	2.68	1.08
	1.48	**1.56**	1.49	0.78	2.72	1.18
(200,0.8)	0.68	2.02	**2.04**	0.48	3.72	1.63
	1.56	2.14	**2.54**	1.42	3.44	1.68
(400,0.5)	3.40	**4.26**	4.24	2.12	5.12	1.62
	3.52	**4.4**	4.32	1.26	5.12	1.64
(400,0.8)	3.02	**5.21**	4.67	1.26	5.98	1.48
	3.52	**5.22**	5.15	1.92	5.96	1.68
(800,0.5)	6.08	**6.45**	6.41	2.02	6.21	1.68
	7.21	**7.61**	7.42	1.52	6.14	1.42
(800,0.8)	7.12	**7.62**	7.41	2.2	6.28	2.02
	7.28	**7.60**	7.42	2.36	6.22	1.41

lncRNA-disease association. If there is more than one peak within 10 cM of the true lncRNA-disease association, only one is considered correct. Table 2 represents the average correct and extraneous (incorrect) lncRNA-disease detections for the six situations and for all three methods for y1 and y2, respectively.

As shown in Fig. 1, the results in the real data show that our proposed method performs better than other methods.

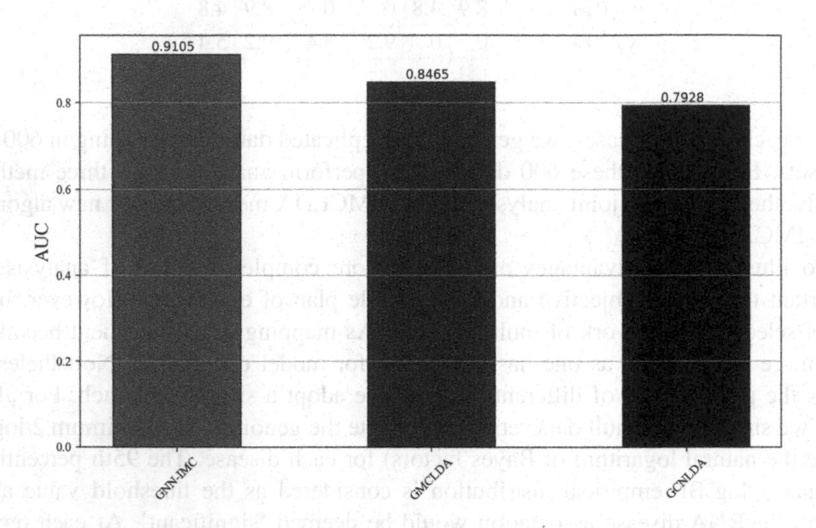

Fig. 1. Comparison of AUC values of three methods on real data

5 Conclusion

In this work, we propose a new machine learning framework for lncRNA-disease association task by combining Graph Neural Networks and nonconvex inductive matrix completion, named GNN-IMC. With the structure of GNN, we could generate subgraphs based on (lncRNA, disease) pairs from the observed association matrix and maps these subgraphs to their corresponding associations. In addition, GNN-IMC is inductive–it can generalize to lncRNAs/diseases unseen during the training (given that their associations exist), and can even transfer to new tasks. We concatenate all the features as the input feature. Dense similarities matrices are constructed by the inner product of features. Once we got embeddings of genes and diseases by GCNs, we use the inner product of them to get the preference score matrix. An ϵ-Insensitive loss is proposed to reduce the effect of data sparsity. We use nonconvex inductive matrix completion approach to optimize this loss function. From our experiments, we could see that GNN-IMC outperforms all other state-of-the-art methods a lot on most of metrics.

In the analysis of our results, we could see that extra information is the key point to improve the performance in disease-gene association task. In the future, we would try to mine more accurate correlations by learning method. Moreover, we will apply our method on more applications that include similarity information and contain sparse data, such as POI recommendation.

Acknowledgement. This work was supported by the National Key R&D Program of China [No.2019YFB1404700], the Natural Science Foundation of Shandong Province [No. ZR2017-LF019].

References

1. Hobert, O.: Gene regulation by transcription factors and microRNAs. Science **319**(5871), 1785–1786 (2008)
2. Swami, M.: Transcription factors: MYC matters. Nat. Rev. Cancer **10**(12), 812 (2010)
3. Collins, F.S., Morgan, M., Patrinos, A.: The human genome project: lessons from large-scale biology. Science **300**(5617), 286–290 (2003)
4. Yuan, L., Guo, L.H., Yuan, C.A., et al.: Integration of multi-omics data for gene regulatory network inference and application to breast cancer. IEEE/ACM Trans. Comput. Biol. Bioinf. **16**(3), 782–791 (2019)
5. International Human Genome Sequencing Consortium. Initial sequencing and analysis of the human genome. Nature **409**(6822), 860 (2001)
6. Louro, R., Smirnova, A.S., Verjovski-Almeida, S.: Long intronic noncoding RNA transcription: expression noise or expression choice? Genomics **93**(4), 291–298 (2009)
7. Yuan, L., Zhu, L., Guo, W.L., Huang, D.S.: Nonconvex penalty based low-rank representation and sparse regression for eQTL mapping. IEEE/ACM Trans. Comput. Biol. Bioinf. **14**(5), 1154–1164 (2017)
8. Geisler, S., Coller, J.: RNA in unexpected places: long non-coding RNA functions in diverse cellular contexts. Nat. Rev. Mol. Cell Biol. **14**(11), 699–712 (2013)
9. Xing, Z., Lin, A., Li, C., et al.: lncRNA directs cooperative epigenetic regulation downstream of chemokine signals. Cell **159**(5), 1110–1125 (2014)
10. Yuan, L., Yuan, C.A., Huang, D.S.: FAACOSE: a fast adaptive ant colony optimization algorithm for detecting SNP epistasis. Complexity **1**, 1–10 (2017)
11. Yuan, L., Huang, D.S.: A network-guided association mapping approach from DNA methylation to disease. Sci. Rep. **9**(1), 1–16 (2019)
12. Chen, X., Yan, C.C., Zhang, X., et al.: Long non-coding RNAs and complex diseases: from experimental results to computational models. Brief. Bioinf. **18**(4), 558–576 (2016)
13. Gao, Y., Meng, H., Liu, S., et al.: LncRNA-HOST2 regulates cell biological behaviors in epithelial ovarian cancer through a mechanism involving microRNA let-7. Hum. Mol. Genet. **24**(3), 841–852 (2014)
14. Yuan, L., Zheng, C.H., Xia, J.F., Huang, D.S.: Module based differential coexpression analysis method for type 2 diabetes. Biomed. Res. Int. **1**, 1–8 (2015)
15. Chen, G., Wang, Z., Wang, D., et al.: LncRNADisease: a database for long-non-coding RNA-associated diseases. Nucl. Acids Res. **41**(D1), D983–D986 (2012)
16. Lan, W., Li, M., Zhao, K., et al.: LDAP: a web server for lncRNA-disease association prediction. Bioinformatics **33**(3), 458–460 (2016)
17. Wang, J., Ma, R., Ma, W., et al.: LncDisease: a sequence based bioinformatics tool for predicting lncRNA-disease associations. Nucl. Acids Res. **44**(9), e90–e90 (2016)
18. Zheng, C.H., Yuan, L., Sha, W., et al.: Gene differential coexpression analysis based on biweight correlation and maximum clique. BMC Bioinf. 15 Suppl **15**(S15), S3 (2014)
19. Lin, Y., Han, K., Huang, D.S.: Novel algorithm for multiple quantitative trait loci mapping by using bayesian variable selection regression. In: International Conference on Intelligent Computing, pp. 862–868 (2016)

Prediction of lncRNA-miRNA Interactions via an Embedding Learning Graph Factorize Over Heterogeneous Information Network

Ji-Ren Zhou, Zhu-Hong You(✉), Li Cheng, Xi Zhou, and Hao-Yuan Li

The Xinjiang Technical Institute of Physics and Chemistry, Chinese Academy of Sciences, Urumqi 830011, China
zhuhongyou@ms.xjb.ac.cn

Abstract. An increasing number of studies show that identification of lncRNA-miRNA interactions (LMIs) helps the researchers to understand lncRNAs functions and the mechanism of involved complicated diseases. However, biological techniques for detecting lncRNA-miRNAs interactions are costly and time-consuming. Recently, many computational methods have been developed to predict LMIs, but only a few can perform the prediction from a network-based point of view. In this article, we propose a novel computational method to predict potential interactions between lncRNA and miRNA via an embedding learning graph factorize over a heterogeneous information network. Specifically, a large-scale heterogeneous information network is built by combing the associations among proteins, drugs, miRNAs, diseases, and lncRNAs. Then, a graph embedding model Graph Factorization is employed to learn vector representations for all miRNA and lncRNA in the heterogeneous network. Finally, the integrated features are fed to a classifier to predict new lncRNA-miRNA interactions. In the experiment, the proposed method performed good prediction results with AUC of 0.9660 under five-fold cross-validation. The experimental results demonstrate our method as an outperform way to predict potential associations between lncRNAs and miRNAs.

Keywords: lncRNA-miRNA interactions · Network biology · Network embedding · Random forest

1 Introduction

Molecular biology indicates that ncRNA is critical in living cells. They participated in amount biology process such as transcription, translation, epigenetic regulation [1–3]. The ncRNAs can be divided into two categories based on lengths, to be specific, those ncRNAs which are shorter than 200 nucleotides were named as small ncRNA including miRNA, siRNA, and so on. Otherwise, long non-coding RNAs are ncRNAs with transcripts longer than 200 nucleotides [4].

From the accumulating research, the interactions of miRNA and lncRNA were uncovered. More specifically, lncRNAs play the role of a sponge, which has inhibition to some miRNAs in multiple biologic processes through reducing regulatory effect on

D.-S. Huang and K.-H. Jo (Eds.): ICIC 2020, LNCS 12464, pp. 270–278, 2020.
https://doi.org/10.1007/978-3-030-60802-6_24

mRNAs. The biology method is time-consuming and cost unaffordable. It is unrealistic to verify the potential interactions through the traditional biological methods such as highlight and knockdown targeted genes. On the other hand, high-throughput technologies including qPCR and yeast two-hybrid screens bought us massive data, which is collected and published by online databases including LncRNADisease [5], HMDD [6], and STRING [7]. These databases give us a possibility to construct a prediction network to find more potential associations through the extensive data from a holistic perspective.

In the accumulating research, massive computational methods have been put forward to give the insight to predict potential associations between small molecules [8–17]. These methods can be divided into different categories according to the associations involved. Ye et al. constructed a regulatory network and use the vote-counting strategy to predict the regulation mechanism in pancreatic cancer among the miRNA-lncRNA-mRNA-TF [18]. In the protein-protein interactions (PPI) field, Ronalf Jansen et al. had proposed a method using Bayesian networks to predict PPI genome-wide in yeast [19]. You et al. used a novel hierarchical model, principal component analysis-ensemble extreme learning machine to predict PPI only through the information of protein sequences [20]. In the field of the ncRNA-protein interactions(RPI), Yi et al. put forward an approach used robust deep learning framework for predicting the associations between ncRNA and protein [21]. V. Suresh et al. used the SVM method to predict RPI through the information of both sequences and structures [22]. In the field of drug-target interactions (DTI), Wang et al. fed rotation forest with the information of drug structure and protein sequences to predict drug-target interactions [23]. Several computational methods have been introduced to the ncRNA-disease problem, especially in miRNA-disease associations. Chen et al. have proposed a model, WBSMDA, within and between the score for miRNA-disease associations prediction. Shi et al. have constructed a bipartite miRNA-disease network to identify more potential associations between two nodes [24]. Recently, an increasing number of methods consider predicting the associations through the intermediary instead of the direct link. Peng et al. captured similarity in a three-layer network involved miRNA, protein, and disease, in which it trained the model as classifier [25]. Chen et al. carried on label propagation algorithms and used lncRNA as an intermediary to predict miRNA-disease associations. The growing number of researchers consider taking more small molecules into the network and address this issue from a global perspective. We are inspired by these works, therefore, add more nodes in our framework considering various molecules in biology progression.

In this study, we selected a series of molecules, including miRNA, lncRNA, disease, drug, and protein. We can have a more comprehensive understanding of associations between small molecules with diseases. Although our work still has some limitations, it still provides us a new perspective through constructing a framework among the ncRNAs, disease, drug, and protein.

2 Materials and Methods

2.1 Selection of Attributes of Nodes Datasets

The attributes information is including the sequences of ncRNA and protein, the molecular fingerprint of the drug, and the semantics of diseases. The information of miRNA, lncRNA, protein, such as nomenclature, and sequence data was downloaded from the miRbase [26], NONCODE [27], STRING [7]. DrugBank was served as the smile of the drug database, after downloaded, we transformed the smile into the corresponding Morgan molecular fingerprint [28]. We downloaded the disease information from the medical subject headings, the MeSH program from the national library of medicine [29].

2.2 The Encoder of the Diseases

The disease information in the MeSH program was presented by the directed acyclic graph, (DAG) approach for showing the character and reduce the bias. In the DAG approach, the directed arrow connecting father node and son node in the graph shows causation, on the opposite, the nodes without causal associations are left unconnected. The disease in MeSH can be described as formula below:

$$DAG(C) = (C, N(C), E(C)) \tag{1}$$

In this formula, C is any disease collected in MeSH. The N(C), E(C) is the set of all the diseases and all corresponding associations related to disease C in the DAG(C) respectively.

2.3 Calculation of Semantic Similarity of Diseases

For calculating the similarity of the semantics of two diseases in DAG, we use the distance to measure the weight of ancestor nodes to son nodes. We define the weight of one specific disease as 1, and the contribution of farther ancestor nodes are less to the semantic of this specific disease. The calculation is shown below:

$$\begin{cases} D_C(C) = 1 \\ D_C(a) = max\{\Delta * D_C(a') | a' \in children\, of\, a\}\, if\, a \neq C \end{cases} \tag{2}$$

The Δ is the semantic contribution factor of corresponding directed arrows of disease with its child disease C. Thus, the Δ should between 0 to 1 as a decreasing factor, which was set as 0.5 for the better correlation with expression similarity. Then we define the semantic value of disease C, DV(A) as

$$DV(C) = \sum\nolimits_{a \in N(c)} D_C(a) \tag{3}$$

For instance, the DV value of Asthma is 1.0 (Asthma) + 0.5 (Lung Disease, Obstructive) + 0.5 (Bronchial Diseases) + 0.5 (Respiratory Hypersensitivity) + 0.5 × 0.5 (Lung Diseases) + 0.5 × 0.5 (Respiratory Tract Diseases) + 0.5 × 0.5 (Hypersensitivity, Immediate) + 0.5 × 0.5 × 0.5 (Hypersensitivity) + 0.5 × 0.5 × 0.5 × 0.5(Immune System Diseases) = 3.4375

The semantic similarity of two diseases are according to distance in DAG, in which the diseases with larger share part are more similar in semantic. The calculation is defined as

$$S(C,D) = \frac{\sum_{a \in N(C) \cap N(D)} (D_C(a) + D_D(a))}{DV(C) + DV(D)} \tag{4}$$

Where $D_C(a)$ and $D_D(a)$ is semantic of disease a to disease A, B respectively, and the semantic similarity based on ancestor diseases and sharing part in DAG.

2.4 Selection of the Associations' Database

The associations appeared in our network including ncRNA-disease, PPI, ncRNA-protein, DTI, and so on. For gaining the proper results, we had to use the data after identifier unification, de-redundancy, simplification, and deletion. The databased involved are shown in Table 1.

Table 1. The sources and number of associations involved in network

Type of associations	Data resources	Number of associations
miRNA-lncRNA	lncRNASNP2 [30]	8,374
miRNA-disease	HMDD [31]	16,427
miRNA-protein	miRTarBase [32]	4,944
lncRNA-disease	LncRNADisease [6] lncRNASNP2 [30]	1,264
lncRNA-protein	LncRNA2Target [33]	690
Protein-disease	DisGeNET [34]	25,087
Drug-protein	DrugBank [35]	11,107
Drug-disease	CTD [36]	18,416
Protein-protein	STRING [7]	19,237
Total	N/A	105,546

After aggregating the database, the final number of nodes that appeared in our network are shown in Table 2.

Table 2. The number of small molecules nodes in the network

Node	Number of nodes
Disease	2062
LncRNA	769
Mirna	1023
Protein	1649
Drug	1025
Total	6528

2.5 Sparse Auto-Encoder

Sparse Auto-Encoder (SAE), an unsupervised feature learning model method, was employed in our model for avoiding the labor-intensive and handcraft feature design. The hidden layer in SAE can extract the feature of the input layer and make the output equal to the input. In the process of mapping the input layer x to the hidden h the encoder function is $h = \sigma(Wx + p)$. When recovering the output layer x from the hidden layer, the decoder layer is $y = \sigma(Wh + p)$. W is the connection parameters, p is the bias, and the σ is $\sigma(x) = max(0, Wx + p)$, which is set as non-linear mapping factor and the ReLU function was chosen.

2.6 Node Representation

Each node was represented by combining the intrinsic attributes and the behavior simultaneously. For constructing the biomolecular network, we employed the Graph Factorization to represent the relationships to the edges in the network globally.

For scalability, some embedding methods had been put forward, even though they all highly need computational complexity. In recent years, some methods based on the undirected network were raised for reducing computational complexity or generating without an explicit objective function. Graph Factorization [39] was the first approach to generate the network in $O(|E|)$ time. For factorizing the adjacency matrix of the graph, gf minimize the loss function:

$$\phi(X, \lambda) = 0.5 \sum_{(i,j) \in E} (W_{ij} - \langle X_i, X_j \rangle)^2 + 0.5\, \lambda \sum_i \|Y_i\|^2 \quad (5)$$

where λ is a regularization coefficient. The total number is over the observed edges instead of possible edges. It is for consideration of scalability and may bring the noise in the solution. The minimum of the loss function is better than zero even if the dimensionality of embedding is $|V|$, because of the negative semidefinite of the adjacency matrix.

2.7 Classifier

Random forest was introduced to the experiment for training the model [37]. Random Forest was proposed by Breiman in 2001 and added more details by Breiman and Culter in 2003. This classifier consists of many decision trees. Each tree is called a base classifier. When a test case was given to the classifier, each tree assigns it a class according to its classifier. Then the most frequent class was from the overall result.

3 Results and Discussion

3.1 Evaluate the Five-Fold Cross-Validation Performance of Our Method

Five-fold cross-validation would divide the dataset into five subsets randomly and in an equal number. Each subset would be the test set in turn, and the remain four sets would be for training the model. For avoiding leaking the test results to the classifier, only 80% of edges would be embedded in the network. Although the GF would discard some connected associations in this situation because of the threshold value mentioned. It could simulate the real situation and have a better performance to predict the potential associations in manual experiments.

For better revealing the performance, we select serial index including accuracy (Acc.), sensitivity (Sen.), specificity (Spec.), precision (Prec.), and Matthews Correlation Coefficient (MCC) to measure. As shown in Table 3 and Fig. 1, the results of our method achieved average Acc., Sen., Spec., Prec., MCC, and area under the curve (AUC) of 91.32%, 91.41%, 91.22%, 91.24%, 82.65%, 96.60% when the model trained on five-fold cross-validation. Receiver operating Characteristic curve (ROC) is a

Table 3. Five-Fold cross-validation results of our methods.

Fold	Acc. (%)	Sen. (%)	Spec. (%)	Prec. (%)	MCC (%)	AUC (%)
0	78.87	76.49	81.25	80.31	57.8	84.86
1	75.15	70.83	79.46	77.52	50.49	79.63
2	81.55	77.68	85.42	84.19	63.28	85.78
3	78.12	74.40	81.85	80.39	56.41	82.84
4	79.76	75.89	83.63	82.26	59.70	83.28
Average	78.69 ± 2.36	75.06 ± 2.64	82.32 ± 2.28	80.93 ± 2.48	57.54 ± 4.71	83.28 ± 2.36

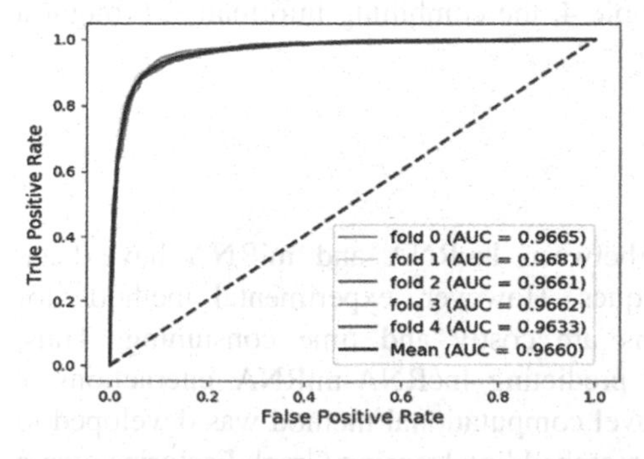

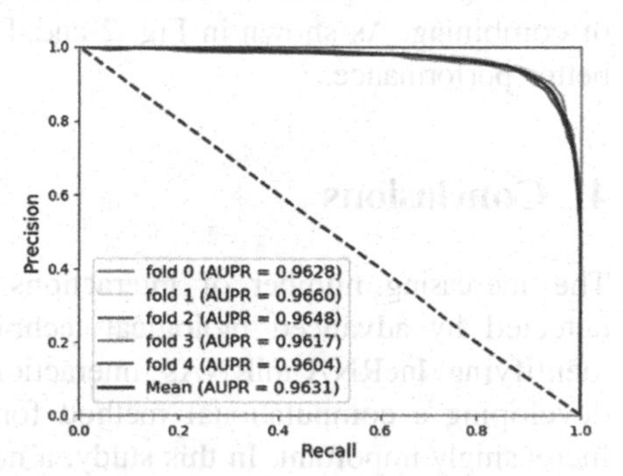

Fig. 1. The ROC and PR curves of our methods. The AUC is the area under the receiver operating characteristic curves (ROC). The AUPR is the area under the precision-recall curves (PR).

common index for measuring performances, and the area under this curve is (AUC), where the abscissa false positive rate(FPR), and the ordinate true positive rate (TPR). Simultaneously, we also use the Precision-Recall (PR) curve, whose abscissa is the recall, and the ordinate is the precision for measuring the performance.

Table 4. Comparison of different features respectively and simultaneously

Feature	Acc. (%)	Sen. (%)	Spec. (%)	Prec. (%)	MCC (%)	AUC (%)
Attribute	88.30 ± 0.81	90.55 ± 0.98	86.04 ± 0.72	86.64 ± 0.71	76.67 ± 1.64	95.02 ± 0.84
Behavior	91.02 ± 0.88	90.67 ± 1.00	91.38 ± 0.88	91.32 ± 0.88	82.06 ± 1.76	96.25 ± 0.27
Both	91.32 ± 0.58	91.41 ± 1.29	91.22 ± 0.56	91.24 ± 0.47	82.65 ± 1.16	96.60 ± 0.17

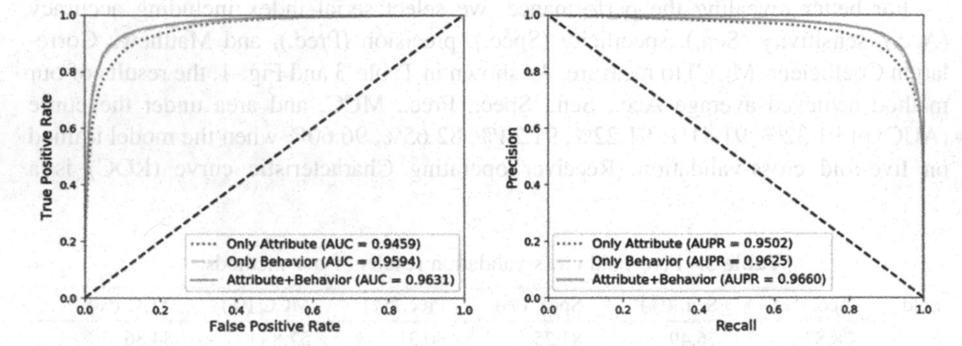

Fig. 2. Comparison of different features under five-fold cross-validation respectively and simultaneously

3.2 Comparison of Different Feature Extraction Methods

In our method, each node was presented by combining the attribute and behavior. For evaluating the impact of different features, we use the information respectively instead of combining. As shown in Fig. 2 and Table 4, the combining information brought a better performance.

4 Conclusions

The increasing number of interactions between lncRNA and miRNA have been detected by advanced biological techniques. However, experimental methods for identifying lncRNA-miRNAs interactions are costly and time consuming. Thus, developing a computational method for predicting lncRNA-miRNA interactions is increasingly important. In this study, a novel computational method was developed to predict lncRNA-miRNA interactions via an embedding learning Graph Factorize over a heterogeneous information network. The experimental results demonstrated that the proposed method achieved an outstanding performance for predicting lncRNA-miRNA interactions. We hope that our work can inspire more research to predict potential associations among small molecules from a more global perspective.

Conflict of Interest. The authors declare that they have no conflict of interest.

Funding. This work is supported by the Xinjiang Natural Science Foundation under Grant 2017D01A78.

References

1. Guttman, M., et al.: Chromatin signature reveals over a thousand highly conserved large non-coding RNAs in mammals. Nature **458**, 223–227 (2009)
2. Mattick, J.S.: The genetic signatures of noncoding RNAs. PLoS Genet. **5**, e1000459 (2009)
3. Lander, E.S., et al.: Initial Sequencing and Analysis of the Human Genome (2001)
4. Kapranov, P., et al.: RNA maps reveal new RNA classes and a possible function for pervasive transcription. Science **316**, 1484–1488 (2007)
5. Chen, G., et al.: LncRNADisease: a database for long-non-coding RNA-associated diseases. Nucl. Acids Res. **41**, D983–D986 (2012)
6. Li, Y., et al.: HMDD v2. 0: a database for experimentally supported human microRNA and disease associations. Nucl. Acids Res. **42**, D1070–D1074 (2014)
7. Szklarczyk, D., et al.: The STRING database in 2017: quality-controlled protein–protein association networks, made broadly accessible. Nucl. Acids Res. gkw937 (2016)
8. Chen, X., Yan, C.C., Zhang, X., You, Z.-H.: Long non-coding RNAs and complex diseases: from experimental results to computational models. Brief. Bioinf. **18**, 558–576 (2017)
9. Chen, Z.-H., Li, L.-P., He, Z., Zhou, J.-R., Li, Y., Wong, L.: An improved deep forest model for predicting self-interacting proteins from protein sequence using wavelet transformation. Front. Genet. **10**, 90 (2019)
10. Chen, Z.-H., You, Z.-H., Li, L.-P., Wang, Y.-B., Qiu, Y., Hu, P.-W.: Identification of self-interacting proteins by integrating random projection classifier and finite impulse response filter. BMC Genom. **20**, 1–10 (2019)
11. Chen, Z.-H., You, Z.-H., Li, L.-P., Wang, Y.-B., Wong, L., Yi, H.-C.: Prediction of self-interacting proteins from protein sequence information based on random projection model and fast Fourier transform. Int. J. Mol. Sci. **20**, 930 (2019)
12. Huang, Y.-A., Chan, K.C., You, Z.-H.: Constructing prediction models from expression profiles for large scale lncRNA–miRNA interaction profiling. Bioinformatics **34**, 812–819 (2018)
13. You, Z.-H., et al.: PBMDA: A novel and effective path-based computational model for miRNA-disease association prediction. PLoS Comput. Biol. **13**, e1005455 (2017)
14. You, Z.-H., Zhou, M., Luo, X., Li, S.: Highly efficient framework for predicting interactions between proteins. IEEE Trans. Cybern. **47**, 731–743 (2016)
15. Zheng, K., You, Z.-H., Li, J.-Q., Wang, L., Guo, Z.-H., Huang, Y.-A.: iCDA-CGR: Identification of circRNA-disease associations based on Chaos Game Representation. PLoS Comput. Biol. **16**, e1007872 (2020)
16. Zheng, K., You, Z.-H., Wang, L., Zhou, Y., Li, L.-P., Li, Z.-W.: DBMDA: a unified embedding for sequence-based miRNA similarity measure with applications to predict and validate miRNA-disease associations. Mol. Therap. Nucl. Acids **19**, 602–611 (2020)

17. Li, S., You, Z.-H., Guo, H., Luo, X., Zhao, Z.-Q.: Inverse-free extreme learning machine with optimal information updating. IEEE Trans. Cybern. **46**, 1229–1241 (2015)
18. Ye, S., Yang, L., Zhao, X., Song, W., Wang, W., Zheng, S.: Bioinformatics method to predict two regulation mechanism: TF–miRNA–mRNA and lncRNA–miRNA–mRNA in pancreatic cancer. Cell Biochem. Biophys. **70**, 1849–1858 (2014)
19. Jansen, R., et al.: A Bayesian networks approach for predicting protein-protein interactions from genomic data. Science **302**, 449–453 (2003)
20. You, Z.-H., Lei, Y.-K., Zhu, L., Xia, J., Wang, B.: Prediction of protein-protein interactions from amino acid sequences with ensemble extreme learning machines and principal component analysis. BMC Bioinf. **14**, S10 (2013). https://doi.org/10.1186/1471-2105-14-S8-S10
21. Yi, H.-C., You, Z.-H., Huang, D.-S., Li, X., Jiang, T.-H., Li, L.-P.: A deep learning framework for robust and accurate prediction of ncRNA-protein interactions using evolutionary information. Mol. Therap. Nucl. Acids **11**, 337–344 (2018)
22. Suresh, V., Liu, L., Adjeroh, D., Zhou, X.: RPI-Pred: predicting ncRNA-protein interaction using sequence and structural information. Nucl. Acids Res. **43**, 1370–1379 (2015)
23. Wang, L., You, Z.-H., Chen, X., Yan, X., Liu, G., Zhang, W.: RFDT: a rotation forest-based predictor for predicting drug-target interactions using drug structure and protein sequence information. Curr. Protein Peptide Sci. **19**, 445–454 (2018)
24. Shi, H., et al.: Walking the interactome to identify human miRNA-disease associations through the functional link between miRNA targets and disease genes. BMC Syst. Biol. **7**, 101 (2013)
25. Peng, J., et al.: A learning-based framework for miRNA-disease association identification using neural networks. Bioinformatics **35**, 4364–4371 (2019)
26. Griffiths-Jones, S., Saini, H.K., Van Dongen, S., Enright, A.J.: miRBase: tools for microRNA genomics. Nucl. Acids Res. **36**, D154–D158 (2007)
27. Fang, S., et al.: NONCODEV5: a comprehensive annotation database for long non-coding RNAs. Nucl. Acids Res. **46**, D308–D314 (2018)
28. Wishart, D.S., et al.: DrugBank: a knowledgebase for drugs, drug actions and drug targets. Nucl. Acids Res. **36**, D901–D906 (2008)
29. Lipscomb, C.E.: Medical subject headings (MeSH). Bull. Med. Libr. Assoc. **88**, 265 (2000)
30. Miao, Y.-R., Liu, W., Zhang, Q., Guo, A.-Y.: lncRNASNP2: an updated database of functional SNPs and mutations in human and mouse lncRNAs. Nucl. Acids Res. **46**, D276–D280 (2018)
31. Huang, Z., et al.: HMDD v3. 0: a database for experimentally supported human microRNA–disease associations. Nucl. Acids Res. **47**, D1013–D1017 (2019)
32. Chou, C.-H., et al.: miRTarBase update 2018: a resource for experimentally validated microRNA-target interactions. Nucl. Acids Res. **46**, D296–D302 (2018)
33. Cheng, L., et al.: LncRNA2Target v2. 0: a comprehensive database for target genes of lncRNAs in human and mouse. Nucl. Acids Res. **47**, D140–D144 (2019)
34. Piñero, J., et al.: DisGeNET: a comprehensive platform integrating information on human disease-associated genes and variants. Nucl. Acids Res. gkw943 (2016)
35. Wishart, D.S., et al.: DrugBank 5.0: a major update to the DrugBank database for 2018. Nucl. Acids Res. **46**, D1074–D1082 (2018)
36. Davis, A.P., et al.: The comparative toxicogenomics database: update 2017. Nucl. Acids Res. **45**, D972–D978 (2017)
37. Breiman, L.: Random forests. Mach. Learn. **45**, 5–32 (2001)

Inferring Drug-miRNA Associations by Integrating Drug SMILES and MiRNA Sequence Information

Zhen-Hao Guo[1,2], Zhu-Hong You[1,2(✉)], Li-Ping Li[1], Zhan-Heng Chen[1,2], Hai-Cheng Yi[1,2], and Yan-Bin Wang[3]

[1] Xinjiang Technical Institute of Physics and Chemistry, Chinese Academy of Sciences, Urumqi 830011, China
zhuhongyou@ms.xjb.ac.cn
[2] University of Chinese Academy of Sciences, Beijing 100049, China
[3] College of Information Science and Engineering, Zaozhuang University, Zaozhuang 277100, Shandong, China

Abstract. The accumulated evidences indicate that drugs not only interact with proteins, but also regulate a wide variety of biomarkers such as miRNAs. Hence, uncovering potential drug-miRNA associations plays significant roles in disease prevention, diagnosis and treatment as well as drug development. In this paper, we discuss how this problem is formulated as a link prediction task in a bipartite graph and construct a computational model to infer unknown drug-miRNA associations. Specifically, the drug SMILES (Simplified molecular input line entry specification) or miRNA sequences can be regarded as a kind of biology language described by distributed representation. The experiment verified associations are treated as positive samples and the same number unlabeled associations are randomly selected as negative samples. Finally, Random Forest classifier is applied to perform the prediction task. In the experiment, the proposed method achieves AUROC of 91.16 and AUPR of 89.21 under 5-fold cross-validation. It demonstrates the great potential of seamless integration of deep learning and biological big data. We hope that this research with great expectations can be used as a practical guidance tool to bring useful inspiration to relevant researchers.

Keywords: Drug · miRNA · Association prediction · Biology language processing

1 Introduction

Despite the massive accumulation of pharmacological and pathological knowledge and the rapid development of data disciplines in few decades, the Research and Development (R&D) costs of pharmaceutical companies remain exorbitant [1]. Most drug candidates are failing in late stage clinical trials, largely due to inappropriate selection

Z. Guo, Z. You—These authors contributed equally to this work

D.-S. Huang and K.-H. Jo (Eds.): ICIC 2020, LNCS 12464, pp. 279–289, 2020.
https://doi.org/10.1007/978-3-030-60802-6_25

of early targets [2]. Therefore, Drug-Target Identification (DTI) plays an irreplaceable role in the drug development process as the first and most important step [3]. Recently, the research focus of precision medicine and individual therapy has changed from the traditional drug response variance to the modern molecular biology development [4, 5]. Evidences point out that many types of biomarkers such as miRNA are regulated by Small Molecule (SM) drugs, which in turn affects disease prevention, diagnosis, treatment and recovery [6, 7]. Generally speaking, uncovering potential drug-miRNA associations plays a significant role in understanding disease mechanism, drug pharmacology and clinical response [8–14].

Considering that wet experiments are labor-intensive, time-consuming, cost-ineffective and expert-dependent, there still exists a huge gap between drug-miRNA association identification and practical therapy [15]. To address this problem, in silico methods are satisfactory choices to speed up this process [16–21]. The task of predicting drug-miRNA associations can be formalized as a link prediction problem in attribute bipartite graphs [22–28]. Many previous computational models can be easily transferred to this field [29–32]. These methods can be mainly divided into 2 categories based on calculation pattern including network-based method and feature-based method [33]. The core idea of the former is to infer potential relationships through known associations, that is, the guilty by association. For example, Emig *et al.* presented a network-based approach for DTIs based on a given disease [34]. Chen *et al.* proposed a method to predict DTIs based on a heterogeneous network through random walk with restart [35]. Cheng et al. developed a computational method for drug discovery based on 3 similarity network including drug-based similarity inference (DBSI), target-based similarity inference (TBSI) and network-based inference (NBI) [36]. The core idea of the latter is to summarize the chemical structure and sequence components to expose unknown relationships. For example, Öztürk *et al.* discover DTIs by adapting and evaluating various SMILES-based similarity methods [37]. You *et al.* detect uncovered protein-protein interactions based on amino acid sequences by combining extreme learning machine and principal component analysis [11]. Wang *et al.* predict DTIs by integrating drug molecular structure and protein sequence information [24]. Wang *et al.* predict the PPIs by using Zernike moments to extract the protein sequence feature from a position specific scoring matrix (PSSM) [17].

How to effectively represent drug and miRNA in the attribute bipartite graph is the first and one of the most important problem [38–45]. Drug SMILES (Simplified Molecular Input Line Entry System) and miRNA sequences are both string descriptors to characterize how drug and miRNA are composed. Just as humans use language to communicate with each other, cells use biological language to convey information between them. In addition, the impressive progress of deep learning in the field of human language processing has reason to convince us that it can achieve the same success in biological language processing [46–55]. The proposal of distributed word vector representation has changed the research paradigm, and how to use cheap data without labels has become a core issue [18, 56–67].

In this paper, we developed a novel machine learning-based method to identify potential drug-miRNA associations by integrating drug SMILES and miRNA sequences. Specifically, we collected the known drug-miRNA associations from ncDR database in March 2020 [68]. Then, drug and miRNA can be represented as continuous vectors by embedding algorithm. After that, the experimental verified drug-miRNA pairs are treated as positive samples and we randomly select the same number unlabeled pairs as negative samples. Finally, Random Forest classifier is applied to carry out the association prediction task. The presented method achieves mean AUROC (Area under Receiver Operating characteristic Curve) of 91.16 and AUPR (Area under Precision – Recall Curve) of 89.21. In addition, we perform the classifier comparison experiment. All results prove the effectiveness and stability of the presented method. We hope that the preliminary exploration of combining deep learning with biological big data can bring useful inspiration to relevant researchers (Fig. 1).

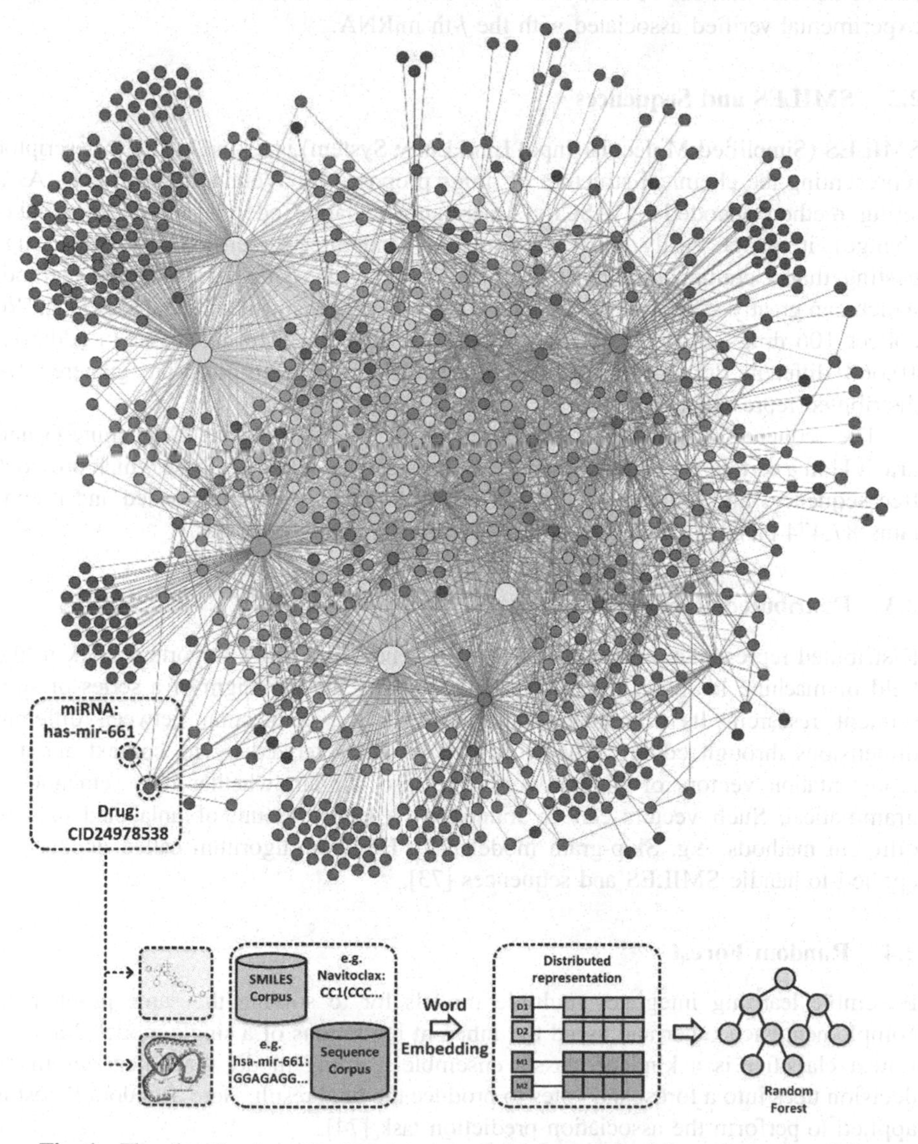

Fig. 1. The visualization of ncDR dataset and the flowchart of the proposed model.

2 Materials and Methods

2.1 Drug-miRNA Associations Collection

Known drug-miRNA associations were downloaded from the ncDR database in March 2020. After eliminating duplication and strictly limiting species to humans, we obtained 106 different drugs and 702 different miRNAs form 3239 independent association pairs. The identifiers of drug and miRNA are uniquely determined by PubChem and miRbase, respectively [69, 70]. All the information can be stored in an adjacency matrix A with d rows and m columns, where d denotes the number of drugs and m denotes number of miRNAs. A_{ij} is defined as 1 if and only if the i-th drug is experimental verified associated with the j-th miRNA.

2.2 SMILES and Sequences

SMILES (Simplified Molecular Input Line Entry System) is a kind of linear descriptor representing the chemical structure of drugs proposed by Weininger *et al.* [71]. As a string method encoded by ASCII (American Standard Code for Information Interchange), it can be easily converted into two- and three-dimensional structure, suggesting that it contains sufficient chemical information. This representation method, which can greatly save storage space, is widely used in mainstream databases [71]. We collect 106 drugs and corresponding SMILES from pubchem database. In addition, 10,668 different drugs and corresponding SMILES in DrugBank are gathered for distributed representation.

The sequence of miRNA is composed of adenine A, cytosine C, guanine G and uracil U in a certain order. miRbase is a comprehensive online database which provides free sequences and annotation query. The whole database is downloaded and it contains 87,474 pieces of miRNAs and their corresponding sequences.

2.3 Distributed Representation

Distributed representation is considered to be one of the most important work in the field of machine learning in recent years, changing the paradigm of a series of subsequent research. Its core idea is to establish the connection between different dimensions through coding. The word can be characterized by its context and the representation vectors of similar words are also similar, whether it is semantic or grammatical. Such vectors can be trained by a large amount of unlabeled data in different methods, e.g. Skip-gram model [72]. Here, an algorithm called doc2vec is applied to handle SMILES and sequences [73].

2.4 Random Forest

Ensemble learning integrates multiple models for to solving the same problem to complement each other and avoid the inherent limitations of a single model. Random Forest classifier is a kind of classic ensemble learning model that integrates many decision trees into a forest and votes to produce the final result. Here, Random Forest is applied to perform the association prediction task [74].

3 Results

3.1 Network Analysis

As mentioned in drug-miRNA associations collection chapter, we obtained 106 different drugs and 702 different miRNAs form 3239 independent drug-miRNA pairs. In addition, we collected the corresponding SMILES and sequences. The length distribution is shown in Fig. 2.

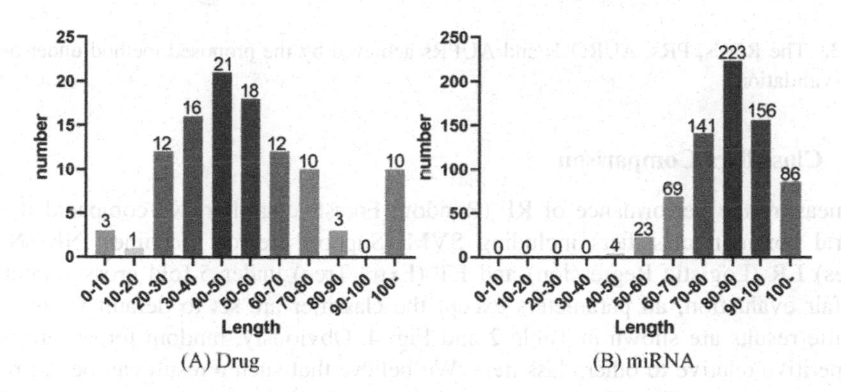

Fig. 2. The length distribution of drug SMILES and miRNA sequences in ncDR.

3.2 Performance Under 5-Fold Cross-Validation

To measure the capability of the presented method, cross-validation is applied. Cross-validation is a frequently used strategy for evaluation. k approximately equal mutually exclusive subsets are divided from the entire data set, one of which is used as the test set to estimate the performance of the model, and the held-out is used as the training set to construct the framework. The average performance is regarded as the results to avoid the randomness and bias. In order to visualize the results, ROC and PR are drawn as well as AUROC and AUPR are calculated, respectively. In addition, a wide range of evaluation criteria including Acc. (Accuracy), Sen. (Sensitivity), Spec. (specificity), Prec. (Precision) and MCC (Matthews correlation coefficient) have also been adopted. The results can be seen in Table 1 and Fig. 3.

Table 1. The Acc., Sen., Spec., Prec. and MCC obtained by the proposed method under 5-fold cross-validation.

Fold	Acc. (%)	Sen. (%)	Spec. (%)	Prec. (%)	MCC (%)
0	83.72	81.79	85.65	85.07	67.49
1	85.19	82.25	88.12	87.38	70.49
2	83.49	81.79	85.19	84.66	67.01
3	82.48	82.25	82.72	82.64	64.97
4	86.32	83.31	89.34	88.65	72.78
Average	**84.24 ± 1.51**	**82.28 ± 0.62**	**86.20 ± 2.60**	**85.68 ± 2.36**	**68.55 ± 3.08**

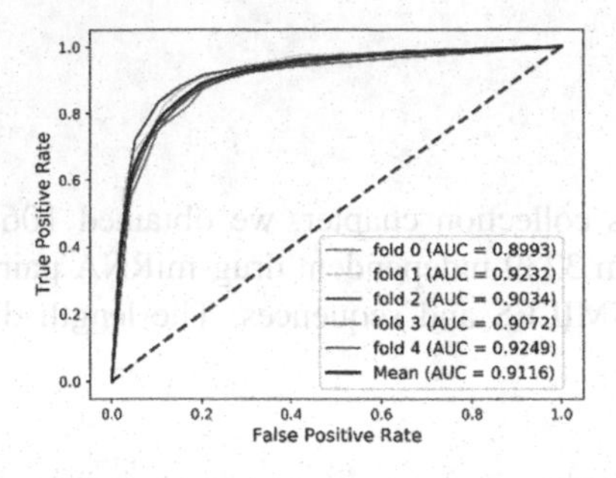

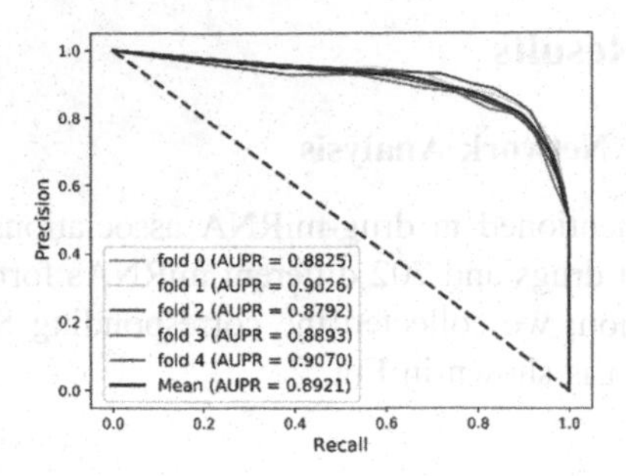

Fig. 3. The ROCs, PRs, AUROCs and AUPRs achieved by the proposed method under 5-fold cross-validation.

3.3 Classifiers Comparison

To measure the performance of RF (Random Forest) classifier, we compared it with several common classifiers including SVM (Support Vector Machine), NB (Naïve Bayes) LR (Logistic Regression) and ET (Extra Tree) under 5-fold cross-validation. For fair evaluation, all parameters except the classifier are set to default values. The specific results are shown in Table 2 and Fig. 4. Obviously, random forests are more competitive relative to other classifiers. We believe that such a result can be attributed to several factors. For SVM, the training speed is extremely slow, and the default parameter settings may affect the performance of the classification. For NB, its assumption is that the attributes are independent of each other, which may not match in this data set. For LR, the dimension of the representation vector is large, so the model is under-fitting and the prediction accuracy is low. For ET and RF, substantial performance achieved by them demonstrate the efficiency of ensemble learning classifier.

Table 2. The Acc., Sen., Spec., Prec. and MCC obtained by different classifiers under 5-fold cross-validation.

Classifier	Acc. (%)	Sen. (%)	Spec. (%)	Prec. (%)	MCC (%)
SVM	50.63 ± 0.22	**99.91 ± 0.08**	1.36 ± 0.48	50.32 ± 0.11	7.40 ± 1.42
NB	57.01 ± 1.10	88.61 ± 1.20	25.41 ± 2.74	54.31 ± 0.77	18.07 ± 2.30
LR	70.38 ± 1.37	67.89 ± 2.66	72.86 ± 1.38	71.44 ± 1.18	40.82 ± 2.71
ET	84.02 ± 1.09	82.53 ± 1.27	85.52 ± 1.50	85.09 ± 1.38	68.09 ± 2.19
RF	**84.24 ± 1.51**	82.28 ± 0.62	**86.20 ± 2.60**	**85.68 ± 2.36**	**68.55 ± 3.08**

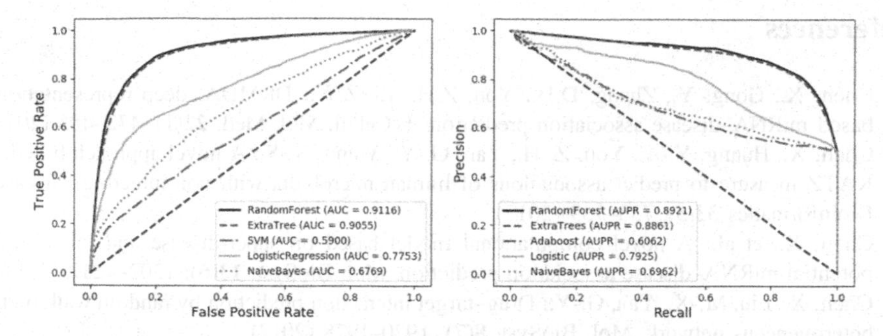

Fig. 4. The ROCs, PRs, AUROCs and AUPRs achieved by different classifiers under 5-fold cross-validation.

4 Conclusion

A large number of biomarkers can be regulated by different drugs, which provides novel insights into individual therapy and drug development. However, there is an insurmountable gap between the researchers and the numerous records scattered in the massive literature. To tackle this challenge, we proposed a computational method to uncover the potential drug-miRNA associations based on a kind of biology language processing. The drug and miRNA can be represented as continues vectors by word embedding algorithm. Then, the Random Forest classifier is applied to carry out the prediction task. Briefly, we implemented network analysis, performed 5-fold cross-validation experiment and executed classifier comparison. All results demonstrated the effectiveness and stability of the presented method. We hope that this preliminary exploration combining deep learning and biological big data can bring useful inspiration to relevant researchers.

Declaration

Acknowledgements. The authors would like to thank all the editors and anonymous reviewers for their constructive advices.

Authors' Contributions. Z-H. G. and Z-H. Y. considered the algorithm, arranged the datasets, and performed the analyses. H-C. Y., Y-B. W. and Z-H. C. wrote the manuscript. All authors read and approved the final manuscript.

Funding. This work is supported in part by the National Natural Science Foundation of China (Grant nos. 61722212, 61902342). The authors would like to thank the editors and anonymous reviewers for their constructive advice.

Competing Interests. The authors declare that they have no competing interests.

References

1. Chen, X., Gong, Y., Zhang, D.H., You, Z.H., Li, Z.W.: DRMDA: deep representations-based miRNA–disease association prediction. J. Cellul. Mol. Med. **22**(1), 472–485 (2018)
2. Chen, X., Huang, Y.-A., You, Z.-H., Yan, G.-Y., Wang, X.-S.: A novel approach based on KATZ measure to predict associations of human microbiota with non-infectious diseases. Bioinformatics **33**(5), 733–739 (2017)
3. Chen, X., et al.: A novel computational model based on super-disease and miRNA for potential miRNA–disease association prediction. Mol. BioSyst. **13**(6), 1202–1212 (2017)
4. Chen, X., Liu, M.-X., Yan, G.-Y.: Drug–target interaction prediction by random walk on the heterogeneous network. Mol. BioSyst. **8**(7), 1970–1978 (2012)
5. Chen, X., Wang, C.-C., Yin, J., You, Z.-H.: Novel human miRNA-disease association inference based on random forest. Mol. Therap. Nucl. Acids **13**, 568–579 (2018)
6. Chen, X., Xie, D., Wang, L., Zhao, Q., You, Z.-H., Liu, H.: BNPMDA: bipartite network projection for MiRNA–disease association prediction. Bioinformatics **34**(18), 3178–3186 (2018)
7. Chen, X., Xie, D., Zhao, Q., You, Z.-H.: MicroRNAs and complex diseases: from experimental results to computational models. Brief. Bioinf. **20**(2), 515–539 (2019)
8. Chen, X., Yan, C.C., Zhang, X., You, Z.-H.: Long non-coding RNAs and complex diseases: from experimental results to computational models. Brief. Bioinf. **18**(4), 558–576 (2016). https://doi.org/10.1093/bib/bbw060
9. Chen, X., et al.: WBSMDA: within and between score for MiRNA-disease association prediction. Sci. Rep. **6**(1), 21106 (2016). https://doi.org/10.1038/srep21106
10. Chen, X., Yan, C.C., Zhang, X., You, Z.-H., Huang, Y.-A., Yan, G.-Y.: HGIMDA: heterogeneous graph inference for miRNA-disease association prediction. Oncotarget **7**(40), 65257 (2016)
11. Chen, X., You, Z.-H., Yan, G.-Y., Gong, D.-W.: IRWRLDA: improved random walk with restart for lncRNA-disease association prediction. Oncotarget **7**(36), 57919 (2016)
12. Chen, X., Zhang, D.-H., You, Z.-H.: A heterogeneous label propagation approach to explore the potential associations between miRNA and disease. J. Transl. Med. **16**(1), 348 (2018)
13. Chen, Z.-H., You, Z.-H., Guo, Z.-H., Yi, H.-C., Luo, G.-X., Wang, Y.-B.: Prediction of drug-target interactions from multi-molecular network based on deep walk embedding model. Front. Bioeng. Biotechnol. **8**, 338 (2020)
14. Chen, Z.-H., You, Z.-H., Li, L.-P., Guo, Z.-H., Hu, P.-W., Jiang, H.-J.: Combining LSTM network model and wavelet transform for predicting self-interacting proteins. In: Huang, D.-S., Bevilacqua, V., Premaratne, P. (eds.) ICIC 2019. LNCS, vol. 11643, pp. 166–174. Springer, Cham (2019). https://doi.org/10.1007/978-3-030-26763-6_16
15. Cheng, F., et al.: Prediction of drug-target interactions and drug repositioning via network-based inference. PLoS Comput. Biol. **8**(5) (2012)
16. Cheng, L., et al.: LncRNA2Target v2. 0: a comprehensive database for target genes of lncRNAs in human and mouse. Nucl. Acids Res. **47**(D1), D140–D144 (2018)
17. Chung, S., Nakagawa, H., Uemura, M., Piao, L., Ashikawa, K., Hosono, N., Morizono, T.: Association of a novel long non-coding RNA in 8q24 with prostate cancer susceptibility. Cancer Sci. **102**(1), 245–252 (2011)
18. Cummings, J., Ward, T.H., Greystoke, A., Ranson, M., Dive, C.: Biomarker method validation in anticancer drug development. Br. J. Pharmacol. **153**(4), 646–656 (2008)
19. Dai, E., et al.: ncDR: a comprehensive resource of non-coding RNAs involved in drug resistance. Bioinformatics **33**(24), 4010–4011 (2017)

20. Emig, D., et al.: Drug target prediction and repositioning using an integrated network-based approach. PloS ONE **8**(4) (2013)
21. Guo, Z.-H., Yi, H.-C., You, Z.-H.: Construction and comprehensive analysis of a molecular association network via lncRNA–miRNA–disease–drug–protein graph. Cells **8**(8), 866 (2019)
22. Guo, Z.-H., You, Z.-H., Huang, D.-S., Yi, H.-C., Chen, Z.-H., Wang, Y.-B.: A learning based framework for diverse biomolecule relationship prediction in molecular association network. Commun. Biol. **3**(1), 1–9 (2020)
23. Guo, Z.-H., et al.: MeSHHeading2vec: a new method for representing MeSH headings as vectors based on graph embedding algorithm. Brief. Bioinf. (2020). https://doi.org/10.1093/bib/bbaa037
24. Guo, Z.-H., You, Z.-H., Li, L.-P., Wang, Y.-B., Chen, Z.-H.: Combining high speed ELM with a CNN feature encoding to predict LncRNA-Disease Associations. In: Huang, D.-S., Jo, K.-H., Huang, Z.-K. (eds.) ICIC 2019. LNCS, vol. 11644, pp. 406–417. Springer, Cham (2019). https://doi.org/10.1007/978-3-030-26969-2_39
25. Guo, Z.-H., et al.: Bioentity2vec: Attribute-and behavior-driven representation for predicting multi-type relationships between bioentities. GigaScience **9**(6), giaa032 (2020)
26. Guo, Z.-H., You, Z.-H., Wang, Y.-B., Yi, H.-C., Chen, Z.-H.: A learning-based method for LncRNA-disease association identification combing similarity information and rotation forest. iScience **19**, 786–795 (2019). https://doi.org/10.1016/j.isci.2019.08.030
27. Guo, Z.-H., You, Z.-H., Yi, H.-C.: Integrative construction and analysis of molecular association network in human cells by fusing node attribute and behavior information. Mol. Therap. Nucl. Acids **19**, 498–506 (2020)
28. Hafner, M., Niepel, M., Sorger, P.K.: Alternative drug sensitivity metrics improve preclinical cancer pharmacogenomics. Nat. Biotechnol. **35**(6), 500–502 (2017)
29. Hay, M., Thomas, D.W., Craighead, J.L., Economides, C., Rosenthal, J.: Clinical development success rates for investigational drugs. Nat. Biotechnol. **32**(1), 40–51 (2014)
30. Hu, P., Huang, Y.-A., Chan, K.C., You, Z.-H.: Learning multimodal networks from heterogeneous data for prediction of lncRNA-miRNA interactions. IEEE/ACM Trans. Comput. Biol. Bioinf. (2019)
31. Huang, Y.-A., Chan, K.C., You, Z.-H.: Constructing prediction models from expression profiles for large scale lncRNA–miRNA interaction profiling. Bioinformatics **34**(5), 812 819 (2017)
32. Huang, Y.-A., Chan, K.C., You, Z.-H.: Constructing prediction models from expression profiles for large scale lncRNA–miRNA interaction profiling. Bioinformatics **34**(5), 812–819 (2018)
33. Huang, Y.-A., Hu, P., Chan, K.C., You, Z.-H.: Graph convolution for predicting associations between miRNA and drug resistance. Bioinformatics **36**(3), 851–858 (2020)
34. Huang, Y.-A., You, Z.-H., Chen, X.: A systematic prediction of drug-target interactions using molecular fingerprints and protein sequences. Curr. Protein Peptide Sci. **19**(5), 468–478 (2018)
35. Huang, Y.-A., You, Z.-H., Chen, X., Chan, K., Luo, X.: Sequence-based prediction of protein-protein interactions using weighted sparse representation model combined with global encoding. BMC Bioinf. **17**(1), 184 (2016)
36. Huang, Z.-A., et al.: PBHMDA: path-based human microbe-disease association prediction. Front. Microbiol. **8**, 233 (2017)
37. Huang, Z.-A., Huang, Y.-A., You, Z.-H., Zhu, Z., Sun, Y.: Novel link prediction for large-scale miRNA-lncRNA interaction network in a bipartite graph. BMC Med. Genom. **11**(6), 113 (2018)

38. Jin, X., Feng, C.-Y., Xiang, Z., Chen, Y.-P., Li, Y.-M.: CircRNA expression pattern and circRNA-miRNA-mRNA network in the pathogenesis of nonalcoholic steatohepatitis. Oncotarget **7**(41), 66455 (2016)
39. Kim, S., et al.: PubChem 2019 update: improved access to chemical data. Nucl. Acids Res. **47**(D1), D1102–D1109 (2019)
40. Kozomara, A., Birgaoanu, M., Griffiths-Jones, S.: miRBase: from microRNA sequences to function. Nucl. Acids Res. **47**(D1), D155–D162 (2018)
41. Le, Q., Mikolov, T.: Distributed representations of sentences and documents. In: Paper presented at the International Conference on Machine Learning (2014)
42. Lee, C.Y., Chen, Y.-P.: Prediction of drug adverse events using deep learning in pharmaceutical discovery. Brief. Bioinf. (2020)
43. Li, J.-Q., Rong, Z.-H., Chen, X., Yan, G.-Y., You, Z.-H.: MCMDA: Matrix completion for MiRNA-disease association prediction. Oncotarget **8**(13), 21187 (2017)
44. Li, J.-Q., You, Z.-H., Li, X., Ming, Z., Chen, X.: PSPEL: in silico prediction of self-interacting proteins from amino acids sequences using ensemble learning. IEEE/ACM Trans. Comput. Biol. Bioinf. (TCBB) **14**(5), 1165–1172 (2017)
45. Li, S., You, Z.-H., Guo, H., Luo, X., Zhao, Z.-Q.: Inverse-free extreme learning machine with optimal information updating. IEEE Trans. Cybern. **46**(5), 1229–1241 (2015)
46. Liaw, A., Wiener, M.: Classification and regression by randomForest. R News **2**(3), 18–22 (2002)
47. Lin, Y., et al.: RNAInter in 2020: RNA interactome repository with increased coverage and annotation. Nucl. Acids Res. **48**(D1), D189–D197 (2020)
48. Liu, X., et al.: SM2miR: a database of the experimentally validated small molecules' effects on microRNA expression. Bioinformatics **29**(3), 409–411 (2013)
49. Ma, L., et al.: Multi-neighborhood learning for global alignment in biological networks. IEEE/ACM Trans. Comput. Biol. Bioinf. (2020)
50. Mikolov, T., Sutskever, I., Chen, K., Corrado, G. S., Dean, J.: Distributed representations of words and phrases and their compositionality. In: Paper Presented at the Advances in Neural Information Processing Systems (2013)
51. Öztürk, H., Ozkirimli, E., Özgür, A.: A comparative study of SMILES-based compound similarity functions for drug-target interaction prediction. BMC Bioinf. **17**(1), 128 (2016)
52. Pammolli, F., Magazzini, L., Riccaboni, M.: The productivity crisis in pharmaceutical R&D. Nat. Rev. Drug Discov. **10**(6), 428–438 (2011)
53. Santos, R., et al.: A comprehensive map of molecular drug targets. Nat. Rev. Drug Discov. **16**(1), 19 (2017)
54. Wang, L., et al.: An ensemble approach for large-scale identification of protein-protein interactions using the alignments of multiple sequences. Oncotarget **8**(3), 5149 (2017)
55. Wang, L., et al.: LMTRDA: using logistic model tree to predict MiRNA-disease associations by fusing multi-source information of sequences and similarities. PLoS Comput. Biol. **15**(3), e1006865 (2019)
56. Wang, L., et al.: A computational-based method for predicting drug–target interactions by using stacked autoencoder deep neural network. J. Comput. Biol. **25**(3), 361–373 (2018)
57. Wang, L., You, Z.-H., Chen, X., Yan, X., Liu, G., Zhang, W.: RFDT: a rotation forest-based predictor for predicting drug-target interactions using drug structure and protein sequence information. Curr. Protein Peptid. Sci. **19**(5), 445–454 (2018)
58. Wang, L., You, Z.-H., Huang, Y.-A., Huang, D.-S., Chan, K.C.: An efficient approach based on multi-sources information to predict circRNA–disease associations using deep convolutional neural network. Bioinformatics **36**(13), 4038–4046 (2020)

59. Wang, M.-N., You, Z.-H., Wang, L., Li, L.-P., Zheng, K.: LDGRNMF: LncRNA-disease associations prediction based on graph regularized non-negative matrix factorization. Neurocomputing (2020)
60. Wang, Y.-B., et al.: Predicting protein–protein interactions from protein sequences by a stacked sparse autoencoder deep neural network. Mol. BioSyst. **13**(7), 1336–1344 (2017)
61. Wang, Y.-B., You, Z.-H., Yi, H., Chen, Z.-H., Guo, Z.-H., Zheng, K.: Combining evolutionary information and sparse bayesian probability model to accurately predict self-interacting proteins. In: Huang, D.-S., Jo, K.-H., Huang, Z.-K. (eds.) ICIC 2019. LNCS, vol. 11644, pp. 460–467. Springer, Cham (2019). https://doi.org/10.1007/978-3-030-26969-2_44
62. Wang, Y., You, Z.-H., Yang, S., Li, X., Jiang, T.-H., Zhou, X.: A high efficient biological language model for predicting protein-protein interactions. Cells **8**(2), 122 (2019)
63. Weininger, D.: SMILES, a chemical language and information system. 1. Introduction to methodology and encoding rules. J. Chem. Inf. Comput. Sci. **28**(1), 31–36 (1988)
64. Wishart, D.S., et al.: DrugBank 5.0: a major update to the DrugBank database for 2018. Nucl. Acids Res. **46**(D1), D1074–D1082 (2017)
65. Wong, L., You, Z.-H., Guo, Z.-H., Yi, H.-C., Chen, Z.-H., Cao, M.-Y.: MIPDH: A Novel Computational Model for Predicting microRNA–mRNA Interactions by DeepWalk on a Heterogeneous Network. ACS Omega (2020)
66. Yi, H.-C., et al.: Learning distributed representations of RNA and protein sequences and its application for predicting lncRNA-protein interactions. Comput. Struct. Biotechnol. J. **18**, 20–26 (2020)
67. Yi, H.-C., You, Z.-H., Huang, D.-S., Li, X., Jiang, T.-H., Li, L.-P.: A deep learning framework for robust and accurate prediction of ncRNA-protein interactions using evolutionary information. Mol. Therap. Nucl. Acids **11**, 337–344 (2018)
68. Yi, H.-C., You, Z.-H., Wang, Y.-B., Chen, Z.-H., Guo, Z.-H., Zhu, H.-J.: *In Silico* identification of anticancer peptides with stacking heterogeneous ensemble learning model and sequence information. In: Huang, D.-S., Jo, K.-H., Huang, Z.-K. (eds.) ICIC 2019. LNCS, vol. 11644, pp. 313–323. Springer, Cham (2019). https://doi.org/10.1007/978-3-030-26969-2_30
69. You, Z.-H., Chan, K.C., Hu, P.: Predicting protein-protein interactions from primary protein sequences using a novel multi-scale local feature representation scheme and the random forest. PLoS ONE **10**(5), e0125811 (2015)
70. You, Z.-H., et al.: PBMDA: A novel and effective path-based computational model for miRNA-disease association prediction. PLoS Comput. Biol. **13**(3), e1005455 (2017)
71. You, Z.-H., Lei, Y.-K., Zhu, L., Xia, J., Wang, B.: Prediction of protein-protein interactions from amino acid sequences with ensemble extreme learning machines and principal component analysis. BMC Bioinf. **14**(8), S10 (2013). https://doi.org/10.1186/1471-2105-14-S8-S10
72. You, Z.-H., Yu, J.-Z., Zhu, L., Li, S., Wen, Z.-K.: A MapReduce based parallel SVM for large-scale predicting protein–protein interactions. Neurocomputing **145**, 37–43 (2014)
73. You, Z.-H., Zhou, M., Luo, X., Li, S.: Highly efficient framework for predicting interactions between proteins. IEEE Trans. Cybern. **47**(3), 731–743 (2017)
74. Zheng, K., You, Z.-H., Li, J.-Q., Wang, L., Guo, Z.-H., Huang, Y.-A.: iCDA-CGR: Identification of circRNA-disease associations based on Chaos Game Representation. PLoS Comput. Biol. **16**(5), e1007872 (2020)

Identification of Rice Drought-Resistant Gene Based on Gene Expression Profiles and Network Analysis Algorithm

Yujia Gao, Yiqiong Chen, Zhiyu Ma, Tao Zeng, Iftikhar Ahmad, Youhua Zhang(✉), and Zhenyu Yue(✉)

School of Information and Computer, Laboratory of Anhui Beidou Precision Agriculture Information Engineering, Anhui Agricultural University, Hefei 230036, Anhui, China
{zhangyh, zhenyuyue}@ahau.edu.cn

Abstract. Drought is a primary cause of grain yield reduction that reduces agricultural production and seriously damage food security worldwide. Drought tolerance has a complex quantitative characteristic with a complicated phenotype that affects different developmental stages of plants. The level of susceptibility or tolerance of rice to several drought conditions is coordinated by the action of different drought-resistant genes. This study presents a bioinformatics approach to identify candidate rice drought-resistant genes based on other known related rice genes. By using the sub-network extraction algorithm with gene co-expression profile, we obtained the integrated network comprising of the known rice drought-resistant related genes (denoted as seed genes) and putative genes (denoted as linker genes). These genes are ranked according to the newly proposed rating scores. Some of the discovered candidate genes were validated by the previous scientific literature and gene set enrichment analysis. The results offer useful gene information that serve as guidance for the researchers and rice breeders. In addition, the proposed approach is sufficiently effective to be applied on other crops via biological network analysis.

Keywords: Drought resistance · Sub-network extraction · Rice gene

1 Introduction

Rice is a major staple food consumed by more than one-third of the world's population especially in Asia [1]. In future it is expected that water shortage will be significantly impacted the production of rice. Drought is a toughest abiotic stress that causes the annual loss of global rice production leading to huge economic losses [2]. The damage in plant growth caused by drought condition is irreversible. As one of the most drought-susceptible plants, rice has characteristics of small root system, thin cuticular wax, and swift stomata closure [3]. In recent years, drought has become major abiotic stress that seriously affects rice yield and their quality.

Y. Gao and Y. Chen—These authors contributed equally to this work

D.-S. Huang and K.-H. Jo (Eds.): ICIC 2020, LNCS 12464, pp. 290–301, 2020.
https://doi.org/10.1007/978-3-030-60802-6_26

Previously, plenty of work has been done on drought resistance in rice which revealed various physiological mechanisms. These studies provided different point of views in plant biotechnology. It has been confirmed that the drought tolerance mechanism are controlled by genetic factors at different growth stages which generally include three types: morphological adaptations, physiological acclimation and cellular adjustments [4]. The resistance to drought stress is positively associated with a series of known and unknown signal transduction pathways, which would eventually lead to physiological and biochemical changes in rice [5, 6]. Genomic analysis of drought-resistant genes reveals a complex molecular landscape. The latest advances of genetic and genomics techniques identify candidate genes and metabolic pathways underlying drought resistance. Various stress-responsive genes have been identified and applied to plants engineering process with improved drought tolerance such as *WRKY*, *DUF221*, *SNAC1*, *OSCA*, *bZIP*, *NAC* (*NAM*, *ATAF1/2*, and *CUC2*), *AP2/ERF*, and *MYB* transcription factor family [6–9].

Discovering candidate genes of rice is an efficient method to improve drought resistance. In other words, the mining of key drought-resistant genes resources may provide the genetic basis for rice production in water-limited conditions and region. At present, despite extensive and collective researches on the mechanism of drought resistance have been made, there are very few methods for identifying rice gene related to drought resistance. However, the mechanism underlying the early response to osmotic stress in plants remains unclear [9]. More candidate genes identification and allele mining needs to be studied urgently [10].

Computational approach is a common way to cope with the biological problems effectively and been utilized to clarify complicated biological network [11, 12] as well as identify candidate genes [13]. There are many successful methods to identify unclear genes and modules in the vast search space of biological network inference and analysis. Among them, the network analysis method is effectively used. Previously, Scott *et al.* proposed a geometric approach for modeling and searching for sub-networks in protein-protein and protein–DNA interaction networks [13]. Noirel *et al.* also presented a sub-network extraction method to proteomics data and identified the metabolic pathways. The extracted result information could be taken as a starting point for in-depth analysis [14]. Smita S *et al.* described abiotic stress tolerant (ASTR) genes and networks under different abiotic stress conditions by transcriptome meta-analysis. About 7,000 ASTR genes, key hub regulatory genes, transcription factors, and functional modules were identified [15]. Yanmei *et al.* identified 517 core drought responsive genes and different modules through the analysis of large amount of transcriptomic datasets [16]. Wang *et al.* proposed a method based on gray BP neural network to improve the effectiveness of the network prediction for rice protein interaction [17]. However, these studies with traditional methods always focused on identifying differentially expressed genes but these methods showed some difficulties in the analytics capabilities with the massive growth of data.

In this study, a sub-network extraction method was used to identify drought-resistant candidate genes by known related genes retrieved from National BioResource Project: Rice Database (NBRP, https://shigen.nig.ac.jp/rice/oryzabase/). The sub-network extraction algorithm was used to analyze the rice gene co-expression data and obtained a subnetwork comprising of the known drought-resistance related genes (indicated as seed genes) and the putative genes (designated as linker genes). The results of our experiment could be used to identify unknown drought-resistant genes, as well as to infer the gene associations which are good candidates for experimental analysis. Overall, our study further provides an insight and guidance for researchers and rice producers, which is essential for developing transgenic crops with enhanced drought stress tolerance and will help to obtain the stress-tolerant crops in modern agriculture.

2 Materials and Methods

2.1 Dataset

In this study the gene expression data was obtained from Plant Express [18], which is widely available microarray-based transcriptome database involving rice and Arabidopsis data (http://bioinf.mind.meiji.ac.jp/OryzaExpress/network/script/index.html). Oryza Express provides information of rice which is integrated omics annotations and gene expression networks (GENs). The omics annotations are collected from main public databases and integrated into OryzaExpress. The rice datasets include 1893 samples. The rice gene expression data (V 3.0) of 153 experimental series was obtained from the web database PlantExpress on November 10, 2019 (list GEO 1893 RMA. TAR.GZ). Then 110 sets therein annotated as "drought" of "tolerance" which were included in the used dataset. Based on the above mentioned dataset, we retrieved another dataset by comparing the gene expression levels of the selected candidate genes, which also contained 110 samples those are not under non-abiotic stress conditions. Next, we reviewed and merged the duplicate gene data by Database for Annotation, Visualization and Integrated Discovery (DAVID, http://david.abcc.ncifcrf.gov).

2.2 Gene Co-expression Network Building

The biological analysis of gene co-expression network has been widely performed to identify significant genes or modules in a specific condition [19]. As a rule, the expression levels of two co-expressed genes change synchronously [4, 20]. It has been proved that the genes sharing the same function are involved in the same regulatory pathway will always tend to present similar expression profiles and hence form clusters or modules in the network [21].

We constructed a gene co-expression network using Python 3.6. The Pearson correlation coefficient (PCCs) and confidence coefficients were used as the co-expression measure. Then the correlation network analysis and network visualization were performed by Cytoscape platform (version 3.8.0). The number of nodes and edges in the co-expression network under different threshold conditions were illustrated in Table 1. Considering the computational complexity and the network sizes, we used the network 2 in Table 1 for subsequent analysis.

Table 1. Numbers of nodes and edges in networks.

Co-expression network	PCCs interval	Confidence interval	Number of Nodes/Edges		
			Node	Edge	Seed gene
1	[0.7,1.0]	[0.0,0.01]	37228	2833372	63
2	[0.8,1.0]	[0.0,0.01]	23306	599952	45
3	[0.9,1.0]	[0.0,0.01]	6327	49696	16

2.3 Centralities Based Analysis of Gene Nodes

After the network construction, the input genes are filtered by network topological features. Topological centrality was presented to verify its validity and feasibility of characterizing the biological importance of gene signatures, pathways and modules [22, 23]. In order to describe the gene signatures, a statistical analysis was performed on the gene co-expression network data by applying the local and global scale. We sorted all the node genes according to 4 topological features (degree, stress centrality, betweenness centrality and closeness centrality) and obtained the first 10000 genes datasets respectively. Then, the intersection of the 4 datasets was used as the input genes. Figure 1 showed a summary workflow for identifying genes related to rice drought-resistant.

2.4 Seed Genes

We retrieved the rice seed genes related to drought stress tolerance from publicly China Rice Data Center (Rice Gene Database, http://www.ricedata.cn) by using specific keywords "TO.0000237" and "water abiotic", i.e. response by the plant in terms of sensitivity to water stress. This database is an authoritative literature knowledge-based rice breeding related biological repository. We used 32 drought-resistance related genes (seed genes) with a high confidence level to conduct follow-up experiments (Table 2).

2.5 Sub-network Extraction Algorithm

The goal is to extract a sub-network initiated by seed genes with a minimum score till all the seeds connected.

Three inputs are described in its basic form which is as follows

1. A rice gene co-expression dataset—It's corresponding to the rice gene co-expression network, where the edges denoted the correlation between pair of genes. The Pearson's correlation coefficients and P-values were used to obtain a sub-network. The weights of every node were assigned to be 1 and the values of Pearson's correlation coefficient were assigned as an edge weight.
2. A set of seeds—It's also defined as terminal in the network.
3. A score function—It was defined to calculate a score to each constructing sub-network in each iterative. The score of a sub-network is the summation of all the weights of edges in network.

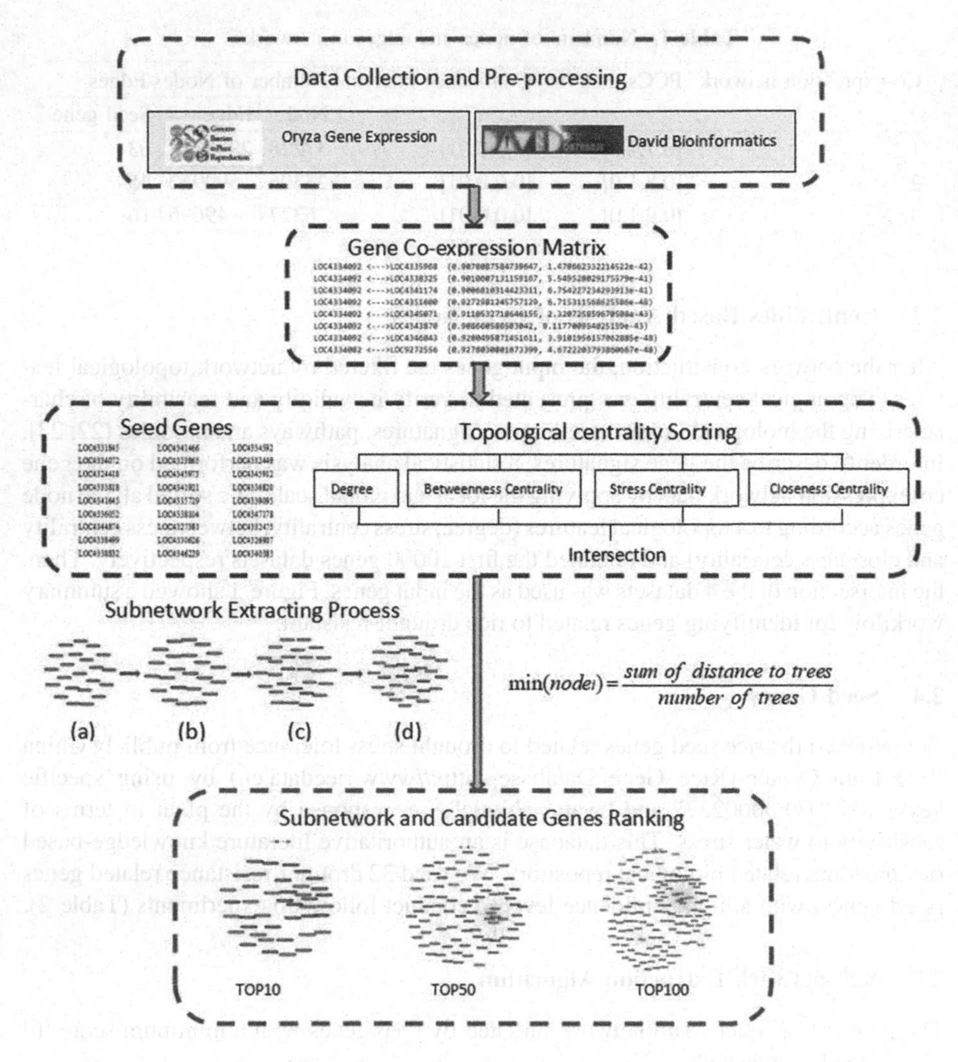

Fig. 1. Summary workflow for identifying candidate rice drought-resistant genes

The sub-network extraction is typical instance of the Steiner tree problem. The normalization was processed to the origin rice gene co-expression data. The Klein-Ravi algorithm was used to solve the subnet extracting problem [24]. A sub-network with minimum score which connected all the given seeds was constructed step by step. In each iterative step of the algorithm, a non-tree node was selected to minimize the following ratio:

$$Min(node_i) = \frac{sum\ of\ distance\ to\ trees}{number\ of\ trees} \tag{1}$$

Finally, an optimal sub-network which is indeed a Steiner tree was obtained.

Table 2. 32 seed genes used in the current study.

Gene symbols	Gene symbols	Gene symbols	Gene symbols
LOC4334072	LOC4336783	LOC4338832	LOC4341445
LOC4339095	LOC4334592	LOC4345912	LOC4326907
LOC4323865	LOC4332449	LOC4341466	LOC4327381
LOC4324418	LOC4343572	LOC4338104	LOC4333818
LOC4332475	LOC9268873	LOC4342317	LOC4336052
LOC4340383	LOC4341645	LOC4332145	LOC4344217
LOC4333432	LOC4330203	LOC4338499	LOC4347178
LOC4341921	LOC4330838	LOC4339653	LOC4349778

We improved the algorithm and solved sub-nctwork-extract problem by objective: the total cost of the edges in the network was minimum target [25]. A detailed account of the algorithm process is given: We create an undirected graph G with a non-negative integral function d that specifies the non-negative PCCS as the cost of each edge $e \in E$ in the Steiner tree was constructed. A subset of seed nodes is called terminals. The goal is to construct a Steiner tree T containing all the terminals such that the score of the total cost of the edges in T is minimum target. Here, T is the sub-network. This algorithm is one of three sub-network extraction algorithms in GenRev, which can run randomly in the network by constructing a Steiner tree and extract relevant sub-network connecting seed nodes. For more details please refer to the original work [26].

2.6 Sorting of Candidate Genes

After the extraction of sub-network, the sum of all edge costs linked with every gene in the sub-network was calculated as the ratings score (RS), which is a prioritization and ranking strategy of candidate genes. Because of the importance of difference between a seed gene and a linked gene, the edge weights of linker and seed genes were doubled when RSs was calculated. A higher rating score of a gene means that it might be more significant in rice drought-resistant. The candidate genes were ranked according to their rating score using by:

$$RS_i = \sum C_{ij} + \sum 2C_{ik} \quad (2)$$

Where RS_i is the rating score of candidate gene i, C_{ij} is the weight of edge (v_i, v_j) linked a non-seed node j, and C_{ik} is the weight of edge (v_i, v_k) linking a seed gene and a linked gene. We sorted all the linked genes according to the rating scores.

2.7 Gene Set Enrichment Analysis

To evaluate the function of the rice drought-resistant genes, we explored the functional enrichment analysis by using DAVID, which is a real-time analysis software and a functional annotation tool, mainly used to analyze gene lists in order to explore the molecular mechanism [27]. We obtained the GO terms with the enrichment FDR value less than 0.05.

3 Results and Discussion

3.1 Subnetwork and Top-100 Candidate Genes

Our algorithm takes a group of seed genes and a rice co-expression network as input, and produces a subnetwork connecting high-confident rice drought-resistant related genes. In the actual process of algorithm, all the relevant data of genes and networks is imported into Cytoscape software platform in order to better demonstrate the experimental results.

The extracted subnetwork contained 122 nodes and 2172 edges. We sorted all the linker genes and obtained the top 100 candidate genes by using the rating score. The top 100 of candidate genes with their rating scores are listed in Table 3. And the subnetwork contained the top 100 genes are illuminated in Fig. 1. Then we performed Wilcox rank sum test for both drought-resistant related and not related samples. Among the 100 genes, there are 51 genes obtained, where the P value is less than 0.05. The results further indicate the two groups shows a different gene expression levels in these genes (Fig. 2).

3.2 Comparing the Proposed Rating Score (RS) and Traditional Degree

The literature analysis of core candidate genes is critical to further understanding of the molecular mechanisms governing abiotic stress, and finally enhancing stress tolerance in crops through genetic manipulation.

The previous studies on identification of candidate genes in biology networks had made remarkable achievements, but the research mainly focused on considering the topological properties just like the degree of the nodes. Here, we performed two experiments and compared the results. One group was performed to filter the gene expression microarray data by using the topological features: degree, eigenvector, betweenness and closeness centrality, while the other group did not. In both experiments we ranked the linker genes from the extracted subnetwork, two groups of genes were selected from the list: the top 20 and the top 50 genes, and analyzed the experimental results. According to the experimental results of two groups, the topological features are more helpful in optimizing datasets for identifying the candidate genes.

We can find 13 candidate rice drought-resistant genes in the RS-top20 group, but not in the Degree-top20 group in Fig. 3(a). Furthermore, 21 candidate rice drought-resistant genes in the RS-top50 group, but not in the Degree-top50 group in Fig. 3(b). Succeeding analysis manifest that no matter among the 13 or the 21 novel genes, we can find literature evidences that report the associations between these gene and rice drought resistant or other abiotic stress, such as the geneLOC4352722, LOC4329036, LOC4331532, and LOC4334710.

Table 3. The top 100 genes with their rating scores.

Gene symbols	RS	Gene symbols	RS	Gene symbols	RS	Gene symbols	RS
LOC4349122	7.752	LOC4341569	4.304	LOC4352486	4.128	LOC4336266	3.409
LOC4332887	7.523	LOC4332466	4.296	LOC4342453	4.082	LOC4342139	3.409
LOC4346910	5.996	LOC4344752	4.294	LOC4333802	4.062	LOC4335818	3.407
LOC4325381	5.928	LOC4330131	4.289	LOC9270480	3.571	LOC4348154	3.405
LOC4352722	5.845	LOC4337711	4.283	LOC4324320	3.518	LOC9268269	3.400
LOC4325581	5.264	LOC4330767	4.275	LOC4339170	3.517	LOC4324537	3.399
LOC4330828	5.248	LOC4336609	4.269	LOC4334710	3.482	LOC4325486	3.398
LOC4325645	5.248	LOC4334845	4.258	LOC4328000	3.476	LOC4330477	3.394
LOC4341847	5.228	LOC4331874	4.254	LOC4336922	3.471	LOC4336511	3.394
LOC4348792	5.197	LOC4327631	4.247	LOC4348854	3.470	LOC4346971	3.390
LOC4325914	5.161	LOC4349889	4.239	LOC4348347	3.459	LOC4325823	3.386
LOC4332263	5.139	LOC4335864	4.236	LOC4336421	3.456	LOC4344021	3.381
LOC4334672	5.118	LOC4349217	4.228	LOC4323993	3.442	LOC4340748	3.378
LOC4352408	5.101	LOC4332479	4.224	LOC4325791	3.440	LOC4342397	3.376
LOC4325517	5.030	LOC4334739	4.223	LOC4347773	3.439	LOC4332925	3.374
LOC4328431	5.007	LOC4339664	4.220	LOC4330797	3.438	LOC4330811	3.373
LOC4331533	4.993	LOC4341032	4.212	LOC4350827	3.436	LOC4352216	3.372
LOC4339201	4.918	LOC4349100	4.205	LOC4331761	3.435	LOC4332348	3.371
LOC9266204	4.459	LOC4344630	4.199	LOC4336288	3.435	LOC4324472	3.365
LOC4334632	4.406	LOC4327786	4.172	LOC4335786	3.435	LOC4328368	3.365
LOC4324908	4.370	LOC9269617	4.156	LOC4331275	3.433	LOC4338582	3.364
LOC4330875	4.362	LOC4338267	4.155	LOC4348208	3.427	LOC4328085	3.363
LOC4342001	4.358	LOC4324990	4.150	LOC4339486	3.419	LOC4328040	3.361
LOC4337135	4.325	LOC4332025	4.143	LOC4342069	3.412	LOC4325738	3.360
LOC4325808	4.313	LOC9266201	4.135	LOC4325563	3.410	LOC4336334	3.358

3.3 The Analysis of Experiment Results

In the sub-network extraction and network topology analysis that are mentioned above, 100 candidate genes of rice drought-resistant are identified. Through literature search and analysis, we obtain some genes with evidence from the gene set. The protein encoded by LOC4352722 is described as auxin-responsive protein IAA31-like, which present the uncover roles of conserved lncRNAs associated with agriculture traits by analysis of non-coding transcriptome in rice and maize [28]. The LOC4329036, described as salt stress root protein RS1that have important role in salt tolerance [29]. As one of developmentally regulated plasma membrane polypeptide homologous genes, it shows a definite effect of abiotic stress on physiological and functional properties. The protein encoded by LOC4331532 is a member of the rice glutathione S-transferase gene family which has been approved the highlighted potential contributions of sense and natural antisense transcript (NAT) to the specification of agriculture traits of rice and maize in various organs at different developmental stages [28–30].

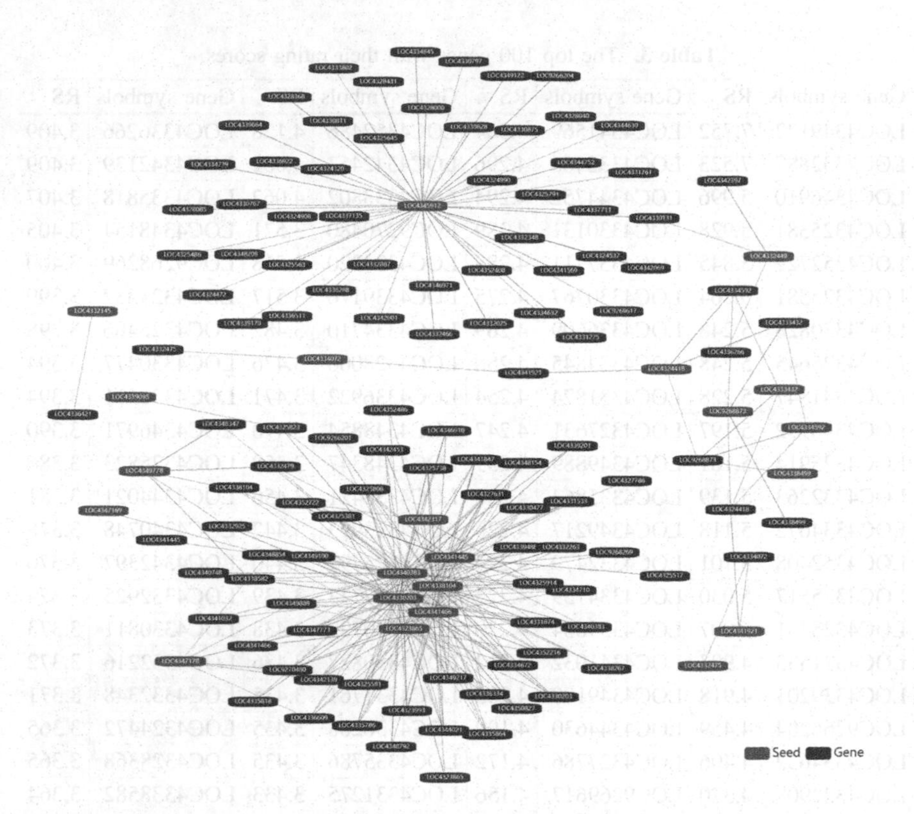

Fig. 2. The subnetwork extracted by Klein-Ravi algorithm. Red nodes represent 22 seed genes, dark nodes represent the other candidate genes. (Color figure online)

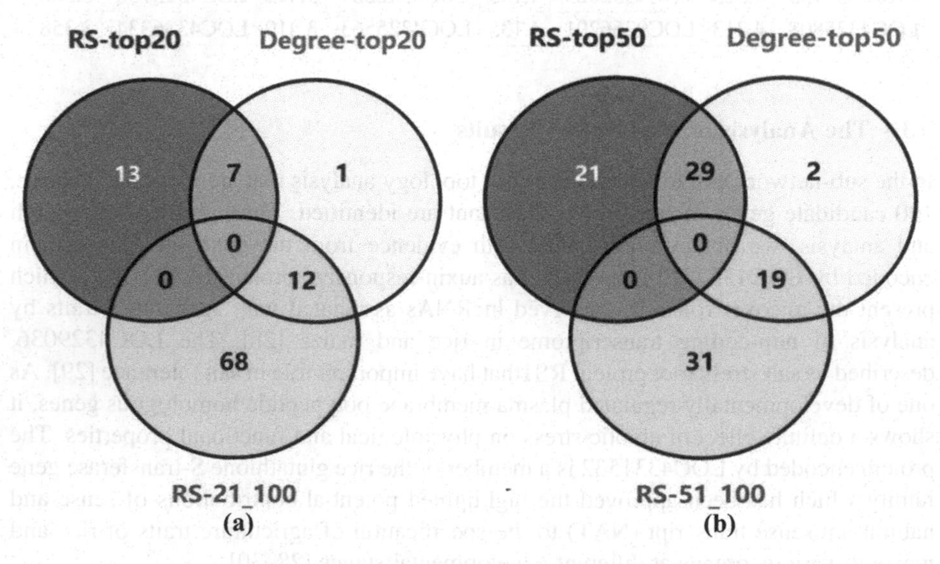

Fig. 3. Venn diagrams demonstrating different group of genes.

The analysis shows that the identified candidate genes play a direct or indirect role in rice drought resistance. Genes that are not analyzed might have relation with rice drought resistance and worth study for further experiment.

We performed the enrichment tests on top-100 candidate genes by using DAVID server to obtain comprehensive biological features. The item in the output of DAVID for GO term named as "GO: 0009535 ~ chloroplast thylakoid membrane" with the FDR <0.05 was selected. From the result of enrichment analysis, most genes represents drought tolerance was related to chloroplast activities. These results highlight the importance and the increase in cellular ABA concentrations and drought tolerance in rice [31].

Funding. This work was supported by the National Key Research and Development Project (2017YFD0301303), the Natural Science Young Foundation of Anhui Agricultural University (2019zd12) and the Introduction and Stabilization of Talent Project of Anhui Agricultural University (yj2019-32).

References

1. Sahebi, M., Hanafi, M.M., Rafii, M.Y., Mahmud, T.M.M., Azizi, P., Osman, M., et al.: Improvement of drought tolerance in rice (Oryza sativa L.): genetics, genomic tools, and the WRKY gene family. Biomed. Res. Int. **2018**, 3158474 (2018)
2. Ashraf, M.: Inducing drought tolerance in plants: recent advances. Biotechnol. Adv. **28**(1), 169–183 (2010)
3. Ji, K., Wang, Y., Sun, W., Lou, Q., Mei, H., Shen, S., et al.: Drought-responsive mechanisms in rice genotypes with contrasting drought tolerance during reproductive stage. J. Plant Physiol. **169**(4), 336–344 (2012)
4. van Noort, V., Snel, B., Huynen, M.A.: The yeast coexpression network has a small-world, scale-free architecture and can be explained by a simple model. EMBO Rep. **5**(3), 280–284 (2004)
5. Xiong, L.M., Schumaker, K.S., Zhu, J.K.: Cell signaling during cold, drought, and salt stress. Plant Cell. **14**, S165–S183 (2002)
6. Li, X., Chang, Y., Ma, S., Shen, J., Hu, H., Xiong, L.: Genome-wide identification of SNAC1-targeted genes involved in drought response in rice. Front. Plant Sci. **10**, 982 (2019)
7. Hu, H.H., Xiong, L.Z.: Genetic engineering and breeding of drought-resistant crops. In: Merchant, S.S. (ed.) Annual Review of Plant Biology, vol. 65, pp. 715–741 (2014)
8. Ganie, S.A., Pani, D.R., Mondal, T.K.: Genome-wide analysis of DUF221 domain containing gene family in Oryza species and identification of its salinity stress-responsive members in rice. PLoS ONE **12**(8) (2017)
9. Li, Y., Yuan, F., Wen, Z., Li, Y., Wang, F., Zhu, T., et al.: Genome-wide survey and expression analysis of the OSCA gene family in rice. BMC Plant Biol. **15**, 261 (2015)
10. Swamy, B.P.M., Kumar, A.: Genomics-based precision breeding approaches to improve drought tolerance in rice. Biotechnol. Adv. **31**(8), 1308–1318 (2013)
11. Ulitsky, I., Shamir, R.: Identifying functional modules using expression profiles and confidence-scored protein interactions. Bioinformatics **25**(9), 1158–1164 (2009)

12. Vandin, F., Clay, P., Upfal, E., Raphael, B.J.: Discovery of mutated subnetworks associated with clinical data in cancer. In: Pacific Symposium on Biocomputing Pacific Symposium on Biocomputing, pp. 55–66 (2012)
13. Scott, M.S., Perkins, T., Bunnell, S., Pepin, F., Thomas, D.Y., Hallett, M.: Identifying regulatory subnetworks for a set of genes. Mol. Cellul. Proteom. **4**(5), 683–692 (2005)
14. Noirel, J., Ow, S.Y., Sanguinetti, G., Jaramillo, A., Wright, P.C.: Automated extraction of meaningful pathways from quantitative proteomics data. Brief. Funct. Genom. Proteom. **7** (2), 136–146 (2008)
15. Smita, S., Katiyar, A., Lenka, S.K., Dalal, M., Kumar, A., Mahtha, S.K., et al.: Gene network modules associated with abiotic stress response in tolerant rice genotypes identified by transcriptome meta-analysis. Funct. Integr. Genom. **20**(1), 29–49 (2020)
16. Lv, Y., Xu, L., Dossa, K., Zhou, K., Zhu, M., Xie, H., et al.: Identification of putative drought-responsive genes in rice using gene co-expression analysis. Bioinformation **15**(7), 480–489 (2019)
17. Wang, X., Wu, Y.J., Wang, R.J., Wei, Y.Y., Gui, Y.M.: Gray BP neural network based prediction of rice protein interaction network. Cluster Comput. J. Netw. Softw. Tools Appl. **22**(2), S4165-S4171 (2019)
18. Kudo, T., Terashima, S., Takaki, Y., Tomita, K., Saito, M., Kanno, M., et al.: PlantExpress: a database integrating OryzaExpress and ArthaExpress for single-species and cross-species gene expression network analyses with microarray-based transcriptome data. Plant Cell Physiol. **58**(1) (2017). https://doi.org/10.1093/pcp/pcw208
19. Serin, E.A.R., Nijveen, H., Hilhorst, H.W.M., Ligterink, W.: Learning from co-expression networks: possibilities and challenges. Front. Plant Sci. **7**, 444 (2016)
20. Lee, H.K., Hsu, A.K., Sajdak, J., Qin, J., Pavlidis, P.: Coexpression analysis of human genes across many microarray data sets. Genome Res. **14**(6), 1085–1094 (2004)
21. Wolfe, C.J., Kohane, I.S., Butte, A.J.: Systematic survey reveals general applicability of "guilt-by-association" within gene coexpression networks. BMC Bioinf. **6**, 227 (2005)
22. Zhuang, D.-Y., Jiang, L., He, Q.-Q., Zhou, P., Yue, T.: Identification of hub subnetwork based on topological features of genes in breast cancer. Int. J. Mol. Med. **35**(3), 664–674 (2015)
23. Chin, C.H., Chen, S.H., Wu, H.H., Ho, C.W., Ko, M.T., Lin, C.Y.: cytoHubba: identifying hub objects and sub-networks from complex interactome. BMC Syst. Biol. **8**, S11 (2014)
24. Klein, P.N., Ravi, R.J.J.A.: A nearly best-possible approximation algorithm for node-weighted Steiner trees. J. Algorithms **19**(1), 104–115 (1995)
25. Ravi, R., Marathe, M.V., Ravi, S.S., Rosenkrantz, D.J., Hunt, H.B.: Approximation algorithms for degree-constrained minimum-cost network-design problems. Algorithmica **31** (1), 58–78 (2001)
26. Zheng, S., Zhao, Z.: GenRev: exploring functional relevance of genes in molecular networks. Genomics **99**(3), 183–188 (2012)
27. Huang, D.W., Sherman, B.T., Lempicki, R.A.: Systematic and integrative analysis of large gene lists using DAVID bioinformatics resources. Nat. Protocols **4**(1), 44–57 (2009)
28. Wang, H., Niu, Q.-W., Wu, H.-W., Liu, J., Ye, J., Yu, N., et al.: Analysis of non-coding transcriptome in rice and maize uncovers roles of conserved lncRNAs associated with agriculture traits. Plant J. **84**(2), 404–416 (2015)

29. Yamada, N., Theerawitaya, C., Kageyama, H., Cha-um, S., Takabe, T.: Expression of developmentally regulated plasma membrane polypeptide (DREPP2) in rice root tip and interaction with Ca2+/CaM complex and microtubule. Protoplasma **252**(6), 1519–1527 (2015)
30. Soranzo, N., Gorla, M.S., Mizzi, L., De Toma, G., Frova, C.: Organisation and structural evolution of the rice glutathione S-transferase gene family. Mol. Genet. Genom. **271**(5), 511–521 (2004)
31. Wang, C., Chen, S., Dong, Y., Ren, R., Chen, D., Chen, X.: Chloroplastic Os3BGlu6 contributes significantly to cellular ABA pools and impacts drought tolerance and photosynthesis in rice. New Phytol. **226**(4), 1042–1054 (2020)

Identification of Human LncRNA-Disease Association by Fast Kernel Learning-Based Kronecker Regularized Least Squares

Wen Li[1], Shu-Lin Wang[1](✉), Junlin Xu[1], and Jialiang Yang[2]

[1] College of Computer Science and Electronic Engineering, Hunan University, Changsha 410082, China
smartforesting@163.com
[2] Geneis Beijing Co., Ltd., Beijing 100102, China

Abstract. As the function of lncRNA is gradually understood, they have been found regulating the expression of target genes at the post-transcriptional level, and their abnormal functions may lead to so many diseases. Then, identifying the lncRNA-disease associations (LDA) can help to better understand its pathogenesis, promote the search for biomarkers of disease diagnosis, and effectively prevent disease. To break through the limitations of the existing computational models, we put forward a novel computational method of lncRNA-disease association identification by employing Fast Kernel Learning with Kronecker Regularized Least Squares (FKL-KronRLS-LDA). This model first extracts three different similarity kernels in disease and lncRNA space respectively. Next, it fuses these distinct kernels into an integrated kernel with the optimized combining weightings indicating their importance. It then combines lncRNA kernel and disease kernel into one larger kernel by Kronecker product kernel. Finally, it adopts the regularization least squares to identify potential associations. In experiments of Leave one out cross validation (LOOCV) and 5-fold cross validation (5-fold CV), FKL-KronRLS-LDA respectively obtains an AUC of 0.917 and 0.856, which outperform other excellent computational models. Furthermore, in the case studies, 9, 8 and 8 out of top 10 identified lncRNAs are successfully confirmed by recent published literature for lung cancer, breast cancer and gastric cancer, respectively. In a word, FKL-KronRLS-LDA can effectively identify potential lncRNA-disease associations for human beings.

Keywords: LncRNA-disease association identification · Similarity kernel fusion · Fast kernel learning · Kronecker regularized least squares

1 Introduction

Long non-coding RNAs' (lncRNAs') transcripts are no less than 200 nucleotides. For a long period of time, lncRNAs have initially been considered as useless, having no biological function. Because they do not encode proteins [1]. In recent years, a large amount of research evidence show that the abnormal expression or dysfunction of long non-coding RNA are important causes of human complex diseases, including

D.-S. Huang and K.-H. Jo (Eds.): ICIC 2020, LNCS 12464, pp. 302–315, 2020.
https://doi.org/10.1007/978-3-030-60802-6_27

neurological degenerative diseases and cancers. Identifying the lncRNA-disease association (LDA) can help to better understand the pathogenesis, facilitate the search for biomarkers of disease diagnosis and prevention, promote research for drug non-coding RNA targets and design specialized molecular tools for disease treatment [2]. In order to overcome the limitation of high cost and labor intensity of traditional biological wet experiments, many computational models have been proposed, such as identification of miRNA-mRNA regulatory modules [3], the discovery of lncRNA and mRNA co-expression modules [4], predicting miRNA-target associations [5], and miRNA-disease association inferring [6, 7].

In recent years, computational methods can roughly be classified into two categories. The first category is biological network-based model. For example, RWRlncD is performed to identify the lncRNA-disease associations by random walk with restart on a heterogeneous network [8]. Since RWRLDA cannot predict infer related lncRNAs for de novo diseases which didn't have any known associations, various types of biological network-based approaches have been proposed [9]. The second category is machine learning-based method [10]. LDAs confirmed by biological experiments are considered as positive labeled samples, while the unknown associations are considered as unlabeled samples. For example, Chen et al. put forward an excellent computational model of Laplacian Regularized Least Squares for LDA prediction (LRLSLDA), which is the first model based on a semi-supervised learning framework to infer LDA [11]. Besides, there are various machine learning-based algorithm, such as bagging support vector machine (SVM) for LDA prediction [12, 13], matrix factorization-based methods [14], matrix completion method [15], multiple kernel fusion-based methods [16], Naive Bayesian classifier-based methods, and recently lots of deep learning methods [17].

In this work, we have proposed a novel computational model of multiple kernel learning for LDA prediction, which is one of machine learning-based methods. Several kernel methods have been used in the previous studies such as in predicting associations of disease-gene pairs, interactions between protein and protein [18], associations of miRNA-disease pairs [19], lncRNA-disease associations [20], drug-target interactions [21], lncRNA-protein interaction [22], and protein-DNA binding residues [23]. The advantage of kernel-based methods is that their optimization is independent of the number of features, known as dimensionless. FKL is an intermediate integration technique which automatically optimizes the combination weightings according to the importance of each base kernels [24]. Since Kronecker RLS outperforms SVM in DTI prediction [25], we chose to apply Kronecker RLS for lncRNA-disease association prediction. Finally, the results of experiments shows the powerful capability of LDA identification.

2 Materials and Methods

2.1 Review

This study is based on the hypothesis that similar diseases are more likely to be associated with similar lncRNAs, and vice versa. We have proposed an extended

Kronecker Regularized Least Square (KronRLS) algorithm to identify possible LDA in the framework of FKL. The flowchart of our work is demonstrated in Fig. 1.

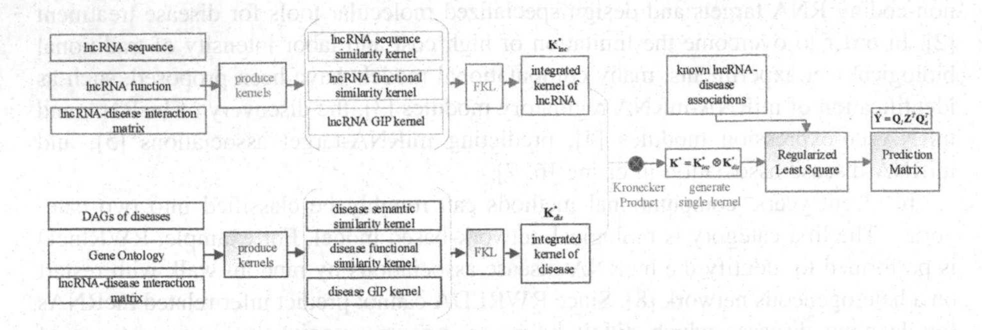

Fig. 1. The workflow of FKL-KronRLS-LDA.

2.2 Benchmark Dataset

The experimentally supported LDAs were downloaded from the database LncRNA-Disease v2.0 [26]. After correcting names of lncRNAs (uniformly mapping to hgnc symbols) and diseases names (uniformly mapping to Mesh names and DO ID), we deleted all the duplicated records and those not human beings. The benchmark dataset was obtained from the state-of-the-art work by Bao et al. [15]. The Dataset consists of 621 known LDAs, including 285 lncRNAs and 226 diseases. The set of lncRNAs is represented as $L = \{l_1, \ldots, l_m\}$, and the set of diseases is represented as $D = \{d_1, \ldots, d_n\}$. The adjacency matrix of LDA is denoted as $Y \in R^{m \times n}$. $Y(i,j)$ is set 1 if the association between lncRNA l_i and disease d_j is experimentally confirmed, otherwise it is set 0.

2.3 Calculation of Similarity Kernels for Diseases and LncRNAs

Disease Semantic Similarity Kernel. There are many existing methods to measure the disease semantic similarity. We downloaded disease MeSH descriptors which provide the classification and associations of diseases from the National Library of Medicine (http://www.nlm.nih.gov/). Then we calculated the semantic similarity through the method proposed by study [27] and implemented it by an R package named DOSE [28]. It is a widely accepted that the more DAGs the two diseases share, the more semantically similar they are. Each disease can be represented by a directed acyclic graph (DAG), such as $DAG(d_i) = (d_i, T(d_i), E(d_i))$. $T(d_i)$ is the set of nodes d_i and all his ancestors. $E(d_i)$ is the set of edges linked to disease d_i. The semantic score of disease d_i is calculated as Eq. (1):

$$DV(d_i) = \sum_{t \in T(d_i)} D1_{d_i}(t) \tag{1}$$

where the $D1_{d_i}(t)$ can be calculated as Eq. (2):

$$D1_{d_i}(t) = \begin{cases} 1 & if\ t = d_i \\ \max\{\Delta * D1_{d_i}(t')\} t' \in child\ of\ t\} & if\ t \neq d_i \end{cases} \tag{2}$$

where the semantic contribution factor Δ is assigned to 0.5. Then, the semantic similarity kernel of disease $\mathbf{K}^{d}_{sem} \in R^{q \times p}$ can be defined as Eq. (3):

$$\mathbf{K}^{d}_{sem}(d_i, d_j) = \frac{\sum_{t \in T(d_i) \cap T(d_j)} (D1_{d_i}(t) + D1_{d_j}(t))}{DV(d_i) + DV(d_j)} \tag{3}$$

Disease Functional Similarity Kernel. Associations of Disease-gene were downloaded from the database DisGeNET (www.disgenet.org/ds/DisGeNET/) [29]. Interactions between genes were downloaded from database BioGrid through a R package biomaRt_2.30.0 [30]. And gene ontology terms of human beings were derived from database Ensemble [31]. We finally calculated the functional similarity kernel of diseases by Jaccard coefficient with gene-gene ontology terms and disease-gene interactions as Eq. (4):

$$\mathbf{K}^{d}_{func}(d_i, d_j) = \frac{|GO_{d_i} \cap GO_{d_j}|}{|GO_{d_i} \cup GO_{d_j}|} \tag{4}$$

where operator $|\cdot|$ denotes the size of the set, and GO_{d_i} represents the gene ontology terms associated with d_j.

LncRNA Sequence Similarity Kernel. We can mark a sequence S with length l as an ordered string of letters $S = s_1 s_2 \cdots s_i \cdots s_l$. The sequences of lncRNAs are obtained from the database LncRNADisease. Referring to the previous studies, we calculate the sequence similarity kernel $\mathbf{K}^{lnc}_{seq}$ by the normalized Smith-Waterman (SW) score as in Eq. (5) [32].

$$\mathbf{K}^{lnc}_{seq}(l_i, l_j) = \frac{SW(S_{l_i}, S_{l_j})}{\sqrt{SW(S_{l_i}, S_{l_j}) SW(S_{l_i}, S_{l_j})}} \tag{5}$$

where S_{l_i} and S_{l_j} are sequences of lncRNA l_i and l_j, respectively. SW is a normalized Smith-Waterman alignment score.

LncRNA Functional Similarity Kernel. Based on the disease semantic similarity aforementioned, we extend the previous computational method of miRNA functional similarity to measure lncRNA functional similarity kernel, implementation details of which can be referred to previous study [33]. It is supposed that lncRNA l_1 is associated with the set of diseases $D1 = \{d_{11}, d_{12}, \ldots, d_{1m}\}$, and lncRNA l_2 is associated with the set of diseases $D2 = \{d_{21}, d_{22}, \ldots, d_{2n}\}$. The functional similarity kernel $\mathbf{K}_{func}^{\ln c}$ between lncRNA$_1$ and lncRNA$_2$ can be calculated as Eq. (6):

$$\mathbf{K}_{func}^{\ln c}(\ln cRNA_1, \ln cRNA_2) = \frac{\sum_{1 \le i \le m} SIM(d_{1i}, D_2) + \sum_{1 \le j \le n} SIM(d_{2j}, D_1)}{m+n} \quad (6)$$

where $SIM(d, D) = \max_{1 \le i \le k}(SIM(d, d_i))$, D_1 and D_2 denote disease group which are related with lncRNA$_1$ and lncRNA$_2$ respectively. Disease d_{1i} belongs to group D_1, and d_{2j} belongs to group D_2.

Gaussian Interaction Profile Kernel. Supposed that the associations are easily influenced by the network topology. We calculated the topology similarity kernel $\mathbf{K}_{GIP}^{\ln c}$ and $\mathbf{K}_{GIP}^{d}$ by virtue of Gaussian interaction profile (GIP) as Eq. (7) [34].

$$\mathbf{K}_{GIP}^{\ln c}(\ln cRNA_i, \ln cRNA_j) = \exp(-\gamma_{\ln c}\|\mathbf{IP}(l_i) - \mathbf{IP}(l_j)\|^2) \quad (7)$$

where $\gamma_{\ln c} = \gamma'_{\ln c} \Big/ \left(\frac{1}{nl}\sum_{i=1}^{nl}\|\mathbf{IP}(l_i)\|^2\right)$, $\gamma_{\ln c}$ is the kernel bandwidth, which can be obtained by normalizing a bandwidth parameter $\gamma'_{\ln c}$. The binary vector $\mathbf{IP}(\cdot)$ are the interaction profile, which denotes whether there is an interaction between lncRNAs and diseases. In the same way, the topology similarity kernel for diseases $\mathbf{K}_{GIP}^{d}$ also can be calculated.

2.4 Kernel Fusion

It is believed that if we can properly fuse multi-view similarity data more precisely, we could infer more potential LDAs. In our work, we attempted to find an optimal weighting strategy to integrate three base kernels $\mathbf{K}_{seq}^{lnc}$, $\mathbf{K}_{func}^{\ln c}$ and $\mathbf{K}_{GIP}^{\ln c}$ into one integrated kernel $\mathbf{K}_{\ln c}^{*}$ in lncRNA space, and integrate three base kernels $\mathbf{K}_{sem}^{d}$, $\mathbf{K}_{func}^{d}$ and $\mathbf{K}_{GIP}^{d}$ into one integrated kernel $\mathbf{K}_{dis}^{*}$ in disease space, respectively. Inspired by previous study, the similarity kernel fusion for lncRNAs can be defined as Eq. (8):

$$\mathbf{K}_{\ln c}^{*} = \sum_{i=1}^{3} w_i^l \mathbf{K}_i^l, \ \mathbf{K}_i^l \in \Re^{m \times m} \quad (8)$$

where $\mathbf{K}_i^l$ represents different base kernel in lncRNA space. w_i^l is element in vector $\mathbf{w}^l \in \Re^{3 \times 1}$. In the same way, $\mathbf{K}_{dis}^{*}$ can also be defined as Eq. (8).

Fast kernel learning attempts to find an optimal weight vector $\mathbf{w}$ by minimizing the distance between the integrated similarity kernel $\mathbf{K}$ and ideal similarity kernel $\mathbf{K}^{ideal}$, which can be defined as $\mathbf{K}^{ideal} = \mathbf{Y}\mathbf{Y}^T$, where $\mathbf{Y}$ is the interaction matrix between lncRNA and disease. For unified description, in the following sections, $\mathbf{w}$ represents $\mathbf{w}^l$

or $\mathbf{w}^d$, and $\mathbf{K}$ represents $\mathbf{K}^*_{\ln c}$ or $\mathbf{K}^*_{dis}$. Therefore, $\mathbf{w}$ can be optimized by the object function $\min_{\mathbf{w}} \|\mathbf{K} - \mathbf{K}^{ideal}\|_F^2$. To avoid overfitting, a regularization term $\|\mathbf{w}\|^2$ is added, as shown in Eq. (9):

$$\min_{\mathbf{w},\mathbf{K}} \|\mathbf{K} - \mathbf{K}^{ideal}\|_F^2 + \lambda\|\mathbf{w}\|^2$$
$$s.t., \sum_{i=1}^{3} w_i = 1, \ (w_i \geq 0, \ i = 1, 2, 3) \tag{9}$$

where regularization coefficient λ is a tradeoff between prediction error and model complexity. $\|\mathbf{X}\|_F$ Represents Frobenius norm of a matrix $\mathbf{X}$. i.e. $\|\mathbf{X}\|_F^2 = tr(\mathbf{X}\mathbf{X}^T)$. Hence, the weights vector $\mathbf{w}$ can be optimized by the minimization of the object function as Eq. (10):

$$\min_{\mathbf{w}} \mathbf{w}^T(\mathbf{A} + \lambda\mathbf{I})\mathbf{w} - 2\mathbf{b}^T\mathbf{w} \quad s.t. \sum_{i=1}^{3} w_i = 1$$
$$\begin{cases} A_{u,v} = tr(\mathbf{K}_u^T \mathbf{K}_v) \\ b_v = tr(\mathbf{Y}^T \mathbf{K}_v) \end{cases} \tag{10}$$

where $tr(\cdot)$ is the trace operator. This optimization problem can be solved by CVX version2.1 program in MATLAB. At last, the optimal weighting parameter vector $\mathbf{w}^l$ and $\mathbf{w}^d$ can be finalized. Above all, the integrated similarity kernel $\mathbf{K}^*_{\ln c}$ and $\mathbf{K}^*_{dis}$ which are the input matrix for subsequent Kronecker RLS have be constructed.

2.5 Kronecker Regularized Least Square

Given $P = \{(x_1, y_1), (x_2, y_2), \ldots, (x_N, y_N)\}$ is a set of training samples, where x_i denotes a pair of lncRNA and disease, $y_i \in \Re$ denotes its binary label, where 1 represents a known association and 0 represents unknown, and $1 < i < N$. Assuming that $|L| = m$ and $|D| = n$ are the size of lncRNA space and disease space respectively, and $N = m \times n$.

According to the representor theorem and the kernel RLS regression, the solution of the objective function of regularized least square (RLS) can solved as Eq. (11):

$$\hat{\mathbf{Y}} = \mathbf{S}(\mathbf{S} + \sigma\mathbf{I})^{-1}\mathbf{Y}_{train} \tag{11}$$

where $\hat{\mathbf{Y}}$ is the predicted matrix, $\mathbf{Y}_{train}$ is the training matrix. $\sigma > 0$ is a regularization coefficient. $\mathbf{S}$ is the kernel matrix with multiple kernel fusion. Then, we combined lncRNA integrated kernel $\mathbf{K}^*_{\ln c}$ and disease integrated kernel $\mathbf{K}^*_{dis}$ into one final kernel $\mathbf{K}^*$ by Kronecker product $\mathbf{K}^* = \mathbf{K}^*_{\ln c} \otimes \mathbf{K}^*_{dis}$, where $\mathbf{K}^* \in \Re^{N \times N}$. At last, the predicted association matrix $\hat{\mathbf{Y}}$ with the kernel $\mathbf{K}^*$ can be obtained by Eq. (12):

$$vec(\hat{\mathbf{Y}}^T) = \mathbf{K}^*(\mathbf{K}^* + \sigma\mathbf{I})^{-1} vec(\mathbf{Y}_{train}^T) \tag{12}$$

Unfortunately, the cost of time is $O((|L||D|)^3)$, which is high even if the size of the lncRNA and disease set are moderate. Because Kronecker RLS needs to implement the inverse of the matrix, whose size is $N \times N = (m \times n) \times (m \times n)$. To speed up calculation, Raymond and Kashima improved Kronecker RLS by two algebraic properties: one is so called '*vec-trick*', and the other is the Eigen-decomposition of Kronecker product. The kernel $\mathbf{K}^*_{\ln c}$ and $\mathbf{K}^*_{dis}$ can be decomposed as $\mathbf{K}^*_{\ln c} = \mathbf{Q}_l \mathbf{\Lambda}_l \mathbf{Q}_l^T$ and $\mathbf{K}^*_{dis} = \mathbf{Q}_d \mathbf{\Lambda}_d \mathbf{Q}_d^T$. The Kronecker product kernel $\mathbf{K}^*$ can be decomposed as $\mathbf{K}^* = \mathbf{K}^*_{\ln c} \otimes \mathbf{K}^*_{dis} = \mathbf{Q} \Lambda \mathbf{Q}^T$, where $\mathbf{Q} = \mathbf{Q}_l \otimes \mathbf{Q}_d$, $\mathbf{\Lambda} = \mathbf{\Lambda}_l \otimes \mathbf{\Lambda}_d$.

Finally, the lncRNA-disease association prediction by Kronecker RLS can be resolved as Eq. (13):

$$\hat{\mathbf{Y}} = \mathbf{Q}_l \mathbf{Z}^T \mathbf{Q}_d^T \tag{13}$$

where $vec(\mathbf{Z}) = (\mathbf{\Lambda}_l \otimes \mathbf{\Lambda}_d)(\mathbf{\Lambda}_l \otimes \mathbf{\Lambda}_d + \sigma \mathbf{I})^{-1} \times vec(\mathbf{Q}_d^T \mathbf{Y}_{train}^T \mathbf{Q}_l)$

Kronecker RLS just needs to implement some matrix multiplication and two Eigen decompositions operations. Therefore, the running time is sharply cut down from $O((|L||D|)^3)$ to $O(|L|^3 + |D|^3)$.

3 Results

3.1 Evaluation Metrics

To evaluate the accuracy of FKL-KronRLS-LDA, we carried out a lot of experiments by global LOOCV and 5-fold CV on the benchmark dataset. The receiver operating characteristic (ROC) and the area under ROC curve (AUC) can be obtained. Besides, the area under the Precision-Recall curve (AUPR) is also calculated. When dealing with unbalanced data, AUPR is a more reliable evaluation metric than AUC, which indicates the performance of the algorithm more accurately.

4 Parameter Optimization

We adjusted the value of λ_l and λ_d in $[2^{-15}, 2^{20}]$ to search for the optimal value through grid search. As shown in Fig. 2, when $\lambda_l = 2^{18}$ and $\lambda_d = 2^{14}$, FKL achieved the best AUPR value of 0.5476. Besides, we tested 11 different values of parameter σ and γ' varying from 2^{-5} to 2^5. According to the best value of AUPR, we picked up the optimal choice. We found that FKL-KronRLS-LDA obtains the best AUPR value of 0.548 when $\sigma = 2^{-1}$. Similarly, when $\gamma' = 1$, AUPR reached the best value of 0.544.

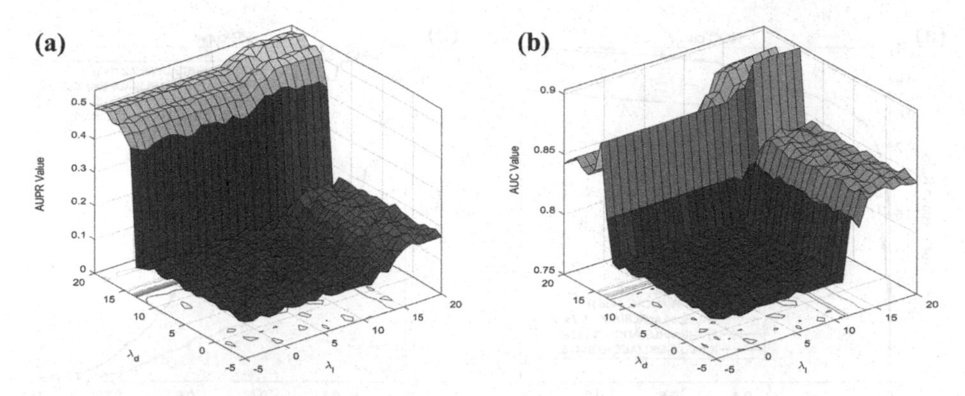

Fig. 2. Parameter optimization for λ_l and λ_d by grid search on benchmark dataset. (A) AUPR values with different λ_l and λ_d. (B) AUC values with different λ_l and λ_d.

4.1 Performance Analysis

Comparison with Different Kernels. To illustrate the advantages of our multiple similarity kernel fusion in LDA, we carried out global LOOCV on benchmark dataset with permutation of diverse kernels. The AUC and AUPR values of different kernel permutation are shown in Table 1. Obviously, our method with integrated similarity kernel $\mathbf{K}^*_{\ln c}$ and $\mathbf{K}^*_{dis}$ by FKL has the best performance. The AUC value and AUPR value achieve 0.8848 and 0.5476 respectively, which are higher than all other single kernel models and average kernel model.

Table 1. The AUC and AUPR values of different kernels by LOOCV

LncRNA similarity	Disease similarity	AUC	AUPR
$\mathbf{K}^{lnc}_{seq}$	$\mathbf{K}^{d}_{func}$	0.7773	0.4265
$\mathbf{K}^{\ln c}_{func}$	$\mathbf{K}^{d}_{sem}$	0.8569	0.4381
$\mathbf{K}^{\ln c}_{GIP}$	$\mathbf{K}^{d}_{GIP}$	0.6914	0.1156
$Avg(\mathbf{K}^{\ln c}_{seq} + \mathbf{K}^{\ln c}_{func} + \mathbf{K}^{\ln c}_{GIP})$	$Avg(\mathbf{K}^{d}_{func} + \mathbf{K}^{d}_{sem} + \mathbf{K}^{d}_{GIP})$	0.8806	0.5239
$\mathbf{K}^{*}_{\ln c}$	$\mathbf{K}^{*}_{dis}$	0.8848	0.5476

Comparison of Performance with Methods of Multiple Kernel Fusion. Several multiple kernel fusion methods are widely used for association prediction, including lncRNA-protein interactions (LPI) prediction, drug-target interaction (DTI) prediction, microRNA-disease association (MDA) prediction, and so on.

We compared our FKL-KronRLS with three multiple kernel fusion-based methods, including FKL-KRR [35], FKL-Spa-LapRLS [36] and KronRLS-MKL [37]. FKL-KronRLS-LDA obtains the AUC value of 0.917, which is higher than FKL-KRR (0.744), FKL-Spa-LapRLS (0.785) and KronRLS-MKL (0.914). The AUPR value of FKL-KronRLS-LDA is 0.521, which is also higher than FKL-KRR (0.201), FKL-Spa-LapRLS (0.222) and KronRLS-MKL (0.509). As shown in Fig. 3, FKL-KronRLS-LDA has the best performance compared with other multiple kernel fusion methods.

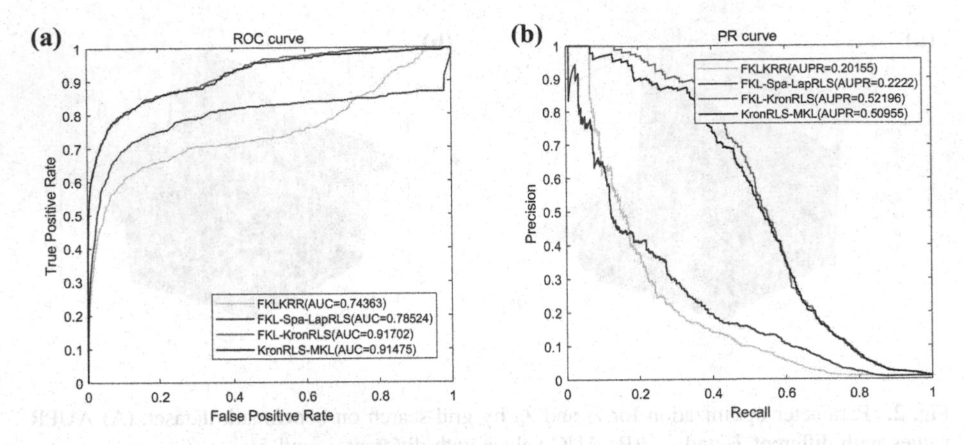

Fig. 3. The results of comparison with other multiple kernel fusion methods by 20 times 5-fold CV. (A) ROC curves and average AUC values. (B) PR curves and average AUPR values.

Comparison of Performance with Other LDA Methods. In order to further verify the prominent advantages of FKL-KronRLS-LDA, we compared it with four existing machine learning-based methods through 20 times 5-fold CV. The results of comparisons are shown in Fig. 4 FKL-KronRLS-LDA gets an AUC value of 0.856, while LRLSLDA, SIMCLDA, SKF-LDA and NCPLDA get the AUC values of 0.705, 0.703, 0.832 and 0.824, respectively. Additionally, the AUPR value of FKL-KronRLS-LDA is 0.465, which is much better than those of LRLSLDA (0.048), SIMCLDA (0.07), SKF-LDA (0.345) and NCPLDA (0.036). Therefore, FKL-KronRLS-LDA is obviously superior to other excellent LDA prediction methods.

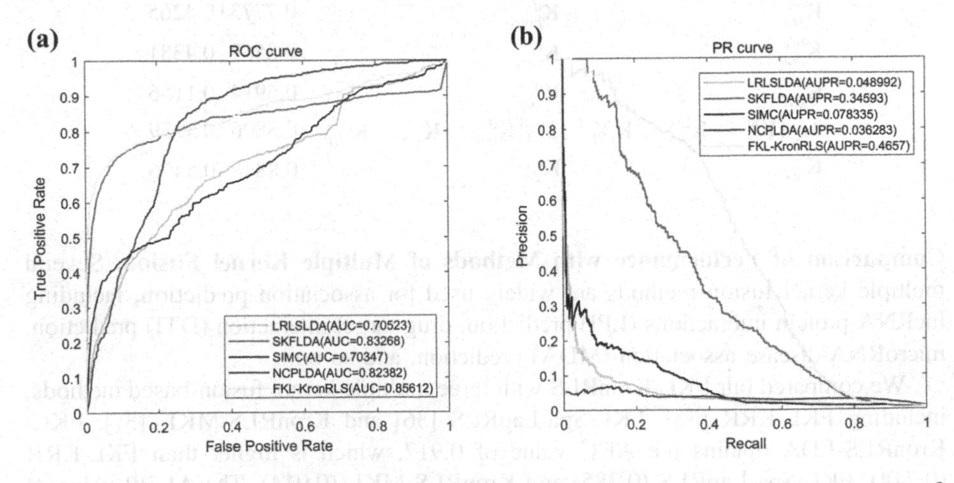

Fig. 4. The results of comparison between four machine learning-based methods by 20 times of 5-fold cross validation. (A) Average AUC values. (B) Average AUPR values.

Comparison of Running Time on Benchmark Dataset. We have conducted a comparison of running time between the methods of SKF, MKL-KronRLS and FKL-KronRLS, which all have outstanding performance. The results of running time are displayed in Table 2.

Table 2. Running time in 20 times.

Rank	Average running time (s)	Standard deviation (s)
FKL-KronRLS	0.6656	0.1011
SKF	2.3109	0.1912
MKL-KronRLS	6.1391	0.3230

As shown above, FKL-KronRLS-LDA obtains a shorter running time 0.6656 s to complete the prediction task, which is much faster than SKF-LDA 2.3109 s and MKL-KronRLS 6.1391 s (as shown in Table 2). In conclusion, FKL-KronRLS-LDA not only have higher AUC and AUPR values but also have shorter running time. So, FKL-KronRLS-LDA is effective and suitable for scaling to larger matrices.

4.2 Case Study

We conducted case studies for three common diseases, lung cancer, breast cancer and gastric cancer, to confirm the validity of FKL-KronRLS-LDA. All of known interactions in benchmark dataset are used to train, while the unknown associations are considered as candidates for prediction. For every queried disease, the candidate lncRNAs are sorted based on their scores of predictions. Among them, top 10 predicted lncRNAs for a queried disease are verified by literature mining from the latest databases: MNDR [38], Lnc2Cancer [39] and LncRNADisease [40]. The results of case study are illustrated in Table 3.

Table 3. Top 10 Identified LncRNAs predicted for Lung Cancer, breast cancer and gastric cancer in descending order of scores.

Lung cancer	PMID	Breast cancer	PMID	Gastric cancer	PMID
ACTA2-AS1	28219204	lincRNA-RoR	27855392	GAS5	26278580
PVT1	28731781	TUG1	28950664	IGF2-AS	31183590
NPTN-IT1	29416684	UCA1	26464647	MALAT1	29162158
MIAT	29487526	MIAT	29100300	SRA1	Unconfirmed
CDKN2B-AS1	25964559	BANCR	29565494	LincRNA-p21	28969031
GAS5	29207621	IGF2-AS	Unconfirmed	HLA-AS1	Unconfirmed
HIF1A-AS1	26339353	DNM3OS	27693451	TUG1	27261864
anti-NOS2A	Unconfirmed	LincRNA-p21	22487937	CASC2	27648142
TUG1	28069000	HIF1A-AS1	Unconfirmed	ESCCAL-1	28146436
CCAT2	28088736	DLEU2	15072547	SNHG16	29081409

Lung cancer is one of most dangerous malignant tumors to human health and life, whose morbidity and mortality rates are rising fastest. It is necessary to identify the associations between abnormal regulation of lncRNAs and lung cancer. In results of our work, 9 out of 10 predicted lncRNAs are successfully confirmed by literature mining or public databases searching. For example, the top 1 ranked lncRNA ACTA2-AS1 is identified to be overexpressed in lung cancer, which is closely associated with lymph node metastasis and tumor differentiation [41].

Breast cancer is the most common cancer among women, with a global incidence of about 24.2%. There is evidence that lncRNAs is one of the key factors involved in cell cycle regulation, and it greatly affects the occurrence and development of breast cancer. In our work, 8 out of 10 candidate lncRNAs predicted by FKL-KronRLS-LDA are confirmed immediately by literature mining. For example, lincRNA-RoR has been dysregulated in many types of cancers, including breast cancer (BC) [42]. The overexpression of TUG1 has been shown to cause cell cycle arrest in breast cancer, induce apoptosis, and significantly inhibit cell proliferation in vitro experiments [43]. Particularly, some associations of prediction are directly supported by the latest literature, such as lncRNA WRAP53, LINC01133, TCL6 and TUSC8.

Gastric cancer is one of the leading causes of cancer deaths globally and is also the second most common cancer in China. Many studies have shown that coding RNA and noncoding RNA (ncRNA) are closely related to gastric cancer, although the mechanism of gastric cancer remains to be elucidated. In our work, 8 out of 10 predicted lncRNAs are immediately confirmed associated with gastric cancer. For example, GAS5 induces gastric cancer cell growth stagnation by restraining G1-S phase translation [44]. IGF2-AS is up-regulated in gastric adenocarcinoma, which is associated with poor prognosis [44].

The results of case study for three selected diseases have further proved that the prediction performance of our method is outstanding. In addition, although some predicted lncRNAs cannot be confirmed immediately, they also rank high in other prediction methods. It is suggested that these lncRNAs have potential and need to be further studied in future experiments.

5 Conclusion

The outstanding contribution of our method is mainly as follows: First, FKL-KronRLS-LDA can avoid biased prediction. Because similarities are not only relied on the lncRNA-disease interaction but also obtained from multi-source data from different views. For the same sake, it also can predict associated lncRNAs for de novo diseases. Second, the prediction of FKL-KronRLS-LDA is more accurate than single kernel model, because of more dimensional information involvement. Third, Kronecker RLS takes advantage of kronecker product algebraic properties to speed up model training without calculating the pairwise function. In conclusion, FKL-KronRLS-LDA has superior performance and can be extended easily to the field of link prediction of large-scale.

Acknowledgement. This work is supported by the National Nature Science Foundation of China (Grant Nos. 61472467 and 61672011), and the National Key Research and Development Program (Grant Nos. 2017YFC1311003).

References

1. Crick, F.H.C., Barnett, L., Brenner, S., Watts-Tobin, R.J.: General nature of the genetic code for proteins. Nature **192**(4809), 1227–1232 (1961)
2. Bian, E., Li, J., Xie, Y., Zong, G., Li, J., Zhao, B.: LncRNAs: new players in gliomas, with special emphasis on the interaction of lncRNAs With EZH2. J. Cell. Physiol. **230**(3), 496–503 (2015)
3. Liang, C., Li, Y., Luo, J.W.: A novel method to detect functional microRNA regulatory modules by bicliques merging. IEEE/ACM Trans. Comput. Biol. Bioinform. (2016)
4. Xiao, Q., et al.: Identifying lncRNA and mRNA co-expression modules from matched expression data in ovarian cancer. In: IEEE/ACM Transactions on Computational Biology and Bioinformatics, Institute of Electrical and Electronics Engineers Inc., August 2018
5. Liu, Y., Luo, J., Ding, P.: Inferring MicroRNA targets based on restricted Boltzmann machines. IEEE J. Biomed. Heal., Informatics (2019)
6. Luo, J., Ding, P., Liang, C., Cao, B., Chen, X.: Collective prediction of disease-associated miRNAs based on transduction learning. IEEE/ACM Trans. Comput. Biol. Bioinform. **14**, 1468–1475 (2017)
7. Luo, J., Long, Y.: NTSHMDA: prediction of human microbe-disease association based on random walk by integrating network topological similarity. IEEE/ACM Trans. Comput. Biol. Bioinform. **17**, 1341–1351 (2018)
8. Sun, J., et al.: Inferring novel lncRNA-disease associations based on a random walk model of a lncRNA functional similarity network. Mol. BioSyst. **10**(8), 2074–2081 (2014)
9. Chen, X.: KATZLDA: KATZ measure for the lncRNA-disease association prediction. Sci. Rep. **5**, 1–11 (2015)
10. Zitnik, M., Nguyen, F., Wang, B., Leskovec, J., Goldenberg, A., Hoffman, M.M.: Machine learning for integrating data in biology and medicine: principles, practice, and opportunities. Inf. Fusion **50**(June 2018), 71–91 (2019)
11. Chen, X., Yan, G.Y.: Novel human lncRNA-disease association inference based on lncRNA expression profiles. Bioinformatics **29**(20), 2617–2624 (2013)
12. Lan, W., et al.: LDAP : A Web Server for lncRNA-Disease Association Prediction, vol. 33, no. October 2016, pp. 458–460 (2017)
13. Li, W., Wang, S., Xu, J., Mao, G., Tian, G., Yang, J.: Inferring latent disease-lncRNA associations by faster matrix completion on a heterogeneous network. Front. Genet. (2019)
14. Fu, G., Wang, J., Domeniconi, C., Yu, G.: Matrix factorization-based data fusion for the prediction of lncRNA-disease associations. Bioinformatics **34**(9), 1529–1537 (2018)
15. Lu, C., et al.: Prediction of lncRNA-disease associations based on inductive matrix completion. Bioinformatics **34**(19), 3357–3364 (2018)
16. Peng, L., Liao, B., Zhu, W., Li, Z., Li, K.: Predicting drug-target interactions with multi-information fusion. IEEE J. Biomed. Heal. Inf. (2017)
17. Zeng, X., et al.: Network-based prediction of drug–target interactions using an arbitrary-order proximity embedded deep forest. Bioinformatics **36**, 1–8 (2020)
18. Ding, Y., Tang, Y., Guo, F.: Identification of protein-protein interactions via a novel matrix-based sequence representation model with amino acid contact information. Int. J. Mol. Sci. (2016)

19. Chen, X., Niu, Y.W., Wang, G.H., Yan, G.Y.: MKRMDA: multiple kernel learning-based Kronecker regularized least squares for MiRNA-disease association prediction. J. Transl. Med. **15**(1), 1–14 (2017)
20. Xie, G., Meng, T., Luo, Y., Liu, Z.: SKF-LDA: similarity kernel fusion for predicting lncRNA-disease association. Mol. Ther. Nucl. Acids **18**(December), 45–55 (2019)
21. Kuang, Q., et al.: A kernel matrix dimension reduction method for predicting drug-target interaction. Chemom. Intell. Lab. Syst. **162**(November 2015), 104–110 (2017)
22. Shen, C., Ding, Y., Tang, J., Jiang, L., Guo, F.: LPI-KTASLP: prediction of LncRNA-protein interaction by semi-supervised link learning with multivariate information. IEEE Access **7**, 13486–13496 (2019)
23. Hu, J., Li, Y., Zhang, M., Yang, X., Shen, H., Yu, D.: Predicting Protein-DNA Binding Residues by Weightedly Combining Sequence-Based Features and Boosting Multiple SVMs, vol. 14, no. 6, pp. 1389–1398 (2017)
24. He, J., Chang, S.F., Xie, L.: Fast kernel learning for spatial pyramid matching. In: 26th IEEE Conference on Computer Vision and Pattern Recognition, CVPR (2008)
25. Ding, H., Takigawa, I., Mamitsuka, H., Zhu, S.: Similarity-based machine learning methods for predicting drug-target interactions: a brief review. Brief. Bioinform. (2013)
26. Bao, Z., Yang, Z., Huang, Z., Zhou, Y., Cui, Q., Dong, D.: LncRNADisease 2.0: an updated database of long non-coding RNA-associated diseases. Nucl. Acids Res. **47**(D1), D1034–D1037 (2019)
27. Wang, D., Wang, J., Lu, M., Song, F., Cui, Q.: Inferring the human microRNA functional similarity and functional network based on microRNA-associated diseases. Bioinformatics **26**(13), 1644–1650 (2010)
28. Package, T., Disease, T., Semantic, O., Annotationdbi, I.: Package. In: DOSE (2019)
29. Piñero, J., et al.: DisGeNET: a comprehensive platform integrating information on human disease-associated genes and variants. Nucl. Acids Res. **45**(D1), D833–D839 (2017)
30. Durinck, S., et al.: BioMart and bioconductor: a powerful link between biological databases and microarray data analysis. Bioinformatics **21**, 3439–3440 (2005)
31. Aken, B.L., et al.: The ensemble gene annotation system. In: Database (Oxford) (2016)
32. Mott, R.: Smith-Waterman algorithm. In: Encyclopedia of Life Sciences (2005
33. Yu, G., Wang, L.G., Yan, G.R., He, Q.Y.: DOSE: an R/Bioconductor package for disease ontology semantic and enrichment analysis. Bioinformatics **31**, 608–609 (2015)
34. van Laarhoven, T., Nabuurs, S.B., Marchiori, E.: Gaussian interaction profile kernels for predicting drug-target interaction. Bioinformatics **27**, 3036–3043 (2011)
35. Shen, C., Ding, Y., Tang, J., Guo, F.: Multivariate information fusion with fast kernel learning to Kernel Ridge Regression in predicting lncRNA-protein interactions. Front. Genet. **10**(Jan), 1–12 (2019)
36. Jiang, L., Xiao, L., Ding, Y., Tang, J., Guo, F.: FKL-Spa-LapRLS: an accurate method for identifying human microRNA-disease association. BMC Genom. **19**(Suppl 10) (2018)
37. Kuang, Q., et al.: A kernel matrix dimension reduction method for predicting drug-target interaction. Chemom. Intell. Lab. Syst. **162**(November 2016), 104–110 (2017)
38. Cui, T., et al.: MNDR v2.0: an updated resource of ncRNA-disease associations in mammals. Nucl. Acids Res. (2018)
39. Gao, Y., et al.: Lnc2Cancer v2.0: updated database of experimentally supported long non-coding RNAs in human cancers. Nucleic Acids Res. **47**(D1), D1028–D1033 (2019)
40. Jia, K., Gao, Y., Shi, J., Zhou, Y., Zhou, Y., Cui, Q.: Annotation and curation of the causality information in LncRNADisease. In: Database (Oxford) (2020)
41. Pan, G.F., Zhou, X.F., Zhao, J.P.: Correlation between expression of long non-coding RNA ZXF1 and prognosis of lung adenocarcinoma and its potential molecular mechanism. Zhonghua Zhong Liu Za Zhi **39**(2), 102–108 (2017)

42. Pan, Y., et al.: The emerging roles of long noncoding RNA ROR (lincRNA-ROR) and its possible mechanisms in human cancers. In: Cellular Physiology and Biochemistry, vol. 40, no. 1–2, pp. 219–229. S. Karger AG, November 2016
43. Fan, S., et al.: Downregulation of the long non-coding RNA TUG1 is associated with cell proliferation, migration, and invasion in breast cancer. Biomed. Pharmacother. **95**, 1636–1643 (2017)
44. Guo, X., et al.: GAS5 inhibits gastric cancer cell proliferation partly by modulating CDK6. Oncol. Res. Treat. **38**(7–8), 362–366 (2015)

A Gaussian Kernel Similarity-Based Linear Optimization Model for Predicting miRNA-lncRNA Interactions

Leon Wong[1,2], Zhu-Hong You[1(✉)], Yu-An Huang[1], Xi Zhou[1], and Mei-Yuan Cao[3]

[1] The Xinjiang Technical Institute of Physics and Chemistry, Chinese Academy of Sciences, Urumqi 830011, China
huangliguang18@mails.ucas.ac.cn, zhuhongyou@ms.xjb.ac.cn
[2] University of Chinese Academy of Sciences, Beijing 100049, China
[3] Guang Dong Polytechnic College, Zhaoqing 526100, Guangdong, China

Abstract. MicroRNAs (miRNAs) and long non-coding RNAs (lncRNAs) are two main functional regulation non-coding RNAs, which involves many important pathological and physiological procedures. Accumulating evidences demonstrated that the interactions between miRNAs and lncRNAs have great impact on modulations of gene expression that are related to many *Human* diseases. However, identification of miRNA-lncRNA interactions via bio-experimental methods suffers from high cost and time consuming. Thus, it is more and more popular for researchers to utilize computational methods in miRNA-lncRNA interactions prediction because of their high-performance. In this study, we propose a gaussian kernel similarity-based linear optimization model for predicting miRNA-lncRNA interactions. Specifically, gaussian kernel similarity method is employed to learn the miRNAs and lncRNAs similarities based on the observed heterogeneous network. Then, an integrated network is constructed by combining the observed heterogeneous network and the constructed similarities. Finally, a linear optimization model is trained to obtain the rating matrix for the unobserved links in the integrated network. To evaluate the performance of our proposed method, *k*-fold cross-validation (CV) and leave-one-out cross-validation (LOOCV) are implemented on the collected dataset. The experimental results show that the proposed model yields high AUCs of 0.8624, 0.9053, 0.9152 and 0.9236 in 2-fold, 5-fold, 10-fold CV and LOOCV, respectively. It is anticipated that our proposed method is promising and reliable to inferring the interactions between miRNAs and lncRNAs for further biological researches.

Keywords: miRNA-lncRNA interaction · Gaussian kernel similarity · Link prediction · Matrix completion

D.-S. Huang and K.-H. Jo (Eds.): ICIC 2020, LNCS 12464, pp. 316–325, 2020.
https://doi.org/10.1007/978-3-030-60802-6_28

1 Introduction

With the rapid development of sequencing techniques, a huge number of gene data has been collected, which is not just involving human species and makes it possible for developing more gene therapy with deep knowledge on the mechanism of life activity. Non-coding RNA (ncRNA), unlike coding RNA, cannot be directly translated into protein. However, more and more studies found that ncRNAs play important roles in regulating gene expression. Especially, microRNA (miRNA), a kind of ~20 nucleotides ncRNA, can be discovered in many essential pathological procedures with respect to tumorigenesis, apoptosis, cell proliferation and development [1, 2]. Long non-coding RNA (lncRNA), a loosely classified class of ncRNA (>200 nucleotides), also has strong regulation functions but only a small fraction of them have been well studied, like XIST, HOTAIR and TERC [3–5]. There are still many efforts should be made for investigation of lncRNAs. Both of miRNAs and lncRNAs have engaged attention and interest of researches [6].

As two main ncRNA, miRNA and lncRNA can interact with each other via various manners. Recently, more and more studies investigate the interaction among different types of ncRNAs [7]. A competing endogenous RNA (ceRNA) hypothesis was proposed to demonstrate how RNAs trigger crosstalk with each other [8]. Specifically, it leads to inhabitation of gene expression when RNAs compete for binding to miRNAs that act as miRNA response elements (MREs) [9]. With respect to lncRNAs, they can be regulated by competing MREs with other RNAs. Some specific lncRNAs' structures are similar to mRNA, and they can be targeted by miRNAs [10]. Moreover, such activities can have great impacts on the pathological processes of disease like vasculature, gastric cancer and other tumors [3, 11–13]. The knowledge of regulation network mediated by miRNAs and lncRNAs can shed light on developing new therapeutic methods [14–16].

Since the interactions between miRNAs and lncRNAs can critically mediate the biological processes, the potential complex mechanisms should be understood well [17, 18]. Due to the evolving bioengineering technology, it can facilitate to collect the experimental data forming the conducive database [19–22]. These cumulative experimental data can be used for further investigation. Especially for computational method, it can offer much beneficial prediction results for researchers to guide their further biological experiments, by which a more ideal outcome may be yielded [23–26]. Based on the limited observed biodata, computational methods at low cost can be the indispensable complement for bio-experiments that always suffer from high cost of time and money [27, 28]. By using compactional method, certain lncRNAs have been discovered but most of them are with underlying functions. By the means of investigating the interactions mediated by lncRNAs and miRNAs, researchers can make progress on the potential biological function annotation of lncRNAs [29–38].

There are many computational methods that are proposed to predict potential interactions between RNAs and targets. Most of them incorporate the basic concepts of RNAs' operating mechanism such as site accessibility, free energy, seed match and conservation [39–43]. Unfortunately, the sophisticated mechanisms probably make these basic concepts invalid especially on human species. Although the existing methods based on general concepts are still widely used for prediction, such as PITA, STarMir and LncTar, their high false-positive rates should be noticed [39, 44–47]. These approaches cannot be faultlessly utilized for predicting miRNA-lncRNA interactions. Therefore, more novel computational approaches for predicting miRNA-lncRNA interactions are urged to be developed.

Currently, the important roles of miRNA-lncRNA interactions attract more and more researchers. Some powerful prediction tools were developed, and evoked strong responses. Huang *et al.* proposed a method EPLMI that was the first technique for inferring miRNA-lncRNA interactions. The model achieved high performance by using a graph-based method based on two-way diffusion, in which two effective matrixes were constructed via weighting the known bipartite graph with similarity matrixes of miRNAs/lncRNAs [48]. In addition, three kinds of bio-similarity matrixes were introduced with respect to expression profile, functional profile and sequence information. Yu *et al.* proposed resource allocation-based algorithm to predict miRNA-lncRNA interactions, combining the sequence information [49]. Hu *et al.* proposed a method LMNLMI that combined with multiple similarity matrixes [50]. Huang *et al.* proposed GBCF that adopted a group-preference Bayesian collative filtering method on the fusion network [51].

In this study, we proposed a novel link prediction model to predict potential miRNA-lncRNA interactions. In detail, effective similarities of miRNAs and lncRNAs are learned by using gaussian kernel similarity method based on the given observed interactions [52]. Combined these similarity matrixes and the known miRNA-lncRNA interactions, an integrated network was constructed. Based on the integrated network, a linear optimization algorithm was employed to do the link prediction task [53]. To evaluate our proposed method, *k*-fold cross validation and leave-one-out cross validation were implemented, in which our proposed method can yield the reliable prediction results. Compared with the existing novel model, our proposed method achieved the best result in 5-fold cross validation. In conclusion, our proposed method is promising for revealing the meaning mechanism of life activity.

2 Materials and Methodology

2.1 Construction of Dataset

In this study, we collected 8091 records of known miRNA-lncRNA interactions from the lncRNASNP database (February 2017 version, http://bioinfo.life.hust.edu.cn/lncRNASNP) [54]. These records are validated by laboratory studies, which can better facilitate the investigation of potential miRNA-lncRNA interactions. After performing data de-duplication on the collection, the final dataset contains 5348 records involving in 275 miRNAs and 780 lncRNAs.

2.2 Construction of Gaussian Kernel Similarity Based on Interaction Profile

In this study, we employ gaussian kernel to construct the efficient similarity as side information based on the interaction profile of miRNAs and lncRNAs, deriving from the assumption that two individuals of miRNAs having resembled profile may interact with more common lncRNAs. The definition of gaussian kernel similarity method is as follow:

$$GSM(r(i), r(j)) = \exp\left(-\gamma_r \|P(r(i)) - P(r(j))\|^2\right) \tag{1}$$

where $r(i)$ indexes i-th RNA of interaction network (i.e. adjacency matrix) by column or by row. $P(r(i))$ denotes the relevant interaction profile of RNA $r(i)$, and it is computed by two-norm of vector $r(i)$. The parameter γ_r is defined as the gaussian kernel bandwidth:

$$\gamma_r = \frac{\gamma_r'}{\left(\sum_{i=1}^{nr} \|P(r(i))\|^2\right)/nr} \tag{2}$$

where new bandwidth γ_r' is set to 1 for simple calculation. Therefore, two gaussian kernel similarity matrixes of GSM_{miRNA} and GSM_{lncRNA} are generated.

2.3 Linear Optimization Method for Link Prediction

In our proposed method, linear optimization method is employed to do the link prediction task on miRNA-lnRNA interaction network.

First, a bilayer network M is constructed by using two gaussian kernel similarity matrixes of miRNAs and lncRNAs and the known interaction network. The definition is as follow:

$$M = \begin{bmatrix} GSM_{miRNA} & MLN \\ MLN^{\top} & GSM_{lncRNA} \end{bmatrix} \tag{3}$$

where MLN is a given miRNA-lncRNA interaction network. Then a rating matrix S can be calculated by using a weighted matrix and the bilayer network M:

$$S = M \cdot C \tag{4}$$

where C is a contribution matrix. Each element s_{ij} can be represented by a linear summation of contributions as follow:

$$s_{ij} = \sum_k m_{ik} c_{kj} \tag{5}$$

Based on self-consistence principle, rating matrix S is highly consistent with matrix M. Therefore, to solve the rating matrix S, C with the small magnitude can be computed as an optimization problem:

$$\min_{C} \alpha\|M - MC\| + \|C\| \tag{6}$$

where α is a regularization parameter. Frobenius-2 norm is applied to solve the minimum optimization problem:

$$\begin{aligned} FN &= \alpha\|M - MC\|^2 + \|C\|^2 \\ &= \alpha \mathrm{Tr}\left[(M - MC)^{\top}(M - MC)\right] + Tr\left(C^{\top}C\right) \\ &= \alpha \mathrm{Tr}\left(M^{\top}M - M^{\top}MC - C^{\top}M^{\top}M + C^{\top}M^{\top}MC\right) + Tr\left(C^{\top}C\right) \end{aligned} \tag{7}$$

To solve the formula, the partial derivative of FN is set to zero with respect to C:

$$\frac{\partial FN}{\partial C} = \alpha\left(2M^{\top}MC - 2M^{\top}M\right) + 2C \tag{8}$$

Then Eq. (8) can be represented as follow:

$$C^{*} = \alpha\left(\alpha M^{\top}M + E\right)^{-1}M^{\top}M \tag{9}$$

where E is an identity matrix. The target rating matrix S can be computed as:

$$S = MC^{*} \tag{10}$$

3 Results and Discussion

3.1 Performance Evaluation of Our Proposed Method

In this section, k-fold cross-validation (CV) is implemented, in which receiver operating characteristic (ROC) curve and the corresponding area under the curve (AUC) can obtained for performance evaluation. If an AUC is higher than 0.5, the model is valid. If an AUC is close to 1, the performance is near perfection. 2-fold CV, 5-fold CV and 10-fold CV are applied to test whether more interaction profile can yield a better result. Each of them is carried out for 20 times so as to reduce the bias caused by the imbalanced learning samples. In addition, leave-one-out CV (LOOCV) is employed.

Further, the average AUCs of 2-fold CV, 5-fold CV and 10-fold CV are 0.8624, 0.9053 and 0.9152 with standard deviation of 0.0031, 0.0017 and 0.0010, respectively. AUC of LOOCV is 0.9236. The best AUCs of k-fold CV are 0.8670, 0.9079 and 0.9170, and the corresponding ROCs are plotted in Fig. 1. These results demonstrated that the model performance can be improved by using more interaction profile.

Meanwhile, the high AUCs yielded by using our proposed method show that our proposed method can reliably detect potential miRNA-lncRNA interactions. By using only interaction profile of miRNA-lncRNA, our proposed method yielded a best AUC at 0.8884 in 5-fold CV, which demonstrated the side information is effective.

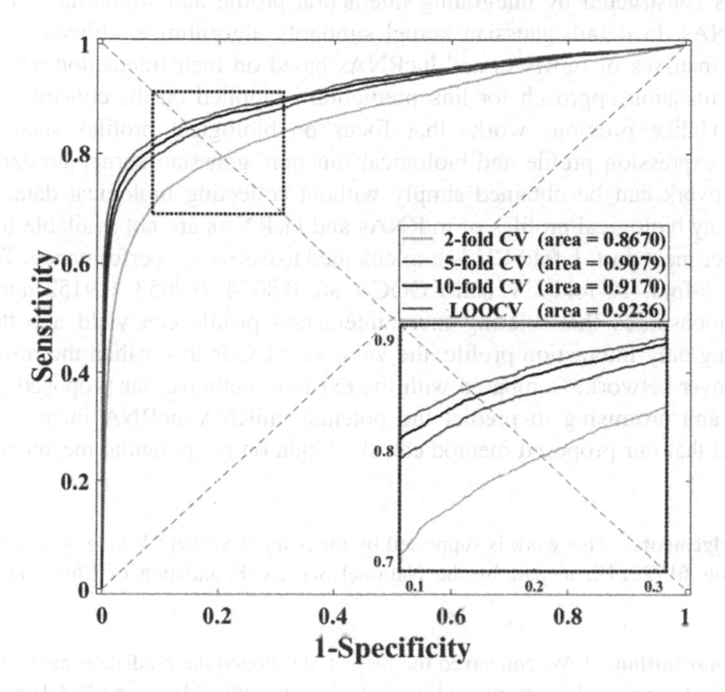

Fig. 1. Performance results of our proposed method by using 2-fold, 5-fold, 10-fold CV and LOOCV

Additionally, there are some novel method developed for miRNA-lncRNA interaction prediction. By using the same constructed dataset, we compared our proposed method with these methods and the 5-fold CV results are listed in Table 1. As a result, our proposed method reaches the highest AUC.

Table 1. Performance comparison among different methods

Method	AUC
EPLMI [55]	0.8447
LMNLMI [50]	0.8926
LNRLMI [56]	0.8960
LCBNI [49]	0.8982
Proposed method	**0.9053**

4 Conclusion

In this study, we propose a novel method for miRNA-lncRNA interaction prediction, in which a matrix factorization algorithm is employ for link prediction and a bilayer network is constructed by integrating interaction profile and similarities of miRNAs and lncRNAs. In detail, gaussian kernel similarity algorithm is utilized to yield the similarity matrixes of miRNAs and lncRNAs based on their interaction network, and linear optimization approach for link prediction is applied on the constructed bilayer network. Unlike previous works that focus on biological profiles such as RNA sequence, expression profile and biological function, gaussian kernel similarity based on the network can be obtained simply without collecting biological data. Unfortunately, many biological profiles of miRNAs and lncRNAs are not available for access. In the experiment part, *k*-fold CV is implemented to assess the performance. The AUCs of 2-fold, 5-fold, 10-fold CV and LOOCV are 0.8624, 0.9053, 0.9152 and 0.9236, which demonstrates that offering more interaction profile can yield a better result. When using only interaction profile, the value of AUC is lower than the results based on the bilayer network. Compared with the existing methods, our proposed method is powerful and promising to predict the potential miRNA-lncRNA interactions. It is anticipated that our proposed method can shed light on the potential mechanism of life activities.

Acknowledgements. This work is supported by the NSFC Excellent Young Scholars Program, under Grants 61722212, in part by the National Science Foundation of China under Grants 61873212.

Author Contribution. L.W. conceived the project, developed the prediction method, designed the experiments, analyzed the result and wrote the manuscript. Z.H.Y. and Y.A.H analyzed the result and revised the manuscript. X.Z and M.Y.C. analyzed the result. All authors read and approved the manuscript.

Conflict of Interest. The authors declare that they have no conflict of interest.

References

1. Bartel, D.P.: MicroRNAs: genomics, biogenesis, mechanism, and function. Cell **116**, 281–297 (2004)
2. Alvarez-Garcia, I., Miska, E.A.: MicroRNA functions in animal development and human disease. Development **132**, 4653–4662 (2005)
3. Baena-Del Valle, J.A., et al.: MYC drives overexpression of telomerase RNA (hTR/TERC) in prostate cancer. J. Pathol. **244**, 11–24 (2018)
4. Gupta, R.A., et al.: Long non-coding RNA HOTAIR reprograms chromatin state to promote cancer metastasis. Nature **464**, 1071–1076 (2010)
5. Clemson, C.M., McNeil, J.A., Willard, H.F., Lawrence, J.B.: XIST RNA paints the inactive X chromosome at interphase: evidence for a novel RNA involved in nuclear/chromosome structure. J. Cell Biol. **132**, 259–275 (1996)

6. Gong, J., Liu, W., Zhang, J., Miao, X., Guo, A.-Y.: lncRNASNP: a database of SNPs in lncRNAs and their potential functions in human and mouse. Nucleic Acids Res. **43**, D181–D186 (2015)
7. Marín, R.M., Vaníček, J.: Efficient use of accessibility in microRNA target prediction. Nucleic Acids Res. **39**, 19–29 (2011)
8. Salmena, L., Poliseno, L., Tay, Y., Kats, L., Pandolfi, P.P.: A ceRNA hypothesis: the Rosetta Stone of a hidden RNA language? Cell **146**, 353–358 (2011)
9. Yan, Y., Zhang, F., Fan, Q., Li, X., Zhou, K.: Breast cancer-specific TRAIL expression mediated by miRNA response elements of let-7 and miR-122. Neoplasma **61**, 672–679 (2014)
10. Tan, J.Y., et al.: Extensive microRNA-mediated crosstalk between lncRNAs and mRNAs in mouse embryonic stem cells. Genome Res. **25**, 655–666 (2015)
11. Xia, T., et al.: Long noncoding RNA associated-competing endogenous RNAs in gastric cancer. Sci. Rep. **4**, 1–7 (2014)
12. Guo, L.-H., Li, H., Wang, F., Yu, J., He, J.-S.: The tumor suppressor roles of miR-433 and miR-127 in gastric cancer. Int. J. Mol. Sci. **14**, 14171–14184 (2013)
13. Ballantyne, M., McDonald, R., Baker, A.: lncRNA/MicroRNA interactions in the vasculature. Clin. Pharmacol. Ther. **99**, 494–501 (2016)
14. Ma, G., Tang, M., Wu, Y., Xu, X., Pan, F., Xu, R.: LncRNAs and miRNAs: potential biomarkers and therapeutic targets for prostate cancer. Am. J. Transl. Res. **8**, 5141 (2016)
15. Beermann, J., Piccoli, M.-T., Viereck, J., Thum, T.: Non-coding RNAs in development and disease: background, mechanisms, and therapeutic approaches. Physiol. Rev. **96**, 1297–1325 (2016)
16. Kumar, M., Goyal, R.: LncRNA as a therapeutic target for angiogenesis. Curr. Top. Med. Chem. **17**, 1750–1757 (2017)
17. Huang, Y.: The novel regulatory role of lnc RNA-mi RNA-mRNA axis in cardiovascular diseases. J. Cell Mol. Med. **22**, 5768–5775 (2018)
18. Wang, L., et al.: LMTRDA: using logistic model tree to predict MiRNA-disease associations by fusing multi-source information of sequences and similarities. PLoS Comput. Biol. **15**, e1006865 (2019)
19. Coordinators, N.R.: Database resources of the national center for biotechnology information. Nucleic Acids Res. **45**, D12 (2017)
20. Li, J.-H., Liu, S., Zhou, H., Qu, L.-H., Yang, J.-H.: starBase v2.0: decoding miRNA-ceRNA, miRNA-ncRNA and protein–RNA interaction networks from large-scale CLIP-Seq data. Nucleic Acids Res. **42**, D92–D97 (2014)
21. Lipscomb, C.E.: Medical subject headings (MeSH). Bull. Med. Libr. Assoc. **88**, 265 (2000)
22. Johnson, M., Zaretskaya, I., Raytselis, Y., Merezhuk, Y., McGinnis, S., Madden, T.L.: NCBI BLAST: a better web interface. Nucleic Acids Res. **36**, W5–W9 (2008)
23. Huang, Y.-A., Chen, X., You, Z.-H., Huang, D.-S., Chan, K.C.: ILNCSIM: improved lncRNA functional similarity calculation model. Oncotarget **7**, 25902 (2016)
24. You, Z.-H., Lei, Y.-K., Gui, J., Huang, D.-S., Zhou, X.: Using manifold embedding for assessing and predicting protein interactions from high-throughput experimental data. Bioinformatics **26**, 2744–2751 (2010)
25. Zheng, K., You, Z.-H., Wang, L., Zhou, Y., Li, L.-P., Li, Z.-W.: DBMDA: a unified embedding for sequence-based miRNA similarity measure with applications to predict and validate miRNA-disease associations. Mol. Ther.-Nucleic Acids **19**, 602–611 (2020)
26. Wang, M.-N., You, Z.-H., Li, L.-P., Wong, L., Chen, Z.-H., Gan, C.-Z.: GNMFLMI: graph regularized nonnegative matrix factorization for predicting LncRNA-MiRNA interactions. IEEE Access **8**, 37578–37588 (2020)

27. You, Z.-H., et al.: PBMDA: a novel and effective path-based computational model for miRNA-disease association prediction. PLoS Comput. Biol. **13**, e1005455 (2017)
28. Wang, L., You, Z.-H., Huang, D.-S., Zhou, F.: Combining high speed ELM learning with a deep convolutional neural network feature encoding for predicting protein-RNA interactions. IEEE/ACM Trans. Comput. Biol. Bioinform. **17**, 972–980 (2018)
29. Yi, H.-C., et al.: ACP-DL: a deep learning long short-term memory model to predict anticancer peptides using high-efficiency feature representation. Mol. Ther.-Nucleic Acids **17**, 1–9 (2019)
30. Zheng, K., You, Z.-H., Wang, L., Zhou, Y., Li, L.-P., Li, Z.-W.: MLMDA: a machine learning approach to predict and validate MicroRNA–disease associations by integrating of heterogenous information sources. J. Transl. Med. **17**, 260 (2019)
31. You, Z., Wang, S., Gui, J., Zhang, S.: A novel hybrid method of gene selection and its application on tumor classification. In: Huang, D.-S., Wunsch, D.C., Levine, D.S., Jo, K.-H. (eds.) ICIC 2008. LNCS (LNAI), vol. 5227, pp. 1055–1068. Springer, Heidelberg (2008). https://doi.org/10.1007/978-3-540-85984-0_127
32. Wang, L., You, Z.-H., Li, L.-P., Zheng, K., Wang, Y.-B.: Predicting circRNA-disease associations using deep generative adversarial network based on multi-source fusion information. In: 2019 IEEE International Conference on Bioinformatics and Biomedicine (BIBM), pp. 145–152. IEEE (2019)
33. Wang, Y., et al.: Predicting protein interactions using a deep learning method-stacked sparse autoencoder combined with a probabilistic classification vector machine. Complexity (2018)
34. Chen, Z.-H., You, Z.-H., Li, L.-P., Wang, Y.-B., Wong, L., Yi, H.-C.: Prediction of self-interacting proteins from protein sequence information based on random projection model and fast Fourier transform. Int. J. Mol. Sci. **20**, 930 (2019)
35. Li, J., et al.: Using weighted extreme learning machine combined with scale-invariant feature transform to predict protein-protein interactions from protein evolutionary information. IEEE/ACM Trans. Comput. Biol. Bioinform. (2020)
36. Zhan, Z.-H., You, Z.-H., Zhou, Y., Li, L.-P., Li, Z.-W.: Efficient framework for predicting ncRNA-protein interactions based on sequence information by deep learning. In: Huang, D.-S., Jo, K.-H., Zhang, X.-L. (eds.) ICIC 2018. LNCS, vol. 10955, pp. 337–344. Springer, Cham (2018). https://doi.org/10.1007/978-3-319-95933-7_41
37. Wang, L., You, Z.-H., Li, Y.-M., Zheng, K., Huang, Y.-A.: GCNCDA: a new method for predicting circRNA-disease associations based on graph convolutional network algorithm. PLoS Comput. Biol. **16**, e1007568 (2020)
38. Wang, M.-N., You, Z.-H., Wang, L., Li, L.-P., Zheng, K.: LDGRNMF: LncRNA-disease associations prediction based on graph regularized non-negative matrix factorization. Neurocomputing (2020)
39. Yue, D., Liu, H., Huang, Y.: Survey of computational algorithms for microRNA target prediction. Curr. Genomics **10**, 478–492 (2009)
40. Guttman, M., Rinn, J.L.: Modular regulatory principles of large non-coding RNAs. Nature **482**, 339–346 (2012)
41. Lewis, B.P., Burge, C.B., Bartel, D.P.: Conserved seed pairing, often flanked by adenosines, indicates that thousands of human genes are microRNA targets. Cell **120**, 15–20 (2005)
42. Betel, D., Koppal, A., Agius, P., Sander, C., Leslie, C.: Comprehensive modeling of microRNA targets predicts functional non-conserved and non-canonical sites. Genome Biol. **11**, R90 (2010)
43. Wang, X.: Composition of seed sequence is a major determinant of microRNA targeting patterns. Bioinformatics **30**, 1377–1383 (2014)
44. Kertesz, M., Iovino, N., Unnerstall, U., Gaul, U., Segal, E.: The role of site accessibility in microRNA target recognition. Nat. Genet. **39**, 1278–1284 (2007)

45. Rennie, W., et al.: STarMir: a web server for prediction of microRNA binding sites. Nucleic Acids Res. **42**, W114–W118 (2014)
46. Li, J., et al.: LncTar: a tool for predicting the RNA targets of long noncoding RNAs. Brief. Bioinform. **16**, 806–812 (2015)
47. Ab Mutalib, N.-S., Sulaiman, S.A., Jamal, R.: Computational tools for microRNA target prediction. In: Computational Epigenetics and Diseases, pp. 79–105. Elsevier (2019)
48. Huang, Y.-A., Chan, K.C., You, Z.-H.: Constructing prediction models from expression profiles for large scale lncRNA–miRNA interaction profiling. Bioinformatics **34**, 812–819 (2018)
49. Yu, Z., Zhu, F., Tianl, G., Wang, H.: LCBNI: link completion bipartite network inference for predicting new lncRNA-miRNA interactions. In: 2018 IEEE International Conference of Safety Produce Informatization (IICSPI), pp. 873–877. IEEE (2018)
50. Hu, P., Huang, Y.-A., Chan, K.C., You, Z.-H.: Learning multimodal networks from heterogeneous data for prediction of lncRNA-miRNA interactions. IEEE/ACM Trans. Comput. Biol. Bioinform. (2019)
51. Huang, Z.-A., Huang, Y.-A., You, Z.-H., Zhu, Z., Sun, Y.: Novel link prediction for large-scale miRNA-lncRNA interaction network in a bipartite graph. BMC Med. Genomics **11**, 17–27 (2018)
52. Huang, Y.-A., You, Z.-H., Chen, X., Huang, Z.-A., Zhang, S., Yan, G.-Y.: Prediction of microbe–disease association from the integration of neighbor and graph with collaborative recommendation model. J. Transl. Med. **15**, 209 (2017)
53. Pech, R., Hao, D., Lee, Y.-L., Yuan, Y., Zhou, T.: Link prediction via linear optimization. Phys. A: Stat. Mech. Appl. **528**, 121319 (2019)
54. Gong, J., Liu, W., Zhang, J., Miao, X., Guo, A.-Y.: lncRNASNP: a database of SNPs in lncRNAs and their potential functions in human and mouse. Nucleic Acids Res. **43**, D181–D186 (2014)
55. Huang, Y.-A., Chan, K.C., You, Z.-H.: Constructing prediction models from expression profiles for large scale lncRNA–miRNA interaction profiling. Bioinformatics **34**, 812–819 (2017)
56. Wong, L., Huang, Y.A., You, Z.H., Chen, Z.H., Cao, M.Y.: LNRLMI: linear neighbour representation for predicting lncRNA-miRNA interactions. J. Cell Mol. Med. **24**, 79–87 (2020)

Identification of Autistic Risk Genes Using Developmental Brain Gene Expression Data

Zhi-An Huang[1](✉), Yu-An Huang[2], Zhu-Hong You[2], Shanwen Zhang[2], Chang-Qing Yu[2], and Wenzhun Huang[2]

[1] Department of Computer Science, City University of Hong Kong, Kowloon 999077, Hong Kong SAR
zahuang2-c@my.cityu.edu.hk

[2] Department of Information Engineering, Xijing University, Xi'an 710000, China

Abstract. Recently, the serious impairments of ASD cause a series of pending issues to increase a major burden of health and finance globally. In this work, we propose an effective convolutional neural network (CNN) - based model to identify the potential autistic risk genes based on the developmental brain gene expression profiles. Based on the 10-fold cross validations, the simulation experiments demonstrate that the proposed model shows supreme classification results as compared to the other state-of-the-art classifiers. In such an imbalanced dataset, the proposed CNN model achieves the F1-score of 63.07 ± 3.9 and the area under ROC curve of 0.6940. In case study, 70% out of the top-10 predicted risk genes have been confirmed to increase the risk of developing ASD via published literatures. The effectiveness enables our model to serve as a candidate tool for accelerating the identification of autistic genetic abnormalities.

Keywords: Autism spectrum disorders (ASD) · Gene prioritization · Autistic biomarkers · Developmental brain gene expression data

1 Background

Autism spectrum disorder (ASD) is a range of common childhood neurodevelopmental disorders including autism, Asperger's syndrome, pervasive developmental disorder-not otherwise specified (PDD-NOS), and other related conditions. ASD individuals typically present with variable deficits of clinical syndromes in restrictive interests, social communication, and repetitive behaviors. It is reported that U.S. Autism Prevalence Rate Soars to 16.8 per 1000 (1 in 59) Children aged eight in 2018, representing an unprecedented spurt of 150% since 2000 [1]. The accepted wisdom in ASD research strongly suggests that individuals with ASD should be diagnosed clinically at the earliest age possible for prompt access to effective treatment [2]. However, no additional supports can be currently used as diagnostic gold standards for ASD such as biochemical markers [3, 4], laboratory tests [5], or neuroimaging analysis [6]. Therefore, the current diagnosis approaches could suffer from the significant limitations in observational strategies.

D.-S. Huang and K.-H. Jo (Eds.): ICIC 2020, LNCS 12464, pp. 326–338, 2020.
https://doi.org/10.1007/978-3-030-60802-6_29

Over the past decades, the key role that genetic variants play in ASD has been implicated in accumulated evidences [7, 8]. In fact, the modern concept of autism defined as a biomedical disorder was primarily coined by its high heritability with several genetic syndromes (e.g. Rett syndrome and Fragile X) [7]. Increasing known genetic mutations causing ASD have been identified. For example, language impairment and delay are extremely common in children with ASD. A gene-expression analysis in developing human brain identified *CNTNAP2* gene as an autism-susceptibility gene to affect their language acquisition [8]. The identified genetic mutations can provide important clues to strengthen our knowledge of molecular pathophysiology and genetic basis with respect to ASD. Nevertheless, each identified genetic mutation/variant merely represents less than 1–2% of the probands. In other words, no single high-risk genetic biomarker has been confirmed for the onset of ASD so far [9].

With increasing high-throughput sequencing data generated [10], it is imperative to develop computer-aided discovery model to identify seminal autistic biomarkers on a large scale [11]. Unlike the routine wet-lab experiments, computational models for system-level biomarker screening is rather efficient and labor-saving to guide the further experimental validation [12]. Generally speaking, statistics-based models and machine learning (ML) – based models are two main classes for the current computational models. As regards to statistics-based models, the expression changes associated with various molecules between different biological states are mostly considered to understand the mechanisms of the alteration that disturb the system's stability [13]. Statistical tests and proxy measures are widely used. For example, eleven microRNA candidates were screened to have significantly high autism gene percentage (AGP) values (p-value < 0.5) via Wilcoxon signed-rank test, where AGP is a novel proxy measure to represent the percentage of autism-associated genes targeted by a single microRNA in order to reveal the specificity of these genes [14]. However, the statistical evidences used as principles have to be collected, concluded, and distilled from the accumulated reports.

By contrast, ML-based models can automatically perform self-evaluation from inputs to outputs based on experience without being explicitly programmed [15, 16]. Most ML-based models are supervised learning algorithms where the instances in both training and testing sets are given the known labels [17, 18]. Given the rules learned from training sets, the instances in testing sets can be assigned the inferred labels [19]. For example, Cogill and Wang utilized the brain developmental gene expression profiling to propose a support vector machine (SVM) approach for prioritization of autistic risk genes and long non-coding RNAs [20]. Following [20], Gök employed feature extraction with Haar wavelet transform as well as discretization methods and classification with Bayes network learning method for the sake of performance improvement [21]. Unfortunately, considering the fact that the negative autistic risk genes are quite deficient, previous works tend to "produce" negative ones in the light of certain hypotheses. For instances, two hundred genes were randomly selected to serve as the negative autism risk genes according to various criteria [22]. [20] and [21] are based on the assumption that if a target gene being relevant to certain human diseases has no associations with ASD or intellectual disability (ID), it could be regarded as negative autism risk genes. These criteria to "produce" non-ASD gene are not sufficiently rigorous. Therefore, a high rate of false-positive results may cause and mislead the self-evaluation by amplifying the systematic bias. In this work, we established stricter criteria to infer qualified non-ASD genes so that the prediction result could be more convincing. The ultimate goal of this work is to accelerate the identification of promising autistic biomarkers for early diagnosis, risk assessment, drug development, and etc. [23–26]. We develop a convolutional neural network (CNN) [27] - based identification model of autistic risk genes by using developmental brain gene expression data.

2 Results and Discussion

2.1 Experiment Setting

To access the classification performance, we implement ten-fold cross validation (10-fold CV) by dividing the whole dataset into 10 equal subsets (folds). Each subset is regards as the test set while the rest is the training set. This process is repeated ten times until all subsets are used to test the model in turns. We conducted 10 times 10-fold CV to reduce the potential bias caused by random sampling. In order to demonstrate the effectiveness of the proposed CNN model, the traditional ML-based classifiers SVM and RF are used for the performance comparison. As a baseline in deep learning (DL), artificial neural network (ANN) is also included. For a fair comparison, the hyperparameters of SVM and RF are optimized by the grid search with CVs. Considering the limited number of iterations (i.e., 30), it is more suitable to find their promising hyperparameter sets based on every combination derived from the provided default values and suggested values as prior knowledge. Unlike the traditional ML-based classifiers, the hyperparameter settings of CNN and ANN model are relatively sophisticated with more tunable arguments. Therefore, we perform the Gaussian processes (GPs)-based Bayesian optimization (BO) algorithm to automatically tune their hyperparameter settings.

Computing confusion matrix is often used to describe the performance of binary classification based on the following four categories: true positives (TP), true negatives (TN), false positives (FP), and false negatives (FN). Accordingly, we evaluate the performance comparison by five common metrics, i.e., accuracy (ACC), specificity (SPE), recall (REC), precision (PRE), and F1-score (F1).

2.2 Performance Comparison

Based on the 10-fold CV, the performance comparison between CNN and other classifiers is tabulated in Table 1. All compared classifiers can show decent results in terms of accuracy and specificity. However, in such an imbalanced dataset, these two metrics could be misleading since the negative samples (i.e., non-ASD genes) occupy the majority of the dataset. It is easier to label the test instances as negative samples.

Therefore, recall and precision seem to be more reasonable to concentrate on the hard classified positive instances. F1-score is also useful to seek a balance between recall and precision. It is shown that the DL-based classifiers tend to outperform the traditional ML-based classifiers. It could be attributed to the fact that the DL classifiers can take advantage of layered structures to capture different levels of feature information by non-linear transformations. Especially on such a high-dimensional biological dataset, the irrelevant, redundant features could overwhelm the training of the traditional ML classifiers. Furthermore, the ML classifiers perform worse in recall, precision, and F1-score. Since there is no balance mechanism to put more focus on the minority class.

The CNN model obtains the supreme overall results to achieve the accuracy of 78.16%, specificity of 83.41%, recall of 56.36%, precision of 48.59%, and F1-score of 63.07%. Thanks to the automatic feature extraction in CNN, the developmental brain gene expression features can be represented as a tensor where local elements are linked to one another. By contrast, the ANN model embodies much more trainable parameters and then requires much more data for training. Consequently, the ANN model suffers from the overfitting problem achieving the inferior performance, as compared to the CNN model.

Table 1. Performance comparison between CNN and other classifiers via 10-fold CV.

Classifier	ACC (%)	SPE (%)	REC (%)	PRE (%)	F1 (%)
CNN	**78.16** ± 2.3	**83.41** ± 3.5	**56.37** ± 3.2	**48.59** ± 2.8	**63.07** ± 3.9
ANN	67.61 ± 3.4	72.88 ± 3.8	43.55 ± 4.1	42.68 ± 4.1	57.21 ± 4.7
RF	63.21 ± 2.3	76.96 ± 1.4	32.46 ± 3.6	29.76 ± 4.3	48.21 ± 3.8
SVM	58.91 ± 2.9	70.89 ± 3.0	23.93 ± 4.3	21.30 ± 3.8	42.41 ± 3.9

In order to further evaluate the performance comparison, the receiver operating characteristic (ROC) curve and the area under ROC curve (AUC) are also used in this work. The ROC curve is plotted based on the true positive rate (TPR, recall) versus false positive rate (FPR, 1-specificity) as varying the thresholds. The AUC value represents a numerical evaluation for model performance. Generally speaking, AUC = 0.5 represents a purely random classification while AUC = 1 represents a perfect classification. As we can see in Fig. 1, the CNN, ANN, RF, and SVM models achieve the AUCs of 0.6940, 0.6521, 0.6018, and 0.5460, respectively. As expected, the CNN model reaches the highest value among the compared classifiers. The CNN model and ANN model tend to manage a trade-off between TPR and FPR, namely achieving a high TPR while maintaining a relatively low FPR.

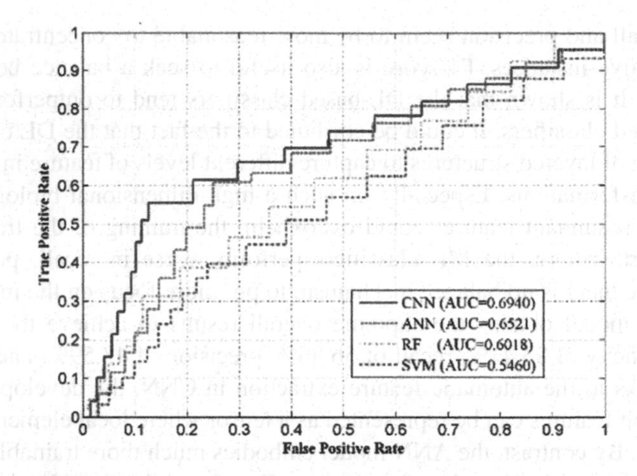

Fig. 1. The performance comparison is evaluated by the ROC curves and AUCs in 10-fold CV.

2.3 Case Study

Aimed at evaluating the classification performance in practice, we attempt to verify the most possible predicted risk genes via published literatures. Based on the assumption that functionally related genes tend to get involved with the pathologically similar human diseases and vice versa [28], the mental disorders-related genes could share the similar abnormal patterns to cause the onset of ASD. It needs to note that the whole dataset for training is not overlapped in the prediction list. As we observe in Table 2, seven out of the top-10 predicted risk genes have been confirmed to increase the risk of developing ASD. For examples, Merner *et al.* found a higher functional expression of *SLC12A5* (1st in the prediction list) variants in ASD cases compared to controls. Anitha *et al.* [29] concluded that the expression of *DNM1L* (2nd in the prediction list) is consistently reduced in autism patients. This case study demonstrates that these predicted risk genes are seminal to act as potential autistic biomarkers for detecting the aberrances in autistic neurological functions.

Table 2. Manual validation for the predicted risk genes via published literatures.

	Gene	Weight	Evidence		Gene	Weight	Evidence
1	*SLC12A5*	0.6303	PMID:26528127	6	*TAF4*	0.5779	DOI: https://doi.org/10.15406/jabb.2017.02.00022
2	*DNM1L*	0.6014	PMID:23116158	7	*NRXN3*	0.5752	PMID:22209245
3	*EN2*	0.5841	PMID:16935268	8	*MAP2*	0.5603	PMID:15541002
4	*SHANK3*	0.5836	PMID:22749736	9	*BRCA2*	0.5599	**Unconfirmed**
5	*DCX*	0.5819	**Unconfirmed**	10	*C12orf65*	0.5557	**Unconfirmed**

3 Conclusions

Although no existing gold standards can provide a definitive validation for ASD so far, leveraging a range of sequencing data and omics data could be a good solution to accelerate the systematic investigation of significant autistic biological autistics [30]. The objective of this work is to develop an effective computer-aided discovery model for identifying the seminal autistic risk gene candidates. As one kind of important biological information, developmental brain gene expression profiles are used to reveal the remarkable features for discriminating between ASD risk genes and non-ASD genes. Based on seven versatile databases, we propose the more rigorous assumptions to screen out 2,831 qualified non-ASD genes from the total 52,376 pathogenic genes. The classification framework can be divided into three steps: 1) Feature selection by Extra-trees; 2) over-sampling by Borderline-SMOTE; 3) CNN-based classification with focal loss (FL) function and automated hyperparameter tuning of BO. The simulation experiments demonstrate the superior classification performance of the proposed model as compared against the ANN, RF, and SVM models. In order to evaluate the practical prediction results, a case study is implemented for manual validation via published literatures. As a result, seven out of the top-10 predicted risk genes have been confirmed to get involved in developing ASD. The current work is expected to provide insights into facilitating the discovery of autistic biomarkers and understanding the pathomechanism of ASD.

4 Materials

The data we used in this work include gene feature sets and label sets. To acquire gene feature sets, we downloaded the developmental brain gene expression data from BrainSpan database [31]. BrainSpan provides the most comprehensive transcriptome of the human developing brain. For each gene instance, there are 524 features to represent the expression values of RNA-sequencing reads in the units of Reads Per Kilobase of transcript per Million mapped reads (RPKM). These features refer to the gene expression levels with different developmental stages on 26 brain structures. We conducted a transformation of $\log_2(RPKM + 1)$ to normalize the data to a narrow range.

For solving a binary classification problem of autistic risk genes, both positive and negative gene instances are requisite. There are several versatile databases compiling such two types of data information. We obtained 754 positive gene instances from the Simons Foundation Autism Research Initiative Gene database (SFARI Gene 3.0) [32], De Rubeis's work [33], and AutismKB [34].

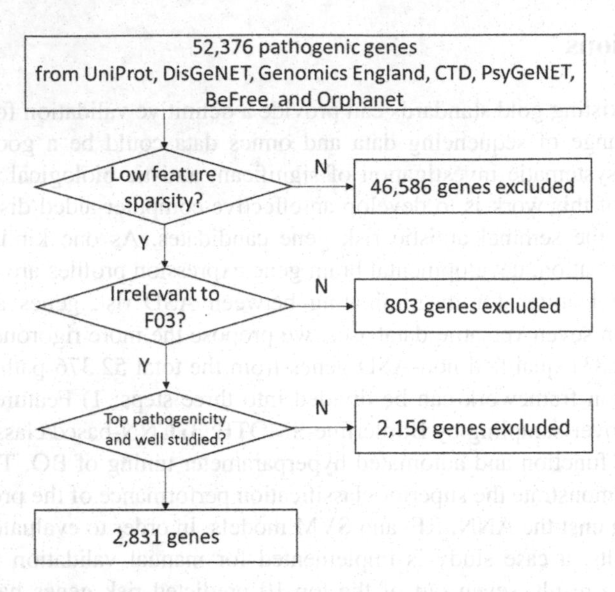

Fig. 2. Data preprocessing of qualified non-ASD genes.

As shown in Fig. 2, to infer qualified negative gene instances, three enhanced considerations are proposed for the assessment based on [20]: 1) The gene instances with high sparsity (more than 50%) in features are excluded; 2) the qualified non-ASD genes should be irrelevant to any diseases in *F03* (i.e., mental disorders) archived in Medical Subject Headings (MeSH) [35] and whilst associated with other human diseases; 3) the qualified non-ASD candidates should be highly pathogenic and well-studied over the long term. The remainders reported to have top 85% associations with 7,186 types of human diseases are considered to serve as non-ASD genes in this work. As shown in Fig. 2, we harvest non-ASD genes from seven versatile databases. Based on the above criteria, 2,831 gene instances are considered as non-ASD genes in this study.

5 Methods

5.1 Model Design

In this section, we describe the details of specific methods and procedures. As shown in Fig. 3, the whole procedure can be divided into three steps. First of all, remarkable feature sets are selected by the Extra-Trees method. Then Borderline-SMOTE is applied for over-sampling on training sets. Finally, focal loss function is compiled in the CNN-based classification model. We utilized the automated hyperparameter tuning of BO to iteratively optimize the model hyperparameters, the best of which is deployed on testing sets.

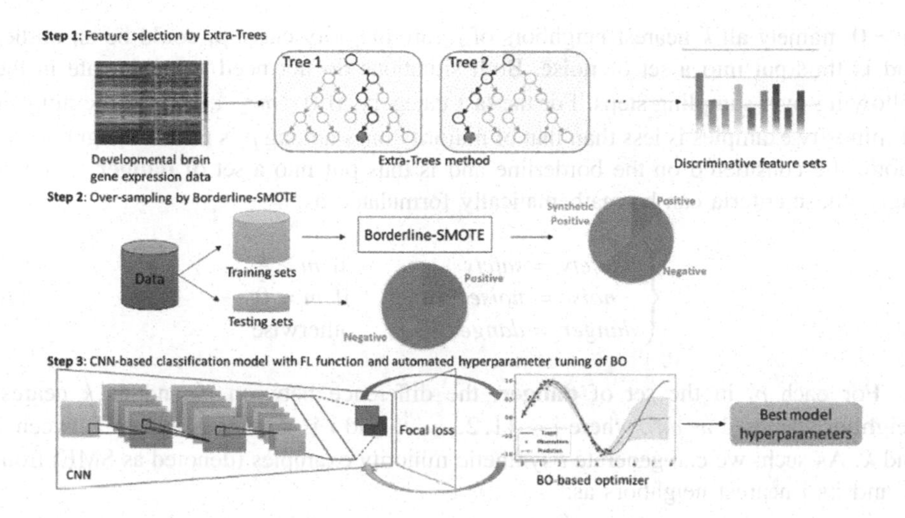

Fig. 3. The flowchart of the proposed model.

5.2 Feature Selection by Extra-Trees

Like RF, the Extra-Trees method follows the same principle of ensemble learning, where a group of weak learners come together to build a strong learner. First, a sample from training sets is randomly chosen with replacement. Then a subset of features are also taken randomly. According to the randomly-generated thresholds randomly drawn for each candidate feature, the best threshold is picked as the best split to split the node iteratively until the tree is grown to the largest. Finally, the above steps are repeated until the predefined number of trees are formed. In this way, we can evaluate the importance of each feature by computing how much each feature contributes to decreasing the weighted Gini impurity.

5.3 Over-Sampling by Borderline-SMOTE

As mentioned above, the dataset is imbalanced with a ratio of 1: 3.75. The positive gene instances are the minority class. In order to achieve the effective learning signal in training process, we perform the Borderline-SMOTE method to over-sample the minority instances for providing more resolution only where it would help. Borderline-SMOTE [36] is a popular extension to SMOTE [37]. The main idea of Borderline-SMOTE is to solve the misclassified examples which lie on the border or edge of the decision boundary. The major procedure of Borderline-SMOTE1 we used can be summarized into three steps.

At first, for each minority example denoted as p_i, we compute its k nearest neighbors from training sets. We denote the number of minority examples among the k nearest neighbors as m. The minority examples are divided into three categories, i.e., safety, danger, and noise based on the following assumption. If $m \geq \frac{k}{2}$, namely the number of minority examples is larger than that of majority ones among p_i's nearest neighbors, p_i is supposed to be easily-classified and is thus put into a set of **safety**. If

$m = 0$, namely all k nearest neighbors of p_i are majority class. p_i could be an outlier and is thus put into a set of **noise**. Both situations do not need to participate in the following over-sampling steps. For the last category, if $0 < m \leq \frac{k}{2}$, namely the number of minority examples is less than that of majority ones among p_i's nearest neighbors, p_i should be considered on the borderline and is thus put into a set of **danger**. Accordingly, these criteria can be mathematically formulated as:

$$\begin{cases} safety = safety \cup p_i, & \text{if } m \geq \frac{k}{2} \\ noise = noise \cup p_i, & \text{if } m = 0 \\ danger = danger \cup p_i, & \text{otherwise} \end{cases} \quad (1)$$

For each p_i' in the set of **danger**, the difference between p_i' and its k nearest neighbors denoted as dif_j, where j = $\{1, 2, \ldots, t\}$ and t is a random integer between 1 and k. As such, we can generate t synthetic minority examples (denoted as SME) from p_i' and its t nearest neighbors as:

$$SME_j = p_i' + r_j \times dif_j \quad (2)$$

where r_j is a random control factor between 0 and 1. Given the number of minority examples in **danger** denoted as n, the above procedure is repeated until each p_i' in **danger** is counted and that $n \times t$ SME are generated. Of course, if $n \times t$ is less than the number of SME we require, more rounds of over-sampling can be invoked as needed.

5.4 CNN-Based Classification Model with FL Function and Automated Hyperparameter Tuning of BO

The synthetic minority examples are incorporated into training sets for the following classification. Inspired by the concept of receptive fields in the visual cortex of animals, CNN has become one of the most efficient neural networks. We propose a relatively small and compact variant of the VGGNet network [38], a well-known CNN architecture, for identification of autistic risk genes. A stack of two 3 × 3 convolutional layers is used, instead of larger size filters (such as 7 × 7 and 5 × 5). To reduce the volume size, a 2 × 2 max-pooling is attached in the following module. At the end of the network, we place the fully connected (FC) layer to give the final probabilities of each label by flattening the output of previous layers. The output activation function is *sigmoid*. For all hidden layers, *ReLU* activation function is utilized followed by the batch normalization. The whole model is trained using Adam optimizer with a learning rate between 0.001 and 0.0001.

A decent loss function plays a key role in the training of a CNN model for properly measuring the resulting error of a run. Since this is a binary classification, binary cross entropy (BCE) loss is widely applied to evaluate how good (or bad) are the estimated probabilities. Given p represents the estimated probability with respect to a positive label, BCE loss can be computed as follows.

$$\mathrm{BCE}(p_t) = -\log(p_t) \tag{3}$$

$$\text{where } p_t = \begin{cases} p & \text{if label is positive} \\ 1-p & \text{otherwise} \end{cases} \tag{4}$$

However, BCE loss tends to degenerate models on class imbalance issues. The majority class tend to appear several times more frequently than the minority class. In other words, the majority class dominates the major loss and that the minority class fails to gain useful learning signal. Therefore, class imbalance issues can overwhelm the BCE loss because of the inefficient training. Here, we replace the BCE loss with a focal loss function (FL) [39] so as to alleviate the class imbalance. FL is developed with a simple extension to BCE loss. The main goal of FL is to put more penalty or focus on those hard, misclassified instances. First, based on Eq. (3), a modulating factor $(1-p_t)^{\omega}$ is integrated to the BCE as:

$$\mathrm{FL}(p_t) = -(1-p_t)^{\omega}\log(p_t) \tag{5}$$

where ω is a tunable decay factor between 0 and 5. Based on the experiment results in [39], ω is set to 2 in this work. It makes sense to smoothly down-weight the loss for the well-classified examples and turn to call for more attention to the misclassified examples. Furthermore, the intuitive idea to alleviate the class imbalance is inverting their class frequency. A weighting factor μ is added to form a μ-balanced variant of FL as:

$$\mathrm{FL}(p_t) = \mu_t(1-p_t)^{\omega}\log(p_t). \tag{6}$$

Here we set $\mu = 0.25$ for majority class and $\mu = 0.75$ for minority class.

Unlike the traditional ML-based classifiers, CNN model has more tunable hyper-parameters for the initial set-up. It is quite challengeable to manually find a near-optimal solution for such sophisticated hyperparameter settings, related to network structure (e.g. filter size, the number of feature maps, and dropout rate) and training algorithm (e.g. momentum, learning rate, and batch size). Therefore, we propose to apply the GPs-based BO algorithm [40] for automatically optimizing hyperparameter settings in CNN model. Since there is no way to take derivatives or have an analytical expression for such an optimization problem, the GPs-BO algorithm has natural advantages to globally query a distribution over functions. We mathematically consider this problem as searching a global maximizer of an unknown objective function f as follows:

$$x^* = argma_{x \in \mathcal{X} \subseteq \mathbb{R}^d} f(x) \tag{7}$$

where $\mathcal{X}$ is the predefined space of interest in global optimization. The GPs-based BO algorithm proposes sampling points in $\mathcal{X}$ and then gains the corresponding observations/responses to define a prior over f, which is utilized for proposing a next sampling point. In this way, the GP posterior is updated iteratively to better approximate f. For building a relatively cheap surrogate model, the GPs-based BO algorithm offers a principled way to gradually reduce the prediction uncertainty by the balanced combinations of exploration and exploitation. Exploration performs random sampling at the locations where the prediction uncertainty is high. Exploitation targets at promising areas where a high objective is expected. To evaluate its efficiency within finite time, the number of iterations is fixed at 30. Given the predefined ranges over hyperparameter settings, four continuous values and four discrete values are included for automated tuning (see Table 3). The validation set is used to obtain the best hyperparameters and that the testing set is tested on the CNN model equipped with the best hyperparameters for the final results.

Table 3. Hyperparameter settings for automated tuning.

For network structure		For training algorithm	
# of feature maps in the 1st convolutional layer	[64, 128]	Batch size	[8, 32]
# of feature maps in the 2nd convolutional layer	[128, 256]	Learning rate	[10^{-3}, 10^{-4}]
# of units in FC layer	[64, 256]	Decay rate	[0.001, 0.02]
Dropout rate	[0.1, 0.3]	Momentum rate	[0.1, 0.99]

References

1. Baio, J., et al.: Prevalence of autism spectrum disorder among children aged 8 years—autism and developmental disabilities monitoring network, 11 sites, United States, 2014. MMWR Surveill. Summar. **67**(6), 1 (2018)
2. Dawson, G., et al.: Early behavioral intervention is associated with normalized brain activity in young children with autism. J. Am. Acad. Child Adolescent Psychiatry **51**(11), 1150–1159 (2012)
3. Huang, Y.-A., Chan, K.C., You, Z.-H., Hu, P., Wang, L., Huang, Z.-A.: Predicting microRNA–disease associations from lncRNA–microRNA interactions via Multiview Multitask Learning. Brief. Bioinf. (2020). https://doi.org/10.1093/bib/bbaa133
4. You, Z.-H., Huang, W.-Z., Zhang, S., Huang, Y.-A., Yu, C.-Q., Li, L.-P.: An efficient ensemble learning approach for predicting protein-protein interactions by integrating protein primary sequence and evolutionary information. IEEE/ACM Trans. Comput. Biol. Bioinf. **16**(3), 809–817 (2018)
5. Grayson, P.C., et al.: Value of commonly measured laboratory tests as biomarkers of disease activity and predictors of relapse in eosinophilic granulomatosis with polyangiitis. Rheumatology **54**(8), 1351–1359 (2015)
6. Huang, Z.-A., Zhu, Z., Yau, C.H., Tan, K.C.: Identifying autism spectrum disorder from resting-state fMRI using deep belief network. IEEE Trans. Neural Netw. Learn. Syst. (2020). https://doi.org/10.1109/TNNLS.2020.3007943
7. Geschwind, D.H.: Advances in autism. Ann. Rev. Med. **60**, 367–380 (2009)
8. Alarcón, M., et al.: Linkage, association, and gene-expression analyses identify CNTNAP2 as an autism-susceptibility gene. Am. J. Hum. Genet. **82**(1), 150–159 (2008)

9. Jeste, S.S., Geschwind, D.H.: Disentangling the heterogeneity of autism spectrum disorder through genetic findings. Nat. Rev. Neurol. **10**(2), 74 (2014)
10. Huang, Z.-A., Wen, Z., Deng, Q., Chu, Y., Sun, Y., Zhu, Z.: LW-FQZip 2: a parallelized reference-based compression of FASTQ files. BMC Bioinf. **18**(1), 179 (2017)
11. Wang, T., Li, L., Huang, Y.-A., Zhang, H., Ma, Y., Zhou, X.: Prediction of protein-protein interactions from amino acid sequences based on continuous and discrete wavelet transform features. Molecules **23**(4), 823 (2018)
12. Huang, Z.-A., Huang, Y.-A., You, Z.-H., Zhu, Z., Sun, Y.: Novel link prediction for large-scale miRNA-lncRNA interaction network in a bipartite graph. BMC Med. Genom. **11**(6), 17–27 (2018)
13. Huang, Y.-A., Chan, K.C., You, Z.-H.: Constructing prediction models from expression profiles for large scale lncRNA–miRNA interaction profiling. Bioinformatics **34**(5), 812–819 (2018)
14. Shen, L., Lin, Y., Sun, Z., Yuan, X., Chen, L., Shen, B.: Knowledge-guided bioinformatics model for identifying autism spectrum disorder diagnostic MicroRNA biomarkers. Sci. Reports. **6**, 39663 (2016)
15. Hu, P., Huang, Y.-A., Chan, K.C., You, Z.-H.: Learning multimodal networks from heterogeneous data for prediction of lncRNA-miRNA interactions. IEEE/ACM Trans. Comput. Biol. Bioinf. (2019). https://doi.org/10.1109/TCBB.2019.2957094
16. Huang, Z.-A., et al.: PBHMDA: path-based human microbe-disease association prediction. Front. Microbiol. **8**, 233 (2017)
17. Wang, L., You, Z.-H., Li, Y.-M., Zheng, K., Huang, Y.-A.: GCNCDA: a new method for predicting circRNA-disease associations based on graph convolutional network algorithm. PLoS Comput. Biol. **16**(5), e1007568 (2020)
18. Huang, Z.-A., et al.: Predicting lncRNA-miRNA interaction via graph convolution auto-encoder. Front. Genet. **10**, 758 (2019)
19. You, Z.-H., et al.: PBMDA: A novel and effective path-based computational model for miRNA-disease association prediction. PLoS Comput. Biol. **13**(3), e1005455 (2017)
20. Cogill, S., Wang, L.: Support vector machine model of developmental brain gene expression data for prioritization of Autism risk gene candidates. Bioinformatics **32**(23), 3611–3618 (2016)
21. Gök, M.: A novel machine learning model to predict autism spectrum disorders risk gene. Neural Comput. Appl. **31**(10), 6711–6717 (2018). https://doi.org/10.1007/s00521-018-3502-5
22. Kou, Y., Betancur, C., Xu, H., Buxbaum, J.D., Ma'Ayan, A.: Network-and attribute-based classifiers can prioritize genes and pathways for autism spectrum disorders and intellectual disability. Am. J. Med. Genet. Part C: Seminar. Med. Genet. **160**, 130–142 (2012)
23. Huang, Y.-A., Hu, P., Chan, K.C., You, Z.-H.: Graph convolution for predicting associations between miRNA and drug resistance. Bioinformatics **36**(3), 851–858 (2020)
24. Huang, Y.-A., et al.: EPMDA: an expression-profile based computational model for microRNA-disease association prediction. Oncotarget **8**(50), 87033 (2017)
25. Jiang, H.-J., Huang, Y.-A., You, Z.-H.: Predicting drug-disease associations via using gaussian interaction profile and kernel-based autoencoder. BioMed Res. Int. (2019). https://doi.org/10.1155/2019/2426958
26. Jiang, H.-J., Huang, Y.-A., You, Z.-H.: SAEROF: an ensemble approach for large-scale drug-disease association prediction by incorporating rotation forest and sparse autoencoder deep neural network. Sci. Rep. **10**(1), 1–11 (2020)
27. Jiang, H.-J., You, Z.-H., Huang, Y.-A.: Predicting drug-disease associations via sigmoid kernel-based convolutional neural networks. J. Transl. Med. **17**(1), 382 (2019)

28. Sun, Y., Zhu, Z., You, Z.-H., Zeng, Z., Huang, Z.-A., Huang, Y.-A.: FMSM: a novel computational model for predicting potential miRNA biomarkers for various human diseases. BMC Syst. Biol. **12**(9), 121 (2018)
29. Anitha, A., et al.: Brain region-specific altered expression and association of mitochondria-related genes in autism. Mol. Autism **3**(1), 12 (2012)
30. Huang, Y.-A., You, Z.-H., Chen, X., Huang, Z.-A., Zhang, S., Yan, G.-Y.: Prediction of microbe–disease association from the integration of neighbor and graph with collaborative recommendation model. J. Transl. Med. **15**(1), 1–11 (2017)
31. Hawrylycz, M.J., et al.: An anatomically comprehensive atlas of the adult human brain transcriptome. Nature **489**(7416), 391 (2012)
32. Abrahams, B.S., et al.: SFARI Gene 2.0: a community-driven knowledgebase for the autism spectrum disorders (ASDs). Mol. Autism **4**(1), 36 (2013)
33. De Rubeis, S., et al.: Synaptic, transcriptional and chromatin genes disrupted in autism. Nature **515**(7526), 209–215 (2014)
34. Xu, L.-M., Li, J.-R., Huang, Y., Zhao, M., Tang, X., Wei, L.: AutismKB: an evidence-based knowledgebase of autism genetics. Nucleic Acids Res. **40**(D1), D1016–D1022 (2012)
35. Lipscomb, C.E.: Medical subject headings (MeSH). Bull. Med. Libr. Assoc. **88**(3), 265–266 (2000)
36. Han, H., Wang, W.-Y., Mao, B.-H.: Borderline-SMOTE: a new over-sampling method in imbalanced data sets learning. In: Huang, D.-S., Zhang, X.-P., Huang, G.-B. (eds.) ICIC 2005. LNCS, vol. 3644, pp. 878–887. Springer, Heidelberg (2005). https://doi.org/10.1007/11538059_91
37. Chawla, N.V., Bowyer, K.W., Hall, L.O., Kegelmeyer, W.P.: SMOTE: synthetic minority over-sampling technique. J. Artif. Intell. Res. **16**, 321–357 (2002)
38. Simonyan, K., Zisserman, A.: Very deep convolutional networks for large-scale image recognition. arXiv preprint arXiv:14091556 (2014)
39. Lin, T.-Y., Goyal, P., Girshick, R., He, K., Dollár, P.: Focal loss for dense object detection. In: Proceedings of the IEEE international conference on computer vision, pp. 2980–2988 (2017)
40. Brochu, E., Cora, V.M., De Freitas, N.: A tutorial on Bayesian optimization of expensive cost functions, with application to active user modeling and hierarchical reinforcement learning. arXiv preprint arXiv:10122599 (2010)

A Unified Deep Biological Sequence Representation Learning with Pretrained Encoder-Decoder Model

Hai-Cheng Yi[1,2], Zhu-Hong You[1,2(✉)], Xiao-Rui Su[1,2], De-Shuang Huang[3], and Zhen-Hao Guo[1,2]

[1] Xinjiang Technical Institute of Physics and Chemistry, Chinese Academy of Sciences, Urumqi 830011, China
zhuhongyou@ms.xjb.ac.cn
[2] University of Chinese Academy of Sciences, Beijing 100049, China
[3] Institute of Machine Learning and Systems Biology, School of Electronics and Information Engineering, Tongji University, Shanghai 201804, China

Abstract. Machine learning methods are increasingly being applied to model and predict biomolecular interactions, while efficient feature representation plays a vital role. To this end, a unified biological sequence deep representation learning framework BioSeq2vec is proposed to extract discriminative features of any type of biological sequence. For arbitrary-length sequence input, the BioSeq2vec produces fixed-length efficient feature representation, which can be applied to various learning models. The performance of BioSeq2vec is evaluated on lncRNA-protein interaction prediction tasks. Experimental results reveal the superior performance of BioSeq2vec in biological sequence feature representation and broad prospects in various genome informatics and computational biology studies.

Keywords: Representation learning · Deep learning · Sequence analysis · Pre-trained model · Seq2Seq

1 Introduction

Biological sequences are the "source code" of life, and the analysis of sequence data provides insights into how genes and proteins work [1]. Due to advances in next-generation sequencing, RNA sequencing, single-cell sequencing, microarrays, and related omics technologies, a vast amount of biological sequences have rapidly accumulated. These rapidly generated sequence data contain a wealth of information but put a great deal of pressure on traditional statistics analytical methods [2–10].

To address this issue, more and more biomedical researches have moved towards applying machine learning methods [11–25]. Among them, deep learning methods have achieved impressive success in many areas where traditional modelling methods seem to have reached their limits [26–37]. The key reason why deep learning can achieve such great success is that deep learning algorithms can automatically learn high-level hidden features from raw data [38–40]. Moreover, pre-training is a commonly used paradigm for deep learning, which first succeeds in the field of image

D.-S. Huang and K.-H. Jo (Eds.): ICIC 2020, LNCS 12464, pp. 339–347, 2020.
https://doi.org/10.1007/978-3-030-60802-6_30

processing. For example, pre-train the convolutional neural network (CNN) model on ImageNet [41], and then fine-tune it for a specific task or data set. Recently, many pre-trained language models also show significant advantages in many tasks in natural language understanding field, including Transformer [42], BERT [43] and so far.

In this study, we propose a deep learning approach for broad-spectrum biological sequence representation using long short-term memory (LSTM) Encoder-Decoder model, named BioSeq2vec. For arbitrary-length nucleic acids or amino acids sequence input, BioSeq2vec outputs an effective fixed-length feature vector, which can be applied to a wide range of downstream machine learning methods in computational biology field. To comprehensive verify the effectiveness of BioSeq2vec, we compared it with widely used sequence feature extraction methods. Then, the performance of BioSeq2vec was verified on lncRNA-protein interaction prediction tasks. The experimental results prove that BioSeq2vec can learn efficient features from biological sequence and can be applied to various bioinformatics and biological studies.

2 Materials and Methods

2.1 BioSeq2vec

The main idea of the Encoder-Decoder structure was first elaborated by Cho *et al.* and Sutskever *et al.* independently [44, 45], also known as sequence-to-sequence (seq2seq), has made a lot of progress in natural language processing field such as machine translation, text summarization, chatbots, etc. Due to the ability to handle variable length sequences, we design BioSeq2vec with Encoder-Decoder structure to learn efficient representation of amino acids and nucleic acids biological sequences.

The input of the model is a sequence of amino acids or nucleic acids. An embedding layer converts these biological sequences into initial encodings with hash as input. First, the sequence is divided into $L - k + 1$ sequence fragment by sliding from left to right with a k window ($k \in [1, L]$). When k is equal to 1, it can be regarded as char-level tokenization (amino acid or nucleic acid). In other cases, it can be seen as word-level encoding. Then, the Encoder map the input into low-dimensional hidden semantic feature vector. The Decoder reproduce a target sequence from this representation. The final intermediate hidden state is produced as the efficient deep representation of biological sequence.

2.2 Model Training

The proposed Encoder-Decoder model was trained by using backpropagation. The Root Mean Square Prop (RMSProp) algorithm is used as an optimization algorithm. The convergence speed of RMSProp is much faster than the well-adjusted stochastic gradient descent (SGD) method with momentum [46, 47], a schematic as shown in Fig. 1.

We considered two training strategies of BioSeq2vec including pre-training with biological sequence corpus and Plug-and-Play without pre-training, which enables more flexible usages of the proposed method. As a unified method, BioSeq2vec can be applied to different types of biological sequences such as miRNA, lncRNA, circRNA, protein and peptide.

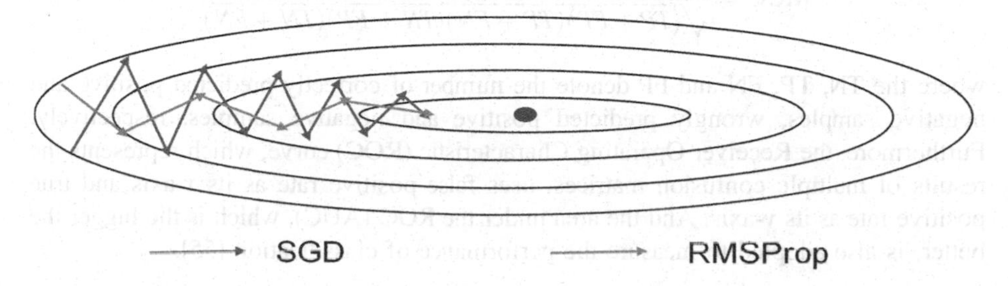

Fig. 1. Schematic diagram of gradient descent comparison between SGD with momentum and RMSProp optimizers.

2.3 Datasets

In order to evaluate the effectiveness of the proposed representation learning method, we trained BioSeq2vec using *human* lncRNA sequence and protein sequence corpus from GENCODE release 33 [48]. Then we verified the performance of the BioSeq2vec feature on lncRNA protein interaction prediction task using RPI488 dataset [49]. The RPI488 data set is a non-redundant data set proposed by Pan *et al.* based on the PDB database [50] and lncRNA-protein structure complexes (least atom distance cutoff <5Å). It contains 488 lncRNA-protein pairs, of which 243 interactive pairs and 245 non-interactive pairs.

2.4 Performance Evaluation Metrics

In the experiments to verify the proposed method, we followed the five-fold cross-validation specification [51–54]. To evaluate our method more fairly, a range of performance evaluation measures were computed including accuracy (Acc.), sensitively (Sens.), specifically (Spec.), precision (Prec.) and Matthews correlation coefficient (MCC), which can be defined as:

$$\text{Acc.} = \frac{TN + TP}{TN + TP + FN + FP} \tag{1}$$

$$\text{Sens.} = \frac{TP}{TP + FN} \tag{2}$$

$$\text{Spec.} = \frac{TN}{TN + FP} \quad (3)$$

$$\text{Prec.} = \frac{TP}{TP + FP} \quad (4)$$

$$\text{MCC} = \frac{TP \times TN - FP \times FN}{\sqrt{(TP + FP)(TP + FN)(TN + FP)(TN + FN)}} \quad (5)$$

where the TN, TP, FN and FP denote the number of correctly predicted positive and negative samples, wrongly predicted positive and negative samples, respectively. Furthermore, the Receiver Operating Characteristic (ROC) curve, which represents the results of multiple confusion matrices, uses false positive rate as its *x*-axis and true positive rate as its *y*-axis. And the area under the ROC (AUC), which is the bigger the better, is also adopted to measure the performance of classification [55].

3 Results and Discussion

3.1 Performance Evaluation on lncRNA-Protein Interaction Prediction Task

In this section, we compared the proposed sequence feature representation method with the *k*-mer feature on lncRNA-protein interaction prediction task under five-fold cross-validation. Three different machine learning classifiers including support vector machine (SVM) [56, 57], Adaptive Boosting (AdaBoost) [58] and Random Forest (RF) [59] are executed to evaluate the performance. The *k*-mer is widely used in biological sequence feature extraction and has achieved good performance in many tasks [14, 19, 21]. The performance comparison of *k*-mer and BioSeq2vec features is performed under five-fold cross-validation. For *k*-mer feature, RNA sequence is represented using 4-mer feature, and protein sequence is represented using conjoint triad feature (3-mer) with reduced amino acid alphabet [60–62].

Table 1. Comparison of *k*-mer and the proposed BioSeq2vec features.

Feature	Classifier	Acc. (%)	Sens. (%)	Spec. (%)	Prec. (%)	MCC (%)	AUC
k-mer	AdaBoost	82.71	85.25	78.84	86.53	65.93	0.877
	RF	87.04	89.28	84.22	89.80	74.28	0.904
	SVM	86.62	91.68	80.48	92.65	73.85	0.896
BioSeq2vec	AdaBoost	83.12	86.98	78.00	88.16	66.80	0.888
	RF	88.06	90.05	85.47	90.61	76.26	0.912
	SVM	89.50	94.52	83.81	95.10	79.55	0.911

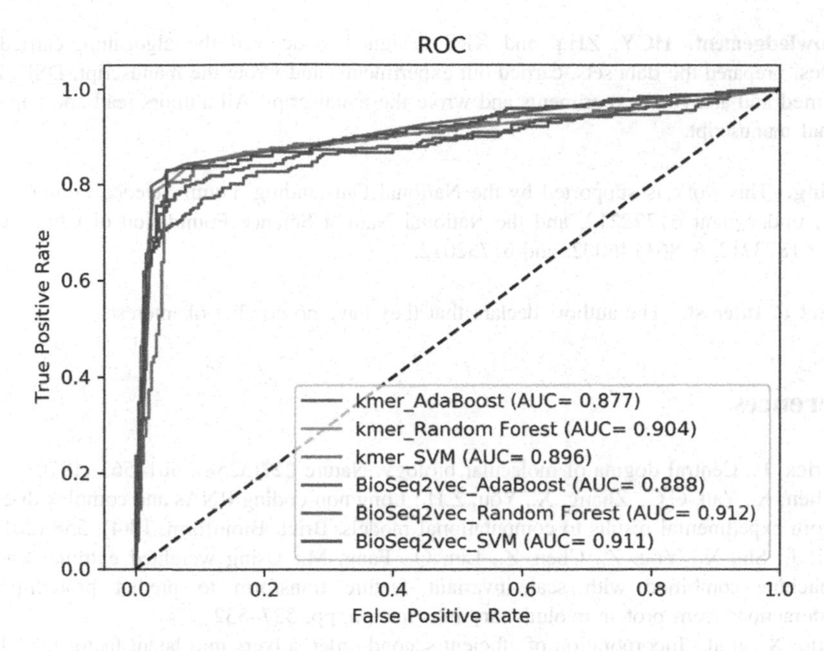

Fig. 2. Comparison of proposed BioSeq2vec and widely used k-mer feature representation methods.

As shown in Table 1 and Fig. 2, BioSeq2vec obtains better performance than widely used *k*-mer feature on all three different classifiers. SVM achieves the best performance with a high accuracy of 89.50%, and improved the accuracy by 3%, the sensitivity by 3%, the specifically by 3.5%, the precision by 2.5%, the MCC about 5% than using k-mer feature. RF and AdaBoost also have better performance when using BioSeq2vec feature. Considering that the effectiveness of *k*-mer has been verified in numerous biological sequence analysis tasks, the experimental results reveal the prospect of the proposed BioSeq2vec method in the representation of biological sequence features.

4 Conclusion

In this work, we propose a deep Encoder-Decoder model to learn unified representation of biological sequences. It combines the advantages of pre-training and deep learning, which can use rich biological sequences to train model with efficient feature representation capabilities. As a unified representation learning method, it can handle any type of biological sequence. For any length sequence input, the BioSeq2vec outputs a fixed-length efficient feature representation, which can be applied to various downstream computational biology or bioinformatics tasks.

Acknowledgement. HCY, ZHY and XRS designed, conceived the algorithm, carried out analyses, prepared the data sets, carried out experiments, and wrote the manuscript; DSH, ZHG performed and analyzed experiments and wrote the manuscript; All authors read and approved the final manuscript.

Funding. This work is supported by the National Outstanding Youth Science Foundation of NSFC, under grant 61722212, and the National Natural Science Foundation of China, under grants 61873212, 61861146002, and 61732012.

Conflict of Interest. The authors declare that they have no conflict of interest.

References

1. Crick, F.: Central dogma of molecular biology. Nature **227**(5258), 561–563 (1970)
2. Chen, X., Yan, C.C., Zhang, X., You, Z.H.: Long non-coding RNAs and complex diseases: from experimental results to computational models. Brief. Bioinform. **18**(4), 558 (2016)
3. Li, J., Shi, X., You, Z., Chen, Z., Lin, Q., Fang, M.: Using weighted extreme learning machine combined with scale-invariant feature transform to predict protein-protein interactions from protein evolutionary information, pp. 527–532
4. Luo, X., et al.: Incorporation of efficient second-order solvers into latent factor models for accurate prediction of missing QoS data. IEEE Trans. Cybern. **48**(4), 1216–1228 (2018)
5. Wang, L., et al.: An improved efficient rotation forest algorithm to predict the interactions among proteins. Soft. Comput. **22**(10), 3373–3381 (2017). https://doi.org/10.1007/s00500-017-2582-y
6. Wang, Y., et al.: Predicting protein interactions using a deep learning method-stacked sparse autoencoder combined with a probabilistic classification vector machine. Complexity **2018** (2018)
7. Li, S., Zhou, M., Luo, X., You, Z.-H.: Distributed winner-take-all in dynamic networks. IEEE Trans. Autom. Control **62**(2), 577–589 (2017)
8. Huang, Y.-A., Chan, K.C., You, Z.-H.: Constructing prediction models from expression profiles for large scale lncRNA–miRNA interaction profiling. Bioinformatics **34**(5), 812–819 (2017)
9. Hu, L., Hu, P., Yuan, X., Luo, X., You, Z.: Incorporating the coevolving information of substrates in predicting HIV-1 protease cleavage sites. IEEE/ACM Trans. Comput. Biol. Bioinf 1 (2019)
10. Li, J.-Q., You, Z.-H., Li, X., Ming, Z., Chen, X.: PSPEL: in silico prediction of self-interacting proteins from amino acids sequences using ensemble learning. IEEE/ACM Trans. Comput. Biol. Bioinf. (TCBB) **14**(5), 1165–1172 (2017)
11. Yi, H.-C., You, Z.-H., Huang, D.-S., Guo, Z.-H., Chan, K.C.C., Li, Y.: Learning representations to predict intermolecular interactions on large-scale heterogeneous molecular association network. iScience **23**(7), 101261 (2020)
12. Chen, X., Xie, D., Zhao, Q., You, Z.H.: MicroRNAs and complex diseases: from experimental results to computational models. Brief. Bioinf. **20**, 515–539 (2017)
13. Luo, X., Zhou, M., Li, S., You, Z., Xia, Y., Zhu, Q.: A nonnegative latent factor model for large-scale sparse matrices in recommender systems via alternating direction method. IEEE Trans. Neural Netw. Learn. Syst. **27**(3), 579–592 (2016)

14. You, Z.-H., Yin, Z., Han, K., Huang, D.-S., Zhou, X.: A semi-supervised learning approach to predict synthetic genetic interactions by combining functional and topological properties of functional gene network. BMC Bioinf. **11**(1), 343 (2010)
15. Chen, X., Huang, Y.-A., You, Z.-H., Yan, G.-Y., Wang, X.-S.: A novel approach based on KATZ measure to predict associations of human microbiota with non-infectious diseases. Bioinformatics **33**(5), 733–739 (2016)
16. Huang, Y.-A., You, Z.-H., Chen, X., Chan, K., Luo, X.: Sequence-based prediction of protein-protein interactions using weighted sparse representation model combined with global encoding. BMC Bioinf. **17**(1), 184 (2016)
17. Wang, L., et al.: Advancing the prediction accuracy of protein-protein interactions by utilizing evolutionary information from position-specific scoring matrix and ensemble classifier. J. Theor. Biol. **418**, 105–110 (2017)
18. You, Z.-H., Huang, W., Zhang, S., Huang, Y.-A., Yu, C.-Q., Li, L.-P.: An efficient ensemble learning approach for predicting protein-protein interactions by integrating protein primary sequence and evolutionary information. IEEE/ACM Trans. Computat. Biol. Bioinf. **16**, 809–817 (2018)
19. You, Z.H., Zhu, L., Zheng, C.H., Yu, H.J., Deng, S.P., Ji, Z.: Prediction of protein-protein interactions from amino acid sequences using a novel multi-scale continuous and discontinuous feature set. BMC Bioinf. **15**(S15), S9 (2014)
20. You, Z.-H., Chan, K.C., Hu, P.: Predicting protein-protein interactions from primary protein sequences using a novel multi-scale local feature representation scheme and the random forest. PLoS Onc **10**(5), e0125811 (2015)
21. You, Z.-H., Lei, Y.-K., Gui, J., Huang, D.-S., Zhou, X.: Using manifold embedding for assessing and predicting protein interactions from high-throughput experimental data. Bioinformatics **26**(21), 2744–2751 (2010)
22. You, Z.-H., Lei, Y.-K., Zhu, L., Xia, J., Wang, B.: Prediction of protein-protein interactions from amino acid sequences with ensemble extreme learning machines and principal component analysis. BMC Bioinf. **14**(Suppl 8), S10 (2013)
23. You, Z.-H., et al.: PRMDA: personalized recommendation-based MiRNA-disease association prediction. Oncotarget **8**(49), 85568 (2017)
24. You, Z.-H., Yu, J.-Z., Zhu, L., Li, S., Wen, Z.-K.: A MapReduce based parallel SVM for large-scale predicting protein-protein interactions. Neurocomputing **145**, 37–43 (2014)
25. An, J.-Y., You, Z.-H., Zhou, Y., Wang, D.-F.: Sequence-based prediction of protein-protein interactions using gray wolf optimizer-based relevance vector machine. Evol. Bioinf. **15**, 1176934319844522 (2019)
26. Yi, H.-C., et al.: ACP-DL: a deep learning long short-term memory model to predict anticancer peptides using high-efficiency feature representation. Mol. Ther. - Nucleic Acids **17**, 1–9 (2019)
27. Yi, H.-C., et al.: Learning distributed representations of RNA and protein sequences and its application for predicting lncRNA-protein interactions. Computat. Struct. Biotechnol. J. (2019)
28. Min, S., Lee, B., Yoon, S.: Deep learning in bioinformatics. Brief. Bioinform. **18**(5), 851–869 (2016)
29. You, Z.-H., et al.: PBMDA: A novel and effective path-based computational model for miRNA-disease association prediction. PLoS Comput. Biol. **13**(3), e1005455 (2017)
30. You, Z.-H., Zhou, M., Luo, X., Li, S.: Highly efficient framework for predicting interactions between proteins. IEEE Trans. Cybern. **47**(3), 731–743 (2017)
31. Guo, Z.-H., You, Z.-H., Wang, Y.-B., Huang, D.-S., Yi, H.-C., Chen, Z.-H.: Bioentity2vec: attribute- and behavior-driven representation for predicting multi-type relationships between bioentities. GigaScience **9**(6), giaa032 (2020)

32. Guo, Z.-H., et al.: MeSHHeading2vec: a new method for representing MeSH headings as vectors based on graph embedding algorithm. Brief. Bioinf. (2020)
33. Guo, Z.-H., You, Z.-H., Huang, D.-S., Yi, H.-C., Chen, Z.-H., Wang, Y.-B.: A learning based framework for diverse biomolecule relationship prediction in molecular association network. Commun. Biol. **3**(1), 118 (2020)
34. Huang, Y.-A., Hu, P., Chan, K.C.C., You, Z.-H.: Graph convolution for predicting associations between miRNA and drug resistance. Bioinformatics **36**(3), 851–858 (2019)
35. Wong, L., Huang, Y.-A., You, Z.-H., Chen, Z.-H., Cao, M.-Y.: LNRLMI: linear neighbour representation for predicting lncRNA-miRNA interactions. J. Cell Mol. Med. **24**(1), 79–87 (2020)
36. Wang, L., You, Z.-H., Huang, Y.-A., Huang, D.-S., Chan, K.C.C.: An efficient approach based on multi-sources information to predict circRNA-disease associations using deep convolutional neural network. Bioinformatics **36**(13), 4038–4046 (2019)
37. Wang, Y., You, Z.-H., Yang, S., Li, X., Jiang, T.-H., Zhou, X.: A high efficient biological language model for predicting protein-protein interactions. Cells **8**(2), 122 (2019)
38. Guo, Z.-H., You, Z.-H., Yi, H.-C.: Integrative construction and analysis of molecular association network in human cells by fusing node attribute and behavior information. Mol. Ther.-Nucleic Acids **19**, 498–506 (2019)
39. Peng, W., Chan, K.C.C., You, Z.: Large-scale prediction of drug-target interactions from deep representations, pp. 1236–1243 (2016)
40. Hu, P., Huang, Y., Chan, K.C.C., You, Z.: Learning multimodal networks from heterogeneous data for prediction of lncRNA-miRNA interactions. IEEE/ACM Trans. Comput. Biol. Bioinf. 1 (2019)
41. Deng, J., Dong, W., Socher, R., Li, L.-J., Li, K., Fei-Fei, L.: ImageNet: a large-scale hierarchical image database, pp. 248–255 (2009)
42. Vaswani, A., et al.: Attention is all you need, pp. 5998–6008 (2017)
43. Devlin, J., Chang, M.-W., Lee, K., Toutanova, K.: Bert: Pre-training of deep bidirectional transformers for language understanding. arXiv preprint arXiv:1810.04805 (2018)
44. Cho, K., et al.: Learning phrase representations using RNN encoder-decoder for statistical machine translation. arXiv preprint arXiv:1406.1078 (2014)
45. Sutskever, I., Vinyals, O., Le, Q.V.: Sequence to sequence learning with neural networks, pp. 3104–3112 (2014)
46. Bottou, L., Curtis, F.E., Nocedal, J.: Optimization methods for large-scale machine learning. SIAM Rev. **60**(2), 223–311 (2018)
47. Nemirovski, A., Juditsky, A., Lan, G., Shapiro, A.: Robust stochastic approximation approach to stochastic programming. SIAM J. Optim. **19**(4), 1574–1609 (2009)
48. Frankish, A., et al.: GENCODE reference annotation for the human and mouse genomes. Nucleic Acids Res. **47**(D1), D766–D773 (2019)
49. Pan, X., Fan, Y.X., Yan, J., Shen, H.B.: IPMiner: hidden ncRNA-protein interaction sequential pattern mining with stacked autoencoder for accurate computational prediction. BMC Genom. **17**(1), 582 (2016)
50. Berman, H.M., et al.: The protein data bank. Nucleic Acids Res. **28**(1), 235–242 (2000)
51. Yi, H., You, Z., Guo, Z., Huang, D., Chan, K.C.C.: Learning representation of molecules in association network for predicting intermolecular associations. IEEE/ACM Trans. Comput. Biol. Bioinf. 1 (2020)
52. Yi, H.-C., You, Z.-H., Huang, D.-S., Li, X., Jiang, T.-H., Li, L.-P.: A Deep learning framework for robust and accurate prediction of ncRNA-protein interactions using evolutionary information. Mol. Ther. Nucleic Acids **11**, 337–344 (2018)

53. Yi, H.-C., You, Z.-H., Wang, M.-N., Guo, Z.-H., Wang, Y.-B., Zhou, J.-R.: RPI-SE: a stacking ensemble learning framework for ncRNA-protein interactions prediction using sequence information. BMC Bioinf. **21**(1), 60 (2020)
54. Wang, L., You, Z.-H., Huang, D.-S., Zhou, F.: Combining high speed ELM learning with a deep convolutional neural network feature encoding for predicting protein-RNA interactions. IEEE/ACM Trans. Comput. Biol. Bioinf. (2018)
55. Yi, H.-C., You, Z.-H., Guo, Z.-H.: Construction and analysis of molecular association network by combining behavior representation and node attributes. Front. Genet. **10**, 1106 (2019)
56. Chang, C.-C., Lin, C.-J.: LIBSVM: a library for support vector machines. ACM Trans. Intell. Syst. Technol. **2**(3), 1–27 (2011)
57. Cortes, C., Vapnik, V.: Support-vector networks. Mach. Learn. **20**(3), 273–297 (1995)
58. Freund, Y., Schapire, R.E.: A decision-theoretic generalization of on-line learning and an application to boosting, pp. 23–37
59. Breiman, L.: Random forests. Mach. Learn. **45**(1), 5–32 (2001)
60. Shen, J., et al.: Predicting protein–protein interactions based only on sequences information. Proc. Natl. Acad. Sci. **104**(11), 4337–4341 (2007)
61. Guo, Z.-H., Yi, H.-C., You, Z.-H.: Construction and comprehensive analysis of a molecular association network via lncRNA–miRNA–disease–drug–protein graph. Cells **8**(8), 866 (2019)
62. Lei, H., et al.: Protein–protein interactions prediction via multimodal deep polynomial network and regularized extreme learning machine. IEEE J. Biomed. Health Inf. **23**(3), 1290–1303 (2019)

Predicting Drug-Target Interactions by Node2vec Node Embedding in Molecular Associations Network

Zhan-Heng Chen[1,2], Zhu-Hong You[1,2(✉)], Zhen-Hao Guo[1,2], Hai-Cheng Yi[1,2], Gong-Xu Luo[1,2], and Yan-Bin Wang[3,4]

[1] The Xinjiang Technical Institute of Physics and Chemistry, Chinese Academy of Sciences, Urumqi 830011, China
zhuhongyou@ms.xjb.ac.cn, zhuhongyou@gmail.com
[2] University of Chinese Academy of Sciences, Beijing 100049, China
[3] School of Cyber Science and Technology, Zhejiang University, Hangzhou 310058, China
[4] College of Information Science and Engineering, Zaozhuang, Shandong 277100, China

Abstract. Accurate identification of drug-target interactions (DTIs) is essential for drug development. It not only helps the researchers to understand the mechanism of drug action, but also contributes to the innovative drug discovery and repositioning. However, due to the limitation the high cost and long time, the traditional experimental methods are difficult to be widely applied for DTIs prediction. In this study, we propose an *in silico* method for predicting drug-target interactions by Node2vec node embedding in molecular associations network (MAN). Specifically, the MAN is constructed by integrating the associations among drug, protein, disease, lncRNA and miRNA. Then, the node2vec embedding method is employed to obtain a behavior feature vector of each node in the network. The traditional attribute feature vector comes from the drug molecular fingerprint and protein sequences. Finally, a random forest (RF) classifier is performed on these features to predict potential drug-target pairs. The experimental results show that the behavior feature could obtain 87.37% accuracy, which is obviously better than the traditional attribute feature. This work is not only more robust and reliable for predicting DTIs, but also provides an alternative way for other biomolecules associations prediction.

Keywords: Drug-target interactions · Multi-molecular network · Node2vec

1 Introduction

Most drug molecules usually perform their functions through the interaction with target proteins in human body. The discovery for drug targets has become the significant focus of innovative drugs research [1, 2]. Hence, prediction of drug-target interactions (DTIs) is one of the most important steps in genomic drug discovery pipeline and drug repurposing [3–5], the purpose is to discover putative new drugs and new uses of existing drugs. Nevertheless, due to the limitation of throughput and cost, the

D.-S. Huang and K.-H. Jo (Eds.): ICIC 2020, LNCS 12464, pp. 348–358, 2020.
https://doi.org/10.1007/978-3-030-60802-6_31

traditional experimental methods are difficult to be widely applied for DTIs prediction. It is of great significance to develop effective calculation methods to predict the interaction between drugs and targets.

Recently, there are already a variety of calculation methods used to identify molecular associations [6–16], especially the interaction between drugs and targets [17–21]. Luo *et al.* developed a computational pipeline to detect novel DTIs from a constructed heterogeneous network, which achieves substantial performance improvement over other state-of-the-art methods [22]. Van Laarhoven *et al.* proposed a simple machine learning method that uses Gaussian interaction profile kernel and regularized least squares for predicting drug-target interactions [23]. Chen *et al.* proposed a drug-target interaction prediction method by random walk on a large-scale heterogeneous network, which combines drug-drug similarity network, protein-protein similarity network and known drug-target interaction network [24]. Ezzat *et al.* provide a comprehensive overview and empirical evaluation on the computational DTIs prediction, which helps understanding the advantages and disadvantages of the state-of-the-art methods [25]. Based on these methods, we proposed a multi-molecular network, also called molecular associations network (MAN) [26] to detect the interactions between drug candidate and related target proteins.

2 Materials and Methods

2.1 Datasets Construction

In the multi-molecular network, the high quality data mainly from nine open source database, which obtained nine known relationships (shown in Table 1) and five types of molecules (shown in Table 2). MAN contained topological relationships and distributions among all the molecules in the heterogeneous network. The drug molecular data and target protein sequences can be collected from DrugBank database and STRING database.

Table 1. Nine known relationships in the molecular associations network

Relationship	Database	Number
Drug-target	DrugBank [27]	11107
Drug-disease	CTD [28]	18416
Protein-disease	DisGeNET [29]	25087
lncRNA-target	LncRNA2Target [30]	690
lncRNA-disease	LncRNADisease [31] lncRNASNP2 [32]	1264
miRNA-target	miRTarBase [33]	4944
miRNA-disease	HMDD [34]	16427
miRNA-lncRNA	lncRNASNP2 [32]	8374
Protein-protein	STRING [35]	19237
Total	N/A	105546

Table 2. The number of 5 types of biomolecules from the nine known relationships

Biomolecule	Number
Drug	1025
Target/Protein	1649
MiRNA	1023
LncRNA	769
Disease	2062
Total	6528

2.2 Molecular Association Network

From the collection of nine known relationships between five types of biomolecules annotated in many famous databases which mentioned above, we constructed a multi-molecular network, also called Molecular Associations Network (MAN) by linking arbitrary two association nodes. The complex molecular associations network is shown in Fig. 1. Based on the known associations, some biomolecules are suggested to interact with each other. In tfhe network graph, the heterogeneous nodes correspond to five types of biomolecules (drug, protein, disease, miRNA and lncRNA), and edges correspond to associations among them. The construction of systematic and MAN network provides a new perspective for predicting interactions between drug and target.

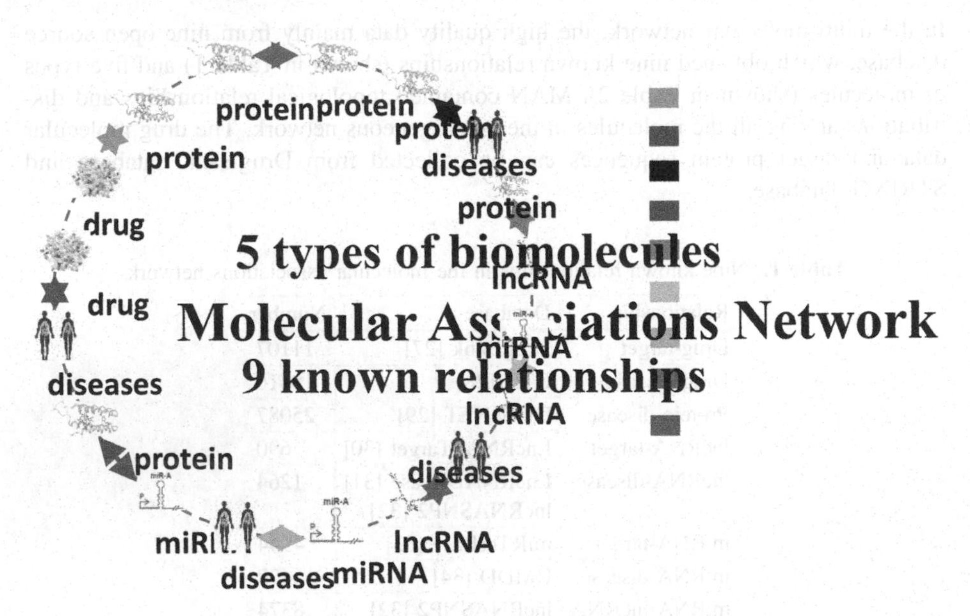

Fig. 1. Construction of multi-molecular network

2.3 Network Embedding: Node2Vec

Grover Aditya and Jure Leskovec proposed a graph embedding method, called Node2Vec, an algorithmic framework for learning continuous feature representations for nodes in graphs [36]. Different from the traditional graph embedding model, it can be seen as an extension of DeepWalk [37]. On the basis of DeepWalk, Node2Vec introduces two biased random walk methods: breadth-first search (BFS) [38] and depth-first search (DFS) [39], to characterize the structural equivalence and homophily of the network. Compared with random walk without any guidance, this method achieves the purpose of biased random walk by introducing Return Parameter and In-out Parameter, that is, the whole random walk process moves between BFS and DFS by setting different offsets. Take node V11 as an example, two search strategies as shown in Fig. 2.

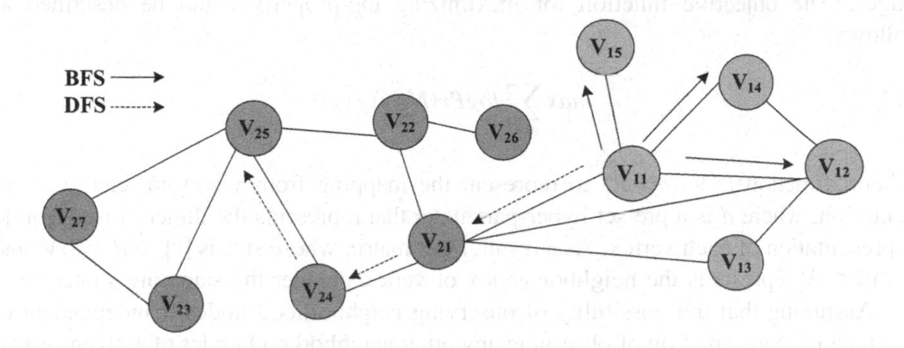

Fig. 2. Two search strategies from node V11 (step = 3)

2.3.1 Biased Random Walk

Suppose that a random walk started with node M and end with node N. Here, due to the use of two different search strategies (BFS and DFS), the selection of strategy will directly affect the result of random walk. The unnormalized transition probability algorithm is introduced to solve this problem. The transfer probability between the two nodes can be described as follow:

$$\pi_{NX} = \alpha_{pq}(M, X) \bullet w_{NX} \tag{1}$$

where, X represents the next position. w is the weight of the edge of the two nodes, which is based on the scenario. α is search bias.

$$\alpha_{pq}(M, X) = \begin{cases} \frac{1}{p} & if\ d_{MX} = 0 \\ 1 & if\ d_{MX} = 1 \\ \frac{1}{q} & if\ d_{MX} = 2 \end{cases} \tag{2}$$

where, d_{MX} is the shortest distance between M and N; p is return parameter, which controls the probability of returning to the original node; q represents in-out parameter, which controls the relationship between BFS and DFS. The setting of different p and q determine the priority of node sequence. When training the model, to find the best p and q by according to the needs of the scene and grid search.

(1) When $d = 0$, it means to return to node M from N. At this time, the search bias is $1/p$, which can be understood as returning to the previous step with a probability of $1/p$;
(2) When $d = 1$, X is the direct neighbor of M, which is equivalent to BFS, then the bias is 1;
(3) When $d = 2$, X is the neighbor's neighbor of M, which is equivalent to DFS, then the bias is $1/q$.

2.3.2 Feature Learning

Now, suppose there is a graph $G = (V, E)$, where V is the set of nodes and E is the set of edges. The objective function for maximizing log-property could be described as follows:

$$\max_f \sum_{v \in V} logPr(N_s(v)|f(v)) \tag{3}$$

where, function f: $V \longrightarrow R^d$ to represent the mapping from vertex to feature representation, where d is a pre-set hyper-parameter that represents the dimension of feature representation of each vertex. As a result, f is a matrix whose size is $|V| \times d$. $v \in V$, and $Ns(v) \subset V$ represents the neighbor vertex of vertex v under the sampling strategy s.

Assuming that the possibility of observing neighborhood nodes is independent of the feature representation of observing any other neighborhood nodes of a given vertex, so as to decompose this conditional probability, then

$$Pr(Ns(v)|f(v)) = \prod_{n_i \in N_S(v)} Pr(n_i|f(v)) \tag{4}$$

Assuming that the influence between two vertices in the feature space is symmetrical, then

$$Pr(n_i|f(v)) = \frac{exp(f(n_i) \bullet f(v))}{\sum_{x \in V} exp(f(x) \bullet f(v))} \tag{5}$$

The purpose of these two assumptions is to better handle the optimization problem. Based on the above assumptions, the objective function can be simplified as follow:

$$max_f \sum_{v \in V} \left[-logZ_v + \sum_{v_i \in N_S(v)} f(v_i) \bullet f(v) \right] \tag{6}$$

For each node,

$$Z_v = \sum_{x \in V} exp(f(v) \bullet f(x)) \tag{7}$$

2.4 Random Forest

Random forest (RF) is an ensemble algorithm that includes a number of decision trees [40]. Focus on the problem of classification, each decision tree is treated as a classifier. Each sample is input into each tree for classification, and the category with the largest number of votes is designated as the final output.

In the process of feature importance assessment using random forest, it depends on the contribution of each feature to each tree in the RF. The contribution usually measured by *Gini* index or error rate of out-of-bag (OOB) data. Assuming that there is n features $f_1, f_2, f_3, \ldots, f_n$, the Gini variable importance measures (VIM) of each feature f_i can be described as follow:

$$Gini_n = \sum_{m=1}^{|M|} \sum_{m' \neq m} p_{nm} p_{nm'} = 1 - \sum_{m=1}^{|M|} p_{nm}^2 \tag{8}$$

where, m represents m classes. p_{nm} is the proportion of class k in node n.

2.5 Performance Measurement Tools

In our study, in order to size up the effectiveness and steadiness of our constructed model, we counted the results of five parameters [41–43]: Accuracy (Acc), recall (sensitivity, hit rate, or true positive rate (TPR)), specificity(selectivity, or true negative rate (TNR)),precision (positive predictive value (PPV)) and Matthews's Correlation Coefficient (MCC), respectively. These parameters can be represented as follows:

$$Acc = \frac{TP + TN}{TP + FP + TN + FN} \tag{9}$$

$$TPR = \frac{TP}{TP + FN} \tag{10}$$

$$TNR = \frac{TN}{FP + TN} \tag{11}$$

$$PPV = \frac{TP}{FP + TP} \tag{12}$$

$$MCC = \frac{(TP \times TN) - (FP \times FN)}{\sqrt{(TP + FN) \times (TN + FP) \times (TP + FP) \times (TN + FN)}} \tag{13}$$

3 Results and Discussion

In this paper, a deep learning method was derived from the idea of molecular association network, and proposed for predicting DTIs. Then, the node2vec embedding method was employed to obtain a behavior feature vector of each node from the multi-molecular network, which is constructed by integrating the associations among drug, protein, disease, lncRNA and miRNA. The good performance was obtained for this method based on behavior feature than traditional attribute feature on the collected datasets. Here, random forest classifier model was used to fulfill the experiment. During this experiment, we set the same parameters to compare the performances of the two different features on the model, the results as shown in Table 3 and 4. From the two tables, it is obvious that the accuracy on behavior features is 5% higher than the accuracy on attribute features under five-fold cross validation.

Table 3. Performance evaluation with RF on attribute features

5-folds	Acc (%)	TPR (%)	TNR (%)	PPV (%)	MCC
1	81.91	79.97	83.84	83.19	0.6386
2	83.12	82.04	84.2	83.85	0.6626
3	82.2	80.6	83.8	83.26	0.6443
4	81.77	80.02	83.53	82.93	0.6359
5	81.75	79.95	83.55	82.94	0.6354
Average	**82.15 ± 0.57**	**80.52 ± 0.89**	**83.78 ± 0.27**	**83.23 ± 0.37**	**0.6434 ± 0.0113**

Table 4. Performance evaluation with RF on behavior features

5-folds	Acc (%)	TPR (%)	TNR (%)	PPV (%)	MCC
1	87.38	82.49	92.26	91.42	0.7511
2	87.89	83.3	92.48	91.72	0.7611
3	87.4	83.44	91.36	90.62	0.7503
4	86.72	81.37	92.08	91.13	0.7387
5	87.47	82.47	92.47	91.64	0.7532
Average	**87.37 ± 0.42**	**82.61 ± 0.83**	**92.13 ± 0.46**	**91.31 ± 0.45**	**0.7509 ± 0.0080**

The ROC curve of random forest classifier based on attribute feature and behavior feature with 5-fold cross-validation that is shown in Fig. 3 and Fig. 4, respectively. It is obvious that the average of AUC is 0.8957 by using attribute information, the average of AUC is 0.9396 by using behavior information based on MAN network. So the behavior information of nodes play an important role in the DTIs predictions.

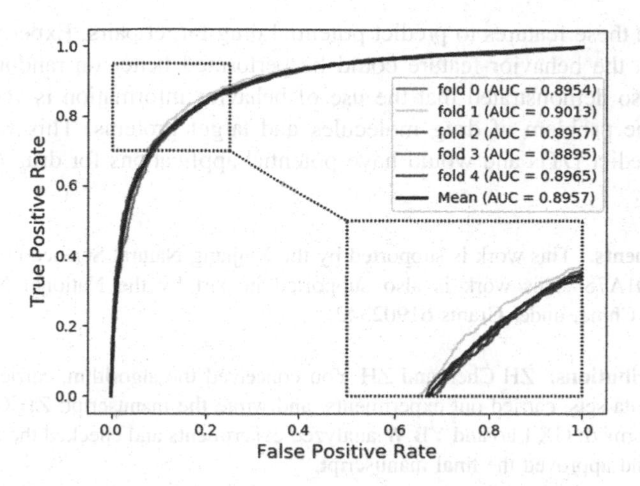

Fig. 3. The ROC curve of random forest on attribute feature

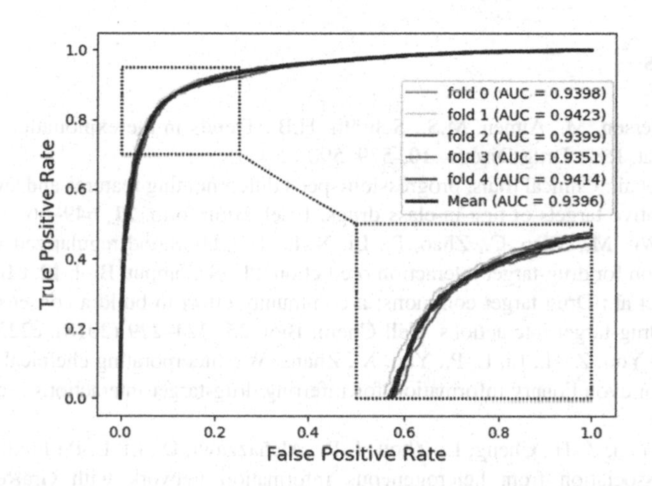

Fig. 4. The ROC curve of random forest on behavior feature

4 Conclusion

In this study, we developed a deep learning method to discover the potential interaction between drugs and target proteins on a large scale by investigating the relationship among five molecules (drug, protein, miRNA, lncRNA and disease). And we construct a novel scheme based on above five molecules and nine relationships between arbitrarily two molecules, which called MAN network. Focus on this network, each node can obtain a feature vector by using node behavior information (the relationship of each node with others could be described by node2vec graph embedding method). The traditional attribute feature vector comes from the drug molecular fin-gerprint and protein sequences on integrated dataset. Finally, a random forest (RF) classifier is performed on these features to predict potential drug-target pairs. Experimental results indicated that the behavior feature could be performed better on random forest classifier. It is also demonstrated that the use of behavior information is very helpful for addressing the problem of drug molecules and target proteins. This work is a new attempt to predict DTIs and would have potential applications for drug discovery and repositioning.

Acknowledgments. This work is supported by the Xinjiang Natural Science Foundation under Grant 2017D01A78. This work is also supported in part by the National Natural Science Foundation of China, under Grants 61902342.

Author Contributions. ZH Chen and ZH You conceived the algorithm, carried out analyses, prepared the data sets, carried out experiments, and wrote the manuscript; ZH Guo and HC Yi designed, performed; GX Luo and YB W analyzed experiments and checked the manuscript; All authors read and approved the final manuscript.

Conflicts of Interest. The authors declare no conflict of interest.

References

1. Rask-Andersen, M., Almén, M.S., Schiöth, H.B.: Trends in the exploitation of novel drug targets. Nat. Rev. Drug Discov. **10**, 579–590 (2011)
2. Li, Y.H., et al.: Clinical trials, progression-speed differentiating features and swiftness rule of the innovative targets of first-in-class drugs. Brief. Bioinform. **21**, 649–662 (2020)
3. Liu, Y., Wu, M., Miao, C., Zhao, P., Li, X.-L.: Neighborhood regularized logistic matrix factorization for drug-target interaction prediction. PLoS Comput. Biol. **12**, e1004760 (2016)
4. Tang, J., et al.: Drug target commons: a community effort to build a consensus knowledge base for drug-target interactions. Cell Chem. Biol. **25**, 224–229 (2018). e222
5. Wang, L., You, Z.-H., Li, L.-P., Yan, X., Zhang, W.: Incorporating chemical sub-structures and protein evolutionary information for inferring drug-target interactions. Sci. Rep. **10**, 1–11 (2020)
6. Ji, B.-Y., You, Z.-H., Cheng, L., Zhou, J.-R., Alghazzawi, D., Li, L.-P.: Predicting miRNA-disease association from heterogeneous information network with GraRep embedding model. Sci. Rep. **10**, 1–12 (2020)
7. You, Z.-H., et al.: PBMDA: a novel and effective path-based computational model for miRNA-disease association prediction. PLoS Comput. Biol. **13**, e1005455 (2017)
8. Chen, X., Yan, C.C., Zhang, X., You, Z.-H.: Long non-coding RNAs and complex diseases: from experimental results to computational models. Brief. Bioinform. **18**, 558–576 (2017)
9. Wang, M.-N., You, Z.-H., Wang, L., Li, L.-P., Zheng, K.: LDGRNMF: LncRNA-disease associations prediction based on graph regularized non-negative matrix factorization. Neurocomputing (2020)
10. Wang, M.-N., You, Z.-H., Li, L.-P., Wong, L., Chen, Z.-H., Gan, C.-Z.: GNMFLMI: graph regularized nonnegative matrix factorization for predicting LncRNA-MiRNA interactions. IEEE Access **8**, 37578–37588 (2020)
11. Wong, L., Huang, Y.A., You, Z.H., Chen, Z.H., Cao, M.Y.: LNRLMI: linear neighbour representation for predicting lncRNA-miRNA interactions. J. Cell Mol. Med. **24**, 79–87 (2020)

12. Guo, Z.-H., You, Z.-H., Wang, Y.-B., Huang, D.-S., Yi, H.-C., Chen, Z.-H.: Bioentity2vec: attribute-and behavior-driven representation for predicting multi-type relationships between bioentities. GigaScience **9**, giaa032 (2020)
13. Guo, Z.-H., et al.: MeSHHeading2vec: a new method for representing MeSH headings as vectors based on graph embedding algorithm. Brief. Bioinf. (2020)
14. You, Z.-H., Zhou, M., Luo, X., Li, S.: Highly efficient framework for predicting interactions between proteins. IEEE Trans. Cybern. **47**, 731–743 (2016)
15. Wang, Y., You, Z.-H., Yang, S., Li, X., Jiang, T.-H., Zhou, X.: A high efficient biological language model for predicting protein–protein interactions. Cells **8**, 122 (2019)
16. Chen, Z.-H., You, Z.-H., Li, L.-P., Wang, Y.-B., Wong, L., Yi, H.-C.: Prediction of self-interacting proteins from protein sequence information based on random projection model and fast Fourier transform. Int. J. Mol. Sci. **20**, 930 (2019)
17. Chen, Z.-H., You, Z.-H., Guo, Z.-H., Yi, H.-C., Luo, G.-X., Wang, Y.-B.: Prediction of drug-target interactions from multi-molecular network based on deep walk embedding model. Front. Bioeng. Biotechnol. **8**, 338 (2020)
18. Wang, Y.-B., You, Z.-H., Yang, S., Yi, H.-C., Chen, Z.-H., Zheng, K.: A deep learning-based method for drug-target interaction prediction based on long short-term memory neural network. BMC Med. Inform. Decis. Mak. **20**, 1–9 (2020)
19. Wang, L., et al.: Identification of potential drug–targets by combining evolutionary information extracted from frequency profiles and molecular topological structures. Chem. Biol. Drug Design (2019)
20. Huang, Y.-A., You, Z.-H., Chen, X.: A systematic prediction of drug-target interactions using molecular fingerprints and protein sequences. Curr. Protein Pept. Sci. **19**, 468–478 (2018)
21. Li, Z., et al.: In silico prediction of drug-target interaction networks based on drug chemical structure and protein sequences. Sci. Rep. **7**, 1–13 (2017)
22. Luo, Y., et al.: A network integration approach for drug-target interaction prediction and computational drug repositioning from heterogeneous information. Nat. Commun. **8**, 1–13 (2017)
23. Van Laarhoven, T., Nabuurs, S.B., Marchiori, E.: Gaussian interaction profile kernels for predicting drug–target interaction. Bioinformatics **27**, 3036–3043 (2011)
24. Chen, X., Liu, M.-X., Yan, G.-Y.: Drug–target interaction prediction by random walk on the heterogeneous network. Mol. BioSyst. **8**, 1970–1978 (2012)
25. Ezzat, A., Wu, M., Li, X.-L., Kwoh, C.-K.: Computational prediction of drug–target interactions using chemogenomic approaches: an empirical survey. Brief. Bioinform. **20**, 1337–1357 (2019)
26. Guo, Z.-H., Yi, H.-C., You, Z.-H.: Construction and comprehensive analysis of a molecular association network via lncRNA–miRNA–disease–drug–protein graph. Cells **8**, 866 (2019)
27. Wishart, D.S., et al.: DrugBank 5.0: a major update to the DrugBank database for 2018. Nucleic Acids Res. **46**, D1074–D1082 (2018)
28. Davis, A.P., et al.: The comparative toxicogenomics database: update 2019. Nucleic Acids Res. **47**, D948–D954 (2019)
29. Piñero, J., et al.: DisGeNET: a comprehensive platform integrating information on human disease-associated genes and variants. Nucleic Acids Res. gkw943 (2016)
30. Cheng, L., et al.: LncRNA2Target v2.0: a comprehensive database for target genes of lncRNAs in human and mouse. Nucleic Acids Res. **47**, D140–D144 (2019)
31. Chen, G., et al.: LncRNADisease: a database for long-non-coding RNA-associated diseases. Nucleic Acids Res. **41**, D983–D986 (2012)

32. Miao, Y.-R., Liu, W., Zhang, Q., Guo, A.-Y.: lncRNASNP2: an updated database of functional SNPs and mutations in human and mouse lncRNAs. Nucleic Acids Res. **46**, D276–D280 (2018)
33. Chou, C.-H., et al.: miRTarBase update 2018: a resource for experimentally validated microRNA-target interactions. Nucleic Acids Res. **46**, D296–D302 (2018)
34. Huang, Z., et al.: HMDD v3.0: a database for experimentally supported human microRNA–disease associations. Nucleic acids research **47**, 1013–1017 (2019)
35. Szklarczyk, D., et al.: The STRING database in 2017: quality-controlled protein–protein association networks, made broadly accessible. Nucleic Acids Res. gkw937 (2016)
36. Grover, A., Leskovec, J.: node2vec: Scalable feature learning for networks. In: Proceedings of the 22nd ACM SIGKDD International Conference on Knowledge Discovery and Data Mining, pp. 855–864 (2016)
37. Perozzi, B., Al-Rfou, R., Skiena, S.: Deepwalk: online learning of social representations. In: Proceedings of the 20th ACM SIGKDD International Conference on Knowledge Discovery and Data Mining, pp. 701–710 (2014)
38. Kurant, M., Markopoulou, A., Thiran, P.: On the bias of BFS (breadth first search). In: 2010 22nd International Teletraffic Congress (ITC 22), pp. 1–8. IEEE (2010)
39. Cui, L., et al.: A novel artificial bee colony algorithm with depth-first search framework and elite-guided search equation. Inf. Sci. **367**, 1012–1044 (2016)
40. Liaw, A., Wiener, M.: Classification and regression by randomForest. R News **2**, 18–22 (2002)
41. Wong, L., You, Z.-H., Guo, Z.-H., Yi, H.-C., Chen, Z.-H., Cao, M.-Y.: MIPDH: a novel computational model for predicting microRNA–mRNA interactions by DeepWalk on a heterogeneous network. ACS Omega **5**, 17022–17032 (2020)
42. Chen, Z.-H., Li, L.-P., He, Z., Zhou, J.-R., Li, Y., Wong, L.: An improved deep forest model for predicting self-interacting proteins from protein sequence using wavelet transformation. Front. Genet. **10**, 90 (2019)
43. Wang, Y., You, Z., Li, L., Chen, Z.: A survey of current trends in computational predictions of protein-protein interactions. Front. Comput. Sci. **14**, 144901 (2020)

Intelligent Computing in Drug Design

Prediction of Drug-Target Interactions with CNNs and Random Forest

Xiaoli Lin[1(✉)], Minqi Xu[1], and Haiping Yu[2]

[1] Hubei Key Laboratory of Intelligent Information Processing and Real-Time Industrial System, School of Computer Science and Technology, Wuhan University of Science and Technology, Wuhan 430065, China
aneya@163.com, minqi.xu@foxmail.com

[2] School of Computer, Huanggang Normal University, Huanggang 438000, China
seapingyu@outlook.com

Abstract. Drug molecules interact with target proteins to influence the pharmacological action of the target to achieve the phenotypic effect, which can facilitate the identification of novel targets for current drug. Traditional biological experiments for discovering new drug-target interactions are expensive and time-consuming. Therefore, it is crucial to develop new prediction methods for identifying potential drug-target interactions. Computing methods have been increasing developed which can quickly and effectively predict drug-target interactions. In particular, machine learning methods have been widely used due to high predictive performance and computational efficiency. This paper first uses MACCS substructure fingerings to encode the drug molecules, then uses CNNs to extract the biological evolutionary information of target protein sequences, and finally uses random forest algorithm to predict drug-target interactions. Four datasets of drug-target interactions including Enzymes, Ion Channels, GPCRs and Nuclear Receptors, are independently used for building models with random forest. The results demonstrate our proposed method has a general compatibility, which is effective and feasible to predict drug-target interactions.

Keywords: Drug-target interactions · CNNs · Random forest · Feature extraction

1 Introduction

Accurate and effective identification of drug-target interactions can accelerate the drug development process and reduce the cost [1, 2]. In addition, the recognition of drug-target interactions can reveal the hidden function of drug molecules or target proteins [3]. Biological organism is a very complex system, and the effect of drugs on biological system is multi-level. In the process of drug development, it is a very challenging work to explore a compound that selectively binds potential targets. Only a small number of candidate chemical molecules have been identified as available drugs, while a large number of compounds still have unknown interaction with proteins. With the development of drug discovery technology, existing methods can identify drug-target

D.-S. Huang and K.-H. Jo (Eds.): ICIC 2020, LNCS 12464, pp. 361–370, 2020.
https://doi.org/10.1007/978-3-030-60802-6_32

interactions to a certain extent, but there are still many undetected interactions. Therefore, the prediction of drug-target interactions is very important for the discovery of new drugs [4], which provides a new idea for drug design and drug discovery.

Over the years, researchers have put forward many intelligent computing methods to predict drug-target interactions [5, 6], which accelerate the development of drug-target interaction prediction technology. A semi supervised method was proposed to integrate the similarity between drugs and the local correlation between targets into a robust PCA model [7]. Deep learning strategy has also been used to predict drug-target interactions [8, 9]. Pliakos [10, 11] proposed a multi label classification framework using label division to predict drug-target interactions. Zhang [12] introduced the label propagation method with linear neighborhood information. Ezzat [13] incorporated the regularization of graphs into the matrix decomposition, and the graph regularization is applied to drive the matrix decomposition method. Olayan [14] extracted features based on the graph and used machine learning classifier to predict interaction. Li [15] used the topological structure to extract features for classification. Rayhan [16] used boosting method to identify drug-target interactions. Mousavian [17] predicted the drug-target interactions by using the features extracted from the Position Specific Scoring Matrix (PSSM) of proteins. CNNs has also been widely applied to the bioinformatics field. Ozkirimli et al. [18] proposed a deep learning model based on protein and ligand sequences to predict the binding affinity of protein-ligand interactions. They used Davis Kinase binding affinity dataset [19] and the KIBA large-scale kinase inhibitors bioactivity data [20, 21] to evaluate the performance of the model. These machine learning models can help to reveal the potential connection of large-scale drug molecules and target proteins. However, the accuracy of prediction is still the problem that need to be solved.

According to studies of Yamanishi et al. [22, 23], the structural similarity of compounds and the sequence similarity of targets have strong correlation with drug-target interactions. Therefore, it is feasible to predict interaction of drugs and targets based on the one-dimensional sequences. This paper uses MACCS substructure fingerings to encode the drug molecules, then uses CNNs to extract the biological evolutionary information of protein sequences, and finally uses random forest to predict drug-target interactions.

2 Method

2.1 Feature Extraction and Feature Selection

The most important problem in drug-target interactions is how to extract features that can effectively represent drugs and targets. The features of drug molecules and target proteins are different. If the same feature extraction model is used in drug feature extraction and target feature extraction, the final prediction results may be affected. Therefore, we use two independent models to extract the features of drug molecules and target proteins.

Studies show that molecule activity class characteristic substructure (MACCS) can also be used to describe drug molecules. Therefore, in order to extract the features of drug molecule structure, drug molecules are encoded with MACCS fingerings [24]. Target protein sequences are usually stored in the form of one-dimensional sequences. To effectively express the characteristics of target proteins, the convolution layer and pooling layer are used to complete the feature extraction of the target protein sequence, and then the extracted features are input into the random forest classifier to obtain the classification results. The number of filters in convolution layer will greatly affect the generalization ability of the model, and the number of trees in random forest will also affect the classification accuracy.

Feature selection is the process of finding the optimal feature subset, which can be regarded as the problem of combinatorial optimization. The final subset is a part of the original features, and the optimal feature subset has the greatest impact on the performance of the classification model. In this paper, we use random forest to calculate the importance of feature variables, which ensures the interpretability of the importance of original features.

According to a certain deletion ratio, a new feature set is obtained. A new random forest is established by using the new feature set, the importance of each feature is calculated, and then it is sorted in descending order. Each feature set corresponds to a random forest, and the corresponding out of bag (OOB) error rate is calculated. At the same time, using the importance ranking algorithm based on random forest to remove the irrelevant features for improving the classification accuracy. The top features of drug and target protein with high importance can be selected by importance ranking algorithm, which are shown in Fig. 1 and Fig. 2 respectively.

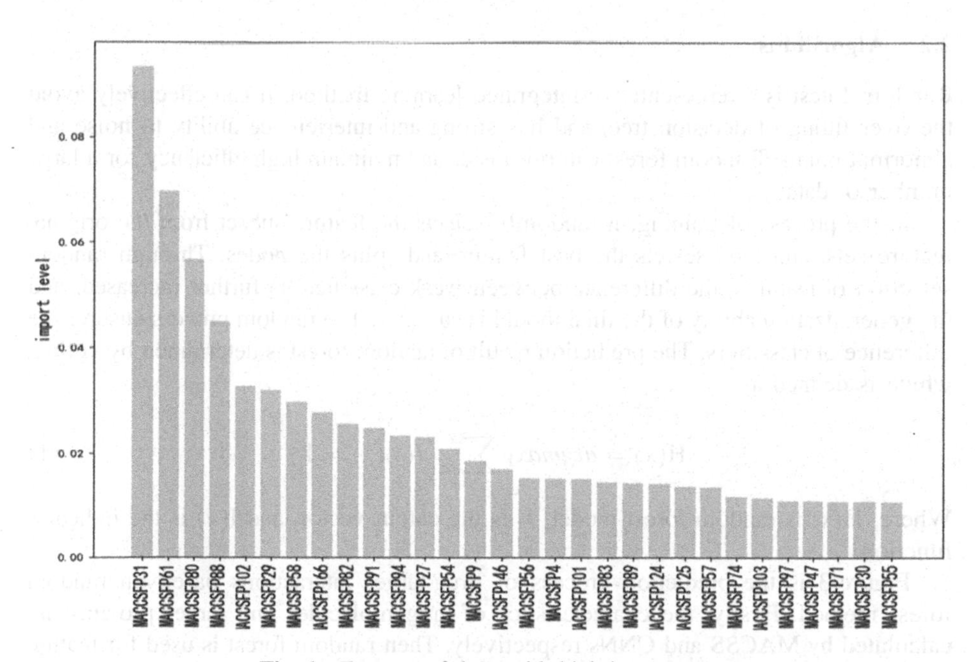

Fig. 1. Features of drug with high importance

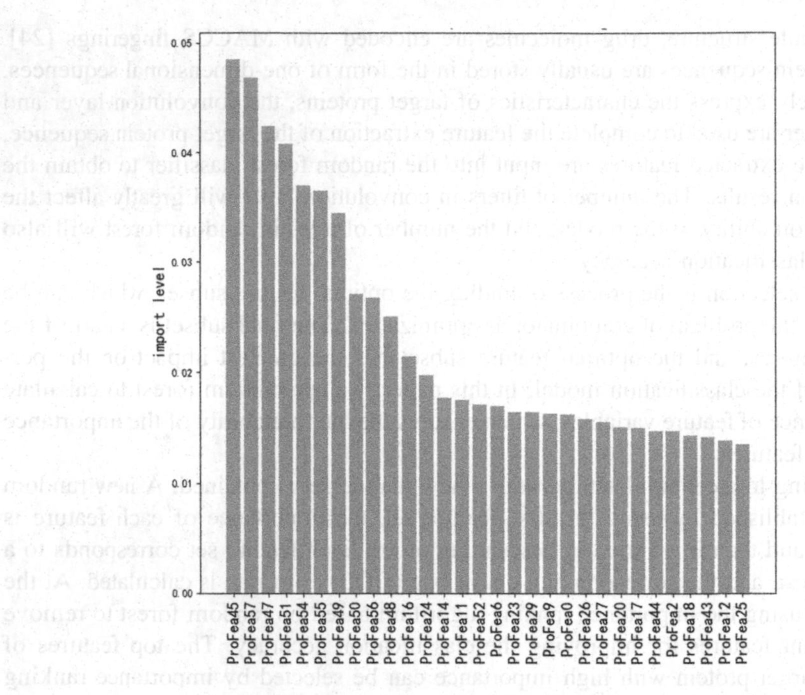

Fig. 2. Features of target protein with high importance

2.2 Algorithms

Random forest is a representative integrated learning method. It can effectively avoid the over fitting of decision tree, and has strong anti-interference ability to noise and abnormal points. Random forest can run faster and maintain high efficiency for a large number of data.

In the process of training, it randomly selects the feature subset from the original feature sets, and then selects the best feature and splits the nodes. Through random selection of features, the difference between weak classifiers is further increased, and the generalization ability of the final model is stronger. The random process ensures the difference of classifiers. The prediction result of random forest is determined by voting, which is defined as

$$\mathrm{H(x)} = argmax_Y \sum\nolimits_{i=1}^{k} I(h_i(x) = Y) \tag{1}$$

Where, H(x) is random forest model, Y is the output vector, and I(∗) is the indicator function.

Figure 3 is the prediction process of drug-target interactions based on random forest method. Firstly, the characteristics of drug molecules and target proteins are calculated by MACSS and CNNs respectively. Then random forest is used for feature selection, and features from drug and target protein space is integrated into a global random forest model for classification.

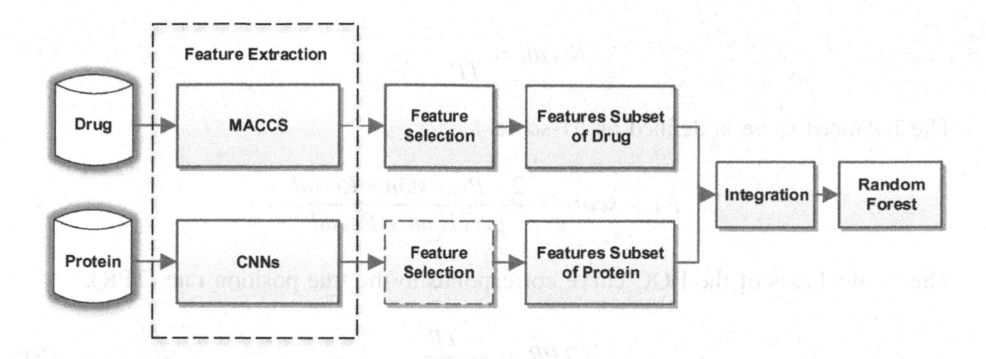

Fig. 3. Prediction process of drug-target interactions

3 Experiments

3.1 Dataset

This paper uses the gold standard dataset introduced by Yamanishi et al. [22] for comparing with previous methods. This dataset includes the information about drug-target interactions which comes from SuperTarget [25], KEGG [26], DrugBank [27] and BRENAD [28]. There are 2926 positive samples, 664 amino acid sequences and 445 SMILES sequences in drug and enzymes interactions. And there are 635 positive samples, 95 amino acid sequences and 223 SMILES sequences in drug and GPCRs interactions. And there are 1476 positive samples, 204 amino acid sequences and 210 SMILES sequences in the interactions between drug and ion channels interactions. In addition, there are 90 positive samples, 54 amino acid sequences and 26 SMILES sequences in the interactions between drug and nuclear receptors interactions. The detail information of datasets is shown in Table 1.

Table 1. The datasets of the drug-target interactions

	Enzymes	GPCRs	Ion channels	Nuclear receptors
Amino acid sequences	664	95	204	54
SMILES sequences	445	223	210	26
Interaction pairs	2926	635	1476	90

3.2 Evaluation Criteria

There are several measures used to evaluate the performance. Precision and Recall are defined as

$$Precision = \frac{TP}{TP + FP} \tag{2}$$

$$Recall = \frac{TP}{TP + FN} \tag{3}$$

The balanced score is defined as F1-score.

$$F1 - score = \frac{2 * Precision * Recall}{precision + Recall} \tag{4}$$

The vertical axis of the ROC curve corresponds to the true position rate (TPR).

$$TPR = \frac{TP}{TP + FN} \tag{5}$$

The horizontal axis of the ROC curve corresponds to the false positive rate (FPR).

$$FPR = \frac{FP}{FP + TN} \tag{6}$$

The true positive (*TP*) represents the number of positive samples that can be correctly identified, and the true negative (*TN*) represents the number of negative samples that can be correctly identified. The false positive (*FP*) represents the number of positive samples that cannot be correctly identified. The false negative (*FN*) represents the number of negative samples that cannot be correctly identified.

3.3 Experimental Results

The experiments have been carried out. The different kinds of drug-target pairs were separately predicted to fully exploit the general characteristics of the SMILES sequences and amino acid sequences. There have been many state-of-the-art methods achieved good results of predicting drug-target interactions. The results are shown in Table 2 for compare with these state-of-the-art methods. Our method achieves the best performance in GPCRs, Ion Channels and Nuclear Receptors datasets, which is a little better than iDTI-ESBoost [16], but is better than other methods. In Enzymes dataset, AUROC of our method is worse than those of Bigram-PSSM and iDTI-ESBoost, but is better than those of other methods.

To verify the prediction ability of our method for drug-target interactions, four kinds of datasets are integrated. Then, all positive samples are used as training sets to train the model, then, unknown drug-target pairs are predicted. The drug-target pairs with the higher prediction score is verified by SuperTarget database. Table 3 lists the prediction performance for the whole dataset, and ROC curve of prediction is shown in Fig. 4. Figure 5 shows the change of the accuracy of the random forest classifier on the train set and cross validation as the number of trees increases. When the number of trees is small, the accuracy scores of training set and cross validation are very low, indicating that the model fits. When the number of trees increases to 250, the two accuracy scores did not change much.

Experiment results shows that our method can significantly improve the drug-target interaction prediction level. In addition, the average prediction rate of the classification

Table 2. Comparision of the performance based on four benchmark gold datasets in AUROC

Methods	Enzymes	GPCRs	Ion channels	Nuclear receptors
Yamanishi [22]	0.904	0.851	0.899	0.843
Yamanishi [23]	0.892	0.812	0.827	0.835
DBSI [29]	0.807	0.802	0.802	0.757
NetCBP [30]	0.825	0.803	0.823	0.839
Bigram-PSSM [17]	0.948	0.889	0.872	0.869
iDTI-ESBoost [16]	**0.968**	0.932	0.936	0.928
Our Method	0.938	**0.942**	**0.948**	**0.935**

Table 3. Prediction performance for the whole dataset

	Precision	Recall	F1-score
0	0.91	0.94	0.92
1	0.81	0.72	0.76
Macro avg	0.86	0.83	0.84
Weighted avg	0.88	0.88	0.88

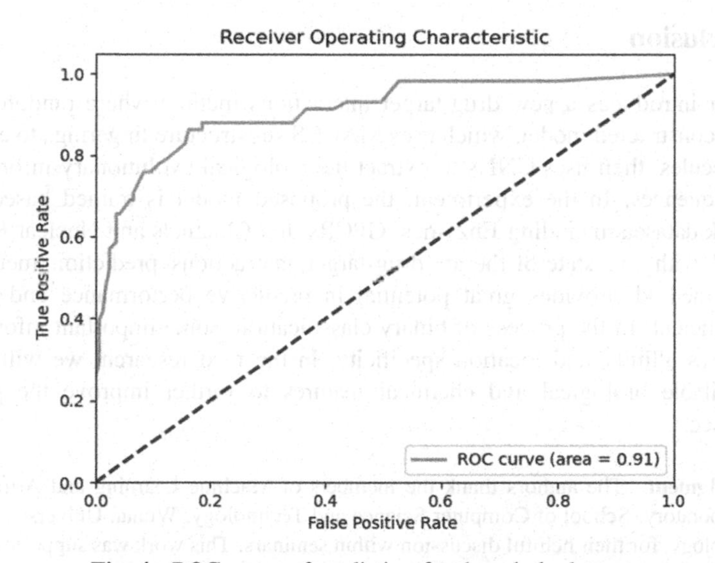

Fig. 4. ROC curve of prediction for the whole dataset

model still needs to be improved, which makes it more difficult to predict the unknown interaction. For Enzymes, GPCRs, Ion Channels and Nuclear Receptors datasets, our method has greatly reduced the prediction range and improved the prediction accuracy, which plays an important role in reducing the cost of drug development and improving the efficiency of drug-target interaction prediction.

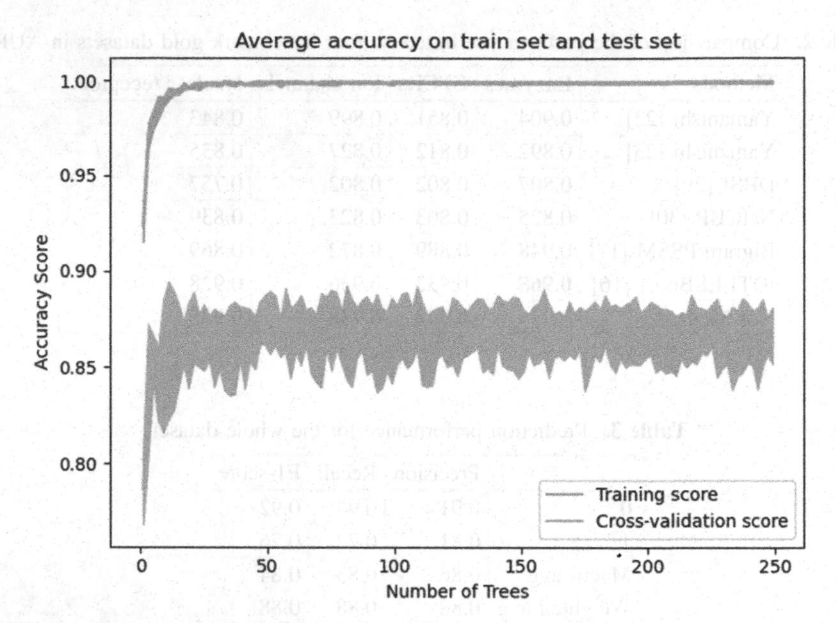

Fig. 5. Average accuracy on train set and test set

4 Conclusion

This paper introduces a new drug-target interactions method where random forest is built on reconstructed model, which uses MACCS substructure fingerings to encode the drug molecules, then uses CNNs to extract the biological evolutionary information of protein sequences. In the experiment, the proposed model is trained based on four benchmark datasets including Enzymes, GPCRs, Ion Channels and Nuclear Receptors. Compared with the state-of-the-art drug-target interactions prediction methods, the proposed method provides great potential in predictive performance and computationally efficient. In the process of binary classification, some important information is lost, such as affinity and location specificity. In the next research, we will focus on more available biological and chemical features to further improve the prediction performance.

Acknowledgment. The authors thank the members of Machine Learning and Artificial Intelligence Laboratory, School of Computer Science and Technology, Wuhan University of Science and Technology, for their helpful discussion within seminars. This work was supported in part by Hubei Province Natural Science Foundation of China (No. 2018CFB526, 2019CFB797), by National Natural Science Foundation of China (No. 61502356, 61972299, 61702385).

References

1. Kim, I.W., Jang, H., Hyunkim, J., et al.: Computational drug repositioning for gastric cancer using reversal gene expression profiles. Sci. Rep. **9**, 2660 (2019)
2. Ganotra, G.K., Wade, R.C.: Prediction of drug-target binding kinetics by comparative binding energy analysis. ACS Med. Chem. Lett. **9**(11), 1134–1139 (2018)
3. Kingsmore, K.M., Grammer, A.C., Lipsky, P.E.: Drug repurposing to improve treatment of rheumatic autoimmune inflammatory diseases. Nat. Rev. Rheumatol. **16**, 32–52 (2020)
4. Williams, G., Gatt, A., Clarke, E., et al.: Drug repurposing for Alzheimer's disease based on transcriptional profiling of human iPSC-derived cortical neurons. Transl. Psychiatry **9**, 220 (2019)
5. Stokes, J.M., Yang, K., Swanson, K., et al.: A deep learning approach to antibiotic discovery. Cell **180**(4), 668–702 (2020)
6. Zhang, W., Lin, W., Zhang, D., Wang, S., Shi, J., Niu, Y.: Recent advances in the machine learning-based drug-target interaction prediction. Curr. Drug Metab. **20**(3), 194–202 (2019)
7. Peng, L., Liao, B., Zhu, W., Li, Z., Li, K.: Predicting drug-target interactions with multi-information fusion. IEEE J Biomed. Health Inf. **21**(2), 56–72 (2017)
8. Zong, N., Kim, H., Ngo, V., Harismendy, O.: Deep mining heterogeneous networks of biomedical linked data to predict novel drug-target associations. Bioinformatics **33**(15), 2337–2344 (2017)
9. Wen, M., et al.: Deep-learning-based drug-target interaction prediction. J. Proteome **16**(4), 1401–1409 (2017)
10. Pliakos, K., Vens, C., Tsoumakas, G.: Predicting drug-target interactions with multi-label classification and label partitioning. IEEE/ACM Trans. Comput. Biol. Bioinf. (2019, early access). https://doi.org/10.1109/TCBB.2019.2951378. https://ieeexplore.ieee.org/document/8890853
11. Pliakos, K.: Mining biomedical networks exploiting structure and background information. KU Leuven, Belgium (2019)
12. Zhang, W., Chen, Y., Li, D.: Drug-target interaction prediction through label propagation with linear neighborhood information. Molecules **22**(12), 2056 (2017)
13. Ezzat, A., Zhao, P., Wu, M., Li, X., Kwoh, C.: Drug-target interaction prediction with graph regularized matrix factorization. IEEE/ACM Trans. Comput. Biol. Bioinf. **14**(3), 646–656 (2017)
14. Olayan, R.S., Ashoor, H., Bajic, V.B.: DDR: efficient computational method to predict drug-target interactions using graph mining and machine learning approaches. Bioinformatics **34** (7), 1164–1173 (2017)
15. Li, Z., et al.: Identification of drug-target interaction from interactome network with 'guilt-by-association' principle and topology features. Bioinformatics **32**(7), 1057–1064 (2016)
16. Rayhan, F., Ahmed, S., Shatabda, S., et al.: iDTI-ESBoost: identification of drug target interaction using evolutionary and structural features with boosting. Sci. Rep. **7**(1), 17731 (2017)
17. Mousavian, Z., Khakabimamaghani, S., Kavousi, K., et al.: Drug-target interaction prediction from PSSM based evolutionary information. J. Pharmacol. Toxicol. Methods **78**, 42–51 (2016)
18. Öztürk, H., Özgür, A., Ozkirimli, E.: DeepDTA: deep drug-target binding affinity prediction. Bioinformatics **34**(17), i821–i829 (2018)
19. Davis, M.I., Hunt, J.P., Herrgard, S., et al.: Comprehensive analysis of kinase inhibitor selectivity. Nat. Biotechnol. **29**(11), 1046–1051 (2011)

20. He, T., Heidemeyer, M., Ban, F., et al.: SimBoost: a read-across approach for predicting drug-target binding affinities using gradient boosting machines. J. Cheminf. **9**(1), 24 (2017)
21. Tang, J., Szwajda, A., Shakyawar, S., et al.: Making sense of large-scale kinase inhibitor bioactivity data sets: a comparative and integrative analysis. J. Chem. Inf. Model. **54**(3), 735–743 (2014)
22. Yamanishi, Y., Araki, M., Gutteridge, A., et al.: Prediction of drug-target interaction networks from the integration of chemical and genomic spaces. Intell. Syst. Mol. Biol. **24** (13), 232–240 (2008)
23. Yamanishi, Y., Masaaki, K., Minoru, K., et al.: Drug-target interaction prediction from chemical, genomic and pharmacological data in an integrated framework. Bioinformatics **26** (12), 246–254 (2010)
24. Cao, D., Liu, S., Xu, Q., et al.: Large-scale prediction of drug-target interactions using protein sequences and drug topological structures. Anal. Chim. Acta **752**, 1–10 (2012)
25. Gunther, S., Kuhn, M., Dunkel, M., et al.: SuperTarget and matador: resources for exploring drug-target relationships. Nucleic Acids Res. **36**, 919–922 (2007)
26. Kanehisa, M., Goto, S., Hattori, M., et al.: From genomics to chemical genomics: new developments in KEGG. Nucleic Acids Res. **34**(90001), 354–357 (2006)
27. Wishart, D.S., Knox, C., Guo, A.C., et al.: Drugbank: a knowledgebase for drugs, drug actions and drug targets. Nucleic Acids Res. **36**(suppl 1), D901–D906 (2008)
28. Jeske, L., Placzek, S., Schomburg, I., et al.: BRENDA in 2019: a European ELIXIR core data resource. Nucleic Acids Res. **47**, 542–549 (2019)
29. Cheng, F., Liu, C., Jiang, J., et al.: Prediction of drug-target interactions and drug repositioning via network-based inference. PLoS Comput. Biol. **8**(5), e1002503 (2012)
30. Chen, H., Zhang, Z.: A semi-supervised method for drug-target interaction prediction with consistency in networks. PLoS ONE **8**(5), e62975 (2013)

DTIFS: A Novel Computational Approach for Predicting Drug-Target Interactions from Drug Structure and Protein Sequence

Xin Yan[1,2], Zhu-Hong You[3(✉)], Lei Wang[2,3(✉)], Li-Ping Li[3], Kai Zheng[1], and Mei-Neng Wang[4]

[1] School of Computer Science and Technology, China University of Mining and Technology, Xuzhou, China
[2] College of Information Science and Engineering, Zaozhuang University, Zaozhuang, China
leiwang@ms.xjb.ac.cn
[3] Xinjiang Technical Institute of Physics and Chemistry, Chinese Academy of Sciences, Ürümqi, China
zhuhongyou@ms.xjb.ac.cn
[4] School of Mathematics and Computer Science, Yichun University, Yichun 336000, China

Abstract. Identification and prediction of Drug-Target Interactions (DTIs) is the basis for screening drug candidates, which plays a vital role in the development of innovative drugs. However, due to the time-consuming and high cost constraints of biological experimental methods, traditional drug target identification technologies are often difficult to develop on a large scale. Therefore, *in silico* methods are urgently needed to predict drug-target interactions in a genome-wide manner. In this article, we design a novel *in silico* approach, named DTIFS to predict the DTIs by combining Feature weighted Rotation Forest (FwRF) classifier with protein amino acids information. This model has two outstanding advantages: a) using the fusion data of protein sequence and drug molecular fingerprint, which can fully carry information; b) using the classifier with feature selection ability, which can effectively remove noise information and improve prediction performance. More specifically, we first use Position-Specific Score Matrix (PSSM) to numerically convert protein sequences and utilize Pseudo Position-Specific Score Matrix (PsePSSM) to extract their features. Then a unified digital descriptor is formed by combining molecular fingerprints representing drug information. Finally, the FwRF is applied to implement on *Enzyme*, *Ion Channel*, *GPCR*, and *Nuclear Receptor* datasets. The results of the 5-fold CV experiment show that the prediction accuracy of this approach reaches 91.68%, 88.11%, 84.72% and 78.33% on four benchmark datasets, respectively. To further validate the performance of the DTIFS, we compare it with other excellent methods and Support Vector Machine (SVM) model. The experimental results of cross-validation indicated that DTIFS is feasible in predicting the relationship among drugs and target, and can provide help for the discovery of new candidate drugs.

Keywords: Drug-target interaction · Feature selection · Rotation forest · Pseudo Position-Specific Score Matrix

D.-S. Huang and K.-H. Jo (Eds.): ICIC 2020, LNCS 12464, pp. 371–383, 2020.
https://doi.org/10.1007/978-3-030-60802-6_33

1 Introduction

Identifying the interaction ship between drugs and targets is a crux area in genomic drug discovery, which not only helps to understand various biological processes, but also contributes to the development of new drugs [1, 2]. The emergence of molecular medicine and the completion of the Human Genome Project provide better conditions for the identification of new drug target proteins. Although the researchers have made a lot of efforts, only a small number of candidate drugs can be approved by the Food and Drug Administration (FDA) to enter the market so far [3–5]. An important reason for this situation is due to the inherent defects of the experimental methods. As is known to all, biological laboratory methods to identify DTIs are usually expensive, time-consuming, and are limited to small-scale studies. *In silico* methods can narrow the scope of candidate targets and provide supporting evidence for the drug target experiments, thus speeding up drug discovery. Therefore, *in silico*-based methods are urgently required to improve efficiency and reduce time in identifying potential DTIs across the genome [6–8].

In recent years, researchers have developed a variety of *in silico*-based methods to analyze and predict DTIs [9]. For example, Wu *et al.* [10] proposed the SDTBNI model in 2016, which searches for unknown DTIs through new chemical entity-substructure linkages, drug-substructure linkages and known DTIs networks. Based on biomedical related data and Linked Tripartite Network (LTN), Zong *et al.* [11] used the target-target and drug-drug similarities calculated by DeepWalk to predict DTIs. In addition, Peng *et al.* [12] combines the biological information of targets and drugs with PCA-based convex optimization algorithms to predict new DTIs using semi-supervised inference method. Ezzat *et al.* [13] used ensemble learning algorithm to predict DTIs by decrease features with subinterval features through three dimensionality reduction models. Generally speaking, drugs with chemical similarity also have similar biochemical activity, that is, they can bind to similar target proteins. Based on the above assumptions, the use of medicinal chemical molecular structure information and protein sequence information to predict the DTIs model has achieved good results. For example, Wen *et al.* [14] extracted drug and target features from their chemical substructure and sequence information, and used deep belief network (DBN) to predict potential DTIs.

In this article, according to the assumption that the interaction between drugs and target proteins largely depend on the information of target protein sequences and drug molecular sub-structure fingerprints, a novel *in silico*-based model is proposed to infer potential DTIs. Our feature combines the fingerprint of the drug molecule structure and the protein sequence encoded by a feature extraction method called Pseudo Position-Specific Score Matrix (PsePSSM). In the experiment, we adopt the FwRF classifier to predict the results on the four DTIs benchmark datasets, including *Enzyme*, *Ion Channel*, *GPCR* and *Nuclear Receptor*. In order to verify the performance of the proposed model, we compared with SVM classifier model, different feature extraction models and existing excellent methods. The promoting experimental results show that DTIFS has excellent performance and can effectively predict potential DTIs. The flow chart of the proposed model is shown in Fig. 1.

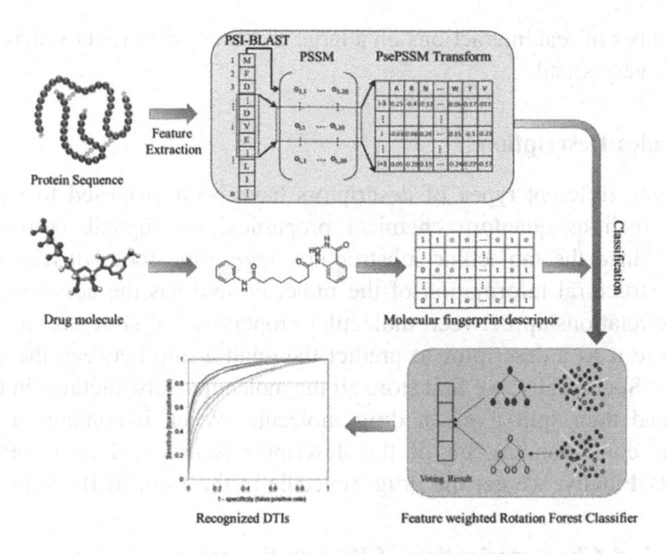

Fig. 1. The flow chart of the proposed model

2 Materials and Methods

2.1 Benchmark Datasets

In this article, we applied four protein targeting datasets, including *Enzyme*, *Ion Channel*, *GPCR* and *Nuclear Receptor*. These datasets are applied as the benchmark datasets by Yamanishi *et al.* [15] and collected from the BRENDA [16], DrugBank [17], SuperTarget & Matador [18] and KEGG BRITE [19]. The number of drugs was 445, 210, 233 and 54, and the number of target proteins was 664, 204, 95 and 26 in these benchmark datasets, respectively. Among these data, 5127 pairs of drug-target were confirmed to interact with each other, corresponding to 2926, 1476, 635 and 90 pairs in four datasets, respectively [20–22].

The DTIs network can be expressed by a bipartite graph in which nodes represent drugs or targets, and edges represent their interactions. If there is a relationship between the nodes, connect them with edges, otherwise they do not connect. The edges in the initial bipartite graph represent the real DTIs have been detected by biological experiments. The number of initial edges is relatively small compared to a completely connected bipartite graph. For example, there are totally $54 \times 26 = 1404$ connections in the bipartite graph of the *Nuclear Receptor* dataset. However, the experiment detected representatives known drug-target interactions only 90. Therefore, the number of positive drug-target pairs (e.g., 90) accounted for only 6.41% of the total number of drug-target pairs (e.g., 1404), much less than the number of negative drug-target pairs (e.g., $1404 - 90 = 1314$). The same problem also appears in the other three datasets. In order to solve the problem of data imbalance, we randomly select negative drug-target pairs with the same number of positive drug-target pairs. In fact, such negative samples may contain drug-target pairs with interactions. To reduce this possibility, we randomly selected ten times negative sample sets in the experiment. From a statistical point of view, the number of real interactions on a large bipartite graph is chosen as the negative sample set is very small.

2.2 Molecules Description

In recent years, different types of descriptors have been proposed to represent drug compounds, such as quantum chemical properties, topological, constitutional and geometrical. Since the molecular substructure fingerprint does not require the three-dimensional structural information of the molecule and has the advantage of directly reflecting the relationship between molecular properties and structure, more and more researchers use it as a descriptor to predict the relationship between the drug and the target protein. Specifically, we first store all the molecular substructures in the form of a dictionary, and then split a given drug molecule. When it contains a certain substructure, the corresponding bit of the descriptor is assigned to 1; otherwise it is assigned to 0. Finally, we get the drug molecule in the form of Boolean vectors.

2.3 Numerical Characterization of Protein Sequences

In the experiment, we used position-specific scoring matrix proposed by Gribskov *et al.* [23] to convert protein sequence numerically. PSSM is widely used in protein binding site prediction [24, 25], protein secondary structure prediction [26], and prediction of disordered regions [27]. PSSM is an $L \times 20$ matrix that can be expressed as $S = \{\partial_{i,j} : i = 1 \cdots L and j = 1 \cdots 20\}$, where 20 represents the number of the amino acids and L denotes the length of the protein sequence. PSSM matrix $S(i,j)$ can be expressed as follows:

$$S = \begin{bmatrix} \partial_{1,1} & \partial_{1,2} & \cdots & \partial_{1,20} \\ \partial_{2,1} & \partial_{2,2} & \cdots & \partial_{2,20} \\ \vdots & \vdots & \vdots & \vdots \\ \partial_{L,1} & \partial_{L,2} & \cdots & \partial_{L,20} \end{bmatrix} \tag{1}$$

where $\partial_{i,j}$ denotes the probability that the ith residue being mutated into the *jth* amino acid during the evolutionary process of protein multiple sequence alignment. In the experiment, we use Position-Specific Iterated BLAST (PSI-BLAST) tool [28–31] to generate PSSM based on *SwissProt* dataset.

2.4 Feature Extraction Algorithm

Effective protein feature descriptors can not only mine useful information, but also improve the performance of the approach. In this study, we introduce the feature extraction algorithm Pseudo Position-Specific Score Matrix (PsePSSM), which concept from Chou *et al.* [32]. The PsePSSM is expressed by formula as follows:

$$e_{i,j} = \frac{e^0_{i,j} - \frac{1}{20}\sum_{k=1}^{20} e^0_{i,k}}{\sqrt{\frac{1}{20}\sum_{v=1}^{20}\left(e^0_{i,v} - \frac{1}{20}\sum_{k=1}^{20} e^0_{i,k}\right)^2}} \quad i = 1..20, j = 1..20 \tag{2}$$

where $e^0_{i,j}$ denotes the raw score generated by PSI-BLAST, which is typically a positive or negative integer. This is not the final score, because if it exceeds 20 amino acids, the score may contain 0; if the same conversion procedure continues, the score may remain unchanged. The positive number signifies that the frequency of corresponding mutations in the alignment is higher than that of accidental expectations. Conversely, the negative number signifies that the frequency of corresponding mutations in the alignment is lower than that of accidental expectations. However, based on the PSSM formula, proteins of different lengths will produce a matrix of different numbers of rows. Therefore, Eq. 2 is used to convert the PSSM matrix into a uniform pattern.

$$\bar{M}_{PSSM} = [\bar{e}_1 \bar{e}_2 \cdots \bar{e}_{20}] \tag{3}$$

and

$$\bar{e}_j = \frac{1}{L}\sum_{i=1}^{L} e_{i,j} \quad j = 1..20 \tag{4}$$

where $\bar{e}_j$ indicates the average score of P protein when amino acid residues evolve into j-type amino acids. However, if only $\bar{M}_{PSSM}$ is used to indicate protein P, all information about sequence order will be lost during evolution. In order to prevent this from happening, we introduce the idea of pseudo amino acid to improve Eq. 2. Therefore, according to the formula 4, we can get the features of segmented PsePSSM:

$$\bar{e}_j = \begin{cases} \frac{1}{L}\sum_{i=1}^{L} e_{i,j} & j = 1..20,\ \lambda = 0 \\ \frac{1}{L-\lambda}\sum_{i=1}^{L-\lambda}\left(e_{i,j} - e_{i+\lambda,j}\right)^2 & j = 1..20,\ \lambda < L \end{cases} \tag{5}$$

where e_j is a related factor for j-type amino acid, whose contiguous distance is along each segmented protein sequence.

2.5 Feature Weighted Rotation Forest Classifier

In this paper, we use the feature weighted rotation forest (FwRF) [33–35] to accurately predict DTIs. Compared with the original rotation forest, FwRF adds the function of weight selection. Through FwRF we can remove the noise features with small weights, thus increasing the content of useful information and improving the accuracy of prediction. The weights of the features are calculated by χ^2 statistical method. The feature F for the class can be obtained by the following formula.

$$\chi^2 = \sum_{i=1}^{n}\sum_{j=1}^{2}\frac{\left(\Upsilon_{ij} - \beta_{ij}\right)^2}{\beta_{i,j}} \tag{6}$$

where n is the number of values in feature F, Υ_{ij} is the number of features in the f_j class with a value of v_i, which can be expressed as a formula:

$$\Upsilon_{ij} = count(F = v_i \ and \ C = f_j) \tag{7}$$

$\beta_{i,j}$ is the expected value of v_i and f_j, which is defined as follows:

$$\beta_{i,j} = \frac{count(F = v_i) \times count(C = f_j)}{N} \tag{8}$$

where $count(C = f_j)$ is the number of samples with the value of f_j in class C, $count(F = v_i)$ is the number of samples whose value of feature F is v_i, and N is the number of samples.

The implementation steps of feature weighted rotation forest are as follows: Firstly, the weights of all features are calculated by Eq. 6; secondly, the features are sorted according to the weights; finally, the desired features are selected according to the given feature selection rate r. After performing these steps, we get a new dataset and send it to rotation forest.

Assuming $\{x_i, y_i\}$ contains S training samples, where in $x_i = (x_{i1}, x_{i2}, \ldots, x_{in})$ be an n-dimensional feature vector. Let X is the training sample set, Y is the corresponding labels and F is the feature set. Then X is $S \times n$ matrix, which is composed of n observation feature vector composition. Assuming that the number of decision trees is N, then the decision trees can be expressed as $D_1, D_2, \ldots, D_N$. The algorithm is executed in the following steps.

(1) Using the appropriate parameter K to randomly divide F into K independent and uncrossed subsets, the number of each subset feature is n/k.
(2) A corresponding column of features in the subset $D_{i,j}$ is selected from the training set X to form a new matrix $X_{i,j}$. Then, 75% of the data is extracted from X in the form of bootstrap to form a new set $X'_{i,j}$.
(3) Use matrix $X'_{i,j}$ as the feature transform to generate coefficients in matrix $M_{i,j}$.
(4) Using the coefficients obtained from the matrix $M_{i,j}$ to form a sparse rotation matrix R_i, the expression of which is as follows:

$$R_i = \begin{bmatrix} e_{i,1}^{(1)}, \ldots, e_{i,1}^{(N_1)} & 0 & \cdots & 0 \\ 0 & e_{i,2}^{(1)}, \ldots, e_{i,2}^{(N_2)} & \cdots & 0 \\ \vdots & \vdots & \ddots & \vdots \\ 0 & 0 & \cdots & e_{i,k}^{(1)}, \ldots, e_{i,k}^{(N_k)} \end{bmatrix} \tag{9}$$

In classification, the test sample x is determined to belong to the y_i class by the x generated by the classifier D_i of $d_{i,j}(XR_i^e)$. Then calculate the confidence class by the following average combination formula:

$$\lambda_j(x) = \frac{1}{N}\sum\nolimits_{i=1}^{N} d_{i,j}(XR_i^e) \tag{10}$$

Finally, the class with the largest value $\lambda_j(x)$ is discriminated as.

3 Results and Discussion

3.1 Evaluation Criteria

In this paper, accuracy (Accu.), sensitivity (Sen.), precision (Prec.), and Matthews correlation coefficient (MCC) are used to estimate the performance of DTIFS. Their formulas are as follows:

$$Accu. = \frac{TP + TN}{TP + TN + FP + FN} \tag{11}$$

$$\text{Sen.} = \frac{\text{TP}}{\text{TP} + \text{FN}} \tag{12}$$

$$\text{Prec.} = \frac{\text{TP}}{\text{TP} + \text{FP}} \tag{13}$$

$$\text{MCC} = \frac{\text{TP} \times \text{TN} - \text{FP} \times \text{FN}}{\sqrt{(\text{TP}+\text{FP})(\text{TP}+\text{FN})(\text{TN}+\text{FP})(\text{TN}+\text{FN})}} \tag{14}$$

where *TP* is the number of drug-target pairs that are related to each other to be correctly identified; *FP* is the number of drug-target pairs that are related to each other to be incorrectly identified; *TN* is the number of drug-target pairs that are not related to each other to be correctly identified; *FN* is the number of drug-target pairs that are not related to each other to be incorrectly identified. Moreover, the receiver operating characteristic (ROC) curve [36–39] and area under the ROC curve (AUC) are used to visually display the performance of the classifier.

3.2 Evaluation of Model Prediction Ability

After finding the optimal parameters of the DTIFS, we put them in benchmark datasets, including *Enzyme*, *Ion Channel*, *GPCR* and *Nuclear Receptor*. In order to avoid over-fitting of the model, we use 5-fold cross-validation (CV) method to evaluate the performance of the model. More specifically, we split the dataset into five subsets, one of which is taken as the test set, and the remaining four are used as the training set. Then, the cross-validation process will be repeated five rounds. The results from the 5 times are then averaged to produce the final result.

Tables 1, 2, 3 and 4 list the predicted results by DTIFS on four benchmark datasets. In *Enzyme* dataset, we gained the average of accuracy, sensitivity, precision, MCC, and AUC were 91.68%, 90.84%, 92.39%, 83.39%, and 91.72%. Their standard deviations were 0.84%, 1.68%, 1.37%, 1.68%, and 1.06%. In *Ion Channel* dataset, we achieved these evaluation criteria were 88.11%, 90.30%, 86.57%, 79.02%, and 88.27%. Their standard deviations were 1.01%, 1.61%, 2.29%, 1.55%, and 1.36%. In *GPCR* dataset, we yielded the average of these evaluation criteria were 84.72%, 84.73%, 84.73%, 74.06%, and 85.57%. Their standard deviations were 1.94%, 3.45%, 4.21%, 2.68%, and 2.28%. In *Nuclear Receptor* dataset, we gained the average of these evaluation criteria were 78.33%, 81.97%, 78.08%, 65.56%, and 75.31%. Their standard deviations were 5.34%, 7.85%, 12.56%, 6.05%, and 5.87%. Figures 2, 3, 4 and 5 draw the ROC curve generated from DTIFS on the four benchmark datasets.

Table 1. Experimental results of 5-fold CV of DTIFS on *Enzyme* dataset

Test set	Accu. (%)	Sen. (%)	Prec. (%)	MCC (%)	AUC (%)
1	90.51	89.20	91.27	81.03	90.04
2	92.82	93.22	92.59	85.64	92.96
3	91.62	90.19	92.74	83.28	92.09
4	91.97	89.68	94.40	84.05	91.79
5	91.47	91.90	90.96	82.94	91.73
Average	91.68 ± 0.84	90.84 ± 1.68	92.39 ± 1.37	83.39 ± 1.68	91.72 ± 1.06

3.3 Comparison Between DTIFS and SVM Classifier Model

As the most versatile Support Vector Machine (SVM) classifier has been widely used by various problems. In order to estimate DTIFS clearly, we compare the results of DTIFS and SVM classifier model on the same dataset. From the Table 5 we can see that DTIFS has achieved excellent results on the four benchmark datasets. Among the evaluation parameters accuracy, sensitivity, MCC and AUC, DTIFS have achieved the highest results, and DTIFS on precision is only slightly lower than that of SVM classifier model in *Enzyme* and *Ion Channel* datasets. This result indicates that the FwRF classifier is suitable for the proposed model and can effectively improve the performance of the model.

3.4 Comparison with Existing Methods

The prediction of the relationship between drugs and targets has drawn increasing interest of researchers. So far, a lot of excellent computational approaches have been designed. To better verify the proposed approach, we compare it with other existing methods using 5-fold CV on the same benchmark datasets. Table 6 lists the details of other excellent methods and DTIFS on four benchmark datasets in terms of the AUC. It can be seen that the results obtained by DTIFS on *Enzyme* and *Ion Channel* datasets are significantly higher than those of other existing methods, and the results achieved on *GPCR* datasets by DTIFS only lower than the highest result 1.13%. The performance of DTIFS on *Nuclear Receptor* dataset is not very good, it may be because the sample number of the *Nuclear Receptor* dataset is too small, and the training of the classifier is not sufficient.

Table 2. Experimental results of 5-fold CV of DTIFS on *Ion Channel* dataset

Test set	Accu. (%)	Sen. (%)	Prec. (%)	MCC (%)	AUC (%)
1	86.61	90.38	83.76	76.76	86.16
2	87.63	91.92	84.78	78.22	87.83
3	88.98	91.61	87.22	80.36	89.68
4	88.31	89.67	87.62	79.33	89.07
5	89.02	87.93	89.47	80.43	88.59
Average	88.11 ± 1.01	90.30 ± 1.61	86.57 ± 2.29	79.02 ± 1.55	88.27 ± 1.36

Table 3. Experimental results of 5-fold CV of DTIFS on *GPCR* dataset

Test set	Accu. (%)	Sen. (%)	Prec. (%)	MCC (%)	AUC (%)
1	82.28	86.21	77.52	70.73	82.86
2	87.01	88.62	85.16	77.38	88.82
3	86.22	86.52	88.41	76.00	86.72
4	84.63	80.33	86.73	73.83	84.37
5	83.46	81.95	85.83	72.37	85.11
Average	84.72 ± 1.94	84.73 ± 3.45	84.73 ± 4.21	74.06 ± 2.68	85.57 ± 2.28

Table 4. Experimental results of 5-fold CV of DTIFS on *Nuclear Receptor* dataset

Test set	Accu. (%)	Sen. (%)	Prec. (%)	MCC (%)	AUC (%)
1	69.44	83.33	65.22	55.90	72.22
2	77.78	85.00	77.27	64.34	69.69
3	80.56	92.31	66.67	67.47	74.25
4	83.33	77.78	87.50	72.05	75.31
5	80.56	71.43	93.75	68.03	85.08
Average	78.33 ± 5.34	81.97 ± 7.85	78.08 ± 12.56	65.56 ± 6.05	75.31 ± 5.87

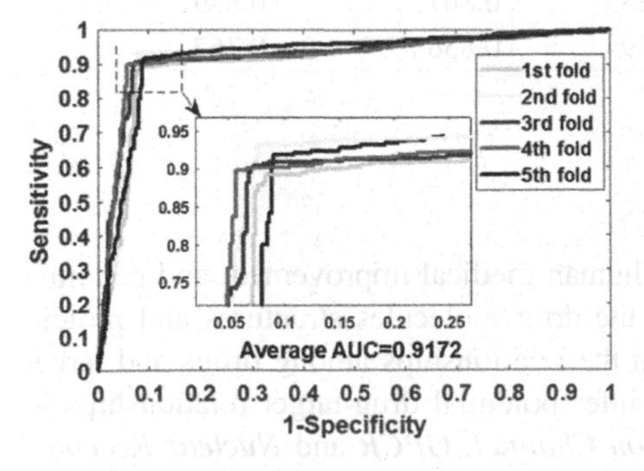

Fig. 2. ROC curves obtained by DTIFS on *Enzyme* dataset.

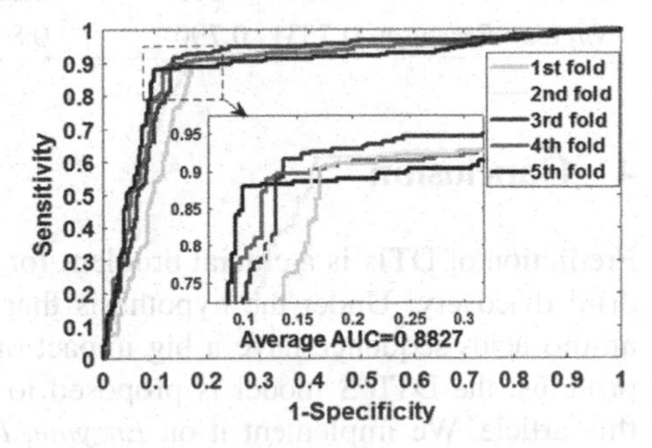

Fig. 3. ROC curves obtained by DTIFS on *Ion Channel* dataset.

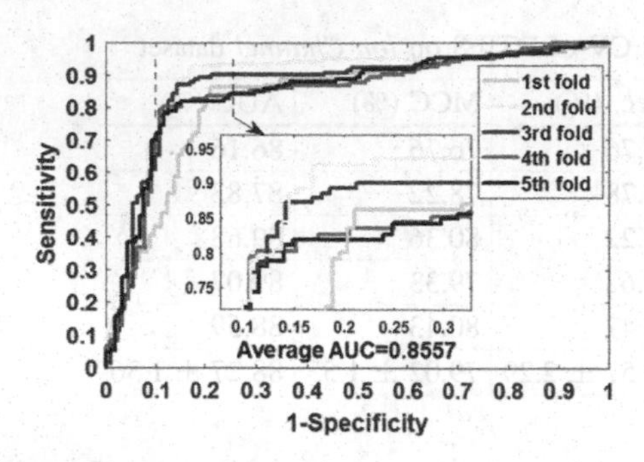

Fig. 4. ROC curves obtained by DTIFS on *GPCR* dataset.

Fig. 5. ROC curves obtained by DTIFS on *Nuclear Receptor* dataset.

Table 5. Results of comparison experiments between DTIFS and SVM classifier model on four benchmark datasets

Dataset	Method	Accu. (%)	Sen. (%)	Prec. (%)	MCC (%)	AUC (%)
Enzyme	PsePSSM+SVM	84.20 ± 0.60	69.90 ± 1.70	98.00 ± 0.50	71.50 ± 1.00	84.30 ± 1.20
	DTIFS	91.68 ± 0.84	90.84 ± 1.68	92.39 ± 1.37	83.39 ± 1.68	91.72 ± 1.06
Ion Channel	PsePSSM+SVM	81.90 ± 1.20	69.70 ± 3.70	92.40 ± 2.20	66.00 ± 1.90	81.70 ± 1.20
	DTIFS	88.11 ± 1.01	90.30 ± 1.61	86.57 ± 2.29	79.02 ± 1.55	88.27 ± 1.36
GPCR	PsePSSM+SVM	70.00 ± 2.10	50.40 ± 7.80	82.30 ± 3.30	42.80 ± 4.90	70.10 ± 2.70
	DTIFS	84.72 ± 1.94	84.73 ± 3.45	84.73 ± 4.21	74.06 ± 2.68	85.57 ± 2.28
Nuclear Receptor	PsePSSM+SVM	63.30 ± 3.60	57.60 ± 7.90	67.50 ± 14.60	29.60 ± 7.40	61.80 ± 5.80
	DTIFS	78.33 ± 5.34	81.97 ± 7.85	78.08 ± 12.56	65.56 ± 6.05	75.31 ± 5.87

Table 6. Performances of other excellent methods and DTIFS on four benchmark datasets in terms of the AUC

Dataset	DTIFS	MLCLE [40]	NetCBP [41]	SIMCOMP [42]	AM-PSSM [43]
Enzyme	0.9172	0.842	0.8251	0.863	0.843
Ion Channel	0.8827	0.795	0.8034	0.776	0.722
GPCR	0.8557	0.850	0.8235	0.867	0.839
Nuclear Receptor	0.7531	0.790	0.8394	0.856	0.767

4 Conclusion

Prediction of DTIs is a crucial problem for human medical improvement and genomic drug discovery. Under the hypothesis that the drug molecules structures and protein amino acids sequence have a big impact on the relationships among drugs and target proteins, the DTIFS model is proposed to infer potential drug-target relationships in this article. We implement it on *Enzyme*, *Ion Channel*, *GPCR* and *Nuclear Receptor* datasets, and obtained excellent results. To further evaluate the performance of the proposed approach, we compared it with the SVM classifier model and other existing methods on the same datasets. In these comparative experiments, DTIFS showed superior performance. These results demonstrated that DTIFS has high reliability and stability and can be used as an effective tool for predicting DTIs.

Acknowledgements. This work is supported in part by the National Natural Science Foundation of China, under Grants 61702444, in part by the West Light Foundation of The Chinese Academy of Sciences, under Grant 2018-XBQNXZ-B-008, in part by the Chinese Postdoctoral Science Foundation, under Grant 2019M653804, in part by the Tianshan youth - Excellent Youth, under Grant 2019Q029, in part by the Qingtan scholar talent project of Zaozhuang University. The authors would like to thank all anonymous reviewers for their constructive advices.

References

1. Xia, Z., Wu, L.-Y., Zhou, X., et al.: Semi-supervised drug-protein interaction prediction from heterogeneous biological spaces. BMC Syst. Biol. **4** (2010)
2. Wang, L., You, Z.-H., Chen, X., et al.: A computational-based method for predicting drug–target interactions by using stacked autoencoder deep neural network. J. Comput. Biol. **25**, 361–373 (2018)
3. Landry, Y., Gies, J.-P.: Drugs and their molecular targets: an updated overview. Fundam. Clin. Pharmacol. **22**, 1–18 (2008)
4. Wang, L., et al.: Computational methods for the prediction of drug-target interactions from drug fingerprints and protein sequences by stacked auto-encoder deep neural network. In: Cai, Z., Daescu, O., Li, M. (eds.) ISBRA 2017. LNCS, vol. 10330, pp. 46–58. Springer, Cham (2017). https://doi.org/10.1007/978-3-319-59575-7_5
5. Wang, L., Yan, X., Liu, M.-L., et al.: Prediction of RNA-protein interactions by combining deep convolutional neural network with feature selection ensemble method. J. Theor. Biol. **461**, 230–238 (2019)
6. Wang, L., You, Z.H., Chen, X., et al.: An ensemble approach for large-scale identification of protein-protein interactions using the alignments of multiple sequences. Oncotarget **8**, 5149 (2017)
7. Chen, Z.-H., Yi, H.-C., Guo, Z.-H., et al.: Prediction of drug-target interactions from multi-molecular network based on deep walk embedding model. Front. Bioeng. Biotechnol. **8**, 338 (2020)
8. Gao, Z.G., Wang, L., Xia, S.X., et al.: Ens-PPI: a novel ensemble classifier for predicting the interactions of proteins using autocovariance transformation from PSSM. Biomed. Res. Int. **8** (2016)
9. Wang, L., You, Z.-H., Li, L.-P., et al.: incorporating chemical sub-structures and protein evolutionary information for inferring drug-target interactions. Sci. Rep. **10**, 1–11 (2020)
10. Wu, Z., Cheng, F., Li, J., et al.: SDTNBI: an integrated network and chemoinformatics tool for systematic prediction of drug–target interactions and drug repositioning. Brief. Bioinform. **18**, 333–347 (2017)
11. Zong, N., Kim, H., Ngo, V., et al.: Deep mining heterogeneous networks of biomedical linked data to predict novel drug-target associations. Bioinformatics **33**, 2337–2344 (2017)
12. Peng, L., Liao, B., Zhu, W., et al.: Predicting drug-target interactions with multi-information fusion. IEEE J. Biomed. Health Inf. **21**, 561–572 (2017)
13. Ezzat, A., Wu, M., Li, X.L., et al.: Drug-target interaction prediction using ensemble learning and dimensionality reduction. Methods **129**, 81 (2017)

14. Wen, M., Zhang, Z., Niu, S., et al.: Deep-learning-based drug-target interaction prediction. J. Proteome Res. **16**, 1401 (2017)
15. Yamanishi, Y., Araki, M., Gutteridge, A., et al.: Prediction of drug-target interaction networks from the integration of chemical and genomic spaces. Bioinformatics **24**, I232–I240 (2008)
16. Schomburg, I., Chang, A., Ebeling, C., et al.: BRENDA, the enzyme database: updates and major new developments. Nucleic Acids Res. **32**, D431–D433 (2004)
17. Wishart, D.S., Knox, C., Guo, A.C., et al.: DrugBank: a knowledgebase for drugs, drug actions and drug targets. Nucleic Acids Res. **36**, D901–D906 (2008)
18. Gunther, S., Kuhn, M., Dunkel, M., et al.: SuperTarget and matador: resources for exploring drug-target relationships. Nucleic Acids Res. **36**, D919–D922 (2008)
19. Kanehisa, M., Goto, S., Hattori, M., et al.: From genomics to chemical genomics: new developments in KEGG. Nucleic Acids Res. **34**, D354–D357 (2006)
20. Wang, L., You, Z.H., Chen, X., et al.: RFDT: a rotation forest-based predictor for predicting drug-target interactions using drug structure and protein sequence information. Curr. Protein Pept. Sci. **19**, 445–454 (2018)
21. Jiang, H.-J., Huang, Y.-A., You, Z.-H.: SAEROF: an ensemble approach for large-scale drug-disease association prediction by incorporating rotation forest and sparse autoencoder deep neural network. Sci. Rep. **10**, 4972 (2020)
22. Wang, L., Wang, H.-F., Liu, S.-R., et al.: Predicting protein-protein interactions from matrix-based protein sequence using convolution neural network and feature-selective rotation forest. Sci. Rep. **9**, 9848 (2019)
23. Gribskov, M., McLachlan, A.D., Eisenberg, D.: Profile analysis: detection of distantly related proteins. Proc. Natl. Acad. Sci. U.S.A. **84**, 4355–4358 (1987)
24. Wang, L., You, Z.-H., Yan, X., et al.: Using two-dimensional principal component analysis and rotation forest for prediction of protein-protein interactions. Sci. Rep. **8**, 12874 (2018)
25. Zheng, K., You, Z.-H., Wang, L., et al.: MLMDA: a machine learning approach to predict and validate MicroRNA–disease associations by integrating of heterogenous information sources. J. Transl. Med. **17**, 260 (2019)
26. Jones, D.T.: Protein secondary structure prediction based on position-specific scoring matrices. J. Mol. Biol. **292**, 195–202 (1999)
27. Jones, D.T., Ward, J.J.: Prediction of disordered regions in proteins from position specific score matrices. Proteins-Struct. Funct. Bioinf. **53**, 573–578 (2003)
28. Altschul, S.F., Madden, T.L., Schaffer, A.A., et al.: Gapped BLAST and PSI-BLAST: a new generation of protein database search programs. Nucleic Acids Res. **25**, 3389–3402 (1997)
29. Wang, L., You, Z.-H., Xia, S.-X., et al.: Advancing the prediction accuracy of protein-protein interactions by utilizing evolutionary information from position-specific scoring matrix and ensemble classifier. J. Theor. Biol. **418**, 105–110 (2017)
30. Wang, L., You, Z.-H., Huang, D.-S., et al.: Combining high speed ELM learning with a deep convolutional neural network feature encoding for predicting protein-RNA interactions. IEEE/ACM Trans. Comput. Biol. Bioinf. **1**, 1 (2018)
31. Wang, L., You, Z.-H., Chen, X., et al.: LMTRDA: using logistic model tree to predict MiRNA-disease associations by fusing multi-source information of sequences and similarities. PLoS Comput. Biol. **15**, e1006865 (2019)
32. Chou, K.C.: Prediction of protein cellular attributes using pseudo-amino acid composition. Proteins-Struct. Funct. Genet. **43**, 246–255 (2001)
33. Rodriguez, J.J., Kuncheva, L.I.: Rotation forest: A new classifier ensemble method. IEEE Trans. Pattern Anal. Mach. Intell. **28**, 1619–1630 (2006)

34. Wang, L., et al.: An improved efficient rotation forest algorithm to predict the interactions among proteins. Soft. Comput. **22**(10), 3373–3381 (2017). https://doi.org/10.1007/s00500-017-2582-y
35. Zheng, K., You, Z.-H., Wang, L., et al.: Dbmda: A unified embedding for sequence-based mirna similarity measure with applications to predict and validate mirna-disease associations. Mol. Ther.-Nucleic Acids **19**, 602–611 (2020)
36. Zweig, M.H., Campbell, G.: Receiver-operating characteristic (ROC) plots: a fundamental evaluation tool in clinical medicine. Clin. Chem. **39**, 561–577 (1993)
37. Zheng, K., Wang, L., You, Z.-H.: CGMDA: an approach to predict and validate MicroRNA-disease associations by utilizing chaos game representation and LightGBM. IEEE Access **7**, 133314–133323 (2019)
38. Chen, Z.-H., You, Z.-H., Li, L.-P., et al.: Identification of self-interacting proteins by integrating random projection classifier and finite impulse response filter. BMC Genom. **20**, 1–10 (2019)
39. Chen, Z.-H., Li, L.-P., He, Z., et al.: An improved deep forest model for predicting self-interacting proteins from protein sequence using wavelet transformation. Front. Genet. **10**, 90 (2019)
40. Pliakos K., Vens, C., Tsoumakas, G.: Predicting drug-target interactions with multi-label classification and label partitioning. IEEE/ACM Trans. Comput. Biol. Bioinf. (2019)
41. Chen, H., Zhang, Z.: A semi-supervised method for drug-target interaction prediction with consistency in networks. PLoS One **8**, e62975 (2013)
42. Öztürk, H., Ozkirimli, E., Özgür, A.: A comparative study of SMILES-based compound similarity functions for drug-target interaction prediction. BMC Bioinf. **17**, 1–11 (2016)
43. Mousavian, Z., Khakabimamaghani, S., Kavousi, K., et al.: Drug-target interaction prediction from pssm based evolutionary information. J. Pharmacol. Toxicol. Methods **78**, 42–51 (2015)

HGAlinker: Drug-Disease Association Prediction Based on Attention Mechanism of Heterogeneous Graph

Xiaozhu Jing[1], Wei Jiang[1], Zhongqing Zhang[1], Yadong Wang[1,2](✉), and Junyi Li[1](✉)

[1] School of Computer Science and Technology, Harbin Institute of Technology (Shenzhen), Shenzhen 518055, Guangdong, China
{ydwang,lijunyi}@hit.edu.cn

[2] Center for Bioinformatics, School of Computer Science and Technology, Harbin Institute of Technology, Harbin 150001, Heilongjiang, China

Abstract. Drug repositioning raises great research interests because it saves a lot of time and economic cost in drug development. Predicting drug-disease associations based on integrated multi-dimensional biological data and networks will have better biological interpretation. At present, there are analysis challenges in scalability and information fusion across biological heterogeneous networks. In this paper, we construct a drug-disease-protein three-layer heterogeneous network, and propose a model HGAlinker to predict the links of drug and disease nodes based on the attention mechanism of heterogeneous network graph. HGAlinker has excellent performance compared with other related research methods. We analyze the model parameters and prove the robustness of the model. Through the case study, we demonstrate the biological effectiveness of the model. HGAlinker's method can also be extended to other prediction of heterogeneous network.

Keywords: Drug repositioning · Heterogeneous network · Attention mechanism · Link prediction

1 Introduction

Drug molecules are generally designed for specific targets of diseases. However, due to the complexity of diseases, a disease may be the result of abnormal expression of multiple genes, and drug molecules may have effects on targets of multiple diseases. Therefore, it is of great practical significance to explore the potential use of drugs, that is, drug repositioning, which can replace drug development to some extent.

At present, the emergence of new biomedical technology makes use of large-scale and multi perspective measurement results [1, 2]. However, based on the complexity of disease generation mechanism, no single data type can capture all the factors related to understanding every phenomena of the disease. It is imperative to explore drug repositioning in an appropriate integrated environment.

Most studies focus on the use of similarity to find common characteristics of drug or disease data. For example, Ha et al. learned drug relocation by the similarity of drug

D.-S. Huang and K.-H. Jo (Eds.): ICIC 2020, LNCS 12464, pp. 384–396, 2020.
https://doi.org/10.1007/978-3-030-60802-6_34

chemical structure [3]. With more and more researches recognizing the complexity of pathological process, clinical data, genomic data and others have been included in the scope of research. These research categories include graph model, matrix decomposition and deep learning. Graph methods construct heterogeneous network by using pharmacological process related dimension data, then they use walking or propagation strategy to mine node relevance [4–8]. Matrix decomposition technology decomposes the known drug-disease treatment relationship matrix into two low-dimensional spaces to explore the correlation between them [9, 10]. However, the model is easily limited for matrix sparsity.

Network representation learning (NRL) is applied to encode the potential information of the local and overall structure in the network. It is often combined with the downstream network tasks (such as link prediction). It will eventually project the graphs or nodes into the low dimension space. Mainstream methods, such as node2vec [11], deepwalk [8, 12], generally use random-walk-based strategics. In recent years, there are also matrix decomposition methods [13, 14] and deep-learning-based methods [15, 16]. These are all models developed for homogeneous networks. However, most of the real problems are modeled as heterogeneous networks with multiple nodes or edges.

In recent years, a large number of algorithm frameworks combine NRL and link prediction. The link prediction effect under this framework is better than the traditional methods. However, these methods can only evaluate the embedding quality when they are combined with the downstream tasks. These algorithms also lack scalability and expressibility [11, 12].

Recently, a large number of researches transfer deep learning technology to network representation learning, so as to learn the low-dimensional representation of network information, which is conducive to the research of link prediction. A lot of work has been done to transfer the concept of convolution to graph data. Furthermore, attention mechanism is added to the network of the graph convolution network. Velikovic [17] et al. proposed graph attention network (GAT). GAT adjusts the weight of neighborhood nodes in an adaptive way when aggregating node information. Wang [18] et al. proposed a hierarchical heterogeneous graph attention network and achieved good results in multiple data sets. In addition, success in predicting microRNA-disease associations promotes researchers to study drug-disease link prediction [19].

We start from the problem of drug-disease link prediction to the field of drug repositioning, and apply the mainstream framework of network representation learning combined with link prediction. Deep learning model for node embedding are chosen. Other kinds of dimensional information, such as target protein, and Semantic differences between different types of association nodes are also taken into account.

In this paper, we propose HGAlinker, which is a link prediction model based on the heterogeneous graph attention, to predict novel drug-disease associations. Since our goal is to make predictions between heterogeneous nodes, we mine a variety of heterogeneous graph patterns for drug and disease nodes, and fuse the structural information and semantic information in the graphs layer by layer. Our contributions can be summarized as follows: (i) As far as we know, this is the first attempt to solve the problem of drug repositioning based on link prediction of heterogeneous graphs attention. And it enriches the research of drug repositioning. (ii) We propose a framework for link prediction across heterogeneous networks. For research nodes, we mine different heterogeneous graph patterns and fuse different semantic information. Our model performs well and can be applied to large-scale heterogeneous graphs. (iii) Through experiments, our model is superior to other models and has high robustness. And we prove the biological effectiveness of the model by the case study.

All additional files are available at: https://github.com/sharpwei/HGAlinker.

2 Dataset

We select three heterogeneous networks from the study of Luo [15] et al., including: drug-protein interaction network (DrugBank version 3.0 [20]), drug-disease association network, protein-disease association network (the Comparative Toxicogenics Database 2013 [21]). We also collected drug-drug isomorphic network data (DrugBank version 3.0).

There are 708 drug nodes and 5603 disease nodes in our data set (see Table 1 for detailed statistics). We analyze the number of diseases that can be treated in each drug node, among which the drug can treat at least one disease and at most 2918 diseases. Therefore, we divide 0-3000 into 20 intervals, record the number of drugs which can treat the number in the interval of diseases, and draw a histogram in Fig. 1(a).

In any network, if there is an edge between nodes, the edge weight is set to 1, otherwise it is 0. Based on these datasets, we build a three-layer heterogeneous network of drugs, diseases and proteins (Fig. 1(b)). Each node is allowed to carry eigenvectors.

3 Methods

HGAlinker predicts the heterogeneous edges of drug and disease nodes by integrating multiple heterogeneous graph patterns hierarchically, based on the attention mechanism of heterogeneous network. The working process is in Fig. 2.

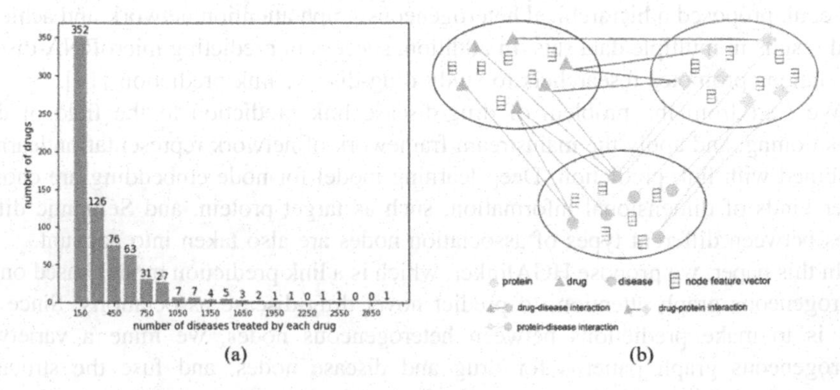

Fig. 1. Data overview. (a) drug-disease association (b) drug-disease-protein three-layer heterogeneous network

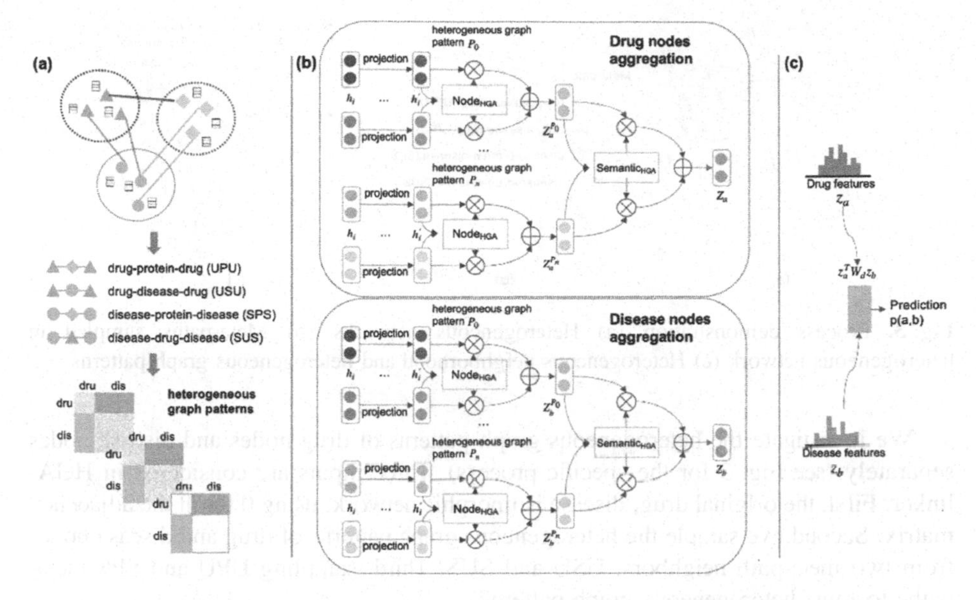

Fig. 2. Overall framework of HGAlinker (a) Mining heterogeneous graph patterns of similar nodes (b) Graph pattern fusion based on heterogeneous graph attention network (c) Link prediction for drug-disease links

3.1 Mining Heterogeneous Graph Patterns of Similar Nodes

Our target is to get drug-disease predictions. In the whole heterogeneous network, the same kind of nodes can be associated not only through the existing edge links, but also through the heterogeneous nodes. Considering the multiple relationships between the same kind of nodes, we can get better prediction results. Therefore, based on meta-path, HGAlinker mines the heterogeneous graph patterns of the two nodes. Sun [22] et al. defined a meta-path as $P = A_1 \xrightarrow{R_1} A_2 \xrightarrow{R_2} \ldots \xrightarrow{R_{n-2}} A_{n-1} \xrightarrow{R_{n-1}} A_n$ (simply recorded as $A_1A_2 \ldots A_n$), where $A_i \in A$ represents the object of type i and $R_j \in R$ represents the relationship type of type j. In our model, a two-step meta-path of drug1-protein-drug2 (UPU) indicates that two drugs can interact with the same target protein. Drug1-disease-drug3 (USU) indicates that both drugs can treat the same disease. These two meta-paths contain different semantic information and heterogeneous network structure information. And they also reveal that drug2 and drug3 are neighbors of drug1 based on meta-path.

In our work, we define heterogeneous neighbor and heterogeneous graph pattern. For any node i, the node that can be reached via meta-path is called heterogeneous neighbor of node i. Node i itself is also included. For any two nodes, all their heterogeneous neighbors and existing heterogeneous associations form a heterogeneous graph pattern.

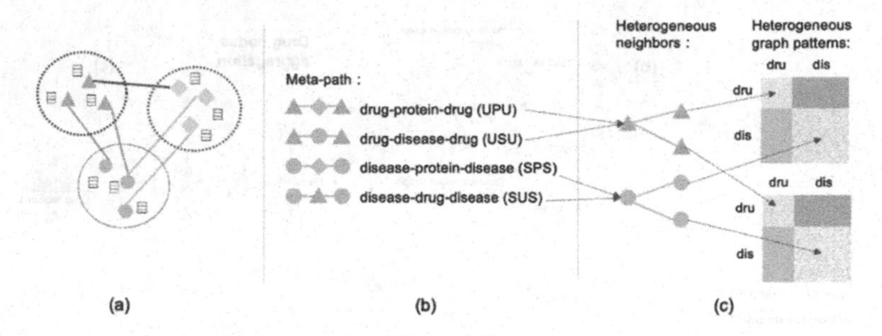

Fig. 3. Process demonstration (a) Heterogeneous networks (b) Meta-paths sampled in heterogeneous network (c) Heterogeneous neighborhood and heterogeneous graph patterns

We investigate the heterogeneous graph patterns of drug nodes and disease nodes separately (see Fig. 3 for the specific process). Three inputs are considered in HGA-linker. First, the original drug, disease isomorphic network, using 0 to fill the adjacency matrix. Second, we sample the heterogeneous graph patterns of drug and disease nodes from two meta-path neighbors: USU and SUS. Third, sampling UPU and SPS meta-paths to mine heterogeneous graph patterns.

3.2 Graph Pattern Fusion Based on Heterogeneous Graph Attention Method

In order to get high quality node embedding from heterogeneous networks, we follow the algorithm proposed by Wang [5] et al., which improves the graph attention mechanism to adapt to heterogeneous networks. We apply attention mechanism to the heterogeneous graph pattern and semantic fusion of heterogeneous graph pattern layer by layer. Then we can extract the rich potential information in heterogeneous network.

Step 1. Aggregation of node information in heterogeneous graph pattern

We use the attention mechanism to learn the neighborhood information of nodes differently. Due to the characteristics of heterogeneous graph pattern, node information in different heterogeneous graph patterns have different biological semantics.

We project the two different kinds of nodes into the same feature space to eliminate the differences. Different from Grover [4] et al., our specific type of transformation matrix is applied to different types. Transformation formula is as follows:

$$h_i' = M_{P_i} \cdot h_i \tag{1}$$

Where h_i' is the original feature vector of the node. We choose one-hot vector as the original feature vector of the node. h_i' is the feature vector after the node projection.

Then we learn the weights of nodes to (i, j) through self-attention [23]. Because different nodes have different neighborhood, the weight e_{ij}^{P} is asymmetric.

$$e_{ij}^{P} = node_{HGA}\left(h_i^{'}, h_j^{'}, P\right) \tag{2}$$

Among them, $node_{HGA}$ is just a algorithm specification symbol, P refers to the set of heterogeneous graph patterns.

Next, we normalize the weight coefficient by softmax, also asymmetrically:

$$\alpha_{ij}^{P} = softmax\left(e_{ij}^{P}\right) = \frac{exp\left(\sigma\left(\partial_P^T \cdot \left[h_i^{'} \| h_j^{'}\right]\right)\right)}{\sum_{k \in N_i^P} exp\left(\sigma\left(\partial_P^T \cdot \left[h_i^{'} \| h_j^{'}\right]\right)\right)} \tag{3}$$

Where $j \in N_i^P$, N_i^P represents the neighborhood of the heterogeneous graph patterns of node i (including itself), $\|$ represents the connection operation, and ∂_P is the attention vector of nodes on the heterogeneous graph pattern P.

To stabilize the learning process of self-attention and solve the problem of large variance of heterogeneous graph data, the model adopts multi-head-attention. Thus, we repeatedly aggregate the node information K times:

$$z_i^P = \|_{k=1}^{K} \sigma\left(\sum_{j \in N_i^P} \alpha_{ij}^{P} \cdot h_j^{'}\right) \tag{4}$$

We can get node embedding sets $\{Z_{P_0}, Z_{P_1}, \ldots, Z_{P_n}\}$ with different semantics.

Step 2. Semantic information fusion of heterogeneous graph patterns

For the heterogeneous graph pattern set, by applying the node information aggregation process, we can get n groups of heterogeneous graph pattern node information embedding with different semantics. To get node embedding with both network structure information and semantic information, the model continues to use attention mechanism to fuse semantic information differently:

$$\left(\beta_{P_0}, \beta_{P_1}, \ldots, \beta_{P_n}\right) = semantic_{HGA}(Z_{P_0}, Z_{P_1}, \ldots, Z_{P_n}) \tag{5}$$

Where $semantic_{HGA}$ is the algorithm specification symbol, $\left(\beta_{P_0}, \beta_{P_1}, \ldots, \beta_{P_n}\right)$ is the weight coefficient set of heterogeneous graph pattern fusion.

Here's how to get the coefficient. First, a layer of MLP is used to embed nodes on any heterogeneous graph pattern for nonlinear transformation. Set up a semantic fusion vector λ, and use it to measure the similarity between multiple groups of heterogeneous graph pattern nodes. Here, we equally consider the importance of node embedding on all heterogeneous graph patterns as follows:

$$\omega_{P_i} = \frac{1}{|v|} \sum_{i \in v} \lambda^T \cdot \tan h\left(W \cdot z_i^P + \theta\right) \tag{6}$$

Where W is the weight matrix and θ is the bias vector. All nodes embedded in heterogeneous graph mode share all the above parameters. We mentioned the usefulness of using softmax normalization earlier. Similarly, we use softmax to normalize all ω_P to obtain the weight of heterogeneous graph patterns P_i, which is expressed as β_{P_i}.

$$\beta_{P_i} = \frac{exp(\omega_{P_i})}{\sum_{i=1}^{n} exp(\omega_{P_i})} \tag{7}$$

Thus, we can fuse different heterogeneous graph patterns together:

$$Z = \sum_{i=1}^{n} \beta_{P_i} \cdot Z_{P_i} \tag{8}$$

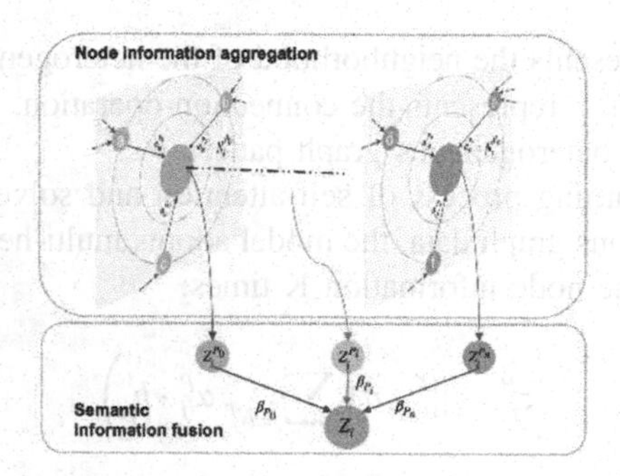

Fig. 4. Information aggregation

The whole information aggregation process is shown in Fig. 4.

3.3 Drug-Disease Link Prediction

Next, HGAlinker uses the previously learned node embedding to reconstruct the edge of the drug-disease heterogeneous network. Similar to the edge decoder in Decagon [7], our decoder is modified as follows:

$$P(i,j) = \sigma\left(z_i^T W_d z_j\right) \tag{9}$$

Where $z_i^T \in R^c$ is learning embedding of drug node i, $W_d \in R^{c*c}$ is training parameter matrix, which simulates the interaction between z_i^T and z_j, and σ is sigmoid function. To reduce the risk of over fitting, we make the parameters of bilinear decoder model be shared among different drug disease pairs.

HGAlinker uses cross entropy loss as a loss function to train the whole model:

$$L(i,j) = -log\,P(i,j) - E_{k\sim P(j)} log(1 - P(i,k)) \tag{10}$$

Due to the lack of negative samples, HGAlinker uses the negative sampling strategy [7] to solve this problem, and the last term of Eq. (10) is also reflected. For each training sample (i, j), HGAlinker randomly samples a non-linked node k to form a link (i, k) as a negative example, and the second node k follows the sampling distribution P. Considering all edges, the cross entropy loss of the final model is as follows:

$$L = \sum\nolimits_{(i,j) \in E_{ud}} L(i,j) \tag{11}$$

Where E_{ud} represents all the connections between drugs and diseases. As discussed earlier, HGAlinker trains the model in an end-to-end manner, where the loss function gradient is propagated back to the parameters in the heterogeneous graph attention model and edge decoding model.

4 Results

4.1 HGAlinker Yields Superior Performance in Predicting New Drug–Disease Interactions

Table 1 shows the heterogeneous network data statistics. Here we demonstrate the performance advantages of our model. We first combine some other network representation learning methods with the link prediction part of HGAlinker to compare the advantages of our model and the embedded part of nodes in heterogeneous network data fusion. At the same time, we also repeat some other research work on this issue for comparison.

GAT [17]. The attention mechanism is applied to the neighborhood information of the aggregation node to learn the neighborhood information differently.

Graph factorization (GF) [24]. Use matrix decomposition to reduce the dimension of node adjacency matrix, with low time complexity.

Node2vec [11]. The random walk strategy is improved, and the network nodes are embedded by processing word vectors.

Metapath2vec [25]. Using random walk based on meta-path to build heterogeneous neighborhood of nodes, learning heterogeneous network node embedding.

Decagon [26]. A new model of graph convolutional network is developed, which can predict the link prediction of multiple relationships in heterogeneous networks.

Table 1. Data statistics

Relations (A-B)	Number of A	Number of B	Number of A-B	Meta-paths	Heterogeneous neighbor
Drug-protein	708	1512	1923	UPU	3110
Drug-disease	708	5,603	199214	USU	217488
Disease-protein	5603	1512	1,596,745	SPS	4302164
Disease-drug	5603	708	199214	SUS	5576179

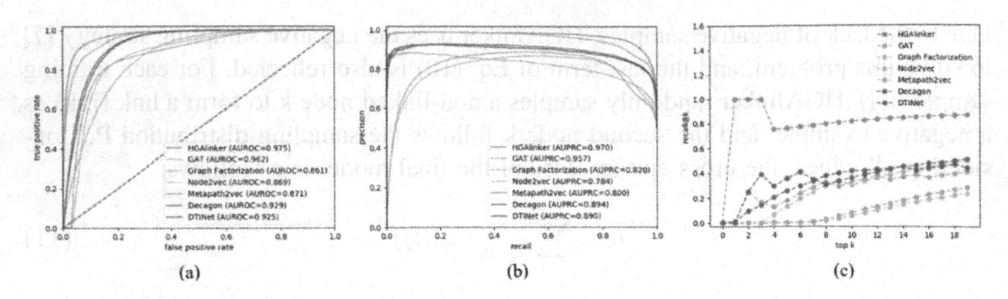

Fig. 5. Performance comparison of different methods (a) ROC curves of different methods (b) PR curve of different methods (c) recall of top k predictions of HGAlinker and other six methods

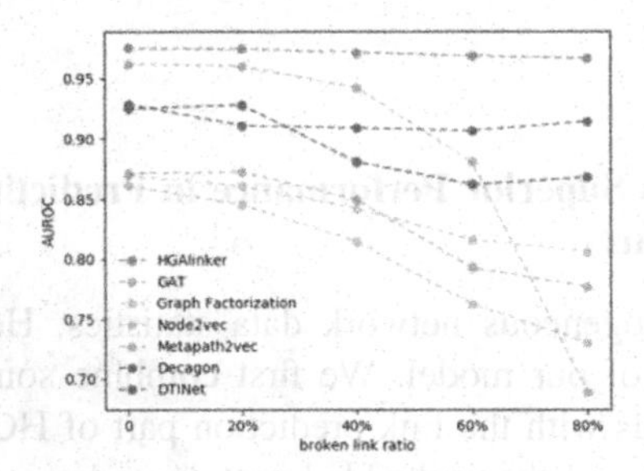

Fig. 6. Model robustness test

DTInet [15]. Using the dimension reduction strategy of DCA, the nodes are projected into the same low-dimensional space, with the help of similarity between nodes.

A certain sample size is enough to verify the model performance and test the model performance. Due to the large scale of the edge in our data sample, considering the training time of the model, we adjust the proportion of training set, verification set and test set to 90:5:5. Then we test the performance of different methods in our dataset. As shown in Fig. 5, HGAlinker is superior to other models in three evaluation indexes. On the one hand, the model can not only mine topological information in the network, but also make full use of specific information such as drug-disease and drug- protein. On the other hand, HGAlinker has excellent learning ability, and can better adapt to the complex heterogeneous network. In terms of ROC and PR curves, the values of AUROC and AUPRC of HGAlinker are higher than those of other models, showing excellent performance. At the same time, we also tested the performance of different models on recall of K, which has more biological research value. According to Fig. 5 (c), our model is much higher than the second-best model.

4.2 Robustness of HGAlinker

In order to test the robustness of the model, we randomly destroy 20%, 60% and 80% edges between existing links (Fig. 6). HGAlinker has always outperformed all other models and remained stable.

4.3 Parameter Sensitivity Analysis

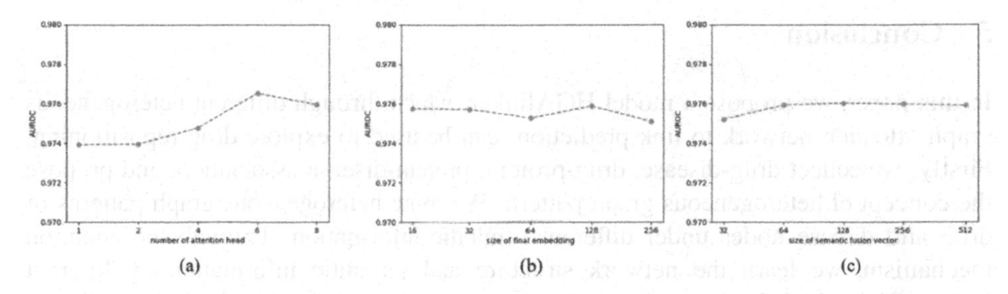

Fig. 7. Sensitivity analysis of parameters

In this paper, we discuss the influence of the size of attention head K, the dimension of semantic fusion vector and the size of final embedded Z of nodes on the model results. Here, we use AUROC as the evaluation index.

Attention Head K: In Fig. 7(a), we test different values of K. We found that the best result is obtained when k = 6.

The size of Final Embedding Z: As shown in Fig. 7(b), when the dimension of Z is less than 128, the model effect basically tends to be stable, and reaches the best effect at 128. After 128, the performance of the model declines obviously, which may be caused by over fitting.

Semantic Fusion Vector: We test different sizes of values, which will affect the ability to learn the semantic information of different heterogeneous graph patterns. As shown in Fig. 7(c), in the 32–512 range, the model has good robustness.

4.4 Case Study

In order to further explore the validity of HGAlinker results in biological interpretation, we selected two diseases for detailed case analysis. Among them, Alzheimer disease is a brain degenerative disease, characterized by the hidden hair of dementia. Anemia, hemolytic is a condition of insufficient circulating red blood cells (anemia) or hemoglobin due to premature destruction of red blood cells (RBC).

We first run the HGAlinker model on the complete data set to get all the drugs for each disease and rank them according to the possibility. Then we delete the existing drug-disease association and select the remaining top 20 possible drugs for each disease. Table that contains the remaining top 20 possible drugs for Alzheimer disease

(MESH: D000544) and Anemia, Hemolytic (MESH: D000743) can be seen in additional files in github.

Sulfasalazine, Indomethacin, Docetaxel are the top3 drug related to Alzheimer. Sulfasalazine, Indomethacin, Docetaxel are the top3 drug related to Anemia, Hemolytic. All our prediction results of Alzheimer disease can be found in the database CTD [27], and 80% of the predictions of Anemia, Hemolytic can also be verified. Thus, the biological validity of HGAlinker can be proved.

5 Conclusion

In this paper, we propose a model HGAlinker, which through different heterogeneous graph attention network to link prediction, can be used to explore drug repositioning. Firstly, we collect drug-disease, drug-protein, protein-disease association, and propose the concept of heterogeneous graph pattern. We mine heterogeneous graph patterns of drug and disease nodes under different semantic information. Through the attention mechanism, we learn the network structure and semantic information in different layers. We embed the learned nodes as the input downstream of the link prediction task, and find new insights into the relationship between drugs and diseases.

In order to prove the adaptability of our model on heterogeneous networks, we compare other representation learning algorithms with our link prediction tasks. At the same time, we also choose some of the previous work of this problem, which fully proves the superiority of our model. Through parameter analysis, robustness experiment and case analysis, we prove the robustness and effectiveness of our model.

Although we have made great efforts in the experiment, our work is still possible to carry out further. First, we have a small number of drug nodes, especially the number of specific drugs. Due to the complexity of the disease mechanism, our three-layer heterogeneous network may not describe the whole biological process completely. After that, we can further screen and add more dimensional data, such as side effects and disease genes, and enrich the drug data set. Secondly, due to the limitation of time, we did not collect the auxiliary information of drugs and diseases. In the next step, we can do text mining for biomedical related data.

In conclusion, HGAlinker has excellent learning ability, and can easily expand the dataset, which has a good explanation in the research of drug relocation. In the future, we can do further work in mining the auxiliary information of drugs and disease nodes.

Acknowledgements. This work was supported by the grants from the National “863” Key Basic Research Development Program (2014AA021505), the National Key Research Program (2017YFC1201201) and the startup grant of Harbin Institute of Technology (Shenzhen).

Authors’ Contributions. XJ designed the study, performed bioinformatics analysis and drafted the manuscript. All of the authors performed the analysis and participated in the revision of the manuscript. JL and YW conceived of the study, participated in its design and coordination and drafted the manuscript. All authors read and approved the final manuscript.

Additional Files. All additional files are available at: https://github.com/sharpwei/HGAlinker.

Competing Interests. The authors declare that they have no competing interests.

References

1. Sun, M.Y., Zhao, S.D., et al.: Graph convolutional networks for computational drug development and discovery. Brief. Bioinform. **21**(3), 919–935 (2020)
2. Zhao, T., Hu, Y., et al.: Identifying drug-target interactions based on graph convolutional network and deep neural network. Brief. Bioinf. (2020)
3. Ha, S., Seo, Y.-J., et al.: IDMap: facilitating the detection of potential leads with therapeutic targets. Bioinformatics **24**(11), 1413–1415 (2008)
4. Chen, H., Zhang, H., et al.: Network-based inference methods for drug repositioning. Comput. Math. Methods Med. **2015**, 1–7 (2015)
5. Li, Z.-C., Huang, M.-H., et al.: Identification of drug–target interaction from interactome network with 'guilt-by-association' principle and topology features. Bioinformatics **32**(7), 1057–1064 (2015)
6. Martinez, V., Navarro, C., et al.: DrugNet: network-based drug-disease prioritization by integrating heterogeneous data. Artif. Intell. Med. **63**(1), 41–49 (2015)
7. Luo, H., Wang, J., et al.: Drug repositioning based on comprehensive similarity measures and Bi-Random walk algorithm. Bioinformatics **32**(17), 2664–2671 (2016)
8. Zong, N., Kim, H., et al.: Deep mining heterogeneous networks of biomedical linked data to predict novel drug–target associations. Bioinformatics **33**(15), 2337–2344 (2017)
9. Zhang, P., Wang, F., et al.: Towards drug repositioning: a unified computational framework for integrating multiple aspects of drug similarity and disease similarity, pp. 1258–1267 (2014)
10. Regenbogen, S., Wilkins, A.D., et al.: Computing therapy for precision medicine: collaborative filtering integrates and predicts multi-entity interactions (2016)
11. Grover, A., Leskovec, J.: node2vec: scalable feature learning for networks (2016)
12. Perozzi, B., Al-Rfou, R., et al.: DeepWalk: online learning of social representations (2014)
13. Wang, D., Cui, P., et al.: Structural deep network embedding (2016)
14. Zeng, X., Zhu, S., et al.: deepDR: a network-based deep learning approach to in silico drug repositioning. Bioinformatics **35**(24), 5191–5198 (2019)
15. Luo, Y., Zhao, X., et al.: A network integration approach for drug-target interaction prediction and computational drug repositioning from heterogeneous information. Nat. Commun. **8**(1), 573 (2017)
16. Gong, Y., Niu, Y., et al.: A network embedding-based multiple information integration method for the MiRNA-disease association prediction. BMC Bioinf. **20**(1), 468 (2019)
17. Velikovi, P., Cucurull, G., et al.: Graph attention networks (2017)
18. Wang, X., Ji, H., et al.: Heterogeneous graph attention network (2019)
19. Luo, J., Xiao, Q.: A novel approach for predicting microRNA-disease associations by unbalanced bi-random walk on heterogeneous network. J. Biomed. Inf. **66**, 194–203 (2017)
20. Knox, C., Law, V., et al.: DrugBank 3.0: a comprehensive resource for 'omics' research on drugs. Nucleic Acids Res. **39**, D1035–D1041 (2010)
21. The comparative toxicogenomics database: update 2013. J. Nucleic Acids Res. (2013)
22. Sun, Y., Han, J., et al.: PathSim: meta path-based top-K similarity search in Heterogeneous information networks. Proc. VLDB Endow. **4**, 992–1003 (2011)
23. Vaswani, A., Shazeer, N., et al.: Attention is all you need. In: 31st Conference on Neural Information Processing Systems (NIPS 2017) (CA), pp. 5998–6008 (2017)

24. Ahmed, A., Shervashidze, N., et al.: Distributed large-scale natural graph factorization. In: Proceedings of the 22nd International Conference on World Wide Web, Rio de Janeiro, Brazil, pp. 37–48 (2013)
25. Dong, Y., Chawla, N.V., et al.: metapath2vec: scalable representation learning for heterogeneous networks. In: Proceedings of the 23rd ACM SIGKDD International Conference on Knowledge Discovery and Data Mining, Halifax, NS, Canada, pp. 135–144 (2017)
26. Marinka, Z., Monica, A., et al.: Modeling polypharmacy side effects with graph convolutional networks. Bioinformatics **34**(13), i457–i466 (2018)
27. Allan, P., et al.: The comparative toxicogenomics database: update 2017 (2016)

Computational Genomics

A Probabilistic Matrix Decomposition Method for Identifying miRNA-Disease Associations

Keren He, Ronghui Wu, Zhenghao Zhu, Jinxin Li, and Xinguo Lu(✉)

Hunan University, Lushan Nan Rd. No. 2, Changsha 410082, China
hnluxinguo@126.com

Abstract. MicroRNA (miRNA) is a non-coding RNA molecule whose length is about 22 nucleotides. The growing evidence shows that miRNA makes critical regulations in the development of complex diseases, such as cancers and cardiovascular diseases. Predicting potential miRNA-disease associations can provide a new perspective to achieve a better schemes of disease diagnosis and prognosis. However, predicting some potential essential miRNAs only with few known associations has proved challenging. Here we propose a novel method of probabilistic matrix decomposition for identifying miRNA-disease associations in heterogeneous omics data. First, we construct disease similarity network and miRNA similarity network to preprocess the miRNAs with none available associations. Then, we apply probabilistic factorization to obtain two feature matrices of miRNA and disease. Finally, we utilize obtained feature matrixes to identify potential associations for all diseases. The results indicate that PMDA is superior over other methods in predicting potential miRNA-disease associations in sparse and unbalance data. Moreover, we further estimate the performance of novel interactions in three typical diseases, and simulation results illustrate that PMDA could also achieve satisfactory predictive performance for novel diseases and miRNAs.

Keywords: Probabilistic matrix decomposition · Disease similarity · miRNA similarity · miRNA-disease associations

1 Introduction

Complex diseases, such as cancers, cardiovascular and cerebrovascular diseases, and respiratory diseases, have become continuing global threats for their increasing prevalence and high mortality rate [1, 2]. The occurrence of disease is an extremely complicated process, and the cause of disease has always been focused by medical research. Thereinto, many computational methods [3, 4] have been proposed to identify the driver mutations and understand the functional impact of these mutations. MiRNA is a small non-coding RNA molecule (containing about 22 nucleotides) found in human, animals and some viruses, that can inhibit the expression of post-transcriptional genes by binding to target messenger ribonucleic acid (mRNA) [5–7]. Recently, there is growing evidence shows that microRNA (miRNA) play a critical role in the development of complex diseases. Existing research reveals that miRNA infect in regulating the cell cycle, the developmental timing of organisms and cause abnormal

D.-S. Huang and K.-H. Jo (Eds.): ICIC 2020, LNCS 12464, pp. 399–410, 2020.
https://doi.org/10.1007/978-3-030-60802-6_35

physiological responses in individuals [8, 9]. It provides a new perspective to achieve a better schemes of disease diagnosis and prognosis, as well as therapeutic targets. However, expensiveness and high failure rate of the biological verification experiment make obtaining miRNA-disease associations still challenging.

With the development of various high throughput analysis techniques, miRNA activities in physiologic and pathologic conditions is constantly undergoing researches. Meanwhile, a vast amount of omics data like miRNA-target interaction, transcription factor (TF) and miRNA co-regulatory motif identification have been generated in recent years. Many computational methods based on machine learning have been proposed to predict potential miRNA related associations with disease. Xuan et al. proposed a supervised learning approach to predict potential disease-associated miRNAs by searching for the most similar miRNA neighbors [10]. Pasquier et al. summarized a vector space model with singular value decomposition (SVD) to estimate the association of miRNAs with diseases by considering multiple miRNA-related information sources [11]. However, these analytical approaches inefficiently explore the interaction between miRNA and disease, so these prediction datasets have relatively high false-negative and false-positive rates. In addition, many other algorithms have been proposed to solve this problem. Lu et al. distinguished the subtype-specific drivers from normal genes by applying a module-network analysis [12]. Chen et al. proposed an approach IMCMDA that integrates functional similarity of miRNAs, disease semantic similarity, and nu- clear similarity of Gaussian interaction spectra of disease and miRNAs [13]. IMCMDA can effectively solve the problem of negative sample selection, and obtain satisfactory performance for miRNA-disease associations. Ding et al. proposed a prediction method for MFSP based on path computational miRNA functional similarity [14]. Dong et al. proposed a model of edge trace perturbation, named as EPMDA [15]. Its theory is based on an edge of graph will affect the whole structure, so the characteristics of each edge plays an important basis in miRNA-disease associations. However, these approaches could not efficaciously deal with miRNA which have very few associations with disease due to the distribution is unbalanced. Consequently, it remains a challenge to achieve significant performance for predicting interaction between miRNA and disease.

Hence, we propose a method, named as probabilistic matrix decomposition for identifying miRNA-disease associations (PMDA), to predict potential interaction between miRNA and disease. And we apply PMDA on near 500 miRNAs with 328 diseases from HMDD 2.0 [9, 16, 17]. To quantify the similarities for miRNAs and diseases, we fully exploit the semantic associations between diseases to quantify the similarities for diseases and miRNAs [18]. These processes improve the miRNAs with none available associations in the preprocessor. Moreover, to extend our method to miRNA or disease with few associations, we construct a weighted neighbor learning step based on prior information to assist both novel diseases or miRNAs and those with sparse known associations for prediction of potential miRNA-disease associations. Then, we decompose the adjacency matrix of miRNA-disease into low-order characteristic matrices U and V of the miRNAs and diseases based on the probability matrix decomposition. Finally, we utilize two feature matrices of miRNA and disease to identify potential associations for all diseases. The experimental results show that PMDA can more precisely predict the novel interaction than others (Fig. 1).

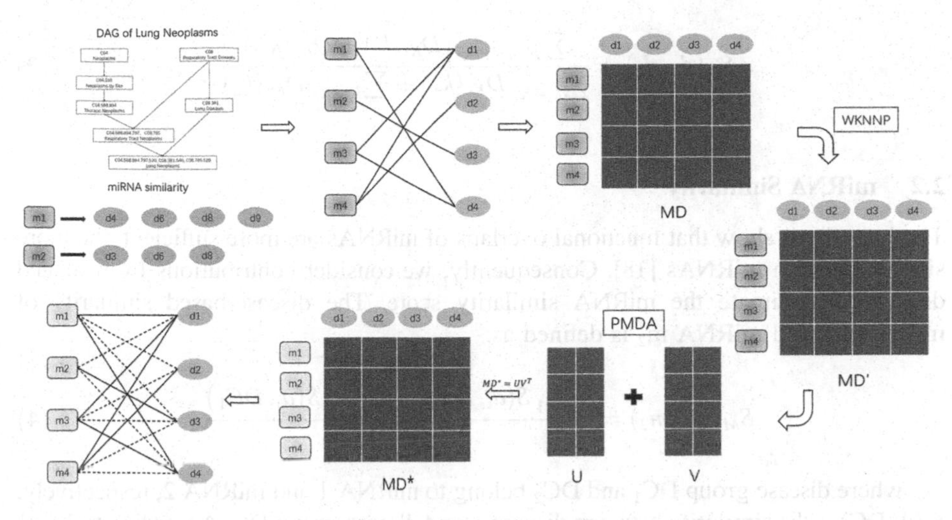

Fig. 1. PMDA workflow

2 Methods and Materials

2.1 Disease Similarity

We download directed acyclic graphs (DAGs) of category C to obtain the various relationships between diseases [19–21]. In DAGs, each node, which points to its children, represents a disease. And the edge represents a parent node point to its children. For example, lung Diseases, as the father node, point to Lung tumors and each node has a MeSH number in graph of lung neoplasms. For example, The MeSH number of Lung Diseases is C08.381, its' MeSH Numbers of children diseases is C04.588.894.797.520, C08.381.540 and C08.785.520.

Here we define the semantic contribution of the child disease k in the DAG relative to the disease t:

$$\begin{cases} D_d(k) = max\{\Delta * D_d(k') | k' \in children\ of\ k\}, k \neq d \\ D_d(k) = 1, k = d \end{cases} \tag{1}$$

where Δ is the semantic contribution factor of the edge E t connecting disease d and its child disease k, and Δ is set to 0.5 to better measure the semantic contribution according to the distance. In the DAG, disease d is the most specific disease, so we define contribution to its semantic value as 1. Subsequently, we can get the semantic value of disease d as follow:

$$D(d) = \sum \{D_d(k_i)\}, k_i \in K \tag{2}$$

where k_i is a disease subset of disease d. From this, the disease pair has more of the same ancestral diseases, the more similarity they are. Therefore, we can derive the following objective function for the similarity between the disease pair:

$$S_D(d_1, d_2) = \frac{\sum_{k' \in K_1 \cap K_2} (D_{K_1}(k') + D_{K_2}(k'))}{\sum_{k_1 \in K_1} D_{K_1}(k_1) + \sum_{k_2 \in K_2} D_{K_2}(k_2)} \quad (3)$$

2.2 miRNA Similarity

The hypothesis show that functional overlaps of miRNAs are more sufficient, the more similar between miRNAs [18]. Consequently, we consider contributions from related diseases to estimate the miRNA similarity score. The disease-based similarity of miRNA m_1 and miRNA m_2 is defined as:

$$S_M(m_1, m_2) = \frac{\sum_{i=1}^{m} S(d_i, DC_2) + \sum_{j=1}^{n} S(d_j, DC_1)}{m + n} \quad (4)$$

where disease group DC_1 and DC_2 belong to miRNA 1 and miRNA 2, respectively. S (d, DC) is the similarity between disease d and disease group DC. According to Eq. 1 and Eq. 2, S (d, DC) is defined as:

$$S(d, DC) = \max_{1 \le i \le l} S_D(d, d_i),\ d_i \in DC \quad (5)$$

2.3 Infrequent miRNAs with Neighbors Learning

We construct a miRNA-disease adjacency matrix MD, which includes 5186 associations involving 328 diseases and 490 miRNAs, from the HMDD 2.0 database [22, 23]. Let M $\{m_1, m_2, \ldots, m_n\}$ and D $\{d_1, d_2, \ldots, d_m\}$ denote the set of miRNAs and diseases, respectively. MM (m_n) is denoted as the associations of miRNA m n in nth row of matrix MD and DM (d_m) is denoted in mth column of matrix MD. It is obvious that adjacency matrix is sparse. Especially, some infrequent miRNAs are blank that the associations are unknown. It may lead to poor performance in predicting the potential association between miRNA and disease. Here, we utilize a preprocessed step [24] to solve the above problem. According to Eq. 4, the miRNA-related interactions MM (m_n) is defined as follows:

$$\begin{cases} M(m_n) = \frac{\sum_{i=1}^{K} \theta_i * MD(m_i)}{\sum_{i=1}^{K} SM(m_i, m_n)} \\ \theta_i = \mu^{i-1} * SM(m_i, m_n) \end{cases} \quad (6)$$

where MD (m_i) is the disease-related associations of miRNA m_i. MiRNA m_i are sequential neighbors of miRNA m_n. And they are obtained by arranging in descending order of similarity with miRNA m_i; θ_i is the weight coefficient, and $\mu \in [0, 1]$. In the same way, the disease-related interactions of disease d_j is defined as follows:

$$\begin{cases} DM(d_m) = \frac{\sum_{j=1}^{K} \theta_j * MD(d_j)}{\sum_{j=1}^{K} SD(d_j, d_m)} \\ \theta_j = \mu^{j-1} * SD(d_j, d_m) \end{cases} \tag{7}$$

where $MD(d_j)$ is the miRNA-related associations of disease d_j.

In the end, we combine the above two matrices to obtain the novel miRNA-related associations:

$$MD' = \frac{\lambda_1 MM(m_n) + \lambda_2 DM(d_m)}{\lambda_1 + \lambda_2} \tag{8}$$

where λ_i is the weight coefficient and we assign $\lambda_1 = \lambda_2 = 1$. Finally, the updated miRNA-disease matrix MD is defined as:

$$MD = max\left(MD, MD'\right) \tag{9}$$

where the new association is obtained to replace the original matrix (MD nm = 0).

2.4 Probabilistic Matrix Decomposition for Disease-Associated MiRNA

Probabilistic Matrix Decomposition

Probabilistic matrix decomposition (PMD) is applied to recommending tasks in existing approaches of collaborative filtering [25]. And PMD is a factor-based model for recommending information. It aims to model the original matrix as the product of two low-rank user matrices and item matrices. Given the miRNA-disease matrix MD and scores in the (0–1) range, $U \in R\ N * K$ and $V \in R\ M * K$ are potential miRNAs and disease signature matrices, and column vectors represent feature vectors. Here, we mathematically formulate the problem of conditional distribution of observed scores as the following objective function:

$$p\left(MD|U, V, \sigma^2\right) = \prod_{i=1}^{N} \prod_{j=1}^{M} \left[\mathrm{N}\left(MD_{ij}|U_i^T V_j, \sigma^2\right)\right]^{I_{ij}} \tag{10}$$

where $N(x|\mu,\sigma^2)$ is a probability density function satisfying the Gaussian distribution with mean μ and variance σ^2, I_{ij} is an index function, and $I_{ij} = 1$ if miRNA i is associated with disease j, otherwise 0. Moreover, the miRNAs and disease feature vectors U and V also satisfy the zero-mean spherical Gaussian prior:

$$\begin{cases} p\left(U|\sigma_U^2\right) = \prod_{i=1}^{N} \mathrm{N}\left(U_i|0, \sigma_U^2 I\right) \\ p\left(V|\sigma_V^2\right) = \prod_{j=1}^{M} \mathrm{N}\left(V_j|0, \sigma_V^2 I\right) \end{cases} \tag{11}$$

To generate a PMD for miRNA-disease associations, the posterior distribution of characteristics with miRNA and disease is calculated as follows:

$$\begin{aligned} & p\left(U, V \mid MD, \sigma^2, \sigma_U^2, \sigma_V^2\right) \\ & = p\left(MD \mid U, V, \sigma^2, \sigma_U^2, \sigma_V^2\right) * p(U, V) / p\left(MD \mid \sigma^2, \sigma_U^2, \sigma_V^2\right) \\ & \sim p\left(MD \mid U, V, \sigma^2, \sigma_U^2, \sigma_V^2\right) * p(U, V) \\ & = p\left(MD \mid U, V, \sigma^2, \sigma_U^2, \sigma_V^2\right) * p(U) * p(V) \end{aligned} \tag{12}$$

where the value of $p(MD|\sigma^2, \sigma_U^2, \sigma_V^2)$ is close to 1.

Optimization

To maximize the probability p(MD|U,V,σ ²), it is satisfied the conditions of maximum likelihood estimation (MLE) and maximum posterior probability (MAP), the log-posterior is represented as:

$$\operatorname{Ln}\left(p\left(U, V \mid MD, \sigma^2, \sigma_U^2, \sigma_V^2\right)\right)$$

$$= -\frac{1}{2\sigma^2}\sum\nolimits_{i=1}^{N}\sum\nolimits_{j=1}^{M} I_{ij}\left(MD_{ij} - U_i^T V_j\right)^2 - \frac{1}{2\sigma_U^2}\sum\nolimits_{i=1}^{N} U_i^T U_i - \frac{1}{2\sigma_V^2}\sum\nolimits_{j=1}^{M} V_j^T V_j \tag{13}$$

$$-\frac{1}{2}(K * N * ln\sigma_U^2 + K * M * ln\sigma_V^2$$

where C is a constant that does not depend on parameters. And maximizing the log-posterior on U and V with hyperparameters being kept fixed will be equivalent to minimizing the following objective function:

$$E = \frac{1}{2}\sum\nolimits_{i=1}^{N}\sum\nolimits_{j=1}^{M} I_{ij}\left(MD_{ij} - U_i^T V_j\right)^2 + \frac{\lambda_U}{2}\sum\nolimits_{i=1}^{N}\|U_i\|_{Fro}^2 + \frac{\lambda_V}{2}\sum\nolimits_{j=1}^{M}\|V_j\|_{Fro}^2 \tag{14}$$

As a result, we update each miRNA and disease for using a stochastic gradient method until convergence. According to Eq. 14, the derivative can be calculated as:

$$\begin{cases} \frac{\partial E}{\partial U} = -\left(MD_{ij} - U_i^T V_j\right)V_j + \lambda_U U_i \\ \frac{\partial E}{\partial V} = -\left(MD_{ij} - U_i^T V_j\right)U_i + \lambda_V V_j \end{cases} \tag{15}$$

Then we obtained the updated miRNA and disease characteristics U,V based on the gradient descent algorithm.

$$U_i \leftarrow U_i - \varepsilon\frac{\partial E}{\partial U} = U_i + \varepsilon\left(MD_{ij} - U_i^T V_j\right)V_j + \lambda_U U_i \tag{16}$$

$$V_j \leftarrow V_j - \varepsilon\frac{\partial E}{\partial V} = V_j + \varepsilon\left(MD_{ij} - U_i^T V_j\right)U_i + \lambda_V V_j$$

where ε is the learning rate. The minimum value of the function is reached by training convergence.

Finally, we obtain the predicted miRNA-disease correlation matrix MD * and rank the disease-associated miRNAs according to the entities in the matrix MD *. In principle, the top miRNAs in each column of MD * are more likely to be associated with the corresponding disease.

3 Experimental Results

3.1 Evaluation Metrics

We apply the root mean squared error (RMSE) as one of the performance evaluation metrics to validate if the estimated value calculated by PMDA is approached to the real value from the data. According to above computational methods, we define the root mean squared error, denoted RMSE (R, h), of PMDA as the deviation between estimated value and real value from the data, which is given by:

$$RMSE(R,h) = \sqrt{\frac{1}{m}\sum_{i=1}^{m}\left[\widehat{R}(m_i) - R(m_i)\right]^2} \tag{17}$$

where n is the number of miRNA. $\widehat{R}(m_i)$ and $R(m_i)$ is the estimated value and the real value at miRNA m_i respectively. The deviation indicates the root mean squared error of estimated value to the real value. High bias represents poor accurate of estimated result.

To compare the superiority of PMDA over other existing methods, we also use receiver operating characteristic (ROC) as the performance evaluation:

$$TPR = \frac{TP}{TP + FN} \tag{18}$$

where TPR and FPR is the true positive rate and false positive rate. TPR is represented the proportion of actual positive instances in the positive class. FPR is represented the proportion of actual false instances in the false class and FPR = 1 − TPR. TP and TN represent the number of positive samples and negative samples identified correctly, respectively. FP and FN denote the number of positive samples and negative samples identified incorrectly, respectively.

3.2 Performance for Comparing with Other Methods

To systematically evaluate PMDA performance, we firstly perform 5-fold cross validation and compare it with the following methods: IMCMDA [26], GRNMF [24], DMpred [27], IMDAILM [28], EPMDA [29]. The parameter values of the comparison method are selected according to the recommended values in the original text and the data sets used in these methods the same as in this article. As demonstrated in Fig. 2, the AUC values of PMDA with five methods are 0.880, 0.839, 0.865, 0.875, 0.855 and 0.886, respectively. we can intuitively see that PMDA achieves the best performance,

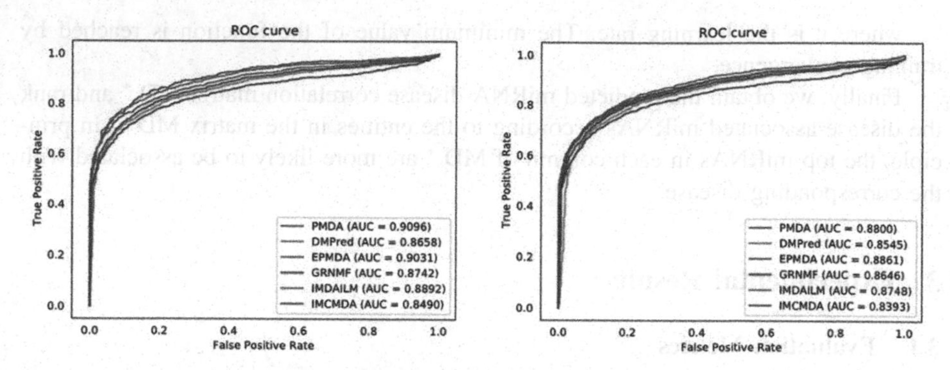

Fig. 2. ROC curves for PMDA and other methods in miRNA-disease association prediction

and its average AUC values are 4.1%, 0.5%, 1.5%, and 2.5% higher than IMCMDA, IMDAILM, GRNMF, and DMPred, respectively.

we perform 10-fold cross validation and compare with these methods. Due to increase training data in each set, miRNA and disease characteristics can be better reflected. The figure shows that the PMDA algorithm performs better, with an average AUC value of 0.91 which is 2.94% higher than other algorithms.

Table 1. Performance comparison with AUC for 10 diseases between PMDA and other methods

Disease name	PMDA	IMCMDA	GRNMF	DMPred	IMDAILM	EPMDA
Breast neoplasms	0.933	0.798	0.885	0.877	0.901	0.916
Hepato-cellular carcinoma	0.915	0.721	0.844	0.865	0.929	0871
Stomach neoplasms	0.862	0.744	0.837	0.802	0.824	0.847
Prostatic neoplasms	0.853	0.804	0.898	0.854	0.867	0.924
Pancreatic neoplasms	0.876	0.858	0.936	0.891	0.857	0.886
Ovarian neoplasms	0.904	0.868	0.909	0.906	0.870	0.849
Melanoma	0.920	0.801	0.885	0.876	0.886	0.919
Lung neoplasms	0.868	0.852	0.905	0.921	0.902	0.889
Heart failure	0.869	0.942	0.803	0.895	0.894	0.867
Gliob-lastoma	0.899	0.769	0.868	0.868	0.844	0.878
Average AUC	0.8899	0.8157	0.8770	0.8755	0.8774	0.8806

Then, we also test 10 common diseases associated with at least 50 verified associations under cross validation setting and five widely applied miRNA-disease prediction algorithms: (a) IMCMDA, (b) GRNMF, (c) DMPred, (d)IMDAILM, (e)EPMDA is compared with the PMDA. As shown in Table 1, PMDA is outperformed with other methods for the ten diseases [30]. For instance, the average AUC values for IMCMDA, GRNMF, DMPred, IMDAILM, EPMDA are 0.8157,0.8770,0.8755,0.8774,0.8806, respectively. The average AUC of PMDA is respectively higher than 7.4%, 1.2%, 1.4%,

1.2% and 8.5% of the other five methods. The highest AUC value of Breast Neoplasms is 0.9. AUC value of PMDA for these 10 diseases all exceed 0.85, and five common diseases were higher than other methods, such as: breast tumor, liver tumor, stomach tumor, lung tumor and heart disease.

We also used the paired t-test to measure whether PMDA algorithm in 10 diseases was significantly higher than the other methods. The t-test results are shown in Table 2. The statistical results show that, at the significance level of p-value < 0.05, the effect of PMDA is obviously better than that of other methods (Fig. 3).

Table 2. p-value between PMDA and others

	IMCMDA	GRNMF	DMPred	IMDAILM	EPMDA
P-values	4.32e−04	6.78e−05	1.51e−03	3.31e−04	8.57e−06

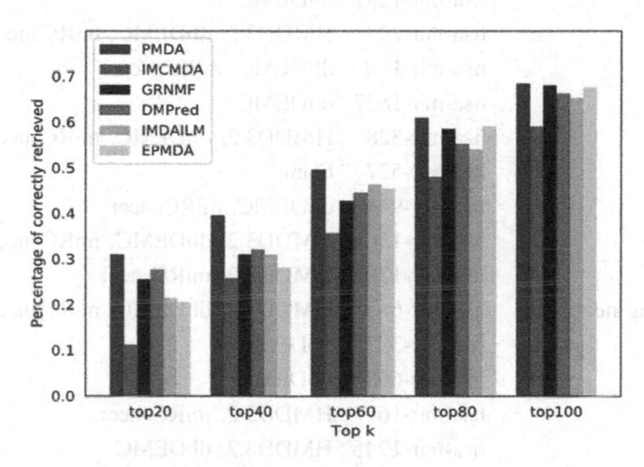

Fig. 3. Percentage of correctly retrieved known associations between miRNAs and diseases for various ranking thresholds

To future verify PMDA algorithm's performance to predict miRNA-disease interactions, experiments were carried out on HMDD3.2 database, miRCancer database and dbDEMC database respectively. MiRCancer provides a comprehensive collection of miRNA expression profiles in a variety of human cancers with an accuracy rate of 100%. In dbDEMC, 209 newly published datasets were collected from GEO (Gene Expression Omnibus) and TCGA (Cancer Genome Atlas). And in HMDD 3.2, they manually collected 35547 miRNA-disease association entries which include 1206 miRNA genes, 893 diseases from 19280 papers. The incidence of breast, lung and colon cancers are among the top 10. Therefore, miRNAs may play a key role in these diseases. PMDA algorithm predicts new miRNA-disease interactions and evaluates its effectiveness in predicting miRNA-disease interactions in these three diseases.

As show in Table 3, top 10 potential candidate miRNA predicted by PMDA algorithm in breast, lung and colon can be confirmed to be effective. Among the candidate miRNAs, more than half of them were confirmed to be related to breast, lung and colon cancer in HMDD 3.2 database. In addition, some candidate genes of different tumors play an important role in the regulation process. For example, the overexpression of hsa-mir-1287 in breast cancer affects cell proliferation, and is also expressed in lung cancer and colon cancer abnormally; the expression of hsa-mir-302f was abnormal in lung cancer and colon cancer; hsa-mir-668 has regulatory effects in all three diseases. Therefore, miRNAs may have regulatory effects on a variety of diseases.

Table 3. Top 10 potential candidate miRNA predicted by PMDA in three disease

Disease	miRNA	Evidence dataset
Breast neoplasms	hsa-mir-668	HMDD3.2, dbDEMC
	hsa-mir-1287	dbDEMC
	hsa-mir-223	HMDD3.2, dbDEMC, miRCancer
	hsa-mir-150	dbDEMC, miRCancer
	hsa-mir-1827	dbDEMC
	hsa-mir-328	HMDD3.2, dbDEMC, miRCancer
	hsa-mir-527	blank
	hsa-mir-944	dbDEMC, miRCancer
	hsa-mir-130a	HMDD3.2, dbDEMC, miRCancer
	hsa-mir-1207	HMDD3.2, miRCancer
Lung neoplasms	hsa-mir-663	HMDD3.2, dbDEMC, miRCancer
	hsa-mir-302f	dbDEMC
	hsa-mir-668	dbDEMC
	hsa-mir-16	HMDD3.2, miRCancer
	hsa-mir-1245	HMDD3.2, dbDEMC
	hsa-mir-26	HMDD3.2, miRCancer
	hsa-mir-527	dbDEMC
	hsa-mir-539	dbDEMC
	hsa-mir-127	HMDD3.2, miRCancer
	hsa-mir-1287	dbDEMC
Colon neoplasms	hsa-mir-1290	HMDD3.2, dbDEMC, miRCancer
	hsa-mir-138	HMDD3.2
	hsa-mir-375	HMDD3.2, dbDEMC, miRCancer
	hsa-mir-527	blank
	hsa-mir-155	HMDD3.2, miRCancer
	hsa-mir-328	HMDD3.2, dbDEMC, miRCancer
	hsa-mir-584	HMDD3.2, dbDEMC
	hsa-mir-1287	dbDEMC
	hsa-mir-668	dbDEMC
	hsa-mir-302f	dbDEMC

4 Conclusion

Many previous methods could not efficaciously predict miRNAs only with few known associations. In this study, we firstly integrate two kinds of similarity networks to update the miRNA-disease association network. Then, we apply probabilistic factorization to obtain two feature matrices of miRNA and disease. Meanwhile, we formulate a similarity feature matrix as constraints in the factorization process. To demonstrate the advantages of PMDA, we compare it with other state-of-the-art methods. In addition, the potential associations predicted by PMDA is verified. The experiment results indicate that PMDA can effectively predict the potential associations of its elements in unknown dataset, especially in imbalance and sparse datasets. However, there are many other kinds of data including genes, lncRNAs, microbiota and so on related to disease. Therefore, combining more biological data to improve the prediction abilities of our model is a direction in our future study.

Acknowledgements. This work was supported by Natural Science Foundation of China (Grant No. 61972141) and Natural Science Foundation of Hunan Province, China (Grant No. 2018JJ2053)

References

1. Schork, N.J.: Genetics of complex disease. Am. J. Respirat. Crit. Care Med. **156**(4), S103–S109 (1997)
2. Lu, X., Lu, J., Liao, B., Li, X., Qian, X., Li, K.: Driver pattern identification over the gene co-expression of drug response in ovarian cancer by integrating high throughput genomics data. Sci. Rep. **7**, 1–17 (2017)
3. Lu, X., Qian, X., Li, X., Miao, Q., Peng, S.: DMCM: a data-adaptive mutation clustering method to identify cancer-related mutation clusters. Bioinformatics **35**(3), 389–397 (2018)
4. Lu, X., Wang, X., Liu, P., Zhu, Z., Ding, L.: A novel method to predict protein regions driving cancer through integration of multi-omics data. In: Huang, D.-S., Jo, K.-H., Huang, Z.-K. (eds.) ICIC 2019. LNCS, vol. 11644, pp. 303–312. Springer, Cham (2019). https://doi.org/10.1007/978-3-030-26969-2_29
5. Ambros, V.: The functions of animal microRNAs. Nature **431**(7006), 350–355 (2004)
6. Bartel, D.P.: MicroRNAs: genomics, biogenesis, mechanism, and function. Cell **116**(2), 281–297 (2004)
7. Bartel, D.P.: Metazoan MicroRNAs. Cell **173**(1), 20–51 (2018)
8. Chou, C.H., Chang, N.W., Shrestha, S., et al.: MiRTarBase 2016: updates to the experimentally validated miRNA-target interactions database. Nucleic Acids Res. **44** (Database issue), D239–D247 (2015)
9. Li, Y., Qiu, C., Tu, J., et al.: HMDD v2.0: a database for experimentally supported human microRNA and disease associations. Nucleic Acids Res. **42**(Database issue), D1070–D1074 (2014)
10. Xuan, P., Han, K., Guo, M., Guo, Y., Li, J., et al.: Prediction of microRNAs associated with human diseases based on weighted k most similar neighbors. PLOS One **8**(8), e70204 (2013)
11. Pasquier, C., Gardès, J.: Prediction of miRNA-disease associations with a vector space model. Sci. Rep. **6**, 27036 (2016)

12. Lu, X., Li, X., Liu, P., Qian, X., Miao, Q., Peng, S.: The integrative method based on the module-network for identifying driver genes in cancer subtypes. Molecules. **23**(2), 183 (2018). https://doi.org/10.3390/molecules23020183
13. Chen, X., Yan, C.C., Zhang, X., et al.: WBSMDA: within and between score for mirna-disease association prediction. Sci. Rep. **6**(1), 21106 (2016)
14. Ding, P., Luo, J., Xiao, Q., et al.: A path-based measurement for human miRNA functional similarities using miRNA-disease associations. Sci. Rep. **6**, 32533 (2016)
15. Yadong, D., Yongqi, S., Chao, Q., Weiguo, Z.: EPMDA: edge perturbation based method for miRNA-disease association prediction. IEEE/ACM Trans. Comput. Biol. Bioinform. (2019). https://doi.org/10.1109/TCBB.2019.2940182
16. Chih-Hung, C., Sirjana, S., et al.: miRTarBase update 2018: a resource for experimentally validated microRNA-target interactions. Nucleic Acids Res. **46**(D1), D296–D302 (2018)
17. Hwang, S., Kim, C.Y., Lee, I., et al.: HumanNet v2: human gene networks for disease research. Nucleic Acids Res. **47**(D1), D573–D580 (2019)
18. Wang, D., Wang, J., Lu, M., et al.: Inferring the human microRNA functional similarity and functional network based on microRNA-associated diseases. Bioinformatics **26**(13), 1644–1650 (2010)
19. Gregori, E.: Mesh networks: commodity multihop ad hoc networks. Commun. Mag. IEEE **43**(3), 123–131 (2005)
20. Cheng, L., Li, J., Ju, P., et al.: SemFunSim: a new method for measuring disease similarity by integrating semantic and gene functional association. PLoS One **9**(6), e99415 (2014)
21. Schriml, L.M., Mitraka, E., Munro, J., et al.: Human Disease Ontology 2018 update: classification, content and workflow expansion. Nucleic Acids Res. **47**(D1), D955–D962 (2018)
22. Zeng, X., Zhang, X., Zou, Q.: Integrative approaches for predicting microRNA function and prioritizing disease-related microRNA using biological interaction networks. Brief. Bioinform. **17**(2), 193 (2016)
23. Xia, L., Qianghu, W., Yan, Z., el al.: Prioritizing human cancer microRNAs based on genes' functional consistency between microRNA and cancer. Nucleic Acids Res. **39**(22) , e153 (2011)
24. Xiao, Q., Luo, J., Liang, C., et al.: A graph regularized non-negative matrix factorization method for identifying microRNA-disease associations. Bioinformatics **34**(2), 239–248 (2017)
25. Salakhutdinov, R., Mnih, A.: Probabilistic matrix factorization advances in neural information processing systems 21 (NIPS 21). Vancouver, Canada (2008)
26. Chen, X., Yan, G.: Semi-supervised learning for potential human microRNA-disease associations inference. Sci. Rep. **4**(1), 5501 (2015)
27. Yingli, Z., Ping, X., Xiao, W., et al.: A non-negative matrix factorization based method for predicting disease-associated miRNAs in miRNA-disease bilayer network. Bioinformatics **34**(2), 267–277 (2018)
28. Yuhua, Y., Binbin, J., Sihong, S., et al.: IMDAILM: inferring miRNA-disease association by integrating lncRNA and miRNA data. IEEE Access **8**, 16517–16527 (2020)
29. Xinguo, L., Xinyu, W., Li, D., Jinxin, L., Yan, G., Keren, H.: frDriver: a functional region driver identification for protein sequence. IEEE/ACM Trans. Comput. Biol. Bioinform. https://doi.org/10.1109/TCBB.2020.3020096
30. Ping, X., Ke, H., Yahong, G., Jin, L., Xia, L., Yingli, Z., Zhaogong, Z., Jian, D.: Prediction of potential disease-associated microRNAs based on random walk. Bioinformatics **31**(11), 1805–1815 (2015)

Artificial Intelligence in Biological and Medical Information Procession

CT Scan Synthesis for Promoting Computer-Aided Diagnosis Capacity of COVID-19

Heng Li[1], Yan Hu[1(✉)], Sanqian Li[1], Wenjun Lin[1], Peng Liu[2], Risa Higashita[3], and Jiang Liu[1,4,5(✉)]

[1] School of Computer Science and Engineering, Southern University of Science and Technology, Shenzhen 518055, China
{huy3,liuj}@sustech.edu.cn

[2] Big Data Research Center, University of Electronic Science and Technology of China, Chengdu 611731, China

[3] Tomey Corporation, Nagoya 451-0051, Japan

[4] Department of Computer Science and Engineering, Guangdong Provincial Key Laboratory of Brain-Inspired Intelligent Computation, Southern University of Science and Technology, Shenzhen 518055, China

[5] Ningbo Institute of Industrial Technology Chinese Academy of Sciences, Ningbo, China

Abstract. Nowadays, with the rapid spread of Corona Virus Disease 2019 (COVID-19), this epidemic has become a threatening risk for global public health. Medical workers and researchers all over the world are struggling against the novel coronavirus in the front line. Because the computed tomography (CT) images from infected patients exposure characteristic abnormalities, automatic CT analyzers based on AI-based algorithms are extensively employed as effective weapons to aid clinicians. However, unbalanced data and lack of annotations obstruct AI-based algorithms applying in aided diagnosis because of their low performance. Therefore, in order to solve the above problems, a general-purpose solution is proposed to synthesize COVID-19 CT scans from non-COVID-19 data for providing high-quality negative-positive paired CT scans. Particularly, we introduce an elastic registration algorithm of CT images to manufacture paired training data. Then, a conditional Generative Adversarial Networks (GANs) based image-to-image translation model is implemented to synthesize COVID-19 CT scans from non-COVID-19 data. The effectiveness of our proposed algorithm used in COVID-19 aided diagnosis is verified in the experiments, and the identification and detection capacities of the classification models have been enhanced with the generated CT scans. Specifically, the precise lesion location is achieved by the generated data with a weakly supervised algorithm of class activation mapping (CAM). The model and code of this paper are publicly available at https://github.com/lihengbit/Synthesis-of-COVID-19-CT-Scan.

Keywords: CT scan synthesis · Image-to-image translation · Weakly supervised · Lesion location

D.-S. Huang and K.-H. Jo (Eds.): ICIC 2020, LNCS 12464, pp. 413–422, 2020.
https://doi.org/10.1007/978-3-030-60802-6_36

1 Introduction

Since the outbreak of the novel coronavirus infection named COVID-19 in December 2019, the virus has quickly spread around the world in the past few months. Until now, more than three million cases of COVID-19 have been reported in over 200 countries and territories, resulting in approximately 250,000 deaths (with a fatal rate of 6.90%). This has led to great public health concern in the international community, and governments strengthen the control of the virus spread.

Comprehensive screening is one of the top priority measures to deter the spreading of COVID-19. With the accumulation of findings on COVID-19 infection's radiological features, ground-glass opacification with occasional consolidation in the peripheries has been proved to be the predominant imaging pattern of COVID-19. Chest X-ray and non-contrast thoracic CT have been thus adopted as a significant tool in the screening, detection, and quantification of COVID-19 infection [9, 12]. To aid clinicians on screening and diagnosis with radiological images, rapid automatic evaluation is necessary to interpret large numbers of X-ray and CT data. In various medical studies, AI-based algorithms have been developed to assist in analyzing potentially large numbers of clinical image data [5]. Therefore, researchers are making efforts to develop AI solutions to analyze multiple radiological images in parallel to detect any lung abnormalities caused by COVID-19 [1, 5, 14].

In recent years, due to the strong ability of nonlinear modeling, the algorithms based on DNNs have become the most popular AI pipeline for CAD. Considering a large amount of data samples is the foundation of developing DNNs algorithms, COVID-19 image data are collected by researchers to construct the public platform for fighting with the epidemic [2, 14]. By assembling medical images from websites and publications, Cohen et al. created a dataset including X-ray and CT images to develop AI-based approaches to predict and understand the infection of COVID-19 [2]. Zhao et al. built a COVID-CT-Dataset, which at present contains 349 CT images with clinical findings of COVID-19 and is continuously expanding. And a baseline method has also been adopted to diagnose the infection of COVID-19 on the dataset [14]. Some AI-based solutions to assist detection and diagnosis of COVID-19 has been presented based on public COVID-19 dataset [1, 5]. Abbas et al. adopted a previously developed convolutional neural networks (CNNs) model for the classification of COVID-19 chest X-ray images [1]. Through modifying and adapting existing DNNs models, Gozes et al. developed an AI-based automated CT image analysis tool, which achieves high accuracy in the detection of COVID-19 positive patients [5].

However, despite the limited volumes, the public datasets also suffer from data imbalance, such as the number and quality between the negative and positive data are always asymmetries. Inspired by the strategies, which improve the segmentation performances of deep neural networks with synthesized image data in the medical field [3, 4], in this paper, we propose a synthesis model to generated COVID-19 CT scans from non-COVID-19 ones with elastic registration and conditional Generative Adversarial Networks (cGAN). Subsequently, weakly supervised CNN models and CAM are implemented with the synthesized data to identify and detect the infection of COVID-19. The contribution of our approach is concentrated in two aspects: first, the proposed synthesis model mitigates the limitation of data volume and quality; second, the performances of DNNs have been improved by the synthesized paired data through promoting the training stage.

2 Methodology

In this paper, in order to solve the problems of data imbalance and lack of annotation, we propose to synthesize COVID-19 CT scans from negative ones to establish high-quality data source for CAD model training and further promote the capacity of the models applied in aided diagnosis. As illuminated in Fig. 1, the proposed scheme synthesis of the COVID-19 CT scan consists of two portions, which are described in this section. The proposed solution not only mitigates the lack of high-quality training data, but also addresses the problem of negative-positive data imbalance.

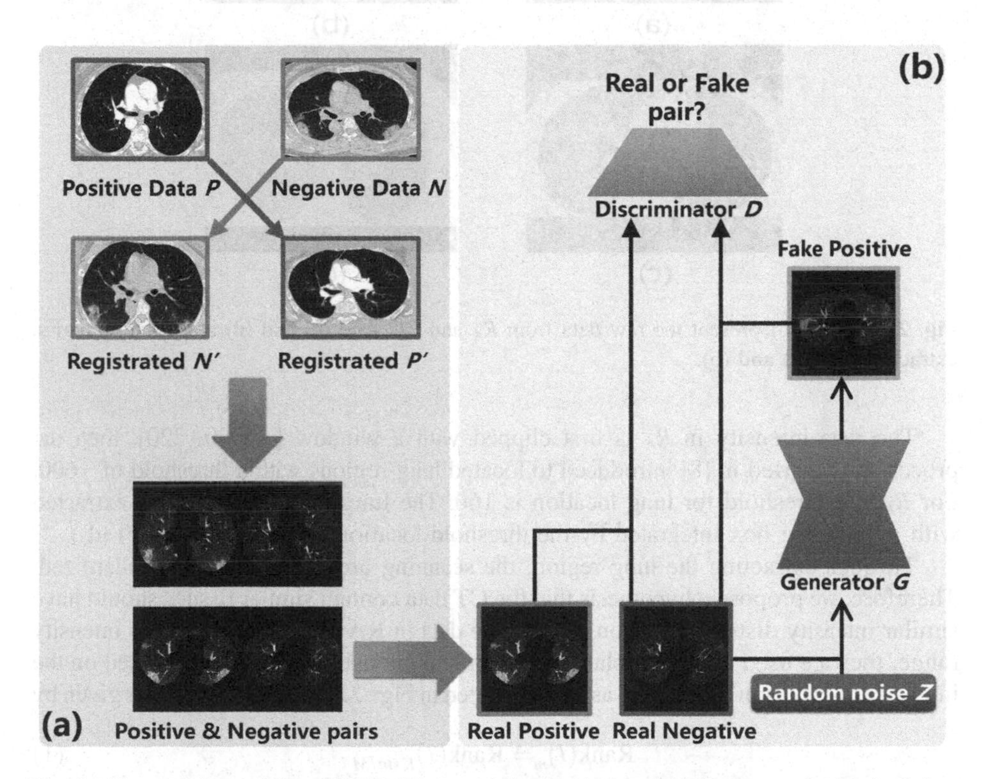

Fig. 1. The workflow of the COVID-19 scan synthesis strategy. (a) The approach of paired data construction. (b) The synthesis of COVID-19 scans from the negative data.

2.1 Data Preprocessing

The raw data of this study are acquired from covid-chestxray-dataset [2], which contains 20 CT cases. Considering the data obtained from different resources, significant differences in imaging appears between the data. To prepare the data for the following study, a normalization pipeline is presented in this section.

The dataset is composed of the data from two resources, which refer to R_A (resource A) and R_B (resource B) in this study. As displayed in Fig. 2(a) (b), in R_A, the intensity of data ranges from −1100 to 3000, while in R_B the range is [0, 255]. Additionally, the diversity of the scanning area is observed between the resources.

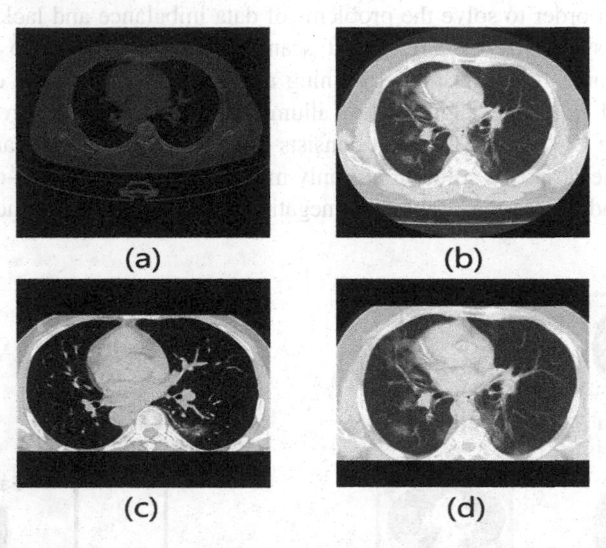

Fig. 2. (a) and (b) present the raw data from R_A and R_B. And (c) and (d) are the lung region extracted from (a) and (b).

The data intensity in R_A is first clipped with a window [−1000, 220], then the processing reported in [8] introduced to located lung regions with a threshold of −600. For R_B, the threshold for lung location is 160. The lung regions are finally extracted with a bounding box integrated by the threshold location (shown in Fig. 2(c) (d)).

Through extracting the lung region, the scanning area of all data is standardized. Therefore, we propose a hypothesis that the CT data contain similar tissues should have similar intensity distribution. Considering the data in RA obey the typical CT intensity range, they are used as the template to perform an intensity normalization based on the intensity distribution of CT data, as demonstrated in Fig. 3. The normalization is given by

$$\mathrm{Rank}(I)_m = \mathrm{Rank}(J)_{K*m/M} \tag{1}$$

where Rank refers to sort the intensity in increasing order, and $\lfloor\cdot\rfloor$ indicates round down. m represents the index of an intensity value in the order. I and J are the source and template images, and K and M are the voxel numbers of the images.

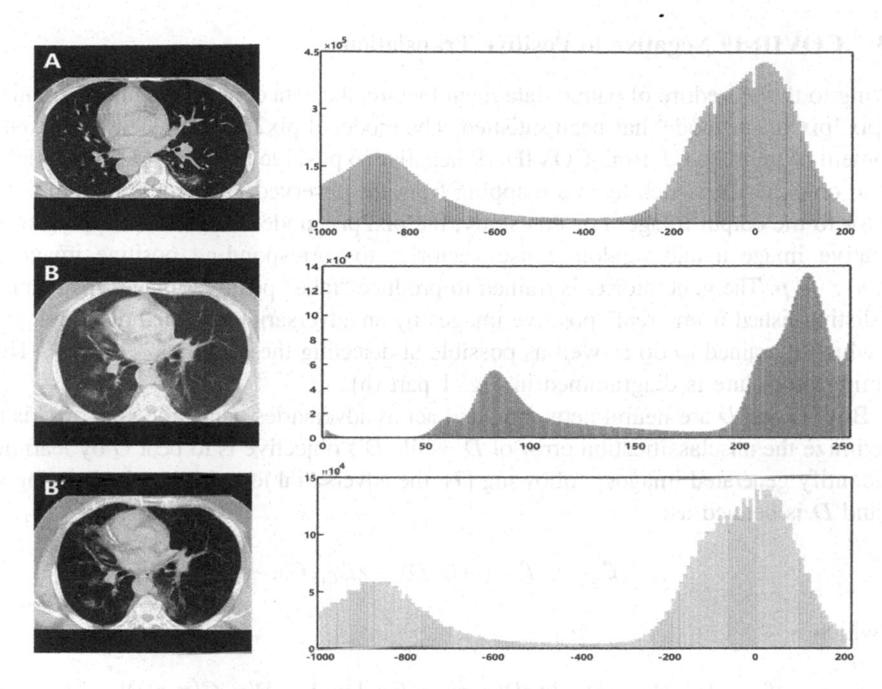

Fig. 3. The first row presents the intensity distribution of CT data in R_A. The original and normalized intensity distributions of CT data in R_B are displayed in the second and third rows.

2.2 Paired Data Manufacture

In this study, the image-to-image translation model named pix2pix [7] is employed to conduct the generation of COVID-19 CT scans. Though the models constructed by pix2pix always provide the "upper bound" performance in the missions of image generation [16], the demanding of paired training data deters it from been extensively implemented in many scenarios.

With the intention to generate high-quality COVID-19 CT scans, we proposed a paired data manufacture scheme by the flexible registration of CT. The above-mentioned COVID-19 data are treated as positive data, while the negative ones are acquired from LUNA16 dataset [11]. As presented in Fig. 1 part (a), the flexible registration algorithm and Advanced Normalization Tools (ANTs) proposed by NJ Tustison et al. [13] is applied to bilaterally register the positive and the negative data. The positive data P and the negative one N are registered to output P', which is P registered to N, as well as N'. The paired data of positive and negative COVID-19 scans with the unified structure are hence built by combining the slices of P and N', P' and N. Finally, the lung region is segmented by the segmentation model provided by [6].

2.3 COVID-19 Negative to Positive Translation

Owing to the procedure of paired data manufacture, the data demanding for the training of pix2pix image model has been satisfied. The model of pix2pix is implemented to the problem of translate CT from COVID-19 negative to positive [7]. The pix2pix model is based on a cGAN, which learn a mapping from the observed image and random noise vector to the output image. For this study, the pix2pix model learns the mapping from negative image n and random noise vector z, to corresponding positive image p, $G : n, z \rightarrow p$. The generator G is trained to produce "fake" positive images that cannot be distinguished from "real" positive images by an adversarially trained discriminator, D, which is trained to do as well as possible at detecting the generator's "fakes". This training procedure is diagrammed in Fig. 1 part (b).

Both G and D are neural networks, and act as adversaries, since the goal of G is to maximize the misclassification error of D, while D's objective is to beat G by learning to identify generated images. Following [7], the adversarial loss driving the learning of G and D is defined as:

$$\mathcal{L}_{n2p} = \mathcal{L}_{cGAN}(G, D) + \lambda \mathcal{L}_{\mathcal{L}_1}(G), \tag{2}$$

in which

$$\mathcal{L}_{cGAN}(G, D) = E_{n,p}[\log D(n, p)] + E_{n,z}[\log(1 - D(n, G(n, z)))], \tag{3}$$

$$\mathcal{L}_{\mathcal{L}_1}(G) = E_{n,p,z}[||p - G(n, z)||_1], \tag{4}$$

where G tries to minimize the adversarial loss against an adversarial D that tries to maximize it, i.e. $G^* = arg \min_G \max_D \mathcal{L}_{cGAN}(G, D) + \lambda \mathcal{L}_{\mathcal{L}_1}(G)$.

3 Experiments and Results

With the intent to evaluate the proposed synthesis model and understand the capacity promotion of DNNs, the diagnosis of COVID-19 is performed in the experiments, including infection identification and lesion detection.

3.1 Dataset

Through the mentioned data processing, the entire dataset of 20 COVID-19 cases from CORONACASES.ORG [2] are introduced to this study, while 20 non-COVID-19 cases are randomly selected from LUNA16 [11] as the negative dataset. By registering the positive and negative datasets, 400 paired CT data are obtained, and 10,254 paired scans are acquired from the paired data.

To promote the robustness of the translation model, data augmentation is performed with horizontal and vertical flip, 90° to 270° rotation, random move, random intensity and contrast. After the augmentation, the dataset contains 82,032 paired scans, and 49,220 scans (60%) are used as the training set, 16,406 scans (20%) as the validating set, and the remaining 16,406 scans (20%) as the test set. It should be noticed that scans from an identical CT must be put into the same set, otherwise the slice similarity of an identical CT case will lead to overfitting.

3.2 Performance Promotion with the Synthesized Images

Evaluating the quality of synthesized images is an open and difficult problem [10]. To holistically evaluate the quality of the translation results, we adopt pre-trained DNNs as a pseudo-metric to measure the discriminability of the synthesis. The intuition is that if the synthesized images are realistic, networks trained on real images will be able to classify the synthesized image correctly, vice versa. Further, the performance of the DNNs will be improved by training on the integrated dataset of the real and fake ones.

Therefore, the results not only exhibit the generated examples, but also quantitatively evaluate the quality with CAD of COVID-19. The experimental results verified the quality of the synthesized images, and illuminated the capacity of performance promotion with the synthesized images. By mixing real data with the synthesized images, models are superiorly trained that remarkable identification and lesion detection performances are provided by the classification models.

Synthesized Image Gallery. Six scans of synthesized images are exhibited in Fig. 4. According to the appearance of the synthesized images, it is a particularly thorny problem to manually distinguish the real and fake positive scans. The synthesized images enjoy diverse pathological traits of COVID-19.

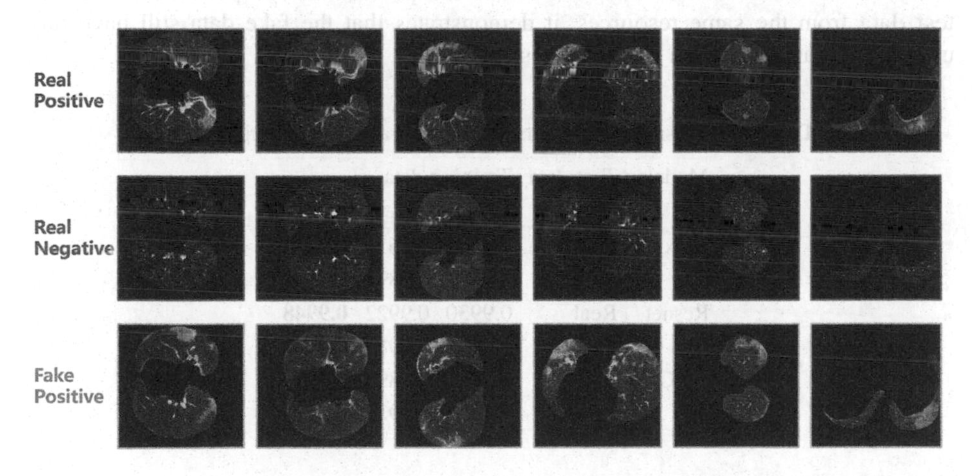

Fig. 4. A gallery of synthesized images. Each column shows a pair of synthesis case. The corresponding real and negative scans are shown in the first and second rows. The synthesized images are presented in the last row.

Specifically, the synthesis diversity is easily inferred from the first two columns, which presents scans derived from the same scans with different augmentation processing. Despite the same training pairs, the synthesized pathological appearances vary greatly from each other for the first two columns. This verifies that the model generalizes properly and did not trivially memorize the examples in the training set. The last columns display the scans at the top and bottom of the lung region. The robustness of the synthesis model is demonstrated by generating positive scans with special lung regions.

Classification Models Promotion. To quantitatively evaluate the capacity promotion and the quality of the synthesized scans, classic convolutional neural networks, including Alexnet, Resnet, VGG, are introduced to handle real and fake data. Specifically, Resnet and VGG are implemented with Resnet-18 and VGG16.

The classification models are trained to identify positive and negative COVID19 scans. The negative data are constantly from real cases, while positive data are respectively selected from real, fake and mixed ones. In the fake and mixed cases, the total number of data is the same as that of the real case. And real and fake data separately composes half of the total volume.

Subsequently, for each individual networks, three models are trained with the three datasets, as well as tested on real and fake data. The classification results are provided in Table 1. No matter the models are trained on data from which resources, decent performances are provided on both test datasets. Therefore, the quality of the synthesis images preserves a large fraction of the image quality present in the training data. In addition, the dataset mixed the real and fake data provides the most significant information that for the three networks the premier performances are all come from the models tuned on the mixed dataset. This also proved data synthesis is a meaningful study. As the models trained on real and fake data provide the best performances on the test data from the same resources, it demonstrates that the fake data still have tiny underlying variances from the real ones.

Table 1. Classification results of models trained with variant data.

Models	Test data	Training data		
		Real	Fake	Mixed
Alexnet	Real	0.9831	0.9610	**0.9838**
	Fake	0.9800	0.9894	0.9843
Resnet	Real	0.9930	0.9922	**0.9948**
	Fake	0.9863	0.9989	0.9971
VGG	Real	0.9907	0.9894	**0.9914**
	Fake	0.9864	0.9928	0.9928

Lesion Localization with CAM. CAM is a weakly supervised localization algorithm to visualize the lesion findings of classification models [15]. It provides heat maps that visualizing the importance distribution of input image region for neural network decisions. In CAD studies, CAM has been widely used to locate lesions with only image-level annotations and provide doctors with diagnostic proof.

With the above trained classification models, CAM is implemented here with the models of ResNet-18 to visualize the COVID-19 lesions, as well as verify the feature extract capacity of the models. The feature map of the last convolution layer is employed to conduct CAM.

The heat maps plotted by CAM based on the features of real data are exhibited in Fig. 5. It is surprising that even though the models trained with the real data outperform those with fake ones, the fake data endow the models with a better locate ability. Furthermore, the models trained with mixed data, are equipped with the top locate the ability to the lesions.

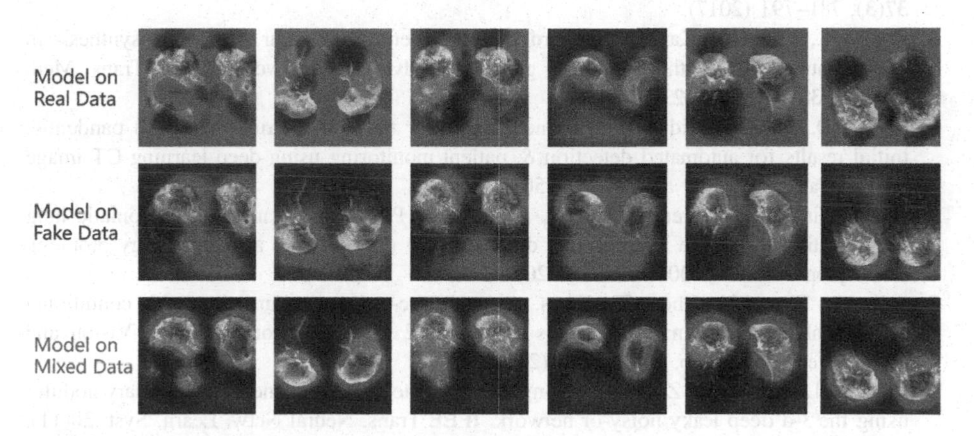

Fig. 5. The important distribution of models trained with real, fake and mixed data are successively displayed in the first, second and third rows.

This phenomenon could be interpreted as the result of the discriminator D in GAN. The discriminator D concentrates on screening the most typical pathological traits in images. The fake data consequently lead the models to accurately find the lesions. And the diversity and the typicality of the lesions are simultaneously fed to the models by mixing the real and fake data.

4 Conclusions

In this paper, we proposed to improve the performances of COVID-19 aided diagnosis algorithms with data synthesis. A synthesis scheme for COVID-19 CT scans was introduced with elastic registration and cGAN, and the capacity promotion and the quality of the synthesized data were verified by CAD with DNNs in the experiments. The neural networks trained with the synthesized data provided superior performances that the capacities for infection identification and lesion location were discernibly improved.

References

1. Abbas, A., Abdelsamea, M.M., Gaber, M.M.: Classification of covid-19 in chestx-ray images using detrac deep convolutional neural network. arXiv preprint arXiv:2003.13815 (2020)
2. Cohen, J.P., Morrison, P., Dao, L.: Covid-19 image data collection. arXiv 2003.11597 (2020). https://github.com/ieee8023/covid-chestxray-dataset
3. Costa, P., et al.: End-to-end adversarial retinal image synthesis. IEEE Trans. Med. Imaging **37**(3), 781–791 (2017)
4. Dar, S.U., Yurt, M., Karacan, L., Erdem, A., Erdem, E., C¸ukur, T.: Image synthesis in multi-contrast MRI with conditional generative adversarial networks. IEEE Trans. Med. Imaging **38**(10), 2375–2388 (2019)
5. Gozes, O., et al.: Rapid ai development cycle for the coronavirus (covid-19) pandemic: initial results for automated detection & patient monitoring using deep learning CT image analysis. arXiv preprint arXiv:2003.05037 (2020)
6. Hofmanninger, J., Prayer, F., Pan, J., Rohrich, S., Prosch, H., Langs, G.: Automatic lung segmentation in routine imaging is a data diversity problem, not a methodology problem. arXiv preprint arXiv:2001.11767 (2020)
7. Isola, P., Zhu, J.Y., Zhou, T., Efros, A.A.: Image-to-image translation with conditional adversarial networks. In: Proceedings of the IEEE Conference on Computer Vision and Pattern Recognition, pp. 1125–1134 (2017)
8. Liao, F., Liang, M., Li, Z., Hu, X., Song, S.: Evaluate the malignancy of pulmonary nodules using the 3-d deep leaky noisy-or network. IEEE Trans. Neural Netw. Learn. Syst. **30**(11), 3484–3495 (2019)
9. Ng, M.Y., et al.: Imaging profile of the covid-19 infection: radiologic findings and literature review. Radiol. Cardiothorac. Imaging **2**(1), e200034 (2020)
10. Salimans, T., Goodfellow, I., Zaremba, W., Cheung, V., Radford, A., Chen, X.: Improved techniques for training gans. In: Advances in Neural Information Processing Systems, pp. 2234–2242 (2016)
11. Setio, A.A.A., et al.: Validation, comparison, and combination of algorithms for automatic detection of pulmonary nodules in computed tomography images: the luna16 challenge. Med. Image Anal. **42**, 1–13 (2017)
12. Shi, H., et al.: Radiological findings from 81 patients with covid-19 pneumonia in wuhan, china: a descriptive study. Lancet Infect. Dis. **20**, 425–434 (2020)
13. Tustison, N.J., et al.: Large-scale evaluation of ants and freesurfer cortical thickness measurements. Neuroimage **99**, 166–179 (2014)
14. Zhao, J., Zhang, Y., He, X., Xie, P.: Covid-ct-dataset: a CT scan dataset aboutcovid-19. arXiv preprint arXiv:2003.13865 (2020)
15. Zhou, B., Khosla, A., Lapedriza, A., Oliva, A., Torralba, A.: Learning deep features for discriminative localization. In: Proceedings of the IEEE Conference on Computer Vision and Pattern Recognition, pp. 2921–2929 (2016)
16. Zhu, J.Y., Park, T., Isola, P., Efros, A.A.: Unpaired image-to-image translation using cycle-consistent adversarial networks. In: Proceedings of the IEEE International Conference on Computer Vision, pp. 2223–2232 (2017)

Traffic Data Prediction Based on Complex-Valued S-System Model

Bin Yang and Wei Zhang(✉)

School of Information Science and Engineering, Zaozhuang University, Zaozhuang, China
zz_zhangwei@163.com

Abstract. To predict traffic data accurately could make an important role in network management. In order to improve forecasting accuracy, this paper proposes complex-valued S-system model (CVSS) forecast small-scale traffic data. According to the form of CVSS model, complex-valued restricted gene expression programming (CVRGEP) is utilized to search the optimal the representation of CVSS. Complex-valued differential evolution (CVDE) is proposed to evolve the parameters of model. The small-scale traffic data is utilized to test our method. Our method has better prediction performances than neural network (NN), radial basis function neural network (RBF), flexible neural tree (FNT), ordinary differential equation (ODE) and S-system.

Keywords: Traffic data · Complex-valued · S-system · Differential evolution

1 Introduction

Prediction is the pre-estimation and prediction of things that have not yet happened or are not yet clear. Certain methods or techniques are utilized to simulate the unknowable and complex intermediate processes, and infer the future situations [1, 2]. The prediction of network traffic data could provide the future data accurately, and can be applied for congestion control, network bandwidth allocation and the design of high-performance router [3, 4]. Traffic data is a non-stationary time series signal, which contains noise, irregularity and chaos. To forecast traffic data accurately has been difficult issue in the past decades.

S-system model is a kind of special ordinary differential equation (ODE) system, and utilized for modeling the molecular biological systems [5]. S-Tree model was proposed to represent S-system mode [6]. However S-tree consists of all the representations of differential equations in S-system, which leads to overwhelming computational complexities when the number of variables is large. Palafox et al. proposed dissipative particle swarm optimization and $L1$ regularizer to optimize S-system model in order to infer large and complex gene regulatory networks [7]. Zhang et al. utilized restricted additive tree and cuckoo search to optimize S-system model for forecasting financial data [8].

In the past decade, complex-valued methods have been proposed to solve problem problems. Goh and Mandic proposed a complex-valued recurrent neural network (CVRNN) to predict the complex data from real-world and synthetic problems [9].

D.-S. Huang and K.-H. Jo (Eds.): ICIC 2020, LNCS 12464, pp. 423–431, 2020.
https://doi.org/10.1007/978-3-030-60802-6_37

Shamima et al. proposed complex-valued version of relaxation network to improve the prediction accuracy of the secondary structure of proteins [10]. Scardapane et al. presented a novel complex-valued neural network based on fully complex, non-parametric activation function and kernel expansion in order to solve the prediction and channel equalization problems [11]. Yang and Chen proposed a novel complex-value model, namely complex-valued polynomial model (CPM) [12].

In order to predict traffic data more accurately, this paper proposes the complex-valued version of S-system (CVSS) in order to solve traffic prediction problem. CVSS contains complex-valued functions and parameters. Complex-valued restricted gene expression programming and complex-valued differential evolution are utilized to evolve the structures and parameters. The small-scale traffic data is utilized to test our method. The results reveal that our method has better prediction performances than neural network, radial basis function neural network, flexible neural tree, ordinary differential equation and S-system.

2 Methods

2.1 Complex-Valued S-System Model

Complex-valued S-system (CVSS) contains complex coefficients and functions. In a CVSS model, the $i-th$ complex-valued ODE is described as follows.

$$Z_i^{'}(t) = \alpha_i \prod_{j=1}^{N} Z_j^{g_{ij}}(t) - \beta_i \prod_{j=1}^{N} Z_j^{h_{ij}}(t) \tag{1}$$

Where N is the number of variables, Z_i and Z_j are the $i-th$ and $j-th$ the complex-valued input variables, respectively. α_i and β_i are complex-valued rate constants of the $i-th$ variables. h_{ij} and g_{ij} are real-valued kinetic orders.

2.2 Complex-Valued Restricted Gene Expression Programming

In order to optimize the structure of CVSS model, complex-valued restricted expression programming (CVRGEP) is proposed. In CVRGEP algorithm, the number of genes is set as 2. The relationship between genes is subtraction. Each gene contains two part: head and tail, which are created by variable set ($V = \{z_1, z_2, \ldots, z_n\}$) and function set ($F = \{{}^*2, {}^*3, \ldots, {}^*n\}$). The lengths of head part and tail part are set as the same as gene expression programming. Suppose that variable set is given as $V = \{z_1, z_2, \ldots, z_5\}$ and function set is given as $F = \{{}^*2, {}^*3, {}^*4\}$. An example of chromosome of CVRGEP is depicted in Fig. 1. In order to represent the parameters of CVSS model, a real-valued parameter (h_{ij} or g_{ij}) is assigned to each variable and a complex-valued parameter (α_i or β_i) is assigned to each gene. The corresponding parse tree is depicted in Fig. 2. The CVSS model obtained is $\frac{dz_i}{dt} = \alpha_i z_1^{g_{i2}} z_2^{g_{i3}} z_3^{g_{i1}} - \beta_i z_1^{h_{i3}} z_2^{h_{i1}} z_3^{h_{i4}} z_4^{h_{i2}}$. The genetic operators of CVRGEP are as the same as gene expression programming, which are introduced detailedly in Ref [13].

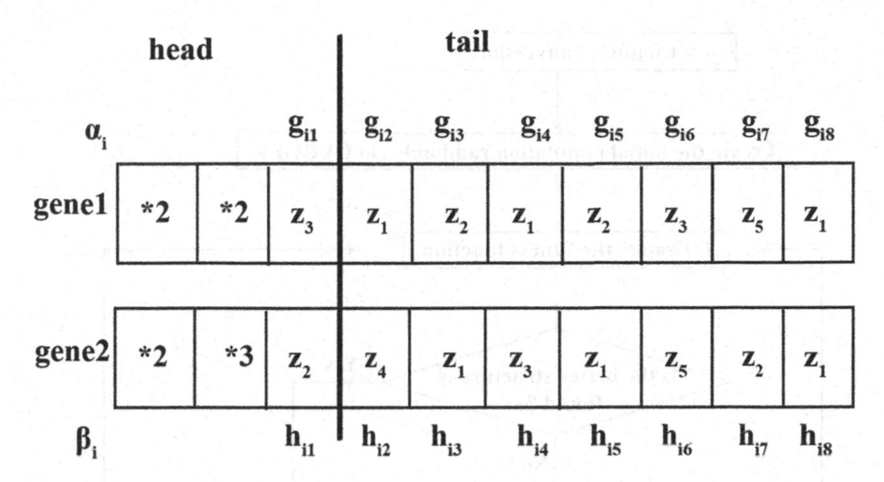

Fig. 1. An example of complex-valued restricted gene expression programming model.

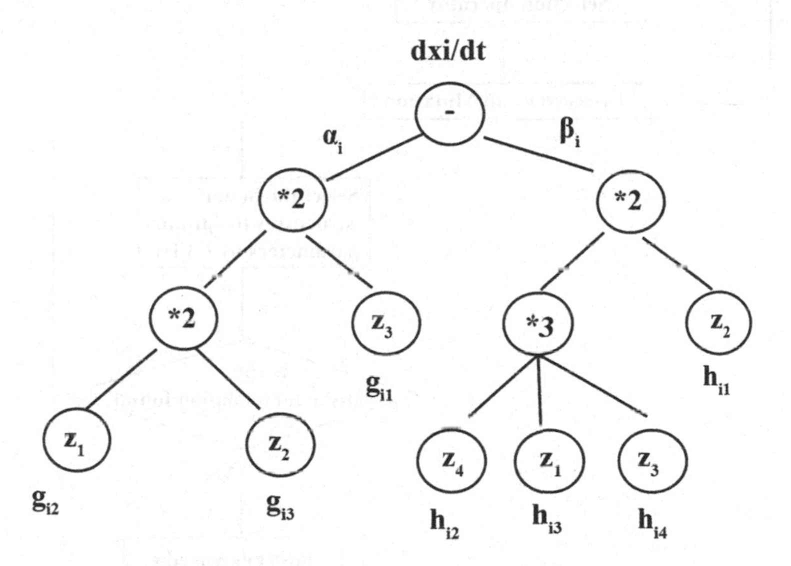

Fig. 2. The corresponding parse tree.

2.3 Complex-Valued Differential Evolution

Complex-valued differential evolution (CVDE) is the complex-valued version of differential evolution (DE). In CVDE, each complex-valued solution contains real and imaginary parts, which could improve the diversity of population and premature convergence of DE. The CVDE is described as follows.

1) In the complex-valued solution space, m individuals $(X_1, X_2, \ldots, X_m)$ are created randomly and uniformly. Each individual (X_i) is an $n-$dimensional complex-valued vector $[X^R_{i,1} + X^I_{i,1}i,\ X^R_{i,2} + X^I_{i,2}i,\ \ldots,\ X^R_{i,n} + X^I_{i,n}i]$, where $X^R_{i,j}$ and $X^I_{i,j}$ are real and imaginary parts of the $j-th$ gene of the $i-th$ individual.

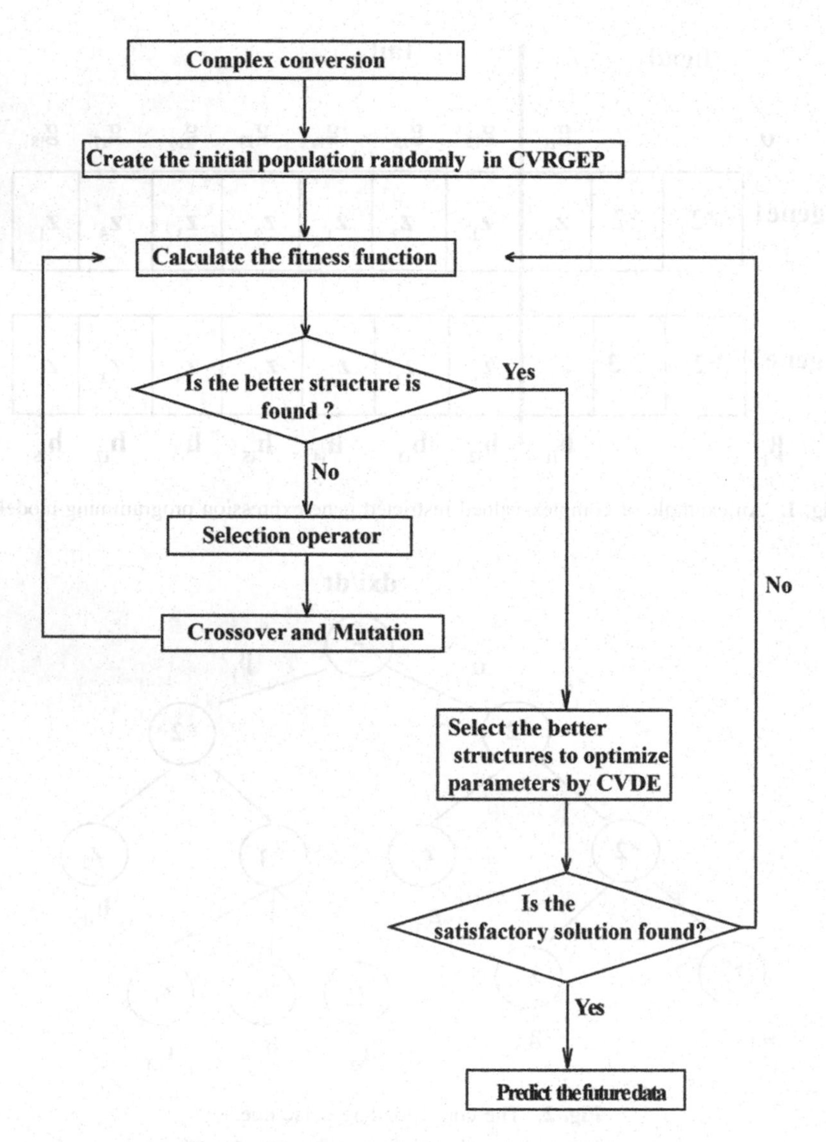

Fig. 3. The flowchart of traffic data prediction.

2) Mutation operator. The real part and imaginary part of each individual could be mutated separately. For individual X_i, select three individuals X_{p1}, X_{p2} and X_{p3} randomly. The mutation operator is defined as follows.

$$Y_{i,j}^{R} = X_{p1,j}^{R} + a \times (X_{p2,j}^{R} - X_{p3,j}^{R}). \tag{2}$$

$$Y_{i,j}^{I} = X_{p1,j}^{I} + a \times (X_{p2,j}^{I} - X_{p3,j}^{I}). \tag{3}$$

Where a is a scale factor, which could control the influence of difference vector.

3) Crossover operator could strength the diversity of population, which is implemented as follows.

$$D_{i,j}^{R} = \begin{cases} Y_{i,j}^{R}, & rand\,(0,1) \leq p_c \\ X_{i,j}^{R} \end{cases}. \tag{4}$$

$$D_{i,j}^{I} = \begin{cases} Y_{i,j}^{I}, & rand\,(0,1) \leq p_c \\ X_{i,j}^{I} \end{cases}. \tag{5}$$

4) Selection operator is utilized to select the individuals with the higher fitness values to the next generation. If the fitness value of the mutated individual D_i is larger than individual X_i, D_i will be select to replace the individual X_i.

3 Traffic Data Prediction

Complex-valued S-system model and the hybrid evolutionary algorithm based on CVRGEP and CVDE are utilized to predict the traffic data and the flowchart is depicted in Fig. 3. Firstly real-valued traffic data need to be converted into complex-valued data according to Ref [14]. Then the proposed hybrid evolutionary algorithm is utilized to optimize the structure and parameters of CVSS model. Lastly the optimal CVSS model is utilized to predict the future traffic data.

4 Experimental Results and Analysis

The used small-time scale traffic data are corrected from Ref [15], which contain 3600 time points. The first 3300 sample data are utilized as training set and the rest data are utilized for testing. Five criterions (*RMSE*, *MAP*, *MAPE*, R^2 and *ARV*) are used to evaluate the performance of our method, which are defined as follows.

$$RMSE = \sqrt{\frac{1}{m}\sum_{k=1}^{m}\left(y_k^t - y_k^f\right)^2}. \tag{6}$$

$$MAP = \max\left(\frac{|y_k^t - y_k^f|}{y_k^f} \times 100\right). \tag{7}$$

$$MAPE = \frac{1}{m}\sum_{k=1}^{m}\left(\frac{|y_k^f - y_k^f|}{y_k^f}\right) \times 100. \tag{8}$$

$$R^2 = 1 - \frac{\sum_{k=1}^{m} (y_k^t - y_k^f)^2}{\sum_{k=1}^{m} (y_k^t - \overline{y})^2}. \tag{9}$$

$$ARV = \frac{\sum_{k=1}^{m} (y_k^t - y_k^f)^2}{\sum_{k=1}^{m} (y_k^f - \overline{f})^2}. \tag{10}$$

Where y_k^t is the target value at $k - th$ time point and y_k^f is the forecasting value at $k - th$ time point. $\overline{f}$ is the mean of dataset.

The prediction results of CVSS model for small-time scale traffic data are depicted in Fig. 4, which could show that the predicted curve of CVSS is very close to the real one. In order show the predicted the results clearly, the predicted errors are depicted in Fig. 5, whose distribution is displayed in Fig. 6. From Fig. 5 and Fig. 6, we can see that the predicted errors of CVSS model are extremely small, which mainly concentrate around zero.

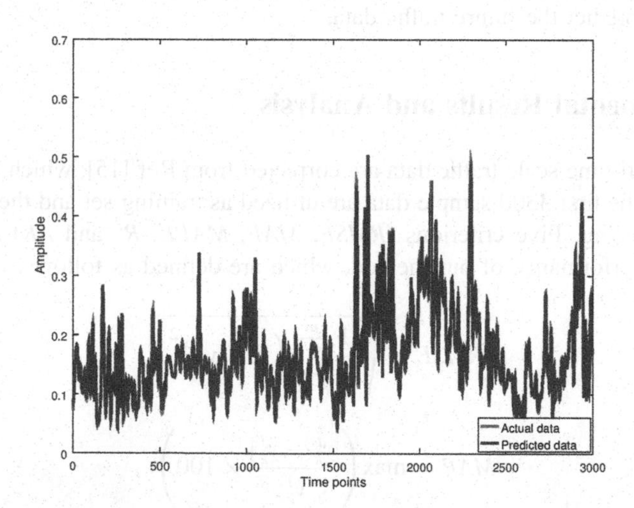

Fig. 4. The prediction results of CVSS model for small-time scale traffic data.

NN-PSO, RBF-NN, ODE, FNT and S-system have been also utilized to predict the traffic data. The predicted performances of six methods are listed in Table 1. Table 1 shows that CVSS model has the best performances in terms of RMSE, MAPE, ARV and R^2. In terms of MAP, NN-PSO method has the smallest performance, while our

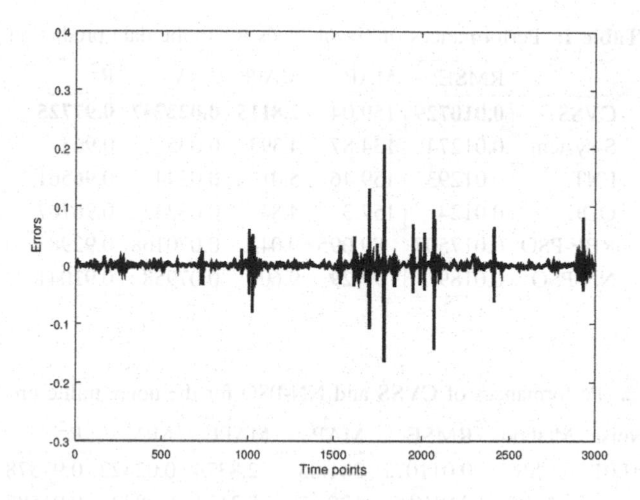

Fig. 5. The predicted errors.

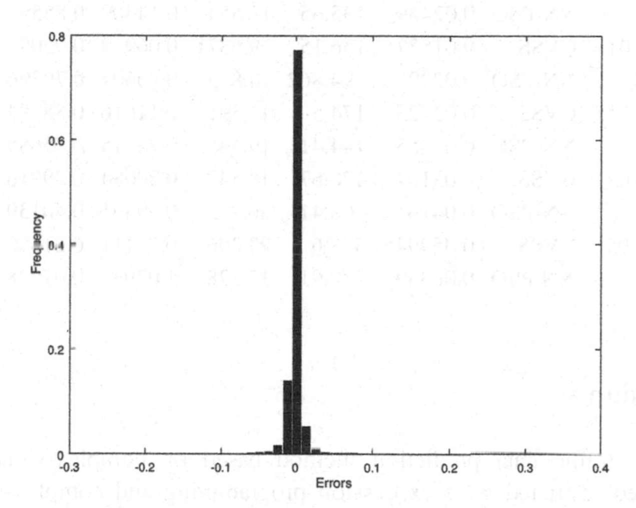

Fig. 6. The distribution of the predicted errors.

method performs worse, which proves that our method has bad prediction at some sample points.

In order to test the robustness to noise for our proposed method, we add 0.001, 0.005, 0.01, 0.015, 0.02 and 0.05 white Gaussian noise to the traffic data. The predicted performances of our method and NN-PSO with noisy data are listed in Table 2. The predicted performances prove that CVSS performs better than NN-PSO for noisy data prediction. When the noise rate is less than 0.02, CVSS could obtain the good predicted performances. When noise rate reaches to 0.05, our method has the bad performances.

Table 1. Performances of six methods for traffic data prediction.

	RMSE	MAP	MAPE	ARV	R^2
CVSS	**0.010729**	159.04	**2.8115**	**0.022747**	**0.97725**
S-system	0.01274	134.87	4.393	0.035	0.965
FNT	0.01293	159.46	5.01	0.0344	0.96561
ODE	0.0124	157.3	4.84	0.03233	0.96767
RBF-PSO	0.017507	89.095	9.042	0.070168	0.92983
NN-PSO	0.0189	**77.29**	9.609	0.07958	0.92041

Table 2. Performances of CVSS and NN-PSO for the noisy traffic prediction.

Noise	Methods	RMSE	MAP	MAPE	ARV	R^2
0.001	CVSS	0.011071	201.05	2.8374	0.02422	0.97578
	NN-PSO	0.019156	125.57	8.3671	0.08422	0.91587
0.005	CVSS	0.013653	151.95	5.7881	0.03694	0.9631
	NN-PSO	0.02439	135.65	13.552	0.14409	0.8559
0.01	CVSS	0.01877	156.15	9.6871	0.06919	0.9308
	NN-PSO	0.0299	94.862	14.875	0.20604	0.79396
0.015	CVSS	0.02423	174.5	13.391	0.11616	0.88384
	NN-PSO	0.03655	144.41	19.445	0.24315	0.75685
0.02	CVSS	0.03107	176.67	18.542	0.20084	0.79916
	NN-PSO	0.04161	108.41	26.745	0.59861	0.40139
0.05	CVSS	0.059945	558.6	22.006	0.78111	0.21889
	NN-PSO	0.06339	230.91	17.328	1.0793	0.07928

5 Conclusion

In this paper, traffic data prediction method based on complex-valued S-system, complex-valued restricted gene expression programming and complex-valued differential evolution is proposed. The small-scale traffic data is utilized to test our method. The predicted errors of our method are extremely small. CVSS model has the better forecasting performances than NN-PSO, RBF-PSO, FNT, ODE and S-system in terms of RMSE, MAPE, ARV and R^2. In terms of MAP, NN-PSO method has the smaller performance than our method, which shows that the predicted errors of our method may be large at some sample points. We also test the robustness to noise for our proposed method and find that CVSS could obtain the good predicted performances when the noise rate reaches to 0.02.

Acknowledgment. This work was supported by the talent project of "Qingtan scholar" of Zaozhuang University, the PhD research startup foundation of Zaozhuang University (No. 2014-BS13), and foundation of Zaozhuang University (No. 2015YY02).

References

1. Bittanti, S., Colaneri, P., Nicolao, G.D.: An algebraic riccati equation for the discrete-time periodic prediction problem. Syst. Control Lett. **14**(1), 71–78 (1990)
2. Grenander, U.: A prediction problem in game theory. Arkiv fr Matematik **3**(4), 371–379 (1957)
3. Liu, J., Yang, O.W.W.: Using fuzzy logic control to provide intelligent traffic management service for high-speed networks. IEEE Trans. Netw. Serv. Manage. **10**(2), 148–161 (2013)
4. Chrysostomou, C., Pitsillides, A., Hadjipollas, G., Polycarpou, M., Sekercioglu, A.: Congestion control in differentiated services networks using fuzzy logic. In: 43rd IEEE Conference on Decision and Control, Nassau, Bahamas, pp. 549–556. IEEE (2004)
5. Yang, B., Zhang, W., Wang, H., Song, C., Chen, Y.: TDSDMI: Inference of time-delayed gene regulatory network using S-system model with delayed mutual information. Comput. Biol. Med. **72**, 218–225 (2016)
6. Cho, D.Y., Cho, K.H., Zhang, B.T.: Identification of biochemical networks by S-tree based genetic programming. Bioinformatics **22**(13), 1631–1640 (2006)
7. Palafox, L., Noman, N., Iba, H.: Reverse engineering of gene regulatory networks using dissipative particle swarm optimization. IEEE Trans. Evol. Comput. **17**(4), 577–587 (2013)
8. Zhang, W., Yang, B.: Stock market forecasting using S-system model. In: Xhafa, F., Patnaik, S., Zomaya, Albert Y. (eds.) IISA 2017. AISC, vol. 686, pp. 397–403. Springer, Cham (2018). https://doi.org/10.1007/978-3-319-69096-4_55
9. Goh, S.L., Mandic, D.P.: Nonlinear adaptive prediction of complex-valued signals by complex-valued PRNN. IEEE Trans. Signal Process. **53**(5), 1827–1836 (2005)
10. Shamima, B., Savitha, R., Suresh, S., Saraswathi, S.: Protein secondary structure prediction using a fully complex-valued relaxation network. In: 2013 International Joint Conference on Neural Networks (IJCNN), Dallas, TX, USA, pp. 3015–3022. IEEE (2013)
11. Scardapane, S., Van Vaerenbergh, S., Hussain, A., Uncini, A.: Complex-valued neural networks with nonparametric activation functions. IEEE Trans. Emerg. Topics Comput. Intell. **4**(2), 140–150 (2020)
12. Yang, B., Chen, Y.: A new complex-valued polynomial model. Neural Process. Lett. **50**(3), 2609–2626 (2019). https://doi.org/10.1007/s11063-019-10042-8
13. Ferreira, C.: Gene expression programming: a new adaptive algorithm for solving problems. Complex Syst. **13**(2), 87–129 (2001)
14. Saoud, L.S., Rahmoune, F., Tourtchine, V., Baddari, K.: Fully complex valued wavelet network for forecasting the global solar irradiation. Neural Process. Lett. **45**(2), 475–505 (2016). https://doi.org/10.1007/s11063-016-9537-7
15. Meng, Q.F., Chen, Y.H., Peng, Y.H.: Small-time scale network traffic prediction based on a local support vector machine regression model. Chin. Phys. B **18**(6), 2194–2199 (2009)

Classification of Protein Modification Sites with Machine Learning

Jin Sun[1], Wenzheng Bao[2], Yi Cao[1], and Yuehui Chen[1](✉)

[1] School of Information Science and Engineering, University of Jinan, Jinan 250024, China
yhchen@ujn.ed.cn
[2] School of Information and Electrical Engineering, Xuzhou University of Technology, Xuzhou 221018, China

Abstract. Lysine malonylation is a newly discovered type of protein post-translational modification, which plays an essential role in many biological activities. A good knowledge of malonylation sites can serve as guidance in solving a large number of biological problems, such as disease diagnosis and drug discovery. There have already been several experimental approaches to identify modification sites, but they are relatively expensive. In this work, we propose three novel machine learning models and utilizes several effective feature description methods. The model is trained based on the cross validation method named Split to Equal Validation (SEV). The experiments show that our model outperforms the others considerably.

Keywords: Lysine malonylation · Feature description · Machine learning · Split to equal validation

1 Introduction

Protein post-translational modification (PTM) is an essential mechanism to regulate protein functions by the covalent and generally enzymatic modification. Hundreds of types of PTMs have been discovered and reported in the field of biology [1]. They played vital roles in influencing almost all aspects of cell biology and pathogenesis, e.g., gene expression, cell division and cell signaling. As one of a newly identified PTM type in both eukaryotic and prokaryotic [2], Lysine malonylation (Kmal) is connected with various biological processes, and some Lysine residues may even have something to do with cancer [3]. Therefore, it is very critical to identify and understand Lysine residues in the studies of biology and disease.

Different from traditional experimental methods, computational approaches for PTMs prediction are more effective and cost-saving. Usually, as the PTM site prediction is abstracted as a typical classification problem, a series of machine learning approaches have been successfully utilized in this field. For example, Logistic Regression (LR) has been applied in ModPred [4] for predicting 23 different modifications. It is sequence-based, evolutionary, and has physicochemical properties. Besides, Musite [5], which is a general and kinase-specific protein phosphorylation site prediction method, applies Support Vector Machines (SVMs) and bears three types of

D.-S. Huang and K.-H. Jo (Eds.): ICIC 2020, LNCS 12464, pp. 432–445, 2020.
https://doi.org/10.1007/978-3-030-60802-6_38

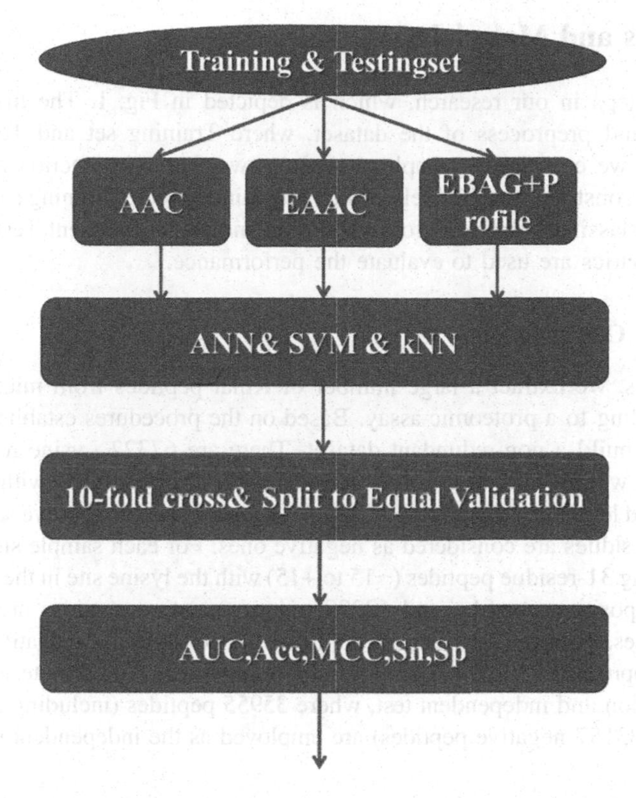

Fig. 1. The working flow of our reaserch.

features: K nearest neighbor score, disorder scores and amino acid frequencies. Wei et al. have proposed PhosPred-RF [6] for predicting phosphorylation sites, which exploits the evolutionary information features from position specific scoring matrices. Deep Learning techniques have also been applied in this area, such as the recently published tool MusiteDeep [7], which is used for general and kinase-specific phosphorylation site prediction.

In this work, we first employ artificial neural network (ANN) classifier based on Stochastic Gradient Descent (SGD) algorithm for protein Kmal site prediction. Then, we investigate a wide range of feature extraction methods and finally choose EBAG + Profile and EAAC methods to train our predictors. In view of the fact that the Kmal prediction problem can be regarded as a binary classification one, we adopt the original SEV method to solve the inherent imbalance problem of positive and negative samples in the training set. The result of our experiment shows that ANN performs better than SVM and kNN predictors. In conclusion, ANN can be a useful tool for identifying Lysine residues.

2 Methods and Materials

There are 4 steps in our research, which is depicted in Fig. 1. The first step is the construction and preprocess of the dataset, where Training set and Testing set are formed. Then we encode the samples based on two feature extraction methods. The next step is to construct three models, which are trained by the Training set. Finally, we tested all the classifiers through cross-validation and an independent Testing set. Five assessment metrics are used to evaluate the performance.

2.1 Dataset Construction

In this process, we extract a large number of Kmal peptides from mice and human species according to a proteomic assay. Based on the procedures established by Chen et al. [8], we build a non-redundant dataset. There are 67322 Lysine residues in the Training set, where sites with high confidence (Kmal peptides with Andromeda scores >50 and localization probability >0.75) are considered as positive samples, while other lysine residues are considered as negative ones. For each sample site, we extract the surrounding 31-residue peptides (−15 to +15) with the lysine site in the center. Then, 5023 unique positive peptides and 62299 unique negative peptides are retained for further analyses. It is obvious that the ratio of the positive and negative samples in training set approaches 1 to 12. Therefore, the train models would be tested by Split to Equal Validation and independent test, where 35955 peptides (including 2798 positive peptides and 33157 negative peptides) are employed as the independent test set.

2.2 Feature Encodings

EBAG + Profile Encoding

EBAG + Profile encoding is an integration scheme consisting of two different feature encoding methods proposed by Han [9]. One of them is Encoding Based on Attribute Grouping (EBAG) [10], which divide 20 types of amino acids into 5 groups by their physical and chemical properties. Table 1 shows the derived groups by on EBAG.

Table 1. Groups of amino acid residues according to EBAG encoding

Group	Amino acid residue	Label
C1	*A, F, G, I, L, M, P, V, W*	Hydrophobic
C2	*C, N, Q, S, T, Y*	Polar
C3	*D, E*	Acidic
C4	*H, K, R*	Basic
C5	*X*	Gaps

The other encoding method is Profile encoding, which calculates the frequency of each amino acid residue occurring in the protein peptides. The frequency is held as the representation of the residue in the sequence so that each peptide with 31 residues can be transformed into a 31-dimensional vector. The way to combine EBAG and Profile is to replace source amino acid peptides with EBAG sequence and then to encode the sequence according to the Profile method. By this operation, a peptide with 31 residues is converted to a vector of 31-dimensional vector using the EBAG + Proifle encodings.

EAAC Encoding

A typical predictive encoding scheme for PTMs is AAC encoding [11], which can reflect the frequency of 20 amino acid residues surrounding the modification site. In this process, we code each amino acid using the EAAC method proposed by chen et al. [12], which is an improved version of the AAC encoding. As the 8-size window of each peptide sliding from the N-terminus to C-terminus of a sample, the EAAC method calculated the frequency of the 20 amino acid residues. Accordingly, the dimension of features can be calculated as follows:

$$N\_s = L\_p - L\_s + 1 \tag{1}$$

$$D\_eaac = N\_s \times 20 \tag{2}$$

where L_p refers to the length of each peptide, L_s stands for the length of sliding windows and D_eaac represents the dimension of feature vector. If we set L_s to 8, a peptide with 31 residues will correspond to 24(31−8 + 1) sliding windows and be converted to a 24 × 20 matrix.

3 Construction of classifiers

Artificial Neural Network

ANN is a traditional machine learning model, which has been widely used in the prediction of lysine PTMs. In this work, we construct an ANN model with four layers, i.e. input layer, output layer and two hidden layers. The input layer receives the feature sequences generated by different encoding methods; The two hidden layers both contain 100 neurons and adopt "ReLu" as their activation function. The output layer contains one single unit, producing the probabilistic score for each site. The architecture is depicted in Fig. 2.

Support Vector Machine

SVM is a well-established and commonly employed algorithm based on the structural risk minimization of the statistical learning theory [13]. SVM can transform the samples into a high-dimensional feature space, and then construct an Optimal Separating Hyperplane (OSH) to maximize the distance between two classes of training samples. In this process, based on Tensorflow [14] and Scikit-learn [15], we employ Support Vector Classify (SVC) as our SVM model, where the kernel function applied was linear kernel.

k-Nearest Neighbor algorithm

kNN algorithm is another widely used algorithm that calculates the distances between samples in order to cluster them [16]. Suppose we have a training set D and a new

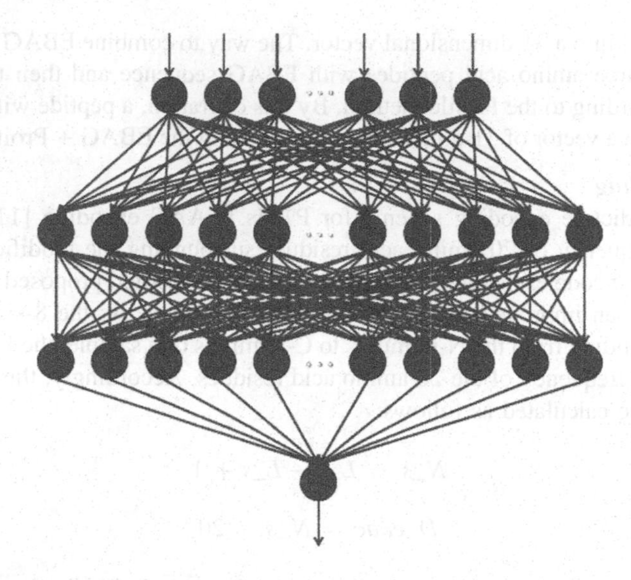

Fig. 2. Architecture of the proposed ANN model

sample with an unknown label, this new sample will be labeled according to the labels of the k samples that are most close to it. In this process, we build a kNN model implemented by Tensorflow and Scikit-learn.

3.1 Cross Validation Methods

Generally, to training an ANN model, the data-set is divided into two parts: a Training set and a Testing set. To make full use of the Training set samples, we train the model through 10-fold cross-validation. The samples in the Training set are further divided into training set. Then, the training set is used to adjust the model parameters, while the validating set is used to measure how the model is trained and to obtain the validation scores. After the cross-validation, the trained model is evaluated on the independent Testing set to get the testing scores (Fig. 3).

In view of the fact that the classifier is always more sensitive to the category containing more samples in binary classification problems, it is necessary to preprocess the training set. In the previous studies, we proposed a new cross validation strategy named SEV (Split to Equal Validation), which can well solve the problem of imbalanced training samples in PTM sites prediction. Here, we adopt the SEV method together with the 10-fold cross-validation, to conduct comparative experiments. The working flow of SEV is shown as follows:

As shown in Fig. 4, SEV consists of five steps. Assuming that the ratio of negative samples to positive samples in the training set is to n:1, (1) divide the negative samples into n groups; (2) combine the positive samples with each group of negative samples and build n new balanced subsets. (3) train model 1 with subset 1 and verify it with

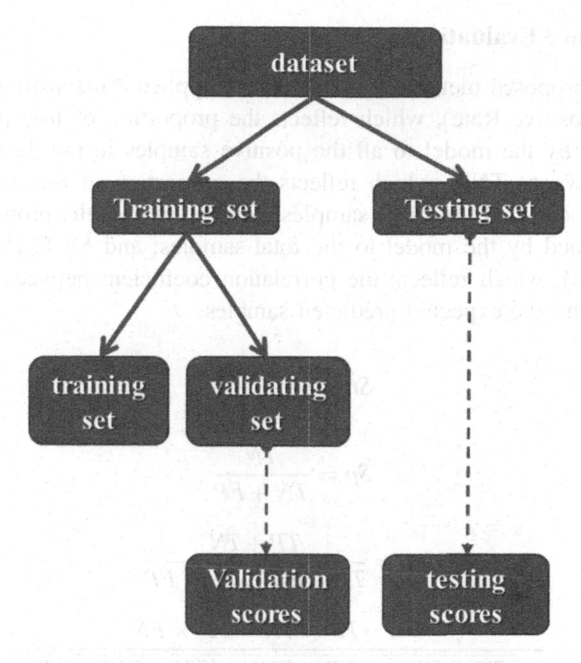

Fig. 3. Process of dataset usage partition

subset 2. Similarly, train model 2 with subset 2 and verify it with subset, and so on. (4) repeat step 3 to train and validate n models. (5) evaluate the n models by the independent test set.

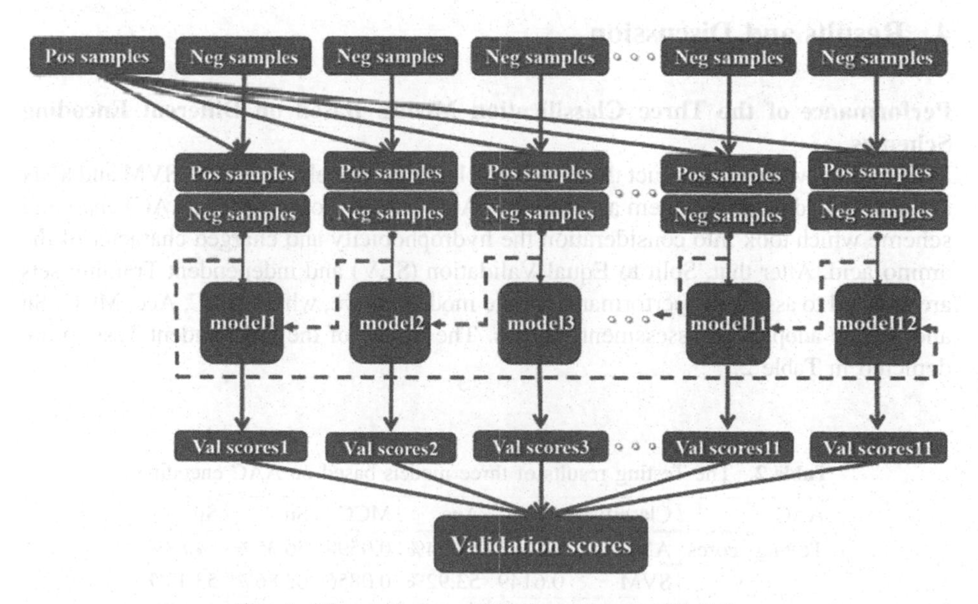

Fig. 4. The work flow of Split to Equal validation

3.2 Performance Evaluation

To evaluate the proposed method, four criteria are applied: Sn(sensitivity), also known as TPR (True Positive Rate), which reflects the proportion of true positive samples (TP) determined by the model to all the positive samples in the data set; Sp (specificity), also known as TNR, which reflects the proportion of true negative samples judged by the model in all negative samples. Acc (accuracy), the proportion of correct samples determined by the model to the total samples; and MCC (Mathew's Correlation Coefficient), which reflects the correlation coefficient between the actual predicted samples and the expected predicted samples.

$$Sn = \frac{TP}{TP + FN} \tag{3}$$

$$Sp = \frac{TN}{TN + FP} \tag{4}$$

$$Acc = \frac{TP + TN}{TP + FN + TN + FP} \tag{5}$$

$$MCC = \frac{TP \times TN - FP \times FN}{\sqrt{(TP + FP) \times (TP + FN) \times (TN + FN) \times (TN + FP)}} \tag{6}$$

As shown in the above, *TP*, *FP*, *TN*, and *FN* represent true positives, false positives, true negatives and false negatives respectively. Besides, ROC curves and AUC value are given for evaluation.

4 Results and Discussion

Performance of the Three Classification Models Based on Different Encoding Schemes

In this study, we first construct three machine learning models, i.e. ANN, SVM and kNN algorithm, and then train them according to Amino Acid Composition (AAC) encoding scheme which took into consideration the hydrophobicity and charged character of the amino acid. After that, Split to Equal Validation (SEV) and independent Training sets are utilized to assess the performance of the models above, where AUC, Acc, MCC, Sn and Sp are adopted as assessment metrics. The results of the independent Testing are depicted in Table 2.

Table 2. The Testing results of three models based on AAC encoding

AAC	Classifier	AUC	Acc	MCC	Sn	Sp
Testing scores	ANN	0.5833	54.94%	0.0598	56.35%	54.83%
	SVM	0.6149	53.92%	0.0856	62.86%	53.17%
	kNN	**0.6224**	47.64%	0.0906	71.20%	45.67%

From the results, we speculate that feature extraction schemes are very important to the final classification accuracy. Therefore, we adopt EBAG + Profile encoding method, which utilizes the physical and chemical properties of amino acids. EAAC encoding method, which is based on AAC encoding and conforms to the probability of specific amino acids in the peptide sequence, is also employed. The Testing scores are depicted in the Table 3 and Table 4 respectively.

Table 3. The Testing results of three models based on EBAG + Profile encoding

EBAG + Pofile	Classifier	AUC	Acc	MCC	Sn	Sp
Testing scores	ANN	**0.6552**	56.57%	0.1226	67.23%	55.68%
	SVM	0.5041	82.71%	0.0056	12.06%	88.61%
	kNN	0.5874	64.69%	0.0705	46.38%	66.21%

As we know, the larger AUC value is, the more likely that the current classification algorithm rank positive samples ahead of negative ones so as to get better classification results. Therefore, it is obvious that the classifiers under EAAC encoding scheme performs better than those under the other two schemes in obtaining higher AUC values. Also, for MCC and Acc value, EAAC encoding get higher scores and exhibits similar advantages to others.

In addition, it is certain that ANN outperforms SVM and kNN under EAAC encoding scheme. As for independent test, when EAAC is taken, the AUC value of ANN is **0.7471**, while that of SVM and kNN algorithm are 0.6322 and 0.6317 respectively. The results in Table 2, 3 and 4 together show that different types of classifiers have different but equally important impacts on prediction performance. In this research, ANN was the best.

Comparing the Results of SEV with that of 10-Fold Cross-Validation and Scaling the Number of Training Samples by SEV

In this experiment, we use SEV method to preprocess the training samples, and to train the classifier model. Furthermore, on the premise that other conditions remain unchanged, we use 10-fold cross-validation instead of SEV to do the same experiment. The experiment results are shown in Table 5:

The experiment was based on the neural network model, using 10-fold cross-validation and SEV method respectively. It can be seen that although the Acc of the 10-fold cross is higher, its AUC value is not. This is because, in the case of the imbalanced positive and negative samples, the classifier would guess the samples with higher probability in the training set, while prediction capability is not improved. The SEV method could overcome this problem. Although Acc value was not as good as 10-fold cross verification, SEV achieves much better AUC values, implying its prediction capability is much powerful.

In order to further explore the influence of the imbalance between positive samples and negative samples in training set, we use SEV method to scale the training samples. We verify the fact that unbalanced training data would lead to very low Sn and very high Sp of the classifier. Besides, in order to further explore the far-reaching impact of

Table 4. The Testing results of three models based on EAAC encoding

EAAC	Classifier	AUC	Acc	MCC	Sn	Sp
Testing scores	ANN	**0.7471**	63.54%	0.2002	74.16%	62.65%
	SVM	0.6322	56.21%	0.1028	63.61%	55.59%
	kNN	0.6317	43.60%	0.0931	76.19%	40.88%

Table 5. The Testing scores of 10-fold cross-validation and Split to Equal Validation

Scores	Validation methods	AUC	ACC	MCC	SN	SP
Validation scores	10F CV	0.5751	21.11%	0.0776	95.23%	15.14%
	SEV	**0.8465**	64.35%	0.3260	100.00%	61.48%
Testing scores	10F CV	0.6965	90.12%	0.1060	10.91%	96.73%
	SEV	**0.7471**	63.54%	0.2002	74.16%	62.65%

positive and negative sample ratio on classifier performance, we calculate the five metrics i.e. AUC, Acc, MCC, Sn and Sp by randomly adjusting different positive and negative sample ratios from 1:1 to 1:12 under SEV.

It can be seen in Table 6, Fig. 5 and Fig. 6, as the proportion of negative samples in the training set increases, AUC value of the model gradually decreases from 0.7471 to about 0.7033, and MCC value gradually decreases to about 0.1292, but ACC value continuously increases to 89.43%. It consistent with our previous conclusion: when the positive and negative samples of the training set are extremely imbalanced, the classifier tends to guess the class with more samples, instead of inferring the label from features.

Table 6. The Testing scores based on different proportion of positive and negative samples

Scale-up ratio	AUC	Acc	MCC	Sn	Sp
1 : 1	0.7471	63.54%	0.2002	74.16%	62.65%
1 : 2	0.7399	76.67%	0.1982	52.99%	78.65%
1 : 3	0.7324	82.60%	0.1828	38.61%	86.28%
1 : 4	0.7290	84.33%	0.1689	32.65%	88.64%
1 : 5	0.7228	85.54%	0.1623	28.93%	90.26%
1 : 6	0.7205	86.71%	0.1476	24.13%	91.94%
1 : 7	0.7196	87.27%	0.1499	23.12%	92.62%
1 : 8	0.7193	86.64%	0.1539	25.18%	91.77%
1 : 9	0.7012	87.32%	0.1340	20.91%	92.86%
1 : 10	0.6984	88.08%	0.1382	19.61%	93.80%
1 : 11	0.6972	88.96%	0.1337	16.87%	94.97%
1 : 12	0.7033	89.43%	0.1292	15.10%	95.64%

In addition, Table 6 implies that when the ratio of positive and negative samples in the training set reaches 1:9, the AUC value tends to be stable, fluctuating only around 0.7 without further decline. This means that in this experiment, although the ratio of positive and negative samples had been changing in a more imbalanced direction, the performance of the classifier would not decrease continually, but tend to be stable after reaching a certain threshold.

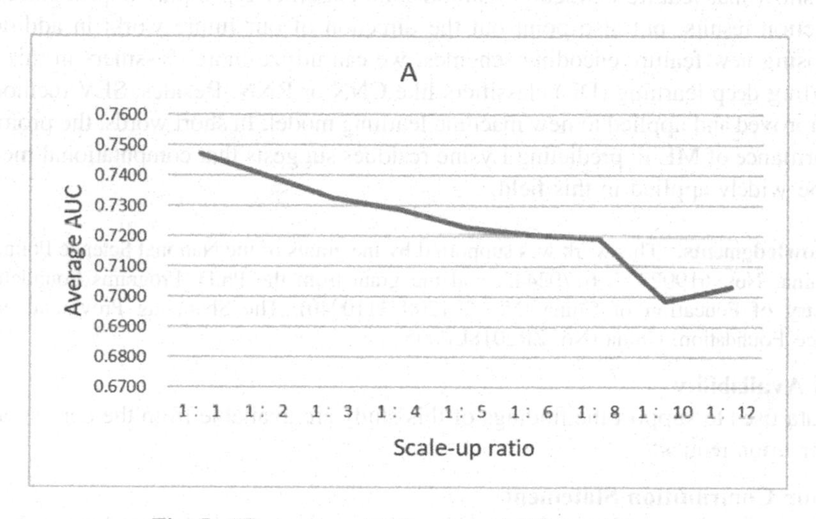

Fig. 5. The average AUC values of different ratios.

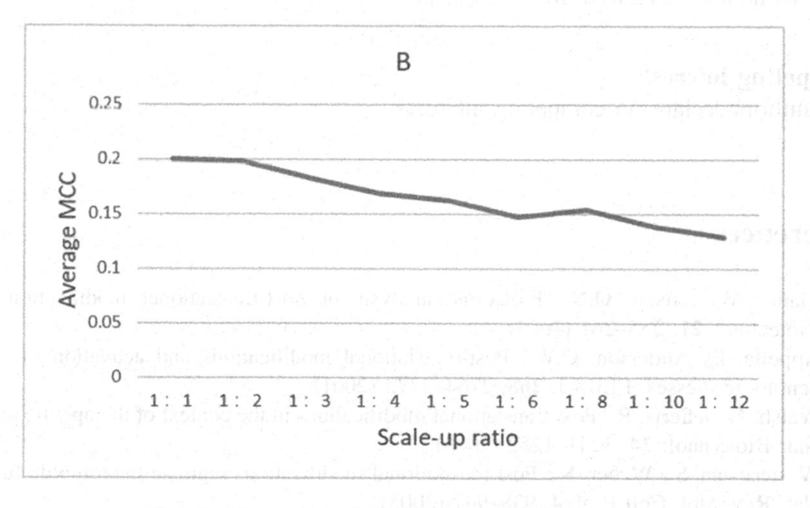

Fig. 6. The average MCC values of different ratios.

5 Conclusions and Future Work

In this work, we first adopt two feature extraction schemes according to different physical and chemical properties as well as occurrence frequency, and construct three ML classifiers. We have found that the application of SEV method could solve the problem of sample imbalance in binary classification cases. The experiment results not only show that feature extraction methods and classifier types play important roles in prediction results, but also point out the direction of our future work: in addition to proposing new feature encoding schemes, we can utilize more classifiers in this field, including deep learning (DL) classifiers like CNN or RNN. Besides, SEV method can be improved and applied to new machine learning model. In short words, the prominent performance of ML in predicting Lysine residues suggests that computational methods can be widely applied in this field.

Acknowledgments. This work was supported by the grants of the National Science Foundation of China, Nos. 61902337, 61702445, and the grant from the Ph.D. Programs Foundation of Ministry of Education of China (No. 20120072110040). The Shandong Provincial Natural Science Foundation, China (No. ZR2018LF005).

Data Availability

To data used to support the findings of this study are available from the corresponding author upon request.

Author Contribution Statement

W.B. conceived the method. Z.L designed the method. B.Y. designed the website of this algorithm. Y.Z. conducted the experiments and W.B. wrote the main manuscript text. All authors reviewed the manuscript.

Competing Interests

The authors declare no competing interests.

References

1. Mann, M., Jensen, O.N.: Proteomic analysis of post-translational modifications. Nat. Biotechnol. **21**, 255–261 (2003)
2. Appella, E., Anderson, C.W.: Post-translational modifications and activation of p53 by genotoxic stresses. FEBS J. **268**, 2764–2772 (2001)
3. Walsh, G., Jefferis, R.: Post-translational modifications in the context of therapeutic proteins. Nat. Biotechnol. **24**, 1241–1252 (2006)
4. Westermann, S., Weber, K.: Post-translational modifications regulate microtubule function. Nat. Rev. Mol. Cell Biol. **4**, 938–947 (2003)
5. Keller, J.N., Hanni, K.B., Markesbery, W.R.: Impaired proteasome function in Alzheimer's disease. J. Neurochem. **75**, 436–439 (2001)

6. Maccioni, R.B., Munoz, J.P., Barbeito, L.: The molecular bases of Alzheimer's disease and other neurodegenerative disorders. Arch. Med. Res. **32**, 367–381 (2001)
7. Ishigami, A., Maruyama, N.: Importance of research on peptidylarginine deiminase and citrullinated proteins in age-related disease. Geriatr. Gerontol. Int. **10**, S53-S58 (2010)
8. Mangat, P., Wegner, N., Venables, P.J., Potempa, J.: Bacterial and human peptidylarginine deiminases: targets for inhibiting the autoimmune response in rheumatoid arthritis? Arthritis Res. Therapy **12**, 209 (2010). https://doi.org/10.1186/ar3000
9. Schwenzer, A., Jiang, X., Mikuls, T.R., Payne, J.B., Sayles, H., Quirke, A.M., et al.: Identification of an immunodominant peptide from citrullinated tenascin-C as a major target for autoantibodies in rheumatoid arthritis. Ann. Rheum. Dis. **75**, 1876–1883 (2016)
10. Brill, A., Fuchs, T.A., Savchenko, A.S., Thomas, G.M., Martinod, K., De Meyer, S.F., et al.: Neutrophil extracellular traps promote deep vein thrombosis in mice. J. Thromb. Haemost. **10**, 136–144 (2012)
11. Van Venrooij, W.J., Pruijn, G.J.M.: Citrullination: a small change for a protein with great consequences for rheumatoid arthritis. Arthritis Res. Therapy **2**, 249–251 (2000)
12. Guo, Q., Bedford, M.T., Fast, W.: Discovery of peptidylarginine deiminasc-4 substrates by protein array: antagonistic citrullination and methylation of human ribosomal protein S2. Mol. BioSyst. **7**, 2286–2295 (2011)
13. Wang, S., Wang, Y.: Peptidylargininc deiminases in citrullination, gene regulation, health and pathogenesis. Biochem. Biophys. Acta. **1829**, 1126–1135 (2013)
14. Bicker, K.L., Subramanian, V., Chumanevich, A.A., Hofseth, L.J., Thompson, P.R.: Seeing citrulline: development of a phenylglyoxal-based probe to visualizc protein citrullination. J. Am. Chem. Soc. **134**, 17015–17018 (2012)
15. Stensland, M., Holm, A., Kiehne, A., Fleckenstein, B.: Targeted analysis of protein citrullination using chemical modification and tandem mass spectrometry. Rapid Commun. Mass Spectrom. **23**, 2754–2762 (2009)
16. Hermansson, M., Artemenko, K.A., Ossipova, E., Eriksson, H., Lengqvist, J., Makrygiannakis, D., et al.: MS analysis of rheumatoid arthritic synovial tissue identifies specific citrullination sites on fibrinogen. Proteomics Clin. Appl. **4**, 511–518 (2010)
17. Bao, W., Yang, B., Huang, D., Wang, D., Liu, Q., Chen, Y., et al.: IMKPse: identification of protein malonylation sites by the key features into general PseAAC. IEEE Access **7**, 54073–54083 (2019)
18. Bao, W., Wang, D., Chen, Y.: Classification of protein structure classes on flexible neutial tree. IEEE/ACM Trans. Comput. Biol. Bioinf. **14**, 1122–1133 (2017)
19. Qiu, W., Xiao, X., Xu, Z., Chou, K.: iPhos-PseEn: identifying phosphorylation sites in proteins by fusing different pseudo components into an ensemble classifier. Oncotarget **7**, 51270–51283 (2016)
20. Qiu, W., Sun, B., Xiao, X., Xu, Z., Jia, J., Chou, K.: iKcr-PseEns: Identify lysine crotonylation sites in histone proteins with pseudo components and ensemble classifier. Genomics **110**, 239–246 (2017)
21. Gao, J., Tao, X., Zhao, J., Feng, Y., Cai, Y., Zhang, N.: Computational prediction of protein epsilon lysine acetylation sites based on a feature selection method. Comb. Chem. High Throughput Screening **20**, 629–637 (2017)
22. Cai, Y., Huang, T., Hu, L., Shi, X., Xie, L., Li, Y.: Prediction of lysine ubiquitination with mRMR feature selection and analysis. Amino Acids **42**, 1387–1395 (2012). https://doi.org/10.1007/s00726-011-0835-0

23. Hasan, M.A.M., Li, J., Ahmad, S., Molla, M.K.I.: predCar-site: carbonylation sites prediction in proteins using support vector machine with resolving data imbalanced issue. Anal. Biochem. **525**, 107–113 (2017)
24. Cheng, X., Xiao, X., Chou, K.: pLoc-mEuk: predict subcellular localization of multi-label eukaryotic proteins by extracting the key GO information into general PseAAC. Genomics **110**, 50–58 (2018)
25. Bao, W., Yuan, C., Zhang, Y., Han, K., Nandi, A.K., Honig, B., et al.: Mutli-features prediction of protein translational modification sites. IEEE/ACM Trans. Comput. Biol. Bioinf. **15**, 1453–1460 (2018)
26. Jia, J., Liu, Z., Xiao, X., Liu, B., Chou, K.: iSuc-PseOpt: Identifying lysine succinylation sites in proteins by incorporating sequence-coupling effects into pseudo components and optimizing imbalanced training dataset. Anal. Biochem. **497**, 48–56 (2016)
27. Xu, Y., Wang, Z., Li, C., Chou, K.: iPreny-PseAAC: identify C-terminal cysteine prenylation sites in proteins by incorporating two tiers of sequence couplings into PseAAC. Med. Chem. **13**, 544–551 (2017)
28. Li, B., Hu, L., Niu, S., Cai, Y., Chou, K.: Predict and analyze S-nitrosylation modification sites with the mRMR and IFS approaches. J. Proteomics **75**, 1654–1665 (2012)
29. Zhang, Q., Sun, X., Feng, K., Wang, S., Zhang, Y., Wang, S., et al.: Predicting citrullination sites in protein sequences using mRMR method and random forest algorithm. Comb. Chem. High Throughput Screening **20**, 164–173 (2017)
30. Bao, W., Yang, B., Bao, R., Chen, Y.: LipoFNT: lipoylation sites identification with flexible neural tree. Complexity **2019**, 1–9 (2019)
31. Bao, W., Yang, B., Li, D., Li, Z., Zhou, Y., Bao, R.: CMSENN: computational modification sites with ensemble neural network. Chemometr. Intell. Lab. Syst. **185**, 65–72 (2019)
32. Shao, J., Xu, D., Tsai, S.N., Wang, Y., Ngai, S.M.: Computational identification of protein methylation sites through bi-profile Bayes feature extraction. PLoS ONE **4**, e4920 (2009)
33. Bao, W., Chen, Y., Wang, D.: Prediction of protein structure classes with flexible neural tree. Biomed. Mater. Eng. **24**, 3797–3806 (2014)
34. Szilágyi, A., Skolnick, J.: Efficient prediction of nucleic acid binding function from low-resolution protein structures. J. Mol. Biol. **358**, 922–933 (2006)
35. Kumar, K.K., Pugalenthi, G., Suganthan, P.N.: DNA-Prot: identification of DNA binding proteins from protein sequence information using random forest. J. Biomol. Struct. Dyn. **26**, 679–686 (2009)
36. Lin, W.Z., Fang, J.A., Xiao, X., Chou, K.C.: iDNA-prot: identification of DNA binding proteins using random forest with grey model. PLoS ONE **6**, e24756 (2011)
37. Song, L., Li, D., Zeng, X. et al.: nDNA-prot: identification of DNA-binding proteins based on unbalanced classification. BMC Bioinform. **15,** 298 (2014). https://doi.org/10.1186/1471-2105-15-298
38. Shi, S.P., Qiu, J.D., Sun, X.Y., Suo, S.B., Huang, S.Y., Liang, R.P.: PLMLA: prediction of lysine methylation and lysine acetylation by combining multiple features. Mol. BioSyst. **8**, 1520–1527 (2012)
39. Florian, G., Shubin, R., Chunaram, C., Jürgen, C., Matthias, M.: Predicting post-translational lysine acetylation using support vector machines. Bioinformatics **26**, 1666 (2010)
40. Li, S., Li, H., Li, M., Shyr, Y., Xie, L., Li, Y.: Improved prediction of lysine acetylation by support vector machines. Protein Peptide Lett. **16**, 977–983 (2009)

41. Xu, Y., Wang, X.B., Ding, J., Wu, L.Y., Deng, N.Y.: Lysine acetylation sites prediction using an ensemble of support vector machine classifiers. J. Theor. Biol. **264**, 130–135 (2010)
42. Suo, S.B., Qiu, J.D., Shi, S.P., Sun, X.Y., Huang, S.Y., Chen, X., et al.: Position-specific analysis and prediction for protein lysine acetylation based on multiple features. PLoS ONE **7**, e49108 (2012)
43. Shao, J., Xu, D., Hu, L., Kwan, Y.W., Wang, Y., Kong, X., et al.: Systematic analysis of human lysine acetylation proteins and accurate prediction of human lysine acetylation through bi-relative adapted binomial score Bayes feature representation. Mol. BioSyst. **8**, 2964–2973 (2012)
44. Li, Y., Wang, M., Wang, H., Tan, H., Zhang, Z., Webb, G.I., et al.: Accurate in silico identification of species-specific acetylation sites by integrating protein sequence-derived and functional features. Sci. Rep. **4**, 5765 (2014)

RFQ-ANN: Artificial Neural Network Model for Predicting Protein-Protein Interaction Based on Sparse Matrix

Wenzheng Ma[1], Wenzheng Bao[2], Yi Cao[1], and Yuehui Chen[1(✉)]

[1] School of Information Science and Engineering, Zaozhuang University, Zaozhuang 277100, China
yhchen@ujn.edu.cn
[2] School of Information and Electrical Engineering, Xuzhou University of Technology, Xuzhou 221018, China

Abstract. Protein is a complex organic substance with a spatial structure, which exists widely in the body of living things. Almost all living things rely on protein to form an important part of the body, perform many physiological function adjustments, and obtain energy. Without protein, there is no life. Most of these functions of proteins are realized by the interaction between proteins. The occurrence of many diseases is also closely related to protein-protein interactions (PPIs). Therefore, research on PPIs is crucial for many fields such as clinical medicine. We propose a new method RFQ-ANN to predict PPIs, and verify the instability of the artificial neural network (ANN) model in making predictions by changing the number of hidden layers. Experiments have shown that it is not that the more hidden layers, the better.

Keywords: Protein-protein interaction · Sparse matrix · Artificial neural network · Prediction

1 Introduction

In biological systems, protein has always been an extremely important substance that exists in the process of the life of every cell. Many phenomena in biology such as translation, transcription, cell cycle regulation, and signal transduction are regulated by protein-protein interactions (PPIs) [1–3]. There are many types of proteins, and the interaction between them is also diverse. Some proteins are tightly bound, while others only briefly bind to cooperate. PPIs act through the motifs or domains of proteins, constitutes a major component of the cell's biochemical reaction network [4, 5]. As an important part of biological systems, PPIs control a large number of cell activity events, including cell proliferation, differentiation and death. In short, only by making PPIs run smoothly can the normal life activities of cells be guaranteed [6, 7].

With the expansion of human knowledge and technological progress, people gradually attach importance to the research of PPIs, and the technology and methods for detecting PPIs have also developed rapidly [8–14]. The first to appear are some high-throughput laboratory biotechnologies, such as yeast two-hybrid [15, 16] and

D.-S. Huang and K.-H. Jo (Eds.): ICIC 2020, LNCS 12464, pp. 446–454, 2020.
https://doi.org/10.1007/978-3-030-60802-6_39

co-immunoprecipitation [17]. But these laboratory techniques are usually not accurate enough and expensive. Later, phylogenetic profiling and other technologies also appeared, but it requires a priori knowledge of protein-related biology. In recent years, with the rapid development of machine learning, researchers have begun to use a combination of machine learning and protein sequence methods to predict PPIs [18–20]. Sun et al. applied deep learning to the prediction of PPIs and achieved a high average accuracy. Wong et al. proposed a rotating forest model based on the PR-LPQ descriptor has achieved a very high prediction accuracy for PPIs on the Saccharomyces cerevisiae dataset [21]. In addition, machine learning methods such as support vector machines are also widely used in the prediction of PPIs [22].

In this article, we propose an artificial neural network (ANN) model based on sparse matrix multiple feature extraction method to verify the instability of the artificial neural network model. We first construct a sparse matrix, and then use the sparse matrix to extract the sequence feature vectors, including reduced sequence feature vector(R), frequency feature vector(F) and quantitative feature vector(Q). After that, we combine them to obtain the final feature vector of a protein sequence. The feature vector is used as the input of the ANN model. According to the evaluation results obtained by changing the number of hidden layers of the artificial neural network classifier, we find that the artificial neural network is unstable. In general, AUC (Area Under Curve) value, accuracy (ACC), specificity (Sp) and Matthew correlation coefficient (MCC) are used to evaluate the performance of our proposed model.

2 Multiple Feature Extraction Based on Sparse Matrix

In recent years, methods of using machine learning to study PPIs have emerged endlessly. Recently Kong et al. proposed the FCT-WSRC model. Their feature extraction method FCT is based on sparse matrix and unit circle, and used PCA method to reconstruct feature subspace [23]. Inspired by Kong et al., We improved their FCT feature extraction method in the FCT-WSRC model, and obtained better prediction accuracy for the same data set. The feature extraction method we proposed is as follows.

2.1 Construction of R Vector

In this part, we consider the physical and chemical properties of protein sequences and divide the 20 amino acids into 6 categories. The classification of the 20 amino acids constituting the protein sequence is shown in Table 1.

For an amino acid sequence, we replace each amino acid in the sequence according to the classification principle of physical and chemical properties. In this way, all protein sequences are reduced to category sequences. Fi is used to indicate the frequency of the i-th element in the simplified sequence. Since we divide amino acids into 6 categories, $i \in \{1, 2, 3, 4, 5, 6\}$. Fi can be described by formula (1).

Table 1. Classification of proteins

Category	Property	Amino acid
C1	Aliphatic	A, C, I, L, M, V
C2	Aromatic	F, H, W, Y
C3	Polar	N, Q, S, T
C4	Positive	K, R
C5	Negative	D, E
C6	Special conformations	G, P

$$F_i = \frac{n_i}{l} \tag{1}$$

Where n_i is the number of occurrences of the i-th amino acid in the simplified sequence, and l is the length of the protein sequence. Then the A vector can be constructed as:

$$\mathrm{R} = (f_1, f_2, \ldots, f_i, \ldots, f_6 \tag{2}$$

Suppose a known protein sequence is 'M, G, P, A, H, A, A, A, K, I, Q, V, G, G, E, R, A, K, M, M, L, N, E, V, P, P, P, A, V, Q, I, Y', then its reduction sequence is 'C1, C6, C6, C1, C2, C1, C1, C1, C4, C1, C3, C1, C6, C6, C5, C4, C1, C4, C1, C1, C1, C3, C5, C1, C6, C6, C6, C1, C1, C3, C1, C2'. According to formula (1), F1 = 15/32 = 0.46, F2 = 2/32 = 0.06, F3 = 3/32 = 0.09, F4 = 2/32 = 0.06, F5 = 2/32 = 0.06, F6 = 0.22. Then the A vector of the protein sequence S can be constructed as A = (0.46, 0.06, 0.09, 0.06, 0.06, 0.22).

2.2 Construction of F and Q Vectors

Before constructing the F and Q vectors, we first construct a 20 * n sparse matrix B, where n is the number of amino acids in the protein sequence. We put 20 kinds of amino acids into O, O = {A, V, L, I, M, C, F, W, Y, H, S, T, N, Q, K, R, D, E, G, P}.

Given the protein sequence S = S1, S2, … Sn, then the sparse matrix corresponding to the sequence S can be expressed by formulas (3) and(4).

$$E_{20\times n} = \begin{pmatrix} a_{11} & a_{12} & a_{13} & \cdots & a_{1n} \\ a_{21} & a_{22} & a_{23} & \cdots & a_{2n} \\ a_{31} & a_{32} & a_{33} & \cdots & a_{3n} \\ \vdots & \vdots & \vdots & \ddots & \vdots \\ a_{20,1} & a_{20,2} & a_{20,3} & \cdots & a_{20,n} \end{pmatrix} \tag{3}$$

$$a_{ij} = \begin{cases} 0, O(i) \neq S(j) \\ 1, O(i) = S(j) \end{cases} \tag{4}$$

When the i-th amino acid in O is the same as the j-th amino acid in S, the corresponding element in the i-th row and j-th column in the sparse matrix takes 1; otherwise, it takes 0.

After that, we finally got a 20 * n sparse matrix about the protein sequence. We first divide each row vector of the sparse matrix into M sub-vectors, and the F and Q vectors are extracted based on them. Among them, the F vector is composed of two parts of the frequency of '0' and '1' in each sub-vector, and the Q vector is composed of the sum of the number of occurrences of '01' and '10', the number of occurrences of '11' and the number of occurrences of '111' in the sub-vector. Assuming M = 4, the protein sequence is 'MGPAHAAAKIQVGGERAKMMLNEVPPPAVQIY', then the first subsequence of this sequence is 'MGPAHAAA'. It can be seen that the first sub-vector of the first row vector of the sparse matrix is {0,0,0,1,0,1,1,1}. Then the F vector is 4 * 100%/8 = 50%, 4 * 100%/8 = 50%, and T is 3, 2, 1. The F and Q vectors are combined to form a 4 * 20 * (2 + 3) = 400-dimensional feature vector.

2.3 Reconstructing Feature Vectors

Each protein sequence can be extracted into its corresponding vectors R, F and Q. Among them, R is a 6-dimensional vector, in the front we combined the vectors F and Q into a 400-dimensional vector FQ. On this basis, we then combine the R vector with the FQ vector to obtain a 406-dimensional feature vector. Finally, each protein is extracted with a 406-dimensional feature vector. When we study PPIs, it is our only way to study each protein separately. However, in the end, we are still exploring the relationship between protein pairs. Therefore, we combine the feature vectors of protein pairs to construct an 812-dimensional feature vector of protein pairs to represent the relationship between two proteins.

3 Build Classifier

In this article, we constructed ANN-based predictors to predict PPIs. The performance of ANN-based predictors of PPIs can prove the instability of ANN. The test results show that, it is not that the more layers, the more accurate the results.

ANN is a complex network structure that mimics the structure of human brain tissue. It uses a mathematical model to simulate the activity of neurons, an information processing system established by simulating the structure and function of neural network organizations in the brain. Generally, ANN consists of an input layer, multiple hidden layers, and an output layer. The common ANN structure diagram is shown in Fig. 1.

Figure 1 is a four-layer ANN structure diagram. When constructing the classifier, we increased the number of hidden layers from 1 to 20. We constructed 20 ANNs with different structures to observe the prediction of ANN. Through continuous trial and error, we finally set the parameter of optimization weight to stochastic gradient descent. In order to improve the efficiency of parallelization, speed up the processing of data, and reduce training shock, we set the "batch_size" parameter to 32.

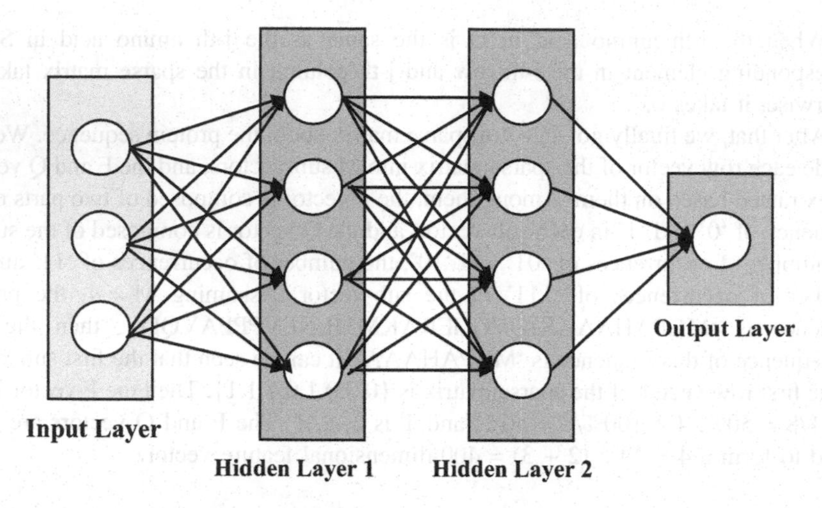

Fig. 1. Artificial neural network structure

4 Results and Discussion

4.1 Evaluation of the Method

In order to verify the reliability of our proposed method, we applied our proposed method to a non-redundant H. pylori dataset. The dataset contains 1458 interacting protein pairs and 4626 non-interacting protein pairs. We use this dataset to train the ANN model for 10-fold cross-validation. The data set is divided into 10 parts. Nine of them is the training set and the other one as the test set. Make sure that each of them has done the test set and training set at the same time. In this way, the average value is obtained after training 10 times, and the result after 10-fold cross-validation is obtained as the model score.

In this paper, we also used accuracy (Acc), sensitivity (Sn) specificity (Sp) and Matthew correlation coefficient (MCC) as the evaluation indicators of our proposed method. Their definition is as follows.

$$Acc = \frac{TP + TN}{TP + TN + FP + FN} \tag{5}$$

$$Sn = \frac{TP}{FN + TP} \tag{6}$$

$$Sp = \frac{TN}{FP + TN} \tag{7}$$

$$MCC = \frac{TP \times TN - FP \times FN}{\sqrt{(TP + FP) \times (TP + FN) \times (TN + FP) \times (TN + FN)}} \tag{8}$$

In addition, we use the AUC value to evaluate the performance of our model. AUC refers to the area under the ROC curve and is a numerical value. The larger the value of AUC, the better the effect of the model.

4.2 Comparison of Feature Extraction Methods

In a model, the feature extraction method is very important, it can often play a decisive role in the performance of the model. This article was inspired by Kong et al., improved their feature extraction method in FCT-WSRC model, and innovated. Kongs' feature extraction method FCT is based on frequency and unit circle. We combine it with ANN to form a new predictor, and then compare it with our model RFQ-ANN. The experimental comparison results are shown in Table 2.

Table 2. Performance comparison of the two methods

Dataset	Classifier	AUC	Acc	MCC	Sn	Sp
H. pylori	FCT-ANN	0.674746	0.615451	0.236099	0.512864	0.718213
	RQT-ANN	**0.742806**	**0.672961**	**0.351125**	**0.588336**	**0.757732**

It can be seen from Table 2 that our feature extraction method is superior to the FCT method proposed by Kong et al. in five indicators. The AUC value of our model reached 0.742806, and the accurate value reached 0.672961. MCC is commonly used to measure the performance of binary classification models, usually returning a value between −1 and +1. From the experimental data, we can see that our model has better classification performance. Sn and Sp respectively reflect the model's ability to recognize positive and negative samples. Obviously, our method is superior to FCT. It can be seen from this that our feature extraction method has better recognition capabilities for sequence features.

4.3 Instability of ANN

The classifier in our model is ANN, and it can be known that ANN has instability by the experimental results of changing the structure of ANN. We take the number of hidden layers of ANN as the only variable, and the value range is from 1 to 20 layers. Each hidden layer contains 100 neurons. It can be seen from the experimental data in Table 3 that, it is not that the more hidden layers, the better.

The bold data in Table 3 is the highest value of each indicator. It can be seen from Table 3 that the with the same dataset and feature extraction method, the performance of the model is the highest when the number of hidden layers is 2. In order to more clearly show the instability of the ANN model, the changes in AUC value, Acc value, and running time are presented in a combined graph. The range of the left ordinate is 0 to 1, corresponding to the values of AUC and Acc. The unit of the right ordinate is seconds, corresponding to the running time of the classifier. Figure 2 is shown below.

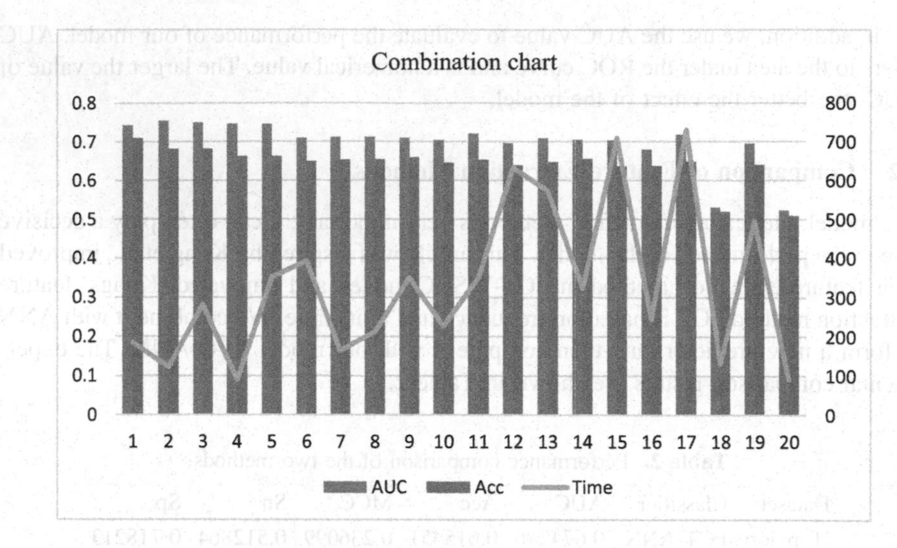

Fig. 2. Changes in AUC, Acc and time

Table 3. Performance of ANN with increasing number of hidden layers from 1 to 20

Number of hidden layers	AUC	Acc	MCC	Sn	Sp
1	0.742806	0.672961	0.351125	0.588336	0.757732
2	**0.753637**	0.681545	**0.369971**	0.586621	0.776632
3	0.749682	**0.682403**	0.367693	0.620926	0.743986
4	0.746468	0.662661	0.329831	0.581475	0.743986
5	0.744478	0.662661	0.326341	**0.624357**	0.701031
6	0.708169	0.649785	0.303304	0.572899	0.726804
7	0.710811	0.652361	0.310312	0.559177	0.745704
8	0.711502	0.654077	0.312617	0.571184	0.737113
9	0.710432	0.659227	0.321660	0.590051	0.728522
10	0.703520	0.645494	0.295568	0.559177	0.731959
11	0.720935	0.653219	0.308369	0.598628	0.707904
12	0.695848	0.638627	0.282547	0.543739	0.733677
13	0.709171	0.647210	0.297393	0.578045	0.716495
14	0.705017	0.654936	0.311589	0.603774	0.706186
15	0.702785	0.636910	0.278674	0.545455	0.728522
16	0.679531	0.637768	0.278829	0.562607	0.713058
17	0.718209	0.648927	0.302729	0.560892	0.737113
18	0.529809	0.519313	0.039614	0.418525	0.620275
19	0.694168	0.639485	0.286801	0.524871	0.754296
20	0.523415	0.509013	0.070853	0.027444	**0.991409**

From Fig. 2, we can clearly see that with the increase of the number of hidden layers, the accuracy of the classifier has not been improved all the time. And even in the end, there is a cliff-like decline. The running time of the model is also irregular. From this we can see that the ANN classifier is unstable, and its performance is not getting better with the increase of the number of hidden layers.

5 Conclusion

With the rapid development of biology, people pay more and more attention to PPIs. In the development history of studying PPIs, many methods have emerged. And now the most widely used by researchers is based on protein sequence and machine learning methods. We proposed a model based on sparse matrix called RFQ-ANN. In RFQ-ANN, we first divide the 20 amino acids that make up the protein sequence into 6 categories. Then the original protein sequence is replaced with a reduced sequence according to the classification, and the reduced feature vector R is extracted according to the reduced sequence. The next step is to construct a sparse matrix and construct frequency-based eigenvectors F and Q based on the sparse matrix and its sub-vectors of row vectors. After performing three feature extractions on the protein sequence, we combine the feature vectors R, F, and Q as a feature vector that ultimately represents the protein sequence. After that, we reconstruct the feature vector and obtain the protein pair feature vector representing the protein pair relationship as the input of the ANN classifier. When constructing the ANN classifier, we considered the influence of the ANN structure on the model, and constructed ANN classifiers with different structures with the number of hidden layers increasing from 1 to 20. It can be known from the experimental results that the ANN classifier is unstable. We will focus on improving the feature extraction method in the future, and gradually optimize our model to provide more reliable support for the prediction of PPIs.

References

1. Zhang, Q.C., et al.: Structure-based prediction of protein–protein interactions on a genome-wide scale. Nature **490**(7421), 556–560 (2012)
2. Wu, J., et al.: Integrated network analysis platform for protein-protein interactions. Nature Methods 6(1), 75 (2009)
3. De Las Rivas, J., Fontanillo, C.: Protein–protein interactions essentials: key concepts to building and analyzing interactome networks. PLoS Comput. Biol. **6**(6), e1000807 (2010)
4. Sugaya, N., Ikeda, K.: Assessing the druggability of protein-protein interactions by a supervised machine-learning method. BMC Bioinform. **10**(1), 263 (2009). https://doi.org/10.1186/1471-2105-10-263
5. Shen, J., et al.: Predicting protein–protein interactions based only on sequences information. In: Proceedings of the National Academy of Sciences, vol. 104, no. 11, pp. 4337–4341 (2007)
6. Zhang, Y.P., Zou, Q.: PPTPP: a novel therapeutic peptide prediction method using physicochemical property encoding and adaptive feature representation learning. Bioinformatics **36**, 3982–3987 (2020)

7. Shen, Z., Lin, Y., Zou, Q.: Transcription factors–DNA interactions in rice: identification and verification. Brief. Bioinform. **21**, 946–956 (2019)
8. Liu, G.-H., Shen, H.-B., Dong-Jun, Yu.: Prediction of protein–protein interaction sites with machine-learning-based data-cleaning and post-filtering procedures. J. Membr. Biol. **249**(1-2), 141–153 (2016). https://doi.org/10.1007/s00232-015-9856-z
9. Chatterjee, P., Basu, S., Kundu, M., Nasipuri, M., Plewczynski, D.: PPI_SVM: Prediction of protein-protein interactions using machine learning, domain-domain affinities and frequency tables. Cell. Mol. Biol. Lett. **16**(2), 264–278 (2011). https://doi.org/10.2478/s11658-011-0008-x
10. You, Z.H., et al.: Prediction of protein-protein interactions from amino acid sequences with ensemble extreme learning machines and principal component analysis. BMC Bioinform. **14** (S8), S10 (2013). BioMed Central. https://doi.org/10.1186/1471-2105-14-S8-S10
11. Wei, L., et al.: Improved prediction of protein–protein interactions using novel negative samples, features, and an ensemble classifier. Artif. Intell. Med. **83**, 67–74 (2017)
12. Patel, S., et al.: DeepInteract: deep neural network based protein-protein interaction prediction tool. Curr. Bioinform. **12**(6), 551–557 (2017)
13. Singh, G., et al.: SPRINGS: prediction of protein-protein interaction sites using artificial neural networks. No. e266v2. PeerJ PrePrints (2014)
14. Jia, C., Zuo, Y., Zou, Q.: O-GlcNAcPRED-II: an integrated classification algorithm for identifying O-GlcNAcylation sites based on fuzzy undersampling and a K-means PCA oversampling technique. Bioinformatics **34**(12), 2029–2036 (2018)
15. Schwikowski, B., Uetz, P., Fields, S.: A network of protein–protein interactions in yeast. Nat. Biotechnol. **18**(12), 1257–1261 (2000)
16. Coates, P.J., Hall, P.A.: The yeast two-hybrid system for identifying protein–protein interactions. J. Pathol. J. Pathol. Soc. Great Brit. Ireland **199**(1), 4–7 (2003)
17. Free, R.B., Hazelwood, L.A., Sibley, D.R.: Identifying novel protein-protein interactions using co-immunoprecipitation and mass spectroscopy. Curr. Protoc. Neurosci. **46**(1), 5–28 (2009)
18. Huang, Y.A, et al.: Sequence-based prediction of protein-protein interactions using weighted sparse representation model combined with global encoding. BMC bioinform. **17**(1), 184 (2016). https://doi.org/10.1186/s12859-016-1035-4
19. Zhang, S.-W., Hao, L.-Y., Zhang, T.-H.: Prediction of protein–protein interaction with pairwise kernel support vector machine. Int. J. Mol. Sci. **15**(2), 3220–3233 (2014)
20. Sun, T., et al.: Sequence-based prediction of protein protein interaction using a deep-learning algorithm. BMC Bioinform. **18**(1), 277 (2017)
21. Wong, L., You, Z.-H., Li, S., Huang, Y.-A., Liu, G.: Detection of protein-protein interactions from amino acid sequences using a rotation forest model with a novel PR-LPQ descriptor. In: Huang, D.-S., Han, K. (eds.) ICIC 2015. LNCS (LNAI), vol. 9227, pp. 713–720. Springer, Cham (2015). https://doi.org/10.1007/978-3-319-22053-6_75
22. Zhang, S.-W., Hao, L.-Y., Zhang, T.-H.: Prediction of protein–protein interaction with pairwise kernel support vector machine. Int. J. Mol. Sci. **15**(2), 3220–3233 (2014)
23. Kong, M., et al.: FCTP-WSRC: protein–protein interactions prediction via weighted sparse representation based classification. Front. Genet. **11**, 18 (2020)

Recent Advances in Swarm Intelligence: Computing and Applications

A Novel Hybrid Algorithm Based on Bacterial Foraging Optimization and Grey Wolf Optimizer

Xiaobing Gan and Baoyu Xiao(✉)

College of Management, Shenzhen University, Shenzhen 518060, China
winsonxiao2019@163.com

Abstract. A novel hybrid algorithm named GMBFO, with the combination between Grey Wolf Optimizer and the modified Bacterial Foraging Optimization, is presented in the paper. To improve the fixed chemotaxis step size in the standard BFO algorithm, the paper incorporates a nonlinear-decreasing adaptive mechanism into BFO. Besides that, an effective swarm learning strategy with the other three current global best individuals is proposed. In the dispersal and elimination step, we adopt the roulette wheel selection and local mutation mechanism to improve the diversity of the whole bacterial population. To testify the optimization performance of the proposed GMBFO, six benchmark functions with 45 dimensions are selected. Compared with BFO and the other three BFO variants, the GMBFO algorithm has an excellent capability in function optimization.

Keywords: Bacterial foraging optimization · Grey wolf optimizer · Roulette wheel selection · Mutation · Function optimization

1 Introduction

In recent years, several swarm optimization algorithms have emerged and developed, such as Particle Swarm Optimization (PSO) [1], Genetic Algorithm (GA) [2], Grey Wolf Optimizer (GWO) [3], and so on. Moreover, as a novel bio-inspired heuristic optimization algorithm, the Bacterial Foraging Optimization (BFO) [4] was proposed by Passino in 2002, mainly mimicking the foraging process and swimming behaviors of E. coli. Due to its great performance of local exploitation and robustness, the BFO algorithm has gradually become a potential research direction in current years, and it has been widely obtained the attention by academics in various optimization fields including vehicle routing problem (VRP) [5], face recognition [6], the structural learning of Bayesian networks [7], and so on.

Furthermore, to enhance the convergence speed and optimization accuracy of the standard BFO algorithm, numerous improved BFO algorithms have sprung up. As the chemotactic capability plays a crucial role in the process of searching for the optimal solution, lots of academics have contributed to improve it. Niu et al. [8, 9] took the lead in proposing adaptive mechanisms respectively with linear and nonlinear-decreasing strategies, which are both beneficial to balance the capability of the local exploitation

D.-S. Huang and K.-H. Jo (Eds.): ICIC 2020, LNCS 12464, pp. 457–468, 2020.
https://doi.org/10.1007/978-3-030-60802-6_40

and global exploration. In [10], Chen et al. incorporated the hyperbolic tangent function into the swarm communication mechanism, aiming at improving the accuracy of the cell-to-cell signal exchange. In [11], Chen et al. proposed an improved BFO variant (CCGBFO) whose Gaussian mutation, as well as chaotic local search operator with a "shrinking" mechanism, were applied to the process of chemotaxis. In [12], Wang et al. improved the fixed step size utilizing strategies of progressive exploration towards the local optimum and adaptive raid followed by the current best individual. After compared with the convergence results of GA, PSO, and BFO, the proposed algorithm had been examined to have an outstanding performance in optimization.

Another direction is to redesign the whole structure of the BFO algorithm so as to decrease the computation complexity and improve the convergence performance. Niu et al. [13] adopted a novel BFO variant that is just involved in one rather than three nested loops. More detaily, the paper integrated the reproduction, dispersal, and elimination process into the chemotaxis process. Besides, several researchers have worked on the hybridization of BFO with other remarkable algorithms. Researchers [14, 15] presented some hybrid methods by combining the main idea of PSO with BFO, prompting the bacterium to learn from the current global best individual. Novel hybrid algorithms [6, 16] integrating GA into BFO for coping with optimization problems were put forward. Sarasiri et al. [17] proposed a new variant involving in BFO and Tabu Search Algorithm (TS), and applied to the identification of nonlinear friction models. Turanoglu et al. [18] carried out a new hybrid algorithm based on BFO and Simulated Annealing (SA). Yildz et al. [19] introduced an efficient method named CDEOA which combined with Differential Evolution Optimization (DE) and BFO.

However, owing to the stationary chemotaxis step size and a lack of effective swarm communication, the standard BFO algorithm has a poor performance of the convergence accuracy and it is quite easy to trap into the local optimum. Furtherly, to enhance the performance of BFO, the paper firstly employs an adaptive mechanism with a nonlinear-decreasing strategy to improve the fixed step length. Moreover, to the best of our knowledge, there are few studies on the hybridization of GWO and BFO. Consequently, the paper tries to introduce a novel hybrid algorithm named GMBFO, which combines GWO with the modified BFO to greatly enhance the comprehensiveness of group communication and effectively improve the convergence performance. More importantly, the bacterium could obtain more valuable information and reasonably adjust its moving location via the comprehensive learning mechanism with multiple superior individuals to find out the optimum solution in an effective way. Besides, to retain the best individual, the paper sorts out the healthier bacterium by assessing their current fitness value to perform the reproduction step rather than by calculating the sum of fitness value in the whole lifecycle of each bacterium. Last but not least, to alleviate the premature problem and increase the diversity of the population, the bacterium is strictly abided by the roulette wheel selection and local mutation mechanism to randomly disperse in a new location.

The rest of the paper will be organized as follows: Sect. 2 and Sect. 3 respectively describes the standard BFO algorithm and Grey Wolf Optimizer (GWO). The details of the proposed algorithm GMBFO is shown in Sect. 4. Numerical experimental results are demonstrated in the next section. The last section presents conclusions and future work.

2 The Standard Bacterial Foraging Optimization

The standard Bacterial Foraging Optimization (BFO) [4] mainly simulates E. coli' s three significant processes during searching for nutrients and avoiding harmful substances, respectively for chemotaxis, reproduction, elimination, and dispersal.

2.1 Chemotaxis

The bacterium i ($i\epsilon SS$) performs tumbling through randomly selecting one direction and moves toward the food with a fixed chemotactic step size $C(i)$. After tumbling and moving, if the bacterium explores a better position (better fitness value), it could continue to swim in terms of the same direction until the searching position gets worse (worse fitness value) or the bacterium reaches to the maximum limit of swimming denoted as N_s. The related equation about updating positions for each bacterium can be shown as follows:

$$\theta^i(j+1,k,l) = \theta^i(j,k,l) + C(i) \times \frac{\Delta(i)}{\sqrt{\Delta^T(i)\Delta(i)}}. \tag{1}$$

where SS is the number of the whole bacterial population. $\theta^i(j,k,l)$ denotes the position of the bacterium i in j th chemotaxis, k th reproduction, l th elimination and dispersal. $\Delta(i)$ represents a direction vector whose all elements range from -1 to 1.

2.2 Reproduction

To enhance the convergence accuracy of BFO, after the process of chemotaxis, the healthier individual ($Sr = SS/2$) would have the capability to perform reproduction while the other half bacterium with poor performance would be replaced. Moreover, the healthy degree f of the bacterium i can be measured by the sum of fitness value in its whole lifecycle. The corresponding equation can be presented as (2).

$$f_{i,health} = \sum\nolimits_{j=1}^{N_c} J(i,j,k,l). \tag{2}$$

where N_c denotes the whole number of chemotaxis while J denotes the fitness value.

2.3 Dispersal and Elimination

The process of dispersal and elimination mirrors that when confronting with unexpected environment changes, some bacteria needs to disperse randomly. Thus, they would local in a new position of the search map.

3 Grey Wolf Optimizer

Grey Wolf Optimizer (GWO) [3] was proposed by Mirjalili, greatly simulating the specific social leadership hierarchy in the colony of grey wolves and their pack hunting behavior during searching for prey.

Complying with the strict social hierarchy, the population of grey wolves prefers to collaborate and cooperate for hunting and living. More detailly, the social hierarchy could be divided into four levels. The top of the hierarchy pyramid represents a leader of the whole population and it can be named as α which is responsible for making a determination about hunting, food distribution, and the sleeping time or location. Moreover, the second level is β that can be regarded as subordinates with the most crucial responsibility to assist the grey wolf α to make decisions. The third hierarchy is represented as σ which plays the role as the advisor to the α and the β.

Generally, in the GWO algorithm, the individuals are guided by the leader α (the global best fitness value), the significant subordinates β (the second-best fitness value) and σ (the third-best fitness value) to continuously update their current positions and effectively search for the prey (the optimal solution).

4 The GMBFO Algorithm

4.1 Nonlinear-decreasing Adaptive Chemotaxis Mechanism

In the standard BFO algorithm, the fixed run length to a large extent makes a negative impact on the convergence rate and the solution accuracy. Inspired by plenty of prior studies [8–12], we recognize the importance of the improvement of chemotactic strep size. In the paper, we also adopt a nonlinear-decreasing mechanism presented as Eq. (3), which helps the bacterium to have a great balance between the local exploitation and the global exploration.

$$C'(i) = C_{max} - (C_{max} - C_{min}) \times \frac{2\sqrt{3}}{3} \times \sin\left(\frac{\pi}{3} \times \frac{current\ iterations}{total\ iterations}\right). \tag{3}$$

Furthermore, as for the basic BFO, if the bacterium locates in a better position than the previous, it could continue to move forward along the same direction. On the contrary, after reaching to a worse position than the last one, the bacterium would retain the current location and go into the next step. However, it is not hard for us to find that the bacterium with the defective position negatively influences the convergence speed of the BFO algorithm. As a result, illuminated by [15], we also implement the self-regulation strategy. Specifically, with the ability to memorize their prior search records, the bacterium could autonomously modify to the last better location when it locates in the worse one.

4.2 The Cooperative Learning Mechanism

As for the BFO algorithm, the bacterium only just relies on randomly choosing one direction to move on and has a limited chance to learn with the searching experience of elites. In other words, the reason why the basic BFO algorithm has a poor performance in the convergence speed and the solution precision might attribute to a lack of swarm communication. To enhance the group learning, we propose a new hybrid algorithm incorporating the GWO [3] into the BFO algorithm.

Referring to the foraging results of the elite, the bacterium could adjust their positions by learning from the current global best position θ^{α}, the second-best location θ^{β}, and the third-best search agent θ^{δ} with nonlinear-decreasing adaptive step size. The corresponding mathematic model about the updated position of the bacterium i is followed as Eq. (4) where ω controls the study capability, and X_1 represents a predicted position towards the current global best individual. Furthermore, Eqs. (6), (7) and (8) respectively demonstrates the distance between the bacterium i and three of the best individuals α, β and δ where A and C are both regarded as convergency factors. Specifically, the vector A denoted as Eq. (9) ranges from $-a$ to a and a decreases from 1 to 0 in a nonlinear way.

$$\begin{aligned}\theta^{i}(j+1,k,l) =& \theta^{i}(j,k,l)+\omega\times\left\{\left[\theta^{i}(j,k,l)-X_{1}\right]\times C^{i}(i)\times\frac{\Delta(i)}{\sqrt{\Delta^{T}(i)\Delta(i)}}\right\} \\ &+(1-\omega)\times\left(\frac{D_{\alpha}+D_{\beta}+D_{\delta}}{3}\right).\end{aligned} \tag{4}$$

$$X_{1} = \theta^{\alpha}(j,k,l)-D_{\alpha}. \tag{5}$$

$$D_{\alpha} = A_1 \left| C_1 \times \theta^{\alpha}(j,k,l) - \theta^{i}(j,k,l) \right|. \tag{6}$$

$$D_{\beta} = A_2 \left| C_2 \times \theta^{\beta}(j,k,l) - \theta^{i}(j,k,l) \right|. \tag{7}$$

$$D_{\delta} = A_3 \left| C_3 \times \theta^{\delta}(j,k,l) - \theta^{i}(j,k,l) \right|. \tag{8}$$

$$A = 2a \times rand() - a;\ A_1, A_2, A_3 \in A. \tag{9}$$

$$a = \exp\left(-\varphi \times \left(\frac{current\ iterations}{total\ iterations} \right)^{\tau} \right).$$

$$C = rand();\ C_1, C_2, C_3 \in C. \tag{10}$$

4.3 The Improved Reproduction Mechanism

In the standard BFO algorithm, the healthy degree of the bacterium i is measured by calculating total fitness values in its whole lifecycle. However, it is likely to discard those bacteria which have a poor performance in initial iterations but have the greatest exploration ability in the current condition. To be considered more reasonable, the paper assesses the healthy condition of bacteria in terms of their current fitness values.

4.4 The Roulette Wheel Selection and Mutation Mechanism

Inspired by GA [2], we try to incorporate the roulette wheel strategy and local mutation mechanism into the BFO algorithm to improve the diversity of the bacterial population, as well as to alleviate the problem of early trapping into the local optimum. In the elimination and dispersal step, after being evaluated the proportion of its fitness value in the whole population by equation $J_i(j,k,l) / \sum_{i=1}^{SS} J(j,k,l)$, the individual with a better fitness value would have a smaller probability to scatter in a new position. By contrast, part of elements of the bacterial position vector will be randomly mutated within a certain probability P_{ed}.

The corresponding pseudo code of the proposed GMBFO is shown in Table 1.

Table 1. The pseudo code of the proposed GMBFO algorithm

```
Initialize parameters and the bacterial population θ^i(j,k,l), i ∈ (1,2,3,...,SS);
Initialize ω,φ,τ,a,A,C; Evaluate the fitness value J_i(j,k,l), i ∈ (1,2,3,...,SS);
Evaluate θ_α, θ_β, θ_δ and their own fitness value J_α,J_β,J_δ;
For l = 1 : N_ed
    For k = 1 : N_re
        For j = 1 : N_c
            For i = 1 : SS
                Update J_i(j,k,l) and let Jlast = J_i(j,k,l), Plast = θ^i(j,k,l);
                The bacterium i updates its position with equation (4);
                Calculate the fitness value J_i(j+1,k,l);
                Let m = 0 (initialize the swimming length)
                While m < N_s
                    m = m + 1
                    If the updated fitness value J_i(j+1,k,l) gets better
                        Update Jlast and Plast;
                        The bacterium i preforms swimming using equation (4);
                    Else
                        m=Ns;
                        J_i(j+1,k,l)=Jlast; θ^i(j+1,k,l) = Plast;
                    End
                End
            Update a,A,C,θ_α, θ_β,θ_δ, J_α,J_β,J_δ;
            End
        End
    Sort out the healthy degree of bacteria based on their current fitness values;
    Perform reproduction process;
    End
Each bacterium disperses based on the roulette wheel mechanism or has a chance to mutate;
Generate a random number P ranging from 0 to 1;
If P(i) < (J_i / Σ_{i=1}^{SS} J)
The bacterium i is randomly located on a new position θ';
Else
Part of elements of the bacterium i perform local mutation within a certain probability P_ed;
End
End
```

5 Experimental Results and Discussion

5.1 Parameter Settings

The performance of GMBFO algorithm is compared with the original BFO [4] and the other BFO variants involved in the basic BFO algorithm with linear-decreasing strategy (BFO-LDC) [8], the algorithm improved by nonlinear-decreasing chemotactic step size (BFO-NDC) [9], as well as the hybrid algorithm with PSO operator (BFO-PSO) [14]. Followed with the corresponding parameter settings in [20], the population of bacteria SS for all involved algorithms are set to 50 and the total number of chemotaxis N_C, reproduction N_{re}, elimination and dispersal N_{ed} respectively are 1000, 5 and 2 in the paper. Thus, the total iteration is $N_C \times N_{re} \times N_{ed} = 10000$. Besides, we also set $N_s = 4$, $P_{ed} = 0.25$. As for specified parameters settings in the GMBFO algorithm, we set $C_{min} = 0.01$, $C_{max} = 1.5, \omega = 0.05, \varphi = 4$, $\tau = 4$. Additionally, other relevant parameters in above improved BFO algorithms are generally followed the previous studies [1, 8, 9, 14, 20] and they are respectively shown as follows. BFO-LDC settings: $C_{min} = 0.01$, $C_{max} = 1.5$; BFO-NDC: $C_{min} = 0.01$, $C_{max} = 1.5$, $\lambda = 4$; BFO-PSO:$C(i) = 0.1$, the inertia weight $\omega = 0.8$, the study rate $c = 1.5$. According to the above parameter settings, six classical benchmark functions in 45 dimensions are chosen to identify the performance of the above five algorithms, respectively for three unimodal functions (Sphere, Schwefel's Problem 2.22, and Quartic) and three multimodal functions (Alpine, Rastrigin, and Griewank). Moreover, owing to the simple constructs just with one local optimum which is the global best solution, unimodal functions are always used to examine the convergence speed of different algorithms. Comparatively, multimodal functions with more than one local optimum largely increase the computation complexity and lots of algorithms easily fall into a local best solution. Meanwhile, the optimal values among all of these functions are both 0. To ensure the objectivity and reliability of findings, each algorithm is fully operated in 10 times for each benchmark function and the average results are taken.

5.2 Experimental Results and Analyses

Experimental data results are illustrated in Table 2. In Table 2, there contains four necessary metrics including the minimum solution ('Best'), the worst fitness value ('Worst'), the average value ('Mean'), and the standard deviation ('Std'). Additionally, the average fitness values about six benchmark functions with 45 dimensions through 10 running times for each algorithm are respectively shown in Fig. 1.

Table 2. Comparison between GMBFO and other algorithms with 45 dimensions in 10 times

Functions	Metrics	BFO	BFO-LDC	BFO-NDC	BFO-PSO	GMBFO
Sphere	Best	4.78e + 03	1.13e − 01	1.78e − 01	5.28e − 05	**3.08e − 33**
	Worst	6.55e + 03	2.04e − 01	2.71e − 01	2.41e − 03	**5.46e − 29**
	Mean	5.88e + 03	1.51e − 01	2.37e − 01	4.72e − 04	**1.39e − 29**
	Std	5.55e + 02	2.78e − 02	3.44e − 02	6.95e − 04	**2.01e − 29**
Schwefel's Problem 2.22	Best	2.48e + 01	1.30e + 04	6.15e + 05	2.56e − 01	**1.81e − 47**
	Worst	1.56e + 02	2.04e + 10	1.17e + 13	5.52e − 01	**6.08e − 46**
	Mean	1.26e + 02	5.07e + 09	1.40e + 12	3.61e − 01	**1.65e − 46**
	Std	3.80e + 01	7.38e + 09	3.64e + 12	9.15e − 02	**1.83e − 46**
Quartic	Best	1.20e + 00	5.25e − 02	8.04e − 02	1.86e − 06	**1.43e − 08**
	Worst	2.10e + 00	1.23e − 01	1.93e − 01	**1.72e − 05**	6.79e − 02
	Mean	1.80e + 00	8.88e − 02	1.38e − 01	**7.39e − 06**	8.57e − 03
	Std	2.61e − 01	2.82e − 02	3.89e − 02	**5.40e − 06**	2.10e − 02

(*continued*)

Table 2. (*continued*)

Functions	Metrics	BFO	BFO-LDC	BFO-NDC	BFO-PSO	GMBFO
Alpine	Best	1.97e + 01	2.37e + 01	3.66e + 01	1.38e + 00	**4.68e − 46**
	Worst	2.28e + 01	3.25e + 01	4.54e + 01	5.43e + 00	**1.02e − 38**
	Mean	2.11e + 01	2.72e + 01	4.02e + 01	3.42e + 00	**1.02e − 39**
	Std	1.03e + 00	3.37e + 00	3.08e + 00	1.22e + 00	**3.23e − 39**
Rastrigin	Best	2.60e + 02	4.76e + 02	2.60e + 02	4.53e + 01	**2.66e − 14**
	Worst	2.86e + 02	6.48e + 02	3.47e + 02	6.61e + 01	**5.54e + 01**
	Mean	2.71e + 02	6.01e + 02	3.11e + 02	5.75e + 01	**2.58e + 01**
	Std	8.46e + 00	4.85e + 01	2.65e + 01	**6.43e + 00**	2.12e + 01
Griewank	Best	6.73e + 02	1.10e + 00	2.22e + 02	9.78e − 05	**0.00e + 00**
	Worst	6.80e + 02	1.13e + 00	2.61e + 02	1.74e − 02	**1.55e − 15**
	Mean	6.77e + 02	1.12e + 00	2.36e + 02	4.07e − 03	**6.22e − 16**
	Std	1.91e + 00	1.12e − 02	1.42e + 01	6.81e − 03	**4.39e − 16**

First, despite being inferior to the BFO-PSO algorithm, the GMBFO has a prominent improvement in the convergence speed compared with the other three algorithms (BFO, BFO-LDC, BFO-NDC). It might be accountable for the nonlinear-adaptive chemotaxis method which improves the balance between the local search and the global search, and the comprehensive learning strategy with efficient guidance from three of the current global best individuals. Second, we could not difficult to observe that the solution accuracy of the GMBFO algorithm generally outperforms than the others in the latter iterations expect for the Quartic benchmark function. The findings could illustrate that the GMBFO algorithm has the worse capability in optimizing the unimodal function like Quartic than the BFO-PSO. Moreover, it might seem that the proposed GMBFO algorithm has a more excellent capability to optimize multimodal functions, particularly for coping with high-dimensional problems. This could partly attribute to the roulette wheel selection and mutation mechanisms which help the bacterium to escape from the local optimum and effectively alleviate the premature problem. However, the improvement of the diversity of the whole population also to some certain extent leads to the increasement of the computation time and affects the convergence accuracy, especially for the unimodal function like Quartic.

To examine whether the findings of the proposed GMBFO algorithm are significantly different from the other algorithms', we employ Friedman-test. The statistical analyses about the average fitness value between GMBFO and the other algorithms in 10 times are presented in Table 3. Specifically, the confidence interval is set to 95%. In other words, if the probability p is less than 0.05, the findings represent that there has a significant difference between pairwise objects. From Table 3, we could observe that almost of statistical results of GMBFO are significant. As a result, the proposed GMBFO is generally superior to the other four algorithms and it has an outstanding performance in function optimization.

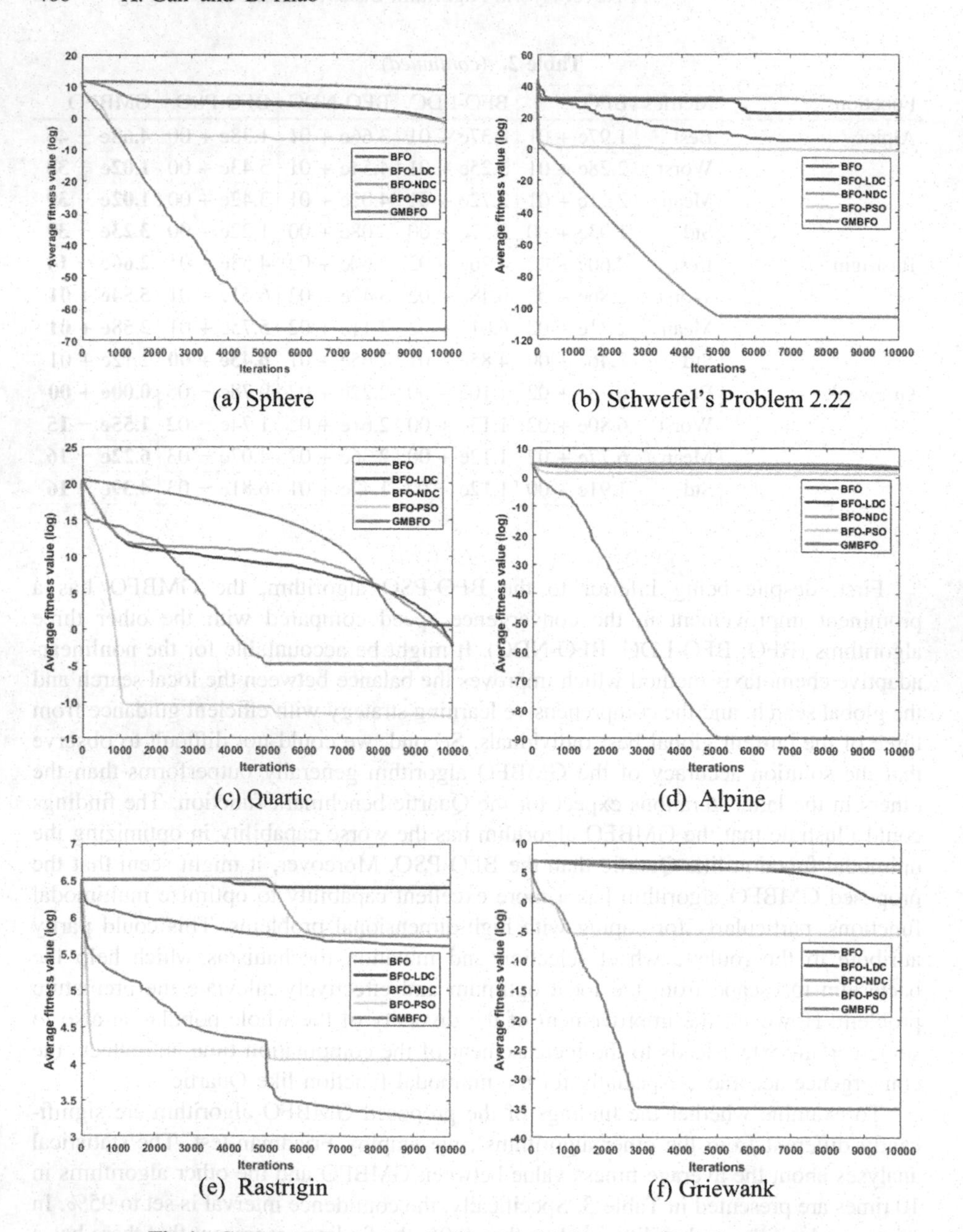

Fig.1. Convergence results of different algorithms with 45 dimensions

Table 3. Comparison between GMBFO' s results and the other algorithms on Friedman-test

Function	The statistic results	Two-tailed P
Sphere	GMBFO ≫ BFO-PSO ≈ BFO-LDC ≈ BFO-NDC ≈ BFO	0.05
Schwefel's Problem 2.22	GMBFO ≫ BFO-PSO ≫ BFO ≈ BFO-LDC ≈ BFO-NDC	0.05
Quartic	BFO-PSO ≈ GMBFO ≈ BFO-LDC ≈ BFO-NDC ≈ BFO	0.05
Alpine	GMBFO ≫ BFO-PSO ≫ BFO-LDC ≈ BFO-NDC ≫ BFO	0.05
Rastrigin	GMBFO ≈ BFO-PSO ≈ BFO ≫ BFO-NDC ≈ BFO-LDC	0.05
Griewank	GMBFO ≫ BFO-PSO ≈ BFO-LDC ≫ BFO-NDC ≈ BFO	0.05

Note: "A ≫ B" represents A is significantly better than B with the significant level of 5%; "A ≈ B" denotes the results of A are better than B, but there is no significant difference.

6 Conclusions and Future Work

In the paper, we put forward a novel hybrid algorithm (GMBFO) that is combined GWO with the modified BFO to effectively enhance the convergence rate and the solution accuracy of the original BFO. More detailly, the bacterium could update its run-length based on an adaptive chemotaxis mechanism. An effective swarm communication mechanism about learning from top-three best individuals is also incorporated into the chemotaxis step. Meanwhile, to reasonably retain the healthier bacteria in the reproduction, we evaluate the healthy degree of each bacterium via considering their current fitness value in the end of lifecycle. Last but not least, according to the roulette wheel selection and local mutation mechanism, some bacteria could adaptively disperse to a new position. Finally, to examine the optimization efficiency of the proposed GMBFO algorithm, six benchmark functions in 45 dimensions are chosen. Experimental results show that the GMBFO algorithm outperforms the other algorithm in general.

In the future, we will furtherly improve the performance of GMBFO and proceed to do more experiments through several exceptional test functions. Besides that, how to apply the proposed GMBFO algorithm into real-world problems like the airline scheduling, and job-shop distribution would become another research direction.

References

1. Eberhart, R., Kennedy, J.: Particle swarm optimization. In: Proceedings of the IEEE International Conference on Neural Networks, pp. 1942–1948. IEEE Press, New York (1995)
2. Holland, J.H.: Adaptation in Natural and Artificial Systems: An Introductory Analysis with Applications to Biology, Control, and Artificial Intelligence. MIT Press, Cambridge (1992)

3. Mirjalili, S., Mirjalili, S.M., Lewis, A.: Grey wolf optimizer. Adv. Eng. Softw. **69**, 46–61 (2014)
4. Passino, K.M.: Biomimicry of bacterial foraging for distributed optimization and control. IEEE Control Syst. Mag. **22**, 52–67 (2002)
5. Tan, L., Lin, F., Wang, H.: Adaptive comprehensive learning bacterial foraging optimization and its application on vehicle routing problem with time windows. Neurocomputing **151**, 1208–1215 (2015)
6. Panda, R., Naik, M.K.: A novel adaptive crossover bacterial foraging optimization algorithm for linear discriminant analysis based face recognition. Appl. Soft Comput. **30**, 722–736 (2015)
7. Yang, C., Ji, J., Liu, J., et al.: Structural learning of bayesian networks by bacterial foraging optimization. Int. J. Approximate Reasoning **69**, 147–167 (2016)
8. Niu, B., Fan, Y., Wang, H., et al.: Novel bacterial foraging optimization with time-varying chemotaxis step. Int. J. Artif. Intell. **7**, 257–273 (2011)
9. Niu, B., Wang, H., Tan, L., et al.: Improved BFO with adaptive chemotaxis step for global optimization. In: 2011 Seventh International Conference on Computational Intelligence and Security, pp. 76–80. IEEE Press, New York (2011)
10. Chen, Y., Li, Y., Wang. G., et al.: A novel bacterial foraging optimization algorithm for feature selection. Expert Syst. with Appl. **83**, 1–17 (2017)
11. Chen, H., Zhang, Q., Luo. J., et al.: An enhanced bacterial foraging optimization and its application for training kernel extreme learning machine. Appl. Soft Comput. **86**, 1–24 (2020)
12. Wang, D., Qian, X., Ban. X., et al.: Enhanced bacterial foraging optimization based on progressive exploitation toward local optimum and adaptive raid. IEEE Access **7**, 95725–95738 (2019)
13. Niu, B., Liu, J., Wu. T., et al.: Coevolutionary structure-redesigned-based bacterial foraging optimization. IEEE-ACM Trans. on Comput. Biol. Bioinform. **15**, 1865–1876 (2018)
14. Biswas, A., Dasgupta, S., Das, S., et al.: Synergy of PSO and bacterial foraging optimization – a comparative study on numerical benchmarks. In: Corchado, E., Corchado, J.M., Abraham, A. (eds.) Innovations in Hybrid Intelligent Systems. ASC, vol. 44, pp. 255–263. Springer, Heidelberg (2007). https://doi.org/10.1007/978-3-540-74972-1_34
15. Pang, B., Song, Y., Zhang. C., et al.: Bacterial foraging optimization based on improved chemotaxis process and novel swarming strategy. Appl. Intell. **49**, 1283–1305 (2019). https://doi.org/10.1007/s10489-018-1317-9
16. Kim, D.H., Abraham, A., Cho. J.H.: A hybrid genetic algorithm and bacterial foraging approach for global optimization. Inf. Sci. **177**, 3918–3937 (2007)
17. Sarasiri, N., Suthamno, K., Sujitjorn, S.: Bacterial foraging-tabu search metaheuristics for identification of nonlinear friction model. J. Appl. Math. **2012**, 1–24 (2012)
18. Turanoglu, B., Akkaya, G.: A new hybrid heuristic algorithm based on bacterial foraging optimization for the dynamic facility layout problem. Expert Syst. Appl. **98**, 93–104 (2018)
19. Yildiz, Y.E., Altun, O.: Hybrid achievement oriented computational chemotaxis in bacterial foraging optimization: a comparative study on numerical benchmark. Soft. Comput. **19**, 3647–3663 (2015)
20. Niu, B.: Bacterial Colony Optimization and Bionic Management. Science Press, China (2014). (in Chinese)

An Analysis of K-Means, Particle Swarm Optimization and Genetic Algorithm with Data Clustering Technique

Maja Gulan[1] and Kaishan Huang[2](✉)

[1] Faculty of Technical Sciences, University of Novi Sad, Novi Sad, Serbia
majica.gulan@gmail.com
[2] Innovation and Entrepreneurship Education Center, Shenzhen University, Shenzhen, China
hks@szu.edu.cn

Abstract. Data clustering is a popular data analysis approach for organizing similar objects without needing to be monitored - a collection of data, into meaningful groups (or clusters). Many approaches from different disciplines have been proposed as solutions for clustering problems and each algorithm has its advantages and drawbacks. This research paper offered a brief overview and performance comparison analysis between different existing data mining clustering strategies based on Particle Swarm Optimization, Genetic Algorithm, K-means and Hybrid PSO Solutions. The reasoning behind choosing these algorithms is that PSO has, due to the applicability of its advanced hybrid variants, reached a remarkable position in this field, and the Genetic Algorithm based unsupervised clustering technique provides a stable clustering performance with less computational time required. However, we found that Hybrid PSO solutions have been shown to produce excellent results in a wide variety of real-world data. In this paper, a brief review of these algorithms applied to data clustering has been described.

Keywords: Data clustering · K-means · Particle swarm optimization · Genetic algorithm · Hybrid PSO solutions

1 Introduction

Clustering analysis is a crucial part of Data Mining research and a group of multivariate techniques whose primary purpose is to group objects (respondents, products, or other objects) based on their characteristics. The grouping of objects is done in such a way that every object is incredibly similar (or connected) to the others within the cluster, and also that the resulting groups have the property of internal homogeneity within each cluster and high external (between clusters) diversity (different from each other or unrelated). Irrespective of what the definition is, it is generally true that the aim of clustering is to spot the "natural" structure in a data set. According to the methodology utilized by the algorithm, we are able to distinguish the following standard categories: Hierarchical, Exclusive, Overlapping and Probabilistic Clustering [15]. Cluster analysis is widely used in many engineering fields, like machine learning, pattern recognition, bioinformatics

D.-S. Huang and K.-H. Jo (Eds.): ICIC 2020, LNCS 12464, pp. 469–480, 2020.
https://doi.org/10.1007/978-3-030-60802-6_41

and image processing [14]. It has also found a purpose in other fields, like medicine, psychology, biology, sociology, astrophysics, statistics, marketing, economics, data mining, machine learning, planning development and our daily life [22].

The basic steps of cluster analysis include the sorting of a range of variables into clusters using a clustering algorithm (which includes a selection of similarity/diversity measures in addition to the clustering criteria), validation and interpretation of results. The basic applications of cluster analysis are:

- **Data exploration.** cluster analysis, we "discover" an unknown structure.
- **Data reduction.** Cluster analysis can play a major role within the compression of the information contained in the data, which is very significant due to the difficulty of working with an oversized data set. Clustering allows the data to be grouped into "interesting" clusters, so rather than working on the entire dataset as a unity, typical representatives of the clusters thus obtained are often considered.
- **Generation of hypotheses.** The results of cluster analysis conducted on datasets of an unknown structure are clusters (groups) whose number and composition can help to define the data structure hypothesis [13].
- **Testing of hypotheses.** Cluster analysis is employed to verify the validity of specific hypotheses, and a technique to verify accuracy is to use cluster analysis on a representative dataset.
- **Forecasting.** The resulting clusters are specific in terms of the characteristics that belong to them. Therefore, an unknown entity will be classified into a particular cluster based on its "similarities" to the characteristics of that cluster [10].

The paper deals with the study of the clustering efficiency based on combining the PSO and the K-means Algorithms, and PSO and GA Algorithms and then comparing it with other existing clustering techniques like standard PSO, K-means and Genetic Algorithm. The inspiration for this idea is the fact that mixing PSO with other algorithms can give excellent results in terms of accuracy and efficiency.

The remainder of this work is organized as follows: Sect. 2 is a brief introduction to the K-means technique. Section 3 is a general insight into Particle Swarm Optimization and Sect. 4 into Genetic Algorithm. Following this, Sect. 5 highlights the Hybrid PSO Solutions. The topic of Sect. 6 deals with the main key points of the details of experiments with result analysis. Section 7 provides the final conclusion.

2 K-Means

One of the best known and most effective non-hierarchical clustering methods is the K-means algorithm, which is known for being a quick algorithm (in terms of time complexity) for applying to large datasets. Clusters are described by utilizing a centroid that represents the arithmetic mean of the objects located in the cluster. It starts from *k* clusters (determined arbitrarily, or based on the previous clustering), and objects are classified into those clusters whose centroid is closest to them.

K-Means Algorithm. Denoted by A is a finite set of points of the *n*-dimensional space R^n:

$$A = \{a^1, \ldots, a^m\} \text{ where is } a^i \in R^n,\ i = 1, \ldots, m \tag{1}$$

Step 1. Select an initial solution consisting of k centers from the points.
Step 2. Assign points $a^i \in A$ to the nearest center to get the k-division of set A.
Step 3. Redefine the centers for this new division and return to step 2, until the cluster centers in the last two iterations are matched [8].

The reasons for the wide application of K-Means algorithm are:

- The time complexity is $O(m \times k \times l)$ where m is the number of cases, k is the number of clusters, and l is the number of iterations of the algorithm. As k and l are fixed beforehand, the time complexity of this algorithm is linear for sample size.
- The spatial complexity is $O(k+m)$.
- The algorithm is sequence independent. For a given initial set of cluster centers, it generates the identical division regardless of the order of cases.
- High-speed performance and measurable efficiency when working with large data sets.
- Ease of implementation in various problems.

3 Particle Swarm Optimization

Swarm Intelligence (SI) belongs to the field of Artificial Intelligence (AI) and is a relatively young interdisciplinary subfield of research that is supported by the collective behavior of decentralized and self-organized systems. The name was introduced by Gerardo Beng and Jing Wang in 1989. regarding the development of cellular (cell) robots [18]. SI systems are composed of a population of individual members (particles), also called agents, who interact with one another locally as well as interact with their environment. Particles or agents in a set or swarm follow simple rules and there is no central control structure to manipulate them. An intelligent agent is part of a system that analyses its surroundings and makes choices that maximize the chance of its success. Numerous algorithms have been created in the past few years for solving issues of numerical and combinatorial optimizations. Most promising among them are swarm intelligence algorithms.

PSO is a population-based stochastic search process method for continuous nonlinear function optimization that simulates the so-called social behaviors, like behavior of an ant colony, bird flocks, fish schools, herds of animals, groups of bacteria, etc. Unlike in other evolutionary computation techniques, each particle in this algorithm is additionally associated with a velocity [21]:

$$v_i(k+1) = v_i(k) + g_{1i}(p_i - x_i(k)) + g_{2i}(G - x_i(k)) \tag{2}$$

and position:

$$x_i(k+1) = x_i(k) + v_i(k+1). \tag{3}$$

where in (2) and (3) the variables are: i – particle index, k – descrete time index, v – velocity of i-th particle, x – position of i-th particle, p – best position found by i-th particle (personal best), G – best position found by a swarm (global best, personal best).

The particle changes the position of each of its centroids at each iteration, by talking into account its and neighbors' experience. As fitness function he distance between the centroid and a document of the collection is used [12]. The air is to evaluate the obtained solution of each particle, and the value of the fitness function can be measured by the following formula:

$$Fitness = \frac{\sum_{i=1}^{k}\left\{\frac{\sum_{j=1}^{p_i} d(o_i, m_{ij})}{p_i}\right\}}{k} \tag{4}$$

where in (4) the variables are: k - the number of clusters, m_{ij} - the jth document in the ith cluster, p_i - the number of documents in the cluster c_i, o_i - the centroid of the ith cluster, $d(o_i, m_{ij})$ - the distance between the document m_{ij} and the centroid o_i.

The main advantages of PSO are very few parameters to deal with and also a large number of processing elements (dimensions), which enables us to fly around the solution space effectively. Also, it is simple to code and has small computational costs, which are also important features.

4 Genetic Algorithm

GA is a metaheuristic optimization technique that works to find an accurate or approximate solution to a given problem by mimicking the process of natural evolution. In the 1970s, John Holland formulates the basic principles of genetic algorithms inspired by Darwin's theory of human evolution in his work [11]. GA is often applied to optimization problems, them being able to find the global optimum and in a set with multiple local extremes. In doing so, the function whose optimum is to be determined need not be continuous and differentiable. However, as a function of GA, one cannot know for sure whether the found solution is a local or global optimum, or whether it is cut off with the desired point, the process of optimizing needs to be repeated in order to increase the reliability of the solution. Also, due to a large number of computational operations, GA is slow and characterized by slow convergence to the found solution.

In the search for a solution to an optimization problem, a population of individuals is creating, so that each unit represents a candidate for solving the given problem. By encoding individuals, chromosomes are formed. Each chromosome is made up of several genes that represent encoded variables of a defined problem. The measure of the quality of an individual (*fitness*) is determined by the application of the adaptation function, which should adequately represent the problem that needs to be solved by applying a GA. The fitness value is the sum of the intra-cluster distances of all clusters. This sum of distance has a profound impact on the error rate [21]:

$$Fitness = \sum |X_j - Z_i|, \ i = 1, \ldots, K \ j = 1, \ldots, n \qquad (5)$$

where in (5) the variables are: K – the number of clusters, n – the number of data sets, Z_i – the cluster center at point i, X_j – the cluster for data point j.

Through several iterations of the algorithm, new generations of chromosomes are formed by simulating the processes of natural selection, mutation and reproduction. By the mechanism of selection, the irregular chromosomes are rejected, while the higher quality individuals are given a greater chance for survival and reproduction while forming the next generation. By crossing the chromosomes, new individuals are created by recombining the genes of the parent chromosomes. The process of mutation involves the random change of a gene, bringing new genetic material into the population. Mutation is the basic mechanism for preventing premature convergence of the genetic algorithm to the local optimum. The described procedure is repeated until the criteria for stopping the optimization process are fulfilled, and the highest quality individual of the last generation represents an approximate solution to the given problem.

GA-Based Clustering Technique. The searching capability of GA is exploited so as to look for appropriate cluster centres in the data set in order that the similarity metric of the resulting clusters is optimized. The chromosomes encode the centres of a fixed number of clusters and they are represented as strings of real numbers. GA has the ability to conduct a global search for solutions and this is the main reason that makes its use appropriate for clustering. The superiority of the GA-clustering algorithm over the commonly used K-means algorithm is in producing a more optimal solution.

5 Hybrid PSO Solutions

The main reason why PSO and GA were selected for analysis in this paper is that in recent years those two are among the most popular Evolutionary computation based algorithms that have received disseminated attention in research. On the one side, the Genetic Algorithm is a computational abstraction of biological evolution that can be used to solve certain optimization problems [19]. It is a probabilistic optimal algorithm that is based on evolutionary theories. On the other side, unlike gradient methods, the application of the PSO technique does not involve determining the derivative of the function, so is therefore characterized by easy implementation and fast convergence. It is often applied in combination with fuzzy logic, or other optimization techniques to create a hybrid optimization algorithm of higher efficiency.

6 Relationship Between PSO and GA

PSO and GA techniques have numerous likenesses: they start with a group of a randomly created population and both use a fitness value to evaluate the population, which is the sum of the intra-cluster distances of all clusters. The aim is to update the population and search for the optimum with random methods. In GA techniques, operators which are included are recombination, mutation and selection, while PSO does not have a direct recombination operator. In PSO the data exchange takes place only among the particle's own experience and therefore the experience of the best particle in the swarm, and in GA, it is being carried from fitness dependent selected parents to descendants.

The information-sharing mechanism of PSO is a one-way information sharing mechanism because only global best gives out the data to others. However, in the GA, chromosomes share information, in the way that the entire population moves like one group towards an optimal area. The benefits of PSO compared to GA are that PSO is simple to implement and there are fewer parameters to regulate [19].

GA and PSO are both well known nature-inspired metaheuristic optimization techniques frequently utilized in nearly all areas of engineering and science. The first is inspired by Darwin's concept of evolution, and the latter is a swarm-based algorithm inspired by the movement and nature of birds. However, the GA has a number of limitations like complex operational parameters including mutation and crossover, generation of the global solution and variable response time. On the other hand, PSO suffers from the problem of slow convergence. So, suitable parameter selection and proper adjustments in the parameters of both methods could also lead to the creation of very efficient cluster centres [16].

Hybrid PSO and GA Clustering Approach (HPSOGA). Optimal location management in mobile computing with Hybrid GA and PSO was proposed by Lipo Wang and Guanglin Si [25] and Priya I. Borkar and Leena H. Patil [6] present a model of hybrid GA and PSO (HGAPSO) for Web Information Retrieval. Two years later, authors Yicong Zhou, Yue-Jiao Gong, Jing-Jing Li et al. (2015) brought that social learning in PSO can help collective efficiency, while individual reproduction in GA facilitates global effectiveness [14].

PSO usually locates optimal solutions at a quick convergence speed, but on the other side, fails to regulate its velocity step size to continue optimization in the binary search space, which results in premature convergence. In contrast, research has shown that GA can adjust its mutation step size dynamically to better reflect the granularity of the local search area. However, GA suffers from a slow convergence speed, and although they have been successfully applied to a great spectrum of various problems, using GA for large-scale optimization might be expensive because of its requirement of a large number of function evaluations for convergence. Therefore, HPSOGA has been proposed to overcome those problems and mix the advantages of PSO and GA. The idea behind this is that such a hybrid approach is predicted to possess merits of PSO with those of GA. The biggest advantage of PSO over GA is its algorithmic simplicity. Information will be swapped between two particles to fly to the new search area, by applying crossover operation. Therefore, in this mixed algorithm, the crossover operation is additionally included, which might improve the diversity of individuals [14].

7 Relationship Between PSO and K-Means

Limitations and drawbacks of classical K-means algorithm can be grouped into four parts:

1. An initial selection of the number of clusters must be previously known and specified by the user.
2. Because of its sensitivity to initial partition, it can only generate a local (neighborhood) optimal solution.
3. Results directly depend upon the initial cluster centroid chosen by the algorithm.

On the other side, the limitations of the PSO are as follows:

1. Due to its nature, it is inclined towards quick and premature convergence in mid optimum points.
2. When the search space is high, the convergence speed slows down considerably.
3. This algorithm does not perform very well when the dataset is complex or large [20].

Even if K-means is one of the most popular clustering algorithms, the benefits of PSO should be emphasized. K-means is well-known, but its primary disadvantages are its sensitivity to the initialization of the beginning K centroids, as it was explained before. However, PSO solves this issue by reducing the impact that starting conditions have because it starts searching from multiple positions in parallel. Due to this, it often yields more precise results, even though PSO tends to converge slower than the standard K-means algorithm [17].

Two different strategies for clustering data utilizing PSO were proposed by Van der Merwe and Engelhrecht [24]. In the first method, PSO was used to discover the centroid of a user-specified number of clusters. On the other hand, the second method propagates extended use of PSO where K-means clustering is employed for seeding the initial swarm. When the two approaches are compared with plain K-means, it turned out that PSO approaches have superior convergence to lower quantization errors and in general larger inter cluster and smaller intra cluster distances.

Hybrid PSO and K-Means Clustering Approach (HPSOKM). It is important to emphasize the fact that the performance of the PSO can be improved even further. Rana, Jasola and Kumar [20] proposed the so-called PSO and K-means algorithm, which is a hybrid of PSO and K-means. The issues related to global and local minimums play an important role when data sets and attributes associated are very large and the classification based on clustering is important and critical. In case of certain data sets like medical, security, finance etc. the error generated because of the K-means method is not acceptable at all. The objective function of the K-means algorithm is not convex and hence it can contain many local minima. In this section, the aim is to present a hybrid clustering algorithm based on combining the K-means and PSO algorithms. The motivation for this idea is the fact that PSO, at the beginning stage of the algorithm, starts the clustering process due to its quick convergence speed, and also the result of PSO is then tuned by the K-means near-optimal solutions.

With this HPSOKM, it has been shown by researchers that it arrives at much better results in the analysis of data and prediction when comparing to the results of each algorithm individually. Although PSO is a good clustering method, it does not perform well when the dataset is too large or complex. K-means is added in sequence to the PSO to obtain better result through further refinement in cluster formation. The PSO algorithm is utilized to assist in finding a vicinity of the optimal solution by a global search during the initial stage. The results from the PSO are used as the initial seed of the K-means algorithm, which is applied for refining and later generating the final result of the data clustering problem [20]. HPSOKM overcomes the disadvantages of both algorithms, uses the advantages of both algorithms to produce the best optimization and improves the clustering technique by avoiding being trapped in a local optimal solution.

8 Experiments and Results Analysis

The following section will illustrate a comparison of the proposed techniques from the point of view of the global fitness parameter. The proposed HPSOKM is tested with four data sets [23] which represent the following document collections: “Banking and Finance”, “Science”, “Programming Languages” and “Sport”. The PSO parameters that are fixed in the program are shown in Table 1 and all parameters which are changed at each test are shown in Table 2. In this experiment the specified fitness function in (4) was used. The algorithm implementation was performed by using MATLAB software and was inspired by the following paper [12].

Table 1. Fixed PSO parameters and their values

Parameter	Value
Number of clusters	5
Confident coefficient at its best position	0.72
Confident coefficient at its neighboring	1.49
Inertia factor	0.3

The results from the comparison of three different clustering methods obtained from the experiment are shown in Table 3. Experimental results show that if PSO is hybridized with other clustering algorithms such as K-means, it then has an advantage over standard PSO while not inheriting its drawbacks. Hybridized PSO generates the lowest values of the fitness function. It is a good indicator that the clusters which are generated by HPSOKM algorithm are the most compact.

Figure 1 illustrates the convergence of the different algorithms on the Banking and Finance document collection [12]. In this figure, it can be noticed that K-means converge quickly and prematurely. The fitness values grow from 11.45 to 8.26 after 10 iterations, and then it does not change after. For PSO and HPSOKM the two curves have the same trajectories until the 25th iteration because those algorithms both use

Table 2. Changeable parameters and their values

Parameter	Value
Number of particles	10–25
Number of iterations in PSO and HPSOKM	5–50
Number of iterations in K-means	10–25
Number of particles	10–25
Dimension of the problem	500–10000

Table 3. Performances of three different clustering methods (fitness values)

Document collection	K-means	PSO	HPSOKM
Banking and Finance	5.093	6.982	3.211
Science	7.245	8.632	5.447
Programming Languages	4.788	4.871	2.922
Sport	9.090	10.093	4.63

PSO and do the same work. However, after the 25th iteration, the trajectories have changed: PSO converges slowly and HPSOKM converges quickly. Finally, we can deduce that the HPSOKM algorithm gives the best results.

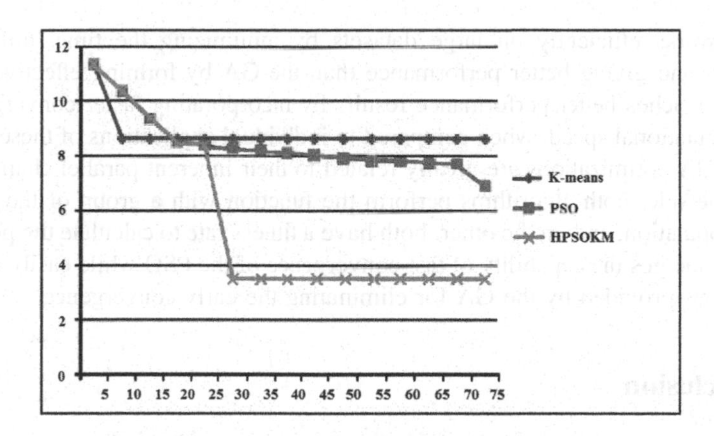

Fig. 1. Convergence graph of the three clustering algorithms (fitness value/iterations)

In next section, the three different optimization techniques are tested on five data sets and the results of standard PSO and GA are compared with the results of HPSOGA in term of optimal fitness values and elapsed time. For the comparison purpose, the stopping criteria, that is the number of maximum generations, is the same for all the algorithms and equals 100 runs. In this experiment the specified fitness function in (5) was used. The algorithm implementation was performed by using MATLAB software and was inspired by the following paper [14].

In Table 4, the HeartDisease dataset is considered as the best example because HPSOGA can produce an optimized result faster than the other two. HPSOGA methodology enhances better performance results by incorporating high computational speed and the faster convergence than the individual comparison. Table 5 present the CPU time, taken to compete for the optimization for a dataset. Even though the HPSOGA gets the best-optimized value, it consumes more time.

Table 4. Result comparisons with optimal fitness absolute values

Datasets	PSO	GA	HPSOGA
HeartDisease	186.7309	3.13E + 04	186.7039
Data	186.7309	1.13E + 04	186.7309
DataScience	185.1554	2.01E + 02	185.1554

Table 5. Result comparisons with time

Datasets	PSO	GA	HPSOGA
HeartDisease	33.659704	49.111702	51.12346
Data	37.381394	169.634856	140.483837
DataScience	1592.40601	9797.553331	10876.302102

PSO works efficiently on large datasets by minimizing the time, utilizing less parameters and giving better performance than the GA by forming effective clusters. HPSOGA reaches better performance results by incorporating faster convergence and high computational speed, when compared to individual applications of these methods. PSO and GA optimizations are greatly related to their inherent parallel characteristics. On the one side, both algorithms perform the function with a group of the randomly created population, and on the other, both have a fitness rate to calculate the population. HPSOGA merges the capability of fast convergence of the PSO while easily exploiting the solutions provided by the GA for eliminating the early convergence.

9 Conclusion

Many algorithms have been devised for clustering in past years because it is one of the most common and helpful data mining techniques which have application in various cases. In this paper, we introduced and evaluated some of the techniques which have been used often in recent years. A review of various research efforts done in the areas of K-means, PSO and GA is presented and it is concluded that when PSO is used in combination with K-means or GA it produces superior results in terms of accuracy and efficiency because the algorithms compensate each other's drawbacks.

Acknowledgement. The work described in this paper was supported by The Natural Science Foundation of China (Grant No.71971143, 71571120), Natural Science Foundation of Guangdong Province (2020A1515010749), Key Research Foundation of Higher Education of Guangdong Provincial Education Bureau (2019KZDXM030), Guangdong Province Soft Science Project (2019A101002075), Guangdong Province Educational Science Plan 2019 (2019JKCY010) and Guangdong Province Postgraduate Education Innovation Research Project (2019SFKC46).

References

1. Abdel-Kadel, R.F.: Genetically improved PSO algorithm for efficient data clustering. In: Second International Conference on Machine Learning and Computing, pp. 71–75 (2010)
2. Acid, S., De Campos, L., Fernandez-Luna, J., Huete, J.: An information retrieval model based on simple bayesian networks. Int. J. Intell. Syst. **18**(2), 251–265 (2003)
3. Ahmadyfard, A., Modares, H.: Combining PSO and k-means to enhance data clustering.In: International symposium on telecommunications, pp. 688–691 (2008)
4. Al-Maqaleh, B., Shahbazkia, H.: A genetic algorithm for discovering classification rules in data mining. Int. J. Comput. Appl. **41**(8), 42–43 (2012)
5. Ballardini, A.L.: PSO Clustering Algorithm (2014). https://github.com/iralabdisco/pso-clustering. Accessed 28 April 2020
6. Borkar, P., Patil, L.: Web information retrieval using genetic algorithm-particle swarm optimization. Int. J. Future Comput. Commun. **2**(6), 595–599 (2013)
7. Cui, X., Potok, T.E.: Document clustering analysis based on hybrid PSO + K-means algorithm. J. Comput. Sci (Special Issue) **27**, 33 (2005)
8. Dragnic, N.: Konstrukcija i analiza klaster algoritma sa primenom u definisanju bihejvioralnih faktora rizika u populaciji odraslog stanovnistva Srbije. Univerzitet u Novom Sadu, Novi Sad (2015)
9. Glisovic, N., Davidovic, T., Raskovic, M.: Clustering when missing data by using the variable neighborhood search. In: XLIV Symposium on Operational Research, Zlatibor, Serbia, pp. 158–159 (2017)
10. Halkidi, M., Bastiakis, Y., Vazirgiannis, M.: On clustering validation techniques. J. Intell. Inf. Syst. **17**(2-3), 108–109 (2001)
11. Holland, J.: Adaptation in natural and artificial systems. an introductory analysis with application to biology, control, and artificial intelligence. University of Michigan Press, Ann Arbour. (1975)
12. Kamel, N., Ouchen, I., Baali, K.: A sampling PSO-K-means algorithm for document clustering. In: Pan, J.S., Krömer, P., Snášel, V. (eds.) Genetic and Evolutionary Computing. Advances in Intelligent Systems and Computing, pp. 45–54. Springer, Cham. (2014). https://doi.org/10.1007/978-3-319-01796-9_5
13. Kovacic, Z.J.: Multivarijaciona analiza. Ekonomski fakultet, Beograd (1994)
14. Malini Devi, G., Lakshmi Prasanna, M., Seetha, M.: Performance analysis of genetic algorithm with PSO for data clustering. Int. J. Adv. Eng. Res. Sci. – IJAERS, 3(11), 115–119 (2016)
15. Matteucci, M.: A tutorial on Clustering Algorithms - Hierarchical Clustering Algorithms. http://home.deib.polimi.it/matteucc/Clustering/tutorial_html/hierarchical.html. Accessed 24 April 2020
16. Nayak, J., Kanungo, D.P., Behera, H.S.: an improved swarm based hybrid k-means clustering for optimal cluster centers. In: Advances in Intelligent Systems and Computing (2015)

17. Omran, M., Salman, A., Engelbrecht, A.: Image classification using particle swarm optimization. In: Proceedings of the 4th Asia-Pacific Conference on Simulated Evolution and Learning (2002)
18. Prascevic, Z., Prascevic, N.: Particle Swarm Optimization. Izgradnja **62**, 339–345 (2008)
19. Premalatha, K., Natarajan, A.M.: Hybrid PSO and GA models for document clustering. Int. J. Adv. Soft Comput. Appl. 2(3) 309–310 (2010)
20. Rana, S., Jasola, S., Kumar, R.: A hybrid sequential approach for data clustering using K-Means and particle swarm optimization algorithm. Int. J. Eng. Sci. Technol. **2**(6) (2010)
21. Sarkar, S., Roy, A., Purkayastha, B.: Application of particle swarm optimization in data clustering. Int. J. Comput. Appl. **65**(25) (2013)
22. Saxena, A, Prasad, M, Gupta, A, et al.: A review of clustering techniques and developments. Neurocomputing **267**, 664–681 (2017)
23. TREC. Text Retrieval Conference (1999). http://trec.nist.gov. Accessed 02 Aug 2020
24. Van der Merwe, D.W., Engelhrecht, A.P.: Data clustering using particle swarm optimization. In: Conference of Evolutionary Computation CEC 2003, Canberra, ACT, Australia, pp. 215–220 (2003)
25. Wang, L., Si, G.: Optimal location management in mobile computing with hybrid Genetic algorithm and Particle Swarm Optimization (GAPSO). In: IEEE Transaction Information Theory, pp. 1877–1886 (2012)

A Novel Computational Approach for Predicting Drug-Target Interactions via Network Representation Learning

Xiao-Rui Su[1,2], Zhu-Hong You[1(✉)], Ji-Ren Zhou[1], Hai-Cheng Yi[1,2], and Xiao Li[1,2]

[1] The Xinjiang Technical Institute of Physics and Chemistry, Chinese Academy of Sciences, Urumqi 830011, China
zhuhongyou@ms.xjb.ac.cn
[2] University of Chinese Academy of Sciences, Beijing 100049, China

Abstract. Detection of drug-target interactions (DTIs) has a beneficial effect on both pathogenesis and drugs discovery. Although a huge number of DTIs have been generated recently, the number of known interactions is still very small. Thus, it is strongly needed to develop computational methods to accurately and effectively predict DTIs. In this paper, a large-scale computational method is proposed to predict potential DTIs via network representation learning. More specifically, known associations among drugs, proteins, miRNA and disease are formulated as a biomolecular association network, and the network representation method Structural Deep Network Embedding (SDNE) is used to extract network-based features of drug and target nodes. Then, the fingerprints of drug compounds and sequence information of proteins are also adopted. Finally, an ensemble Random Forest classifier is used to classify and predict DTIs. Experiment results show that the proposed method achieved a good prediction performance with an accuracy of 83.68% and AUC of 0.9052. It is anticipated that proposed model is feasible and effective to predict DTIs at a global molecule level, which is a new respective for future biomedical researches.

Keywords: Drug-Target interactions · Molecular association network · Combined feature · SDNE · Drug discovery

1 Introduction

Drug-Target Interactions (DTIs) studying is beneficial to understanding various interactions and biological processes [1–3]. Previous studies have shown that there are approximately 6000–8000 targets of pharmacological interest in the human genome [4, 5]. But it is a pity that only a small portion has been confirmed to be related to approved drugs in spite of ongoing efforts have been made by traditional biological experiments. It is well known that the conventional investigation of the interactions between drug and protein requires much energy and material resources. Besides, the development of new drug also is a long and expensive process. Therefore, developing an accurate and efficient method to accelerate the pace of drug development is urgent. Especially with the completion of the Human Genome Project and the emergence of molecular

D.-S. Huang and K.-H. Jo (Eds.): ICIC 2020, LNCS 12464, pp. 481–492, 2020.
https://doi.org/10.1007/978-3-030-60802-6_42

medicine, the rapid development of biological enables researchers to retrieve information and features and study DTIs through system integration [6–8]. Simultaneously, some databases related to DTIs have been established, which are freely available [9, 10]. The generation of these databases provide the vital basis for predicting DTIs on a larger scale.

In line with different study strategies, drug target prediction can be divided into three categories: (1) prediction based on ligand structure features, (2) prediction based on protein structure features and (3) prediction based on data mining methods. Drug target prediction based on ligand structure features mainly includes chemical similarity search and reverse pharmacophore search [11, 12]. However, this method only considers the three-dimensional structure of the macromolecule of the corresponding biological target, so it cannot fully reflect the mutual binding mode of the query molecule and the predicted target. Drug target prediction based on protein structure features is generally referred to as a reverse molecular docking method [13–15]. This method molecularly docks the query molecules with several molecules in the target database, and selects the best candidate for the combination, which is considered as a potential target [16, 17].

Computational methods have been widely applied to predict DTI, which is the third strategy of studying drug targets. Cheng et al. [2] proposed a drug-target network with network-based inference (NBI) to predict new DTIs. This method is derived from the recommendation algorithms of complex network theory. However, this method does not consider the biological information of protein. Wang et al. [1, 18] proposed a computational model for predicting DTIs using drug structure and protein sequences. Both drug structure and protein sequence were encoded as feature vectors, which were fed into a rotation forest-based predictor. Wang et al. [19] proposed a deep learning-based model for DTIs prediction, which utilized the Sparse Principal Component Analysis (SPCA) to compress the features of drugs and proteins into a uniform vector space. Then using deep LSTM network detected the interactions between drugs and targets. All three methods consider the biological information of proteins and drugs. However, due to the extremely complex associations between small biomolecules, predicting DTIs with drug and protein information is limited. Small molecules that are not directly related can be indirectly linked by other molecules to act on drugs or proteins. Therefore, a comprehensive model needs to be built to predict DTIs from a global perspective.

In this paper, we proposed a novel computational model for predicting drug-target interactions using drug molecular fingerprint, protein sequence feature and network embeddings based on Molecular Association Network (MAN). More specifically, the MAN is constructed by collecting nine known associations between five small molecules. Besides, network representation learning method SDNE is applied to extract network embedding of drug and target nodes. SDNE is the first network representation learning method, which introduces deep learning. SDNE uses an auto-encoder structure to optimize both the first-order and second-order similarities. The learned vector representation can retain local and global structures, which is robust to sparse networks. And the fingerprints of drug compounds and sequence information of proteins are also adopted. The positive sample set is constructed by the known interactive drug-target pairs and the same number of negative samples are randomly selected from the unknown pairs. In our experiments, 11107 positive samples and 11107 negative samples are used to train our model. Then, these efficient features are used to train a Random Forest classifier to infer drug-target interactions. Under 5-fold cross-validation, the combined features show better performance than individual attribute feature and behavior feature. Experimental results indicate that the proposed method is promising to infer drug-target interactions. Which can be used as a competitive computational tool for drug discovery.

2 Materials and Method

2.1 Molecular Association Network Dataset

MAN is a dataset based on small molecules levels, containing biological small molecule transcripts (miRNA, lncRNA and protein), disease, drug and their associated relationships. It is also a comprehensive and systematic molecule biological associations network [20–25]. MAN contains five kinds of nodes and nine types of relationship, which are collected from a number of valuable datasets [9, 26–29]. The MAN dataset is obtained after completing database integration, unification of identifiers, redundancy elimination and simplification. The details are shown in Table 1 and Table 2.

Table 1. Nine types of relationships and their quantity contained in MAN

Relationship	Quantity
lncRNA - Protein	690
Drug - Disease	18416
Drug - Protein	11107
Protein - Disease	25087
Protein - Protein	19237
lncRNA - Disease	1264
miRNA - Protein	4944
miRNA - Disease	16427
miRNA - lncRNA	8347
Total	105546

Table 2. The quantity of five kinds of nodes in MAN

Node	Quantity
Protein	1649
Disease	2062
Drug	1025
miRNA	1023
lncRNA	769
Total	6528

2.2 Attribute Feature Extraction Using Molecular Fingerprints and Protein Sequences

Attribute feature describes the biological information of nodes in MAN, which is a 64-demension vector. Drug attribute feature is extracted from the SMILES [30] of drugs downloaded from DrugBank. Current studies show that the use of various molecular substructure fingerprints to represent pharmaceutical compounds is effective [21, 25, 31–36]. In this work, drug fingerprint records information on 881 substructures, so the attribute feature of drug molecule is the 881 binary vector.

Protein attribute feature is extracted from protein sequences downloaded from STRING [37]. According to the polarity of the side chain, 20 amino acids have been divided into four groups by Shen *et al.* [38]. Hence, each amino acid has a label for the category. The 3-mer method uses a sliding window of length of 3 to slide from the beginning of sequence to the end of the sequence in one step. Since there are 64 possible sorts of four groups of amino acid, the target vector is a 64-dimension vector. The target vector is normalized when the slide is over. Finally, attribute feature of protein is obtained (see Fig. 1).

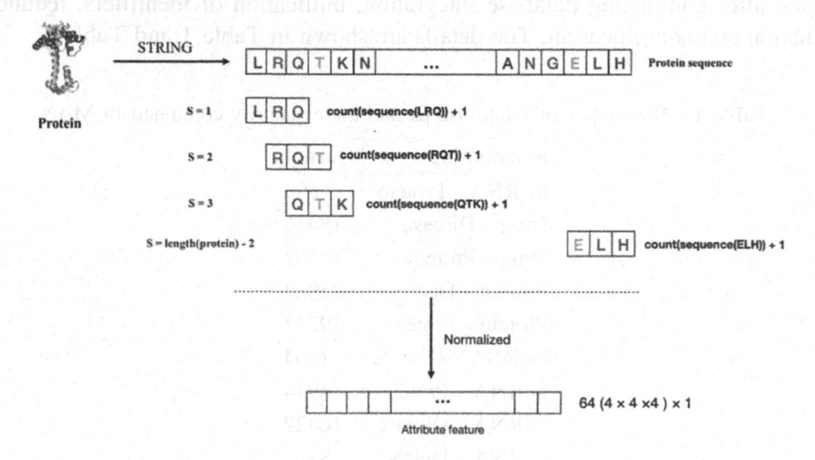

Fig. 1. Schematic diagram of a 3-mer method for extracting protein attribute feature

Due to the different dimension of protein attribute feature and drug attribute feature, stacked autoencoder is adopted to get a suitable subspace from the original feature space. Stacked autoencoder has encoder and decoder two parts. Encoder process can be defined as follow:

$$h = f(x) = s_f(Wx + b) \tag{1}$$

In this formula, W represents the weight matrix, b represents the bias vector and x represents the input. s_f is an activation function.

Decoder process is defined as:

$$\bar{x} = \bar{f}(h) = s_{\bar{f}}(\bar{W}h + \bar{b}) \tag{2}$$

$\bar{W}$ represents the weight matrix of output layer and $\bar{b}$ still represents the bias vector. In this work, ReLU function is selected as the activation function. Finally, a 64-dimension vector is obtained as the attribute feature of drug and proteins.

2.3 Behavior Feature Extraction Based on SDNE

Behavior feature describes the global structure of MAN. Unlike previous studies, we are looking forward to detect potential new associations from complicated network. Therefore, extracting relationship of network is crucial to our work.

In order to extracting behavior feature, network embedding method SDNE [39] is employed to learn the deep structure of MAN. Inspired by deep learning, SDNE is a semi-supervised model and has proven to have the power to learn deep and complex network structures. It captures highly nonlinear network structures through multiple layers of nonlinear functions. In order to fully understand the global network structure and preserve the local structure, SDNE adopts the first-order proximity and second-order proximity. The first-order proximity preserves the network local structure by supervised learning, and the second-order proximity captures the global structure through unsupervised learning. By jointly optimizing them in a semi-supervised depth model, the method can preserve local and global network structures together and is robust to sparse networks.

2.4 Node Representation

Each node in MAN has two features, attribute feature and behavior feature. Attribute feature describes the biological information of nodes, behavior feature describes the network structure of nodes. In order to represent the nodes in MAN more comprehensively and systematically, the combination of attribute feature and behavior feature is used to represent nodes, called combined feature, which is used as the input of our model (see Fig. 2).

2.5 Evaluation Criteria

In order to evaluate the performance of proposed method, a few criteria are used, including accuracy (Acc.), sensitivity (Sen.), specificity (Spec.), precision (Prec.) and Matthews Correlation Coefficient (MCC), Receiver operating characteristic (ROC), Area Under Curve (AUC) and Area Under Precision-Recall (AUPR). These criteria are sufficient to measure the quality, robustness, and predictability of the model from different perspectives. They are defined as follows:

$$Acc. = \frac{TN + TP}{FP + TP + FN + TN} \tag{3}$$

$$Sen. = \frac{TP}{TN + TP} \tag{4}$$

$$Prec. = \frac{TP}{TP + FP} \tag{5}$$

$$MCC = \frac{(TN \times TP) - (FN \times FP)}{\sqrt{(FN + TP) \times (FP + TN) \times (FP + TP) \times (FN + TN)}} \tag{6}$$

Where FP, TP, FN and TN represent false positive, true positive, false negative and true negative, respectively.

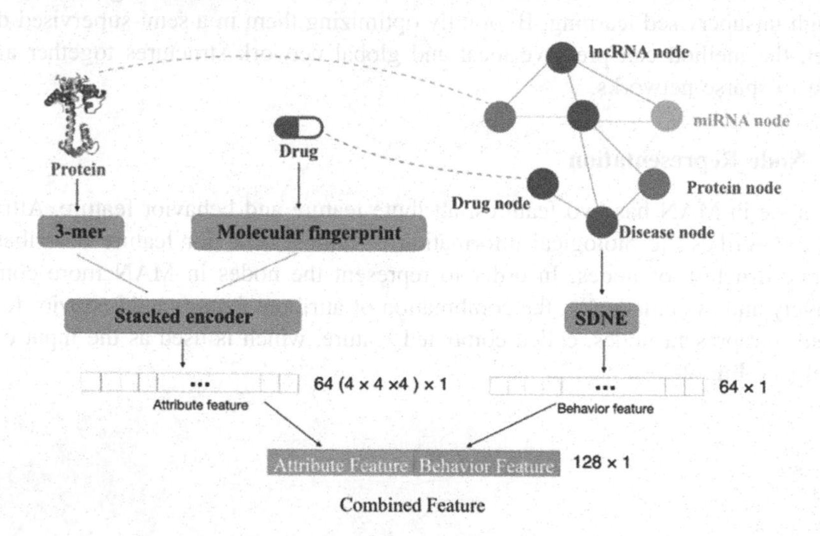

Fig. 2. Combined feature generation process diagram

3 Results and Discussion

3.1 Prediction Performance of Proposed Model

To systematical estimate proposed model performance, we perform 5-fold cross-validation experiments, which not only detects the stability of the model but avoids the overfitting. The proposed model performs well in the MAN dataset. Table 3 lists the prediction performance of Random Forest (RF) classifier, it yields an accuracy of 83.68%, sensitivity of 79%, specificity of 88.35%, precision of 87.15%, and MCC of 67.65%. The standard deviations are 0.25, 0.48, 0.48, 0.44, and 0.50, respectively. It can be observed that proposed method yields good prediction performance with AUC value of 0.9052 and AUPR value of 0.9128 (see Fig. 4 and Fig. 3).

Table 3. The 5-fold cross-validation performance of proposed method

Fold	Acc. (%)	Sen. (%)	Spec. (%)	Prec. (%)	MCC (%)	AUC (%)
0	83.98	78.85	89.11	87.86	68.32	90.91
1	83.60	78.71	88.48	87.23	67.51	90.44
2	83.87	79.84	87.89	86.83	67.95	90.54
3	83.60	78.94	88.25	87.05	67.49	90.37
4	83.35	78.68	88.01	86.78	66.99	90.34
Average	83.68 ± 0.25	79.00 ± 0.48	88.35 ± 0.48	87.15 ± 0.44	67.65 ± 0.50	90.52 ± 0.23

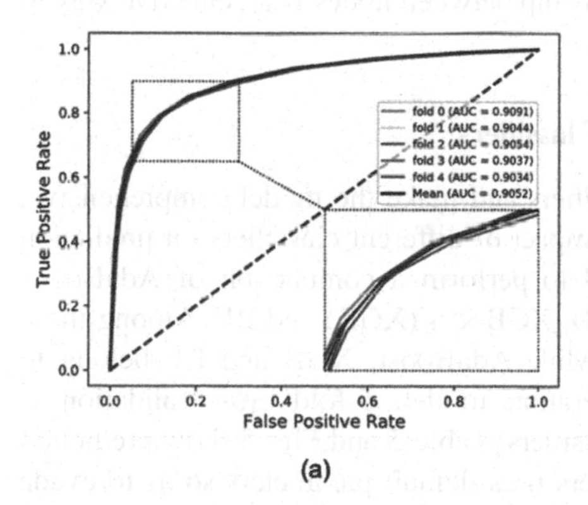

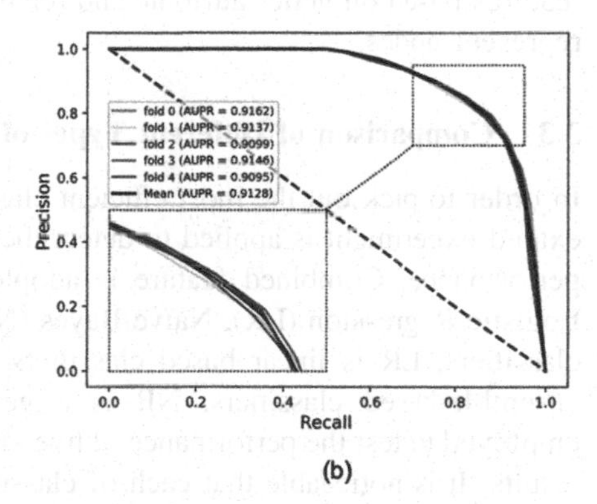

Fig. 3. The 5-fold cross-validation performance of proposed method, (a) shows the ROC curve (b) shows the PR curve

3.2 Comparison of Three Node Representations

In order to better explore the impact of node representations on final performance, we evaluate and compare the three different node representations using RF under 5-fold cross-validation, respectively. Six evaluation metrics are calculated in the study, which are shown in Table 4 and Fig. 4.

It can be obtained that representing nodes by attribute feature achieve the lowest prediction performance with an average accuracy of 81.17%, sensitivity of 76.98%, specificity of 85.37, precision of 84.03%, MCC value of 62.57% and AUC value of 0.8797. Compared with attribute feature, behavior feature has a better performance, it yields an accuracy of 82.72%, sensitivity of 77.78%, specificity of 87.66%, precision of 86.31%, MCC of 65.77% and AUC of 0.8963. Nevertheless, combined feature achieves the best performance. As shown in Table 4, the AUC of combined feature is 0.9052, higher than 0.8963(behavior feature) and 0.8797(attribute feature), respectively. In addition, the other five appraisal metrics are higher than the other node representations with an average accuracy of 83.68%, sensitivity of 79.00%, specificity of 88.35%, precision of 87.15% and MCC of 67.65%.

On the other hand, the results of combined feature have the lowest standard deviation, which of accuracy, sensitivity, specificity, precision, MCC and AUC are 0.25, 0.48, 0.48, 0.44, 0.50 and 0.23, respectively. Attribute feature standard deviation of six evaluation metrics are 0.44, 0.34, 0.96, 0.85, 0.93 and 0.30. Behavior feature standard deviation of six evaluation metrics are 0.57, 0.68, 0.52, 0.59, 1.14 and 0.48. At this point, a verdict can be got that using combined feature to represent nodes makes the model more stable and robust.

Figure 4 shows that representing nodes with combined feature gets the highest value of not only AUC but AUPR, further demonstrating that combined feature performs better in predicting drug-target interactions. This is due to that combined feature considers both node information and associations of nodes, especially in the case of large data volume and complex associations. Therefore, it indicates that extracting features based on nodes attribute and relationship between nodes is an effective way to represent nodes.

3.3 Comparison of Different Types of Classifiers

In order to pick out the most efficient classifiers and make the model comprehensive, extend experiment is applied to detect the impact of different classifiers on prediction performance. Combined feature is adopted to perform a comparison of AdaBoost, Logistic Regression (LR), Naïve Bayes (NB), XGBoost (XGB) and RF. Among these classifiers, LR is linear based classifiers, while AdaBoost, XGB and RF belong to ensemble-based classifiers. NB is a generation model. 5-fold cross-validation is employed to test the performance of five classifiers. Table 5 and Fig. 5 show prediction results. It is noticeable that each of classifiers uses default parameters so as to evade bias.

Generation model NB, which is a series of simple probability classifiers based on Bayesian theorem. The model assumes that the individual attributes are independent of each other. Owing to this situation is not common, and the rule does not apply to MAN dataset. Hence, NB achieves the worst prediction performance with an average accuracy of 64.49%, sensitivity of 39.25%, specificity of 89.72%, precision of 79.86%, MCC of 33.80% and AUC value of 72.70%. It also can be observed from (see Fig. 5) that AUPR of NB is the lowest with 0.7490.

Table 4. The results of different feature extractions method

Feature	Acc. (%)	Sen. (%)	Spec. (%)	Prec. (%)	MCC (%)	AUC (%)
Attribute	81.17 ± 0.44	76.98 ± 0.34	85.37 ± 0.96	84.03 ± 0.85	62.57 ± 0.93	87.97 ± 0.30
Behavior	82.72 ± 0.57	77.78 ± 0.68	87.66 ± 0.52	86.31 ± 0.59	65.77 ± 1.14	89.63 ± 0.48
Combined	83.68 ± 0.25	79.00 ± 0.48	88.35 ± 0.48	87.15 ± 0.44	67.65 ± 0.50	90.52 ± 0.23

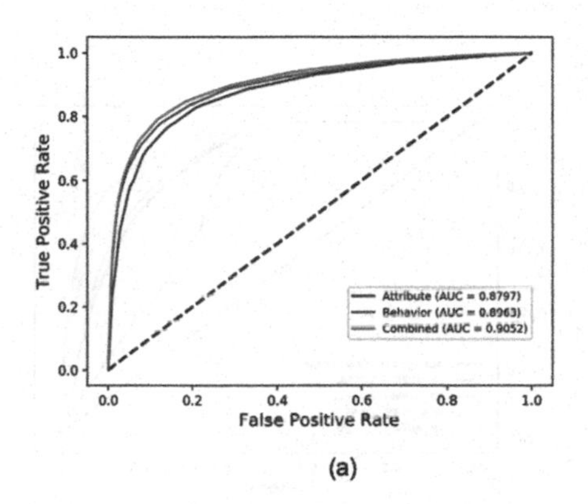

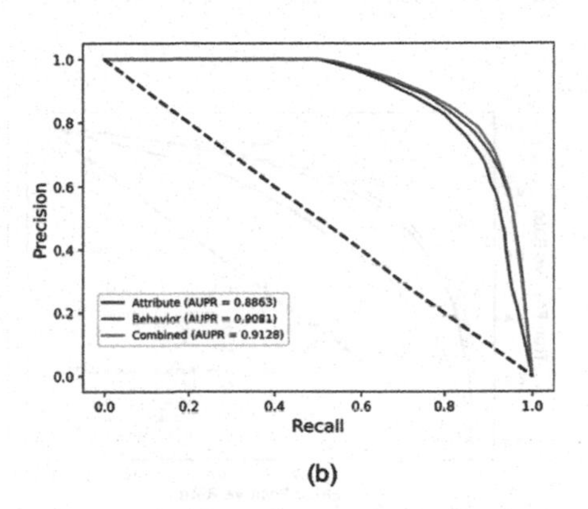

Fig. 4. The 5-fold cross-validation performance of three node representations, (a) shows the ROC curve, (b) shows the PR curve

Linear model LR, which has a better performance than NB, attaining an average accuracy of 78.07%, sensitivity of 74.07%, specificity of 82.07%, precision of 80.51%, MCC of 56.33% and AUC of 0.8503. Compared with NB, both accuracy and AUC have increased by 14% and 13%, respectively. However, the linear model LR is not the most suitable for our model owing to data used in experiment is complex and non-linear, comprising a plenty of associations between five kinds of nodes. This point can be proved by the experiment results in Table 5.

AdaBoost, XGB and RF have quite different classification performance in spite of all of them belong to ensemble-based model. RF performs the best among three classification models, whose results of over-all accuracy, sensitivity, specificity, precision, MCC and AUC are 83.69%, 79.00%, 88.35%, 87.15%, 67.65% and 0.9052. Besides, the standard deviation of RF is the lowest, indicating the model adopting RF is the most stable. Unfortunately, AdaBoost gets the adverse results with the lowest accuracy 76.78%, the lowest AUC 83.71% and the highest standard deviation. Though XGB performs better than AdaBoost, the accuracy and AUC obtained by XGB, which are 81.08 and 0.8831, respectively, are not as high as RF. Simultaneously, XGB gets higher standard deviation of over-all six evaluation metrics than RF. Among these three classifications, RF stands out. Therefore, RF is selected as the final classifiers of our model due to the excellent performance. In conclusion, all the results demonstrate that our model is an effective method to predict associations between drug and proteins.

Table 5. The performance of different classifiers

Classifier	Acc. (%)	Sen. (%)	Spec. (%)	Prec. (%)	MCC (%)	AUC (%)
AdaBoost	76.78 ± 0.68	73.50 ± 0.96	80.06 ± 1.13	78.67 ± 0.94	53.68 ± 1.37	83.71 ± 0.66
LR	78.07 ± 0.71	74.07 ± 0.84	82.07 ± 0.68	80.51 ± 0.74	56.33 ± 1.41	85.03 ± 0.32
NB	64.49 ± 0.96	39.25 ± 4.27	89.72 ± 4.49	79.86 ± 4.85	33.80 ± 2.89	72.70 ± 2.08
XGBoost	81.08 ± 0.62	75.93 ± 0.59	86.22 ± 0.69	84.65 ± 0.74	62.49 ± 1.26	88.31 ± 0.56
RF	83.68 ± 0.25	79.00 ± 0.48	88.35 ± 0.48	87.15 ± 0.44	67.65 ± 0.50	90.52 ± 0.23

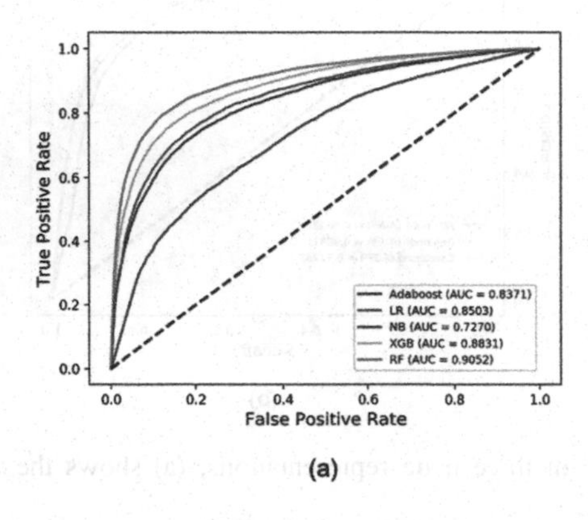

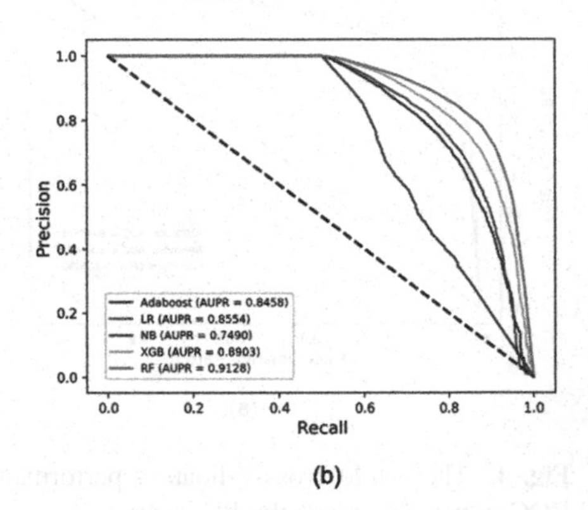

Fig. 5. The 5-fold cross-validation performance of five classifiers, (a) shows the ROC curve of different classifiers and the value of AUC, (b) shows the PR curve and the value of AUPR

4 Conclusions

In this study, an efficient technique base on MAN has been proposed for predicting drug-target using molecular fingerprints and protein sequence information. The results exhibit that the proposed method is significantly promising for predicting drug-target interactions. MAN is constructed by combining nine known valuable datasets in order to detect the potential links between target nodes. 3-mer and molecular fingerprints are used to extract the attribute feature for protein and drug, respectively. SDNE is applied to extract the behavior feature. Combined feature, which is the combination of attribute feature and behavior feature, are adopted to represent the nodes in MAN. Compared with other two node representations, proposed model performs well.

References

1. Wang, L., et al: RFDT: a rotation forest-based predictor for predicting drug-target interactions using drug structure and protein sequence information. Curr. Protein Peptide Sci. **19**, 445–454 (2018)
2. Cheng, F., et al.: Prediction of drug-target interactions and drug repositioning via network-based inference. PLoS Comput. Biol. **8**, e1002503 (2012)

3. You, Z.-H., et al.: PBMDA: A novel and effective path-based computational model for miRNA-disease association prediction. PLoS Comput. Biol. **13**, e1005455 (2017)
4. Drews, J.: Drug discovery: a historical perspective. Science **287**, 1960–1964 (2000)
5. Mabonga, L., Kappo, A.P.: Protein-protein interaction modulators: advances, successes and remaining challenges. Biophys. Rev. **11**(4), 559–581 (2019). https://doi.org/10.1007/s12551-019-00570-x
6. Huang, Y.-A., You, Z.-H., Chen, X.: A systematic prediction of drug-target interactions using molecular fingerprints and protein sequences. Curr. Protein Peptide Sci. **19**(5), 468–478 (2018)
7. Chen, X., You, Z.H., Yan, G.Y., Gong, D.W.: IRWRLDA: improved random walk with restart for lncRNA-disease association prediction. Oncotarget **7**, 57919–57931 (2016)
8. Chen, X., Yan, C.C., Zhang, X., You, Z.H., Huang, Y.A., Yan, G.Y.: HGIMDA: Heterogeneous graph inference for miRNA-disease association prediction. Oncotarget **7**, 65257–65269 (2016)
9. Kanehisa, M., Goto, S.: KEGG: kyoto encyclopedia of genes and genomes. Nucleic Acids Res. **28**, 27–30 (2000)
10. Wishart, D.S., et al.: DrugBank 5.0: a major update to the DrugBank database for 2018. Nucleic Acids Res. **46**, D1074-D1082 (2017)
11. Günther, S., et al.: SuperTarget and Matador: resources for exploring drug-target relationships. Nucleic Acids Res. **36**, D919-D922 (2007)
12. Willett, P., Barnard, J.M., Downs, G.M.: Chemical Similarity Searching (1998)
13. Rognan, D.J.M.I.: Structure-based approaches to target fishing and ligand profiling. Molecular Informat **29**, 176–187 (2010)
14. Chen, Y.Z., Zhi, D.G.: Ligand–protein inverse docking and its potential use in the computer search of protein targets of a small molecule. Proteins Struct. Function Bioinformat. **43**, 217–226 (2001)
15. Li, H., et al.: TarFisDock: a web server for identifying drug targets with docking approach. Nucle Acids Res. **34**, 219–224 (2006)
16. Huang, D.-S., Zhang, L., Han, K., Deng, S., Yang, K., Zhang, H.: Prediction of protein-protein interactions based on protein-protein correlation using least squares regression. Curr. Protein Pept. Sci. **15**, 553–560 (2014)
17. Wang, J.C., Chu, P.Y., Chen, C.M., Lin, J.H.: idTarget: a web server for identifying protein targets of small chemical molecules with robust scoring functions and a divide-and-conquer docking approach. Nucleic Acids Research **40**(1), W393-W399 (2012)
18. Wang, L., You, Z.O., Li, L.I., Yan, X., Song, C.O.: Identification of potential drug–targets by combining evolutionary information extracted from frequency profiles and molecular topological structures. Chemical Biology & Drug Design (2019)
19. Yi, H.-C., You, Z.-H., Huang, D.-S., Li, X., Jiang, T.-H., Li, L.-P.: A deep learning framework for robust and accurate prediction of ncRNA-protein interactions using evolutionary information. Molecular Therapy-Nucleic Acids **11**, 337–344 (2018)
20. Guo, Z.H., Yi, H.C., You, Z.H.: Construction and Comprehensive Analysis of a Molecular Association Network via lncRNA–miRNA –Disease–Drug–Protein Graph (2019)
21. Yi, H., You, Z., Guo, Z., Huang, D., Chan, K.C.C.: Learning representation of molecules in association network for predicting intermolecular associations. In: IEEE/ACM Transactions on Computational Biology and Bioinformatics, p. 1 (2020)
22. Jiang, H.J., Huang, Y.A., You, Z.H.: SAEROF: an ensemble approach for large-scale drug-disease association prediction by incorporating rotation forest and sparse autoencoder deep neural network. Sci. Rep. **10**, 4972 (2020)

23. You, Z.-H., Zhou, M., Luo, X., Li, S.: Highly efficient framework for predicting interactions between proteins. IEEE Trans. Cybern. **47**, 731–743 (2016)
24. You, Z.-H., Yin, Z., Han, K., Huang, D.-S., Zhou, X.: A semi-supervised learning approach to predict synthetic genetic interactions by combining functional and topological properties of functional gene network. BMC Bioinformatics **11**, 343 (2010)
25. You, Z.-H., Lei, Y.-K., Gui, J., Huang, D.-S., Zhou, X.: Using manifold embedding for assessing and predicting protein interactions from high-throughput experimental data. Bioinformatics **26**, 2744–2751 (2010)
26. Cheng, L., et al.: LncRNA2Target v2. 0: a comprehensive database for target genes of lncRNAs in human and mouse. Nucleic Acids Res. **47**, D140–D144 (2018)
27. Chou, C.-H., et al.: miRTarBase update 2018: a resource for experimentally validated microRNA-target interactions. Nucleic Acids Res. **46**, D296–D302 (2017)
28. Dweep, H., Gretz, N.: miRWalk2. 0: a comprehensive atlas of microRNA-target interactions. Nature Methods **12**, 697 (2015)
29. Huang, Z., et al.: HMDD v3. 0: a database for experimentally supported human microRNA–disease associations. Nucleic Acids Res. **47**, D1013–D1017 (2018)
30. Weininger, D.: SMILES, a chemical language and information system. 1. Introduction to methodology and encoding rules. J. Chem. Inf. Comput. Sci. **28**, 31–36 (1988)
31. Wang, R., Li, S., Wong, M.H., Leung, K.S.: Drug-protein-disease association prediction and drug repositioning based on tensor decomposition. In: 2018 IEEE International Conference on Bioinformatics and Biomedicine (BIBM), pp. 305–312. IEEE (2018)
32. Guo, Z.-H., Yi, H.-C., You, Z.-H.: Construction and comprehensive analysis of a molecular association network via lncRNA–miRNA–Disease–Drug–Protein graph. Cells **8**, 866 (2019)
33. Wang, Y.-B., You, Z.-H., Li, L.-P., Huang, Y.-A., Yi, H.-C.: Detection of interactions between proteins by using legendre moments descriptor to extract discriminatory information embedded in pssm. Molecules **22**, 1366 (2017)
34. An, J.-Y., et al.: Identification of self-interacting proteins by exploring evolutionary information embedded in PSI-BLAST-constructed position specific scoring matrix. Oncotarget **7**, 82440 (2016)
35. An, J.Y., You, Z.H., Chen, X., Huang, D.S., Yan, G., Wang, D.F.: Robust and accurate prediction of protein self-interactions from amino acids sequence using evolutionary information. Mol. BioSyst. **12**, 3702 (2016)
36. Chan, K.C., You, Z.-H.: Large-scale prediction of drug-target interactions from deep representations. In: 2016 International Joint Conference on Neural Networks (IJCNN), pp. 1236–1243. IEEE (2016)
37. Szklarczyk, D., et al.: The STRING database in 2017: quality-controlled protein–protein association networks, made broadly accessible. Nucleic Acids Res. gkw937 (2016)
38. Shen, J., et al.: Predicting protein–protein interactions based only on sequences information. Proc. Natl. Acad. Sci. **104**, 4337–4341 (2007)
39. Wang, D., Cui, P., Zhu, W.: Structural deep network embedding. In: Proceedings of the 22nd ACM SIGKDD international conference on Knowledge discovery and data mining, pp. 1225–1234. ACM (2016)

Predicting LncRNA-miRNA Interactions via Network Embedding with Integrated Structure and Attribute Information

Bo-Wei Zhao[1,2], Ping Zhang[3], Zhu-Hong You[1,2(✉)], Ji-Ren Zhou[1], and Xiao Li[1]

[1] The Xinjiang Technical Institute of Physics and Chemistry, Chinese Academy of Sciences, Urumqi 830011, China
zhuhongyou@ms.xjb.ac.cn

[2] University of Chinese Academy of Sciences, Beijing 100049, China

[3] The School of Computer Sciences, BaoJi University of Arts and Sciences, Baoji 721016, China

Abstract. Accumulating evidence demonstrated that microRNAs (miRNAs) and long noncoding RNAs (lncRNAs) are related with some complex human diseases. LncRNA-miRNA interactions (LMIs) play an important role in regulatory of gene networks. However, the biological experiments for detecting lncRNA-miRNA interactions are often expensive and time-consuming. Thus, it is urgent to develop computational method for predicting LMIs. In this paper, we propose a novel computational approach LMMAN to predict potential lncRNA-miRNA associations based on molecular associations network (MAN). More specifically, the known relationships among miRNA, lncRNA, protein, drug and disease are firstly integrated to construct a molecular association network. Then, a network embedding model LINE is employed to extract network behavior features of lncRNA and miRNA nodes. Finally, the random forest classifier is used to predict the potential lncRNA-miRNA interactions. In order to evaluate the performance of the proposed LMMAN approach, five-fold cross-validation tests are implemented on benchmark dataset lncRNASNP2. The proposed LMMAN approach can achieve the high AUC of 0.9644, which is obviously better than the existing methods. The promising results reveal that LMMAN can effectively predict new lncRNA-miRNA interactions and can be a good complement to relevant biomedical fields in the future.

Keywords: Molecular associations · LMMAN · lncRNA · miRNA · LINE

1 Introduction

MicroRNAs (miRNAs) are endogenous small, non-coding, single-stranded and short RNA sequences [1, 2]. More than ten years have passed since the discovery of miRNA. In recent years, researchers have made more and more in-depth studies on miRNA and made progress in many aspects of miRNA, such as biological initiation, function and target gene regulation [3–18]. The past years have evidence shown that lncRNA has been as a participator in various cellular processes, in which cell differentiation, cell

D.-S. Huang and K.-H. Jo (Eds.): ICIC 2020, LNCS 12464, pp. 493–501, 2020.
https://doi.org/10.1007/978-3-030-60802-6_43

growth and death [19–22]. Thus, the total number of lncRNA continues to rise, the number of human lncRNA go beyond the number of protein-coding genes [23, 24]. However, in this case the functions of most lncRNAs are unknown and many lncRNAs may not have obvious functions, the functional roles and mechanism of lncRNA defined in some classical literatures are very clear. For example, XIST, HOTAIR and TERC [24].

Previous studies have shown that the calculation and prediction of miRNA target sites are extensive networks of interaction between miRNA and lncRNA [25–32]. Saakshi Jalali *et al.* have proposed that miRNA interactions with lncRNAs play an important role in regulation mediated by miRNAs, and the miRNA-lncRNA interactions have been verified through zebrafish model experiments, but their view is only on a single data set and the results are too limited [33].

Yoon *et al.* believed that the interaction between miRNA and lncRNA was robust and dynamically controlled in protein expression [34]. Dario Veneziano *et al.* proposed that lncRNA is associated with miRNA in some biological processes in humans, which proves that lncRNA-miRNA interactions are involved in a variety of physiological and pathological processes, especially cancer [35]. In previous, provided the model or the network structure is unitary, usually two molecules, or find a molecule as the medium, in the prediction of correlation between the two molecules, which INLMI network is proposed by Hu *et al.* [36]. Chen *et al.* proposed HGLDA and LFSCM that a novel model is based on semantic similarity of diseases, they are still studying the network structure formed between lncRNA-diseases and miRNA-diseases, and have not achieved a comprehensive study [37]. Although computer-related techniques have been used for a long time to predict biological molecular interactions, there are few machine learning-based lncRNA-miRNA interaction prediction methods [38, 39]. Huang *et al.* proposed a novel bidirectional diffusion model called EPLML, which uses a two-part diagram of known lncRNA-miRNA interactions to predict unknown lncRNA-miRNA interactions [23].

The miRNA-target reasoning tool mentioned by the previous researchers was not comprehensive, and the latest research results of miRNA-lncRNA were not included, so the interaction of miRNA-lncRNA could not be predicted comprehensively [29]. Up to now, a novel network architecture has been established. Guo *et al.* have proposed a network model called MAN [40]. In their experiments, they have proved that among various attributes and associations, there is a great advantage for studying the interaction between biological molecules, which can make the prediction results have good robustness and accuracy. In this study, LMEMAN makes full use of established biomolecular relationship networks, such as miRNA-lncRNA, miRNA-disease, miRNA-protein, lncRNA-disease, lncRNA-protein, protein-disease, drug-protein, drug-disease, protein-protein, and lncRNA sequences and miRNA sequences (themselves attributes) to predict the interaction between lncRNA-miRNA. Experimental results showed that LMMAN was able to predict lncRNA-miRNA interactions more accurately than other advanced methods, helping to discover that did not exist in our data set without any known interactions of lncRNA-miRNA. In addition, the performance of the proposed method is evaluated, and the actual data is tested by using the data set of the latest database [lncRNASNP2] [41].

2 Materials

In order to improve the efficiency of the experiment, we used the lncRNASNP2 database in January 2018 version, this version can be downloaded from http://bioinfo.life.hust.edu.cn/lncRNASNP2 [41]. In addition, in order to validate LMMAN, we specially looking for 9 kinds of database, in which relational database of miRNA-diseases used the HMDD V3.0 (http://www.cuilab.cn/hmdd/), and compared with the old version increased 2-fold more entries and the classification of these associations is more accurate [42]. miRTarBase (http://miRTarBase.mbc.nctu.edu.tw/) is used for the miRNA-protein association database, which contains 4,076 miRNAs and 23,054 gene proteins [43]. After screening, the number of associations of miRNAs and proteins were 4,944. We also found the same lncRNA and diseases database that uses LncRNADisease (http://cmbi.bjmu.edu.cn/lncRNAdisease/) [44] and lncRNASNP2 [41], LncRNA2Target (http://123.59.132.21/lncRNA2target/) [45] for lncRNA and protein database, DisGeNET (http://www.disgenet.org/) [46] for protein and disease, DrugBank5.0 (http://www.drugbank.ca/) [47] for drug and protein, CTD (http://ctdbase.org/) [48] for drug-disease, and STRING (http://string-db.org/) [49] for protein and protein. In this experiment, these data were filtered and the associated pairs were left. After the above database is aggregated, classify different nodes and the final statistical data are obtained respectively, as shown in Table 1.

Table 1. The number of 5 types of nodes in LMMAN.

Node	Number
Protein	1,649
Disease	2,062
Drug	1,025
miRNA	1,023
lncRNA	769
Total	6,528

3 Methods

3.1 miRNA and lncRNA Sequence

In this experiment, miRNA and lncRNA Sequences are extracted from the miRbase [50] and NONCODE [51], respectively. To represent node properties. Firstly, for miRNA and lncRNA, the obtained RNA sequences are k-mer treated, which 3-mer is used here (i.e., the sequences are divided into strings with 3 bases). For instance, “AACTGACTGA” can be divided into “AAC, ACT, CTG, TGA, GAC, ACT, CTG, TGA”. Secondly, considering the characteristics of attributes information, we utilize a 64 ($4 \times 4 \times 4$) dimensional vector to represent the attributes of each node in LMMAN.

3.2 Representation of Node Relationships

LINE [52] is an approach based on neighborhood similarity assumptions, except that, unlike DeepWalk [53], which uses DFS to construct a neighborhood, LINE can be viewed as an algorithm that uses BFS to construct a neighborhood. In addition, LINE can be used on weighted diagrams, but DeepWalk can only be used on unauthorized diagrams.

First-order similarity is used to describe the local similarity between pairs of vertices in the graph. It is formally described as follows: if there are direct connected edges between u and v, then edge weight w_{uv} is the similarity of two vertices, if there are no direct connected edges, first-order similarity is 0. In this paper, second-order similarity is used to describe the relationship between each node, which is formally defined as: $p_u = \left(w_{u,1}, \cdots, w_{u,|V|}\right)$ represent first-order similarity between node u and all other nodes, then second-order similarity of u and v can be expressed by similarity of p_u and p_v. If there is no identical neighbor node between u and v, the degree of similarity is 0.

In this paper, two embedding vectors are maintained for each vertex: one is the vector of the vertex itself; the other is the vector to be represented when the point is the context vertex of other vertices. For directed edges (i,j), the probability of generating context (neighbor) vertex v_j under the given vertex v_i is:

$$p_2\left(v_j|v_i\right) = \frac{\exp\left(\vec{u}_j'^T \cdot \vec{u}_i\right)}{\sum_{k=1}^{|V|} \exp\left(\vec{u}_k'^T \cdot \vec{u}_i\right)} \quad (1)$$

in which $|V|$ is context(neighbor) or the number of vertices. In order to retain the *second-order* similarity, we should use the context conditional distribution $p_2(\cdot|v_i)$ characterized by low dimensions to be close to the empirical distribution $\hat{p}_2(\cdot|v_i)$. Therefore, the optimization objective is defined as:

$$O_2 = \sum\nolimits_{i \in V} \lambda_i d(\hat{p}_2(\cdot|v_i), p_2(\cdot|v_i)) \quad (2)$$

in which λ_i is the factor controlling the importance of nodes, which can be estimated by vertex degree or PageRank and other methods. The empirical distribution is defined as:

$$\hat{p}_2(v_j|v_i) = \frac{w_{ij}}{\mathrm{d}_i} \quad (3)$$

in which w_{ij} is the edge weight of the edge (i,j), and the degree of the d_i vertex v_i. For the weighted graph:

$$\mathrm{d}_i = \sum\nolimits_{k \in N(I)} w_{ik} \quad (4)$$

uses KL divergence and sets λ_i = d_i, ignoring the constant, there is:

$$O_2 = -\sum\nolimits_{(i,j)\in E} w_{ij}\, log\, p_2(v_j|v_i) \quad (5)$$

3.3 Full Feature Construction for Each Node

In the first two sections, we showed how the properties of each node can be represented as vectors and how the relationships between nodes can be transformed into representation vectors. LMMAN have been proposed for computing inference potential lncRNA-miRNA interaction based on similar biological molecules network by network representation, each RNA molecule expressed by a 3-mer that is 64 (4 * 4 * 4) dimensional vector and each corresponding to the dimensions of the vector said in a sequence of 3-mer [54]. Meanwhile, in the network, the interaction between biomolecules will be expressed in terms of 64 (4 * 4 * 4) dimensions. We obtain vectors through certain methods for effective combination, forming a vector set of 128 dimensions for each node, and then splicing with known node correlation relations to form a vector set of 256 dimensions. Finally, machine learning method is used for training and prediction.

3.4 Random Forest Algorithm

Random forest (RF) is an integrated algorithm based on decision trees [55]. It is a machine learning method that combines several independent decision trees to obtain the final prediction results by voting or taking the mean value, which is usually more accurate and stable than a single tree.

This assume the dataset as follows:

$$D = \{(a_1, n_1), (a_2, n_1), L, (a_u, n_u)\} \quad (6)$$

Where, a is the sample feature and n is the sample. Besides, there is $i = 1, 2, \cdots, I$. I is the number of subsets of samples. First, m samples were randomly selected from the original data set to obtain the training subset ($i \in m$). D_i is put into CART decision tree for training. In the training process, the segmentation rules of each node randomly select k features from all the features. Then, select an optimal feature from the k features as the segmentation point to delimit the molecular tree. Finally, the class that voted the most was used as the predicted category.

4 Results and Discussion

4.1 Performance Evaluation for LMMAN

In this experiment, 5-fold cross-validation in cross-validation is adopted. Firstly, data are divided into five subsets of the same size. Secondly, choosing four subsets as training set, and the remaining a subset for the test set. Finally, this process is repeated for five times, so that each data can be tested exactly once and the total error is the sum of all 5 runs. The 5-fold cross-validation results performed by our method are detailed in Table 2.

Table 2. 5-Fold cross-validation results performed by LMMAN.

Fold	Acc. (%)	Sen. (%)	Spec. (%)	Prec. (%)	MCC (%)	AUC (%)
0	90.87	89.19	92.54	92.28	81.78	96.32
1	89.76	86.87	92.66	92.21	79.66	95.85
2	91.07	89.01	93.13	92.84	82.22	96.99
3	90.30	88.42	91.18	91.87	80.65	96.27
4	90.98	88.95	93.01	92.71	82.03	96.76
Average	90.60	88.49	92.70	92.38	81.27	96.44

4.2 Comparison of Different Feature Extraction Methods

For the LMMAN model, different attributes of lncRNA and miRNA are used to compare. In the experimental design, there were three categories. The first method is extracted with pure attribute information as the node feature. The second is extracted with behavior information via LINE network embedding. The third is based on attribute information combined behavior information of global representation, in which the correlation of lncRNA-miRNA and the remaining 8 relationships are processed by LINE to generate vector sets. The 5-fold cross-validation details are offered in Table 3. The results clearly show that after adding the behavior information, the prediction effect of the model is much better than any single feature extraction method.

Table 3. Comparison of different features.

Feature	Acc. (%)	Sen. (%)	Spec. (%)	Prec. (%)	MCC (%)	AUC (%)
Attribute	85.99 ± 0.4	84.76 ± 1.02	87.22 ± 1.15	86.91 ± 0.93	72.02 ± 0.82	92.98 ± 0.34
Behavior	90.13 ± 0.66	87.80 ± 0.99	92.45 ± 0.63	92.09 ± 0.64	80.35 ± 1.31	95.96 ± 0.54
Both	90.60 ± 0.56	88.49 ± 0.95	92.70 ± 0.38	92.38 ± 0.39	81.27 ± 1.09	96.44 ± 0.45

5 Conclusions

The lncRNA-miRNA interactions (LMIs) play an important role in regulatory of gene networks. In this paper, we proposed a novel computational method LMMAN for predicting lncRNA-miRNA interactions based on multi-molecular association network. To evaluate the prediction performance of the proposed LMMAN model, five-fold cross-validation tests are performed on benchmark dataset. The promising results illustrated the effectiveness of our method. We believe that the proposed approach can be a good complement for relevant biomedical research.

References

1. Zhou, S., Xuan, Z., Wang, L., Ping, P., Pei, T.: A novel model for predicting associations between diseases and LncRNA-miRNA pairs based on a newly constructed bipartite network. Comput. Math. Methods Med. **2018**, 6789089 (2018)
2. Holley, C.L., Topkara, V.K.: An introduction to small non-coding RNAs: miRNA and snoRNA. Cardiovasc. Drugs Ther. **25**, 151–159 (2011)
3. Lin, S.L., Miller, J.D., Ying, S.Y.: Intronic microRNA (miRNA). J. Biomed. Biotechnol. **2006**, 26818 (2006)
4. Chen, X., Gong, Y., Zhang, D.H., You, Z.H., Li, Z.W.: DRMDA: deep representations-based miRNA–disease association prediction. J. Cell Mol. Med. **22**, 472–485 (2018)
5. Chen, X., et al.: A novel computational model based on super-disease and miRNA for potential miRNA–disease association prediction. Mol. BioSyst. **13**, 1202–1212 (2017)
6. Chen, X., Xie, D., Wang, L., Zhao, Q., You, Z.-H., Liu, H.: BNPMDA: bipartite network projection for MiRNA–disease association prediction. Bioinformatics **34**, 3178–3186 (2018)
7. Chen, X., et al.: WBSMDA: within and between score for MiRNA-disease association prediction. Sci. Rep. **6**, 21106 (2016)
8. Chen, X., Yan, C.C., Zhang, X., You, Z.-H., Huang, Y.-A., Yan, G.-Y.: HGIMDA: heterogeneous graph inference for miRNA-disease association prediction. Oncotarget **7**, 65257 (2016)
9. Chen, X., Zhang, D.-H., You, Z.-H.: A heterogeneous label propagation approach to explore the potential associations between miRNA and disease. J. Transl. Med. **16**, 348 (2018)
10. Huang, Y.-A., Hu, P., Chan, K.C., You, Z.-H.: Graph convolution for predicting associations between miRNA and drug resistance. Bioinformatics **36**, 851–858 (2020)
11. Bouba, I., Kang, Q., Luan, Y.S., Meng, J.: Predicting miRNA-lncRNA interactions and recognizing their regulatory roles in stress response of plants. Math. Biosci. **312**, 67–76 (2019)
12. Li, J.-Q., Rong, Z.-H., Chen, X., Yan, G.-Y., You, Z.-H.: MCMDA: matrix completion for MiRNA-disease association prediction. Oncotarget **8**, 21187 (2017)
13. Qu, J., et al.: In Silico prediction of small molecule-miRNA associations based on the HeteSim algorithm. Mol. Ther.-Nucleic Acids **14**, 274–286 (2019)
14. Sun, Y., Zhu, Z., You, Z.-H., Zeng, Z., Huang, Z.-A., Huang, Y.-A.: FMSM: a novel computational model for predicting potential miRNA biomarkers for various human diseases. BMC Syst. Biol. **12**, 121 (2018)
15. Wang, L., et al.: LMTRDA: using logistic model tree to predict MiRNA-disease associations by fusing multi-source information of sequences and similarities. PLoS Comput. Biol. **15**, e1006865 (2019)
16. You, Z.-H., et al.: PBMDA: a novel and effective path-based computational model for miRNA-disease association prediction. PLoS Comput. Biol. **13**, e1005455 (2017)
17. You, Z.-H., et al.: PRMDA: personalized recommendation-based MiRNA-disease association prediction. Oncotarget **8**, 85568 (2017)
18. Zheng, K., You, Z.-H., Wang, L., Zhou, Y., Li, L.-P., Li, Z.-W.: Dbmda: a unified embedding for sequence-based mirna similarity measure with applications to predict and validate mirna-disease associations. Mol. Ther.-Nucleic Acids **19**, 602–611 (2020)
19. Chen, X., Yan, C.C., Zhang, X., You, Z.-H.: Long non-coding RNAs and complex diseases: from experimental results to computational models. Brief. Bioinf. **18**, 558–576 (2017)
20. Chen, X., You, Z.-H., Yan, G.-Y., Gong, D.-W.: IRWRLDA: improved random walk with restart for lncRNA-disease association prediction. Oncotarget **7**, 57919 (2016)

21. Guo, Z.-H., You, Z.-H., Wang, Y.-B., Yi, H.-C., Chen, Z.-H.: A learning-based method for lncRNA-disease association identification combing similarity information and rotation forest. iScience **19**, 786–795 (2019)
22. Yi, H.-C., et al.: Learning distributed representations of RNA and protein sequences and its application for predicting lncRNA-protein interactions. Comput. Struct. Biotechnol. J. **18**, 20–26 (2020)
23. Huang, Y.-A., Chan, K.C., You, Z.-H.: Constructing prediction models from expression profiles for large scale lncRNA–miRNA interaction profiling. Bioinformatics **34**, 812–819 (2018)
24. Quinn, J.J., Chang, H.Y.: Unique features of long non-coding RNA biogenesis and function. Nat. Rev. Genet. **17**, 47–62 (2016)
25. Jeggari, A., Marks, D.S., Larsson, E.: miRcode: a map of putative microRNA target sites in the long non-coding transcriptome. Bioinformatics **28**, 2062–2063 (2012)
26. Chen, X., Huang, Y.-A., Wang, X.-S., You, Z.-H., Chan, K.C.: FMLNCSIM: fuzzy measure-based lncRNA functional similarity calculation model. Oncotarget **7**, 45948 (2016)
27. Hu, P., Huang, Y.-A., Chan, K.C., You, Z.-H.: Learning multimodal networks from heterogeneous data for prediction of lncRNA-miRNA interactions. IEEE/ACM Trans. Comput. Biol. Bioinf. (2019)
28. Huang, Y.-A., Chen, X., You, Z.-H., Huang, D.-S., Chan, K.C.: ILNCSIM: improved lncRNA functional similarity calculation model. Oncotarget **7**, 25902 (2016)
29. Huang, Z.A., Huang, Y.A., You, Z.H., Zhu, Z., Sun, Y.: Novel link prediction for large-scale miRNA-lncRNA interaction network in a bipartite graph. BMC Med. Genom. **11**, 113 (2018)
30. Huang, Z.-A., et al.: Predicting lncRNA-miRNA interaction via graph convolution auto-encoder. Front. Genet. **10**, 758 (2019)
31. Wang, M.-N., You, Z.-H., Li, L.-P., Wong, L., Chen, Z.-H., Gan, C.-Z.: GNMFLMI: graph regularized nonnegative matrix factorization for predicting LncRNA-MiRNA interactions. IEEE Access **8**, 37578–37588 (2020)
32. Wong, L., Huang, Y.A., You, Z.H., Chen, Z.H., Cao, M.Y.: LNRLMI: linear neighbour representation for predicting lncRNA-miRNA interactions. J. Cell. Mol. Med. **24**, 79–87 (2020)
33. Jalali, S., Bhartiya, D., Lalwani, M.K., Sivasubbu, S., Scaria, V.: Systematic transcriptome wide analysis of lncRNA-miRNA interactions. PLoS One **8**, e53823 (2013)
34. Yoon, J.-H., Abdelmohsen, K., Gorospe, M.: Functional interactions among microRNAs and long noncoding RNAs. Seminar. Cell Dev. Biol. **34**, 9–14 (2014)
35. Veneziano, D., Marceca, G.P., Di Bella, S., Nigita, G., Distefano, R., Croce, C.M.: Investigating miRNA-lncRNA interactions: computational tools and resources. Methods Mol. Biol. **1970**, 251–277 (2019)
36. Hu, P., Huang, Y.-A., Chan, K.C.C., You, Z.-H.: Discovering an integrated network in heterogeneous data for predicting lncRNA-miRNA Interactions. In: Huang, D.-S., Bevilacqua, V., Premaratne, P., Gupta, P. (eds.) ICIC 2018. LNCS, vol. 10954, pp. 539–545. Springer, Cham (2018). https://doi.org/10.1007/978-3-319-95930-6_51
37. Chen, X.: Predicting lncRNA-disease associations and constructing lncRNA functional similarity network based on the information of miRNA. Sci. Rep. **5**, 13186 (2015)
38. Zhang, W., Tang, G., Wang, S., Chen, Y., Zhou, S., Li, X.: Sequence-derived linear neighborhood propagation method for predicting lncRNA-miRNA interactions. In: 2018 IEEE International Conference on Bioinformatics and Biomedicine (BIBM), pp. 50–55. IEEE (2018)

39. Zhang, W., Qu, Q., Zhang, Y., Wang, W.: The linear neighborhood propagation method for predicting long non-coding RNA–protein interactions. Neurocomputing **273**, 526–534 (2018)
40. Guo, Z.-H., Yi, H.-C., You, Z.-H.: Construction and comprehensive analysis of a molecular association network via lncRNA–miRNA–disease–drug–protein graph. Cells **8**, 866 (2019)
41. Miao, Y.R., Liu, W., Zhang, Q., Guo, A.Y.: lncRNASNP2: an updated database of functional SNPs and mutations in human and mouse lncRNAs. Nucleic Acids Res. **46**, D276–D280 (2018)
42. Huang, Z., et al.: HMDD v3.0: a database for experimentally supported human microRNA-disease associations. Nucleic Acids Res. **47**, D1013–D1017 (2019)
43. Chou, C.H., et al.: miRTarBase update 2018: a resource for experimentally validated microRNA-target interactions. Nucleic Acids Res. **46**, D296–D302 (2018)
44. Chen, G., et al.: LncRNADisease: a database for long-non-coding RNA-associated diseases. Nucleic Acids Res. **41**, D983–D986 (2012)
45. Jiang, Q., et al.: LncRNA2Target: a database for differentially expressed genes after lncRNA knockdown or overexpression. Nucleic Acids Res. **43**, D193–D196 (2014)
46. Piñero, J., et al.: DisGeNET: a comprehensive platform integrating information on human disease-associated genes and variants. Nucleic Acids Res. gkw943 (2016)
47. Wishart, D.S., et al.: DrugBank 5.0: a major update to the DrugBank database for 2018. Nucleic Acids Res. **46**, D1074–D1082 (2017)
48. Davis, A.P., et al.: The comparative toxicogenomics database: update 2019. Nucleic Acids Res. **47**, D948–D954 (2018)
49. Szklarczyk, D., et al.: The STRING database in 2017: quality-controlled protein–protein association networks, made broadly accessible. Nucleic Acids Res. gkw937 (2016)
50. Kozomara, A., Birgaoanu, M., Griffiths-Jones, S.: miRBase: from microRNA sequences to function. Nucleic Acids Res. **47**, D155–D162 (2018)
51. Fang, S., et al.: NONCODEV5: a comprehensive annotation database for long non-coding RNAs. Nucleic Acids Res. **46**, D308–D314 (2017)
52. Tang, J., Qu, M., Wang, M., Zhang, M., Yan, J., Mei, Q.: LINE: large-scale information network embedding. In: Proceedings of the 24th International Conference on World Wide Web, pp. 1067-1077. International World Wide Web Conferences Steering Committee (2015)
53. Perozzi, B., Al-Rfou, R., Skiena, S.: Deepwalk: Online learning of social representations. In: Proceedings of the 20th ACM SIGKDD International Conference on Knowledge Discovery and Data Mining, pp. 701-710. ACM (2014)
54. Kirk, J.M., et al.: Functional classification of long non-coding RNAs by k-mer content. Nat. Genet. **50**, 1474 (2018)
55. Amaratunga, D., Cabrera, J., Lee, Y.-S.: Enriched random forests. Bioinformatics **24**, 2010–2014 (2008)

Machine Learning Techniques in Bioinformatics

A Novel Computational Method for Predicting LncRNA-Disease Associations from Heterogeneous Information Network with SDNE Embedding Model

Ping Zhang[1], Bo-Wei Zhao[2,3], Leon Wong[2,3], Zhu-Hong You[2,3](✉), Zhen-Hao Guo[2,3], and Hai-Cheng Yi[2,3]

[1] The School of Computer, BaoJi University of Arts and Sciences, Baoji 721016, China
jcjyxy@163.com

[2] The Xinjiang Technical Institute of Physics and Chemistry, Chinese Academy of Sciences, Urumqi 830011, China
zhuhongyou@ms.xjb.ac.cn

[3] University of Chinese Academy of Sciences, Beijing 100049, China

Abstract. Recent studies have shown that lncRNAs play a critical role in numerous complex human diseases. Thus, identification of lncRNA and diseases associations can help us to understand disease pathogenesis at the molecular level and develop disease diagnostic biomarkers. In this paper, a novel computational method LDAMAN is proposed to predict potential lncRNA-disease interactions from heterogeneous information network with SDNE embedding model. Specifically, known associations among lncRNA, disease, microRNA, circular RNA, mRNA, protein, drug and microbe are integrated to construct a molecular association network and a network embedding model SDNE is employed to extract network behavior features of lncRNA and disease nodes. Finally, the XGBoost classifier is used for predicting potential lncRNA-disease associations. In the experiment, the proposed method obtained stable AUC of 92.58% using 5-fold cross validation. In summary, the experimental results demonstrate our method provides a systematic landscape and computational prediction tool for lncRNA-disease association prediction.

Keywords: LncRNA-disease associations · Disease · Network embedding · Heterogeneous information network · SDNE

1 Introduction

In recent years, large numbers of evidences have confirmed that non-coding RNAs (ncRNAs), especially lncRNAs as a kind of important heterologous ncRNAs, play significant roles in multiple biological processes and are associated with complex diseases [1–3]. However, it is worth pointing out that, experimentally verified lncRNA-disease associations are still comparatively limited resulting the collected known association data are far from enough. Therefore, it is imperative to construct effective

D.-S. Huang and K.-H. Jo (Eds.): ICIC 2020, LNCS 12464, pp. 505–513, 2020.
https://doi.org/10.1007/978-3-030-60802-6_44

and accurate computational model to predict potential lncRNA-disease associations using computational methods.

In the past few years, to predict lncRNA-disease associations, a variety of computational models have been proposed for further experimental validation [4–7]. Chen *et al.* propose an improved model called IRWRLDA based on random walk, with the initial probability vector of known disease-lncRNA associations, lncRNA expression similarity, and lncRNAs by combining disease semantic similarity [8]. Yu *et al.* adopt the naive Bayesian classifier and present a computational model called NBCLDA to explore potential lncRNA-diseases relationships [9]. Ou-Yang *et al.* utilize the estimated representations of lncRNAs and diseases to predictive lncRNA-disease association via a two-side sparse self-representation algorithm termed TSSR [10]. In contrast to the above models based on machine learning, the network-based approach is now regarded as a perfect method for predicting lncRNA-disease associations owing to its comprehensive view as well as precision of predictive performance [11–15]. For instance, Yang *et al.* make direct use of the propagation algorithm as Random Walker on Protein Complex Network (RWPCN) to construct bipartite network for predicting and prioritizing disease genes ranking the potential candidates for all diseases [16]. When detecting potential links, Sun *et al.* developed a network called RWRlncD. RWRlncD network is derived from the applying to lncRNA-lncRNA functional similarity network and random walks restarting [17]. Afterwards, to identify potential lncRNA-disease associations, Zhou *et al.* propose a rank-based model (RWRHLD) in which the known lncRNA- diseases association network, lncRNA-lncRNA crosstalk network, disease-disease similarity network and miRNA-associated are combined into a heterogeneous network with a random walk with restart [18]. Ding *et al.* propose a TPGLDA model, which originates from the research of tripartite graph. It is proposed as a systematic approach to predictive lncRNA-disease associations by integrating gene-disease associations with lncRNA-disease associations based on lncRNA-disease-gene tripartite graph [19]. To enhance predictive effect, Mori *et al.* add biological sequence information to a disease-target-ncRNA tripartite network to predictive association between ncRNAs and diseases [20]. In term of bipartite network, a model to infer potential lncRNA-disease associations by constructing a bipartite network followed a power-law distribution is proposed by Ping *et al.* [21]. To identify lncRNA-disease associations, Sumathipala *et al.* also build a multi-level complex network (tripartite network) called LION [22].

In July 2019, Yi and Guo *et al.* formally proposed a novel MAN (Molecular Associations Network) [23–30] model by integrating the associations among miRNA, lncRNA, protein, drug, and disease [31–43], in which any kind of potential associations can be predicted. In this paper, a novel computational method LDAMAN is proposed to predict potential lncRNA-disease interactions from heterogeneous information network with SDNE embedding model. In order to estimate the prediction performance of LDAMAN, the 5-fold cross validation was implemented, and simulation results show that the area under the curve (AUC) reached 92.58% by using a XGBoost classifier based on known molecular association datasets.

2 Materials and Methods

2.1 LDAMAN

According to available datasets, the complex molecular associations network of biomolecules for LDAMAN based on MAN (Molecular Associations Network) consists of two parts: nodes and edges. Determining the edges between any two nodes in the whole complex associations network helps us to have a deep and comprehensive understanding of associations between lncRNA and disease. LDAMAN includes the following crucial two parts:

Nodes (molecular): ncRNA (miRNA, lncRNA), protein (target), drug and disease.

Edges (association): lncRNA-disease association, miRNA-disease association, miRNA-lncRNA association, protein-protein interaction, lncRNA-protein interactions, drug-target interaction, protein-disease association, miRNA-protein associations and drug-disease associations.

2.2 LDAMAN Datasets

In order to systematically and holistically establish the LDAMAN, known molecular associations. these data are together to construct a LDAMAN network. In current condition, we consider five object types: lncRNAs, diseases, miRNAs, proteins and drugs. After performing the inclusion of identifier unification, de-redundancy, simplification and deletion of the irrelevant items, we collected nine relational associations between these objects from public databases. The details of the final LDAMAN objects data obtained are shown in the following Table 1.

Table 1. The databases of nine kinds of associations in the LDAMAN

Relationship type	Database
lncRNA-disease	LncRNADisease [44]
	lncRNASNP2 [45]
miRNA-lncRNA	lncRNASNP2 [45]
lncRNA-protein	LncRNA2Target [46]
miRNA-disease	HMDD [47]
protein-disease	DisGeNET [48]
Drug-disease	CTD [49]
miRNA-protein	miRTarBase [50]
Drug-protein	DrugBank [51]
Protein-protein	STRING [52]

2.3 Collection of the Human LncRNA-Disease Associations

To estimate the prediction performance of our newly proposed method, known lncRNA-disease associations data were gathered from the lncRNASNP2 and LncRNADisease database. After processing the data that describe the same lncRNA-disease relationships based on evidence from different experiments, we obtained 1264 independent lncRNA-disease association pairs including 345 different lncRNAs and 295 different diseases.

2.4 LncRNA and Protein Sequence Representation

In LDAMAN, The sequences of lncRNA, miRNA, and protein were downloaded from miRbase [53], NONCODE [54], and STRING [52] databases. In order to represent the attribute of the nodes in LDAMAN, a 64 ($4 \times 4 \times 4$) dimensional vector was used to encode ncRNA sequences. In the RNA sequence, each attribute in nodes represented the normalized frequency of the corresponding K-mer (k = 3) appearing. According to the method from Shen et al. [55], we divided 20 amino acids into 4 types based on polarity of the side chain. Then, by a 3-mer that is 64 ($4 \times 4 \times 4$) dimensional vector, we represented each protein sequence. To simplify the processing, each dimension of the vector represented the normalized frequency of the corresponding K-mer (k = 3) in the protein sequence.

2.5 Directed Acyclic Graph (DAG)

In LDAMAN, each disease including all related annotation terms which can be obtained from MeSH can be represented by a DAG (Directed Acyclic Graph) [46]. Generally speaking, DAG can be expressed as $DAG = (D, N(D), E(D))$, for a given disease D, $N(D)$ denotes D itself together with all its ancestor nodes, while $E(D)$ denotes all relationships connecting between nodes in the $DAG(D)$. The $D_d(t)$ of a disease t in a DAG to the semantics of disease D is defined as follows:

$$\begin{cases} D_d(D) = 1 \\ D_d(t) = max\{0.5 * D_d(t') | t' \in children\ of\ t\}\ if\ t \neq d \end{cases} \quad (1)$$

The semantic similarity score between two diseases i and j can then be calculated by:

$$S(i,j) = \frac{\sum_{t \in T(i) \cap T(j)} (D_i(t) + D_j(t))}{\sum_{t \in T(i)} D_i(t) + \sum_{t \in T(j)} D_j(t)} \quad (2)$$

2.6 Drug Molecular Fingerprint Representation

The DrugBank database [51] is a bioinformatics resource that combines detailed drug data with comprehensive drug target information. In LDAMAN, the smiles of drugs were downloaded from DrugBank database and then transformed the smiles of drugs into corresponding Morgan Fingerprint by RDKit (Open-Source Cheminformatics Software) API.

2.7 SDNE Network Embedding

For large-scale networks, some existing network embedding algorithms require complex computational complexity, resulting in sub-optimal network representations. To obtain the optimal network representation structure and optimal similarity measurement, we adopt SDNE [56] to perform network embedding in LDAMAN. In SDNE, for given the input x^i, for each layer, the hidden representations are shown as follows:

$$y_i^{(1)} = \sigma\left(W^{(1)}x^i + b^{(1)}\right), k = 1 \tag{3}$$

$$\mathrm{y}_i^{(k)} = \sigma\left(W^{(k)}y_i^{(k-1)} + \mathrm{b}^{(k)}\right), k = 2, \ldots, K \tag{4}$$

After obtaining y_i^k, we can obtain the output $\hat{x}_i$ by reversing the calculation process of encoder. The objective function is shown as follows:

$$O = \sum\nolimits_{n=1}^{\infty} \| \hat{x}_i - x_i \|_2^2 \tag{5}$$

2.8 XGBoost Classifier

Compared to the traditional Gradient Boosting algorithm, XGBoost [57], eXtreme Gradient Boosting, can be faster than other integrated algorithms using gradient enhancement. It has been recognized as an advanced evaluator with ultra-high performance in classification. To estimate the prediction performance of proposed LDAMAN, considering that the influence of classifiers on LDAMAN, we choose XGBoost classifier to predict potential lncRNA-disease associations.

3 Results and Discussion

In this section, we presented results and discussions of the proposed LDAMAN method. In the experiment, we choose XGBoost as a classifier using SDNE network embedding algorithm to construct LDAMAN for lncRNA-disea{, #1}se association prediction. Here, the AUC score is used to evaluate the predictive performance of various methods. Through 5-fold cross-validation, by plotting TPR versus FPR at different thresholds, receiver operating characteristics (ROC) curves can be obtained and the areas under ROC curve (AUCs) are calculated. Table 2 shows the results of 5-fold cross validation for LDAMAN based on SDNE. As seen from the Table 2, there exist clear differences between LDAMAN and previous methods. Specifically, it can be easily found that LDAMAN achieved a reliable AUC of 92.58%, which is significantly higher than AUCs of previous methods.

Table 2. Five-fold cross validation results of LDAMAN

Fold	Acc. (%)	Sen. (%)	Spec. (%)	Prec. (%)	MCC (%)	AUC (%)
0	84.58	84.58	84.58	84.58	69.17	93.11
1	81.62	78.26	84.98	83.90	63.38	91.25
2	83.60	81.82	85.38	84.84	67.24	92.39
3	85.57	84.19	86.96	86.59	71.17	93.16
4	84.92	84.13	85.71	85.48	69.85	92.94
Average	**84.06 ± 1.54**	**82.60 ± 2.66**	**85.52 ± 0.91**	**85.08 ± 1.02**	**68.16 ± 3.03**	**92.58 ± 0.80**

4 Conclusions

More and more studies have shown that lncRNA plays an extremely important role in a variety of biological processes and relate to the pathological change of human diseases. Therefore, identifying associations between lncRNAs and diseases can improve the understanding of disease pathogenesis at the lncRNA level. In this paper, we proposed a novel computational method LDAMAN to predict potential lncRNA-disease associations. To evaluate the prediction performance of our method, 5-fold cross validation experiment is implemented and the experiment results demonstrate the effectiveness and prediction performance. Therefore, LDAMAN can be a powerful tool for predicting potential lncRNA-disease associations.

Funding. This research was funded by the National Natural Science Foundation of China, grant number 61772333. This research was funded by the Special Project of Education Department of Shaanxi Provincial Government of china, grant number 16JK1048.

References

1. Lv, J., et al.: Identification and characterization of long intergenic non-coding RNAs related to mouse liver development. Mol. Genet. Genom. **289**(6), 1225–1235 (2014). https://doi.org/10.1007/s00438-014-0882-9
2. Yanofsky, C.: Establishing the triplet nature of the genetic code. Cell **128**, 815–818 (2007)
3. Core, L.J., Waterfall, J.J., Lis, J.T.: Nascent RNA sequencing reveals widespread pausing and divergent initiation at human promoters. Science **322**, 1845–1848 (2008)
4. Chen, X., Yan, C.C., Zhang, X., You, Z.: Long non-coding RNAs and complex diseases: from experimental results to computational models. Brief. Bioinform. **18**, 558–576 (2016)
5. Chen, X., et al.: NRDTD: a database for clinically or experimentally supported non-coding RNAs and drug targets associations. Database 2017 (2017)
6. Huang, Y., Chen, X., You, Z., Huang, D., Chan, K.C.C.: ILNCSIM: improved lncRNA functional similarity calculation model. Oncotarget **7**, 25902–25914 (2016)
7. Guo, Z., You, Z., Wang, Y., Yi, H., Chen, Z.: A learning-based method for LncRNA-disease association identification combing similarity information and rotation forest. iScience **19**, 786–795 (2019)

8. Chen, X., You, Z.-H., Yan, G.-Y., Gong, D.-W.: IRWRLDA: improved random walk with restart for lncRNA-disease association prediction. Oncotarget **7**, 57919 (2016)
9. Yu, J., Ping, P., Wang, L., Kuang, L., Li, X., Wu, Z.: A novel probability model for lncRNA–disease association prediction based on the naïve bayesian classifier. Genes **9**, 345 (2018)
10. Ou-Yang, L., et al.: LncRNA-disease association prediction using two-side sparse self-representation. Front. Genet. **10**, 476 (2019)
11. You, Z., Yin, Z., Han, K., Huang, D., Zhou, X.: A semi-supervised learning approach to predict synthetic genetic interactions by combining functional and topological properties of functional gene network. BMC Bioinf. **11**, 343 (2010)
12. Li, S., Zhou, M., Luo, X., You, Z.: Distributed winner-take-all in dynamic networks. IEEE Trans. Autom. Control **62**, 577–589 (2017)
13. Chen, X., Xie, D., Wang, L., Zhao, Q., You, Z., Liu, H.: BNPMDA: bipartite network projection for MiRNA–disease association prediction. Bioinformatics **34**, 3178–3186 (2018)
14. Wang, M., You, Z., Wang, L., Li, L., Zheng, K.: LDGRNMF: LncRNA-disease associations prediction based on graph regularized non-negative matrix factorization. Neurocomputing (2020)
15. Ma, L., et al.: Multi-neighborhood learning for global alignment in biological networks. IEEE/ACM Trans. Comput. Biol. Bioinf. (2020)
16. Yang, P., Li, X., Wu, M., Kwoh, C.-K., Ng, S.-K.: Inferring gene-phenotype associations via global protein complex network propagation. PloS One **6**, e21502 (2011)
17. Sun, J., et al.: Inferring novel lncRNA–disease associations based on a random walk model of a lncRNA functional similarity network. Mol. BioSyst. **10**, 2074–2081 (2014)
18. Zhou, M., et al.: Prioritizing candidate disease-related long non-coding RNAs by walking on the heterogeneous lncRNA and disease network. Mol. BioSyst. **11**, 760–769 (2015)
19. Ding, L., Wang, M., Sun, D., Li, A.: TPGLDA: novel prediction of associations between lncRNAs and diseases via lncRNA-disease-gene tripartite graph. Sci. Rep. **8**, 1065 (2018)
20. Mori, T., Ngouv, H., Hayashida, M., Akutsu, T., Nacher, J.C.: ncRNA-disease association prediction based on sequence information and tripartite network. BMC Syst. Biol. **12**, 37 (2018)
21. Ping, P., Wang, L., Kuang, L., Ye, S., Iqbal, M.F.B., Pei, T.: A novel method for lncRNA-disease association prediction based on an lncRNA-disease association network. IEEE/ACM Trans. Comput. Biol. Bioinf. **16**, 688–693 (2018)
22. Sumathipala, M., Maiorino, E., Weiss, S.T., Sharma, A.: Network diffusion approach to predict lncRNA disease associations using multi-type biological networks: LION. Front. Physiol. **10**, 888 (2019)
23. Yi, H., You, Z., Guo, Z.: Construction and analysis of molecular association network by combining behavior representation and node attributes. Front. Genet. **10**, 1106 (2019)
24. Guo, Z.-H., Yi, H.-C., You, Z.-H.: Construction and comprehensive analysis of a molecular associations network via lncRNA-miRNA-disease-drug-protein graph. Cells **8**, 866 (2019)
25. Yi, H.-C., You, Z.-H., Huang, W.-Z., Guo, Z.-H., Wang, Y.-B., Cheng, Z.-H.: Construction of large-scale heterogeneous molecular association network and its application in molecular link prediction. In: Basic & Clinical Pharmacology & Toxicology, p. 5. Wiley, Hoboken (2019)
26. Yi, H.C., You, Z.H., Guo, Z.H., Huang, D.S., Kcc, C.: learning representation of molecules in association network for predicting intermolecular associations. IEEE/ACM Trans. Comput. Biol. Bioinf. 1 (2020)
27. Guo, Z., You, Z., Yi, H.: Integrative construction and analysis of molecular association network in human cells by fusing node attribute and behavior information. Mol. Ther. Nucleic Acids **19**, 498–506 (2020)

28. Guo, Z.-H., You, Z.-H., Huang, D.-S., Yi, H.-C., Chen, Z.-H., Wang, Y.-B.: A learning based framework for diverse biomolecule relationship prediction in molecular association network. Commun. Biol. **3**, 1–9 (2020)
29. Guo, Z., You, Z., Yi, H., Zheng, K., Wang, Y.: MeSHHeading2vec: a new method for representing MeSH headings as feature vectors based on graph embedding algorithm. bioRxiv 835637 (2019)
30. Guo, Z.-H., You, Z.-H., Wang, Y.-B., Huang, D.-S., Yi, H.-C., Chen, Z.-H.: Bioentity2vec: attribute-and behavior-driven representation for predicting multi-type relationships between bioentities. GigaScience **9**, giaa032 (2020)
31. Chen, X., Xie, D., Zhao, Q., You, Z.: MicroRNAs and complex diseases: from experimental results to computational models. Brief. Bioinform. **20**, 515–539 (2019)
32. Chen, X., Huang, Y., You, Z., Yan, G., Wang, X.: A novel approach based on KATZ measure to predict associations of human microbiota with non-infectious diseases. Bioinformatics **33**, 733–739 (2016)
33. You, Z., et al.: PBMDA: a novel and effective path-based computational model for miRNA-disease association prediction. PLOS Comput. Biol. **13**, e1005455 (2017)
34. Zheng, K., You, Z.-H., Li, J.-Q., Wang, L., Guo, Z.-H., Huang, Y.-A.: iCDA-CGR: Identification of circRNA-disease associations based on chaos game representation. PLoS Comput. Biol. **16**, e1007872 (2020)
35. Wang, L., You, Z., Li, Y., Zheng, K., Huang, Y.: GCNCDA: a new method for predicting CircRNA-disease associations based on graph convolutional network algorithm. bioRxiv 858837 (2019)
36. Ji, B.-Y., You, Z.-H., Cheng, L., Zhou, J.-R., Alghazzawi, D., Li, L.-P.: Predicting miRNA-disease association from heterogeneous information network with GraRep embedding model. Sci. Rep. **10**, 1–12 (2020)
37. Zheng, K., You, Z.-H., Wang, L., Zhou, Y., Li, L.-P., Li, Z.-W.: Dbmda: A unified embedding for sequence-based mirna similarity measure with applications to predict and validate mirna-disease associations. Mol. Ther.-Nucleic Acids **19**, 602–611 (2020)
38. Zheng, K., Wang, L., You, Z.: CGMDA: an approach to predict and validate MicroRNA-disease associations by utilizing chaos game representation and LightGBM. IEEE Access **7**, 133314–133323 (2019)
39. Wang, M., You, Z., Li, L., Wong, L., Chen, Z., Gan, C.: GNMFLMI: graph regularized nonnegative matrix factorization for predicting LncRNA-MiRNA interactions. IEEE Access **8**, 37578–37588 (2020)
40. Wong, L., Huang, Y.A., You, Z.H., Chen, Z.H., Cao, M.Y.: LNRLMI: linear neighbour representation for predicting lncRNA-miRNA interactions. J. Cell Mol. Med. **24**, 79–87 (2020)
41. Hu, P., Huang, Y., Chan, K.C.C., You, Z.: Learning multimodal networks from heterogeneous data for prediction of lncRNA-miRNA interactions. IEEE/ACM Trans. Comput. Biol. Bioinf. 1 (2019)
42. Huang, Y., et al.: Predicting lncRNA-miRNA Interaction via graph convolution auto-encoder. Front. Genet. **10**, 758 (2019)
43. Huang, Z., Huang, Y., You, Z., Zhu, Z., Sun, Y.: Novel link prediction for large-scale miRNA-lncRNA interaction network in a bipartite graph. BMC Med. Genom. **11**, 113 (2018)
44. Miao, Y.-R., Liu, W., Zhang, Q., Guo, A.-Y.: lncRNASNP2: an updated database of functional SNPs and mutations in human and mouse lncRNAs. Nucleic Acids Res. **46**, D276–D280 (2017)
45. Chen, G., et al.: LncRNADisease: a database for long-non-coding RNA-associated diseases. Nucleic Acids Res. **41**, D983–D986 (2012)

46. Cheng, L., et al.: LncRNA2Target v2.0: a comprehensive database for target genes of lncRNAs in human and mouse. Nucleic Acids Res. **47**, D140–D144 (2018)
47. Huang, Z., et al.: HMDD v3.0: a database for experimentally supported human microRNA–disease associations. Nucleic acids research **47**, D1013–D1017 (2018)
48. Piñero, J., et al.: DisGeNET: a comprehensive platform integrating information on human disease-associated genes and variants. Nucleic Acids Res. gkw943 (2016)
49. Davis, A.P., et al.: Text mining effectively scores and ranks the literature for improving chemical-gene-disease curation at the comparative toxicogenomics database. PLoS One **8**, e58201 (2013)
50. Chou, C.-H., et al.: miRTarBase update 2018: a resource for experimentally validated microRNA-target interactions. Nucleic Acids Res. **46**, D296–D302 (2017)
51. Wishart, D.S., et al.: DrugBank 5.0: a major update to the DrugBank database for 2018. Nucleic Acids Res. **46**, D1074–D1082 (2017)
52. Szklarczyk, D., et al.: The STRING database in 2017: quality-controlled protein–protein association networks, made broadly accessible. Nucleic Acids Res. gkw937 (2016)
53. Kozomara, A., Birgaoanu, M., Griffiths-Jones, S.: miRBase: from microRNA sequences to function. Nucleic Acids Res. **47**, D155–D162 (2018)
54. Fang, S., et al.: NONCODEV5: a comprehensive annotation database for long non-coding RNAs. Nucleic Acids Res. **46**, D308–D314 (2017)
55. Shen, J., et al.: Predicting protein–protein interactions based only on sequences information. Proc. Natl. Acad. Sci. **104**, 4337–4341 (2007)
56. Wang, D., Cui, P., Zhu, W.: Structural deep network embedding. In: Proceedings of the 22nd ACM SIGKDD International Conference on Knowledge Discovery and Data Mining, pp. 1225–1234. ACM (2016)
57. Chen, T., Guestrin, C.: XGboost: a scalable tree boosting system. In: Proceedings of the 22nd ACM SIGKDD International Conference on Knowledge Discovery and Data Mining, pp. 785–794 (2016)

Expression and Gene Regulation Network of *ELF3* in Breast Invasive Carcinoma Based on Data Mining

Chenxia Ren[1], Pengyong Han[1], Chandrasekhar Gopalakrishnan[3], Caixia Xu[2], Rajasekaran Ramalingam[3], and Zhengwei Li[4](✉)

[1] Central Lab, Changzhi Medical College, Changzhi 046000, China
[2] Department of Pharmacy, Changzhi Medical College, Changzhi 046000, China
[3] Bioinformatics Lab, Department of Biotechnology, School of Bio Sciences and Technology, Vellore Institute of Technology (Deemed to be University), Vellore 632014, India
[4] School of Computer Science and Technology, China University of Mining and Technology, Xuzhou 221116, China
zwli@cumt.edu.cn

Abstract. Epigenetic infractions could have a profound impact on the tumorigenicity. E74 like ETS (Erythroblast Transformation Specific) transcription factor 3 (*ELF3*) which encodes an epithelial-restricted member of the ETS transcription factor family, is long suspected to be a prominent player in breast cancer. However, *ELF3* operates differently in various cancer types. Its precise role in breast cancer propagation remains elusive. Hence in this analysis, to resolve the functionality of *ELF3* in breast cancer pathology, we have studied *ELF3* expression, mutations, function networks and regulation in data from patients with breast invasive carcinoma (BRCA), procured from various different public databases. The present study protrudes staggered evidence on the prominence of *ELF3* in BRCA and its potential role as a novel biomarker. *ELF3* expression was assessed using Oncomine and UALCAN analysis, and data concerning *ELF3* genetic alterations and associated survival rates were further analyzed using c-BioPortal. Further, LinkedOmics was used to indicate correlations between *ELF3* and other pertinent differentially expressed genes. Gene enrichment analysis of *ELF3* associated genes was conducted using the Web-based Gene SeT AnaLysis Toolkit. The expression levels of *ELF3* were considerably increased in patients with BRCA and also in various molecular subtypes of breast cancer cells. *ELF3*'s behavior was permuted in 82 of 996 (8%) BRCA patients. Expression of this gene is ascribed to functional networks pertaining to biochemical pathways such as 'cell adhesion molecule binding', 'histone deacetylase binding', 'chromosome segregation', 'DNA-binding transcription activator activity' and 'Cell cycle' pathways. The present study revealed the expression pattern variations and potential changes in bio-molecular pathways of *ELF3* in BRCA, explicating the pernicious nature of *ELF3* and establishing a genomic foundation for further investigation of the role of *ELF3* in breast carcinogenesis.

C. Ren, P. Han and C. Gopalakrishnan—Contributed to the work equally.

D.-S. Huang and K.-H. Jo (Eds.): ICIC 2020, LNCS 12464, pp. 514–523, 2020.
https://doi.org/10.1007/978-3-030-60802-6_45

Keywords: *ELF3* · Breast invasive carcinoma · Data mining · Genome analysis · Biomarker discovery

1 Introduction

Invasive breast carcinoma is the leading cause of all cancer diagnoses in women. It is forayed by the systemic spread of cancer cells and formation of metastases [1]. Although substantial progress has been made in the diagnosis and treatment of breast cancer, a more sensitive prognostic biomarker identification is warranted for the betterment of its mitigation [2, 3].

ELF3 is a member of the epithelial-specific subfamily of ETS transcription factors, which play a role in modulating immune system [4, 5]. *ELF3* was discovered more than 20 years ago, the understanding of its functions in tumor cells has increased substantially. Loss of *ELF3* in epithelial cells could result in cancer and other diseases [6]. However, experiments have demonstrated that *ELF3*'s overexpression suppressed epithelial-mesenchymal transition in ovarian cancer. Although, recent studies have shown that *ELF3* promotes cancer cell proliferation and migration in prostate cancer, colorectal cancer and breast cancer. *ELF3* is posited as a tumor suppressor in many epithelial tumors and yet it displays oncogenic properties in other types. Nevertheless, the role of *ELF3* in BRCA is obscure [7–11].

Hence in the present study, employing robust genomic pipelines, we have investigated the expression levels and functional networks of *ELF3* in BRCA to clarify the precise nature of *ELF3* in BRCA tumorigenicity. Further, we have advocated the potential of *ELF3* as a potential bio-marker in BRCA.

2 Materials and Methods

2.1 Oncomine Analysis

The DNA copy number and mRNA expression of ELF3 in BRCA were investigated inside the Oncomine database (www.oncomine.org) [12]. Oncomine is the world's largest oncogene chip database and integrated data mining platform, contains gene expression data sets and data from TCGA about cancer tissues and normal tissues. This analysis drew on a series of TCGA BRCA and BRCA studies from Curtis research group which used Illumina HumanHT-12 V3.0 R2 Array. *ELF3* expression was involved in evaluating BRCA tissue in respect to its expression in normal tissue, and $P < 0.05$ as the cutoff criterion was considered statistically significant.

2.2 UALCAN Analysis

UALCAN (http://ualcan.path.uab.edu) is a user-friendly web asset for in-depth analyses of TCGA gene expression information [13]. UALCAN can be used to determine the relative expression levels of a target gene across several clinicopathological features. *ELF3* expression was evaluated by UALCAN in various tumor subgroups, including tumor stage, molecular subtype, tumor grade, etc.

2.3 c-BioPortal Analysis

We utilized c-BioPortal (http://cbioportal.org) [14] to analyze the ELF3 genetic alterations in the TCGA BRCA samples and the effect of gene expression dysregulation on the patient's overall survival and disease-free survival using the Kaplan-Meier analysis in c-BioPortal.

2.4 LinkedOmics Analysis

The LinkedOmics database (http://www.linkedomics.org) is a web-based platform for analyzing 32 TCGA cancer-associated multi-dimensional datasets [15]. The 'Link Finder' module in the LinkedOmics was utilized to study differentially expressed genes (DEGs) in TCGA BRCA cohort (n = 526) whose expression levels correlated with those of *ELF3*. The searching and targeting datasets were both of 'RNAseq', and the results were statistically analyzed using Pearson's correlation coefficient. Moreover, the 'Link Interpreter' module was utilized to recruit Web-based Gene SeT Analysis Toolkit to perform analyses of GO and KEGG pathways in which the identified DEGs were involved in the analysis [16]. The entire overview of the pipeline incorporated by us is surmised in Fig. 1.

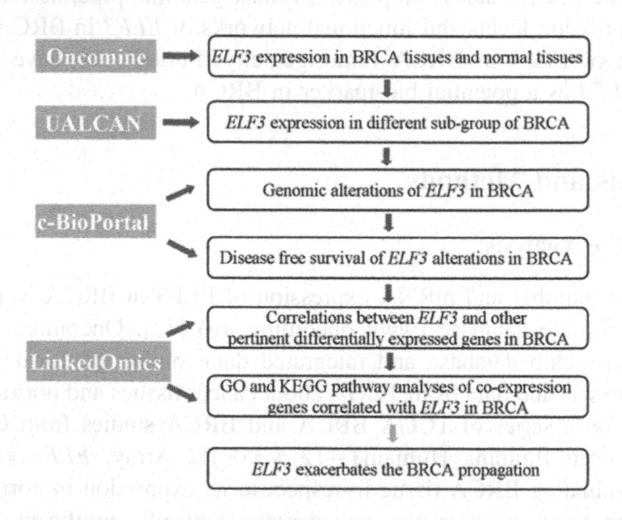

Fig. 1. Chart illustrating the workflow of the pipeline employed in the analysis

3 Results

3.1 *ELF3* Expression in BRCA

We initially evaluated *ELF3* transcription levels in multiple BRCA studies using datasets of Oncomine from TCGA and the Gene Expression Omnibus (GEO). Data revealed that mRNA expression and DNA copy number variation (CNV) of *ELF3* were significantly higher in BRCA tissues than in normal tissues ($p < 0.05$). The fold differences were above 1.5, and *ELF3* ranked almost within the top 5% (Fig. 2). Further sub-group analysis of multiple clinic pathological features of 1097 BRCA samples in the TCGA consistently showed high transcription of *ELF3* (Fig. 3B). In tumor stage subgroup (normal-vs Stage 1, normal-vs-Stage 2 and normal-vs-Stage 3)analysis, the transcription level of *ELF3* was essentially higher in BRCA patients than healthy individuals (Fig. 3A); molecular subtyping subgroups (normal-vs-luminal, normal-vs-HER2 positive and normal-vs-Triple negative) analysis (Fig. 3C); lymph node tumor type subgroup (normal-vs N0, normal-vs-N1, normal-vs-N2 and normal-vs-N3) (Fig. 3D). Data in the Oncomine database revealed that mRNA expression and DNA copy number variation (CNV) of *ELF3* were significantly higher in BRCA tissues than in normal tissues ($p < 0.05$), and *ELF3* ranked within the top 1% based on DNA CNVs.

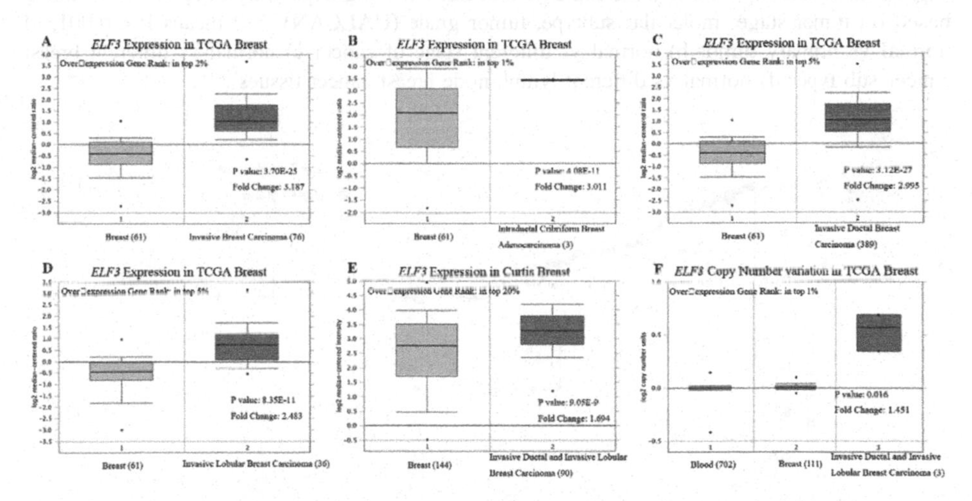

Fig. 2. *ELF3* transcription in BRCA. Levels of *ELF3* mRNA and DNA copy number were significantly higher in *ELF3* than in normal tissue. Shown are fold change, associated p values, and overexpression gene rank, based on Oncomine analysis. (A–E) Box plots show *ELF3* mRNA levels in different types of BRCA. F Box plot shows *ELF3* copy number in TCGA breast.

3.2 Genomic Alterations of ELF3 in BRCA

We then used the c-BioPortal to determine the types and frequency of *ELF3* alterations in BRCA based on sequencing data from BRCA patients in the TCGA database. *ELF3* was altered in 82 of 996 (8%) BRCA patients. These modifications were manifested as

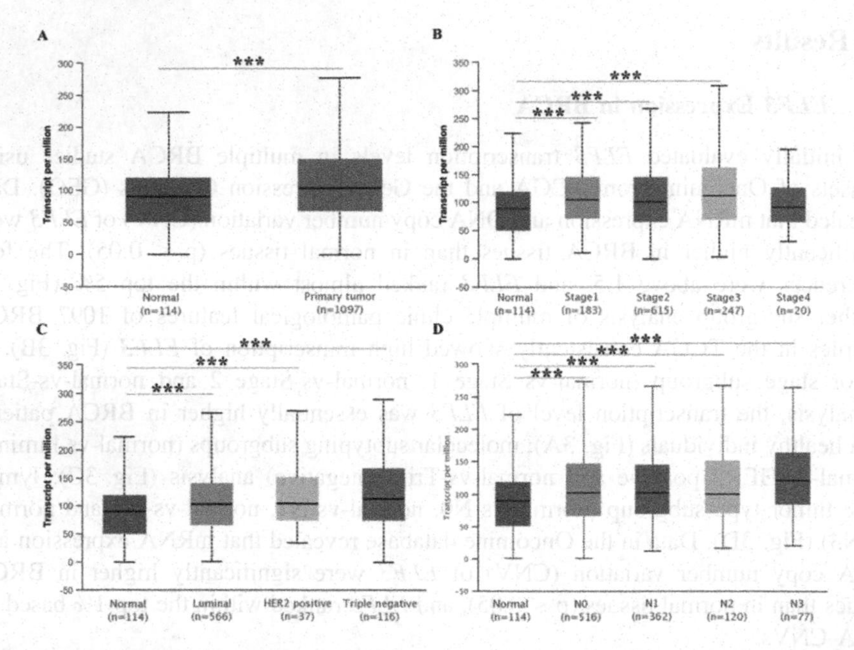

Fig. 3. Boxplots show the relative expression of *ELF3* in subgroups of BRCA patients, stratified based on tumor stage, molecular subtype, tumor grade (UALCAN). *** means P < 0.001. a) normal vs primary tumor b) normal vs different stage of cancer c) normal vs different breast cancer sub types d) normal vs different lymph-node breast cancer tissues

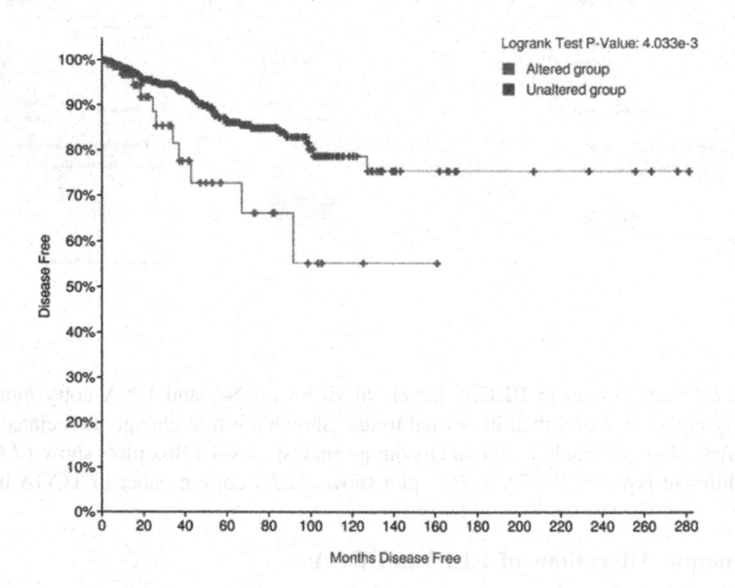

Fig. 4. Disease free survival of *ELF3* alterations in BRCA (cBioPortal). P < 0.05.

mRNA amplification in 79 cases, missense mutation in 2 cases and fusion in 1 case. In addition, Kaplan-Meier survival analysis demonstrated statistically significant findings that *ELF3* alterations was related to poor disease-free survival (P < 0.01) in BRCA (Fig. 4).

3.3 GO and KEGG Pathway Analyses of Co-expression Genes Correlated with *ELF3* in BRCA

The LinkedOmics was used to analyze mRNA sequencing data from BRCA patients in the TCGA. A Spearman test was utilized to analyze connections among *ELF3* and other genes differentially expressed in BRCA (Fig. 5A). The 50 critical gene sets positively connected with *ELF3* is given in the heat map (Fig. 5B). Additionally, significant GO term examination by gene set enrichment analysis demonstrated that genes differentially expressed in connection with *ELF3* were found to be 'Centromere Protein A containing chromatin organization', 'Notch signaling pathway', 'chromosome segregation', 'cell adhesion molecule binding', 'histone deacetylase binding', 'ephrin receptor binding', 'DNA-binding transcription activator activity', 'NF-kappaB binding', 'condensed chromosome', 'midbody', 'nuclear chromatin, chromosomal region' (Fig. 6). KEGG pathway analysis showed 'enrichment in the Cell cycle', 'Pyrimidine metabolism', 'Bladder cancer', 'RNA transport', et al. (Fig. 6D). Figure 6E showed the *ELF3*-associated DEGs involved in Cell cycle. These data indicate that *ELF3* may serve in nucleus to positively formulate cell cycle of BRCA.

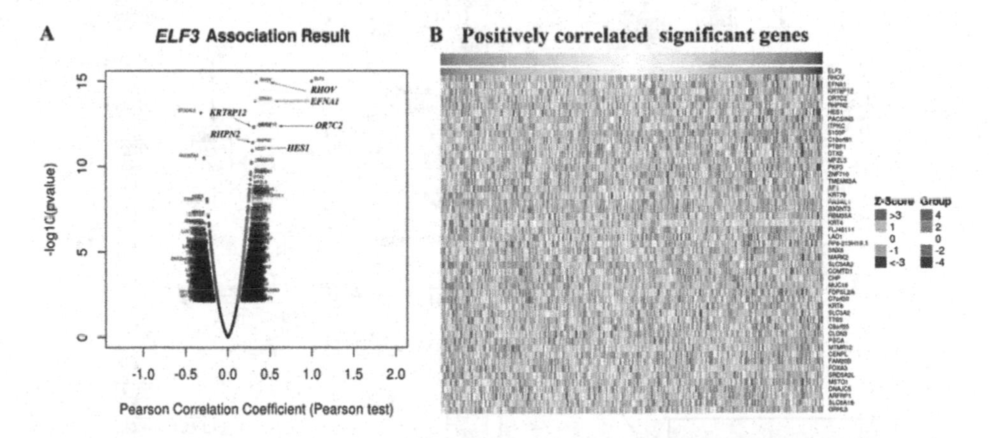

Fig. 5. Genes differentially expressed in correlation with *ELF3* in BRCA (LinkedOmics). (A) A T-test was used to analyze correlations between *ELF3* and genes differentially expressed in BRCA. (B) Heat maps showing genes positively and negatively correlated with *ELF3* in BRCA (TOP 50). Red indicates positively correlated genes and blue indicates negatively correlated genes. (Color figure online)

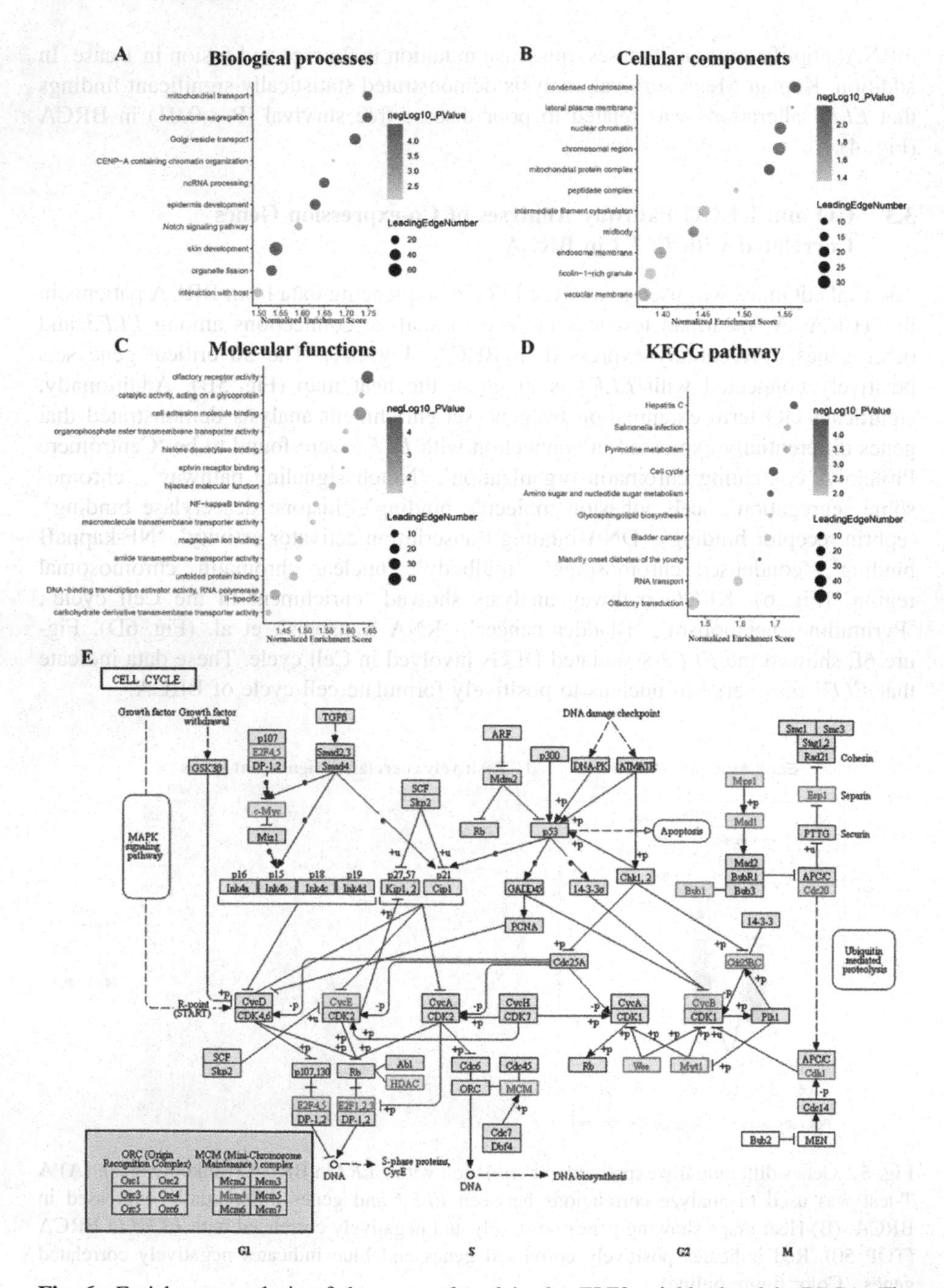

Fig. 6. Enrichment analysis of the genes altered in the *ELF3* neighborhood in BRCA. The bubble diagrams display the enrichment results of the top 50 genes altered in the *RBM8A* neighborhood in LIHC. The Leading-Edge Number: the number of leading-edge genes (A) Cellular components. (B) Biological processes. (C) Molecular functions. (D) KEGG pathway analysis. (E) KEGG pathway annotations of the ribosome pathway. The red star de-notes altered genes. (Color figure online)

4 Discussion

Till now cancer has been an epigenetic enigma. Increasing evidence indicates that transcription factors play an important role in the development and progression of cancers. *ELF3* regulates transcription of many genes that are involved in cellular transformation and inflammation [17–19], and *ELF3* plays different roles in different types of cancer [20]. It has been reported that *ELF3* is highly expressed in the luminal subtype of breast cancer cells and $HER2^{+}$ breast cancer cells, and conversely *ELF3* knockdown can inhibit cyclin D1 resulting in G1 arrest in breast cancer cell lines [21].

Hence, in this study, with the help of a computationally comprehensive and a statistically robust pipeline we have ascertained the definitive role of *ELF3* in BRCA tumor propagation.

Initially, we measured the expression level of *ELF3* in BRCA tissues and normal breast tissues by mining TCGA database and found that *ELF3* expression levels were significantly increased, including all breast cancer subtypes. Our study found that the copy number of *ELF3* was increased in BRCA, and that the major type of alteration was amplification. This congruence of increased *ELF3* with the pathology insinuates a positive relationship between *ELF3* and BRCA. This can be further bolstered by the findings in Fig. 3A and B wherein there is a significant increase of *ELF3* in primary and other stages of breast cancer. The same trends can be observed in different molecular subtypes of breast cancer (Fig. 3C and D). This statistically recurring pattern of increased *ELF3* gene expression could potentially ascribe the causality of BRCA. Moreover, the persistent elevated levels of *ELF3* in breast cancer patients and various other breast cancer subtype cells could render its potential as a viable biomarker in assessing BRCA.

Furthermore, the pathological influence of *ELF3* as visualized by survival analysis (Fig. 4), indicates that patients with *ELF3* mutation showed a considerably decreased survival as compared to the *ELF3* unaltered cohort.

Further calibration of *ELF3* and its co-expressed gene network ascertains a group of biochemical pathways that initiates, fosters and accelerate cell division. The genes that are co-expressed with *ELF3* could be positively regulated by *ELF3*. Also, the set of genes that are expressed with *ELF3* actuates pathways that are very distinctive of aggravating cell proliferation (Fig. 5 and 6).

Compiling the aforementioned findings, it is safe to presume that in BRCA, *ELF3* exacerbates the cancer propagation.

5 Conclusion

In conclusion, this study shows that *ELF3* expression is upregulated in BRCA tissues and the alteration of *ELF3* in BRCA is significantly associated with poor disease-free survival in BRCA patients. In addition, *ELF3* may work in BRCA by regulating cell cycle signaling pathways. This study gives staggered evidence for the significance of *ELF3* in BRCA carcinoma and its potential candidacy as a novel biomarker.

Funding. This study was supported by Science and technology innovation project of Shanxi province universities (2019L0683), Changzhi Medical College Startup Fund for PhD faculty (BS201922), Provincial university innovation and entrepreneurship training programs (2019403).

References

1. Wu, J.M., et al.: Heterogeneity of breast cancer metastases: comparison of therapeutic target expression and promoter methylation between primary tumors and their multifocal metastases. Clin. Cancer Res. **14**, 1938–1946 (2008). https://doi.org/10.1158/1078-0432.CCR-07-4082
2. Turashvili, G., Brogi, E.: Tumor heterogeneity in breast cancer. Front. Med. **4**, 227 (2017). https://doi.org/10.3389/fmed.2017.00227
3. Lindström, L.S., et al.: Clinically used breast cancer markers such as estrogen receptor, progesterone receptor, and human epidermal growth factor receptor 2 are unstable throughout tumor progression. JCO **30**, 2601–2608 (2012). https://doi.org/10.1200/JCO.2011.37.2482
4. Oliver, J.R., Kushwah, R., Hu, J.: Multiple roles of the epithelium-specific ETS transcription factor, ESE-1, in development and disease. Lab. Invest. **92**, 320–330 (2012). https://doi.org/10.1038/labinvest.2011.186
5. Oettgen, P., et al.: Isolation and characterization of a novel epithelium-specific transcription factor, ESE-1, a member of the ETS family. Mol. Cell. Biol. **17**, 4419–4433 (1997). https://doi.org/10.1128/MCB.17.8.4419
6. Oettgen, P., Barcinski, M., Boltax, J., Stolt, P., Akbarali, Y., Libermann, T.A.: Genomic organization of the human ELF3 (ESE-1/ESX) gene, a member of the ETS transcription factor family, and identification of a functional promoter. Genomics **55**, 358–362 (1999). https://doi.org/10.1006/geno.1998.5681
7. Wang, J.-L., et al.: Elf3 drives β-catenin transactivation and associates with poor prognosis in colorectal cancer. Cell Death Dis. **5**, e1263 (2014). https://doi.org/10.1038/cddis.2014.206
8. Longoni, N., et al.: ETS transcription factor ESE1/ELF3 orchestrates a positive feedback loop that constitutively activates NF-κB and drives prostate cancer progression. Cancer Res. **73**, 4533–4547 (2013). https://doi.org/10.1158/0008-5472.CAN-12-4537
9. Shatnawi, A., et al.: ELF3 is a repressor of androgen receptor action in prostate cancer cells. Oncogene **33**, 862–871 (2014). https://doi.org/10.1038/onc.2013.15
10. Nakarai, C., et al.: Significance of ELF3 mRNA expression for detection of lymph node metastases of colorectal cancer. Anticancer Res. **32**, 3753–3758 (2012)
11. Yeung, T.-L., et al.: ELF3 is a negative regulator of epithelial-mesenchymal transition in ovarian cancer cells. Oncotarget **8**, 16951–16963 (2017). https://doi.org/10.18632/oncotarget.15208
12. Rhodes, D.R., et al.: Oncomine 3.0: genes, pathways, and networks in a collection of 18,000 cancer gene expression profiles. Neoplasia **9**, 166–180 (2007). https://doi.org/10.1593/neo.07112
13. Chandrashekar, D.S., et al.: UALCAN: a portal for facilitating tumor subgroup gene expression and survival analyses. Neoplasia **19**, 649–658 (2017). https://doi.org/10.1016/j.neo.2017.05.002
14. Cerami, E., et al.: The cBio cancer genomics portal: an open platform for exploring multidimensional cancer genomics data. Cancer Discov. **2**, 401–404 (2012). https://doi.org/10.1158/2159-8290.CD-12-0095

15. Vasaikar, S.V., Straub, P., Wang, J., Zhang, B.: LinkedOmics: analyzing multi-omics data within and across 32 cancer types. Nucleic Acids Res. **46**, D956–D963 (2018). https://doi.org/10.1093/nar/gkx1090
16. Wang, J., Vasaikar, S., Shi, Z., Greer, M., Zhang, B.: WebGestalt 2017: a more comprehensive, powerful, flexible and interactive gene set enrichment analysis toolkit. Nucleic Acids Res. **45**, W130–W137 (2017). https://doi.org/10.1093/nar/gkx356
17. Kwon, J.H., Keates, S., Simeonidis, S., Grall, F., Libermann, T.A., Keates, A.C.: ESE-1, an enterocyte-specific ETS transcription factor, regulates MIP-3alpha gene expression in Caco-2 human colonic epithelial cells. J. Biol. Chem. **278**, 875–884 (2003). https://doi.org/10.1074/jbc.M208241200
18. Brown, C., et al.: ESE-1 is a novel transcriptional mediator of angiopoietin-1 expression in the setting of inflammation. J. Biol. Chem. **279**, 12794–12803 (2004). https://doi.org/10.1074/jbc.M308593200
19. Grall, F.T., et al.: The Ets transcription factor ESE-1 mediates induction of the COX-2 gene by LPS in monocytes. FEBS J. **272**, 1676–1687 (2005). https://doi.org/10.1111/j.1742-4658.2005.04592.x
20. Li, L., et al.: Epithelial-specific ETS-1 (ESE1/ELF3) regulates apoptosis of intestinal epithelial cells in ulcerative colitis via accelerating NF-κB activation. Immunol. Res. **62**(2), 198–212 (2015). https://doi.org/10.1007/s12026-015-8651-3
21. Sinh, N.D., Endo, K., Miyazawa, K., Saitoh, M.: ETS1 and ESE1 reciprocally regulate expression of ZEB1/ZEB2, dependent on ERK1/2 activity, in breast cancer cells. Cancer Sci. **108**, 952–960 (2017). https://doi.org/10.1111/cas.13214

Embracing Disease Progression with a Learning System for Real World Evidence Discovery

Zefang Tang[1], Lun Hu[2], Xu Min[1], Yuan Zhang[1], Jing Mei[1], Kenney Ng[3], Shaochun Li[1], Pengwei Hu[1(✉)], and Zhuhong You[2(✉)]

[1] IBM Research, Beijing, China
hupengwei@hotmail.com
[2] Technical Institute of Physics and Chemistry, Chinese Academy of Sciences, Urumqi, China
zhuhongyou@ms.xjb.ac.cn
[3] Center for Computational Health and MIT-IBM Watson AI Lab, Cambridge, MA, USA

Abstract. Electronic Health Records (EHRs) have been widely used in healthcare studies recently, such as the analyses for patient diagnostic outcome and understanding of disease progression. EHR is a treasure for researchers who conduct the Real-World study to discovering Real-World Evidence (RWE). In this paper, we design an end-to-end learning system for disease states discovery based on a data-driven strategy. A large-scale proprietary EHR data mart containing about 55 million patients with over 20 billion data records is used for data extraction and analysis. Given a disease of interest, researchers could easily obtain the hidden disease states. Once our system were operational, biomedical researchers could get the results for downstream analyses such as disease prediction, drug design and outcome analyses.

Keywords: Real World Evidence discovery · Disease progression · Biomedical analyses · EHRs

1 Introduction

Electronic Health Records (EHRs) have became the richest source to trace and evaluate the healthcare delivery. EHR contains structured clinical data including demographic factors, vital signs and lab tests, diagnoses, medications and procedures, as well as unstructured data like clinical notes. Not surprisingly, EHR data provides a broader view of patient encounters thus brings tremendous opportunities in healthcare research.

To leverage this huge amount of EHR data, practitioners have been focused on supervised learning tasks like risk prediction. For example, [1] developed Doctor AI on 260k patients to predict the diagnose and medication categories for subsequent visits. [2] leveraged longitudinal data of over 703K patients to predict acute kidney injury

Z. Tang and L. Hu—Contributed equally.

D.-S. Huang and K.-H. Jo (Eds.): ICIC 2020, LNCS 12464, pp. 524–534, 2020.
https://doi.org/10.1007/978-3-030-60802-6_46

(AKI) episodes. The data-driven approaches have shown the powerful prediction capability for secondary analysis of EHR data to provide actionable insights for healthcare stakeholders [3, 4]. Comparing these works, many SOTA supervised learning models have also been widely applied in predicting disease on small data sets [5, 6] and explore the disease mechanism [7]. However, unsupervised learning task for disease-subtype identification is more challenging, especially for the progression associated states which need to be clinically meaningful and be accepted by domain experts. The early identification of the patients at risk of deterioration would help these patients get timely intervention, which could improve disease management in the long term.

As a longitudinal unsupervised learning task, disease progression modeling has drawn much attention. Chronic obstructive pulmonary disease (COPD) has been well studied [8, 9], while the diabetes and its complications are also been analyzed [10, 11]. These efforts open the door of modeling disease progression by using advanced machine learning algorithms. However, when facing up with large-scale EHR database, dedicated developments are urgently required to address the data pre-processing for cohort construction, parameter tuning for modeling and profiling, as well as the evaluation for model efficiency.

This paper offers a glimpse of an end-to-end learning system to provide one-stop disease progression modeling for any given disease of interest. Our main contributions are:

1. We propose an end-to-end learning system for Real-World Evidence (RWE) discovery from electronic health records. This system can be reused in any large-scale EHR database to construct population of interest, conduct versatile analyses as well as provide downstream profiling or evaluations, in an automatic and efficient way.
2. Our system adopts hidden variable models of high complexity and flexibility as data-driven methods for disease progression analysis. The system is able to deal with different types of input data simultaneously, including both continuous data and binary data which are common in the medical domain.
3. We use Type 2 Diabetes Mellitus (T2DM) as a case study, to validate our system. A cohort of T2DM patients was extracted from a proprietary EHR data mart and was used to identify underlying hidden states and corresponding transition patterns. We provide qualitative and quantitative evaluation measurements to demonstrate that our system can achieve clinically meaningful insights.

2 Methods

Our end-to-end learning system consist of cohort construction, disease progression model, disease state profiling and evaluation steps to obtain progression associated hidden states, as shown in Fig. 1. In this section, we first give an overall description of our learning system, and then we provide detailed explanations for each of the modules.

2.1 Overall Architecture

Given a disease of interest, the system first automatically selects a list of categorical and continuous features without expert advice. Next, imputation and dropping out are used in the system to deal with the missing data. Then a time related model is trained on the established cohort data to discover hidden disease states during disease progression. After hidden states were predicted, the system would profile the features and provide the statistic results and evaluation results to delineate the differences among hidden states.

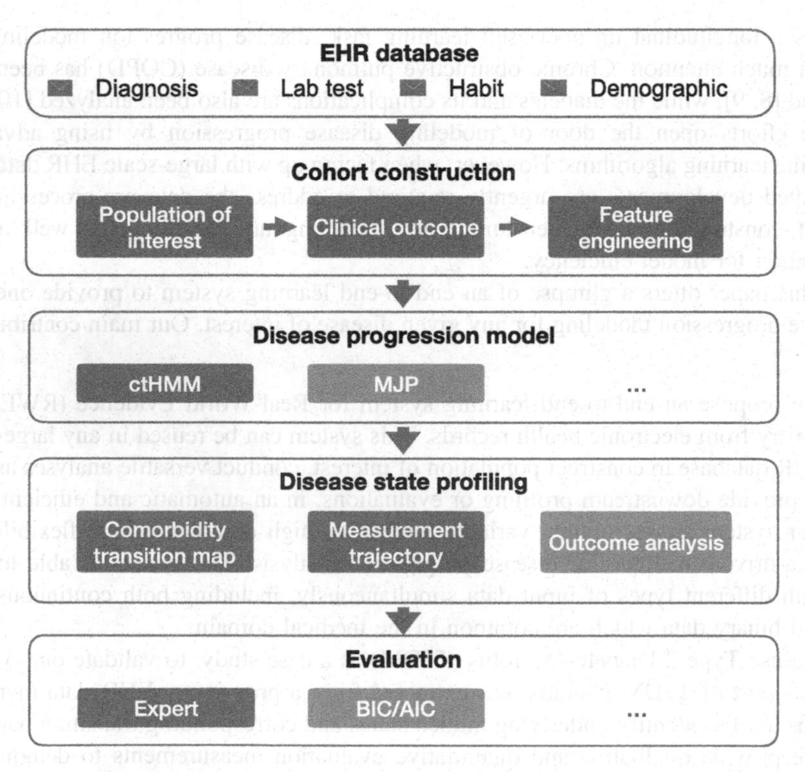

Fig. 1. The proposed end-to-end data driven learning system to predict progression hidden states of a given disease.

2.2 Cohort Construction

2.2.1 Calculation Incidence Odds Ratio (OR)

For a disease of interest i, the system uses a data-driven strategy to identify the comorbidities of disease i among other diseases based on their incidence Odds Ratio (OR). The incidence OR of disease j towards disease i could be formulated as

$$OR_{i,j,l} = \frac{I_{\text{observed}}}{I_{\text{random}}} = \frac{I_{i,j,l}}{\sum_{k=1}^{n_b} I^k_{\cdot,j,l}/n_b}, \forall i,j \in D, \quad (1)$$

where D is the disease space of the EHR database. I_{observed} represents the percentage of patients who get disease j after the diagnosis of disease i within l-year interval. I_{random} indicates the mean value of randomly sampled incidence of disease j within l-year interval using n_b times bootstrap. The randomly sample $I^k_{\cdot,j,l}$ d incidence $I^k_{\cdot,j,l}$ is independent of disease i.

In bootstrap step, the system performs permutations for n_b times to calculate l-year random incidence rate of j disease. Specifically, in each permutation the system collects m patients for analysis. For each patient, the system randomly chooses a time point before disease j happened, and then calculates the percentage of patients who get disease j within l-year intervals. The OR values represent the hazard ratios of disease j incidence after disease i is diagnosed.

2.2.2 Data Extraction

Based on the OR value of comorbidities, the system identifies the comorbidities with high OR values as discrete features. Meanwhile, the system extracts disease correlated clinical measurements together with demographic information as continuous features.

2.2.3 Imputation

To deal with the missing data, the system first conducts case deletion for those patients who have features without any records from beginning to end. Then the blank observations are imputed as the median value of the feature for the corresponding patient.

2.3 Continuous-Time Hidden Markov Model

In the hidden states prediction step, the system would use continuous-time hidden Markov model (ctHMM) [12, 13], Markov jump processes (MJP) [14], or other time-related model to predict disease hidden states. In our case study, we used ctHMM for hidden states prediction.

Continuous-time Hidden Markov Model (ctHMM) is an attractive approach for disease progression modeling, which uses a latent continuous-time Markov Chain (CTMC) to account for complex medical observations arriving irregularly in time. ctHMM is based on a *time-homogeneous* CTMC, which is a Markov chain whose state transitions can happen at any arbitrary time. As a consequence, in contrast to a traditional Hidden Markov Model (HMM), ctHMM has two levels of hidden information: 1) at observation time, the state of Markov chain is hidden, 2) the state transitions are also hidden. We prefer ctHMM because it can not only 1) extract discrete hidden states for patient subtyping or clustering, but also 2) model disease progression trajectories in a continuous timeline.

2.3.1 Formulation of ctHMM

We will describe ctHMM formally below. A ctHMM is defined by a finite and discrete state space S, a state transition rate matrix Q, an initial state probability distribution π, and a distribution $p(o|s)$ of observation o depending on hidden states s. In detail, the elements in the transition rate matrix Q satisfy $q_i = \sum_{j \neq i} q_{ij}$, $q_{ii} = -q_i$, which means the sum of each row is 0. The sojourn time in each state i is modeled as an exponential distribution with *rate parameter* q_i, which is $f(t) = q_i e^{-q_i t}$ with mean $1/q_i$. The q_{ij} describes the transition rate from state i to state j. The observation distribution $p(o|s)$ depends on the type of input data. It can be either Gaussian distribution for continuous data, or Bernoulli distribution for binary data.

A fully observed ctHMM contains four sequences, including: the underlying state transition time $(t'_0, t'_1, \cdots, t_{V'})$, the corresponding state $Y = \{y_0 = s(t'_0), \cdots, y_{V'} = s(t'_{V'})\}$, the observed data $O = (o_1, o_2, \cdots, o_V)$, and the corresponding observation time $T = (t_1, t_2, \cdots, t_V)$. Here $s(t)$ denotes the state at time t. As a result, the complete likelihood of the ctHMM can formulated as:

$$
\begin{aligned}
CL &= \prod_{v'=0}^{V'-1} \frac{q_{y'_{v'}, y_{v'+1}}}{q_{y_{v'}}} q_{y_{v'}} e^{-q_{y_{v'}} \tau_{v'}} \prod_{v=0}^{V} p(o_v|s(t_v)) \\
&= \prod_{v'=0}^{V'-1} q_{y_{v'}, y_{v'+1}} e^{-q_{y_{v'}} \tau_{v'}} \prod_{v=0}^{V} p(o_v|s(t_v)) \\
&= \prod_{i=1}^{|S|} \prod_{\substack{j=1 \\ j \neq i}}^{|S|} q_{ij}^{n_{ij}} e^{-q_i \tau_i} \prod_{v=0}^{V} p(o_v|s(t_v))
\end{aligned}
\qquad (2)
$$

where $\tau_{v'} = t'_{v'+1} - t'_{v'}$ is the time interval between two state transitions, n_{ij} is the total number of transitions from state i to state j, and τ_i is the total time the chain remains in state i.

2.3.2 Learning Algorithm

Estimation of the parameters for a ctHMM is computationally challenging, since we have only O and T observed. We here use an EM algorithm to learn parameters following[9]. In brief, take the estimation of Q for example, in the Expectation step we calculate the expected log-likelihood given the current estimate of $\hat{Q}_0$:

$$
\begin{aligned}
L(Q) = &\sum_{i=1}^{|S|} \sum_{\substack{j=1 \\ j \neq i}}^{|S|} \left\{ \log(q_{ij}) \mathbb{E}\left[n_{ij}|O, T, \hat{Q}_0\right] - q_i \mathbb{E}\left[\tau_i|O, T, \hat{Q}_0\right] \right\} \\
&+ \sum_{v=0}^{V} \mathbb{E}\left[\log p(o_v|s(t_v))|O, T, \hat{Q}_0\right]
\end{aligned}
\qquad (3)
$$

Then in the Maximization step, we can easily obtain an optimal estimate of Q by taking the derivative of $L(Q)$ with respect to Q:

$$\begin{aligned} \hat{q}_{ij} &= \frac{\mathbb{E}\left[n_{ij}|O,T,\hat{Q}_0\right]}{\mathbb{E}\left[\tau_i|O,T,\hat{Q}_0\right]}, i \neq j \\ \hat{q}_{ii} &= -\sum_{j \neq i} \hat{q}_{ij}. \end{aligned} \tag{4}$$

In fact, the challenge lies in the E-step where the expectations of n_{ij} and τ_i need to be computed according to the observed data O and T. Practically, it can be decomposed by computing the posterior state probabilities using the *forward-backward* or *Viterbi* algorithm, and computing the end-state conditioned expectations using the *Expm*, *Unif* or *Eigen* method [12].

2.3.3 Explanation in Disease Progression

The learned parameters of ctHMM have explicit interpretations in disease progression. First of all, the hidden discrete state space S encodes the information of $|S|$ different disease states. For binary observation data, such as diagnosis, modeled with Bernoulli distribution, each state is represented by a vector of $\mathbf{p}$ whose element p_i means the occurrence probability of the i-th disease. For continuous observation data, such as lab tests, modeled with Gaussian distribution, each state is represented by a vector of μ whose element μ_i means the expectation of the i-th measurement.

Second, the transition rate matrix Q contains the information of transitions between different disease states. We can constrain Q to an upper triangular matrix with $q_{ij} = 0$, $\forall j < i$, to force the algorithm to learn a series of states. If $q_i = 0$, we call this state an *absorbing* state where a patient will never leave, e.g., the death state. Otherwise, the time a patient stays in the i-th state, is exponentially-distributed with rate q_i, whose expectation and variance are both equal to $1/q_i$. Then the patient will jump to other states according to the transition proportion q_{ij}/q_i. Besides, given the transition rate matrix Q, we can compute the transition probability matrix $P(t)$ corresponding to a time interval t by

$$P(t) = e^{Qt} = \sum_{n=0}^{\infty} Q^n \frac{t^n}{n!} \tag{5}$$

The element $P_{ij}(t)$ represents the probability that a patient in state i will transit to state j after time t.

Table 1. Statistics of the dataset

# of patients	17,378
# of visits	609,697
Median # of visits per patient	24
# of categorical features	27
# of continuous features	12

2.4 Multiple Comparison and Evaluation

The system used Tukey's HSD (honestly significant difference) test for multiple comparisons of each feature among different hidden states. The HSD could be formulated as:

$$HSD = \frac{M_i - M_j}{\sqrt{MS_w / n_h}} \tag{6}$$

where $M_i - M_j$ is the difference between the pair of means, and M_i should be larger than M_j. MS_w is the Within Mean Square, and n_h is the number of groups or treatments. In addition, outcome analysis based Cox-regression model would be integrated into the system in the future.

To explore the medical meanings of the results, we provide not only qualitative evaluations by depicting an occurrence probability matrix heatmap, but also quantitative evaluations by incorporating clinician knowledge. Specifically, human experts give an expert score *ES* for each comorbidity, according to its chronological order in disease progression. Then a progression score *PS* for each hidden state can be calculated as: $PS = \sum_i ESi \cdot pi$ (7), where N means the number of comorbidities, ES_i indicates the expert score of *i*-th disease, and p_i represents the occurrence probability of the *i*-th disease as mentioned before. Currently, we are also planning to provide numerical evaluations such as Bayesian information criteria (BIC) or Akaike's Information Criteria (AIC) in future versions. Finally, researchers could use our system to obtain a series of results and evaluations and then discuss with experts to discover more insights.

3 Results

In our case study, we applied our system in a proprietary EHR dataset, which contains about 55 million patients with over 20 billion data points collected from participating hospitals and care networks. Type 2 diabetes mellitus (T2DM) was chosen as disease of interest. We used parts of the system functions to obtain the T2DM hidden states.

3.1 Cohort Construction

By using the parameters $l = 2$, $n_b = 100$, $m = 10000$ with the thresholds $I_{\text{observed}} > 4\%$ and OR > 1, the system obtained 24 comorbidities associated with T2DM. Next, the system extracted 200,000 T2DM patients records with comorbidity information, demographic information (gender and age), habits (alcohol and tobacco) as well as T2DM associated clinical measurements (11 T2DM-associated clinical test measurements such as alanine amino transferase, HbA1c and total cholesterol). After imputation and filtering steps, we finally get a 609,697 records dataset with 17,378 patients information for hidden states prediction (Table 1). To determine the number of the hidden states, we tried different values of $|S|$ including 4, 5 and 6. Then by consulting domain expert according to the characteristic of T2DM, we set $|S| = 4$ for further analysis.

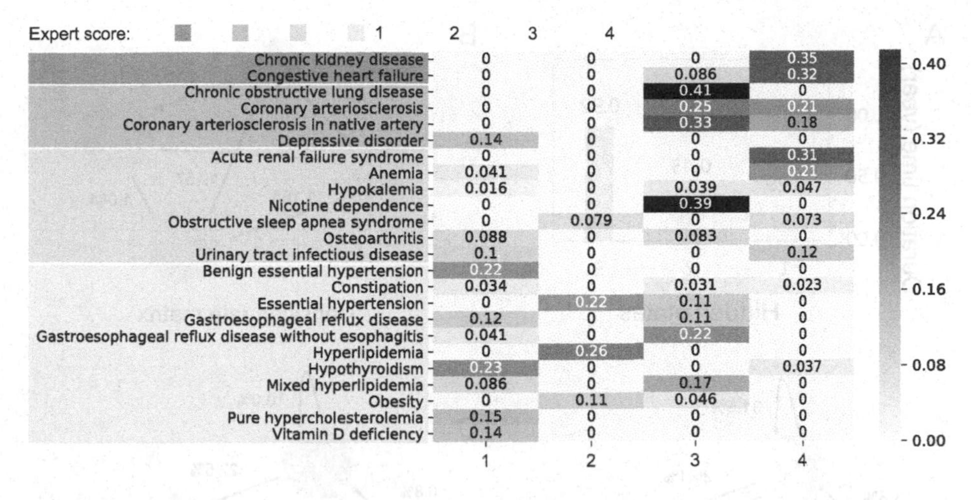

Fig. 2. The heatmap shows comorbidity occurrence probabilities in each hidden state. The comorbidities were ranked by expert scores, which were marked with red panel in the vertical axis. In the heatmap, the blue blocks show the probabilities of each comorbidity in corresponding hidden state. (Color figure online)

3.2 Hidden States Profiling

The heatmap shown in Fig. 2 indicates the different occurrence probabilities of the comorbidities according to the **p** vectors of ctHMM modeling results. The value in each block represents the occurrence probability of each comorbidity in a certain hidden state. Meanwhile, the system ranked the comorbidities based on expert score ($ES \in [1, 2, 3, 4]$) referred to as expert domain knowledge.

The expected duration times (year) of each hidden state are 0.29, 0.43 and 0.92 (state 4 is absorbing state), respectively (Fig. 3A). The transition rate matrix Q learned in ctHMM, is shown in Fig. 3(B), and the associated transition probability matrix $P(t)$ among different hidden states within 6 or 12 months are shown in Fig. 3(C–D). Within 6 months, 82.4% (96.9% within 12 months) of the state 1 patients would move into other state, while 41.8% (66.2% within 12 months) of the state 3 patients would progress towards state 4, which is relatively consistent to the increasing tendency of duration time of each state.

The lab test values of the patients in each hidden state have different tendency. Glomerular filtration rate has a decreasing tendency in the cohorts as the T2DM progresses while glucose and creatinine will increase when T2DM progresses (Fig. 4).

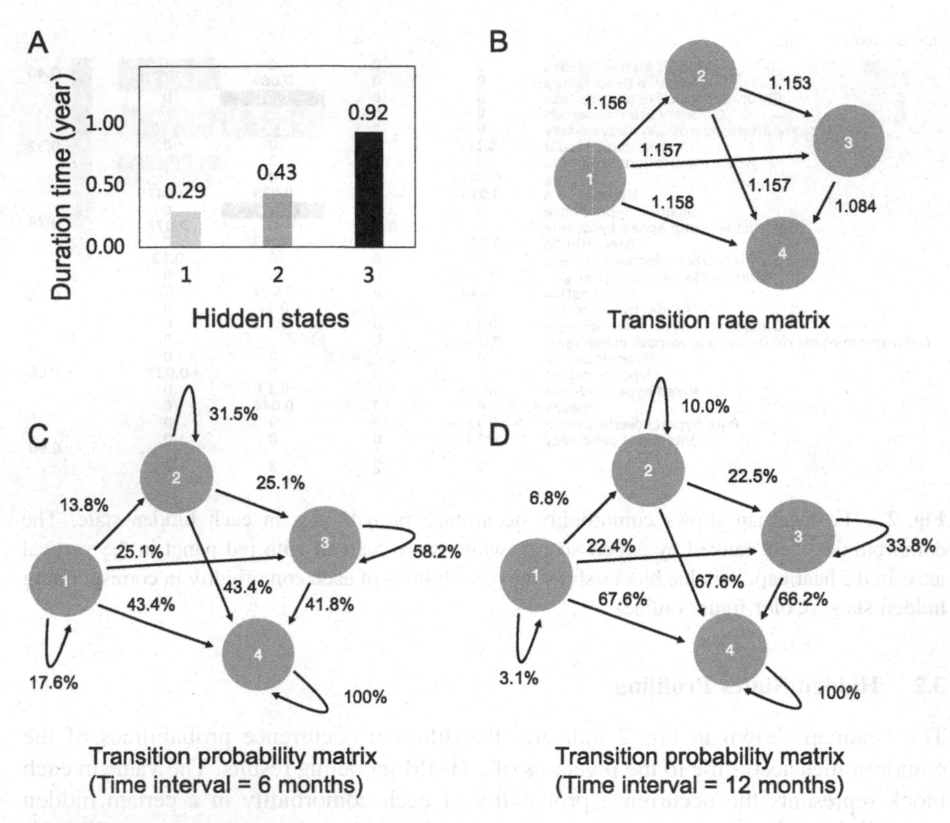

Fig. 3. The hidden states transition profile. (A) Bar plot indicated the expected duration time of each hidden state. (B) The transition plot shows the transition rate matrix Q. (C) The transition plot shows transition probability matrix $P(t)$ in 6 months ($t = 0.5$). (D) The transition plot shows transition probability matrix $P(t)$ in one year ($t = 1.0$).

3.3 Evaluation

Multiple comparison test (Tukey's HSD test) provides the significant tendencies of the glomerular filtration rate (eGFR), glucose and creatinine (Fig. 4). The increasing tendency of glucose is reasonable since the blood glucose is highly associated with the T2DM progression. eGFR increases shortly after T2DM diagnose and tends to decrease in later states with an increase of creatinine, which is in line with previous studies [15, 16] addressing the hyperfiltration in early T2DM and progressive decline in eGFR. This data-drive approach enables evaluation of current findings and inspires further research.

Meanwhile, the heatmap of occurrence probability shows that in the first two states, patients tend to have metabolic disorders such as hypertension, hypothyroidism and hyperlipidemia. These diseases usually happen in a short period after T2DM diagnosis. When the T2DM progresses into state 3, patients are more likely to have chronic obstructive lung disease (COPD) or coronary arteriosclerosis. These diseases need longer period progression. In the final state, patients are more easily to have long-term progressed diseases such as chronic kidney disease and congestive heart failure.

The occurrence probabilities in different hidden states also show a T2DM progression, which is consistent with the domain knowledge of expert.

Furthermore, according to the progression score *PS* of each hidden state (state 1: 1.95; state 2: 0.74; state 3: 5.05; state 4: 5.40), we can demonstrate that the progression tendency learned by our ctHMM is in accordance with human experience. Above all, the evaluation results consist with the T2DM progression, which indicates that our system have a good performance to predict progression-associated disease hidden states.

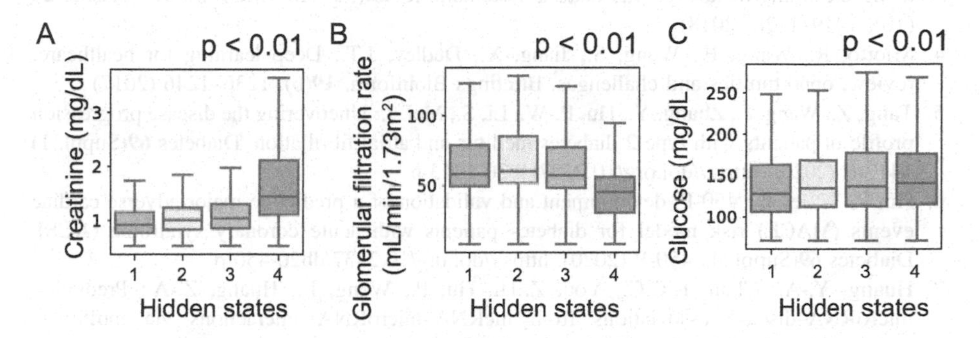

Fig. 4. Box plots show the tendency of creatinine, glomerular filtration rate, glucose in 4 hidden states. Tukey's HSD method was used for multiple comparison.

4 Discussion and Conclusion

EHR data provides enormous treasure for Real-World Evidence discovery. In our project, we designed an end-to-end learning system with purely data-driven strategy to predict the progression hidden states of given disease. Based on this strategy, researchers could easily get the hidden states without domain knowledge. Meanwhile, the system simplify the time-consuming efforts of data processing and accelerate the data construction. Furthermore, statistic methods were implemented for evaluation of the model efficiency. By using this system, we uncovered 4 hidden states of T2DM. The evaluation indicated that the results from our system are highly consistent with expert domain knowledge.

Although the system could provide an efficient way for feature extraction, it is worth noting that the data-driven strategy lacks the feature selection step for improving the model training efficiency, which required us to refine the system in the future. Furthermore, the hidden states prediction of a given disease is just a glance for interpreting the disease. Future studies are still needed to examine the tendencies of the continuous features in our results. Once our system was implemented, fundamental studies could be accelerated since we provide a prediction base- line for any given disease. Oncologists could apply different cancer types into our system to discover cancer subtypes which were associated with the patient diagnosis. Biomedical practitioners could use specific inventions towards different progression-state patients. The industry could exert our system to screen the candidate drug-targeted cohorts. Our system may open another window of hidden states discovery based on EHR data.

References

1. Choi, E., Bahadori, M.T., Schuetz, A., Stewart, W.F., Sun, J.: Doctor AI: predicting clinical events via recurrent neural networks. In: Machine Learning for Healthcare Conference, pp. 301–318 (2016)
2. Tomasev, N., et al.: A clinically applicable approach to continuous prediction of future acute kidney injury. Nature **572**(7767), 116–119 (2019)
3. Xiao, C., Choi, E., Sun, J.: Opportunities and challenges in developing deep learning models using electronic health records data: a systematic review. J. Am. Med. Inform. Assoc. **25** (10), 1419–1428 (2018)
4. Miotto, R., Wang, F., Wang, S., Jiang, X., Dudley, J.T.: Deep learning for healthcare: review, opportunities and challenges. Briefings Bioinform. **19**(6), 1236–1246 (2017)
5. Tang, Z., Wang, J., Zhang, Y., Hu, P.-W., Li, S., Mei, J.: Uncovering the disease progression profile of patients with type 2 diabetes mellitus and atrial fibrillation. Diabetes **69**(Suppl. 1) 1413-P (2020). https://doi.org/10.2337/db20-1413-p
6. Hu, P.-W., et al.: 450-P: development and validation of a predictive major adverse cardiac events (MACE) risk model for diabetes patients with acute coronary syndrome (ACS). Diabetes **69**(Suppl. 1) 450-P (2020). https://doi.org/10.2337/db20-450-p
7. Huang, Y.-A., Chan, K.C.C., You, Z.-H., Hu, P., Wang, L., Huang, Z.-A.: Predicting microRNA–disease associations from lncRNA–microRNA interactions via multiview multitask learning. Briefings Bioinform. (2020)
8. Wang, X., Sontag, D., Wang, F.: Unsupervised learning of disease progression models. In: Proceedings of the 20th ACM SIGKDD International Conference on Knowledge Discovery and Data Mining, pp. 85–94. ACM (2014)
9. Jensen, A.B., et al.: Temporal disease trajectories condensed from population-wide registry data covering 6.2 million patients. Nat. Commun. **5**, 4022 (2014)
10. Pham, T., Tran, T., Phung, D., Venkatesh, S.: Predicting healthcare trajectories from medical records: a deep learning approach. J. Biomed. Inform. **69**, 218–229 (2017)
11. Liu, B., Li, Y., Sun, Z., Ghosh, S., Ng, K.: Early prediction of diabetes complications from electronic health records: a multi-task survival analysis approach. In: Thirty-Second AAAI Conference on Artificial Intelligence (2018)
12. Liu, Y.-Y., Li, S., Li, F., Song, L., Rehg, J.M.: Efficient learning of continuous-time hidden markov models for disease progression. In: Advances in Neural Information Processing Systems, pp. 3600–3608 (2015)
13. Sun, Z., et al.: A probabilistic disease progression modeling approach and its application to integrated Huntington's disease observational data. JAMIA Open **2**(1), 123–130 (2019)
14. Metzner, P., Schutte, C., Vanden-Eijnden, E.: Transition path theory for markov jump processes. Multiscale Model. Simul. **7**(3), 1192–1219 (2009)
15. Tonneijck, L., et al.: Glomerular hyperfiltration in diabetes: mechanisms, clinical significance, and treatment. J. Am. Soc. Nephrol. **28**(4), 1023–1039 (2017)
16. Weldegiorgis, M., et al.: Longitudinal estimated GFR trajectories in patients with and without type 2 diabetes and nephropathy. Am. J. Kidney Diseases **71**(1), 91–101 (2018)

High-Performance Modelling Methods on High-Throughput Biomedical Data

Robust Graph Regularized Extreme Learning Machine Auto Encoder and Its Application to Single-Cell Samples Classification

Liang-Rui Ren[1], Jin-Xing Liu[1(✉)], Ying-Lian Gao[2], Xiang-Zhen Kong[1], and Chun-Hou Zheng[3]

[1] School of Computer Science, Qufu Normal University, Rizhao 276826, China
renliangrui@126.com, sdcavell@126.com, kongxzhen@163.com

[2] Library of Qufu Normal University, Qufu Normal University, Rizhao 276826, China
yinliangao@126.com

[3] College of Computer Science and Technology, Anhui University, Hefei 230601, China
zhengch99@163.com

Abstract. Combined with Auto Encoder (AE), Extreme Learning Machine Auto Encoder (ELM-AE) has attracted the interest of researchers in recent years. Considering the classification tasks of single-cell Ribonucleic Acid sequencing (scRNA-seq) data, in this paper, we propose a novel supervised learning method based on ELM-AE, which is named Robust Graph Regularized Extreme Learning Machine Auto Encoder (RGELMAE). The method introduces $L_{2,1}$-norm minimization on loss function to improve the robustness, and combines with the manifold regularization framework to explore the internal local structure between data points. Finally, RGELMAE is applied to the classification tasks of scRNA-seq data. The experimental results indicate that our method can effectively extract the key information representing the original data, and improve the classification performance of ELM.

Keywords: Single-cell RNA-seq · Extreme learning machine · L_{21}-norm · Auto encoder · Supervised learning · Manifold regularization

1 Introduction

Single-cell RNA sequencing (scRNA-seq) [1–4] has enabled the characterization of highly specific cell types in many human tissues, as well as both primary and stem cell-derived cell lines. An important fact of these studies is the ability to identify the transcriptional signatures that define a cell type or state, and clearly, the ability to accurately predict a cell type and any pathologic-related state will play a critical role in the early diagnosis of disease and decisions around the personalized treatment for patients [5]. Since many single-cell RNA sequencing data have characteristics of high-dimensional-small-sample-size, it is a challenge of how to effectively mine the available information in the data.

D.-S. Huang and K.-H. Jo (Eds.): ICIC 2020, LNCS 12464, pp. 537–545, 2020.
https://doi.org/10.1007/978-3-030-60802-6_47

Recently, Huang et al. proposed an effective learning method, Extreme Learning Machine (ELM) [6–8], to train a Single-Hidden-Layer feedforward neural networks (SLFNs). Compared with the traditional neural network, ELM eliminates the process of iterative parameter adjustment, and avoids the problems of slow convergence speed and local minimum value [9]. Combined with Auto Encoder (AE) [10], Kasun et al. proposed a linear and a nonlinear dimension reduction framework referred to as Extreme Learning Machine Auto-Encoder (ELM-AE) and Sparse Extreme Learning Machine Auto-Encoder (SELM-AE) [11]. ELM-AE maps the original data into a low dimension or high dimension feature space and recovers the original data as much as possible, i.e., the output of ELM-AE is the same as the input, so that the output weight matrix $\boldsymbol{\beta}$ can learn the key information to represent the original data.

Inspired by the previous works, in this paper, we propose a new classification method named Robust Graph Regularized Extreme Learning Machine Auto Encoder (RGELMAE) and apply it to the classification tasks on single-cell RNA sequencing data. In our method, the $L_{2,1}$-norm [12] is introduced to the loss function. Compared with the traditional Frobenius norm (F-norm), the $L_{2,1}$-norm can improve the robustness of the method and reduce the influence of noise and outliers [13]. In addition, the introduction of manifold regularization can make full use of the local structural information between the data, which may further help us extract molecular information that representing different cell types. We evaluate the performance of RGELMAE on four scRNA-seq data sets. During the experiment, it is obvious that our method has better classification performance than other algorithms. At the same time, the experimental results show that the application of our method in the field of bioinformatics will help researchers identify different cell types, which is conducive to better study of cell heterogeneity between single cells.

The rest of the paper is organized as follows: Sect. 2 briefly introduces the main contents of other methods related to our method. And in Sect. 3, our method named RGELMAE is proposed for classifying samples of scRNA-seq data. Section 4 mainly reports and analyzes the experimental results over four scRNA-seq data sets. Section 5 summarizes the paper.

2 Related Work

2.1 Extreme Learning Machine

For a supervised learning problem, $\{\mathbf{P}, \mathbf{T}\} = \{\mathbf{p}_i, \mathbf{t}_i\}_{i=1}^{N}$, where $\mathbf{P} \in \mathbb{R}^{N \times M}$ is the data matrix with N samples, $\mathbf{T} \in \mathbb{R}^{N \times 1}$ is the label matrix. $\mathbf{p}_i$ is the i-th training sample which has M-dimensional features. $\mathbf{t}_i$ is the target output of i-th training sample. The objective function of original ELM is:

$$\mathrm{Min}\frac{1}{2}\|\boldsymbol{\beta}\|^2 + \frac{C}{2}\|\mathbf{H}\boldsymbol{\beta} - \mathbf{T}\|^2, \tag{1}$$

where the first term is the regularization term. C is a trade-off parameter. $\mathbf{T} = [\mathbf{t}_1, \mathbf{t}_2, \ldots, \mathbf{t}_N]^T_{N \times M}$ is the target output matrix of training samples.

$\boldsymbol{\beta} = [\boldsymbol{\beta}_1, \boldsymbol{\beta}_2, \ldots, \boldsymbol{\beta}_{n_h}]^T \in \mathbb{R}^{n_h \times M}$ is the output weight matrix and $\mathbf{H}$ is the hidden layer output matrix. By computing the derivative of $\boldsymbol{\beta}$ and putting it to zero, we have:

$$\boldsymbol{\beta} - C\mathbf{H}^T(\mathbf{H}\boldsymbol{\beta} - \mathbf{T}) = 0. \tag{2}$$

If the number of training samples is larger than the number of hidden nodes, we have:

$$\boldsymbol{\beta} = \left(\mathbf{I}_{n_h} + C\mathbf{H}^T\mathbf{H}\right)^{-1} C\mathbf{H}^T\mathbf{T}, \tag{3}$$

where $\mathbf{I}_{n_h}$ is an identity matrix. And if the number of training samples is less than the number of hidden nodes, we get the solution:

$$\boldsymbol{\beta} = \mathbf{H}^T\left(\mathbf{I}_N + C\mathbf{H}\mathbf{H}^T\right)^{-1} C\mathbf{T}, \tag{4}$$

where $\mathbf{I}_N$ is an identity matrix.

2.2 Extreme Learning Machine Auto Encoder

Based on ELM, ELM-AE is still a single hidden layer feedforward neural network, but it is very effective for information extraction. By random orthogonal weights and bias, ELM-AE maps the original data to the n_h-dimensional feature space. This process can be represented as:

$$\mathbf{H}(\mathbf{P}) = \mathrm{g}(\mathbf{A} * \mathbf{P} + \mathbf{b}), \quad \text{s.t.} \quad \mathbf{A}^T\mathbf{A} = \mathbf{I},\ \mathbf{b}^T\mathbf{b} = \mathbf{I}, \tag{5}$$

where $\mathbf{A} = [\mathbf{a}_1, \mathbf{a}_2, \ldots, \mathbf{a}_{n_h}]$ is the orthogonal random weight matrix, $\mathbf{b} = [\mathbf{b}_1, \mathbf{b}_2, \cdots, \mathbf{b}_{n_h}]$ is the orthogonal random bias matrix. And $\mathrm{g(x)}$ is the activation function. And the hidden output matrix $\boldsymbol{\beta}$ is calculated as follows:

$$\mathrm{Min}\frac{1}{2}\|\boldsymbol{\beta}\|^2 + \frac{C}{2}\|\mathbf{H}\boldsymbol{\beta} - \mathbf{P}\|^2. \tag{6}$$

And the solutions are expressed as:

$$\begin{cases} \boldsymbol{\beta} = \left(\mathbf{I} + C\mathbf{H}^T\mathbf{H}\right)^{-1} C\mathbf{H}^T\mathbf{P}, \ \mathrm{N} > \mathrm{n}_h, \\ \boldsymbol{\beta} = \mathbf{H}^T\left(\mathbf{I} + C\mathbf{H}\mathbf{H}^T\right)^{-1} C\mathbf{P}, \ \mathrm{N} \leq \mathrm{n}_h. \end{cases} \tag{7}$$

2.3 Manifold Regularization

Manifold regularization [14] is a framework proposed to capture the manifold structure based on the smoothness assumption in machine learning [15]. The assumptions are: (1) Both labeled data $\mathbf{X}$ and unlabeled data $\mathbf{X}'$ are derived from the same edge distribution P_χ. (2) If the two data points $\mathbf{x}_1$ and $\mathbf{x}_2$ are close to each other, then their conditional probabilities $\mathrm{P}(\mathbf{y}|\mathbf{x}_1)$ and $\mathrm{P}(\mathbf{y}|\mathbf{x}_2)$ should be very similar. In supervised learning, the conditional probability is approximated by the predicted class indicator $\mathbf{y}_i$ and $\mathbf{y}_j$ with respect to the i-th and j-th data points [15]. Hence, the objective function is:

$$\mathbf{Z} = \frac{1}{2}\sum_{i,j} \mathbf{w}_{ij}\left\|\mathbf{y}_i - \mathbf{y}_j\right\|, \tag{8}$$

and its matrix form is:

$$\mathbf{Z} = \mathrm{Tr}\left(\mathbf{Y}^T\mathbf{Q}\mathbf{Y}\right), \tag{9}$$

where $\mathbf{Y}$ is the predicted label matrix of data points, $\mathbf{Q} = \mathbf{D} - \mathbf{W}$ is the graph Laplacian matrix with $\mathbf{D}$ is a diagonal matrix and its diagonal elements $d_{ii} = \sum_{i=1}^{N} \mathbf{w}_{ij}$ and $\mathbf{W}$ is the weight matrix.

3 Robust Graph Regularized Extreme Learning Machine Auto Encoder

3.1 The Objective Function of RGELMAE

For a training data set $\{\mathbf{P}_{train}, \mathbf{T}_{train}\} = \{\mathbf{p}_i, \mathbf{t}_i\}_{i=1}^{N}$, the objective function of RGELMAE can be expressed as:

$$\begin{aligned} &\mathrm{Min}\tfrac{1}{2}\|\boldsymbol{\beta}\|^2 + \tfrac{C}{2}\|\boldsymbol{\xi}_i\|_{2,1} + \tfrac{\lambda}{2}\mathrm{Tr}\left((\mathbf{H}\boldsymbol{\beta})^T\mathbf{Q}(\mathbf{H}\boldsymbol{\beta})\right), \\ &s.t. \quad \mathbf{h}(\mathbf{p}_i)\boldsymbol{\beta} = \mathbf{p}_i - \boldsymbol{\xi}_i,\ i = 1, 2, \ldots, N, \end{aligned} \tag{10}$$

where $\boldsymbol{\beta}$ is the hidden layer output weight matrix. $\mathbf{H}$ is the hidden layer output matrix. The first term is a regularization item which is constrained by L_2-norm. $\boldsymbol{\xi}_i$ is an error vector of i-th data. $\mathbf{Q}$ is the graph Laplacian matrix. C and λ are penalty coefficients. $\mathrm{Tr}(\cdot)$ is the trace norm and T stands for transpose operation of matrix.

Ding et al. proposed $L_{2,1}$-norm in [12]. The matrix form of $L_{2,1}$-norm is as follows:

$$\|\mathbf{Z}\|_{2,1} = \sum_{i=1}^{m} \|\mathbf{z}_i\|_2 = \sum_{i=1}^{m} \sqrt{\sum_{j=1}^{n} \mathbf{z}_{ij}^2}. \tag{11}$$

It is obviously that the essence of $L_{2,1}$-norm is to first calculate the L_2-norm of each row of the matrix $\mathbf{Z}$, and then calculate the L_1-norm of each column. In the minimization problem, $L_{2,1}$-norm can generate row sparsity. More importantly, the robustness of the method can be improved.

3.2 The Optimization of RGELMAE

To solve the regularization problem of $L_{2,1}$-norm, we introduce the following expression:

$$\|\mathbf{A}\|_{2,1} = \mathrm{Tr}\left(\mathbf{A}^T \mathbf{D} \mathbf{A}\right), \tag{12}$$

where $\mathbf{D}$ is a diagonal matrix and the i-th diagonal element is $d_{ii} = 1 \Big/ \left(2\|\mathbf{A}^i\|_2\right)$, $\mathbf{A}^i$ represents i-th row of $\mathbf{A}$.

According to Eq. (10). An unconstrained optimization problem can be obtained:

$$\mathrm{F} = \frac{1}{2}\|\boldsymbol{\beta}\|^2 + \frac{C}{2}\|\mathbf{P} - \mathbf{H}\boldsymbol{\beta}\|_{2,1} + \frac{\lambda}{2}\mathrm{Tr}\left(\boldsymbol{\beta}^T \mathbf{H}^T \mathbf{Q} \mathbf{H} \boldsymbol{\beta}\right). \tag{13}$$

By calculating the gradient of $\boldsymbol{\beta}$ and setting it equal to zero, we have:

$$\frac{\partial \mathrm{F}}{\partial \beta} = \boldsymbol{\beta} - C\mathbf{H}^T \mathbf{D}(\mathbf{P} - \mathbf{H}\boldsymbol{\beta}) + \lambda \mathbf{H}^T \mathbf{Q} \mathbf{H} \boldsymbol{\beta} = 0. \tag{14}$$

If the number of training samples is larger than the number of hidden nodes, the solution of $\boldsymbol{\beta}$ is:

$$\boldsymbol{\beta} = \left(\mathbf{I}_{n_h} + C\mathbf{H}^T \mathbf{D} \mathbf{H} + \lambda \mathbf{H}^T \mathbf{Q} \mathbf{H}\right)^{-1} C\mathbf{H}^T \mathbf{D} \mathbf{P}, \tag{15}$$

where $\mathbf{I}_{n_h}$ is an identity matrix. If the number of training samples is less than the number of hidden nodes, $\boldsymbol{\beta}$ can be expressed as:

$$\boldsymbol{\beta} = \mathbf{H}^T \left(\mathbf{I}_N + C\mathbf{D}\mathbf{H}\mathbf{H}^T + \lambda \mathbf{Q}\mathbf{H}\mathbf{H}^T\right)^{-1} C\mathbf{D}\mathbf{P}, \tag{16}$$

where $\mathbf{I}_N$ is an identity matrix. In Eq. (15) and Eq. (16), $\mathbf{D}$ is calculated by:

$$\mathbf{D} = \frac{1}{\|\mathbf{P} - \mathbf{H}\boldsymbol{\beta}\|_2}. \tag{17}$$

In particular, the information extraction ability of RGELMAE is reflected by the following equation:

$$\mathbf{P}_{new} = \mathbf{P}\boldsymbol{\beta}^T. \tag{18}$$

To acquire an exact solution of $\boldsymbol{\beta}$, an efficient iterative optimization algorithm is designed. The main flow of the algorithm is outlined as Algorithm 1.

Algorithm 1: RGELMAE

Input:

Training data set $\{\mathbf{P}_{train}, \mathbf{T}_{train}\} = \{\mathbf{p}_i, \mathbf{t}_i\}_{i=1}^{N}$; Testing data set $\{\mathbf{P}_{test}, \mathbf{T}_{test}\} = \{\mathbf{p}_i, \mathbf{t}_i\}_{i=1}^{U}$;

The optimal parameters: C, λ; The number of hidden nodes: n_h.

Output:

The output weight matrix.

Steps:

Step 1: Randomly generate random orthogonal input weights and biases.

Step 2: Calculate the hidden layer output matrix $\mathbf{H}$.

Step 3: Construct the Laplace matrix of the graph $\mathbf{Q}$.

Step 4: Randomly initialize the hidden layer output weight matrix $\boldsymbol{\beta}_1$.

Step 5: for $i = 1, \ldots, k$

Calculate the diagonal matrix $\mathbf{D}$ by Eq. (17);

if $N > n_h$, update $\boldsymbol{\beta}$ by equation Eq. (15);

if $N < n_h$, update $\boldsymbol{\beta}$ by equation Eq. (16);

until the algorithm converges.

4 Experiments

We evaluate the classification performance of our method on four scRNA-seq data sets: Melanoma (Mela2 and Mela6) [16, 17], Multiple myeloma (MM) [18], Hepatocellular Carcinoma (HCC) [19]. The classification performance of RGELMAE also compared with other methods based on ELM: ELM [7], W-ELM [20], SS-ELM [21], ELM-AE (linear), ELM-AE (non-linear), and sparse ELM-AE (SELM-AE) [11].

4.1 Data Sets

In the experiments, we evaluate the performance of our method on four single-cell RNA-sequencing data sets. Table 1 lists the details of the scRNA-seq data sets used in the experiment. All data sets are obtained from Gene Expression Omnibus Database (GEO).

Table 1. Details of the scRNA-seq data sets used in the experiments.

Data sets	Categories	Samples	Training	Testing	Dimensions
HCC	3	138	83	55	19,525
Mela2	2	4,513	2,708	1,805	23,686
Mela6	6	2,887	1,732	1,155	23,686
MM	4	597	358	239	23,398

4.2 Settings

In experiments, the initialization range of orthogonal random weights and biases is (−1, 1). The sigmoid function is selected as the activation function. The optimal number of hidden nodes of RGELMAE for each data set is listed in Table 2.

Table 2. The optimal number of RGELMAE hidden nodes for each data set.

Data sets	HCC	Mela2	Mela6	MM
n_h	80	1000	1000	100

There are two parameters need to be selected in our method: trade-off parameter C and manifold parameter λ. The selection range of them is both between $\{10^{-3}, 10^{-2}, 10^{-1}, 10^{0}, 10^{1}, 10^{2}, 10^{3}\}$. We use 10-fold cross validation to determine the best combination of C and λ.

4.3 Results and Discussion

Each method is performed 30 times on every data set, and Table 3 shows the average results of the 30 experiments, and the best classification results are in bold type.

Table 3. The classification results on single-cell RNA-seq data sets (±variance).

Methods	HCC	Mela2	Mela6	MM
ELM	0.9849 ± 2.85	0.9663 ± 7.01	0.9622 ± 3.91	0.8510 ± 3.69
W-ELM	0.9545 ± 4.28	0.9697 ± 0.00	0.9078 ± 2.95	0.7503 ± 0.29
SS-ELM	0.9333 ± 3.67	0.9699 ± 0.00	0.9542 ± 0.00	0.7987 ± 0.00
ELMAE (linear)	0.9855 ± 0.02	0.9880 ± 0.08	0.9811 ± 0.00	0.8476 ± 0.09
ELMAE (nonlinear)	0.9855 ± 0.00	0.9796 ± 0.13	0.9785 ± 0.00	0.8463 ± 0.10
SELM-AE (linear)	0.9792 ± 0.00	0.9873 ± 0.08	0.9879 ± 0.00	0.8470 ± 0.06
RGELMAE	**0.9884 ± 0.00**	**0.9930 ± 0.00**	**0.9883 ± 0.00**	**0.8594 ± 0.00**

According to the experimental results, we can draw several interesting conclusions:

1) It is clear that the classification performance of RGELMAE is better than other methods. Compared with other methods, RGELMAE introduces $L_{2,1}$-norm minimization to improve the robustness of the method and reduce the influence of noise and outliers. And the introduction of manifold learning enables RGELMAE to learn local structural information between data points, which improves the information extraction ability of the model.

2) All ELM methods related to the Auto Encoder are better than those based on ELM alone. It shows that the ELM method based on Auto Encoder can effectively extract the feature information of the original data set, reduce the redundancy of classification information and improve the performance of the method.
3) RGELMAE shows good classification ability in both binary classification scRNA-seq data sets and multiclass classification scRNA-seq data sets, which provides another applicable framework for binary and multiclass classification problems.

5 Conclusions

In this paper, an interesting method named RGELMAE is proposed and applied to the sample classification of single-cell RNA-seq data. In RGELMAE, the $L_{2,1}$-norm is introduced to the loss function to improve the robustness of the method. We also introduce manifold regularization to learn the local manifold structure of data. The method is applied to the classification of four single-cell RNA sequencing data. Experimental results show the effectiveness of the RGELMAE method compared to other representative methods. In particular, the experimental results on single-cell RNA sequencing data show that ELM and its variants can also be used as a tool for analyzing single-cell RNA sequencing data.

Acknowledgment. This work was supported in part by the NSFC under grant Nos. 61872220, and 61873001.

References

1. Hashimshony, T., Wagner, F., Sher, N., Yanai, I.: CEL-Seq: single-cell RNA-Seq by multiplexed linear amplification. Cell Rep. **2**(3), 666–673 (2012)
2. Jaitin, D., et al.: Massively parallel single-cell RNA-Seq for marker-free decomposition of tissues into cell types. Science **343**(6172), 776–779 (2014)
3. Macosko, E.Z., et al.: Highly parallel genome-wide expression profiling of individual cells using nanoliter droplets. Cell **161**(5), 1202–1214 (2015)
4. Zheng, G.X.Y., et al.: Massively parallel digital transcriptional profiling of single cells. Nat. Commun. **8**(1), 14049 (2017)
5. Stuart, T., Satija, R.: Integrative single-cell analysis. Nat. Rev. Genet. **20**(5), 257–272 (2019)
6. Huang, G., Zhu, Q., Siew, C.K.: Extreme learning machine: theory and applications. Neurocomputing **70**(1), 489–501 (2006)
7. Huang, G.B., Zhou, H.M., Ding, X.J., Zhang, R.: Extreme learning machine for regression and multiclass classification. IEEE Trans. Syst. Man Cybern. Part B Cybern. **42**(2), 513–529 (2012)
8. Huang, G., Huang, G., Song, S., You, K.: Trends in extreme learning machines. Neural Netw. **61**, 32–48 (2015)
9. Peng, Y., Wang, S., Long, X., Lu, B.: Discriminative graph regularized extreme learning machine and its application to face recognition. Neurocomputing **149**, 340–353 (2015)
10. Rumelhart, D.E., Hinton, G.E., Williams, R.J.: Learning representations by back-propagating errors. Nature **323**(6088), 696–699 (1988)

11. Kasun, L.L.C., Yang, Y., Huang, G., Zhang, Z.: Dimension reduction with extreme learning machine. IEEE Trans. Image Process. **25**(8), 3906–3918 (2016)
12. Ding, C., Zhou, D., He, X., Zha, H.: R1-PCA: rotational invariant L1-norm principal component analysis for robust subspace factorization. In: International Conference on Machine Learning, pp. 281–288 (2006)
13. Liu, J.X., et al.: A joint-L-2, L-1-norm-constraint-based semi-supervised feature extraction for RNA-Seq data analysis. Neurocomputing **228**, 263–269 (2017)
14. Belkin, M., Niyogi, P., Sindhwani, V.: Manifold regularization: a geometric framework for learning from labeled and unlabeled examples. J. Mach. Learn. Res. **7**, 2399–2434 (2006)
15. Liu, T., Lekamalage, C.K.L., Huang, G., Lin, Z.: Extreme learning machine for joint embedding and clustering. Neurocomputing **277**, 78–88 (2018)
16. Single-Cell RNA-Seq Reveals Melanoma Transcriptional Heterogeneity. Canc. Discov. **6**(6), 570 (2016)
17. Tirosh, I., et al.: Dissecting the multicellular ecosystem of metastatic melanoma by single-cell RNA-seq. Science **352**(6282), 189–196 (2016)
18. Jang, J.S., et al.: Molecular signatures of multiple myeloma progression through single cell RNA-Seq. Blood Canc. J. **9**(1), 2 (2019)
19. Zheng, H., et al.: Single-cell analysis reveals cancer stem cell heterogeneity in hepatocellular carcinoma. Hepatology **68**(1), 127–140 (2018)
20. Zong, W., Huang, G., Chen, Y.: Weighted extreme learning machine for imbalance learning. Neurocomputing **101**, 229–242 (2013)
21. Huang, G., Song, S., Gupta, J.N.D., Wu, C.: Semi-supervised and unsupervised extreme learning machines. IEEE Trans. Cybern. **44**(12), 2405–2417 (2014)

Intelligent Computing and Swarm Optimization

A Short Survey of Multi-objective Immune Algorithm Based on Clonal Selection

Lingjie Li(✉), Qiuzhen Lin, and Zhong Ming

College of Computer Science and Software Engineering, Shenzhen University, Shenzhen 510860, China
vilitejie@qq.com, {qiuzhlin,mingz}@szu.edu.cn

Abstract. Multi-objective immune algorithm (MOIA) is a heuristic algorithm based on artificial immune system model. Due to its characteristics of antibody clonal selection and automatic antigen recognition in the immune system, immune algorithm has become a research hotspot in the field of multi-objective optimization after the evolutionary algorithm. In this paper, the studies of multi-objective immune algorithm based on clonal selection principle are summarized and discussed. At the beginning, some background about clonal selection principle is introduced. Then, the details of each immune algorithm are expounded, which mainly include its characteristic, mechanism and drawbacks. Moreover, in order to have a visual observation of the performance of immune algorithm on solving multi-objective optimization problems, four state-of-the-art MOIAs based on clonal selection principle are tested on one widely used benchmark problem in the experimental comparison. Finally, some future research directions of MOIAs are briefly discussed.

Keywords: Multi-objective optimization · Multi-objective immune algorithm · Clonal selection principle

1 Introduction

In the field of optimization, multi-objective optimization has become a hot research area, as it involves many fields in the real-life engineering problems. For example, immune algorithms have been proposed for solving flexible job schedule problems [1], power distribution system [2], and network intrusion detection [3]. Generally speaking, we call the optimization problems that include two or more conflicting objectives as multi-objective optimization problems (MOPs) [4]. As a result, there is no single solution that can obtain the optimal result for all the conflicting objectives. That is to say, multi-objective optimization aims to find a set of Pareto-optimal solutions, namely ***PS***, which include the best trade-off solutions for the conflicting objectives. Moreover, the Pareto-optimal front (***PF***) that means the objective function values of ***PS*** is used as the preference information of each objective. Hence, the main goals for tackling a MOP are to obtain a set of final solutions that can approximate the true ***PF*** as closely as possible and cover the true ***PF*** as evenly as possible.

More recently, many multi-objective immune algorithms have been proposed with promising performance on solving various kinds of MOPs, which imitates biological

D.-S. Huang and K.-H. Jo (Eds.): ICIC 2020, LNCS 12464, pp. 549–559, 2020.
https://doi.org/10.1007/978-3-030-60802-6_48

immune mechanism and combines with the evolutionary mechanism of genes. *Farmer* et al. firstly introduced a dynamic model of the immune system based on the immune network theory, and discussed the relationship between the immune system and other artificial intelligence methods [5]. After that, more research studies on the artificial immune system are conducted. In 1996, the first international symposium based on the immune system was held in Japan and the concept of "artificial immune system" (AIS) was first proposed as well.

According to the features inspired from the biologic immune system, most existing MOIAs can be classified into three main categories. They are respectively immune algorithms based on the clonal selection principle, immune algorithms based on the immune network system, and immune algorithms embedding other heuristic operators. In particular, the clonal selection based MOIAs have become the most popular immune algorithms for solving MOPs, as they can achieve a good balance between the diversity and convergence by using a suitable clonal selection mechanism. Hence, in order to have a comprehensive understanding on the principle of clonal selection, most of the clonal selection based MOIAs are summarized and overviewed in this paper.

The remainder of this paper is organized as follows. Section 2 gives some preliminary information, including some definitions in MOP, the principle of clonal selection and the framework of MOIAs. In Sect. 3, the studies of MOIAs based on clonal selection principle are summarized and analyzed. Section 4 shows the performance of some MOIAs on tackling MOPs. Finally, the conclusions and development trends of MOIAs are given in Sect. 5.

2 Preliminary

2.1 Multi-objective Optimization Problems

Normally, a multi-objective optimization problem can be modeled as below,

$$\underset{x \in \Omega}{\mathrm{Min}}\, F(x) = (f_1(x), f_2(x), \ldots, f_m(x))^T \tag{1}$$

where $x = (x_1, \ldots, x_n)^T$ is a decision vector with n dimensions in Ω, and $F: \Omega \rightarrow R^m$ defines the mapping from Ω to R^m (the m-dimensional objective space). However, the objectives often conflict with each other, thus a single solution cannot minimize all the objectives simultaneously. The best trade-offs among the objectives can be properly attained by using the definition of Pareto optimality.

Definition 1 (Pareto-dominance): For two solutions $x_1, x_2 \in \Omega$, we can call that x_1 dominates x_2 (noted by $x_1 \prec x_2$) if and only if $f_i(x_1) \leq f_i(x_2)$ for all $i \in \{1, 2, \ldots, m\}$ and $f_j(x_1) \neq f_j(x_2)$ for at least one $j \in \{1, 2, \ldots, m\}$.

Definition 2 (Pareto-optimal): A solution x is said to be Pareto-optimal if and only if $\neg \exists y \in \Omega$: $y \prec x$.

Definition 3 (Pareto-optimal set): The set ***PS*** is composed by all the Pareto-optimal solutions, defined as $PS = \{x | \neg \exists y \in \Omega: y \prec x\}$.

Definition 4 (Pareto-optimal front): The set ***PF*** includes the values of all the objective functions corresponding to the Pareto-optimal solutions in ***PS***, i.e., $PF = \{F(x) = (f_1(x), f_2(x), \ldots, f_m(x))^T | x \in PS\}$.

2.2 A General Framework of Multi-objective Immune Algorithms

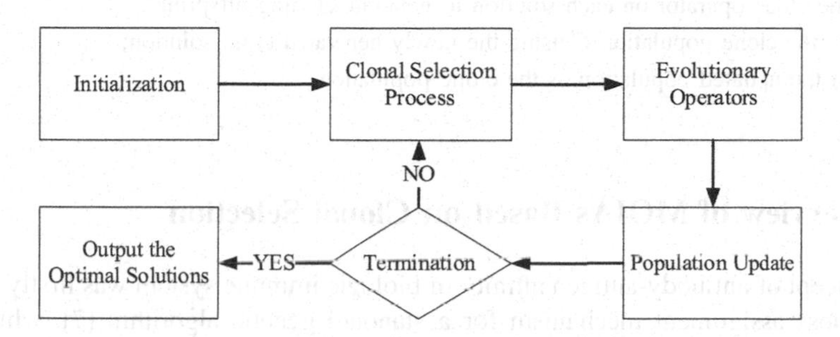

Fig. 1. A general framework of MOIAs.

Figure 1 shows a general framework of MOIAs, which includes six main processes. At the beginning, the population and some relative parameters are initialized during the *Initialization* process. After that, the fitness value of each antibody in the population is calculated for *Clonal Selection*, in which the clone number is assigned for each antibody according to its corresponding fitness value. For most existing MOIAs, the crowding distance value [6] is often used as the metric of clonal selection. After the process of clonal selection, some evolutionary operators, including the cloning, crossover, and mutation operators, are performed successively. Meanwhile, the process of *Population Update* is performed to replace some worse solutions with the promising offspring. The main loop of MOIAs will be terminated when reaching the stop condition. At last, all the antibodies in the final population are outputted as the optimal solutions.

2.3 Clonal Selection Principle in MOIAs

In this part, the principle of clonal selection is introduced. Firstly, n_A solutions with the highest fitness values in the population (namely P) are selected to build the clone population, called C. Then, the clone operator is run on the clone population, which is formulated as follows:

$$P = \sum_{i=1}^{n_A} c_i \otimes a_i, \quad a_i \in C \tag{2}$$

where the symbol $\otimes$ means to copy the individual. c_i indicates the number of clones for solution a_i, which is quantified according to its corresponding fitness value. According to the principle of clonal selection, only a small ratio of individuals with good performance on convergence and diversity are picked up for clonal proliferation. The pseudo-code of clonal selection is given below.

The pseudo-code of clonal selection
1 Select n_A solutions with the highest fitness value to form clone population C;
2 Calculate the clones number c_i for each solution in C;
3 Run the clone operator on each solution to generate cloning offspring;
4 Update the clone population C using the newly generated child solution;
5 Return the updated population as the clone population.

3 Overview of MOIAs Based on Clonal Selection

The concept of antibody-antigen affinity in biologic immune system was firstly applied as a fitness assignment mechanism for a standard genetic algorithm [7], which was considered to be a first attempt to propose a MOIA. On the other hand, the first real-coded MOIA may retrospect to a non-dominated neighbor-based MOIA, namely NNIA [8], in which the crowding distance value is used as the fitness value to assign the clone number for each solution, as defined by,

$$c_i = \left\lceil N \times \frac{CD(a_i)}{\sum_{j=1}^{n_A} CD(a_j)} \right\rceil, \quad a_i \in C \tag{3}$$

where $CD(a_i)$ means the crowding distance value of solution a_i, N indicates the size of population. It should be noticed that the crowding distance values for the boundary solutions are set to be the double of the maximum crowding distance value among n_A solutions. By this way, more clones will be assigned to the less-crowded and boundary areas.

Although the crowding distance based clonal selection principle shows some advantages on the convergence speed, it is normal to encounter the problem regarding the lack of diversity.

To alleviate this issue, a large number of MOIAs based on the clonal selection principle were proposed to enhance the ability of MOIA on maintaining the diversity. According to the part of improvement, most existing MOIAs are proposed with promising performance through designing some novel strategies or operators in the different parts of evolutionary process, which mainly involves the process of clonal selection, the process of evolutionary and the process of population update.

3.1 Improvement in the Process of Clonal Selection

At the aspect of cloning operator, some MOIAs are performed to strengthen the ability of immune algorithm on maintaining diversity, which designs a novel clonal selection principle. For example, NICA [9] proposes a novel clonal selection principle that selects the antibody according to the Pareto-dominance relationship, in which only non-dominated solutions will be selected to proliferation. Moreover, the entire cloning is applied in NICA instead of using the proportional cloning strategy. Similar to NICA, the relationship of Pareto-dominance is also applied in IDSMOA [10] to perform the cloning process. More recently, in order to enhance the ability of MOIA on tackling complicated MOPs with complicated *PS* or *PF*, a novel decomposition-based selection operator is designed in MOIA-DCSS [11], in which the aggregated function values instead of the crowding distance values are used as the fitness value for each solution.

According to the mechanism of the decomposition-based clonal selection, the diversity is guaranteed by using a set of uniformly weight vectors, and the convergence is speeded up by using the aggregated function, which is crucial for solving some complicated MOPs.

In terms of the quantum immune cloning, an adaptive niche quantum-based immune algorithm (ANQICA) [12] is proposed for solving multi-model optimization, in which the quantum coding, immune clonal selection and a niche mechanism are cooperated with each other. In particular, the quantum coding has a strong search ability that can better explore the objective space. The niche mechanism is performed to guarantee that the algorithm can converge to the multi-extremum. This way, ANQICA achieves a better balance between convergence and diversity of the population. Moreover, the quantum immune cloning is also applied in QICCA [13] to tackle the dynamic multi-objective optimization problems (DMOPs). Firstly, a quantum updating operation is designed by adopting the entire cloning and evolving the theory of quantum, which aims to enhance the ability of exploration. Secondly, in order to improve the uniformity and the convergence performance of each solution in population, a co-evolutionary mechanism is incorporated in global operation, including a co-evolutionary competitive operation and a co-evolutionary cooperative operation simultaneously. Similar to ANQICA and QICCA mentioned above, the immune clonal operator and a co-evolutionary mechanism that combine the co-evolutionary competitive operator and cooperative operator are implemented in ICCoA [14], which shows a promising performance on tackling the DMOPs. The main difference between QICCA and ICCoA is that the entire clonal operator is used in the former while the proportional clonal operator is applied in the later.

3.2 Improvement in the Process of Evolution

At the aspect of other heuristic operators, many MOIAs combining with some other heuristic operators are designed to improve the robustness of traditional immune algorithm, which are illustrated below.

In terms of the differential evolution operator, DMMO [15] is proposed with two different evolutionary models, which includes differential evolution based evolutionary model and clonal selection based evolutionary model. In the first model, the differential evolution operator is applied to enhance the objective. For the second model, it uses proportional cloning, recombination and hyper-mutation operators to strengthen the multiple objectives. According to the different evolutionary stage with different characteristics, AIMA [16] divides the whole evolutionary process into three stages, including the early stage, middle stage and later stage, respectively. There are three differential evolution operators with different search characteristics designed for each stage. As the information of search space at the early stage is limited, a DE with a strong global search capability is designed to explore the whole search space effectively. While the main purpose in the later stage is to find the better solutions around the current solutions as much as possible. Hence, a DE with strong local search capability is used in the later stage to concentrate on the exploitation. For the middle stage, a DE is designed to balance exploration and exploitation. Similar to AIMA, a novel adaptive DE operator is proposed in ADE-MOIA [17], which mainly includes a suitable parent strategy and an adaptive parameter control approach. Firstly, in order to provide a correct search and evolutionary direction, two parents are selected from the current population and dominated population, respectively. Then, the crossover rate and scaling factor in ADE-MOIA are adaptively varied according to the evolutionary process and the success rate of offspring, which aims to achieve a better balance between the convergence and the diversity. A uniform distribution selection mechanism is proposed in AUDHEIA [18], in which the individuals in the population are mapped to a hyper-plane firstly. Then, they are clustered to increase the diversity of solutions. Besides that, in order to enhance the distribution of the solutions, the mapped hyper-plane is sectioned evenly and a dynamic threshold is designed as the standard to judge the distribution condition. In case that there is not sufficient solution, the limit optimization variation strategy of the best individual is also applied. A DE inspired recombination strategy that combines a newly designed DE operator and SBX is proposed in IMADE [19].

In terms of the Baldwin effect [20] that can direct the genotypic changes, *Gong et al.* firstly introduce the Baldwinian learning into the artificial immune algorithm, termed as BCSA [21], to deal with MOPs. Based on the Baldwinian learning operator, which simulates the learning mechanism by using information from the antibody population to alter the search space, it can make full use of the exploration performed by the phenotype to facilitate the evolutionary search for good genotypes. After that, two learning mechanisms, including Baldwinian learning operator and orthogonal learning operator [22], are applied into HLCSA [23] during the process of clonal selection. In particular, the former operator aims to direct the genotypic changes based on the Baldwin effect, while the later one is used to search the space defined by one antibody and its best Baldwinian learning vector.

3.3 Improvement in the Process of Population Update

At the aspect of population update, a series of MOIAs are presented to alleviate the drawbacks of traditional immune algorithm through designing the population update. For instance, a dynamic population strategy is proposed in MOIA-DPS [24], in which the size of mating population is adjusted dynamically (enlargement or reduction) according to the status of the external archive. Moreover, in order to balance the diversity and the convergence more properly, a novel differential evolution operator with two search models is also designed in MOIA-DPS to further enhance the exploration ability and the robustness of algorithm. MIMO [25] based on a micro-population is proposed. Firstly, an effective fine-grained selection strategy is performed on the micro-population, which aims to preserve the diversity of population. Then, an adaptive mutation operator is also used according to the fitness value of each solution, which can promote to use relatively large steps for boundary and sparse areas with a high probability. This way, the exploratory capabilities of immune algorithm are improved. A hybrid evolutionary framework for MOIA is proposed in HEIA [26], in which the cloned individuals are divided into several sub-populations. Moreover, in order to prove the effectiveness of the proposed hybrid framework, simulated binary crossover and differential evolution with polynomial mutation are implemented in this framework to solve some complicated MOPs. A co-evolutionary strategy is designed for each sub-population in CONNIA [27], in which the characteristics of each sub-population can be effectively maintained and the diversity of the entire population can be also increased. Moreover, the reference experience from each sub-population is used to improve the efficiency of evolutionary search.

4 Studies on the Performance of MOIAs

In order to show the difference between each immune algorithm and the performance of MOIAs on tackling some MOPs. Four state-of-the-art immune algorithms based on clonal selection principle, including NNIA [8], HEIA [18], AIMA [16] and MOIA-DPS [24], are applied in our experimental comparison. It is easily known from the studies above that HEIA uses the sub-population co-evolutionary strategy to improve the performance, AMIA is enhanced by using three co-operative DE operators, while MOIA-DPS is strengthened by using a dynamic population size as well as a double-module DE search strategy.

Two groups of test problems with different characteristics are used in our experimental comparison, including WFG test problems [28] and UF test problems [29]. It should be pointed out that all the benchmark problems have two conflicting objectives. Besides that, the inverted generation distance (IGD) [30] is used as the metric to estimate the performance of each compared algorithm. It should be noticed that all the results based on IGD are obtained after 30 independent run and the relative parameters for each compared algorithm mentioned above are set according to the corresponding references [8, 16, 18, 24].

Table 1. The average IGD results and standard deviations obtained by each compared algorithm

Problems	HEIA [18]	MOIA-DPS [24]	AIMA [16]	NNIA [8]
WFG2	2.12E−02 (1.26E−03)(2)	**1.00E−02** **(5.18E−04)(1)**	7.82E−02 (8.28E−04)(3)	7.95E−02 (2.40E−03)(4)
WFG3	1.29E−02 (6.34E−04)(3)	**1.15E−02** **(1.19E−04)(1)**	1.18E−02 (1.78E−04)(2)	1.81E−02 (2.70E−03)(4)
WFG4	1.36E−02 (7.38E−04)(3)	**1.08E−02** **(2.78E−04)(1)**	1.29E−02 (5.92E−04)(2)	1.54E−02 (1.49E−03)(4)
WFG5	6.73E−02 (4.43E−04)(3)	**6.55E−02** **(3.87E−05)(1)**	6.65E−02 (2.04E−04)(2)	6.86E−02 (8.44E−04)(4)
UF2	3.50E−02 (6.22E−03)(3)	2.12E−02 (4.77E−03)(2)	**1.50E−02** **(2.28E−03)(1)**	6.63E−01 (0.00E+00)(4)
UF3	1.67E−01 (4.72E−02)(3)	1.12E−01 (4.60E−02)(2)	**8.59E−02** **(1.99E−02)(1)**	1.14E+00 (0.00E+00)(4)
UF4	5.39E−02 (8.02E-03)(3)	4.03E−02 (1.04E−03)(2)	**3.78E−02** **(1.01E−03)(1)**	1.89E−01 (0.00E+00)(4)
UF5	4.95E−01 (2.17E−01)(3)	3.05E−01 (2.14E−01)(2)	**2.95E−01** **(9.90E−02)(1)**	5.06E+00 (0.00E+00)(4)
Parameters setting	1) For WFG problems: N = 100, *MaxGen* = 25 000, k = 4, l = 20; 2) For UF problems: N = 300, *MaxGen* = 150 000;			

Table 1 summarizes the average IGD values and the standard deviations obtained by each compared algorithm, which shows the performance of each algorithm on solving WFG2–WFG5 and UF2–UF5 benchmark problems. It should be pointed out that the symbol of (1), (2), (3) and (4) mean the rank of all the compared algorithms on solving each test problem. Moreover, some relative parameters are summarized in the last row of Table 1, in which N indicates the population size, *MaxGen* is the maximum generation for each compared algorithm, k and l are the position-related and distance-related decision variables for WFG benchmark problems, respectively. The numbers of decision variables for all UF problems are set to 30. Table 2 gives the statistical results for all the compared algorithms based on the IGD results listed in Table 1.

Table 2. The statistical results for all compared algorithms on WFG and UF problems

Test problems	Rank	HEIA	MOIA-DPS	AIMA	NNIA
WFG	(1)	0	4	0	0
	(2)	1	0	3	0
	(3)	3	0	1	0
	(4)	0	0	0	4
UF	(1)	0	0	4	0
	(2)	0	4	0	0
	(3)	4	0	0	0
	(4)	0	0	0	4

Regarding the WFG benchmark problems, they test the convergence pressure of each algorithm. As shown in Table 2, MOIA-DPS can obtain all the best results among five cases, followed by AIMA, HEIA and NNIA. This obvious superiority of MOIA-DPS over others is mainly thanks to the dynamic clonal population size designed in it. According to DPS, the clonal population size is gradually decreased when the external archive is full. This way, it can save the computational resources as well as improve the convergence pressure in MOIA-DPS. Hence, a suitable clonal population size is efficiency for solving such benchmark problems that needs strong convergence pressure during the whole evolutionary process.

Regarding the UF test problems with various complex characteristics and vary complicated ***PS*** shapes, they test the ability on maintaining diversity of each algorithm. We can learn from the results listed in Table 2, AIMA obtains all the best results out of four cases, followed by MOIA-DPS, HEIA and NNIA. The main reason for the superiority of AIMA, MOIA-DPS and HEIA over NNIA is that the differential evolution operators are applied in them, while only SBX operator is used in NNIA. Hence, some evolutionary operators with strong disturbance abilities is crucial for maintaining the diversity. Moreover, three DE operators with different exploration ability are cooperated with each other in AIMA. This way, AIMA achieves a dynamic balance between the diversity and the convergence. The same as AIMA, a novel DE operator with two search modules is designed in MOIA-DPS, which also improves the disturbance ability. On the other hand, HEIA only combines SBX and DE operators in evolutionary process. Hence, AIMA and MOIA-DPS show some advantages over HEIA on solving some complex test problems that need a strong ability on maintaining the diversity during the whole evolutionary process.

To conclude, a suitable clonal population size, the evolutionary operators with strong exploration ability or an efficient clonal selection mechanism can alleviate the problems existing in most traditional immune algorithms that easily encounter premature convergence and lose diversity of the whole population.

5 Conclusions and Future Discussions

This paper presented a short survey of research works of MOIAs based on clonal selection principle. Firstly, we divide most existing clonal selection based MOIAs into several types according to the improvement part, which mainly includes improvement in the process of the clonal selection, evolutionary operators and population update. Then, we describe each immune algorithm in details that includes its design idea, the advantages as well as its drawbacks. Finally, some experimental comparisons are performed to show the difference advantages between different immune algorithms on solving various kinds of MOPs.

Although most existing MOIAs were proposed with some promising performance on solving MOPs, the study on the performance of IAs in a high dimensional space is just getting started. For most existing MOIAs, they may encounter some challenges when solving the optimization problems with many objectives (MaOPs). The first trouble is that the convergence pressure is ineffectiveness using traditional clonal selection principles. The second difficult is that it is easy for most existing IAs to lose the population diversity in a high dimensional space, which has a larger search space. Hence, regarding the future research directions, how to extend the traditional immune algorithm to solve the optimization problems in high dimensions may be a meaningful research filed. They are some suggestions that we can try. 1) Design a suitable and effectiveness clonal selection principle; 2) Design a novel clonal operator that has a strong ability on diversity maintaining in a larger search space. By this way, the ability of IAs in solving MaOPs may be strengthened; 3) Design a hybrid clonal operator that achieves a dynamic combination between different clonal operators with different characteristics. By this way, the robustness of IAs in solving different problems can be enhanced. In addition, the application of immune algorithm in real-life engineering optimization is also one research topic.

Acknowledgement. This work was supported by the National Natural Science Foundation of China (NSFC) under Grants 61876110, 61836005, and 61672358, the Joint Funds of the NNFC under Key Program Grant U1713212, and the Shenzhen Scientific Research and Development Funding Program under grant JCYJ20190808164211203.

References

1. Abdollahpour, S., Rezaeian, J.: Minimizing makespan for flow shop scheduling problem with intermediate buffers by using hybrid approach of artificial immune system. Appl. Soft Comput. **28**, 44–56 (2015)
2. Lima, F.P.A., Lopes, M.L.M., Lotufo, A.D.P., Minussi, C.R.: An artificial immune system with continuous-learning for voltage disturbance diagnosis in electrical distribution systems. Expert Syst. Appl. **56**, 131–142 (2016)
3. Souza, S.S.F., Romero, R., Franco, J.F.: Artificial immune networks Copt-aiNet and Opt-aiNet applied to the reconfiguration problem of radial electrical distribution systems. Electr. Power Syst. Res. **119**, 304–312 (2015)
4. Shivasankaran, N., Kumar, P.S., Raja, K.V.: Hybrid sorting immune simulated annealing algorithm for flexible job shop scheduling. Int. J. Comput. Intell. Syst. **8**(3), 455–466 (2015)
5. Farmer, J.D., Packard, N.H., Perelson, A.S.: The immune system, adaptation, and machine learning. Phys. D **22**, 187–204 (1986)
6. Miettinen, K.: Nonlinear Multiobjective Optimization. Birkhaüser, Basel (1998). https://doi.org/10.1007/978-3-0348-8280-4
7. Yoo, J., Hajela, P.: Immune network simulations in multicriterion design. Struct. Multidisc. Optim. **18**(2–3), 85–94 (1999)
8. Gong, M.G., Jiao, L.C., Du, H.F., Bo, L.F.: Multi-objective immune algorithm with nondominated neighbor-based selection. Evol. Comput. **16**, 225–255 (2008)
9. Shang, R.H., Jiao, L.C., Liu, F., Ma, W.P.: A novel immune clonal algorithm for MO problems. IEEE Trans. Evol. Comput. **16**(1), 35–50 (2012)
10. Xiao, J.K., Li, W.M., Li, X.R., Xiao, X.R., LV, C.Z.: A novel immune dominance selection multi-objective optimization algorithm for solving multi-objective optimization problems. Appl. Intell. **46**(3), 739–755 (2016). https://doi.org/10.1007/s10489-016-0866-z
11. Li, L.J., Lin, Q.Z., Liu, S.B., Gong, D.W., Coello, C.A.C., Ming, Z.: A novel multi-objective immune algorithm with a decomposition-based clonal selection. Appl. Soft Comput. **81**, 1–14 (2019)

12. Liu, J.Y., Wang, H.X., Sun, Y.Y., Li, L.: Adaptive niche quantum-inspired immune clonal algorithm. Nat. Comput. **15**(2), 297–305 (2015). https://doi.org/10.1007/s11047-015-9495-4
13. Shang, R.H., Jiao, L.C., Ren, J.Y., Li, L., Wang, L.P.: Quantum immune clonal coevolutionary algorithm for dynamic multiobjective optimization. Soft. Comput. **18**(4), 743–756 (2013). https://doi.org/10.1007/s00500-013-1085-8
14. Shang, R.H., Jiao, L.C., Ren, Y.J., Li, L., Wang, L.P.: Immune clonal coevolutionary algorithm for dynamic multiobjective optimization. Nat. Comput. **13**(3), 421–445 (2014). https://doi.org/10.1007/s11047-014-9415-z
15. Liang, Z.P., et al.: A double-module immune algorithm for multi-objective optimization problems. Appl. Soft Comput. **35**, 161–174 (2015)
16. Lin, Q.Z., et al.: An adaptive immune-inspired multi-objective algorithm with multiple differential evolution strategies. Inf. Sci. **430–431**, 46–64 (2018)
17. Lin, Q.Z., Zhu, Q.L., Huang, P.Z., Chen, J.Y., Ming, Z., Yu, J.P.: A novel hybrid multi-objective immune algorithm with adaptive differential evolution. Comput. Oper. Res. **62**, 95–111 (2015)
18. Qiao, J.F., Li, F., Yang, S.X., Yang, C.L., Li, W.J., Gu, K.: An adaptive hybrid evolutionary immune multi-objective algorithm based on uniform distribution selection. Inf. Sci. **512**, 446–470 (2020)
19. Qi, Y.T., Hou, Z.T., Yin, M.L., Sun, H.L., Huang, J.B.: An immune multi-objective optimization algorithm with differential evolution inspired recombination. Appl. Soft Comput. **29**, 395–410 (2015)
20. Bereta, M., Burczynski, T.: Immune K-means and negative selection algorithms for data analysis. Inf. Sci. **179**, 1407–1425 (2009)
21. Gong, M.G., Jiao, L.C., Zhang, L.N.: Baldwinian learning in clonal selection algorithm for optimization. Inf. Sci. **180**, 1218–1236 (2010)
22. Wang, Y., Cai, Z.X., Zhang, Q.F.: Enhancing the search ability of differential evolution through orthogonal crossover. Inf. Sci. **185**, 153–177 (2012)
23. Peng, Y., Lu, B.L.: Hybrid learning clonal selection algorithm. Inf. Sci. **296**, 128–146 (2015)
24. Lin, Q.Z., et al.: A multi-objective immune algorithm with dynamic population strategy. Swarm Evol. Comput. **50**, 1–12 (2018)
25. Lin, Q.Z., Chen, J.Y.: A novel micro-population immune multiobjective optimization algorithm. Comput. Oper. Res. **40**, 1590–1601 (2013)
26. Lin, Q.Z., et al.: A hybrid evolutionary immune algorithm for multiobjective optimization problems. IEEE Trans. Evol. Comput. **20**(5), 711–729 (2016)
27. Shi, J., Gong, M.G., Ma, W.P., Jiao, L.C.: A multipopulation coevolutionary strategy for multiobjective immune algorithm. Sci. World J. 1–23 (2014)
28. Huband, S., Barone, L., While, L., Hingston, P.: A scalable multi-objective test problem toolkit. In: Coello Coello, C.A., Hernández Aguirre, A., Zitzler, E. (eds.) EMO 2005. LNCS, vol. 3410, pp. 280–295. Springer, Heidelberg (2005). https://doi.org/10.1007/978-3-540-31880-4_20
29. Zhang, Q.F., Liu, W. D., Li, H.: The performance of a new version of MOEA/D on CEC09 unconstrained MOP test instances. In: 2009 IEEE Congress on Evolutionary Computation, pp. 203–208 (2009)
30. Bosman, P., Thierens, D.: The balance between proximity and diversity in multiobejctive evolutionary algorithms. IEEE Trans. Evol. Comput. **7**(2), 174–188 (2003)

Adaptive Artificial Immune System for Biological Network Alignment

Shiqiang Wang, Lijia Ma[(✉)], and Xiao Zhang

College of Computer Science and Software Engineering, Shenzhen University, Shenzhen 518060, China
wsql920@qq.com, ljma@szu.edu.cn, doudizhangxiao@163.com

Abstract. With the rapid developments of computer technology and biotechnology, a large amount of bioinformatics data such as bioprotein interaction networks have been generated. Protein interactions play an important role in most biochemical functions. In this paper, we propose an adaptive artificial immune system for the alignment of biological protein networks, which adaptively preserves excellent genes to maintain the diversity of individuals in population. Moreover, it adopts a crossover operator to simulate the sexual reproduction while it uses a simple mutation operator to mimic the mutations in the evolutionary process. Extensive experiments on real-world biological networks demonstrate the superiority of the proposed algorithm in maintaining the biological similarity and topological similarity of biological networks over the state-of-the-art algorithms.

Keywords: Bioinformatics data · Protein interaction network · Network alignment · Artificial immune system

1 Introduction

With the rapid development of bioinformatics in the past decades, biological research has produced extensive bio-protein interaction data. These data can be well represented by protein-protein interaction (PPI) networks, and the analysis on those data will greatly help the researchers to understand the potential biological process, such as homology analysis, reconstruction of evolutionary tree among biological objects, and prediction of protein function [1–3].

In recent years, biological network alignments have received great attention, which try to find the most similar regions in topology and function between two or more different bio-protein networks [4–6]. Classical methods include the local network alignment (LNA) and global network alignment (GNA) [7, 8]. LNA aims to find the alignment of one or more similar regions [9]. Note that, it is hard to reveal the potential relationships between biological networks systematically. GNA is the global alignment between two or more different biological networks [1, 10–12], and the studies on GNA are important to identify functional homology among species and explaining the preservation of functional evolution.

D.-S. Huang and K.-H. Jo (Eds.): ICIC 2020, LNCS 12464, pp. 560–570, 2020.
https://doi.org/10.1007/978-3-030-60802-6_49

Nowadays, many meta-heuristic algorithms for global alignment of bio-logical networks have been proposed. For instance, MAGNA [13] uses a genetic algorithm (GA) to directly maximize the accuracy in global network alignment while constructing the alignment, which only takes the edge conservation into consideration. MAGNA++ [14] is the advanced version of MAGNA which also takes node conservation into consideration. PSONA [15] uses a permutation-based particle swarm optimization algorithm (PSO) to search the optimal solution of network alignment. MeAlign [16] uses a memetic algorithm. Optnetalign [17] uses a multiobjective memetic algorithm to find a tradeoff between topological similarity and sequence similarity.

In this paper, we propose an artificial immune algorithm for global alignment in biological networks. To maintain diversity, we propose an adaptive clonal selection. Moreover, to replicate more high-quality individuals, we use the individuals with high fitness values to promote the proliferation of individuals. In addition, to further obtain diversity, we also adopt the concentration of antibody (i.e. individual) in cloning selection. Finally, we propose a local search to improve the performance of the proposed algorithm. Extensive experiments on the real-world biological networks demonstrate the superiority of the proposed algorithm in maintaining the biological similarity and topological similarity of biological networks over state-of-the-art algorithms.

2 Related Works

2.1 Definition of Network Alignment

PPI networks can be well represented by undirected graphs. $G^A(V_A, E_A)$ and $G^B(V_B, E_B)$ represent protein networks of two different species. V_A and V_B represent the proteins, while E_A and E_B represent the interactions between proteins of the G_A and G_B, respectively. The task of network alignment is to find an optimal mapping between two graphs G_A and G_B.

We denote the aforementioned mapping as permutation [15]. A mapping from G_A to G_B can be represented as follows:

$$M\{x_1, x_2, \ldots\ldots, x_{N_B}\}, \tag{1}$$

where, x_i ranges from 1 to N_B, and it represents that the i–th node in G_A is mapped to the x_i–th node in G_B. A simple representation of this mapping can be seen in Fig. 1. It is a NP-hard problem to find the optimal alignment.

2.2 Optimization Model

A good PPI network alignment needs to consider the preservation of both the topological similarity and the biological similarity of biological networks. Here, we adopt the following objective function f to evaluate the fitness of an alignment:

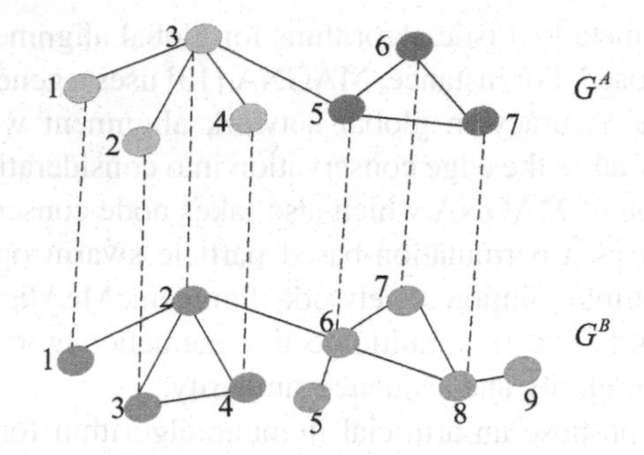

Fig. 1. Above is a simple representation of the mapping from G_A to G_B, in which nodes represent proteins, solid lines represent interactions between proteins in the same network, and dotted lines represent mappings between nodes in different networks. The permutation M represents the mapping from network G_A to network G_B.

$$f = \alpha T_s + (1 - \alpha) N_s, \tag{2}$$

where α is a weight parameter, while T_s and N_s are the topological similarity and biological similarity, respectively. T_s is generally is evaluated as follows:

$$T_s = \frac{|f(E_1)|}{|E_1|}, \tag{3}$$

where $|f(E_1)|$ is the number of edges in target network that are mapped by the edges of the source network. $|E_1|$ represents the number of edges in source network.

N_s is computed as follows:

$$N_s = \sum_{e_{ij} \in M} \frac{I_{ij}}{\max I}, \tag{4}$$

where $e_{ij} \in M$ represents the mapping edge from node $i \in G^A$ to node $j \in G^B$, and I represents the BLAST sequence similarity score matrix. Finally, our task is to optimize this model to get the maximal value of f.

3 Description of the Proposed Algorithm

3.1 ImAlign

In this paper, we apply artificial immune algorithm to global alignment of biological networks. As one of the most important operators of artificial immune system, clonal selection can replicate excellent subjects to keep excellent genes transferred to the next

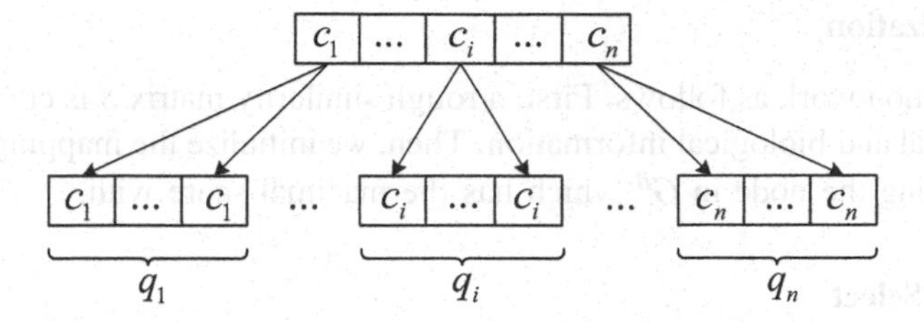

Fig. 2. Schematic illustration of memory cells cloning.

generation [18, 19] and accelerate the speed of solution and convergence [20]. An illustration of the clonal process is vividly depicted in Fig. 2. Memory cell is represented as $C = \{c_1, c_2, \ldots, c_n\}$, in which n is the size of memory bank. C can be cloned to a new population $P^c(c_i) = \{c_i^1 + c_i^2 + \cdots + c_i^{q_i}\}$ which is defined as follows:

$$\begin{aligned} P^c(c_1 + c_2 + \cdots + c_n) &= P^c(c_1) + P^c(c_2) + \cdots + P^c(c_n) \\ &= \{c_1^1 + c_1^2 + \cdots + c_1^{q_1}\} + \{c_2^1 + c_2^2 + \cdots + c_2^{q_2}\} \\ &+ \cdots + \{c_n^1 + c_n^2 + \cdots + c_n^{q_n}\}, \end{aligned} \tag{5}$$

where $P^c(c_i) = \{c_i^1 + c_i^2 + \cdots + c_i^{q_i}\}$, $c_i^j = c_i$, $i = 1, 2, \ldots, n$, $j = 1, 2, \ldots, q_i$. q_i is the number of memory cell c_i being cloned, which is calculated as:

$$q_i = \lceil n_c \times P_i \rceil, \tag{6}$$

where n_c is the size of clone population, and P_i is the adaptive possibility of c_i to be cloned and will be given latter. To enlarge the diversity of population, the concentration of antibody is taken into consideration while cloning. The algorithm framework of the proposed algorithm (called as ImAlign) is shown in **Algorithm 1**.

Algorithm 1 ImAlign

1: **Input**: two networks G^A and G^B, maximum number of iterations N, number of memory cells N_m.
2: Initialize the population.
3: **for** i=0 to N **do**
4: Select the top N_m individuals in population to create memory cells.
5: Clone memory cells to generate clonal population.
6: Crossover clonal population to generate offspring population.
7: Mutate the offspring population.
8: Localsearch the offspring population.
9: Refresh the population.
10: **end for**
11: **Output**: the best individual.

3.2 Initialization

The initialization work as follows. First, a rough similarity matrix S is created based on the topological and biological information. Then, we initialize the mapping node of a in G^A by choosing the node in G^B which has the maximal score with.

3.3 Clone Select

To solve the problem of excessive concentration of antibodies, a concentration-based clonal inhibition method is proposed (see **Algorithm 2**). When an individual's concentration is too high, it will inhibit its probability of being selected. When the concentration of an individual is too low, the probability of its selection will be increased appropriately. As shown in **Algorithm 2**.

Algorithm 2 CloneSelect

1: **Input**: Memory cells.
2: Calculate P_f and P_d of each memory cell.
3: Calculate the probability of each memory cell being cloned P according to (9).
4: **repeat**
5: Generate new individuals based on the P of each memory cell.
6: **Until** the size of cloned population reaches N.
7: Integrating memory cells into population.
8: **Output**: New population.

When calculating the clonal selection probability of memory cells, both the fitness probability and the concentration probability are taken into consideration. Like traditional clonal selection algorithms, high fitness promotes to clone more individuals with high fitness, and better genes will be inherited. However, high concentration inhibits clone to ensure the diversity. Here we set Pf as fitness probability and its formula is as follows:

$$p_f(i) = \frac{fitness(i)}{\sum_{j=1}^{n} fitness(j)}. \tag{7}$$

We define the degree of difference between individual i and individual j as d_{ij}, $d_{ij} = \sum ij_k$, where ij_k represents the difference between individual i and individual j at the k-th position. If the values at the k-th position are same, then the ij_k is set as 0, otherwise 1. d_{ij} is the total degree of difference between individual i and j. And, the greater the d_{ij}, the greater the difference between individual i and j. D_i is the total difference of individual i among the population, which is defined as $D_i = \sum d_{ij}$. Finally the concentration probability is defined as follows:

$$p_d(i) = \frac{D_i}{\sum_{j=1}^{n} D_j}. \tag{8}$$

The greater the difference D_i, the lower the concentration, and the greater the probability of being selected.

Considering the fitness and concentration, we propose an adaptive clonal selection method. The selection probability is defined as follows:

$$P = \gamma * P_f + (1 - \gamma) * P_d. \tag{9}$$

3.4 Crossover Operator

To avoid invalid crossover, memory cells are divided into two elite teams and multiplied into two new populations respectively. As is shown in Fig. 3, different colors represent different individuals. When crossing, one parent comes from elite team 1, and the other parent comes from elite team 2, this strategy can make good use of high quality genes to form higher-quality offspring, and avoid invalid crossing between two same parents.

Here, individuals are represented as permutations from 1 to N_B. The crossover operator proposed by Saraph in 2014 [13] is adopted, which decomposes the permutation into cycles and then crossed.

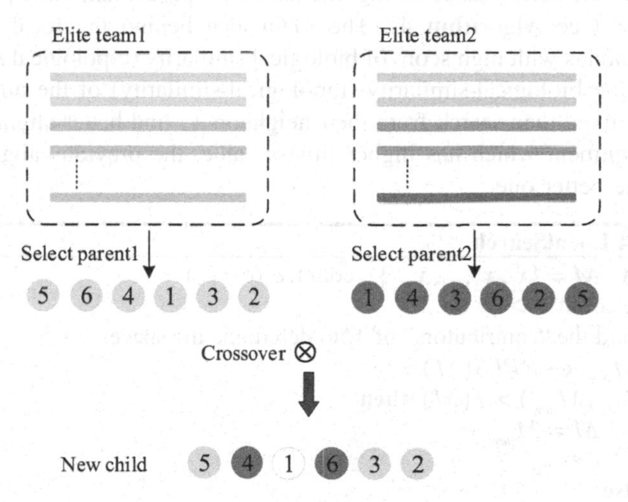

Fig. 3. An illustration of crossover operator. When the crossover operator is executed, one parent is selected from elite team1, and the other is selected from elite team2.

3.5 Mutation Operator

Mutation can increase the diversity of population. Here, in order to ensure the diversity of the population in the evolutionary process, a simple mutation is also proposed to expand the search space, which is given in **Algorithm 3**.

Algorithm 3 Mutation

1: **Input**: A individual.
2: Generate a random number *r*.
3: **if** $r \leq P_{mutate}$ **then**
4: Randomly generate two different numbers i, j in the range of $1\text{-}N_B$.
5: Exchange: $M(i) \rightleftharpoons j$.
6: **else**
7: Do nothing.
8: **end if**
9: **Output**: New individual.

In mutation, two sites *i* and *j* are selected randomly, and then the values of them are exchanged.

3.6 Local Search

After crossover and mutation, a new population is generated, and then it is integrated with memory cells (elite individuals of previous generation).

To improve the performance of alignments, we propose a randomly permuting local search (RPLS) (see **Algorithm 4**). The main idea behind the local search is that mapping two nodes with high score of biological similarity (topological similarity) will contribute to the biological-similarity (topological-similarity) of the alignment. Keep the key nodes fixed then search from their neighbors to find better alignments. If there is a better alignment which has higher fitness value, the previous alignment will be updated by the better one.

Algorithm 4 LocalSearch

1: **Input**: $M = \{x_1, x_2, ..., x_{N_B}\}$, count: $c_0 (c = c_0)$.
2: **while** count $c > 0$ **do**
3: Find the "contributors" of *M* to determine the space.
4: $M_{new} \leftarrow RPLS(M)$.
5: **if** $f(M_{new}) > f(M)$ **then**
6: $M = M_{new}$.
7: $c = c_0$.
8: **else**
9: $c = c - 1$.
10: **end if**
11: **end while**
12: **Output**: A better alignment.

4 Experimental Results

4.1 Evaluation Metric

Edge Correctness (EC). *EC* represents the proportion of the number of conserved edges in the target network to the number of edges in the source network. It is defined as:

$$EC = \frac{|f(E_1)|}{|E_1|}, \tag{10}$$

in which $|f(E_1)|$ is the edges of target network that are mapped by the edges of the source network. $|E_1|$ represents the number of edges in source network.

Functional Coherence (FC). Singh et al. in 2008 proposed the Functional Coherence as evaluation criteria to evaluate the functional consistency of matched protein nodes in alignment [1]. For each protein, there are *GO* terms that belong to it. For each pair of matched proteins in alignment, the more *GO* terms they have in common, the closer the functional correlation between the two proteins.

$$FC(u, v) = \frac{|S_u \cap S_v|}{|S_u \cup S_v|}. \tag{11}$$

In the equation, S_u and S_v are the standard *GO* term sets corresponding to the matched protein nodes *u* and *v* in the alignment. The *FC* value of a network alignment is the average value of *FC*s of all matched protein nodes in the alignment.

4.2 Experimental Setting

Data Set: The experiments are carried out on 10 pairs of protein interaction networks from five eukaryotes in real world. These five species are: mouse, worm, yeast, fruitfly and human. All the data used in the experiments such as the PPI networks of five species of eukaryotes and the BLAST scores of sequence similarity [21], comes from a public data set website IsoBase [22].

Algorithms: In order to show the superiority of our algorithm, we compare with several traditional algorithms with better performance, such as IsoRank [1], MAGNA++ [14], PSONA [15] and MeAlign [16].

Parameters: For all algorithms, the number of population is 100, the number of iterations is 300. For the tradeoff between biological similarity and topological similarity, α is set as 0.1. Other parameters are set default values in the original paper or the recommended values. For the parameters not mentioned, they are set to be the most appropriate.

Table 1. The basic information of real-world PPI network data used in our experiments.

Dataset name	Species	Nodes	Edges	k
M.musculus (mm)	Mouse	290	242	1.669
C.elegansce (ce)	Worm	2805	4495	3.205
S.cerevisiae (sc)	Yeast	5499	31261	11.37
D.melanogaster (dm)	Fruitfly	7518	25635	6.820
H.sapiens (hs)	Human	9633	34327	7.127

4.3 Experimental Results

In this paper, we apply artificial immune algorithm to the global alignment of biological networks. We conduct 10 groups of comparative experiments of five biological protein networks in Table 1. We compare several representative algorithms, which are IsoRank, MAGNA++, PSONA and MeAlign. In order to avoid accidental errors, we take the average of 30 calculation results. We summarize the results of 5 algorithms on 10 pairs of networks. To show the superiority of our algorithm, we make a statistical comparison of the experimental results (see Fig. 4).

More detailed results are given in Table 2.

The maximum value of each row is marked in bold. From the table, we can see that, in terms of topological similarity, ImAlign performs the best and achieves the maximum value for 7 times. At the same time, in terms of biological similarity, ImAlign also performs the best and achieves the maximum value for 8 times.

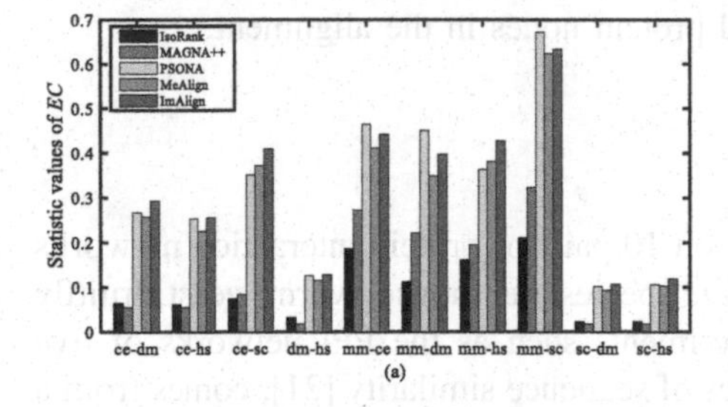

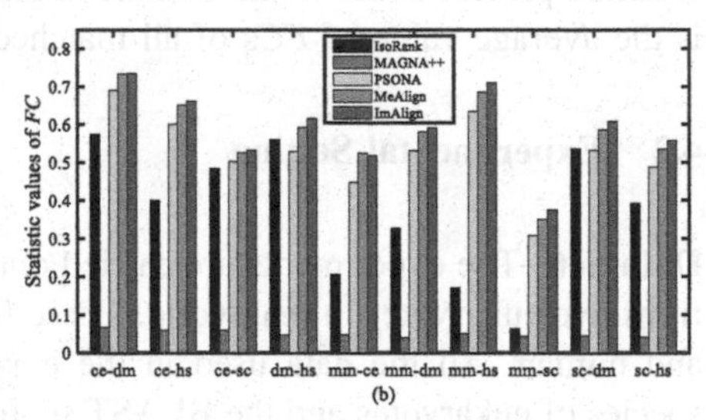

Fig. 4. Statistic values of EC and FC of the five compared algorithms over ten pairs of PPI networks.

Table 2. The averaged *EC* and *FC* ($\alpha = 0.1$) obtained by IsoRank, MAGNA++, PSONA, MeAlign and ImAlign on 10 pairwise networks over 30 independent trials.

Networks	Criteria	IsoRank	MAGNA++	PSONA	MeAlign	ImAlign
mm-ce	*EC*	0.1860	0.2715	**0.4653**	0.4112	0.4420
	FC	0.2060	0.0478	0.4478	**0.5243**	0.5183
mm-sc	*EC*	0.2107	0.3233	**0.6725**	0.6241	0.6332
	FC	0.0628	0.0442	0.3079	**0.3753**	**0.3753**
mm-dm	*EC*	0.1116	0.2218	**0.4514**	0.3495	0.3976

(*continued*)

Table 2. (*continued*)

Networks	Criteria	IsoRank	MAGNA++	PSONA	MeAlign	ImAlign
	FC	0.3278	0.0392	0.5308	0.5807	**0.5907**
mm-hs	*EC*	0.1612	0.1941	0.3635	0.3808	**0.4276**
	FC	0.1702	0.0502	0.6341	0.6849	**0.7106**
ce-sc	*EC*	0.0743	0.0843	0.3509	0.3721	**0.4097**
	FC	0.4840	0.0588	0.5018	0.5259	**0.5321**
ce-dm	*EC*	0.0652	0.0546	0.2660	0.2567	**0.2923**
	FC	0.5737	0.0678	0.6906	0.7336	**0.7356**
ce-hs	*EC*	0.0612	0.0548	0.2526	0.2243	**0.2543**
	FC	0.4001	0.0597	0.6005	0.6510	**0.6616**
sc-dm	*EC*	0.0234	0.0171	0.1010	0.0928	**0.1071**
	FC	0.5357	0.0452	0.5318	0.5855	**0.6087**
sc-hs	*EC*	0.0232	0.0180	0.1005	0.1025	**0.1182**
	FC	0.3927	0.0417	0.4886	0.5349	**0.5567**
dm-hs	*EC*	0.0324	0.0180	0.1256	0.1156	**0.1286**
	FC	0.5597	0.0464	0.5287	0.5925	**0.6157**

5 Conclusion

In this paper, we have proposed an adaptive artificial immune algorithm (ImAlign) to solve the problem of biological network alignment. To obtain a good alignment, ImAlign has proposed an adaptive clonal section method, which combines the adaptive fitness and concentration of individuals. Individuals with high fitness and low concentration are more likely to be cloned, while individuals with low fitness and high concentration are less likely to be cloned. ImAlign has also proposed an efficient local search method to optimize the best alignments. Extensive experiments on ten pairs of PPI networks across five real-word species have demonstrated the effectiveness of ImAlign in preserving the topological and biological information of biological networks. In future work, we will apply our approach to alignment of social networks.

Acknowledgement. This work was supported by the National Natural Science Foundation of China under Grant 61803269, in part by the Natural Science Foundation of Guangdong Province under Grant 2020A1515010790, and in part by the Technology Research Project of Shenzhen City under Grant by JCYJ20190808174801673.

References

1. Singh, R., Xu, J., Berger, B.: Global alignment of multiple protein interaction networks with application to functional orthology detection. Proc. Natl. Acad. Sci. **105**(35), 12763–12768 (2008)
2. Panni, S., Rombo, S.E.: Searching for repetitions in biological networks: methods, resources and tools. Brief. Bioinformatics **12**(1), 118–136 (2013)

3. Todor, A., Dobra, A., Kahveci, T.: Probabilistic biological network alignment. IEEE/ACM Trans. Comput. Biol. Bioinf. (TCBB) **10**(1), 109–121 (2013)
4. Clark, C., Kalita, J.: A comparison of algorithms for the pairwise alignment of biological networks. Bioinformatics **30**, 2351–2359 (2014)
5. Aladağ, A.E., Erten, C.: Spinal: scalable protein interaction network alignment. Bioinformatics **29**(7), 917–924 (2013)
6. Meng, L., Striegel, A., Milenković, T.: Local versus global biological network alignment. Bioinformatics **32**(20), 3155–3164 (2016)
7. Sun, Y., Crawford, J., Tang, J., Milenković, T.: Simultaneous optimization of both node and edge conservation in network alignment via WAVE. In: Pop, M., Touzet, H. (eds.) WABI 2015. LNCS, vol. 9289, pp. 16–39. Springer, Heidelberg (2015). https://doi.org/10.1007/978-3-662-48221-6_2
8. Faisal, F.E., Meng, L., Crawford, J., Milenković, T.: The post-genomic era of biological-network alignment. EURASIP J. Bioinf. Syst. Biol. **2015**(1), 3 (2015)
9. Ciriello, G., Mina, M., Guzzi, P.H., Cannataro, M., Guerra, C.: Alignnemo: a local network alignment method to integrate homology and topology. PloS one **7**(6), e38107 (2012)
10. Lee, D.S., Park, J., Kay, K., Christakis, N., Oltvai, Z., Barabási, A.L.: The implications of human metabolic network topology for disease comorbidity. Proc. Natl. Acad. Sci. **105**(29), 9880–9885 (2008)
11. Aparício, D., Ribeiro, P., Milenković, T., Silva, F.: Temporal network alignment via GoT-WAVE. Bioinformatics **35**, 3527–3529 (2019)
12. Patro, R., Kingsford, C.: Global network alignment using multiscale spectral signatures. Bioinformatics **28**(23), 3105–3114 (2012)
13. Saraph, V., Milenković, T.: Magna: maximizing accuracy in global network alignment. Bioinformatics **30**(20), 2931–2940 (2014)
14. Vijayan, V., Saraph, V., Milenković, T.: Magna++: maximizing accuracy in global network alignment via both node and edge conservation. Bioinformatics **31**(14), 2409–2411 (2015)
15. Huang, J., Gong, M., Ma, L.: A global network alignment method using discrete particle swarm optimization. IEEE/ACM Trans. Comput. Biol. Bioinf. (TCBB) **15**(3), 705–718 (2018)
16. Gong, M., Peng, Z., Ma, L., Huang, J.: Global biological network alignment by using efficient memetic algorithm. IEEE/ACM Trans. Comput. Biol. Bioinf. **13**(6), 1117–1129 (2016)
17. Clark, C., Kalita, J.: A multiobjective memetic algorithm for PPI network alignment. Bioinformatics **31**(12), 1988–1998 (2015)
18. Ulutas, B.H., Kulturel-Konak, S.: A review of clonal selection algorithm and its applications. Artif. Intell. Rev. **36**(2), 117–138 (2011)
19. Chen, D., Lü, L., Shang, M.S., Zhang, Y.C., Zhou, T.: Identifying inuential nodes in complex networks. Physica A **391**(4), 1777–1787 (2012)
20. Mohamed Elsayed, S.A., Ammar, R.A., Rajasekaran, S.: Artificial immune systems: models, applications, and challenges. In: Proceedings of the 27th Annual ACM Symposium on Applied Computing, pp. 256–258. ACM (2012)
21. Altschul, S.F., Gish, W., Miller, W., Myers, E.W., Lipman, D.J.: Basic local alignment search tool. J. Mol. Biol. **215**(3), 403–410 (1990)
22. Park, D., Singh, R., Baym, M., Liao, C.S., Berger, B.: IsoBase: a database of functionally related proteins across PPI networks. Nucleic Acids Res. **39**(suppl1), D295–D300 (2011)

A Novel Decomposition-Based Multimodal Multi-objective Evolutionary Algorithm

Wu Lin(✉), Yuan Li, and Naili Luo

Shenzhen University, Shenzhen 518060, China
{linwu2017,2172272145}@email.szu.edu.cn,
luonaili@163.com

Abstract. Recently, multimodal multi-objective optimization problems (MMOPs) have got widespread attention, which brings difficulties and challenges to current multi-objective evolutionary algorithms in striking a good balance between diversity in decision space and objective space. This paper proposes a novel decomposition-based multimodal multi-objective evolutionary algorithm, which comprehensively considers diversity in both decision and objective spaces. In environmental selection, a decomposition approach is first used to divide union population into K subregions in objective space and the density-based clustering method is used to divide the union population into different clusters in decision space. Then, the nondominated solutions in the same cluster of each subregion are first selected, and then the remaining ones with good convergence in objective space are further selected to form a temporary population with more than N solutions (N is the population size). Next, temporary population is divided into K subregions by a decomposition approach. The pruning process, which deletes one most crowding solution in the most crowding subregion at each time, will be repeatedly run until there are N solutions left. The experimental results demonstrate that our proposed algorithm can better balance diversity in both decision and objective spaces on solving MMOPs.

Keywords: Evolutionary algorithm · Multi-objective optimization · Multimodal optimization · Clustering · Decomposition approach

1 Introduction

Multi-objective optimization problems (MOPs) usually contain more than one objective to be optimized simultaneously, which can be described as follows:

$$\min_{x \in \mathbb{R}^n} F(x) = (f_1(x), \ldots, f_m(x))^T \quad (1)$$

where $\mathbb{R}^n$ is the decision space of a MOP and $x = (x_1, \ldots, x_n)$ is a decision vector with n dimensions in $\mathbb{R}^n$. $f_1(x), \ldots, f_m(x)$ are m objective functions of x, which are conflicting to each other. These Pareto-optimal or non-dominated solutions compose the Pareto optimal set (PS). The set of solutions in the objective space that corresponds to PS is called the Pareto front (PF) [1]. For MOPs, improvement of one objective will

D.-S. Huang and K.-H. Jo (Eds.): ICIC 2020, LNCS 12464, pp. 571–582, 2020.
https://doi.org/10.1007/978-3-030-60802-6_50

always lead to deterioration of the other objectives. Therefore, it is impossible to find the best values of all objectives simultaneously. Multi-objective evolutionary algorithms (MOEAs) aim to find the best trade-off among different objectives, which are uniformly distributed on the Pareto optimal tradeoff surface to provide more choices for decision makers [2].

Recently, multimodal multi-objective optimization problems (MMOPs) are proposed in [3–5], having different PSs for the same PF or even existing local and global PSs simultaneously. For MMOPs, diversity of solutions in decision space is also key factor affecting the performance of final population. Unfortunately, diversity in the decision space is hardly considered in most existing MOEAs. For example, two famous MOEAs, one Pareto-based MOEA (NSGA-II [6]) and another decomposition-based MOEA (MOEA/D [7]), are unable to maintain diversity in decision space because solutions with better convergence in objective space always replace other solutions with worse performance in their environmental selection.

In recent years, there are several attentions focusing on maintaining diversity in decision space. For example, in [8], a concept of crowding distance in decision space was introduced and embedded into NSGA-II as a non-dominated sorting scheme, giving a new MOEA called Omni-optimizer. In DN-NSGA-II [9], a decision space based niching NSGA-II is designed to maintain different solutions corresponding to the same PF point. Moreover, the novel addition and deletion operators are designed in MOEA/D-AD [10], which can obtain multiple non-dominated solutions being far away from each other in decision space. Furthermore, there are other heuristic algorithms having been designed for solving MMOPs, i.e., particle swarm optimization (PSO) algorithms [11, 12]. For example, in [11], a new PSO algorithm using ring topology and special crowding distance (MO_Ring_PSO_SCD) is proposed to solve MMOPs, which helps to maintain multiple PSs. Similarly, in [12], a self-organizing map network is used to construct the neighborhood in decision space, which can help to select good neighborhood leaders. Recently, in TriMOEA-TA&R [13], both the convergence and the diversity archives are used to cooperatively improve the performance, trying to balance diversity in two spaces. However, the above multimodal MOEAs (MMOEAs) [8–13] cannot find all local and global PSs simultaneously for tackling MMOPs with at least one local PS, as global PS always has good convergence that often dominate and replace local PS. When making decision, decision maker wants more options due to some uncontrollable factors. Thus, some local PSs with good performance are also expected to provide some useful information [13]. Based on the above motivation, in our proposed MMOEA/D-M2M, the multi-objective to multi-objective (M2M) decomposition approach [14] is used to maintain diversity in objective space, and the density-based clustering method (DBSCAN) [15] is used to strengthen diversity in decision space, which comprehensively considers diversity in both decision and objective spaces.

Next, Sect. 2 gives the related background in this paper, i.e., the multi-objective to multi-objective decomposition approach and the density-based clustering method. Section 3 introduces the mainframe of our algorithm and the environmental selection. Our simulation results and discussions are given in Sect. 4. Finally, Sect. 5 summarizes our research work and provides some advices.

2 Related Background

2.1 M2M Decomposition

The multi-objective to multi-objective (M2M) decomposition approach decomposes a MOP into multiple constrained multi-objective subproblems by dividing the objective space into multiple subregions with direction vectors [14]. The M2M decomposition approach has K unit vectors $w^1, \ldots, w^K$ in R^m_+. Then, divide R^m_+ into K subregions $\Omega_1, \ldots, \Omega_K$, where $\Omega_k(k = 1, \ldots, K)$ is

$$\Omega_k = \{x \in R^m_+ | \langle x, w^k \rangle \leq \langle x, w^j \rangle \text{ for any } j = 1, \ldots, K\} \quad (2)$$

where $\langle x, w^j \rangle$ is the acute angle between x and w^j. In other words, x belongs to Ω_k if and only if w^k has the smallest angle to x among all the K direction vectors.

2.2 DBSCAN

In the density-based clustering method DBSCAN [15], considering each sample data $x = (x_1, \ldots, x_n)$ in a data set D and the neighborhood parameters ε and *MinPts* (*MinPts* is positive integer), some concepts which are related to DBSCAN are introduced as follows:

ε-neighborhood: The ε-neighborhood $B(x)$ of x is defined as follows:

$$B(x) = \{y \in U | dist(x, y) \leq \varepsilon\} \quad (3)$$

where $d(x, y)$ is Euclidean distance between x and y.
Core objective: A sample data x is said to be core objective if:

$$B(x) \geq MinPts \quad (4)$$

Directly density-reachable: A sample data y is said to be direct density-reachable from x if:

$$y \in B(x) \text{ and } B(y) \geq MinPts \quad (5)$$

Density-reachable: The sample data y is said to be density-reachable from x if there are a set of sample data $p_1, p_2, \ldots, p_n$, where $p_1 = x$ and $p_n = y$ and p_{i+1} is direct density-reachable from p_i.
Density-connected: The sample data x and y are said to be density-connected, if there is a sample data p where x and y are density-reachable from it.

3 The Proposed Algorithm: MMOEA/D-M2M

In our proposed algorithm (i.e., MMOEA/D-M2M), M2M decomposition approach is first used to divide population into multiple subpopulations, which can be helpful to select more promising solutions. Then, it is used again to divide temp population into multiple subpopulations, which aims to find the most crowded solution. In this way, diversity of population in both decision and objective spaces are taken into consideration in MMOEA/D-M2M.

3.1 General Framework

The pseudo-code of the proposed method is shown in Algorithm 1. MMOEA/D-M2M shares a common framework that is employed by many MOEAs. First, a population P with N solutions is randomly initialized in the whole decision space and K uniformly distributed weight vectors are initialized by using the Das and Dennis [16] systematic approach. Then, a set of offspring solutions P' is generated by applying crossover and mutation operations on parents solutions P. Finally, N solutions are selected from the union set of P and P' by adopting an environmental selection procedure. The above steps continue until the stopping criterion is met.

Algorithm 1 General Framework

Input: N (population size), K (number of subregions)
Output: P (final population)

1 initialize a population P with N solutions
2 initialize K weight vectors $W=\{w^1, \ldots, w^K\}$
3 **while** the stopping criterion is not met
4 $\quad P' =$ Variation (P)
5 $\quad U = P \cup P'$
6 $\quad P =$ Environmental Selection (U, N, W)
7 **end while**
8 **return** P

3.2 Environmental Selection

Algorithm 2 shows the general framework of the environmental selection procedure. In line 1, P is initialized as empty sets. In line 2, union population U is normalized. Then, in line 3, U is divided into K subregions in objective space by the M2M decomposition approach. In line 4, the density-based clustering method DBSCAN is used to divide U into multiple clusters in the decision space. Then, the nondominated solutions of the same cluster in each subregion are first selected in lines 5–8. As shown in lines 9–14, the remaining ones with good convergence in objective space are further selected to form a temporary population P with more than N solutions (N is the population size). Next, P is first normalized and then divided into K subregions by the M2M decomposition approach in lines 15–16. In lines 17–21, the pruning process, which deletes one most crowded solution in the most crowded subregion each time, will be repeatedly run until there are N solutions left. Specifically, in line 18, the most crowded subregion P_i is identified by (6) as follows:

$$i = \max_{k}\{|P_k|\}, k = 1, \ldots, K \tag{6}$$

Then, the crowding distances of each solution x in decision space can be computed by the HAD method [17], which is shown as follows:

$$HAD(x) = \frac{|P_i| - 1}{\sum_{y \in P_i \text{ and } y \neq x} 1/d(x, y)} \tag{7}$$

where $d(x, y)$ is the Euclidean distance between normalized solutions x and y in decision space. The smaller the HAD value, the more crowded the solution in P_i. Thus, the solution x with minimal HAD value will be deleted from P_i and P. At last, P is returned in line 22 as the population for next generation.

Algorithm 2 Environmental Selection

Input: U (union population), N (population size), W (a set of weight vectors)
Output: P (updated population)

1 initialize P as empty set
2 normalize U
3 $(P_1,\ldots,P_K)$ = M2M decomposition (U, W)
4 $(C_1,\ldots,C_H)$ = DBSCAN $(U, \varepsilon, MinPts)$
5 **for** $k = 1$ **to** K
6 identify different clusters (C_i) of the solutions in P_k
7 find the non-dominated solutions in each C_i and add them into P
8 **end for**
9 $(F_1,\ldots,F_t)$ = *Non-dominated Sorting* (U)
10 $i = 1$
11 **while** $|P| < N$
12 $i = i + 1$
13 $P = P \cup (F_i \backslash P)$
14 **end while**
15 normalize P
16 $(P_1,\ldots,P_K)$ = M2M decomposition (P, W)
17 **while** $|P| > N$
18 identify the most crowding subregion P_i in $\{P_1,\ldots,P_K\}$ by (6)
19 find the most crowding solution x in P_i by (7)
20 delete x from P_i and P
21 **end while**
22 **return** P

4 Experimental Studies and Discussions

4.1 Parameter Settings

In this paper, 11 test problems were adopted from the benchmark MMOPs in CEC 2019 [4], which have many distinct Pareto optimal sets corresponding to the same PF (i.e., SYM-PART simple, SYM-PART rotated, Omni-test, MMF14 and MMF14_a), and coexistence of local and global PSs (i.e., MMF10-MMF13, MMF15 and MMF15_a).

In this paper, we adopted four MMOEAs (Omni-optimizer [8], DN-NSGA-II [9], MO_Ring_PSO_SCD [11] and TriMOEA-TA&R [13]) for performance comparison, and their parameters settings are suggested in their references. In MMOEA/D-M2M, the number of subregions K is set to be $N/20$ (where N is the size of population). SBX and polynomial-based mutation are used as evolutionary operators [6] in MMOEA/D-M2M, and their parameter settings can be found in [6]. For DBSCAN, the neighborhood parameter *MinPts* is set to be 10, and ε is set by an adaptive strategy as follows:

$$r = \frac{\sum_{i=1}^{n}(x_i^{\max} - x_i^{\min})}{n} \times 0.1 \tag{8}$$

where $x_i^{\min}$ and $x_i^{\max}$ are respectively the minimum and maximum values of i-th decision variable in union population, and n is the number of decision variables.

As suggested in [4], population size N is set to be $100 \times n$ and the maximum function evaluations is set to be $5000 \times n$ for each test problem (n is the number of decision variables of test problem). All the compared algorithms are run independently with 21 times on each test problem. The source codes of all algorithms were written in Matlab, and all experiments were run on PC with Intel Core i7-7700 CPU, 3.60 GHz.

4.2 Performance Metrics

Inverted Generational Distance (IGD [18, 19]) is a common indicator for performance comparison. If P^* is a set of sampling solutions on true PF or PS and P is the final population obtained by an evolutionary algorithm. Thus, IGD can be computed as follows:

$$IGD(P, P^*) = \frac{\sum_{i=1}^{|P^*|} \min_{j=1}^{|A|} d(p^i, x^j)}{|P^*|} \tag{9}$$

To evaluate the performance of the proposed algorithm, two performance metrics are adopted here, i.e., Inverted Generational Distance in decision space (named IGDX), and Inverted Generational Distance in objective space (named IGDF), which have been explained in their corresponding references [9].

4.3 Performance Comparisons

Table 1 and Table 2 give the IGDX and IGDF results of all the compared algorithms on all MMOPs, respectively. In terms of IGDX, MMOEA/D-M2M obtained the best results in 8 out of 11 cases, which are summarized in the second last row of Table 1. MMOEA/D-M2M failed to get the best results only on SYM-PART simple, MMF14, and MMF14_a, because TriMOEA-TA&R performed best on them. However, the other compared algorithms could not perform best in any case. On SYM-PART simple, MMOEA/D-M2M was only beaten by TriMOEA-TA&R, while it was significant better than other algorithms. For MMF14, and MMF14_a, MMOEA/D-M2M only obtained the third best results on them because TriMOEA-TA&R and MO_Ring_PSO_SCD outperformed our algorithm. As summarized in the second last row of Table 2, for the performance in objective space measured by IGDF, MMOEA/D-M2M achieved the best results on MMF10, MMF11, MMF12, MMF13, MMF15, and MMF15_a. Omni-Optimizer and TriMOEA-TA&R respectively performed best in 3 and 2 cases, while the rest competitors could not perform best in any case. Similar, MMOEA/D-M2M only obtained the third best results on MMF14, and MMF14_a, because TriMOEA-TA&R achieved the best results and MO_Ring_PSO_SCD achieved the second best results on

Table 1. Performance comparison of MMOEA/D-M2M and other MMOEAs on IGDX

Problem	Omni-Optimizer	DN-NSGA-II	MO_Ring_PSO_SCD	TriMOEA-TA&R	MMOEA/D-M2M
SYM-PART simple	5.558e+00 (2.61e+00)−	3.810e+00 (9.39e−01)−	1.663e−01 (2.19e−02)−	**2.181e−02 (3.01e−03)**+	8.691e−02 (5.46e−03)
SYM-PART rotated	4.405e+00 (1.22e+00)−	4.613e+00 (1.99e+00)−	1.972e−01 (3.50e−02)−	1.674e+00 (9.48e−01)−	**9.736e−02 (9.89e−03)**
Omni-test	1.501e+00 (2.29e−01)−	1.410e+00 (2.10e−01)−	4.057e−01 (8.74e−02)−	2.536e−01 (1.09e−01)−	**1.629e−01 (3.34e−02)**
MMF10	1.646e−01 (3.31e−02)−	1.617e−01 (4.48e−02)−	1.004e−01 (3.69e−02)−	2.014e−01 (7.52e−05)−	**1.563e−02 (1.90e−03)**
MMF11	2.502e−01 (3.58e−04)−	2.505e−01 (4.85e−04)−	2.169e−01 (2.72e−02)−	2.524e−01 (5.79e−05)−	**1.342e−02 (3.70e−03)**
MMF12	2.396e−01 (1.68e−02)−	2.467e−01 (9.44e−04)−	1.875e−01 (4.53e−02)−	2.477−01 (6.84e−04)−	**1.016e−02 (1.54e−02)**
MMF13	2.858e−01 (9.42e−03)−	2.895e−01 (1.54e−02)−	2.325e−01 (1.43e−02)−	2.714e−01 (6.55e−03)−	**1.109e−01 (8.12e−03)**
MMF14	8.873e−02 (6.29e−03)−	9.794e−02 (9.76e−03)−	5.329e−02 (1.86e−03)+	**3.651e−02 (4.92e−04)**+	5.473e−02 (2.00e−03)
MMF14_a	1.110e−01 (9.66e−03)−	1.196e−01 (5.83e−03)−	5.997e−02 (1.97e−03)+	**5.604e−02 (1.59e−03)**+	8.008e−02 (2.77e−03)
MMF15	2.471e−01 (2.10e−02)−	2.238e−01 (2.59e−02)−	1.562e−01 (1.63e−02)−	2.711e−01 (2.53e−04)−	**8.729e−02 (1.25e−02)**
MMF15_a	2.133e−01 (1.99e−02)−	2.086e−01 (1.46e−02)−	1.641e−01 (1.31e−02) ~	2.194e−01 (2.83e−03)−	**1.606e−01 (1.00e−02)**
Best/all	0/11	0/11	0/11	3/11	8/11
+/−/~	0/11/0	0/11/0	2/8/1	3/8/0	\

"+", "−", and "~" indicate that the results of the compared algorithms are significantly better than, worse than, and similar to that of MMOEA/D-M2M respectively by Wilcoxon's rank sum test with $\alpha = 0.05$.

them. For SYM-PART simple, SYM-PART rotated, Omni-test, MO_Ring_PSO_SCD was worse than Omni-Optimizer and DN-NSGA-II. In addition, MMOEA/D-M2M was beaten by TriMOEA-TA&R on Omni-test, while they achieved the similar results on SYM-PART simple and SYM-PART rotated. MMOEA/D-M2M was only better than MO_Ring_PSO_SCD on these three cases.

In addition, MMOEA/D-M2M is compared with each competitor on all problems, which are shown in Table 1 and Table 2, where "+/−/∼" indicates the competitor is respectively better than, worse than and similar to MMOEA/D-M2M. On IGDX, MMOEA/D-M2M was better than Omni-optimizer and DN-NSGA-II in all cases. MMOEA/D-M2M was better than MO_Ring_PSO_SCD in 8 out of 11 cases, i.e., SYM-PART simple, SYM-PART rotated, Omni-test, MMF10-MMF13, and MMF15. Conversely, it was worse than MO_Ring_PSO_SCD in 2 cases (i.e., MMF14 and MMF14_a). Furthermore, MMOEA/D-M2M was better than TriMOEA-TA&R in 8 out of 11 cases (i.e., SYM-PART rotated, Omni-test, MMF10-MMF13, MMF15, and MMF15_a), while it was worse than TriMOEA-TA&R in 3 cases (i.e., SYM-PART simple, MMF14, and MMF14_a). Furthermore, on IGDF, MMOEA/D-M2M gained better results than Omni-optimizer and DN-NSGA-II in 8 and 8 cases, respectively, while it was only beaten by them on SYM-PART simple, SYM-PART rotated and

Table 2. Performance comparison of MMOEA/D-M2M and other MMOEAs on IGDF

Problem	Omni-Optimizer	DN-NSGA-II	MO_Ring_PSO_SCD	TriMOEA-TA&R	MMOEA/D-M2M
SYM-PART simple	**1.216e−02 (1.72e−03)**+	1.230e−02 (1.78e−03)+	4.042e−02 (5.50e−03)−	3.414e−02 (4.99e−03)∼	3.243e−02 (5.24e−03)
SYM-PART rotated	**1.252e−02 (1.70e−03)**+	1.539e−02 (2.12e−03)+	4.604e−02 (4.32e−03)−	2.843e−02 (3.88e−03)∼	2.672e−02 (3.78e−03)
Omni-test	**6.776e−03 (2.47e−04)**+	7.911e−03 (5.10e−04)+	4.232e−02 (5.35e−03)−	1.841e−02 (4.18e−03)+	3.106e−02 (2.54e−03)
MMF10	1.822e−01 (4.17e−02)−	1.938e−01 (4.11e−02)−	1.297e−01 (1.51e−02)−	2.285e−01 (5.05e−03)−	**2.617e−02 (2.99e−03)**
MMF11	9.565e−02 (9.39e−04)−	9.823e−02 (2.07e−03)−	8.508e−02 (6.16e−03)−	1.633e−01 (7.27e−03)−	**3.511e−02 (7.85e−03)**
MMF12	8.611e−02 (8.13e−03)−	8.404e−02 (2.68e−03)−	6.431e−02 (1.27e−02)−	8.601e−02 (1.47e−03)−	**1.115e−02 (6.83e−03)**
MMF13	1.472e−01 (2.08e−03)−	1.507e−01 (4.40e−03)−	9.329e−02 (1.73e−02)−	2.431e−01 (6.87e−03)−	**5.575e−02 (5.85e−03)**
MMF14	9.765e−02 (3.95e−03)−	1.080e−01 (8.20e−03)−	**8.042e−02 (2.94e−03)**+	8.616e−02 (1.12e−03)+	9.413e−02 (5.94e−03)
MMF14_a	1.049e−01 (4.57e−03)−	1.178e−01 (6.69e−03)−	**7.818e−02 (2.50e−03)**+	7.878e−02 (1.15e−03)+	9.483e−02 (3.80e−03)
MMF15	2.010e−01 (7.37e−03)−	2.162e−01 (1.14e−02)−	1.741e−01 (3.27e−03)−	2.067e−01 (7.10e−04)−	**1.507e−01 (8.40e−03)**
MMF15_a	2.074e−01 (8.92e−03)−	2.259e−01 (1.05e−02)−	1.744e−01 (3.72e−03)−	1.936e−01 (3.67e−03)−	**1.642e−01 (4.85e−03)**
Best/all	3/11	0/11	2/11	0/11	6/11
+/−/∼	3/8/0	3/8/0	2/9/0	3/6/2	\

"+", "−", and "∼" indicate that the results of the compared algorithms are significantly better than, worse than, and similar to that of MMOEA/D-M2M respectively by Wilcoxon's rank sum test with $\alpha = 0.05$.

Omni-test. Moreover, MMOEA/D-M2M outperformed MO_Ring_PSO_SCD in 9 out of 11 cases, while it was only worse than MO_Ring_PSO_SCD on MMF14 and MMF14_a. MMOEA/D-M2M outperformed TriMOEA-TA&R in 6 out of 11 cases, while it was worse than TriMOEA-TA&R on Omni-test, MMF14 and MMF14_a. These comparison results have somehow showed the advantages of MMOEA/D-M2M on tackling most of adopted MMOPs.

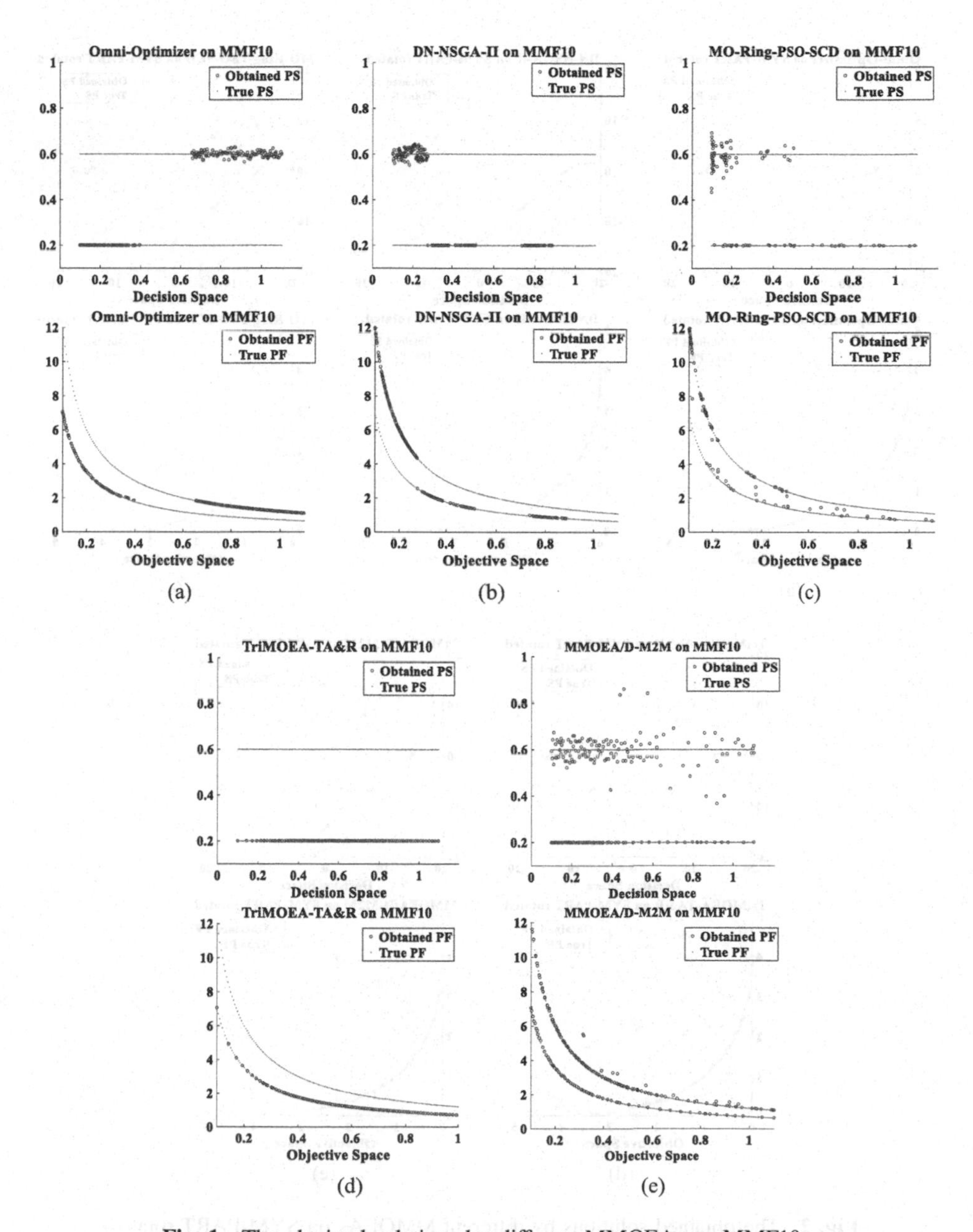

Fig. 1. The obtained solutions by different MMOEAs on MMF10.

Furthermore, in Figs. 1 and 2, the final solution sets obtained by all algorithms on MMF10 and SYM-PART rotated were given, respectively, which show the advantages of MMOEA/D-M2M. As shown in Fig. 1, on MMF10, because Omni-optimizer and DN-NSGA-II used non-dominated sorting to select solutions in objective space, they could not approximate the whole global and local PSs. Due to the use of an index-based ring topology to induce stable niches, MO_Ring_PSO_SCD can find global and local

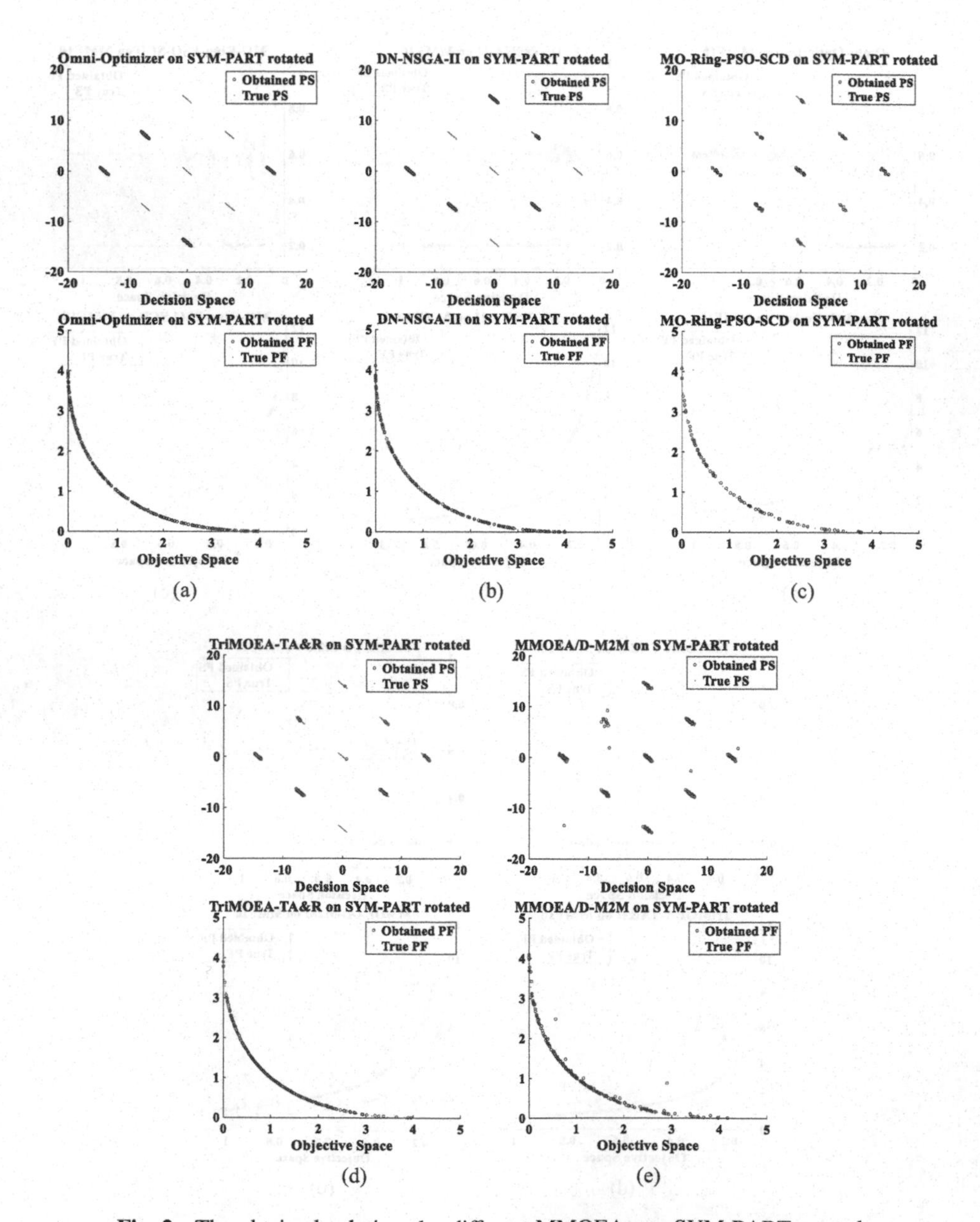

Fig. 2. The obtained solutions by different MMOEAs on SYM-PART rotated.

PSs, while the final solution sets cannot uniformly cover global and local PSs. For TriMOEA-TA&R, it can be seen that only the global PS is found while the local PS is missed because non-dominated sorting is used in environmental selection. It is obvious that MMOEA/D-M2M can find the whole global and local PSs because our proposed algorithm comprehensively considers the diversity in both decision and objective spaces. As shown in Fig. 2, on SYM-PART, the obtained solution sets by Omni-optimizer, DN-NSGA-II and TriMOEA-TA&R could not approximate all PSs because non-dominated sorting is the primary criterion to select solutions in objective space, which will deteriorate diversity in decision space. We can see that MO_Ring_PSO_SCD and MMOEA/D-M2M can find all PSs of SYM-PART rotated, while the final solution set obtained by MMOEA/D-M2M can achieve better performance in both decision and objective spaces.

5 Conclusion

In this paper, a novel decomposition-based multimodal multi-objective evolutionary algorithm, i.e., MMOEA/D-M2M, was proposed for solving MMOPs, in which M2M decomposition approach and DBSCAN are respectively applied on union population, aiming to produce K subregions in objective space and multiple different clusters in decision space. This way, more than N promising solutions are selected to form a temporary population, which is divided into K subregions by M2M decomposition approach, aiming to delete redundant solutions until there are N solutions left. This way, our proposed MMOEA/D-M2M attaches equal importance to diversity in both decision and objective spaces. The experimental results demonstrate that our proposed algorithm is capable of preserving more uniformly and widely distributed solutions in decision space without deteriorating their distributions in objective space, which confirms its superiority on most of MMOPs adopted.

In our future work, we will try to improve the performance of our algorithm on more test problems. In addition, one more reasonable performance metric should be designed to better measure the quality of final solutions obtained by MMOEAs.

References

1. Deb, K.: Multi-Objective Optimization Using Evolutionary Algorithms. Wiley, New York (2001)
2. Fonseca, C.M., Fleming, P.J.: An overview of evolutionary algorithms in multiobjective optimization. Evol. Comput. **7**(3), 205–230 (1995)
3. Yue, C.T., Qu, B.Y., Yu, K.J., Liang, J.J., Li, X.D.: A novel scalable test problem suite for multimodal multiobjective optimization. Swarm Evol. Comput. **48**, 62–71 (2019)
4. Liang, J.J., Qu, B.Y., Gong, D.W., Yue, C.T.: Problem definitions and evaluation criteria for the CEC 2019 special session on multimodal multiobjective optimization. Technical Report, Computational Intelligence Laboratory, Zhengzhou University (2019)
5. Ishibuchi, H., Peng, Y., Shang, K.: A scalable multimodal multi-objective test problem. In: 2019 IEEE Congress on Evolutionary Computation (CEC), Wellington, New Zealand, pp. 310–317. IEEE (2019)

6. Deb, K., Pratap, A., Agarwal, S., Meyarivan, T.: A fast and elitist multiobjective genetic algorithm: NSGA-II. IEEE Trans. Evol. Comput. **6**(2), 182–197 (2002)
7. Zhang, Q.F., Li, H.: MOEA/D: a multiobjective evolutionary algorithm based on decomposition. IEEE Trans. Evol. Comput. **11**(6), 712–731 (2007)
8. Deb, K., Tiwari, S.: Omni-optimizer: a procedure for single and multi-objective optimization. In: Coello Coello, C.A., Hernández Aguirre, A., Zitzler, E. (eds.) EMO 2005. LNCS, vol. 3410, pp. 47–61. Springer, Heidelberg (2005). https://doi.org/10.1007/978-3-540-31880-4_4
9. Liang, J.J., Yue, C.T., Qu, B.Y.: Multimodal multi-objective optimization: a preliminary study. In: 2016 IEEE Congress on Evolutionary Computation (CEC), pp. 2454–2461. IEEE, Vancouver, BC (2016)
10. Tanabe, R., Ishibuchi, H.: A decomposition-based evolutionary algorithm for multi-modal multi-objective optimization. In: Auger, A., Fonseca, C.M., Lourenço, N., Machado, P., Paquete, L., Whitley, D. (eds.) PPSN 2018. LNCS, vol. 11101, pp. 249–261. Springer, Cham (2018). https://doi.org/10.1007/978-3-319-99253-2_20
11. Yue, C., Qu, B., Liang, J.: A multi-objective particle swarm optimizer using ring topology for solving multimodal multi-objective problems. IEEE Trans. Evol. Comput. **22**(6), 805–817 (2018)
12. Hu, Y., et al.: A self-organizing multimodal multi-objective pigeon-inspired optimization algorithm. Sci. China Inf. Sci. **62**(7), 1–17 (2019). https://doi.org/10.1007/s11432-018-9754-6
13. Liu, Y., Yen, G.G., Gong, D.: A multi-modal multi-objective evolutionary algorithm using two-archive and recombination strategies. IEEE Trans. Evol. Comput. **23**(4), 660–674 (2019)
14. Liu, H.L., Gu, F., Zhang, Q.: Decomposition of a multiobjective optimization problem into a number of simple multiobjective subproblems. IEEE Trans. Evol. Comput. **18**(3), 450–455 (2014)
15. Ester, M., Kriegel, H.P., Sander, J., Xu, X.: A density-based algorithm for discovering clusters in large spatial databases with noise. In: International Conference on Knowledge Discovery Data Mining, pp. 226–231 (1996)
16. Das, I., Dennis, J.: Normal-boundary intersection: a new method for generating the Pareto surface in nonlinear multicriteria optimization problems. SIAM J. Optim. **8**(3), 631–657 (1998)
17. Huang, V.L., Suganthan, P.N., Qin, K., Baskar, S.: Differential evolution with external archive and harmonic distance-based diversity measure. https://www.researchgate.net/publication/228967624. Accessed 2008
18. Bosman, P.A., Thierens, D.: The balance between proximity and diversity in multiobjective evolutionary algorithms. IEEE Trans. Evol. Comput. **7**(2), 174–188 (2003)
19. Zhang, Q., Zhou, A., Jin, Y.: RM-MEDA: a regularity model-based multiobjective estimation of distribution algorithm. IEEE Trans. Evol. Comput. **12**(1), 41–63 (2008)

GTCN: Dynamic Network Embedding Based on Graph Temporal Convolution Neural Network

Zhichao Huang, Jingkuan Zhang, Lijia Ma(✉), and Fubing Mao

College of Computer Science and Software Engineering, Shenzhen University, Shenzhen 518060, China
ljma@szu.edu.cn

Abstract. Network embedding aims to learn the low-dimensional node representations from high-dimensional network structures of complex systems. Embedding in dynamic networks is a very difficult but important problem due to the dynamics of network structures in real-world systems and the high computational complexity. In this paper, we propose a novel Graph Temporal Convolution Network (short for GTCN) for the dynamic network embedding. In GTCN, a graph convolution network is used to learn the embedding representations of nodes in each snapshot, while a temporal convolutional network is adopted to parallelly reveal the evolution of node structures in dynamic networks. Extensive experiments on six real dynamic networks show that GTCN has a better performance than the state-art-of-art dynamic network embedding methods in tackling prediction tasks.

Keywords: Dynamic network · Network embedding · Graph convolution · Temporal convolution

1 Introduction

With the development of graphics technology, networks with link structures have become more and more worth studying [1]. In an actual system, nodes and links of a network represent the entities and communications of the system, respectively. However, with the rapid development of the Internet and information systems, communications and entities have grown exponentially. It has been difficult to calculate or express the structural properties of nodes and links in the network [2]. Network embedding has been proved an effective method to learn low-dimension feature of nodes in networks. The low-dimension feature not only preserves the structural features of nodes, but also obtains dense feature representations [3, 4].

Previous embedding methods mainly work on static networks [2, 5–7]. These models are usually designed to preserve local structures by GCN [7–9]. The embedding performance of this encoder-decoder architecture is determined by a loss function which determines the distance of local structures between the original and decoded networks. Classical network embedding methods include the matrix factorization-based approaches (DNR [10]) and the random walk based approaches (such as Deepwalk [5]

D.-S. Huang and K.-H. Jo (Eds.): ICIC 2020, LNCS 12464, pp. 583–593, 2020.
https://doi.org/10.1007/978-3-030-60802-6_51

and Node2vec [2]).The GCN architecture continuously merges information of neighbor nodes by means of message propagation.

The relationship between nodes in social networks and biomolecular networks always changes over time. Compared with static networks, dynamic networks are more suitable for these systems, in which the addition and deletion of nodes and links represent changes in node relationships in the system [11–13]. Moreover, in these networks, network embedding needs to consider both the structure and temporal features of the networks.

In recent years, some dynamic network embedding methods [14–16] have been proposed. For instance, the methods like dynGEM [16] and dynTriad [14] generated dynamic graph embedding based on a simplified assumption that graphs change smoothly. [15] improved classic skip-gram method for dynamic network embedding by updating the embedding of nodes whose links change most dramatic node embedding. dynRNN [17] and dynAERNN [17] used a recursive layer to solve the problem of inconsistent time spans. It is well known that Recurrent Neural Network (RNN) [18] is difficult to train and have high computational cost. Recently, many methods have been proposed for processing temporal data [19–21].

In this paper, we propose a novel model (called GTCN), which preserves both the structural and the temporal feature of dynamic network. GTCN uses GCNto get the embedding vector of each node in each snapshot. Then, it exploits TCN [22] to generate the nodes embedding that evolves over time. Extensive experiments on six real-world networks demonstrate that GTCN outperforms the state-of-the-art methods in link prediction.

2 Problem Definition

A dynamic network can be represented by $G = \{\mathcal{G}^1, \mathcal{G}^2, \ldots .\mathcal{G}^T\}$, where T is the total time step, while G^t is the snapshot of the dynamic graph at time stamp t. Each $\mathcal{G}^t = \{\mathcal{V}^t, \mathcal{E}^t\}$ represents undirected graph, where $\mathcal{V}^t$ denotes node set and $\mathcal{E}^t$ denotes link set at time t. Additionally, we can represent its links as a symmetric adjacency matrix $\mathrm{A}^t = [A_{ij}^t] \in \mathcal{R}^{n \times n}$, where n is the number of nodes, with each entry $\{i, j\}$ of the matrix as

$$A_{ij}^t = \begin{cases} 1 & \text{if } \mathrm{e}_{\mathrm{ij}}^{\mathrm{t}} \in \mathcal{E}_{\mathrm{t}} \\ 0 & \text{otherwise} \end{cases}.$$

Dynamic network embedding aims to obtain the latent representation of each node, so that the latent representation $\mathrm{emb}_i^{t,2} \in \mathcal{R}^{f_2 \times 1}$ (the embedding of node i) can capture both the current time graph structure information and the time information of the graph sequence.

3 Our Solution: GTCN

Our model consists of two parts. First, we apply GCN to deal with current time graph structure feature. Then, we exploit TCN to obtain past time temporal information.

3.1 Current Time Graph Structure Feature

This layer is consisted of 2 GCNs [7], used to grasp the current topology features of the graph.

For the first layer of GCN, the input are the adjacency matrix $\mathrm{A}^t \in \mathcal{R}^{n \times n}$ and the node feature matrix $\mathrm{F}^t \in \mathcal{R}^{n \times n}$ at time t.

Because the dataset used does not have node features, in general, the identity matrix is used as the node feature. Here, we use the following methods to calculate node features.

$$\begin{aligned} \hat{\mathrm{F}}^t &= \mathrm{I}_N + \mathrm{A}^t \\ \mathrm{F}^t &= \left(1 + t_d \cdot \mathrm{F}^{t-1}\right) \odot \hat{\mathrm{F}}^t, \end{aligned} \tag{1}$$

where $t \in \{2, \ldots, T\}$, and $\mathrm{F}^1 = \mathrm{I}_N + \mathrm{A}^1$. $\mathrm{I}_N \in \mathcal{R}^{n \times n}$ is the identity matrix, t_d is time decay value, a hyper parameter and $\odot$ is element-wise multiply. That means the longer time the link remains, the higher the similarity between two nodes the link connected, but once the link disappears, the similarity of the corresponding two nodes become zero and re-accumulated.

The output of the first GCN layer l_1 of a single snapshot is calculated as follows:

$$\mathrm{Z}_s^{t,l_1} = \mathbf{relu}\left(\hat{\mathrm{A}}^t \cdot \mathrm{F}^t \cdot \mathrm{W}_s^{t,l_1}\right), \tag{2}$$

where $\mathrm{Z}_s^{t,l_1} \in \mathcal{R}^{n \times d_1}$ and $\mathrm{W}_s^{t,l_1} \in \mathcal{R}^{n \times d_1}$ is the parameters with dimensionality d_1 of the first GCN layer to be learned and **relu** is defined as:

$$\mathbf{relu}(x) = \begin{cases} 1 & \text{if } \mathrm{x} > 0 \\ 0 & \text{otherwise} \end{cases},$$

and $\hat{\mathrm{A}}^t$ is defined as:

$$\begin{aligned} \tilde{\mathrm{A}}^t &= \mathrm{A}^t + \mathrm{I}_N \\ \tilde{\mathrm{D}}_{ii}^t &= \sum_j \tilde{\mathrm{A}}_{ij}^t \\ \hat{\mathrm{A}}^t &= (\tilde{\mathrm{D}}^t)^{-\frac{1}{2}} \cdot \tilde{\mathrm{A}}^t \cdot (\tilde{\mathrm{D}}^t)^{-\frac{1}{2}}. \end{aligned}$$

For the second GCN layer l_2, the output of the first GCN layer Z_s^{t,l_1} serves as the node feature of the second GCN layer and the output of second GCN layer is calculated as:

$$Z_s^{t,l_2} = \hat{A}^t \cdot Z_s^{t,l_1} \cdot W_s^{t,l_2}, \tag{3}$$

where $Z_s^{t,l_2} \in \mathcal{R}^{n \times d_2}$ and $W_s^{t,l_2} \in \mathcal{R}^{d_1 \times d_2}$ is the parameters with dimensionality d_2 of the second GCN layer.

Finally, we need to use the structure embedding vector [23] of each node Z_s^{t,l_2} to reconstruct the adjacency matrix.

$$\hat{A}_s^t = \mathbf{sigmoid}(Z_s^{t,l_2} \cdot (Z_s^{t,l_2})^T), \tag{4}$$

where $(Z_s^{t,l_2})^T$ means transpose of matrix Z_s^{t,l_2}. **sigmoid** is an activation function, defined as $\mathbf{sigmoid}(x) = \frac{1}{1+e^x}$.

Then, we compare the reconstructed matrix $\hat{A}_s^t$ with the ground truth adjacency matrix A_t at the current moment t as structure loss. Our goal is to minimize the structure loss of time step t.

$$\mathcal{L}_S = \sum_{t=1} norm_s^t \left\| \left(\hat{A}_s^t - A^t \right) \odot B_s^t \right\|_F^2, \tag{5}$$

where $norm_s^t$ is the normalization coefficient, defined as:

$$norm_s^t = \frac{n \times n}{2(n \times n - \|\mathcal{E}^t\|)},$$

$B_s^t \in \mathcal{R}^{n \times n}$ is the penalty matrix

$$B_{s,ij}^t = \begin{cases} \frac{n \times n - \|\mathcal{E}^t\|}{\|\mathcal{E}^t\|} & \text{if } e_{ij}^t \in \mathcal{E}^t \\ 0 & \text{otherwise} \end{cases},$$

where $B_{s,ij}^t$ is the element at the i_{th} row and j_{th} column of matrix B^t and $\odot$ is element-wise multiply.

3.2 Past Time Temporal Feature

In this section, we will use TCN to solve the temporal problem. In order to predict the adjacency matrix of $\mathcal{G}^{t+1}$, we need to consider all the time series information before time $t+1$, that is $\{\mathcal{G}^1, \mathcal{G}^2, \ldots, \mathcal{G}^t\}$.

The input of this layer is the structural feature of each node of the graph at time $\{1, 2, \ldots, t\}$. Here we use the node representation Z_s^{t,l_2} obtained from the GCN above to be the structural feature. We use $Z_{s,i*}^{t,l_2} \in \mathcal{R}^{1 \times d_2}$, the i_{th} row of the matrix, to represent the structural feature of node i in $\mathcal{G}_t$, then our input for the prediction of each node of $\mathcal{G}^t$ is $\{Z_{s,i*}^{1,l_2}, Z_{s,i*}^{2,l_2}, \ldots, Z_{s,i*}^{t,l_2}\}$.

Here we use 2 layers of TCN [22] as a model for processing temporal information. Each layer of TCN consists of a dilated convolution, a causal convolution, and a skip connection.

For convenience, we set $Z^{t,l_2}_{s,i*}$ as:

$$H^{t,0}_i = (Z^{t,l_2}_{s,i*})^T,$$

where $H^{t,0}_i \in \mathcal{R}^{d_2 \times 1}$ means the embedding vector of the node i of $\mathcal{G}^t$ in the layer 0, *i.e.*, the i_{th} row of Z^{t,l_2}_s.

For each layer l, dilated convolution and causal convolution is calculated as follows:

$$\hat{H}^{t,l}_i = \mathbf{relu}\left(\sum_{j=0}^{k-1} W^l_j \cdot H^{(t-dj),(l-1)}_i\right), \tag{6}$$

where $W^l_j \in \mathcal{R}^{f_l \times f_{l-1}}$, f_l is the number of filter (dimension of output) in l_{th} layer, d is the dilation factor and k is the filter size, and $t - dj$ means the past focused step. Therefore, you can consider d as the number of steps between two adjacent columns of the filter. When d is equal to 1, it is equivalent to usual causal convolution. The larger d enables an exponentially large receptive field, *i.e.*, more information of graph dynamic evolution.

For each layer, skip connection (residual connection) [24], add the original input to the output,

$$H^{t,l}_i = \mathbf{relu}\left(W^l \cdot H^{t,(l-1)}_i + \hat{H}^{t,l}_i\right), \tag{7}$$

where $W^l \in \mathcal{R}^{f_l \times f_{l-1}}$, used to make the dimension of original input and output equal.

Then, we use a linear layer and the output of the last TCN layer (here we use two layer TCN, so $emb^{t,3}_i$ is the output of the last TCN layer) to reconstruct the vector representation of the node i at the next time step $t+1$,

$$\hat{V}^{t+1}_i = \mathbf{sigmoid}\left(W_d \cdot H^{t,2}_i + B_d\right), \tag{8}$$

where $W_d \in \mathcal{R}^{n \times f_2}$ and $B_d \in \mathcal{R}^{n \times 1}$ are the parameters of the linear layer.

After obtaining the prediction vector representation of node i at next time step $t+1$, we need to compare the difference between the predicted value and the ground truth, and use this as our temporal loss to optimize,

$$\mathcal{L}_T = norm^t_t \sum_{t=0}^{T} \sum_{i=0}^{n} \left\| \left(\hat{V}^{t+1}_i - V^{t+1}_i\right) \odot B^t_{t,i*} \right\|^2_F, \tag{9}$$

where T is the total time step before $t+1$ and n is the number of nodes.V_i^{t+1} is the i_{th} row of the adjacency matrix A at time $t+1$. $norm_t^t$ and B_t^t are the same as $norm_s^t$ and B_s^t separately except that it used $t+1$ nodes and edges information. And $B_{t,i*}^t \in \mathcal{R}^{n\times 1}$ is the i_{th} row of B_t^t.

3.3 Total Architecture

In this section, we will present the key components ofour overall model.

Current Time Graph Structure Feature Block: For this module, we aim to capture the topological features and the neighbor characteristics of the nodes at each time graph $\mathcal{G}_t$. The parameters at each time step are not shared, the output of this module will be used as input to the next temporal module.

Past Time Temporal Feature Block: In this module, we will capture the temporal features of the graph dynamic changing. We use all the graph structure information before $t+1$ as input, *i.e.*, $\{\mathcal{G}^1, \mathcal{G}^2, \ldots, \mathcal{G}^t\}$, and use TCN to predict $\mathcal{G}^{t+1}$.

Objective: we define the loss for optimization as:

$$\mathcal{L} = \min(\mathcal{L}_\mathcal{S} + \alpha\mathcal{L}_\mathcal{T} + \lambda\mathcal{L}_{reg}), \tag{10}$$

where $\mathcal{L}_\mathcal{S}$ is the graph structure feature loss and $\mathcal{L}_\mathcal{T}$ is the graph dynamic feature loss.α is the parameters to control the contribution of $\mathcal{L}_\mathcal{S}$ and $\mathcal{L}_\mathcal{T}$, which is between 0 and 1. λ is the parameter to control contribution of $\mathcal{L}_{reg}$ and $\mathcal{L}_{reg}$ is an L2-norm regularizer to prevent overfitting, defined as:

$$\mathcal{L}_{reg} = \frac{1}{2}\|W\|_F^2,$$

where W is all the parameters that need to be learned.

Optimization: We use the adam [25] optimizer.

4 Experimental Results

In this section, firstly we introduce 6 real-world dynamic networks as our data set, and compare 5 models including 1 static graph embedding method and 4 dynamic embedding approaches. Then, we analyze the experimental results of the model under different data sets.

Table 1. Statistics of dataset used in our experiments.

Network	Nodes	Links	Time-step
fb-forum	899	33720	7
Hypertext	113	20818	7
Hospital	75	32424	8
School	242	125773	7
Email	167	82927	7
Workplace	92	9827	8

4.1 Experimental Settings

Datasets. It consists of 6 real-life networks that include web forums, networks for medical, school, workplace and citation relationship. The basic properties of the data set are described in Table 1.

Comparison Models. We compare the performance of our model with other state-of-the-art models including one static graph embedding model (SDNE [26]) and four dynamic graph embedding models (dynGEM[16], Dyn2vecAE [17], Dyn2vecRNN [17] and Dyn2vecAERNN [17]).

Task. In real networks, links between nodes tend to follow time changes. Therefore, it is very important to use past node link information to predict changes in node connections at the next moment. Here, we use link prediction as the criterion for judging the pros and cons of our model. In the experiment, we use the information of the past nodes $1, 2, 3, \ldots, t$ to predict the embedding of nodes at the next moment $t+1$, and calculate the relationship between nodes at $t+1$ snapshot with the binary classification model. Finally, we compare the predicted node association with the ground truth connections and adopt Mean Average Precision (mAP) as judging criteria.

We conduct experiments on the tasks of link prediction in dynamic graphs. Firstly, we train the model and acquire the corresponding learned parameters on snapshots $\{\mathcal{G}_1, \mathcal{G}_2, \ldots, \mathcal{G}_t\}$ and apply $\hat{V}_{t+1}$ to predict links in $\mathcal{G}_{t+1}$ during evaluation. At each snapshot of the graph, links could be disappeared or added. We compare the performance of different models in link prediction based on their own abilities.

Metrics. To evaluate the performance of link prediction learned by our model, we use Mean Average Precision (mAP) and precision@k as the metrics.

Detailed Settings. For our datasets with short time steps, when training the model to predict $\mathcal{G}_{t+1}$, we will exploit all the graphs before time $t+1$, *i.e.*, $\{\mathcal{G}_1, \mathcal{G}_2, \ldots, \mathcal{G}_t\}$. In our experiment, we predict the T, $T-1$, $T-2$ and $T-3$ graphs for each data set, and train 20 times for each experiment. Finally, the average of the mAPs for 20 times with 4 time steps is selected as the final assessment metric.

Table 2. Experiment results on dynamic link prediction (mAP)

Network	SDNE		dynGEM		dynAE	
	Average	Max	Average	Max	Average	Max
fb-forum	0.0001	0.0001	0.0001	0.0002	0.0005	0.0005
Hypertext	0.4294	0.4383	0.3690	0.3841	0.3690	0.4211
Hospital	0.4738	0.4865	0.4637	0.4730	0.4515	0.4679
School	0.3619	0.3969	0.3484	0.3548	0.2623	0.3009
Email	0.0003	0.0003	0.0001	0.0002	0.2270	0.2568
Workplace	0.4182	0.4346	0.3735	0.4005	0.1970	0.2118
Network	DynRNN		DynAERNN		GTCN [Ours]	
	Average	Max	Average	Max	Average	Max
fb-forum	0.0002	0.0002	0.0051	0.0054	**0.0274**	**0.0315**
Hypertext	0.3818	0.3994	0.4039	0.4102	**0.4816**	**0.5022**
Hospital	0.4826	0.5111	**0.5195**	**0.5339**	0.4716	0.4885
School	0.4526	0.4712	0.3022	0.3121	**0.5214**	**0.5406**
Email	0.1605	0.2004	0.2475	0.2517	**0.2661**	**0.2812**
Workplace	0.2966	0.3296	0.3816	0.4052	**0.4882**	**0.5033**

Here, we choose 2-layer GCNs to extract the topology structural feature of each snapshot of graphs, and 3-layer TCNs to obtain the dynamic evolution feature.

All the experiments are conducted on Linux platform with 2 GPU cores (Tesla P100) and 32 GB RAM.

4.2 Result and Analysis

In this section we present the performance of different models for link prediction on different datasets.

Table 2 shows the performance (mAP) of each algorithm on each data set. Among them, we take mAP as the criterion for judging the pros and cons of the model. The larger the mAP, the better performance of the model has.

Table 3. Experiment results on dynamic link prediction (Precision@k part-1)

Method	Fb-forum(avg)			Hypertext(avg)			Email(avg)		
	P@100	P@500	P@1000	P@100	P@500	P@1000	P@100	P@500	P@1000
SDNE	0.000	0.001	0.000	0.323	**0.404**	**0.397**	0.000	0.000	0.001
dynGEM	0.000	0.001	0.000	0.065	0.164	0.220	0.000	0.000	0.000
dynAE	0.000	0.000	0.000	0.095	0.237	0.260	0.000	0.011	0.031
dynRNN	0.003	0.001	0.000	0.373	0.340	0.354	0.090	0.174	0.190
dynAERNN	0.003	0.001	0.003	0.395	0.383	0.367	0.068	**0.202**	**0.229**
GTCN [ours]	**0.003**	**0.006**	**0.005**	**0.425**	0.387	0.372	**0.138**	0.122	0.125

Table 4. Experiment results on dynamic link prediction (Precision@k part-2)

Method	School (avg)			Workplace (avg)			Hospital (avg)		
	P@100	P@500	P@1000	P@100	P@500	P@1000	P@100	P@500	P@1000
SDNE	0.313	0.340	0.320	0.270	0.275	0.245	0.350	0.408	0.387
dynGEM	0.315	0.342	0.309	0.183	0.227	0.206	0.208	0.326	0.359
dynAE	0.118	0.136	0.124	0.160	0.131	0.135	0.138	0.269	0.322
dynRNN	0.318	0.242	0.200	0.290	0.311	0.281	0.255	0.423	0.442
dynAERNN	0.330	0.251	0.218	0.258	0.232	0.224	0.350	**0.445**	**0.453**
GTCN [ours]	**0.705**	**0.575**	**0.502**	**0.615**	**0.430**	**0.353**	**0.385**	0.390	0.396

From Table 2, we can see that SDNE, dynGEM, dynAE, dynRNN and dynAERNN perform very poorly on fb-forum data set (smaller than 0.01 and 20 times or smaller than the mAP of GTCN). The reason is that the edges in fb-forum change very sharply at each moment, and both SDNE and dynGEM just consider the adjacency matrix of the graph at the current moment as the training set to predict the graph structure of next time step. Therefore, the violent changes cause SDNE and dynGEM to become very difficult to predict next snapshot of the dynamic network. The dynAE, dynRNN and dynAERNN methods can find long historical data information, but the structure of Auto-encoder makes it difficult for them to mine the characteristics of sparse graph structures. However, GTCN achieves better performance since GTCN uses TCN to discover historical information, and then adopts GCN to cooperate with time feature to some extent to alleviate the impact of the sparse graph structure. However, the degree of the node of fb-forum is very small which results in the adjacency matrix being sparse. Thus, the number of edges that can be trained is small, the training is insufficient and the number of nodes is too large which results in lower accuracy for prediction. Hence, all the algorithms are not very good and even the best performance (mAP) can only reach about 0.02. For email dataset, we also find the same situation that using only the information from the previous moment is not enough to discover the structural characteristics of the dynamic network.

Compared to other dynamic graph embedding algorithms, GTCN performs the best on most of data sets. We attribute this to the GCN and the preprocessing of timing features which contain the node features of each snapshot with a rough summary of temporal information. Compared to dynGEM mentioned above, the model only focuses on the changes in the links of the previous 1 time step graph, which is too short-sighted. Compared with dynAE and dynRNN, the AE and RNN structures are not as good as GCN for capturing the neighbor structure of each node in the graph. Furthermore, the combination of structural features and temporal features are also impossible for AE and RNN.

Table 3 and Table 4 show the precision@ k values for various data sets. The larger the precision@ k is, the better the model performs. As we explained above, fb-forum has many nodes, but links between nodes are sparse and change very sharply, which leads to poor model learning. However, our model can still achieve better performance than other models. For other small data sets, our model can also outperform than other models in most of the datasets in P@100, P@500 and P@1000.

5 Conclusion

In this paper, we have proposed a dynamic network embedding method GTCN. The model first captured the topology properties of the nodes in the graph by the GCN, and then adopted TCN to capture the dynamic temporal changes of the nodes. Experimental results have showed that our model not only captured the topology properties of nodes, but also extracted dynamic temporal changes. They have also indicated that our proposed method outperforms the state-of-the-art methods in link predictions.

Acknowledgement. This work was supported by the National Natural Science Foundation of China under Grant 61803269, in part by the Natural Science Foundation of Guangdong Province under Grant 2020A1515010790, and in part by the Technology Research Project of Shenzhen City under Grant by JCYJ20190808174801673.

References

1. Facchetti, G., Iacono, G., Altafini, C.: Computing global structural balance in large-scale signed social networks. Proc. Natl. Acad. Sci. **108**(52), 20953–20958 (2011)
2. Grover, A., Leskovec, J.: Node2vec: scalable feature learning for networks. In: ACMSIGKDD International Conference on Knowledge Discovery and Data Mining, San Francisco, USA, pp. 855–864 (2016)
3. Hamilton, W.L., Ying, R., Leskovec, J.: Representation learning on graphs: Methods and applications. CoRR**abs/1709.05584** (2017)
4. Zhang, D., Yin, J., Zhu, X., Zhang, C.: Network representation learning: a survey. IEEE Trans. Big Data (2018). https://doi.org/10.1109/tbdata.2018.2850013
5. Perozzi, B., Al-Rfou, R., Skiena, S.: DeepWalk: online learning of social representations. In: ACMSIGKDD International Conference on Knowledge Discovery and Data Mining, NewYork, USA, pp. 701–710 (2014)
6. Tang, J., Qu, M., Wang, M., Zhang, M., Yan, J., Mei, Q.: Line: large-scale information network embedding. In: World Wide Web Conference. Florence, Italy, May 2015
7. Kipf, T.N., Welling, M.: Semi-Supervised Classification with Graph Convolutional Networks. In: International Conference on Learning Representations. Palais des Congrès Neptune, Toulon, France (2017)
8. Wang, X., Cui, P., Wang, J., Pei, J., Zhu, W., Yang, S.: Community preserving network embedding. In: AAAI Conference on Artificial Intelligence, San Francisco, California USA. AAAI (2017)
9. Yang, L., Cao, X., He, D., Wang, C., Wang, X., Zhang, W.: Modularity based community detection with deep learning. In: International Joint Conference on Artificial Intelligence, New York, USA, pp. 2252–2258 (2016)
10. Skrlj, B., Kralj, J., Konc, J., Robnik-Sikonja, M., Lavrac, N.: Deep node ranking: an algorithm for structural network embedding and end-to-end classification. CoRR**abs/1902.03964** (2019)
11. K., E.G., Mirzasoleiman, B., Grosu, R., Leskovec, J.: Dynamic network model from partial observations. In: Advances in Neural Information Processing Systems. Montréal, Canada (2018)
12. Rossetti, G., Cazabet, R.: Community discovery in dynamic networks: a survey. CoRR**abs/1707.03186** (2017)

13. Sekara, V., Stopczynski, A., Lehmann, S.: Fundamental structures of dynamic social networks. Proc. Nat. Acad. Sci. United States Am, p. 201602803 (2016)
14. Du, L., Wang, Y., Song, G., Lu, Z., Wang, J.: Dynamic network embedding by modeling triadic closure process. In: AAAI Conference on Artificial Intelligence, New Orleans, USA (2018)
15. Du, L., Wang, Y., Song, G., Lu, Z., Wang, J.: Dynamic network embedding: an extended approach for skip-gram based network embedding. In: International Joint Conference on Artificial Intelligence (2018)
16. Goyal, P., Kamra, N., He, X., Liu, Y.: Dyngem: deep embedding method for dynamic graphs. CoRR**abs/1805.11273** (2018)
17. Goyal, P., Chhetri, S.R., Canedo, A.: Dyngraph2vec: capturing network dynamics using dynamic graph representation learning. Knowl. Based Syst. **187**, 104816 (2020)
18. Lipton, Z.C.: A critical review of recurrent neural networks for sequence learning. Computer Science (2015)
19. Sankar, A., Wu, Y., Gou, L., Zhang, W., Yang, H.: Dysat: deep neural representation learning on dynamic graphs via self-attention networks. In: International Conference on Web Search and Data Mining, Houson, USA, pp. 519–527 (2020)
20. Vaswani, A., et al.: Attention is all you need. In: Advances in Neural Information Processing Systems, Long Beach, USA, pp. 5998–6008 (2017)
21. Veličković, P., Cucurull, G., Casanova, A., Romero, A., Lio, P., Bengio, Y.: Graph attention networks. In: International Conference on Learning Representation, Vancouver, Canada (2018)
22. Bai, S., Kolter, J.Z., Koltun, V.: An empirical evaluation of generic convolutional and recurrent networks for sequence modeling. CoRR **abs/1803.01271** (2018)
23. Pan, S., Hu, R., Long, G., Jiang, J., Yao, L., Zhang, C.: Adversarially regularized graph autoencoder for graph embedding. In: International Joint Conference on Artificial Intelligence, Stockholm, Sweden, pp. 2609–2615 (2018)
24. He, K., Zhang, X., Ren, S., Sun, J.: Deep residual learning for image recognition. In: IEEE Conference on Computer Vision and Pattern Recognition, Las Vegas, USA, pp. 770–778 (2016)
25. Kingma, D.P., Ba, J.: Adam: a method for stochastic optimization. Computer Science (2014)
26. Wang, D., Cui, P., Zhu, W.: Structural deep network embedding. In: ACM SIGKDD International Conference on Knowledge Discovery and Data Mining, San Francisco, USA, pp. 1225–1234 (2016)

Resource Scheduling Algorithm Based on Evolutionary Computation in Dynamic Cloud Environment

Qiyuan Yu(✉), Shen Zhong, Naili Luo, and Peizhi Huang

Shenzhen University, Shenzhen 518060, China
1800271030@email.szu.edu.cn, 343189068@qq.com,
luonaili@163.com, hpz@szu.edu.cn

Abstract. The growing types of resources and changing user requirements in the cloud environment bring great challenges to the resource scheduling problem. In order to solve the problem of resource scheduling in dynamic cloud environment, this paper constructs a virtual dynamic cloud environment resource scheduling model, which realizes resource scheduling by means of virtual machine migration. In this model, the energy consumption of the cloud environment and the quality of service of the cloud environment after virtual machine migration are taken as two optimization objectives. At the same time, we propose a new dynamic multi-objective optimization algorithm (called DCRS-EA) to solve the resource scheduling problem in dynamic cloud environment. DCRS-EA not only detects whether the environment changes, but also estimates the types of changes, in which different types of change are solved by different response strategies. Finally, the experiments show the superior performance of DCRS-EA when comparing to other dynamic strategies on the built optimization model.

Keywords: Evolutionary algorithm · Dynamic multi-objective optimization · Cloud computing

1 Introduction

Cloud computing is a new computing service model, which can solve complex information management problems in Internet, such as cloud control, cloud evolution and cloud reasoning [1]. How to effectively schedule cloud resources is the core research content of cloud computing. In the process of scheduling, we expect energy consumption to be as low as possible, and virtualization technology is usually used to shut down idle servers to achieve this goal. In addition, a temperature-aware workload migration method is proposed in [2] to reduce energy consumption. However, for a cloud computing center, it is not the market expectations to maximize resource efficiency only by reducing energy consumption. It may be necessary to centrally migrate most virtual machines to multiple servers for maximizing resource utilization, which can cause server latency and subsequently lead to a decline of quality of service (QoS) in the cloud environment. Generally speaking, in the process of resource scheduling, it is necessary to reduce the energy consumption in the cloud environment

D.-S. Huang and K.-H. Jo (Eds.): ICIC 2020, LNCS 12464, pp. 594–606, 2020.
https://doi.org/10.1007/978-3-030-60802-6_52

and ensure the service quality of the cloud environment, which is a conflicting two-objective optimization problem. In order to solve the resource scheduling problem for multiple optimization objectives in cloud environment, Yang et al. proposed an optimized ant colony algorithm based on particle swarm to solve cloud computing environment resources allocation problem [3]. In [4], Ding et al. proposed a multi-objective ant colony optimization method to optimize operating costs and resource waste at the same time. However, it is worth noting that the real cloud computing environment may be constantly changing, such as the types of resources and the needs of users. There is little research on the construction of resource scheduling model in dynamic cloud environment. Therefore, in this paper, we design a virtual resource scheduling model in dynamic cloud environment, which has certain flexibility and expansibility.

When solving the resource scheduling problem in dynamic cloud environment, the algorithms for solving static problems mentioned above are usually difficult to meet the requirements. For the dynamic multi-objective optimization problems (DMOPs), the objective functions of the optimization will change with the environment, so the designed algorithms not only have to optimize the objectives, but also need to detect and respond to the changes in environment. For dynamic multi-objective evolutionary algorithms (DMOEAs), their essence is to add change detection mechanism and change response strategy on the basis of static multi-objective evolutionary algorithms (MOEAs). When the detection mechanism detects a change in the environment, the response strategy is used to initialize the population of the new optimization objectives. Otherwise the static MOEAs will continue to be used. How to design effective change detection mechanism and response strategy is the main research work of DMOEAs. In order to solve the problem of resource scheduling in dynamic cloud environment, this paper proposes a DMOEA with multiple strategies based on the degree of change, and it can work normally when the resource scheduling model changes. The contributions of this paper are mainly reflected in the following two aspects:

2 Related Work

Recently, there has been a lot of research work on cloud resource scheduling algorithm. For the reduction of energy consumption, Mezmaz et al. studied the priority of parallel service in cloud environment and reduced the energy consumption by reschedule of the parallel services [5]. Energy consumption can also be reduced by turning off unnecessary server which is suitable to the cloud environment with live migration of virtual machine [6]. On the other hand, resource utilization is also an important approach in the cloud resource scheduling. A prediction methodology of workload cost was introduced in [7] to afford lower hosting costs while offering equal or better performance. Fehimeh et al. proposed a virtual machine consolidation approach that takes into account both the current and the future utilization of resource, which gets substantial improvement over other heuristic and meta-heuristic algorithms [8]. An adaptive control approach was proposed in [9] for resource allocation that adaptively reacts to dynamic request workloads and resource demands.

When dealing with dynamic multi-objective optimization problems, the focus of the research work is how to deal with dynamic changes as quickly as possible, while ensuring the diversity and convergence of the population. At present, algorithms for dealing with environmental changes can be roughly divided into three categories: the diversity-based DMOEAs, the convergence-based DMOEAs and the prediction strategy based DMOEAs. The diversity-based DMOEAs focus on maintaining the diversity of the population. For example, some hyper-mutation operators are used to disturb some solutions with good convergence abilities in the old environment, which are expected to jump out of their original positions in [10]. In [11], a mechanism of random immigration and the intrinsic diversity maintenance strategy are both designed to adapt the population to the dynamical changes. The major purpose of the convergence-based DMOEAs is to speed up the convergence by using the historical information. For example, a memory-based DMOEA [12] is reported with three models for initialization, exploration and archiving, respectively. In [13], a hybrid memory method is run based on the archive, which reinitializes population by re-evaluating the previous non-dominated solutions or by conducting a local search for the elite solutions preserved in the archive. The prediction approaches obtains the movement rules of the changing PS/PF by analyzing the historical records, which tries to estimate the future PSs/PFs for new environment. For example, a forward prediction strategy is proposed based on autoregressive model [14] and other prediction models are developed by regularity property [15]. They track the movement of the center of the PS manifold in different dynamic environments, thus generating a new population within the neighborhood around the expected location of the new PS.

3 The Resource Scheduling Model for Virtual Dynamic Cloud Environment

In this section, we consider the optimization of energy consumption and the optimization of QoS in cloud environment after migration, and we design a bi-objective dynamic optimization problem for resource scheduling in virtual dynamic cloud environment.

From [16], with the increase of CPU utilization, server energy consumption increases linearly. The lower limit of energy consumption is the energy consumption when CPU is idle, and the upper limit is the energy consumption when CPU utilization reaches 100%. This relationship can be expressed as follows:

$$P(u) = P_{idle} + (P_{busy} - P_{idle}) \times (2u - u^r) \tag{1}$$

where u is the utilization ratio of CPU, P_{busy} and P_{idle} are respectively the energy consumption of CPU utilization at 100% and no-load, and r is a fixed parameter which is set to 1.4 [16]. For the cloud environment, the whole energy consumption is the sum of energy consumption of n hosts, and the first optimization function is set as

$$min\ P_{all} = \sum_{i=1}^{N} P_i(u) \tag{2}$$

where P_{all} is the whole energy consumption and $P_i(u)$ is the energy consumption of ith host (i = 1, 2,..., n).

The second optimization goal is the QoS of cloud services [1]. In this paper, the objective function is designed according to the response time of cloud service and the error probability of cloud service. The response time of the cloud service refers to the time from the request issued by the virtual machines (VMs) to the cloud response, which is calculated as follows:

$$T_j = k_i \times u_i \tag{3}$$

where j = *1*, *2*, ..., m and m is the number of VMs, u_i is the load of the ith host, and k_i is the response time factor of this host. For the error probability of the jth VM, it is proportional to the resources occupied by VM, and it is calculated as:

$$p_j = l_j \times v_j \tag{4}$$

where v_j is the required resource consumption of the jth VM, and l_j is the error probability coefficient which is uniformly set to 1 in this paper. Then the optimization function of QoS is set as:

$$min\ Q = \frac{1}{m}\sum_{j=1}^{m} T_j \times p_j \tag{5}$$

In this optimization problem, we assume that there are m VMs and n hosts, each VM requires different CPU computing resources, and the parameters of VMs are dynamic, so $v_j = v_j(t)$ and Eq. (4) is replaced by:

$$p_j = l_j \times v_j(t) \tag{6}$$

For m user VMs, they are divided into three parts, i.e., V_1, V_2 and V_3. The CPU resources required by the VMs included in V_1 are constant, while the CPU resources required by the VMs in V_2 and V_3 change over time. The required resources for the user VMs are set as:

$$\begin{cases} v_j(t) = v_j(0), \text{ if } j \in V_1 \\ v_j(t) = v_j(0) \times (1 + \sin(\pi \times t)), \text{ if } j \in V_2 \\ v_j(t) = v_j(0) \times (1 - \sin(\pi \times t)), \text{ if } j \in V_3 \end{cases} \tag{7}$$

Here, t is the time index in the time space Ω_t, which is defined by

$$t = \frac{\lfloor \tau/\tau_t \rfloor}{n_t} \tag{8}$$

where τ and τ_t represent the generation counter and the frequency of change, respectively, and n_t is the change severity in a problem at time t.

4 The Dynamic Multi-objective Evolutionary Algorithm

In this part, the dynamic multi-objective resource scheduling algorithm DCRS-EA in dynamic cloud environment will be introduced. If no change is detected in the cloud environment, the static multi-objective algorithm MOEA/D-IRA [17] is used to optimize the population, otherwise the response strategy is used to initialize the population. When the conditions for the termination of evolution are met, the final population of each environment will be outputted. In Sect. 4.1, the coding methods, crossover and mutation operations are introduced. Then, the change detection mechanism and the response strategy are introduced in Sects. 4.2 and 4.3, respectively.

4.1 The Methods of Coding and Evolutionary Operators

In the process of resource scheduling in dynamic cloud environment, the migration process of virtual machine needs to be encoded in order to carry out numerical calculation. Suppose the cloud environment consists of m VMs and n hosts, and an array of m length is used to represent the deployment status of the cloud environment. Each value in the array is used as the host number parasitized by the corresponding VM. Figure 1 shows an example of coding method. In this figure, m and n are respectively equal to 8 and 4, and the current deployment status is encoded as $\{1, 1, 4, 4, 3, 2, 3, 1\}$, which indicates that the first, second and eighth VMs are parasitic on the first host, the third and fourth VMs are parasitic on the fourth host, the fifth and seventh VMs are parasitic on the third host, and the sixth VM is parasitic on the second host.

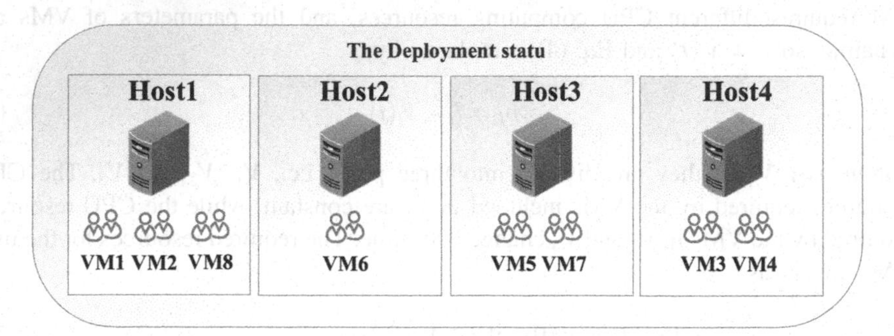

Fig. 1. An example of coding method

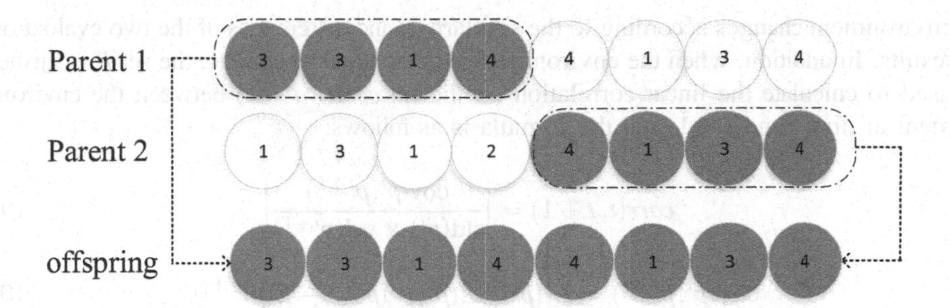

Fig. 2. The crossover operator

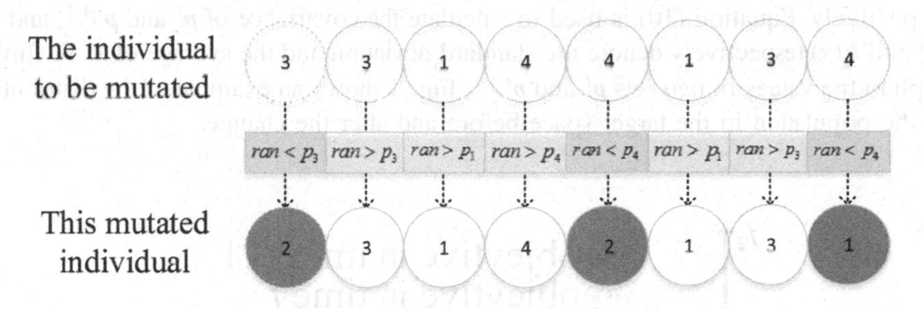

Fig. 3. The mutation operator

Crossover and mutation are the two most important operations in evolutionary algorithms. The crossover method used in this paper is midpoint crossover, as shown in Fig. 2. For two parent individuals, their genes are divided into two parts, and then the front gene of one parent and the latter gene of the other are randomly selected to recombine into a new individual. The mutation strategy of this paper is based on the mutation of gene frequency. In order to preserve the rare genes, we hope that the genes with higher frequency tend to mutate. First of all, the occurrence frequency p_i ($i = 1, 2, \ldots, n$) of each gene in the population is calculated. For an individual to be mutated, a random number *ran* is compared with the gene frequency of each location, and if *ran* is smaller, the gene at the corresponding position is mutated. In Fig. 3, the random number *ran* is less than the first, fifth and eighth gene frequencies, so the corresponding genes are replaced with other low frequency genes.

4.2 The Change Detection Mechanism

In this part, a novel detection strategy is introduced, which not only detects whether the environment has changed, but also sets the changes into three types according to the degree of change. Since the non-dominated solutions can reflect the convergence and diversity of the current population to a certain extent, when detecting whether the environment has changed, it is necessary to use non-dominated sorting to select $N/10$ solutions in the current population to form the set R. After each iteration, the algorithm recalculates the objective values of each solution in R, so as to determine whether the environment changes according to the similarities and differences of the two evaluation results. In addition, when the environment is determined to change, the set R is further used to calculate the linear correlation coefficient $corr(t,\ t + 1)$ between the environment at time t and $t + 1$, and the formula is as follows:

$$corr(t, t+1) = \left| \frac{\mathrm{cov}(\boldsymbol{p}^t, \boldsymbol{p}^{t+1})}{\mathrm{std}(\boldsymbol{p}^t) \times std(\boldsymbol{p}^{t+1})} \right| \tag{9}$$

$$\mathrm{cov}(\boldsymbol{p}^t, \boldsymbol{p}^{t+1}) = E((\boldsymbol{p}^t - E(\boldsymbol{p}^t)) \cdot (\boldsymbol{p}^{t+1} - E(\boldsymbol{p}^{t+1}))) \tag{10}$$

where $\boldsymbol{p}^t$ and $\boldsymbol{p}^{t+1}$ are the set of constant terms of the straight line formed in the objective space by the objective values of the solutions in R at time t and $t + 1$, respectively. Equation (10) is used to calculate the covariance of $\boldsymbol{p}^t$ and $\boldsymbol{p}^{t+1}$, and $std(x)$ and $E(x)$ respectively denote the standard deviation and the average of x. To further explain the values in two sets $\boldsymbol{p}^t$ and $\boldsymbol{p}^{t+1}$, Fig. 4 shows an example of the distribution of the population in the target space before and after the change.

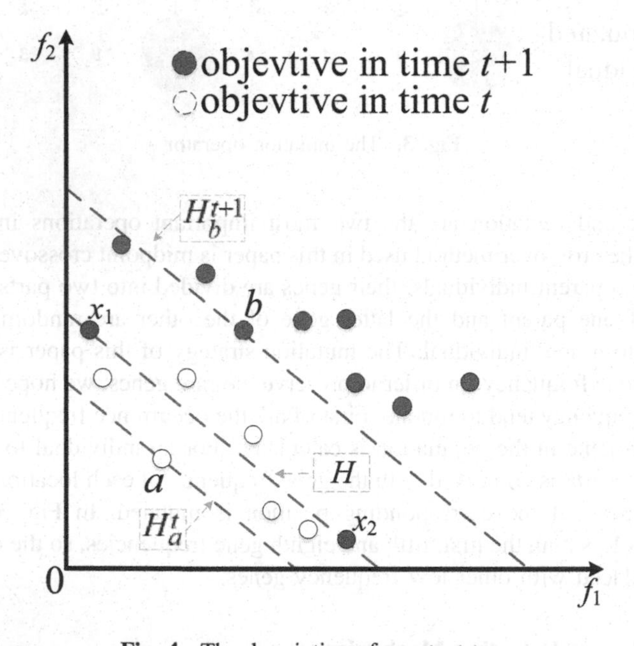

Fig. 4. The description of constant term

In Fig. 4, the eight white points and the eight black points respectively represent the objective values of the solutions in R at time t and time $t + 1$. At time t, For each objective, the solutions with the minimum objective value are selected as the boundary points, i.e., the red points x_1 and x_2, and then a line H is constructed to go through these two points, which is formalized as $k_1 f_1 + k_2 f_2 + p = 0$. After that, this line is translated, so that it passes through each solution in R. For each solution $x_i \in R$ corresponds to a line H_i^t parallel to H and has an constant term $p_i^t \in \boldsymbol{p}^t$, where i = 1, 2,..., N/10. In this figure, H and H_a^t are represented by black solid lines and blue dashed lines, respectively. Similarly, at time t + 1, the same line H is translated and goes through all the black points, and then each solution $x_i \in R$ corresponds to a new line H_i^{t+1} and a constant term $p_i^{t+1} \in \boldsymbol{p}^{t+1}$ at time t + 1.

After calculating the linear correlation coefficient, the interval in which the *corr* is located determines the degree of the change and the three types of changes are set by the following principles:

- Type I: The change is linearly independent if $corr \in [0.0, 0.2)$
- Type II: The change is strong linearly dependent if $corr \in [0.7, 1.0]$
- Type III: The change is weak linearly dependent if $corr \in [0.2, 0.7)$.

4.3 The Response Strategy

The population needs to be reinitialized after a change in the environment. This paper proposes a hybrid strategy including three response schemes to deal with varying degrees of changes in the environment, which is a prediction method based on historical information. In the process of evolution, the final population of the nearest two environments is preserved, and the movement vector of the centroid of the two environments indicates the migration of the new environment. In DCRS-EA, the centroid is used to represent the location information of the population. For each gene locus, we set the gene value of the most frequent occurrence in the population to the gene value of the corresponding locus of the centroid.

The three types of changes mentioned in Sect. 4.2 correspond to three different response schemes. When the changes are considered to be linearly independent, we believe that the change of the environment is drastic and irregular, and the historical population movement information is no longer referential. In the new environment, in order to ensure convergence, the non-dominated solutions are retained as part of the initial population, and the remaining solutions are obtained by randomization method to ensure diversity. The initial population Q^t of time t is represented by the following formula.

$$Q^t = Q_{nd}^{'} \cup X^{'} \tag{11}$$

where $Q_{nd}^{'}$ is the non-dominated solution set of the last population at time t, and $X^{'}$ is the new solution set generated by random initialization, as follows:

$$X^{'} = randomInitialize(N - |Q_{nd}^{'}|) + \varepsilon \tag{12}$$

where the function *randomInitialize*() is used to randomly obtain $N - |Q_{nd}^{'}|$ solutions, N is the population size and ε is the Gaussian noise.

When the change is detected to be strong linearly dependent, we think that the degree of environmental change is small, and the historical population is usually used as the main factor to predict the initial population of the new environment. Assuming that the centroids of the two nearest historical populations are x^{t-1} and x^{t-2}, respectively, the movement vectors v^t of the two populations are calculated by

$$v^t = x^{t-1} - x^{t-2} \tag{13}$$

and the *i*th solution x_i^t of the population Q^t is calculated by

$$x_i^t = x_i^{'} + v^t + \varepsilon \tag{14}$$

where $x_i^{'}$ is the *i*th solution of the last population (i = 1, 2, …, N).

For the week linearly dependent change, we think that its degree is between linearly independent and strong linearly dependent, and the initial population of the new environment will be initialized randomly within a certain range, which is determined by the centroid movement vectors v^t of the last two historical populations. Then, the *i*th solution of initial population Q^t is obtained by

$$x_i^t = x_i^{'} + r \times v^t + \varepsilon \tag{15}$$

where r is a random value in the range (−1,1) and v^t is calculated by Eq. (13).

5 The Experimental Studies

In this part, the mean hypervolume (MHV) [18] is adopted to investigate the performance of convergence and diversity. For testing the performance of the DCRS-EA on the designed dynamic resource scheduling problem, we set up two groups of experiments. The first group of experiments compares DCRS-EA with the classical dynamic multi-objective optimization algorithm DNSGA-II [19] on the MODEL test problem. The second group of experiments replaced other different environmental change response strategies [20] into DCRS-EA and compared with the original method on INS1-2.

For the designed dynamic resource scheduling problem, the energy consumption P_{idle} under no-load of CPU is set in the range [30.0, 50.0], and the energy consumption P_{busy} under fully load of CPU is set to $P_{idle} + ram$, where $ram \in$ [50.0, 120.0]. The response time factor k of hosts is set in the range [1.0, 3.0]. At time t = 0, the required resource of user VM $v(0)$ is set in the range [5.0, 15.0]. In addition, for MODEL1-5, the five groups of setting about (m, n) are (6, 12), (8, 24), (12, 24), (18, 48) and (24, 60), respectively, and the change frequency τ_T is set to 30, 60 and 120. For INS1-2, the two groups of setting about (m, n) are (6, 12) and (12, 36), and the change frequency τ_T is set to 30, 60, 100 and 120. For all the test cases, the change severity n_T is set to 10. Besides, the maximal evolution number *max_gen* is 1200 and the population size N is 100.

Table 1. Comparisonof results of DCRS-EA and DNSGA-II on MODEL test problems using MHV

Instance	τ_T	DNSGA-II	DCRS-EA
MODEL1	30	8.6590E−02 (4.8985E−05)–	**1.0776E−01 (8.2205E−05)**
	60	8.3362E−02 (5.1555E−05)–	**1.1003E−01 (1.4036E−05)**
	120	8.5291E−02(3.8743E−05)–	**1.1873E−01 (3.9157E−05)**
MODEL2	30	5.9899E−02 (1.4146E−05)–	**9.9183E−02 (4.4367E−06)**
	60	5.9957E−02 (2.2845E−05)–	**1.0231E−01 (8.1519E−06)**
	120	6.1863E−02(6.2095E−05)–	**9.4400E−02 (2.7146E−06)**
MODEL3	30	5.6324E−02 (9.6759E−06)–	**9.3947E−02 (5.4080E−06)**
	60	5.4313E−02 (2.8983E−05)–	**9.9462E−02 (9.9468E−06)**
	120	5.5437E−02 (1.5525E−04)–	**1.0312E−01 (3.2922E−05)**
MODEL4	30	5.3054E−02 (1.5929E−05)–	**8.8384E−02 (3.1659E−06)**
	60	5.3162E−02 (1.3925E−05)–	**9.4673E−02 (1.2941E−05)**
	120	5.4646E−02 (1.8944E−05)–	**1.0355E−01 (3.6488E−05)**
MODEL5	30	5.2495E−02 (1.3547E−05)–	**8.3648E−02 (1.0346E−05)**
	60	5.2685E−02 (2.3178E−05)–	**8.8914E−02 (1.5532E−05)**
	120	5.4906E−02 (3.4750E−05)–	**1.0271E−01 (3.1091E−05)**
+/≈/–		0/0/15	

"+", "–", and "≈" indicate that the results of the compared algorithms are significantly better than, worse than, and similar to that of DCRS-EA respectively by Wilcoxon's rank sum test with $\alpha = 0.05$.

Table 2. Comparison of results of DCRS-EA and four response mechanisms on INS test problems using MHV

Instance	n_T	τ_T	PI	PHI	PPHI	HIPPHI	DCES-EA
INS1	10	30	1.026E−01 (3.68E−06)–	1.072E−01 (9.12E−05)–	1.044E−01 (6.06E−06)–	1.069E−01 (3.90E−05)–	**1.078E−01 (8.22E−06)**
		60	1.012E−01 (1.79E−05)–	1.006E−01 (1.11E−04)–	1.061E−01 (8.43E−06)–	**1.101E−01 (3.89E−05)+**	1.100E−01 (1.40E−05)
		100	1.057E−01 (2.43E−05)–	1.056E−01 (3.54E−04)–	1.089E−01 (1.90E−05)–	1.065E−01 (3.94E−05)–	**1.131E−01 (5.89E−05)**
		120	1.006E−01 (2.75E−05)–	1.158E−01 (3.51E−04)–	1.155E−01 (6.68E−05)–	1.111E−01 (7.51E−05)–	**1.187E−01 (3.92E−05)**
	20	30	**1.053E−01 (1.91E−05)+**	1.046E−01 (2.05E−05)+	1.028E−01 (1.12E−05)–	1.026E−01 (1.18E−05)–	1.032E−01 (1.06E−05)
		60	1.033E−01 (2.97E−05)–	1.050E−01 (2.10E−05)–	1.109E−01 (3.01E−05)–	1.108E−01 (4.41E−05)–	**1.142E−01 (2.56E−05)**
		100	1.019E−01 (3.85E−05)–	1.018E−01 (3.83E−05)–	1.201E−01 (4.22E−05)–	1.192E−01 (6.19E−05)–	**1.205E−01 (3.91E−05)**
		120	1.168E−01 (7.74E−05)+	1.161E−01 (6.18E−05)+	1.165E−01 (2.25E−06)+	**1.177E−01 (4.32E−05)+**	1.138E−01 (4.95E−05)

(continued)

Table 2. *(continued)*

Instance	n_T	τ_T	PI	PHI	PPHI	HIPPHI	DCES-EA
INS2	10	30	9.479E−02 (2.41E−06)+	**1.057E−01 (2.82E−05)+**	9.382E−02 (1.54E−06)–	9.921E−02 (1.87E−05)–	9.395E−02 (5.41E−06)
		60	9.827E−02 (5.54E−06)–	9.802E−02 (4.37E−05)–	9.823E−02 (5.19E−06)–	9.2390E−02 (5.19E−06)–	**9.946E−02 (9.95E−06)**
		100	1.024E−01 (9.65E−04)–	9.561E−02 (7.34E−05)–	1.024E−01 (6.17E−06)–	9.427E−01 (6.17E−06)–	**1.031E−01 (2.14E−05)**
		120	**1.064E−01 (2.00E−06)+**	9.980E−02 (1.30E−04)–	1.052E−01 (5.67E−06)+	9.637E−02 (3.27E−05)–	1.031E−01 (3.29E−05)
	20	30	8.971E−02 (3.79E−06)–	9.165E−02 (9.42E−06)+	9.115E−02 (6.55E−06)+	**9.223E−02 (9.75E−06)+**	9.085E−02 (8.51E−06)
		60	9.016E−02 (1.10E−05)–	1.000E−01 (1.07E−05)–	9.919E−02 (1.37E−05)–	9.916E−02 (1.49E−05)–	**1.001E−1 (6.87E−06)**
		100	1.039E−01 (1.19E−05)–	1.031E−01 (3.25E−05)–	1.036E−01 (2.39E−05)–	1.036E−01 (1.82E−05)–	**1.039E−01 (3.79E−05)**
		120	9.862E−02 (3.56E−05)–	9.990E−02 (4.76E−05)+	**1.006E−01 (2.70E−05)+**	9.914E−02 (3.30E−05)+	9.881E−02 (2.78E−05)
+/≈/–			4/0/12	5/0/11	4/0/12	4/0/12	

"+", "–", and "≈" indicate that the results of the compared algorithms are significantly better than, worse than, and similar to that of DCRS-EA respectively by Wilcoxon's rank sum test with $\alpha = 0.05$.

The comparison results of MHV in the first group of experiments are summarized in Table 1. In the experiments, each experiment was run 20 times independently. The best values of each test instance are marked in bold and gray background. Looking at the overall situation, the larger the number of hosts and user VMs, the more difficult it is to optimize the corresponding dynamic problems. The difficulty of optimization is inversely proportional to the change frequency, since the smaller change frequency means that the algorithm has more time to optimize the population with static algorithms. From Table 1, it can be seen that the MHV values of DCRS-EA are better than that of DNSGA-II in all instances. On the whole, under the MHV evaluation indicator, compared with the DNSGA-II, DCRS-EA has better optimization performance in dealing with the designed resource scheduling problem in virtual dynamic cloud environment.

In the second group of experiments, the change response strategies HIPPHI and its three variants [20] are embedded into DCRS-EA and compared with the original DCRS-EA. The three variants of HIPPHI are the Population-Initialized (PI), the Pure Historical Information (PHI) and the Pure Prediction by Historical Information (PPHI). The experimental results are shown in Table 2. Among the 16 simulation experiments, 9 of best results are obtained by DCRS-EA, while PI, PHI, PPHI and HIPPHI obtain 2, 1, 1 and 3 best results, respectively. Besides, from the last row of Table 2, it can be seen that DCRS-EA is better than PI, PHI, PPHI and HIPPHI in 12, 11, 12 and 12 test cases, respectively. In general, according to Table 1 and Table 2, the proposed DCRS-EA shows the excellent performance in dealing with the resource scheduling problem in dynamic cloud environment.

6 Conclusions and Future Work

In this paper, we build a resource scheduling model for dynamic cloud environment, which takes into account the optimization of energy consumption and quality of service. In addition, we design a new dynamic multi-objective optimization algorithm to find an excellent scheduling scheme for the model, which not only detects whether the environment has changed, but also estimates the type of change. For different changes, our response strategy uses different schemes to initialize the population. Finally, by comparing the two groups of experiments, it is obvious that our algorithm has better performance in dealing with the resource scheduling problem in dynamic cloud environment. In real life, there are still many dynamic optimization engineering problems to be solved, which is also one of our main research work in the future.

References

1. Fox, A.M.: Above the clouds: a berkeley view of cloud computing. Eecs Depart. Univ. Calif. Berkeley **53**(4), 50–58 (2009)
2. Tang, Q., Gupta, S.K.S., Varsamopoulos, G.: Energy-efficient thermal-aware task scheduling for homogeneous high-performance computing data centers: a cyber - physical approach. IEEE Trans. Parallel Distrib. Syst. **19**(11), 1458–1472 (2008)
3. Yang, Z., Liu, M., Xiu, J., Liu, C.: Study on cloud resource allocation strategy based on particle swarm ant colony optimization algorithm. In: IEEE International Conference on Cloud Computing & Intelligence Systems, Hangzhou, pp. 488–491(2012)
4. Ding, S., Chen, S.P.: Multi-objective ant colony resource allocation algorithm based on packet cluster mapping in cloud computing. Software **39**(11), 9–14 (2018)
5. Mezmaz, M., Melab, N., Kessaci, Y.: A parallel bi-objective hybrid metaheuristic for energy-aware scheduling for cloud computing systems. J. Parallel Distrib. Comput. **71**(11), 1497–1508 (2011)
6. Lee, Y.C., Zoomaya, A.Y.: Energy efficient utilization of resources in cloud computing systems. J. Supercomput. **60**(2), 268–280 (2012)
7. Lioyd, W.J., et al.: Demystifying the clouds: harnessing resource utilization models for cost effective infrastructure alternatives. IEEE Trans. Cloud Comput. **5**(4), 667–680 (2017)
8. Farahnakian, F., Pahikkala, T., Liljeberg, P., Plosila, J., Hieu, N.T., Tenhunen, T.: Energy-aware VM consolidation in cloud data centers using utilization prediction model. IEEE Trans. Cloud Comput. **7**(2), 524–536 (2019)
9. Gong, S.Q., Yin, B.B., Zheng, Z., Cai, K.Y.: Adaptive multivariable control for multiple resource allocation of service-based systems in cloud computing. IEEE Access **7**, 13817–13831 (2019)
10. Zheng, B.: A new dynamic multi-objective optimization evolutionary algorithm. In: Third International Conference on Natural Computation, ICNC 2007, vol. 5, pp. 565–570. IEEE (2007)
11. Grefenstette, J.J.: Genetic algorithms for changing environments. In: Maenner, R., Manderick, B. (eds.) Parallel Problem Solving from Nature, North Holland, vol. 2, pp. 137–144 (1992)
12. Zhang, Z., Qian, S.: Artificial immune system in dynamic environments solving time-varying non-linear constrained multi-objective problems. Soft. Comput. **15**(7), 1333–1349 (2011)

13. Wang, Y., Li, B.: Investigation of memory-based multi-objective optimization evolutionary algorithm in dynamic environment. In: 2009 IEEE Congress on Evolutionary Computation, pp. 630–637. IEEE (2009)
14. Hatzakis, I., Wallace, D.: Dynamic multi-objective optimization with evolutionary algorithms: a forward-looking approach. In: Proceedings of the 8th Annual Conference on Genetic and Evolutionary Computation, pp. 1201–1208. ACM (2006)
15. Peng, Z., Zheng, J.H., Zou, J., Liu, M.: Novel prediction and memory strategies for dynamic multi-objective optimization. Soft. Comput. **19**(9), 2633–2653 (2015)
16. Fan, X., Weber, W.D., Barroso, L.A.: Power Provisioning for a Warehouse-sized computer. In: 34th ACM International Symposium on Computer Architecture, [S.l.]: [s.n.] (2007)
17. Lin, Q., Jin, G., Ma, Y., Wang, K.C.: A diversity-enhanced resource allocation strategy for decomposition-based multi-objective evolutionary algorithm. IEEE Trans. Cybern. **48**(8), 2388–2401 (2018)
18. Tantar, E., Tantar, A.-A., Bouvry, O.: On dynamic multi-objective optimization, classification and performance measures. In: Proceedings of IEEE CEC, pp. 2759–2766 (2011)
19. Deb, K., Rao N., U.B., Karthik, S.: Dynamic multi-objective optimization and decision-making using modified NSGA-II: a case study on hydro-thermal power scheduling. In: Obayashi, S., Deb, K., Poloni, C., Hiroyasu, T., Murata, T. (eds.) EMO 2007. LNCS, vol. 4403, pp. 803–817. Springer, Heidelberg (2007). https://doi.org/10.1007/978-3-540-70928-2_60
20. Ji, L.Q.: Resource optimization for dynamic migration of virtual machines in cloud environment. Shenzhen University (2018)

An Ensemble Classification Technique for Intrusion Detection Based on Dual Evolution

Qiuzhen Lin(✉), Chao Hu, Shuo Chen, and Peizhi Huang

Shenzhen University, Shenzhen 518060, China
{qiuzhlin,hpz}@szu.edu.cn,
{1810272067,chenshuo2018}@email.szu.edu.cn

Abstract. Intrusion detection systems are becoming increasingly important in network security, which have been widely used to protect our terminal devices against network attacks. However, classification of imbalanced datasets is one of the main challenges in intrusion detection. Traditional methods are biased to the majority classes, which cannot effectively detect the minority classes. This paper presents a novel ensemble classification technique for intrusion detection, which consists of dual evolution. The first evolution is run to select the features from the original training set, which are then used to extract the training set for the basic classifiers. After that, the second one is executed to choose a preferred combination from the base classifiers. When compared to other intrusion detection algorithms, the experimental results on the NSL-KDD and AWID datasets validate the advantages of the proposed algorithm.

Keywords: Intrusion detection · Ensemble learning · Imbalanced classification

1 Introduction

With the rapid development of computer technology, people's life is increasingly inseparable from the Internet, but more and more network attacks are emerging nowadays. Intrusion detection systems (IDSs) play an important role in network security, which analyze the network traffic data and then identify the attack type. As shown in Fig. 1, a generalized framework of IDS is given, where the data in the network will be classified by IDS when passing through the firewall. In this case, only normal visiting data can pass, while attack behavior will be rejected.

According to the deployed detection mechanisms, IDSs can be generally classified into two main categories: misuse detection and anomaly detection. Misuse detection methods [1] will search the signatures of known attacks in network traffic. As a result, they have extremely low false positive rate, but they cannot effectively detect the novel attacks. Anomaly detection methods [2] first establish a baseline with known normal behaviors and then use deviations from this baseline to detect potential intrusions. Thus, they can effectively detect the novel attacks. In this paper, we consider the anomaly detection methods.

D.-S. Huang and K.-H. Jo (Eds.): ICIC 2020, LNCS 12464, pp. 607–618, 2020.
https://doi.org/10.1007/978-3-030-60802-6_53

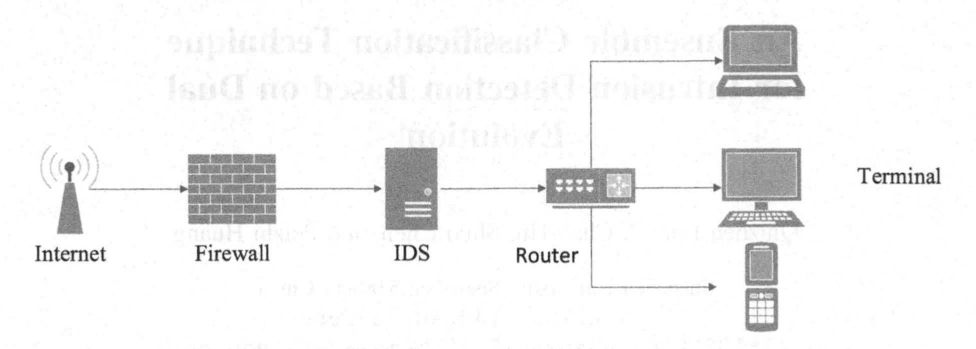

Fig. 1. A generalized framework of IDS

The task of IDSs is to classify the network traffic, where normal behaviors will be allowed while attack behaviors will be intercepted and classified to different attack types. There are two main challenges for IDS, imbalanced samples and redundant features. In the multi-classification problem, imbalanced problem [3] usually refers to the significant difference in the sample size of different categories. Majority classes have much more samples than minority classes. For instance, the NSL-KDD dataset [4] is a benchmark dataset used for evaluating IDS, in which the number of normal samples is more than one thousand times the number of U2r type samples. Traditional intrusion detection methods cannot effectively handle imbalanced problems, as they may classify the samples of minority classes into majority classes. However, in intrusion detection, the minority classes should be detected and marked. On the other hand, there are many redundant and irrelevant features in network traffic data, which will increase the training time of classifiers and reduce the classification accuracy.

In order to overcome the drawback mentioned above, many approaches have been proposed. They can be categorized into three main kinds according to the improvements on the data or algorithms, such as the rebalance of the distribution of data [5], the assignment of appropriate misclassification cost to minority samples on the algorithmic design [6], and the hybridized methods [7] on both data and algorithms. Ensemble learning [8] is a kind of hybridized methods that can improve the performance of classifiers by combining the advantages of several basic classifiers in terms of adaptability and generalization. In this paper, we propose an ensemble classification technique for intrusion detection based on dual evolution (called IDDE), which can handle imbalanced classification problem in intrusion detection. We run the first evolution to select some good features from the original training set to train the basic classifiers and execute the second evolution to select a good combination from the base classifiers. When running the experiments on the NSL-KDD and AWID datasets, the performance of the proposed algorithm is better than other existing intrusion detection algorithms on most cases.

The remainder of this paper is organized as follows. Section 2 introduces some related works. Our ensemble classification method for intrusion detection is described in Sect. 3. In Sect. 4, the experimental results of IDDE and other existing intrusion detection algorithms are compared and discussed. Section 5 provides our conclusions and gives some future directions.

2 Related Work

2.1 Feature Dimension Reduction

In the real world, network traffic data is usually complex and generally has many redundant and irrelevant features. Those high-dimensional data may affect the classification accuracy and classification speed. Thus, feature dimensional reduction is necessary for IDS, which is a common preprocessing step of machine learning algorithms. Dimensionality reduction techniques can be used to improve classification accuracy, reduce irrelevant finite elements and preserve the most important features related to the target classification problems. Generally, there have two kinds of dimensionality reduction approaches: feature extraction [9] and feature selection [10].

Feature extraction methods include principal component analysis (PCA) [11], linear discriminant analysis (LDA) [12] and deep auto-encoder [13]. PCA is the most popular and widely used feature extraction approach, which uses orthogonal transformation to transform a set of features into a set of linearly uncorrelated variable values without losing much information. The principal components are sorted from the largest possible variance to the lowest one, and then the k eigenvectors with largest eigenvalues are selected as new features. However, it is difficult to understand the low-dimensional features extracted from the original features, as these features cannot retain the information of the original data, while feature selection can get a suitable feature subset from original features without any change.

In general, feature selection can be classified into two categories: filter methods and wrapper methods. Filter methods choose the feature subset corresponding to information measures without classifier. Features with higher information gains would be selected and are ranked according to their values of information gains. An approach with relevance boosting and feature interaction enhancing based on mutual information is proposed in [3]. When compared to filter methods, wrapper methods select features subsets by the evaluation under classifiers. Every feature can be selected, so the problem of feature selection is a NP-hard problem. Evolutionary algorithm is an effective method for solving this kind of NP problems. In [14], a wrapper method is proposed, where I-NGSA-III is employed to obtain the optimal feature subsets and then the hierarchical self-organizing maps [15] are used to evaluate the objective value. In general, wrapper methods are mathematically modeled as an optimization problem with multiple objectives. In this paper, wrapper methods for IDS are applied in order to ensure the classification accuracy of both majority and minority classes.

2.2 Ensemble Learning

Ensemble learning is a useful machine learning paradigm with exhibited apparent advantages in many applications, which can be classified into two main types: boosting [16] and bagging [17].

In boosting, every classifier is run in series and depends on the previous one. The next classifier improves the learning sample weight associated to the individual classification error in the previous round, which also reduces the weight of the correct classification for the sample. Therefore, misclassified samples will receive more attention and are expected to be correctly classified in the next round. By this way, the classification problem can be solved by a series of weak classifiers.

Bagging's training data set for individual basic classifiers is obtained by random sampling. *T* sampling sets are obtained through *T* times of random sampling. For these *T* sampling sets, *T* basic classifiers are independently trained, and then the final strong classifier is obtained through a fusion strategy. In [11], an ensemble classifier that uses SVM, IBK, and MLP learners is proposed to enhance the intrusion detection accuracy, and the vote algorithm is based on the average of probabilities. In [7], the training data set is obtained by generating the synthetic data, filling the synthetic data points in minority data space and using those to get ensemble. However, for unbalanced training data sets, some minority classes cannot be sampled by training samples or classified by basic classifiers. To avoid this situation, a novel ensemble method based on bagging is proposed in this paper, which samples the training data sets from different feature subsets.

2.3 Multi-objective Optimization

In the real world, we have to consider more than one objective in many cases. Multi-objective optimization means to optimize multiple objectives in a given problem at the same time. For NSL-KDD, there are five objectives we need to maximize, i.e., the classification accuracy of Normal, Probe, Dos, R2L and U2R. The result of multi-objective optimization is usually a set of trade-off solutions, which are composed of Pareto optimal solutions. In recent years, there are a lot of evolutionary algorithms used to deal with multi-objective optimization problems. In [15], a neural network (GHSOM-pr) is used as the classifier to evaluate the fitness values of some feature subsets and then the NSGA-II algorithm [18] is employed to search the best feature subsets. However, NSGA-II has some limitations in offspring generations; NSGA-III [12] is an effective and state-of-the-art algorithm that can obtain good convergence and diversity. In this paper, we use NSGA-III to get the optimal feature subsets and ensemble selection subsets because of its balanced performance.

2.4 Jaccard's Index

Jaccard's index [19] is a similarity measurement between datasets. It provides a classifier performance measurement for the corresponding class. This index can be used to select the features that fit better for a specific class, since these selected features will get the higher Jaccard's index. Optimizing classifier performance means to maximize a number of objective functions (equal to the number of classes) at the same time. The intrusion detection is a representative imbalanced problem, where 'normal' is a majority class while the 'attack' classes contains four minority subclasses in NSL-KDD, named Dos, Probe, U2R and R2L. Accuracy is not an effective performance measurement for this kind of problems. Thus, we adopt the Jaccard's index as objective functions to be optimized, which can be described as follows:

$$J(k) = \frac{TP}{TP + FN + FP}, \ k \in \{1, 2, \ldots, K\} \tag{1}$$

where K is the total class number, and $J(k)$ represents the Jaccard's index of the k-th class.

3 The Proposed Framework

The proposed IDDE algorithm is an ensemble method with dual evolution on feature selection and basic classifier selection. In Fig. 2, the framework of IDDE is plotted. In this algorithm, a many-objective optimization algorithm (NSGA-III [20]) is used to find the optimal results due to its effective performance. In the feature selection stage, relevant and informative features will be selected to train all the basic classifiers, which can improve the ability of classifier in terms of accuracy and efficiency with less training time. In the basic classifier selection stage, an optimal subset of basic classifiers will be selected and their outputs are combined by using a weighted voting strategy. More details of our algorithm would be introduced in the following subsections.

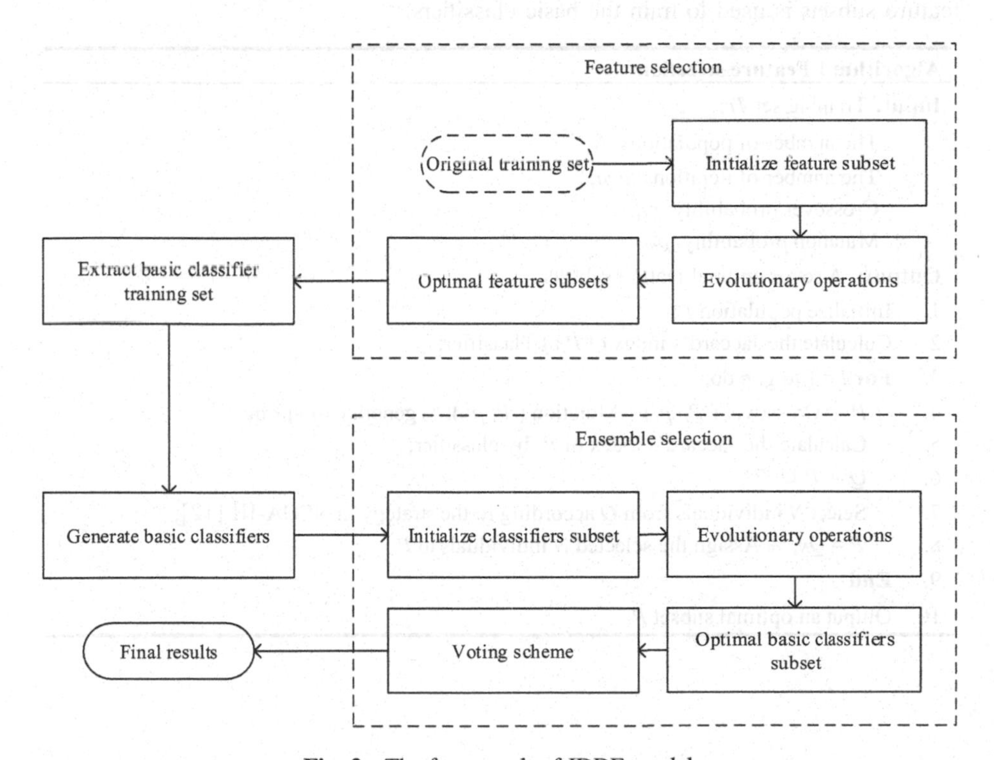

Fig. 2. The framework of IDDE model

3.1 Feature Selection

In the feature selection stage, the individuals in the evolutionary population are defined. The number of features in the original training set is taken as the length of the individual. '0' indicates that the feature at the corresponding position is not selected and '1' indicates that it is selected. Initialize the population by randomly generating a binary sequence with a length of the total number of features. Then calculate the Jaccard's index for each class of base classifier which is trained by training set extracted by feature subsets. Because the individual is a binary string, single-point crossover and bitwise mutation operators are used to generate offspring population. Calculate the Jaccard's index for each class of offspring population and combine offspring population and parent population into a new population. Our method uses the selection strategy in NSGA-III to choose N individuals for this new population at the end of iteration. Finally, we will get a set of optimal feature subsets. The main procedure of the feature selection is presented in Algorithm 1.

The training set of traditional bagging method is obtained by random sampling. However, the ability of basic classifier to correctly classify the minority classes will be reduced when the number of samples in minority classes of the training set is too little. Therefore, our algorithm uses each individual in the optimal feature subsets obtained by Algorithm 1 to extract the corresponding features from the original training set to form a new training set. The training data set obtained by each subset in the optimal feature subsets is used to train the basic classifiers.

Algorithm 1 Feature selection

Input: Training set *Tr*;

The number of populations: N;
The number of iterations: *gen*;
Crossover probability: p_c;
Mutation probability: p_m;

Output: A set of optimal feature subsets;

1. Initialize population P;
2. Calculate the Jaccard's index of P by classifier;
3. **For** $i = 1$ to *gen* do:
4. $P' = \text{Crossover}(P, p_c) + \text{Mutation}(P, p_m)$; % generate offspring
5. Calculate the Jaccard's index of P' by classifier;
6. $Q = P \cup P'$;
7. Select N individuals from Q according to the strategy in NSGA-III [12];
8. $P = Q_N$; % Assign the selected N individuals to P
9. **End**
10. Output an optimal subset P;

3.2 Ensemble Selection

Algorithm 2 Get the selected ensemble classifiers sets

Input: A set of basic classifiers C;
  The number of basic classifiers: S;
  The number of iterations: T;
  Crossover probability: b_c;
  Mutation probability: b_m;

Output: a selected basic classifiers sets and classification results;

1. Initialize population B;
2. Calculate the accuracy of B by the output of the selected basic classifiers;
3. **For** $i = 1$ to T do:
4. $B' = \text{Crossover}(B, b_c) + \text{Mutation}(B, b_m)$; % generate offspring
5.   Calculate the accuracy of B' by the output of the selected basic classifiers;
6. $E = B \cup B'$;
7.   Select S individuals from E according to the strategy in NSGA-III [12];
8. $B = E_S$; % assign the selected E individuals to B
9. **End**
10. Output a selected basic classifiers set and classification result;

A common approach in ensemble learning is to combine all the sets of basic classifiers to obtain better results. However, when the number of basic classifiers is too much, there will be some redundant basic classifiers which reduce classification ability and waste storage space. It is necessary to select the most efficient subset from a set of base classifiers. In ensemble selection stage, the way we define the individual of population is the same as feature selection. '0' indicates that the base classifier is not selected and '1' indicates that it is selected. The main procedure of the ensemble selection is presented in Algorithm 2.

3.3 Voting Scheme

The traditional voting rule is to count all the prediction results of basic classifiers, and the most predicted results will be considered as the final result. Some studies have proposed a weight-based voting method, which uses the classification ability as the standard, and the predicted result of the base classifier with strong classification ability will be multiplied by a higher weight. However, in our model, the difference in the classification ability of the base classifier will not be very large, so we propose a new voting method to adjust the weight based on the type of the predicted result of the basic classifier.

Let X be a test dataset containing J samples and let $x_j \in X$ be the jth test sample in test dataset. Denote $S = \{s_1, \ldots, s_M\}$ as the selected basic classifiers, where s_m is the mth basic selected classifier. The final decision $f(x_j)$ of jth test sample is written as

$$f(x_j) = argmax_y \sum_{m=1}^{M} W_y \times (v_m(x_j) = y) \quad (2)$$

where $v_m(x_j)$ is a predicted label by the mth selected basic classifier for a test sample x_j. w_y is a weight vector based on the predicted label. For example, the weight vector for NSL-KDD is illustrated in Table 1.

Table 1. Weight vector for NSL-KDD.

y	Normal	Dos	Probe	R2L	U2R
W_y	1	1	1	2	3

4 Experimental Results

In this section, the experimental results of IDDE and other ensemble methods in IDS are given. The IDDE is implemented in MATLAB R2018a on a 64-bit Windows 10 PC with an Intel Core 3.60 GHz processor and 24 GB RAM.

The following measures are used to evaluate performance. True Positive (TP) is the number of attack data that is classified as an attack. False Positive (FP) denotes the number of data classified as an attack, but actually it is normal. False Negative (FN) is the number of attack data that is incorrectly classified as normal. True Negative (TN) denotes the number of normal data that is correctly classified as normal.

Accuracy: the percentage of the number of correctly classified records versus the total records, as shown in Eq. (3).

$$Accuracy = \frac{TP + TN}{TP + TN + FP + FN} \quad (3)$$

Precision: the percentage of correct classification in results predicted to be positive samples, as shown in Eq. (4)

$$Precision = \frac{TP}{TP + FP} \quad (4)$$

Recall: the percentage of the number of records identified correctly over the total number of attack records, as shown in Eq. (5)

$$Recall = \frac{TP}{TP + FN} \quad (5)$$

False alarm: the percentage of results that are actually negative samples, which are classified as positive samples, as shown in Eq. (6).

$$False\ Alarm = \frac{FP}{FP + TN} \quad (6)$$

Table 2 presents the percentage of different classes in NSL-KDD. It is obvious that the number of R2L and U2R is significantly less than that of Normal. Table 3 shows the percentage of different classes in AWID [21].

Table 2. Percentage of different classes in NSL-KDD.

Anomaly type	Count	Percentage
Normal	**67342**	53.46%
Dos	**45927**	36.46%
Probe	**11656**	9.25%
R2L	**995**	0.79%
U2R	**52**	0.04%

Table 3. Percentage of different classes in AWID.

Anomaly type	Count	Percentage
Normal	**530785**	92.20%
Flooding	**8097**	1.40%
Injection	**16682**	2.90%
Impersonation	**20079**	3.49%

Table 4. Performance comparison of IDDE and IG-PCA in NSL-KDD

Class	Precision (%)		Recall (%)		False alarm (%)	
	IDDE	IG-PCA-Ensemble	IDDE	IG-PCA-Ensemble	IDDE	IG-PCA-Ensemble
Normal	99.36	**99.42**	**99.50**	95.63	0.72	**0.69**
Dos	**99.71**	95.54	**98.58**	97.34	**0.14**	0.45
Probe	**98.53**	96.22	**99.61**	99.42	**0.15**	2.2
R2l	**60.00**	0	**46.15**	Null	**0.02**	Null
U2r	**94.92**	0	**92.12**	Null	**0.04**	Null

The multi classification performance of our IDDE method and IG-PCA-Ensemble [11] used in NSL-KDD are shown in Table 4. Higher precision and Recall denote better performance, while lower False alarm means that the method is more effective. IG-PCA-Ensemble has only three basic classifiers and the vote algorithm is based on the average of probabilities. For a sample, every basic classifier will get a probability of each classes, the minority classes can easily get a lower probability, so IG-PCA-Ensemble cannot classify the R2l and U2r classes well. We cannot calculate the Recall value and False alarm. Thus, we can conclude that our IDDE method is more effective, especially for R2l and U2r.

Table 5 shows the performance comparison of our IDDE method and IG-PCA-Ensemble. By comparing the results of the two algorithms, we can observe that our IDDE has higher precision, higher Recall and lower False alarm. In addition, the accuracies of binary and multi classification in NSL-KDD and AWID are presented in Table 6. Our IDDE method gets the higher accuracy. In summary, our IDDE method can not only solve the imbalance problem, but also has advantages in terms of the overall accuracy.

To compare the advantages of our algorithm with other ensemble algorithms, Fig. 3 provides the accuracies obtained by J48 [22], AdaBoost [22], Random forest [22] and our IDDE method. In summary, it is obvious that our IDDE method performs better than other ensemble methods.

Table 5. Performance comparison of IDDE and IG-PCA in AWID

Class	Precision (%)		Recall (%)		False alarm (%)	
	IDDE	IG-PCA	IDDE	IG-PCA	IDDE	IG-PCA
Normal	**99.97**	98.84	**99.98**	99.13	**0.09**	4.59
Flooding	**99.94**	97.24	**99.70**	91.43	**0.02**	0.60
Injection	**99.95**	93.58	**99.95**	97.41	**0.01**	0.20
Impersonation	**99.67**	95.28	**99.97**	89.59	**0**	0.26

Table 6. Accuracy comparison of IDDE and IG-PCA in NSL-KDD and AWID

	Data set	IDDE	IG-PCA	Data set	IDDE	IG-PCA
Binary	NSL-KDD	**99.80**	97.86	AWID	**99.59**	98.29
Multi	NSL-KDD	**99.37**	97.13	AWID	**99.96**	98.27

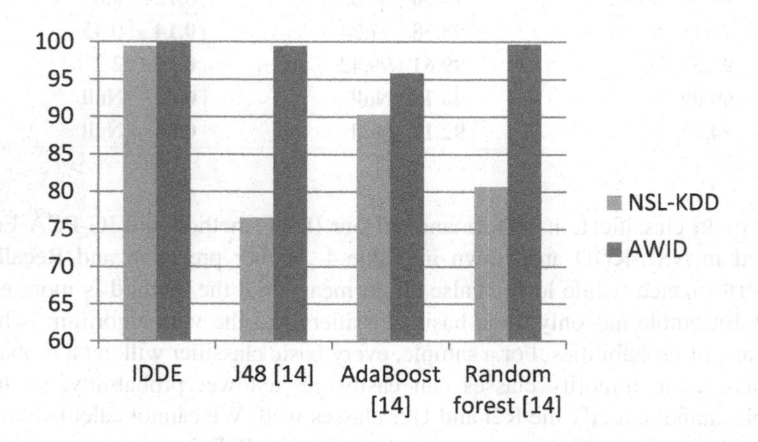

Fig. 3. Comparison of different methods using the accuracy in NSL-KDD and AWID

5 Conclusions

In this paper, we have established a selective ensemble based on dual evolutionary framework for solving the imbalanced problem in IDS. The proposed method extracts the training sets in the feature selection stage, which are used to train the base classifier. Then, the proposed method further adopts ensemble selection to find the most useful subset of basic classifiers. Experimental results on the NSL-KDD dataset and AWID dataset have validated the effectiveness of our proposed method, which can obtain higher accuracy.

In the future work, we plan to use more data sets to verify the feasibility of our approach, and investigate more effective methods to solve the imbalanced problem, such as generating minority classes of training sample to solve the imbalanced problem.

Acknowledgement. This work was supported by the National Natural Science Foundation of China (NSFC) under Grants 61876110, 61836005, and 61672358, the Joint Funds of the NNFC under Key Program Grant U1713212, and the Shenzhen Scientific Research and Development Funding Program under grant JCYJ20190808164211203.

References

1. Ilgun, K., Kemmerer, R.A., Porras, P.A.: State transition analysis: a rule-based intrusion detection system. IEEE Trans. Softw. Eng. **21**(3), 181–199 (1995)
2. Mahadevan, V., Li, W., Bhalodia, V., Vasconcelos, N.: Anomaly detection in crowded scenes. In: 2010 IEEE Computer Society Conference on Computer Vision and Pattern Recognition, San Francisco, CA, pp. 1975–1981 (2010)
3. Zhu, Z., Wang, Z., Li, D., Zhu, Y., Du, W.: Geometric structural ensemble learning for imbalanced problems. IEEE Trans. Cybern. **50**(4), 1617–1629 (2020)
4. Arafat, M., Jain, A., Wu, Y.: Analysis of intrusion detection dataset NSL-KDD using KNIME analytics. In: 13th International Conference on Cyber Warfare and Security, Washington, pp. 573–583 (2018)
5. Wang, S., Yao, X.: Diversity analysis on imbalanced data sets by using ensemble models. In: 2009 IEEE Symposium on Computational Intelligence and Data Mining, Nashville, TN, pp. 324–331 (2009)
6. Saglam, F., Cengiz, M.A.: Noise detection in imbalanced classes using adaptive boosting. In: 2019 4th International Conference on Computer Science and Engineering (UBMK), Samsun, Turkey, pp. 449–452 (2019)
7. Lim, P., Goh, C.K., Tan, K.C.: Evolutionary cluster-based synthetic oversampling ensemble (ECO-ensemble) for imbalance learning. IEEE Trans. Cybern. **47**(9), 2850–2861 (2017)
8. Liu, Z., et al.: Self-paced ensemble for highly imbalanced massive data classification. In: 2020 IEEE 36th International Conference on Data Engineering (ICDE), Dallas, TX, USA, pp. 841–852 (2020)
9. Alipourfard, T., Arefi, H., Mahmoudi, S.: A novel deep learning framework by combination of subspace-based feature extraction and convolutional neural networks for hyperspectral images classification. In: 2018 IEEE International Geoscience and Remote Sensing Symposium, Valencia, pp. 4780–4783 (2018)

10. Peker, M., Arslan, A., Şen, B., Çelebi, F.V., But, A.: A novel hybrid method for determining the depth of anesthesia level: combining ReliefF feature selection and random forest algorithm (ReliefF+RF). In: 2015 International Symposium on Innovations in Intelligent SysTems and Applications (INISTA), Madrid, pp. 1–8 (2015)
11. Salo, F., Nassif, A.B., Essex, A.: Dimensionality reduction with IG-PCA and ensemble classifier for network intrusion detection. Comput. Netw. **148**, 164–175 (2019)
12. Ji, S., Ye, J.: Generalized linear discriminant analysis: a unified framework and efficient model selection. IEEE Trans. Neural Netw. **19**(10), 1768–1782 (2008)
13. Shone, N., Ngoc, T.N., Phai, V.D., Shi, Q.: A deep learning approach to network intrusion detection. IEEE Trans. Emerg. Top. Comput. Intell. **2**(1), 41–50 (2018)
14. Zhu, Y., Liang, J., Chen, J., Ming, Z.: An improved NSGA-III algorithm for feature selection used in intrusion detection. Knowl.-Based Syst. **116**, 74–85 (2017)
15. Yang, Y.H., Huang, H.Z., Shen, Q.N., Wu, Z.H., Zhang, Y.: Research on intrusion detection based on incremental GHSOM. Chin. J. Comput. **37**(5), 1216–1224 (2014)
16. Hui, H.W., Woo, Y.H., Hung, C.H., Leung, K.N.: A double gain-boosted amplifier with widened output swing based on signal-and transient-current boosting technique in CMOS 130-nm technology. In: 2018 IEEE International Conference on Electron Devices and Solid State Circuits (EDSSC), Shenzhen, pp. 1–2 (2018)
17. Yu, Y., Su, H.: Collaborative representation ensemble using bagging for hyperspectral image classification. In: IGARSS 2019 - 2019 IEEE International Geoscience and Remote Sensing Symposium, Yokohama, Japan, pp. 2738–2741 (2019)
18. Deb, K., Pratap, A., Agarwal, S., Meyarivan, T.A.M.T.: A fast and elitist multiobjective genetic algorithm: NSGA-II. IEEE Trans. Evol. Comput. **6**(2), 182–197 (2002)
19. Gupta, A.K., Sardana, N.: Significance of clustering coefficient over jaccard index. In: 2015 Eighth International Conference on Contemporary Computing (IC3), Noida, pp. 463–466 (2015)
20. Deb, K., Jain, H.: An evolutionary many-objective optimization algorithm using reference-point-based nondominated sorting approach, part I: solving problems with box constraints. IEEE Trans. Evol. Comput. **18**(4), 577–601 (2014)
21. Kolias, C., Kambourakis, G., Stavrou, A., Gritzalis, S.: Intrusion detection in 80211 networks: empirical evaluation of threats and a public dataset. IEEE Commun. Surv. Tutor. **18**(1), 184–208 (2016)
22. Abdulhammed, R., Faezipour, M., Abuzneid, A., Alessa, A.: Effective features selection and machine learning classifiers for improved wireless intrusion detection. In: 2018 International Symposium on Networks, Computers and Communications (ISNCC), Rome, pp. 1–6 (2018)

Construction of Large-Scale Heterogeneous Molecular Association Network and Its Application in Molecular Link Prediction

A Novel Stochastic Block Model for Network-Based Prediction of Protein-Protein Interactions

Xiaojuan Wang[1], Pengwei Hu[2], and Lun Hu[1,3](✉)

[1] School of Computer Science and Technology, Wuhan University of Technology, Wuhan, China
[2] IBM Research, Beijing, China
[3] Xinjiang Technical Institute of Physics and Chemistry, Chinese Academy of Science, Urumqi, China
hulun@ms.xjb.ac.cn

Abstract. Proteins interact with each other to play critical roles in many biological processes in cells. Although promising, laboratory experiments usually suffer from the disadvantages of being time-consuming and labor-intensive. Hence, a variety of computational approaches have been proposed to predict protein-protein interactions from an alternative view. However, most of them heavily rest on the biological information of proteins while ignoring the latent structural features in protein interaction networks. In this paper, we propose a novel stochastic block model for network-based prediction of protein interactions. By simulating the generative process of a protein interaction network, our approach can capture the latent structural features of proteins from the perspective of forming protein complexes, thus verifying whether two proteins interact with each other or not. To evaluate the performance of the proposed prediction approach, a series of extensive experiments have been conducted and we have also compared our approach with state-of-the-art network-based prediction model. The experiment results show that our approach has a promising performance when applied to predict protein-protein interactions.

Keywords: Protein interaction network · Protein interaction prediction · Stochastic block model

1 Introduction

Proteins are important component of cells, they provide the material basis for life and undertake various functions in living organisms. Instead of functioning alone, protein interact with each other to form complexes, thus performing their functions. Since protein-protein interactions (PPIs) constitute most of biological processes, such as transcription, replication and translation of DNA, it is of great significance to understand the mechanisms of PPIs. Furthermore, knowing how proteins interact with other not only allows us to get more familiar with cellular mechanisms but also facilitates the design of novel drugs.

D.-S. Huang and K.-H. Jo (Eds.): ICIC 2020, LNCS 12464, pp. 621–632, 2020.
https://doi.org/10.1007/978-3-030-60802-6_54

A lot of efforts have been made to the prediction of PPIs. Several laboratory-based approaches have been developed at early stage, such as two-hybrid [13], TAP-tagging [8, 19], protein chips [25], synthetic lethal analysis [26] and correlated mRNA expression profile [5]. Although promising, laboratory-based approaches are both time-consuming and labor-intensive. Hence, to overcome these disadvantages, a variety of computational approaches have been proposed recently and they are mainly classified into three categories including sequence-based approaches [9–12], evolutionary-based approaches [21] and network-based approaches [1, 17]. Unlike the first two categories that heavily rest on biological information of proteins, network-based approaches are preferred as they only perform their tasks based on protein interaction network data.

As a recent attempt in network-based approaches, L3 [17] follows the intuition that proteins interact not if they are similar to each other, but if one of them is similar to the other's partners. In contrast to the traditional triadic closure principle [16] based on the similarity of neighborhoods [2, 7, 23], L3 relies on network paths of length three and introduces a degree-normalized score to measure the likelihood of being interacting for two proteins. Although the experiment results show that L3 significantly outperforms all existing link prediction methods when applied to predict PPIs, we believe that its performance could be constrained by the assumption of network paths with length three. To address this concern, we intend to propose a more flexible network-based prediction model.

In fact, proteins in the same protein complex are intensively interacted with each other, thus forming a denser structure than those from different complexes. Motivated by this observation, a novel stochastic block model, modified from mixed membership stochastic block model (MMSB), is proposed for network-based prediction of protein interactions. As a well-known community detection model, MMSB analyzes latent structural features by taking into account the memberships of entities. It captures different communities in the network and allows each entity has its own community distribution, which indicates that each node in the network has a K dimensional mixed membership to represent its weight for each community. The introduction of K dimensional community membership allows each node to be assigned to multiple communities instead of only one community, it is reasonable for protein network, because the same protein can participate multiple biological processes under different conditions in real biological system. In generative process, MMSB uses a binary adjacency matrix to represent interaction data and the mixed membership for each node follows Dirichlet distribution. Based on the community membership, there are two interaction indicators for each protein pair to indicate the community they belong to which follows Multinomial distribution. Combining these indicators with a $K \times K$ matrix of Bernoulli parameters which represent the probability of connection between K communities, the pairwise proteins can be classified as interacting or non-interacting.

Considering the huge data of PPIs, we follow the scheme of assortative mixed membership blockmodel (aMMSB) [6] to solve the inference problem of the proposed stochastic block model. Moreover, when determining the complex for each of proteins, we introduce a weighted similarity measure based on the strengths of complexes as well as the distance between the memberships of two proteins. In order to evaluate the performance of our model, we have applied it to three independent PPI datasets and compared its performance with L3. The experiment results show that our prediction approach has a promising performance in terms of several different metrics.

The rest of the paper is organized as follows. The details of our approach are presented in Sect. 2, following which we compare the proposed approach with L3 and also discuss the experiment results in Sect. 3. Finally, the paper ends with a conclusion in Sect. 4.

2 Methods

In aMMSB, protein networks are as input to the model. They are divided into K communities to identify different protein groups and proteins in each group have their mixed memberships, interaction between two proteins are identified by the two elements. To improve the performance of aMMSB, we changed the criterion of original aMMSB and proposed our approach. In this section, we introduce the generative process of aMMSB and the details of our approach.

2.1 The Generative Process of PPI Network

Though aMMSB can capture latent communities and assign each node a mixed membership to describe its weight for each community, the performance of original aMMSB is not good. The interaction indicator z is a K dimensional vector where only one element equals to one, the index of which indicates its corresponding community, and all the other elements equal to zero. In view of the fact that multiple proteins can interact with others in different protein complexes under different conditions, it means one protein may belong to several groups. As a criterion of interacting probability between two proteins, the indicate vectors with only one element equals to one is too strict, a number of interacting protein pairs are predicted to not interact and it leads to many false negative results thus. Due to this severe criterion of original aMMSB, we proposed an improved approach to relax standard and promote performance.

To weaken the impact of harsh criterion, we quantify the difference between two community memberships by Euclidean distance and combine it with community strength to calculate interacting probability. In training set, a protein network is put into aMMSB as input, community strength and community memberships are obtained as evaluation criterion of interacting by the model. The probability that a protein pair can interact is determined by the two elements with normalized function.

Suppose we have observed the links between proteins, each protein can be seen as a node and the link between them can be seen as edge represented by y. $N \times N$ binary matrix is completed by N proteins where y_{ij} is equal to 1 if there is a link between protein i and protein j, y_{ij} is equal to 0 otherwise. For each protein, there is a community membership π to describe its degree of membership for each community which obeys Dirichlet distribution. For the whole network, there is community strength β_k to capture latent communities, β_k is a K dimensional vector and ranges from 0 to 1, the elements in it represent the tightness of each community. Specific process is as follows:

- For each protein i, sample community membership $\pi_i \sim Dirichelet(\alpha)$. The specific equation is: $Dir(\pi_i|\alpha) = \frac{\Gamma\left(\sum_{k=1}^{K} \alpha_k\right)}{\prod_{k=1}^{K} \Gamma(\alpha_k)} \prod_{k=1}^{K} \pi_{i,k}^{\alpha_k - 1}$

- For each community k, sample community strength $\beta_k \sim Beat(\eta)$. The specific equation is: $Beta(\beta_k|\eta_0, \eta_1) = \frac{\Gamma(\eta_0+\eta_1)}{\Gamma(\eta_0)\Gamma(\eta_1)} \beta_k^{\eta_0-1}(1-\beta_k)^{\eta_1-1}$
- For each pair of node i and node j:

  Sample link $y_{ij} = \frac{\gamma - \gamma_{min}}{\gamma_{max} - \gamma_{min}}$, where $\gamma = \sqrt{\sum_{k=1}^{K} \beta_k (\pi_{i,k} - \pi_{j,k})^2}$

2.2 Stochastic Variable Inference

Our approach is a subclass of MMSB model with stochastic variable inference, it is a statistical model that allows nodes to participant in multiple communities. Figure 1 shows corresponding joint distribution of variables applied. The goal of the model is to compute the posterior distribution $p(\pi_{1:N}, \beta_{1:K}, z|y, \alpha, \eta)$ and the strategy adopted is variable inference (VI) [14]. However, traditional variable inference in MMSB deals with all the node pairs each iteration which requires a lot of time. Therefore, stochastic variable inference (SVI) is applied in aMMSB to save time and improve efficiency, aMMSB with SVI can also get comparable results to MMSB with VI.

SVI is a coordinate ascent algorithm that iteratively updates local and global parameters. For each iteration, we first subsample the network and get a subset S, local parameters are optimized given current global parameters and we then update global parameters using stochastic natural gradient with subset S and local parameters. The first step is called local step (L-step) and the second step is called global step (G-step). The specific generative process is described in Algorithm 1.

Algorithm 1 stochastic aMMSB

> Initialize variables randomly: $\gamma = (\gamma_n)_{n=1}^{N}, \lambda = (\lambda_k)_{k=1}^{K}$
> **while** not converging **do**
>   Sample a subset S of node pairs
>   L-step: $\forall (i,j) \in S$, optimize$(\phi_{i\rightarrow j}, \phi_{i\leftarrow j})$
>   Compute natural gradients: $\forall n, \partial\gamma_n^t$; $\forall k, \partial\lambda_k^t$
>   G-step: update γ and λ, $\gamma \leftarrow \gamma + \rho_t \partial\gamma^t$; $\lambda \leftarrow \lambda + \rho_t \partial\lambda^t$
>   Set $\rho_t = (\tau_0 + t)^{-\kappa}$; $t \leftarrow t+1$
> **end while**

The Global Step

The global step updates global parameters community strengths λ and community memberships γ. For a network with N nodes, there is $M = N(N-1)/2$ node pairs, we extract a node pair (i,j) at random. In t-th iteration, the stochastic natural gradients of global parameters are:

$$\partial\gamma_{i,k}^t = \alpha_k + \frac{1}{g(i,j)} \phi_{i\rightarrow j,k}^t - \gamma_{i,k}^{t-1}$$

$$\partial\lambda_{k,m}^t - \eta_{k,m} + \frac{1}{g(i,j)} \phi_{i\rightarrow j,k} \cdot \phi_{i\leftarrow j,k} \cdot y_{ij,m} - \lambda_{k,m}^{t-1}$$

where $y_{ij,0} = y_{ij}$ and $y_{ij,1} = 1 - y_{ij}$, we require that $\sum_t \rho_t^2 < \infty$ and $\sum_t \rho_t = \infty$, we set $\rho_t \triangleq (\tau_0 + t)^{-\kappa}$, where κ is learning rate and $\kappa \in (0.5, 1]$, τ_0 downweights early iterations and $\tau_0 \geq 0$. The time for a G-step is O(*NK*) and the memory required is O(*NK*).

The Local Step

The local step updates local parameters ϕ which represents the posterior approximation of which communities are important in determining whether there is a link. The time for L-step is O(*nK*) where *n* is the number of node pairs sampled in each iteration.

$$\phi^t_{i\rightarrow j,k}|y = 0 \propto \exp\left\{E_q\left[log\pi_{i,k}\right] + \phi^{t-1}_{i\rightarrow j,k}E_q[\log(1-\beta_k)]\right\}$$

$$\phi^t_{i\rightarrow j,k}|y = 1 \propto \exp\left\{E_q\left[log\pi_{i,k}\right] + \phi^{t-1}_{i\rightarrow j,k}E_q[\log\beta_k] + \left(1 - \phi^{t-1}_{i\rightarrow j,k}\right)log \in\right\}$$

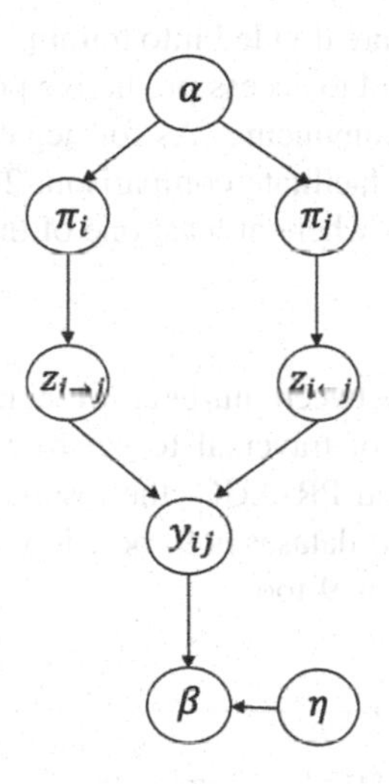

Fig. 1. Graphical model of our approach

3 Results

To evaluate the performance of aMMSB and its improved version experimentally, three protein networks including *yeast* [27], *krogan* [18] and *human* [17, 22] are selected where the human dataset is composed of three human networks containing HI-II-14 [22], HI-III [22] and HI-tested [17]. We adopt 5-fold cross-validation to get more comprehensive results. Except computational experiments in our model, we also applied these datasets to L3 model and compared their results. It shows that aMMSB with improved version acquires remarkable effect and outperforms L3 in numerous evaluation measures.

3.1 Data Structure

Considering the diversity of structure in datasets, the experimental results may be influenced by different protein networks. The datasets in different species are used to get more persuasive results and the detailed analysis of three datasets applied is shown in Table 1.

Table 1. Analysis of three datasets and their main characteristics

Dataset	N	E	k_{av}	ρ	C	SIPs
Yeast	964	3846	7.979	0.008	0.148	0
Krogan	2708	7123	5.261	0.002	0.188	0
Human	6657	32307	9.521	0.001	0.069	616

3.2 Computational Experiments

In experiments, all the datasets are divided into training set and test set, training set is as input network and test set is used to access predictive power of the model, 5-fold cross-validation is used to be more convincing. As for negative dataset, we keep consistent with the strategy used in L3 to facilitate comparison. 244 non-interacting protein pairs are selected including 100 pairs where at least one of the proteins are in the top 500 L3 predictions.

Sensitivity of *K* Value

To explore the relationship between number of communities and performance of aMMSB, we take the method of traversal to get best *K* value. Evaluation measures included are F1-score, AUC and PR-AUC, the results are shown in Fig. 2. Form the figure, the best *K* value in *yeast* dataset is 9, best K value in *krogan* dataset is 10 and best *K* value in *human* dataset is 9 too.

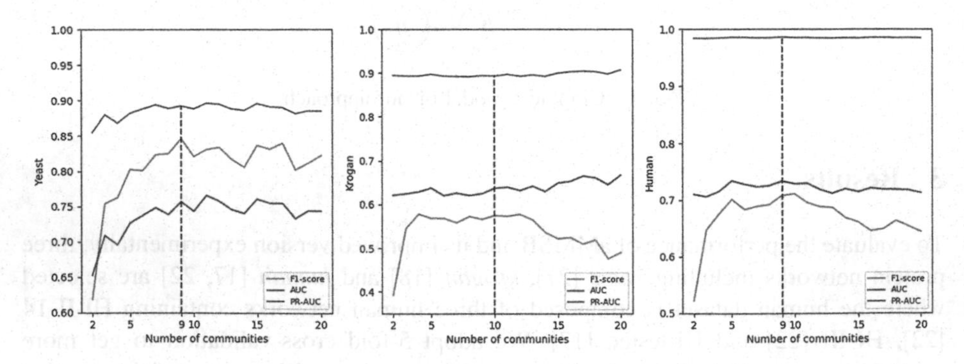

Fig. 2. The performance of different *K* values

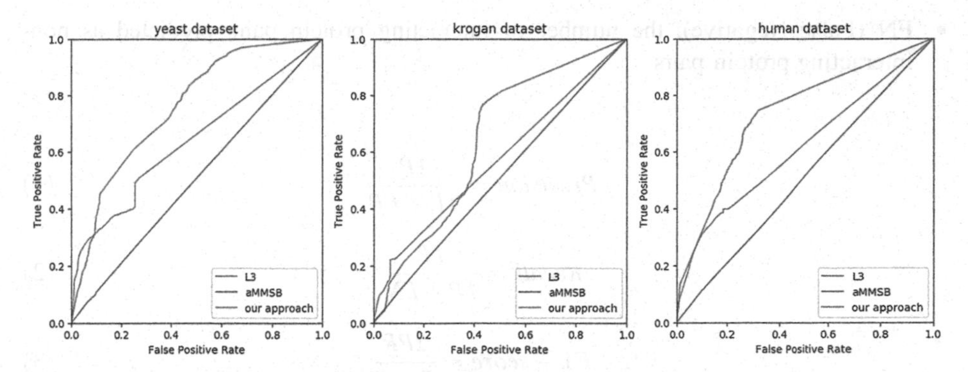

Fig. 3. The performance of AUC

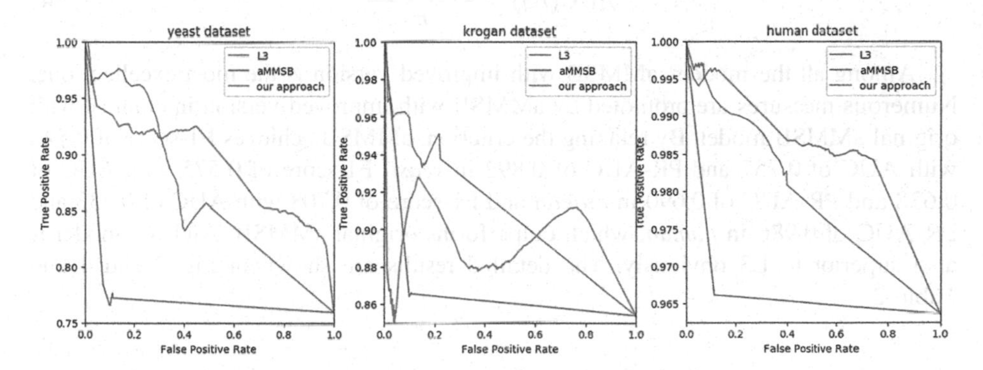

Fig. 4. The performance of PR-AUC

5-Fold Cross-Validation

In 5-fold cross-validation, we split the dataset into five subsets, each of them is as test set and the other four subsets remained are as training set, the test set and training set are rotated five times to access performance. Precision, Recall, F1-score, AUC, PR-AUC, AUC(L3) are used as evaluating measures where AUC(L3) is the measure used in L3 to calculate AUC but there is a little different between it and general AUC calculating method. AUC(L3) follows Eq. 4 where by randomly chooses n pairs of a positive link and negative link, larger score n′ times and equal score n″ times for the positive link are obtained. While AUC is the area under ROC curve [4, 20] which is plotted by the true positive rate against false negative rate and PR-AUC is the area under P-R curve [3] which is plotted by the precision against recall. Several elements used in other evaluation measures are explained as follows:

- TP (True Positive): the number of interacting protein pairs predicted correctly
- TN (True Negative): the number of non-interacting protein pairs predicted correctly
- FP (False Positive): the number of non-interacting protein pairs predicted as interacting protein pairs

- FN (False Negative): the number of interacting protein pairs predicted as non-interacting protein pairs

$$Precision = \frac{TP}{TP + FP} \quad (1)$$

$$Recall = \frac{TP}{TP + FN} \quad (2)$$

$$F1 - score = \frac{2PR}{P + R} \quad (3)$$

$$AUC(L3) = \frac{n' + 0.5n''}{n} \quad (4)$$

Among all the models, aMMSB with improved version is the most excellent one. Numerous measures are promoted by aMMSB with improved version in contrast with original aMMSB model. By relaxing the criterion, aMMSB achieves F1-score of 0.846 with AUC of 0.757 and PR-AUC of 0.892 in *yeast*, F1-score of 0.575 with AUC of 0.638 and PR-AUC of 0.690 in *krogan* and F1-score of 0.708 with AUC of 0.733 and PR-AUC of 0.986 in *human*, which outperforms original aMMSB. And our model is also superior to L3 obviously. The detailed results are shown in Fig. 3 and 4 and Table 2.

Table 2. The performance on three dataset

Dataset	Model	F-measure			AUC	PR-AUC	AUC(L3)
		Precision	Recall	F1-score			
Yeast	L3	0.842	0.393	0.536	0.637	0.866	0.600
	aMMSB	0.773	0.111	0.194	0.504	0.775	0.480
	Our approach	**0.854**	**0.847**	**0.846**	**0.757**	**0.892**	**0.810**
Krogan	L3	**0.942**	0.185	0.309	0.570	**0.909**	0.520
	aMMSB	0.866	0.108	0.191	0.505	0.867	0.505
	Our approach	0.880	**0.427**	**0.575**	**0.638**	0.893	**0.690**
Human	L3	0.982	0.388	0.556	0.612	0.979	0.605
	aMMSB	0.966	0.116	0.207	0.504	0.967	0.505
	Our approach	**0.986**	**0.552**	**0.708**	**0.733**	**0.986**	0.795

From Table 2, we can figure out that recall in *krogan* and *human* is in a low value compared to *yeast*. That is because *krogan* and *human* networks are larger than *yeast* network, when optimize parameters, mixed memberships and community strength in small network are tend to converge and be closer to true value. Thus, there are multiple false negative results in *krogan* and *human*, model in *yeast* dataset performs better. Besides, we assign mixed memberships at random to nodes which appear in test set and not appear in train set, the predicted results involving these nodes are related to a random parameter which is uncontrollable and may leads to false negative results, such uncontrollable nodes in *krogan* accounts for a larger proportion than yeast and human which leads to lower recall value in *krogan*.

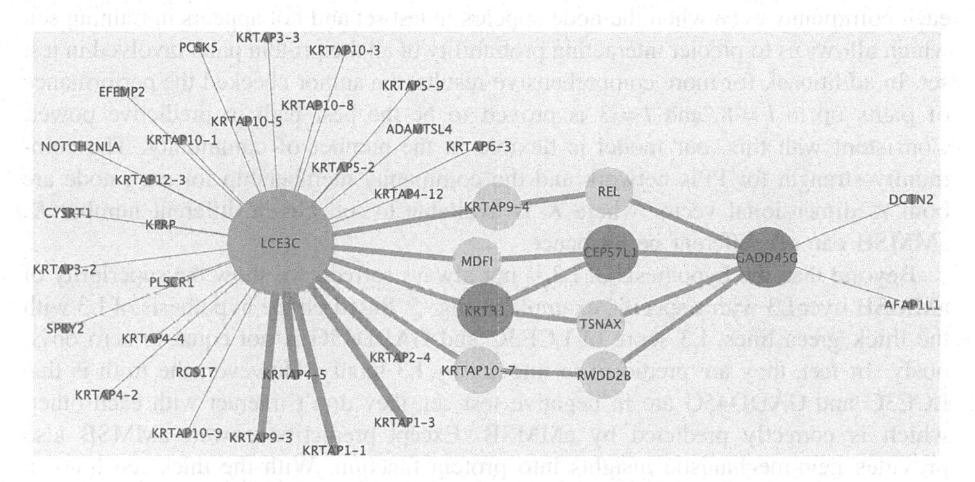

Fig. 5. A part of PPIs identified by our approach

L3 Limitations

L3 is a network-based approach to predict PPIs, it opposes the traditional idea that if two proteins share multiple neighbors, they tend to interact with each other. The author thinks similar proteins are not necessarily interact and interacting proteins are not necessarily similar. Through the study of a variety of protein pairs, the author draws a conclusion that proteins are likely to interact not if they are similar to each other, but if one of them is similar to the other's partners and the probability that two proteins interact is determined by their L3 score which is calculated by Eq. 5. For each protein pair, X and Y:

$$p_{XY} = \sum_{U,V} \frac{a_{XU} a_{UV} a_{VY}}{\sqrt{k_U k_V}} \tag{5}$$

where k_U is the degree of node U and a_{XU} equals to 1 if protein X and protein U interact, and 0 otherwise. In contrast to numerous traditional models based on the similarity of proteins, L3 achieves better results. However, L3 is still not strong enough and have some limitations for PPIs prediction.

L3 dependents on the known interacting protein pairs strongly, it supports the idea that if protein A interact with protein B and protein A is similar with protein C, protein B can interact with protein C, too. This idea requires a complete database, incomplete PPIs dataset limits the performance of L3 vastly. Supposing there is a protein D which appears in test set immensely, many protein pairs containing protein D need to be predicted, but it never appears in training set. Obviously, the probability that protein pairs involving protein D interact can't be predicted. Because we don't know which proteins are able to interact with protein D and which proteins are similar with protein D according to training set. If there are many proteins like protein D in test set, the performance of L3 is greatly compromised. Nevertheless, we describe each protein with a mixed membership in aMMSB model to represent its degree membership for each community even when the node appears in test set and not appears in training set, which allows us to predict interacting probability of all the protein pairs involved in test set. In additional, for more comprehensive results, the author checked the performance of paths up to $l = 8$, and $l = 3$ is proved to be the best path in predictive power. Consistent with this, our model is flexible in the number of community. The community strength for PPIs network and the community membership for each node are both K dimensional vector where K is available to set. Given different number K, aMMSB can get different performance.

Beyond that, the hypothesis of L3 is not always correct, we show the superiority of aMMSB over L3 with a specific example in Fig. 5. Based on the hypothesis of L3 with the thick green lines, L3 score of LCE3C and GADD45G is not equal to zero obviously. In fact, they are predicted to interact by L3 finally. However, the truth is that LCE3C and GADD45G are in negative test set, they don't interact with each other, which is correctly predicted by aMMSB. Except predictive power, aMMSB also provides new mechanistic insights into protein function. With the thick red lines in Fig. 5, LCE3C belongs to LCE family and it is a structural component of the cornified envelope of the stratum corneum involved in innate cutaneous host defense. KRTAP belongs to KRTAP family where KRTAP 1-1 and KRTAP 1-3 belongs to KRTAP type 1 family, KRTAP 2-4 belongs to KRTAP type 2 family, KRTAP 4-5 belongs to KRTAP type 4 family and KRTAP 9-3 belongs to KRTAP type 9 family. Keratin-associated proteins (KRTAP) consists hair keratin intermediate filaments in the hair cortex, they are essential for the formation of a rigid and resistant hair shaft through their extensive disulfide bond cross-linking with abundant cysteine residues of hair keratins. The interactions between LCE3C and KRTAP 1-1, KRTAP 1-3, KRTAP 2-4, KRTAP 4-5, KRTAP 9-3 have been proved in IntAct [15], all of them can participant the biological process of keratinization (GO: 0031424). Their interactions are correctly predicted by aMMSB, but they are predicted to not interact in L3.

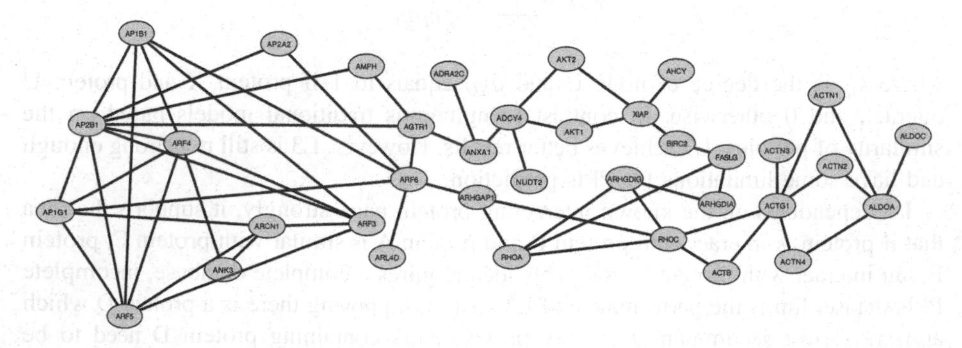

Fig. 6. A part of PPIs predicted by our approach and not included in our dataset

Predicted Protein Pairs Outside the Dataset
Besides the test set, our model can also predict the protein pairs out of the dataset. aMMSB allows each protein to own a mixed membership and assign community strength to each community. With only the two elements, the interacting probability of any protein pairs can be predicted. That is, aMMSB can predict any combinations of protein pairs in dataset only with their mixed memberships and community strength. If there is N proteins in dataset, instead of being limited to test set, $N(N-1)/2$ protein pairs can be predicted by our model. Figure 6 shows a part of protein pairs which is predicted to interact and not belongs to the dataset. All the interactions in Fig. 6 have been proved in STRING [24].

4 Conclusion

We apply a probabilistic model aMMSB to predict PPIs, this model is able to classify protein pairs into interacting pairs and non-interacting pairs with high accuracy and high precision, it is also available to calculate the exact probability of interacting pairs. The basic idea of our approach is that a node may belongs to several communities, it is common with the reality of proteins in multiple biological processes. Besides, the great performance of our approach also provides us a new sight to cellular mechanisms research. The success of the method is based on its ability to capture the structure of whole network and mixed membership of each node. The experimental results show that our approach obtains excellent performance in numerous evaluating measures and outperforms L3 model greatly.

Funding. This work has been supported by the National Natural Science Foundation of China [grant number 61602352], and the Pioneer Hundred Talents Program of Chinese Academy of Sciences.

References

1. Airoldi, E.M., Blei, D.M., Fienberg, S.E., Xing, E.P.: Mixed membership stochastic blockmodels. J. Mach. Learn. Res. **9**(Sep), 1981–2014 (2008)
2. Bass, J.I.F., Diallo, A., Nelson, J., Soto, J.M., Myers, C.L., Walhout, A.J.: Using networks to measure similarity between genes: association index selection. Nat. Methods **10**(12), 1169 (2013)
3. Davis, J., Goadrich, M.: The relationship between precision-recall and roc curves. In: Proceedings of the 23rd International Conference on Machine Learning, pp. 233–240 (2006)
4. Fawcett, T.: An introduction to ROC analysis. Pattern Recogn. Lett. **27**(8), 861–874 (2006)
5. Ge, H., Liu, Z., Church, G.M., Vidal, M.: Correlation between transcriptome and interactome mapping data from saccharomyces cerevisiae. Nat. Genet. **29**(4), 482–486 (2001)
6. Gopalan, P.K., Gerrish, S., Freedman, M., Blei, D.M., Mimno, D.M.: Scalable inference of overlapping communities. In: Advances in Neural Information Processing Systems, pp. 2249–2257 (2012)

7. Granovetter, M.S.: The strength of weak ties. In: Social Networks, pp. 347–367. Elsevier (1977)
8. Ho, Y., et al.: Systematic identification of protein complexes in saccharomyces cerevisiae by mass spectrometry. Nature **415**(6868), 180–183 (2002)
9. Hu, L., Chan, K.C.: Discovering variable-length patterns in protein sequences for protein-protein interaction prediction. IEEE Trans. Nanobiosci. **14**(4), 409–416 (2015)
10. Hu, L., Chan, K.C.: Extracting coevolutionary features from protein sequences for predicting protein-protein interactions. IEEE/ACM Trans. Comput. Biol. Bioinf. **14**(1), 155–166 (2016)
11. Hu, L., Hu, P., Yuan, X., Luo, X., You, Z.H.: Incorporating the coevolving in- formation of substrates in predicting hiv-1 protease cleavage sites. IEEE/ACM Trans. Comput. Biol. Bioinf. (2019, early access)
12. Hu, L., Yuan, X., Hu, P., Chan, K.C.: Efficiently predicting large-scale protein- protein interactions using MapReduce. Comput. Biol. Chem. **69**, 202–206 (2017)
13. Ito, T., Chiba, T., Ozawa, R., Yoshida, M., Hattori, M., Sakaki, Y.: A comprehen- sive two-hybrid analysis to explore the yeast protein interactome. Proc. Natl. Acad. Sci. **98**(8), 4569–4574 (2001)
14. Jordan, M.I., Ghahramani, Z., Jaakkola, T.S., Saul, L.K.: An introduction to variational methods for graphical models. Mach. Learn. **37**(2), 183–233 (1999)
15. Kerrien, S., et al.: The intact molecular interaction database in 2012. Nucleic Acids Res. **40** (D1), D841–D846 (2012)
16. Keskin, O., Tuncbag, N., Gursoy, A.: Predicting protein–protein interactions from the molecular to the proteome level. Chem. Rev. **116**(8), 4884–4909 (2016)
17. Kovács, I.A., et al.: Network-based prediction of protein interactions. Nature Commun. **10** (1), 1–8 (2019)
18. Krogan, N.J., et al.: Global landscape of protein complexes in the yeast saccharomyces cerevisiae. Nature **440**(7084), 637–643 (2006)
19. Mann, M., Pandey, A.: Use of mass spectrometry-derived data to annotate nucleotide and protein sequence databases. Trends Biochem. Sci. **26**(1), 54–61 (2001)
20. Metz, C.E.: Basicprinciplesofrocanalysis. In: Seminarsinnuclearmedicine, vol.8, pp. 283–298. WB Saunders (1978)
21. Mirabello, C., Wallner, B.: InterPred: a pipeline to identify and model protein–protein interactions. Proteins: Struct., Funct., Bioinf. **85**(6), 1159–1170 (2017)
22. Rolland, T., et al.: A proteome-scale map of the human interactome network. Cell **159**(5), 1212–1226 (2014)
23. Simmel, G.: Soziologie: Untersuchungen u ̈ber die formen der vergesellschaftung. BoD–Books on Demand (2015)
24. Szklarczyk, D., et al.: The string database in 2017: quality-controlled protein–protein association networks, made broadly accessible. Nucleic Acids Res. **45**, gkw937 (2016)
25. Tong, A.H.Y., et al.: A combined ex-perimental and computational strategy to define protein interaction networks for peptide recognition modules. Science **295**(5553), 321–324 (2002)
26. Tong, A.H.Y., et al.: Systematic genetic analysis with ordered arrays of yeast deletion mutants. Science **294**(5550), 2364–2368 (2001)
27. Tong, A.H.Y., et al.: Global mapping of the yeast genetic interaction network. Science **303** (5659), 808–813 (2004)

Author Index

Printed in the United States
By Bookmasters